Diffractive Nanophotonics

Diffractive Nanophotonics

Edited by
VICTOR A SOIFER

CISP

CRC Press
Taylor & Francis Group
Boca Raton London New York

CRC Press is an imprint of the
Taylor & Francis Group, an **informa** business

CRC Press
Taylor & Francis Group
6000 Broken Sound Parkway NW, Suite 300
Boca Raton, FL 33487-2742

First issued in paperback 2021
First issued in hardback 2019

ISBN 13: 978-1-03-224238-5 (pbk)
ISBN 13: 978-1-4665-9069-4 (hbk)

Publisher's Note
The publisher has gone to great lengths to ensure the quality of this reprint but points out that some imperfections in the original copies may be apparent.

Visit the Taylor & Francis Web site at
http://www.taylorandfrancis.com

and the CRC Press Web site at
http://www.crcpress.com

Contents

Introduction

Nanophotonics examines the interaction of light with particles of matter or irregularities that are smaller and much less than the wavelength, and devices produced on the basis of the results. Nanophotonics in particular includes the optics of photonic crystals and photonic-crystal waveguides, plasmonics, near-field microscopy, metamaterials and optical micromanipulation.

According to the wave-particle duality, the light during its propagation (except for acts of emission and absorption) can always be considered as a wave. Even if the number of photons is small (very weak light fields) we can observe phenomena of diffraction and interference. Therefore, Maxwell's equations always adequately describe the propagation of light – interference and diffraction in free or homogeneous space.

The interaction of light with matter is described by macroscopic or microscopic electrodynamics. In *macroscopic electrodynamics* Maxwell's equations the matter is described by dielectric and magnetic permeability; the absorption of light is described by the introduction of complex dielectric permittivity material. *Microscopic electrodynamics* is based on Maxwell–Lorentz equations. These equations are a generalization of Maxwell's equations, in which matter is described as a set of moving point charges. According to the electron theory, the Maxwell–Lorentz equations accurately describe the electromagnetic microfields any point in space (including the inter and intratomic fields and even fields inside elementary particles) at any time.

Adequate microscopic description of the emission, absorption, and scattering of light by material is provided by quantum electrodynamics. *Quantum electrodynamics* quantitatively describes the effects of the interaction of radiation with matter, and also consistently describes the electromagnetic interaction between charged particles. *Diffractive nanophotonics*, which is the subject of this book, deals with the processes and devices in which the wave properties of light are predominant. Diffractive nanophotonics suggests the possibility of describing the processes of behaviour of light by the Maxwell equations. The Maxwell equations can be used when working with coherent electromagnetic fields. In this case, the characteristic dimensions of the optical elements and structures should significantly exceed the atomic size. This is necessary to describe the optical properties with the macroscopic characteristics of dielectric permittivity and magnetic permeability. Thus, diffractive nanophotonics studies light diffraction on macro-objects with minimal irregularities of the order of tens of nanometers, up to the quantum dots of about 10 nm in size (this is much larger than the size of individual atoms and simple molecules of the substance). Therefore, the substance is described at the macrolevel, i.e. in the framework of macroscopic electrodynamics. The limitation is that in diffraction of light on free micro- and nanoinhomogeneities (e.g., solid microparticles in fluids), light presses on microparticles with a force of units and tens piconewtons and makes them to move. Thus, strictly speaking, it is required to solve the non-stationary problem of the diffraction of light on a moving heterogeneity and take into account the change in the wavelength of light.

The main purpose of this book is to demonstrate the fruitfulness of the well-established methods of diffractive computer optics in solving nanophotonics tasks. There are 8 chapters and 3 appendices in the book.

In the first chapter the basic equations of the diffractive nanophotonics and related transformation are considered. A system of Maxwell equations is presented and the formulation of the conditions on the interfaces and the Poynting theorem is discussed.

The basic equations that can be obtained from the system of the Maxwell equations are presented. These include the wave equation (with the time dependence of the field), the Helmholtz equation (without such dependence), the Fock–Leontovich equation (approximation of the scalar diffraction theory) and the eikonal equation and transfer equation (ray tracing approach). The integral theorems of optics expressing the field values in the integral form are presented. These include the Green (scalar case) and the Stratton–Chu (vector representation) formulas. The scalar integral transforms are considered for use in optics to calculate the diffraction field in a homogeneous space: the Kirchhoff integral corresponds to the decomposition of the complex amplitude of the field in spherical waves, the Fresnel integral – along parabolic, and the Fourier transform – for plane waves.

The second chapter is devoted to numerical methods for solving diffraction problems under the strict electromagnetic theory. The first section presents difference method for solving Maxwell's equations (FDTD approach). The Yee explicit difference schemes, based on the replacement of the difference derivatives by relationships for grid functions are discussed. In addition, each projection of the electromagnetic field is defined by its grid domain, which improves the order of approximation of the differential problem. The method of transition from grid functions in space and time to complex amplitudes of the field components is developed. Particular attention is given to the imposition of absorbing layers, simulating the free space around the computing domain. The problem of the formation of the electromagnetic wave incident on the optical element is solved.

A method is proposed for the decomposition of the grid area, allowing a large reduction of the duration of calculations by taking into account the structure of the optical elements. The second part of this chapter deals with approaches to numerical solution of the Helmholtz equation. A review of existing approaches is given and their classification in the family of BPM methods (beam propagation method) discussed, and the principal differences between the different versions of the method are shown. The solutions of the scalar Helmholtz equation for paraxial cases (in the approximation of the smooth envelope) and non-paraxial (based on the approximation of the differential operator) distribution are presented. The finite-difference schemes for solving equations and some variants of the boundary conditions are given. Approaches to solving the Helmholtz equation for the vector case are discussed, including the media with an inhomogeneous distribution of the refractive index.

The third chapter examines the diffraction of light on two-dimensional microscopic objects of arbitrary shape. The problem of diffraction of electromagnetic waves with TE or TM-polarization on two-dimensional dielectric objects is solved using the finite element method (FEM) in different conditions. The Helmholtz equation is solved by the combined Galerkin FEM and the boundary element method. Two types of this method are considered for the analysis of diffraction on non-periodic microscopic inhomogeneities and on periodic objects – subwavelength diffractive gratings. As examples, problems of diffraction of plane electromagnetic waves on dielectric and conductive cylinders with the diameter equal to the wavelength are solved. The finite element method is also used to solve the problem of diffraction of a plane wave on a one-dimensional binary dielectric grating with a period of fractions of the wavelength to several wavelengths. Another variation of the FEM is related to the solution of the integral equation of diffraction on a uniform sampling grid. In this case, the problem is reduced to solving a linear system of Gauss algebraic equations. In contrast to the boundary element method, this method does not require the calculation of derivatives of the field, normal to the boundary surface of the local inhomogeneity. A solution of the problem of diffraction of a plane wave on a multilayer dielectric cylinder in the form of a series of cylindrical functions is presented. The coefficients of the series in a general case are found from the recurrence relations. Explicit formulas for the coefficients of the series were derived for a two-layer cylinder. These methods were used to simulate the diffraction of light on Luneburg and Eaton–Lippmann gradient microlenses. This chapter also considers an iterative method for solving integral equations for electromagnetic diffraction of waves with the TE- and TM-polarizations. The conditions under which this method has a relaxation, i.e., is a reduction of the mean error with increasing number of iterations, are determined. The method effectively allows to calculate the diffraction field inside and outside the dielectric inhomogeneity of a size equal to or less than the wavelength.

The fourth chapter is a numerical method for solving the problem of diffraction on periodic diffractive micro- and nanostructures. The method is used to calculate and study the diffraction structures for a number of modern trends in nanophotonics, including plasmonics, metamaterials, nanometrology. The method of rigorous coupled-wave analysis (RCWA) is used to solve the problem of diffraction of a plane wave on two- and three-dimensional diffractive structures and diffractive gratings. This numerical method for solving Maxwell's equations is focused on the analysis of micro- and nanostructures described by a periodic function of the dielectric permittivity. Surface electromagnetic waves (SEW) (surface plasmon–polaritons) are studied, and calculation and study of diffractive structures designed to form interference patterns SEW are carried out. The diffractive structures are composed of a dielectric diffractive grating and a metallic layer deposited on the substrate. The parameters of the diffractive structure are calculated from the excitation conditions at the lower boundary of the metallic layer of a given set of surface electromagnetic waves of different configurations and directions. As a result, directly below the metallic layer there forms a periodic interference pattern of surface electromagnetic waves. Periods of generated interference patterns significantly subwavelength. A promising area of application of the considered structures is nanolithography based on registration of interference patterns of surface electromagnetic waves in the electron resist. The magneto-optical properties of bilayer metal–dielectric heterostructures consisting of a metallic diffractive grating and a dielectric magnetized layer are studied. The calculations show that these structures have resonant magneto-optical effects due to the rotation of the polarization plane of the incident wave and change of the reflectance (transmittance) index of the structure when the magnetization of the layer changes. These structure can be used as magnetic field sensors, gas sensors, light intensity modulation devices, controlled by the external magnetic field.

The fifth chapter describes the simulation of electromagnetic waves in nanophotonic devices. The FullWAVE software is used to calculate the passage of light through two-dimensional photonic crystals in the case in which the light frequency does not fall into the stop band. The results of modelling sharp focusing of light by the two-dimensional gradient photonic–crystal lenses, as well as the results of using these lenses as a coupling device for matching the two planar waveguides with different widths, are considered. It is shown that the width of the focal spot, which is formed near the surface of the photonic crystal (PC) lens, is equal to 0.3 wavelength. The results of experiments with fabrication of two-dimensional photonic-crystal lenses in a film of silicon on fused silica are outlined. The chapter also discussed the radial FDTD-method, which is adapted to solutions of Maxwell's equations for an axially symmetric diffraction laser beam with radial polarization on an axially symmetric optical element. The results of modelling of sharp focusing of laser light with radial polarization using a microaxicon and Mikaelian three-dimensional gradient lenses are presented. The simulation results showed that in both cases focal spots are formed near the surface of the optical element with the inhomogeneous surface waves involved in their formation and it is therefore possible to overcome the diffraction limit. The diameter of the focal spot at half intensity was equal to 0.35 of the wavelength.

The sixth chapter discusses two methods of calculating the spatial modes of microstructured photonic crystal fibres (PCF). This is a relatively new class of optical fibres, which uses the properties of photonic crystals. In the cross section the PCFs have a quartz or glass microstructure with a periodic or aperiodic system of microinclusions, mostly cylindrical microperforations, oriented along the fibre axis. The 'defect' of the microstructure, corresponding to the absence of one or more elements in its centre, is the core of the optical fibre, providing a waveguide mode of propagation of electromagnetic radiation. Two methods for calculating the mode of optical fibres are discussed in detail: the approximate analytical method of matched sinusoidal modes, and grid method based on the use of finite-difference approximations to the stationary wave equations. The basic idea of the method of matched sinusoidal modes (MSM-method), also known as the transverse resonance technique, is based on dividing the PCF section into homogeneous rectangular areas, and the description in each field of a homogeneous area using a system of local sinusoidal modes. The MSM-method is modified by the iterative Krylov method in the most computationally complex stage of the solution of the non-linear problem of eigenvalues of the large matrix to which the problem of finding the propagation constants of modes is reduced. The MSM-method was used to calculate both scalar and vector modes of conventional round step-index fibres and the modes of photonic crystal fibres with a filled core. The basis of the finite-difference method (FD-method) under review in this chapter is the approach based on the use of finite-difference approximations to the stationary vector wave equations for monochromatic light such as the Helmholtz equations. The FD method wins in the speed of the algorithm in the MSM method because the problem of finding the propagation constant and the sampling grid solutions for the transverse components of the electric or magnetic components is directly reduced to a linear matrix problem for the eigenvalues and eigenvectors. The FD method also allows for full vector analysis of modes of photonic crystal fibres with a hollow core. Examples of calculation of modes of Bragg fibres with filled and hollow cores are presented.

The seventh chapter contains the theory of paraxial and non-paraxial laser beams with axial symmetry and an orbital angular momentum. Such beams are called vortex beams, because their energy is propagating in a spiral around the optical axis, forming a 'funnel' as with a wind swirl. In nanophotonics the vortex laser beams are used for optical trapping and rotating micro- and nanoparticles in a ring in the cross-sectional plane of the beam. In near-field diffraction the radius of the ring of the optical vortex is comparable with the wavelength, and considering this radius for the specified intensity level, the radius of the optical vortex can be several times smaller than the wavelength. This property of vortex laser beams is used in modulation nanolithography .

This chapter examines the diffraction of plane, Gaussian and conical waves on a spiral phase plate and a spiral axicon. Explicit analytical expressions are presented for the complex amplitudes of light of vortex beams in the area of Fresnel diffraction and far-field diffraction. The Rayleigh–Sommerfeld integral is used to study paraxial and non-paraxial vector theory of vortex laser beams. It is shown that for the beam waist radius close to the wavelength, the longitudinal component of the vector of the electric field is only a few percent of the transverse component. Considered are the scalar paraxial hypergeometric beams formed by the logarithmic axicon and spiral phase plate. The complex amplitude of the vortex laser beams is proportional to the confluent hypergeometric function or Kummer function. In addition, the family of such hypergeometric beams forms a basis, they all have a ring structure (the intensity on the optical axis is zero), and the thickness of the intensity rings of the transverse diffraction pattern decreases with increasing ring number, tending to zero in the limit. A special form of hypergeometric laser beams are hypergeometric modes that retain their intensity during propagation in space. Non-paraxial hypergeometric beams whose complex amplitude is proportional to the product of two Kummer functions are discussed. Numerical examples of the propagation of such beams and the experimental results on the formation of vortex laser beam using diffractive optical elements are presented.

The eighth chapter discusses methods for calculating the force and torque, exerted by the electromagnetic field focused onto the microparticle of arbitrary form, whose dimensions are comparable with the wavelength of light. There are two ways of calculation of light pressure on the micro-object: the rigorous electromagnetic method (two-dimensional case) and the geometrical optics method (three-dimensional case). The results of both methods of calculation of the pressure force of a Gaussian beam on a dielectric microcylinder are compared. The chapter also describes optical circuits, including diffractive optical elements (DOE) that form the laser vortex beams: Bessel mode, hypergeometric modes of different orders. Such DOEs are produced by optical and electron lithography. In some experiments vortex laser beams were formed by the dynamic liquid crystal microdisplays. The results of the experiments with multiorder optical elements forming multiple vortex beams having different angular orbital angular momentum are discussed. The diameters of the light rings on which microparticles rotated, were tens of microns, and the linear velocity of rotation was equal to a few microns per second.

The book was written by the Image Processing Systems Institute, Russian Academy of Sciences: Chapter 1 – D.L. Golovashkin, V.V. Kotlyar, Chapter 2 – D.L. Golovashkin, A.V. Gavrilov., Chapter 3 – D.V. Nesterenko,

A.G. Nalimov, V.V. Kotlyar, Chapter 4 – L.L. Doskolovich, V.A. Soifer, Chapter 5 – V.V. Kotlyar, V.S. Pavelyev, P.N. Dyachenko, V.A. Soifer, Chapter 6 – V.V. Kotlyar, Y.O. Shuyupova, Chapter 7 – V.V. Kotlyar, A.A. Kovalev, S.N. Khonina, V.A. Soifer, Chapter 8 – R.V. Skidanov, V.V. Kotlyar, A.G. Nalimov, S.N. Khonina, V.A. Soifer.

The book is based on research, performed under the Russian–American program 'Basic Research and Higher Education' (grant CRDF RUXO-014 -Sa-06, PG08-014-1), with the support of RFBR (grants 07-01-96602, 07-02-12134, 07-07-91580, 07-07-97600, 07-0797601, 08-07-99005, 08-07-99007, 10-07-00109, 10-07-00438, 10-07-00453, 10-07-00553, 11-07-00153, 11-07-12036, 12-07-00269, 12-07-00495, 12-07-31115, 12-07-31117, 13-07-97004, 13-07-97005, 12-07-97008), the President of the Russian Federation (MD-5303.2007.9, NSh-3086.2008.9, NSh-7414.2010.9, NSh-4128.2012.9, MD-1929.2013.2, MD-6809.2012.9).

Acknowledgments. The authors are grateful to M.A. Lichmanov for carrying out numerical simulation of light diffraction on multilayer dielectric cylinders (Chapter 3), S.S. Stafeev and J.R. Triandafilov for carrying out the numerical simulation of photonic crystal lens and subwavelength focusing of radially polarized light. Liam O'Faolain (University of St Andrews, Scotland) for their help in making photonic crystal lens (Chapter 5), Jari Turunen (University of Joensuu, Finland) for assistance in the manufacture of diffractive optical elements (Chapter 7).

Sections 4.2 and 4.3 are based on original research papers published in collaboration with A.K. Zvezdin, V.I. Belotelov, D.A. Bykov, E.A. Bezus and V.A. Kotov. The material in section 5.4 is based on the results obtained by the authors in conjunction with Yu.V. Miklyaev.

A.G. Nalimov, V.V. [illegible] N.A. S[illegible] [illegible] V.V. Kotlyar [illegible] Paveyev [illegible] S.S. Stafeev [illegible] Chapter 6 – V.V. Kotlyar [illegible] Chapter 7 – V.V. Kotlyar, A.A. Kovalev, S.N. Khonina [illegible] Chapter [illegible] Kotlyar, A.G. Nalimov [illegible]

The book is based on [illegible] under the Russian–American program "Basic Research and Higher Education" [illegible] RUX0-014-SA-06, PG08-014-1 [illegible] 07-01-[illegible], 07-07-91580, 07-[illegible] 08-07-[illegible] 10-07-00109, 10-07-00438-[illegible] 12-07-00269 [illegible] 2007.9.0056 [illegible] 2010 [illegible] MD-[illegible]/2012 [illegible].

Acknowledgements [illegible] M.A. [illegible] and numerical [illegible] multi-order diffractive [illegible] and [illegible] St. Andrews [illegible] for their help [illegible] optical elements (Chapter 7).

[illegible] 2 and 4 [illegible] based on [illegible] research [illegible] published [illegible] with [illegible] D.L. Rykov [illegible] [illegible] obtained [illegible] MRC [illegible].

Chapter 1

Basic equations of diffractive nanophotonics

There are several books on modern nanophotonics [1–5].The book [1] is devoted only to photonic crystals and does not address other important areas of nanophotonics. Calculation of band gaps in photonic crystals [1] is based on solving Maxwell's equations, rewritten in the form of a problem of the eigenvalues and eigenvectors.

The book [2] deals with almost all areas of nanophotonics, but does not consider the mathematical methods of modelling the diffraction of light. The book [3] focuses on the near-field microscopy for the observation of quantum structures, molecules and biological systems. Analysis of the interaction of light with matter [3] is based on the dipole approximation, which applies to particles of matter with dimensions much smaller than the wavelength of light. The book [4] deals with only one field of nanophotonics and one modelling method. In [4], the authors consider only the localized plasmons as resonance vibrations of metal nanoparticles excited by electromagnetic radiation. Localized plasmons are different from surface plasmons, which are discussed in chapter 4 of this book. Localized plasmons in [4] are analyzed by means of the eigenfunctions of plasmon oscillations, which are the eigenfunctions of the Laplace equation. The book [5] is closest to the present book. The book [5] addresses many aspects of nanophotonics: near-field microscopy, photonic crystals, surface plasmons, quantum emitters, optical trapping. The mathematical modelling methods in nanophotonics problems are discussed: the method of moments, the method of coupled dipoles, the Green's function method. However, the book [5] does not address important areas of nanophotonics, such as as photonic–crystal waveguides and lenses, subwavelength gratings with magnetic and metal layers. Also, the book [5] does not consider the most universal methods for simulation of light diffraction – difference methods for solving Maxwell's equations: FDTD-method and the BPM-method.

Therefore, chapter 1 of this book presents the basic equations of diffractive nanophotonics, which are used in this book: Maxwell's equations in integral and differential forms, and other differential and integral equations derived from Maxwell's equations. Chapter 2 discusses the two main difference methods for solving Maxwell's equations: finite-difference time-domain method (FDTD-method) and the beam propagation method (BPM-method).

1.1. Maxwell equations

1.1.1. Mathematical concepts and notations

In the Cartesian coordinate system with unit vectors $\mathbf{e}_x$, $\mathbf{e}_y$, $\mathbf{e}_z$ we determine the differential operators grad, div, rot and Δ with respect to the scalar f and vector $\mathbf{F}$ functions as follows:

$$\operatorname{grad} f \equiv \nabla f = \mathbf{e}_x \frac{\partial f}{\partial x} + \mathbf{e}_y \frac{\partial f}{\partial y} + \mathbf{e}_z \frac{\partial f}{\partial z},$$

$$\operatorname{div} \mathbf{F} \equiv \nabla \cdot \mathbf{F} = \frac{\partial F_x}{\partial x} + \frac{\partial F_y}{\partial y} + \frac{\partial F_z}{\partial z},$$

$$\operatorname{rot} \mathbf{F} \equiv \nabla \times \mathbf{F} = \left(\frac{\partial F_z}{\partial y} - \frac{\partial F_y}{\partial z} \right) \mathbf{e}_x + \left(\frac{\partial F_x}{\partial z} - \frac{\partial F_z}{\partial x} \right) \mathbf{e}_y + \left(\frac{\partial F_y}{\partial x} - \frac{\partial F_x}{\partial y} \right) \mathbf{e}_z,$$

$$\Delta f \equiv \nabla^2 f = \frac{\partial^2 f}{\partial x^2} + \frac{\partial^2 f}{\partial y^2} + \frac{\partial^2 f}{\partial z^2},$$

$$\Delta \mathbf{F} \equiv \nabla^2 \mathbf{F} = \operatorname{grad} \operatorname{div} \mathbf{F} - \operatorname{rot} \operatorname{rot} \mathbf{F} \cdot$$

For a cylindrical coordinate system with unit vectors $\mathbf{e}_\rho$, $\mathbf{e}_\varphi$, $\mathbf{e}_z$:

$$\operatorname{grad} f = \mathbf{e}_\rho \frac{\partial f}{\partial \rho} + \mathbf{e}_\varphi \frac{1}{\rho} \frac{\partial f}{\partial \varphi} + \mathbf{e}_z \frac{\partial f}{\partial z},$$

$$\operatorname{div} \mathbf{F} = \frac{1}{\rho} \frac{\partial}{\partial \rho} \left(\rho F_\rho \right) + \frac{1}{\rho} \frac{\partial F_\varphi}{\partial \varphi} + \frac{\partial F_z}{\partial z},$$

$$\operatorname{rot} \mathbf{F} = \left(\frac{1}{\rho} \frac{\partial F_z}{\partial \varphi} - \frac{\partial F_\varphi}{\partial z} \right) \mathbf{e}_\rho + \left(\frac{\partial F_\rho}{\partial z} - \frac{\partial F_z}{\partial \rho} \right) \mathbf{e}_\varphi + \left(\frac{1}{\rho} \frac{\partial}{\partial \rho} \left(\rho F_\varphi \right) - \frac{1}{\rho} \frac{\partial F_\rho}{\partial \varphi} \right) \mathbf{e}_z,$$

$$\Delta f = \frac{1}{\rho} \frac{\partial}{\partial \rho} \left(\rho \frac{\partial f}{\partial \rho} \right) + \frac{1}{\rho^2} \frac{\partial^2 f}{\partial \varphi^2} + \frac{\partial^2 f}{\partial z^2}.$$

In a spherical coordinate system with unit vectors $\mathbf{e}_r$, $\mathbf{e}_\theta$, $\mathbf{e}_\varphi$ the following representations apply:

$$\operatorname{grad} f = \mathbf{e}_r \frac{\partial f}{\partial r} + \mathbf{e}_\theta \frac{1}{r} \frac{\partial f}{\partial \theta} + \mathbf{e}_\varphi \frac{1}{r \sin\theta} \frac{\partial f}{\partial \varphi},$$

$$\operatorname{div} \mathbf{F} = \frac{1}{r^2} \frac{\partial}{\partial r} \left(r^2 F_r \right) + \frac{1}{r \sin\theta} \frac{\partial}{\partial \theta} \left(\sin\theta F_\theta \right) + \frac{1}{r \sin\theta} \frac{\partial F_\varphi}{\partial \varphi},$$

$$\mathrm{rot}\,\mathbf{F} = \left(\frac{1}{r\sin\theta} \frac{\partial}{\partial\theta} \left(\sin\theta F_{\varphi} \right) - \frac{1}{r\sin\theta} \frac{\partial F_{\theta}}{\partial\varphi} \right) \mathbf{e}_r$$

$$+ \left(\frac{1}{r\sin\theta} \frac{\partial F_r}{\partial\varphi} - \frac{1}{r} \frac{\partial}{\partial r} \left(rF_{\varphi} \right) \right) \mathbf{e}_{\theta}$$

$$+ \left(\frac{1}{r} \frac{\partial F_r}{\partial r} \left(rF_{\theta} \right) - \frac{1}{r} \frac{\partial F_r}{\partial\theta} \right) \mathbf{e}_{\varphi},$$

$$\Delta f = \frac{1}{r^2} \frac{\partial}{\partial r} \left(r^2 \frac{\partial f}{\partial r} \right) + \frac{1}{r^2 \sin\theta} \frac{\partial}{\partial\theta} \left(\sin\theta \frac{\partial f}{\partial\theta} \right) + \frac{1}{r^2 \sin^2\theta} \frac{\partial^2 f}{\partial\varphi^2}.$$

The most important integral relationships of vector analysis are:

The Gauss–Ostrogradskii theorem:

$$\int_V div\mathbf{F}dv = \oint_S (\mathbf{F}, \mathbf{n})\, ds,$$

where **n** is the unit vector of the external normal; V is a region of space bounded by the surface S.

Stokes' theorem:

$$\int_S \mathrm{rot}\mathbf{F}ds = \oint_L \mathbf{F}dl,$$

where L is the contour bounding the surface S.

1.1.2. Maxwell's equations in differential form

The electromagnetic theory of light is based on a system of Maxwell's equations [1]:

$$\mathrm{rot}\mathbf{H} = \frac{1}{c}\frac{\partial \mathbf{D}}{\partial t} + \frac{4\pi}{c}\mathbf{j} + \frac{4\pi}{c}\mathbf{j}_{cm}, \tag{1.1}$$

$$\mathrm{rot}\mathbf{E} = -\frac{1}{c}\frac{\partial \mathbf{B}}{\partial t}, \tag{1.2}$$

$$\mathrm{div}\mathbf{D} = 4\pi\rho, \tag{1.3}$$

$$\mathrm{div}\mathbf{B} = 0. \tag{1.4}$$

The names of the electromagnetic quantities appearing in (1.1)–(1.4) are given in Table 1.1.

Functions $\mathbf{E} = \mathbf{E}(\mathbf{r}, t)$, $\mathbf{H} = \mathbf{H}(\mathbf{r}, t)$, $\mathbf{D} = \mathbf{D}(\mathbf{r}, t)$, $\mathbf{B} = \mathbf{B}(\mathbf{r}, t)$ describe the electromagnetic field in an environment characterized by parameters $\varepsilon = \varepsilon\,(\mathbf{E}, \mathbf{r}, t)$, $\mu = \mu(\mathbf{H}, \mathbf{r}, t)$, $\rho = (\mathbf{r}, t)$, $\mathbf{j} = \mathbf{j}(\mathbf{E}, \mathbf{r}, t)$ ($\mathbf{r}$ are the spatial coordinates, t is time), and external current $\mathbf{j}_e$, the use of which will be stipulated separately.

Assuming that the processes are local and instantaneous (at each point the state is independent of neighbouring points and at each moment of time of 'prehistory'), we associate the characteristics of the field and the medium by material equations [1]

Table 1.1. Electromagnetic quantities in the Gaussian CGS system

Name	Designation
Charge	q
Current	I
Charge density	ρ
Current density	j
Conductivity	σ
Electric vector	E
Magnetic vector	H
Electric displacement	D
Magnetic induction	B
Permittivity	ε
Magnetic permeability	μ
Speed of light in vacuum	c

$$\mathbf{D} = \varepsilon \mathbf{E}, \tag{1.5}$$

$$\mathbf{B} = \mu \mathbf{H}, \tag{1.6}$$

$$\mathbf{j} = \sigma \mathbf{E}, \tag{1.7}$$

and the law of conservation of charge

$$\text{div}\mathbf{j} = -\frac{\partial \rho}{\partial t}. \tag{1.8}$$

It is also assumed that the paremeters of the medium are independent of the vectors of the field and do not change with time: $\varepsilon = \varepsilon(\mathbf{r})$, $\mu = \mu(\mathbf{r})$ (linear medium), are scalar (isotropic medium), the field does not cause polarization and magnetization of the medium.

If the electric and magnetic vectors can be expressed as $\mathbf{E} = \text{Re}\,(\mathbf{E}\exp(-i\omega t))$, $\mathbf{H} = \text{Re}\,(\mathbf{H}\exp(-i\omega t))$, where $\mathbf{E} = \mathbf{E}\,(\mathbf{r})$, $\mathbf{H} = \mathbf{H}\,(\mathbf{r})$ are the complex functions [1], ω is the cyclic frequency, i is the imaginary unit, we speak of a monochromatic field for which (1.1) and (1.2) take the form:

$$\text{rot}\mathbf{H} = -ik_0\dot{\varepsilon}\mathbf{E}, \tag{1.9}$$

$$\text{rot}\mathbf{E} = ik_0\mu\mathbf{H}, \tag{1.10}$$

where $\dot{\varepsilon} = \varepsilon - i\dfrac{\sigma}{\omega}$, $k_0 = \dfrac{\omega}{c} = \dfrac{2\pi}{\lambda}$ - the wave number.

1.1.3 Maxwell's equations in integral form

Integrating (1.1) (1.2) on the surface S, bounded by L, and applying the Stokes theorem, we obtain the equation:

$$\oint_L \mathbf{H}dl = \frac{1}{c}\frac{d}{dt}\int_S \mathbf{D}ds + \frac{4\pi}{c}\mathbf{I}, \tag{1.11}$$

$$\oint_L \mathbf{E}dl = -\frac{1}{c}\frac{\text{d}}{\text{d}t}\int_S \mathbf{B}ds. \tag{1.12}$$

Equations (1.3) (1.4) are integrated over the volume V, bounded by the surface S. Then, applying the Gauss–Ostrogradskii theorem, we obtain:

$$\oint_S (\mathbf{D},\mathbf{n})\,ds = 2\pi q, \tag{1.13}$$

$$\oint_S (\mathbf{B},\mathbf{n})\,ds = 0. \tag{1.14}$$

The system (1.11)–(1.14) is called the Maxwell equations in integral form.

1.1.4. Fields at interfaces

Applying Maxwell's equations in integral form for an infinitely small contours and volume at the interface between two media, we obtain the following boundary conditions [1] for the electromagnetic fields:

$$((\mathbf{D}_1-\mathbf{D}_2), \mathbf{e}_y) = 4\pi\xi, \tag{1.15}$$
$$((\mathbf{E}_1-\mathbf{E}_2), \mathbf{e}_z) = 0, \tag{1.16}$$
$$((\mathbf{B}_1-\mathbf{B}_2), \mathbf{e}_y) = 0, \tag{1.17}$$
$$((\mathbf{H}_1-\mathbf{H}_2), \mathbf{e}_z) = 4\pi(\boldsymbol{\eta}, \mathbf{i})c, \tag{1.18}$$

where $\xi = \lim\limits_{\Delta S\to 0} \frac{\Delta q}{\Delta S}$ is the surface charge density, $\eta = \lim\limits_{\Delta l\to 0} e_x \frac{\Delta \mathbf{I}}{\Delta l}$ is the density of the surface current (the plane separating media 1 and 2 perpendicular to the vector $\mathbf{e}_y$).

1.1.5. Poynting's theorem

Multiplying (1.1) by **E**, and (1.2) by **H**, we obtain:

$$(\mathbf{E},\mathrm{rot}\,\mathbf{H}) = \frac{1}{c}\left(\mathbf{E},\frac{\partial \mathbf{D}}{\partial t}\right) + \frac{4\pi}{c}(\mathbf{E},\mathbf{j}),$$

$$(\mathbf{H},\mathrm{rot}\,\mathbf{E}) = -\frac{1}{c}\left(\mathbf{H},\frac{\partial \mathbf{B}}{\partial t}\right).$$

Subtracting the second equation from the first, we obtain the Poynting theorem [1], in which

$$\mathrm{div}\left[\mathbf{E},\mathbf{H}\right] = -\frac{1}{c}\left(\left(\mathbf{H},\frac{\partial \mathbf{B}}{\partial t}\right) - \left(\mathbf{E},\frac{\partial \mathbf{D}}{\partial t}\right)\right) - \frac{4\pi}{c}(\mathbf{j},\mathbf{E}). \tag{1.19}$$

In the integral form

$$\frac{c}{4\pi}\oint_S \left(\left[\mathbf{E},\mathbf{H}\right],\mathbf{n}\right)ds = -\frac{1}{4\pi}\int_V \left(\left(\mathbf{H},\frac{\partial \mathbf{B}}{\partial t}\right) + \left(\mathbf{E},\frac{\partial \mathbf{D}}{\partial t}\right)\right)dv - \int_V (\mathbf{j},\mathbf{E})\,dv \tag{1.20}$$

we have the energy balance equation of the electromagnetic field in the volume V. The energy in the volume V is $W = \frac{1}{8\pi}\int_V \left((\mathbf{H},\mathbf{B}) + (\mathbf{E},\mathbf{D})\right)dv$, the consumed power

$P=\int_V (\mathbf{j},\mathbf{E})dv$, and $\mathbf{\Pi}=\frac{c}{4\pi}[\mathbf{E},\mathbf{H}]$ is the Umov–Poynting vector indicating the direction of energy movement and equal in magnitude to the density of its flux.

The monochromatic field is described using the complex Umov–Poynting vector $\Pi=\frac{c}{8\pi}\left[\mathbf{E},\mathbf{H}^*\right]$, where the asterisk denotes complex conjugation, and the average value of the Umov–Poynting vector is equal to the real part of the complex.

1.2. Differential equations of optics

1.2.1. The wave equation

In Maxwell's equations, we eliminate from consideration the currents and charges which usually absent in the problems of optics. Then, equations (1.1) and (1.2) take the form:

$$\operatorname{rot}\mathbf{H}=\frac{\varepsilon}{c}\frac{\partial \mathbf{E}}{\partial t}, \tag{1.21}$$

$$\operatorname{rot}\mathbf{E}=\frac{\mu}{c}\frac{\partial \mathbf{H}}{\partial t}. \tag{1.22}$$

Divide both sides of (1.22) by μ and apply the operator rot

$$\operatorname{rot}\left(\frac{1}{\mu}rot\mathbf{E}\right)+\frac{1}{c}\operatorname{rot}\frac{\partial \mathbf{H}}{\partial t}=0. \tag{1.23}$$

Equation (1.21) is differentiable with respect to time in order to eliminate the second term of equation (1.23):

$$\operatorname{rot}\left(\frac{1}{\mu}\operatorname{rot}\mathbf{E}\right)+\frac{\varepsilon}{c^2}\frac{\partial^2 \mathbf{E}}{\partial t^2}=0.$$

Then, given that

$$\operatorname{rot}\alpha u=\alpha\operatorname{rot}u+\left[\operatorname{grad}\alpha,u\right] \text{ and}$$

we obtain:

$$\nabla^2\mathbf{E}-\frac{\varepsilon\mu}{c^2}\frac{\partial^2 \mathbf{E}}{\partial t^2}+\left(\operatorname{grad}\left(\ln\mu\right)\right)\times\operatorname{rot}\mathbf{E}-\operatorname{grad}\operatorname{div}\mathbf{E}=0 \tag{1.24}$$

To the equation div(εE) = 0 we apply the identity div $\alpha u = \alpha$ div u + (u, grad α),and obtain ε div **E** + (**E**, grad ε) = 0. Expressing from the last equation div**E**, we substitute it into (1.24), writing the wave equation [1] for the electric field in an inhomogeneous dielectric medium

$$\nabla^2\mathbf{E}-\frac{\varepsilon\mu}{c^2}\frac{\partial^2 \mathbf{E}}{\partial t^2}+\left[\operatorname{grad}\left(\ln\mu\right),\operatorname{rot}\mathbf{E}\right]+\operatorname{grad}\left(\mathbf{E},\operatorname{grad}\left(\ln\varepsilon\right)\right)=0. \tag{1.25}$$

Similarly, we obtain the wave equation for the magnetic field vector **H**:

$$\nabla^2\mathbf{H} - \frac{\varepsilon\mu}{c^2}\frac{\partial^2\mathbf{H}}{\partial t^2} + \left[\operatorname{grad}(\ln\varepsilon), \operatorname{rot}\mathbf{H}\right] + \operatorname{grad}\left(\mathbf{H}, \operatorname{grad}(\ln\mu)\right) = 0. \quad (1.26)$$

For a homogeneous medium, electric ε and magnetic μ permeability are constant and the wave equations take the form

$$\nabla^2\mathbf{E} - \frac{\varepsilon\mu}{c^2}\frac{\partial^2\mathbf{E}}{\partial t^2} = 0, \quad (1.27)$$

$$\nabla^2\mathbf{H} - \frac{\varepsilon\mu}{c^2}\frac{\partial^2\mathbf{H}}{\partial t^2} = 0. \quad (1.28)$$

1.2.2. Helmholtz equations

The wave equations written for the complex amplitudes (monochromatic waves), called the Helmholtz equation. For an inhomogeneous medium, they have the form:

$$\nabla^2\mathbf{E} + k_0^2\varepsilon\mu\mathbf{E} + \left[\operatorname{grad}(\ln\mu), \operatorname{rot}\mathbf{E}\right] + \operatorname{grad}\left(\mathbf{E}, \operatorname{grad}(\ln\varepsilon)\right) = 0, \quad (1.29)$$

$$\nabla^2\mathbf{H} + k_0^2\varepsilon\mu\mathbf{H} + \left[\operatorname{grad}(\ln\varepsilon), \operatorname{rot}\mathbf{H}\right] + \operatorname{grad}\left(\mathbf{H}, \operatorname{grad}(\ln\mu)\right) = 0, \quad (1.30)$$

and for a homogeneous one

$$\nabla^2\mathbf{E} + k_0^2\varepsilon\mu\mathbf{E} = 0, \quad (1.31)$$

$$\nabla^2\mathbf{H} + k_0^2\varepsilon\mu\mathbf{H} = 0. \quad (1.32)$$

Equations (1.31) and (1.32) can be solved independently for each projection of the electric and magnetic vectors **E** and **H**, and these projections can be described by a single scalar function U:

$$\nabla^2 U + k_0^2\varepsilon\mu U = 0. \quad (1.33)$$

1.2.3. The Fock–Leontovich equation

We represent the function U as $U = U\exp(ik_0 z)$ and substitute it into equation (1.33) for the vacuum. Assuming that $\left|\frac{\partial^2 U}{\partial z^2}\right| << k_0\left|\frac{\partial U}{\partial z}\right|$, we obtain the Fock–Leontovich parabolic wave equation

$$2ik_0\frac{\partial U}{\partial z} + \Delta_\perp U = 0, \quad (1.34)$$

where $\Delta_\perp U = \frac{\partial^2 U}{\partial x^2} + \frac{\partial^2 U}{\partial y^2}$.

The parabolic equation (1.34) in the scalar optics is used to describe paraxial optical fields, which are distributed mainly along a certain direction in space in a small solid angle.

1.2.4. Eikonal and transport equations

We write the function U as $U = U_0 \exp(ik_0\psi)$, where $\psi = \psi(x, y, z)$ – eikonal, U_0 is the amplitude (real function). Substituting it into (1.33), we obtain:

$$\frac{\partial^2 U_0}{\partial x^2}+\frac{\partial U^2{}_0}{\partial y^2}+\frac{\partial^2 U_0}{\partial z^2}+2ik_0\left(\frac{\partial\psi}{\partial x}\frac{\partial U_0}{\partial x}+\frac{\partial\psi}{\partial y}\frac{\partial U_0}{\partial y}+\frac{\partial\psi}{\partial z}\frac{\partial U_0}{\partial z}\right)+$$

$$ik_0U_0\left(\frac{\partial^2\psi}{\partial x^2}+\frac{\partial^2\psi}{\partial y^2}+\frac{\partial^2\psi}{\partial z^2}\right)-$$

$$-k_0^2U_0\left(\left(\frac{\partial\psi}{\partial x}\right)^2+\left(\frac{\partial\psi}{\partial y}\right)^2+\left(\frac{\partial\psi}{\partial z}\right)^2\right)+k_0^2\varepsilon\mu U_0=0.$$

Equating to zero the imaginary part, we obtain the transport equation:

$$2\left(\frac{\partial\psi}{\partial x}\frac{\partial U_0}{\partial x}+\frac{\partial\psi}{\partial y}\frac{\partial U_0}{\partial y}+\frac{\partial\psi}{\partial z}\frac{\partial U_0}{\partial z}\right)+U_0\left(\frac{\partial^2\psi}{\partial x^2}+\frac{\partial^2\psi}{\partial y^2}+\frac{\partial^2\psi}{\partial z^2}\right)=0. \quad (1.35)$$

The remaining terms amount to the following equation:

$$\frac{1}{k_0^2}\left(\frac{\partial^2 U_0}{\partial x^2}+\frac{\partial U^2{}_0}{\partial y^2}+\frac{\partial^2 U_0}{\partial z^2}\right)-U_0\left(\left(\frac{\partial\psi}{\partial x}\right)^2+\left(\frac{\partial\psi}{\partial y}\right)^2+\left(\frac{\partial\psi}{\partial z}\right)^2\right)+\varepsilon\mu U_0=0,$$

from which, putting $\lambda\to 0$ (geometrical optics approximation), we obtain the eikonal equation,

$$\left(\frac{\partial\psi}{\partial x}\right)^2+\left(\frac{\partial\psi}{\partial y}\right)^2+\left(\frac{\partial\psi}{\partial z}\right)^2=n^2, \quad (1.36)$$

where $n=\sqrt{\varepsilon\mu}$ is the refractive index of the medium.

1.3. Integral theorems of optics

Analysis of the electromagnetic field can be carried out not only by means of differential equations of Maxwell, Helmholtz, and others, but also with the help of equivalent integral equations and transformations. In this case, the Maxwell equations for monochromatic light in a homogeneous region of space are equivalent to the Stratton–Chu vector integral equations. The solution of the differential Helmholtz equation is convenient to study with the help of the Kirchhoff–Helmholtz integral expression (third Green's formula), and the Fock–Leontovich paraxial equation is equivalent to the Fresnel integral transform.

1.3.1. Green's formulas

For two continuous functions unctions $u(x, y, z)$ and $v(x, y, z)$ together with their derivatives in region V, bounded by a piecewise smooth surface S, there is the second Green formula [1]:

$$\int_V (u\Delta v - v\Delta u)dV = \oint_S (u\frac{\partial v}{\partial \mathrm{n}} - v\frac{\partial u}{\partial \mathrm{n}})dS. \tag{1.37}$$

where n is the vector of the outer normal to surface S, $\Delta = \nabla^2 = \frac{\partial^2}{\partial x^2} + \frac{\partial^2}{\partial y^2} + \frac{\partial^2}{\partial z^2}$ is the Laplace operator or Laplacian.

With the help of Green's formula (1.37) the solution of the Helmholtz equation at interior points of a homogeneous region V can be expressed in terms of values of the solution and its derivatives on the boundary S using the third Greens formula (Helmholtz–Kirchhoff integral [2])

$$u(x) = \frac{1}{4\pi}\oint_S \left\{ \frac{\partial u(x')}{\partial n}\frac{e^{ikR}}{R} - u(x')\frac{\partial}{\partial n}\left(\frac{e^{ikR}}{R}\right) \right\} dS, \tag{1.38}$$

where R is the distance between points $\mathbf{x} \in V$ and $\mathbf{x}' \in S$, $u\ (x, y, z)$ is the solution of the Helmholtz equation in a homogeneous space

$$(\Delta + k^2)u(x, y, z) = 0,$$

$k = 2\pi/\lambda$ is the wave number of light with wavelength λ.

The function

$$G = \frac{e^{ikR}}{4\pi R} \tag{1.39}$$

describes a spherical wave, is the Green function of a homogeneous space and satisfies the inhomogeneous Helmholtz equation with a point source

$$(\Delta + k^2)G(x, x') = \delta(x - x'). \tag{1.40}$$

In regions of space with a constant refractive index and without sources, the integral representation (1.38) holds for any Cartesian component of the vector of the strength of the electric field

$$\mathbf{E}(\mathbf{x}) = \frac{1}{4\pi}\oint_S \left\{ \frac{\partial \mathbf{E}(\mathbf{x}')}{\partial \mathbf{n}}\frac{\mathrm{e}^{ikR}}{R} - \mathbf{E}(\mathbf{x}')\frac{\partial}{\partial \mathbf{n}}\left(\frac{\mathrm{e}^{ikR}}{R}\right) \right\} dS. \tag{1.41}$$

Diffraction of scalar waves on a dielectric object

For example, consider the scalar problem of diffraction of electromagnetic waves in a homogeneous dielectric object [3]. Let the function $E_1(x)$ and $E_2\ (x)$ satisfy the two Helmholtz equations inside the region V (inside the object) and on the outside:

$$\begin{aligned} &(\Delta + k_1^2)E_1(\mathbf{x}) = 0, \quad \mathbf{x} \in V, \\ &(\Delta + k_2^2)E_2(\mathbf{x}) = -g, \quad \mathbf{x} \notin V + S \end{aligned} \tag{1.42}$$

the boundary conditions

$$\begin{aligned} &E_1(\mathbf{x})\,|_S = E_2(\mathbf{x})\,|_S, \\ &\frac{\partial E_1(\mathbf{x})}{\partial \mathbf{n}}\,|_S = \frac{\partial E_2(\mathbf{x})}{\partial \mathbf{n}}\,|_S \end{aligned} \tag{1.43}$$

and the Sommerfeld radiation conditions at infinity

$$\frac{\partial E_2(\mathbf{x})}{\partial \mathbf{n}} - ik_2 E_2(\mathbf{x}) = o\left(\frac{1}{r}\right), \quad r \to \infty. \tag{1.44}$$

where $o(x)$ is a function whose order of magnitude is larger than x when $x \to 0$.

In equations (1.42), the function g describes the density of light sources outside the region V, occupied by an object; there are no sources within the object. With the help of Green's theorem (1.37) and (1.38) the solution of (1.42) with the conditions taken into account (1.43) and (1.44) can be reduced to solving a Fredholm integral equation of the second kind

$$E_1(\mathbf{x}) = \frac{k_1^2 - k_2^2}{4\pi} \int_V E_1(\mathbf{x}') \frac{e^{ik_2 R}}{R} dV + \frac{1}{4\pi} \int_{V'} g(\mathbf{y}) \frac{e^{ik_2 R}}{R} dV, \quad \mathbf{x} \in V, \tag{1.45}$$

where $\mathbf{x}$ and $\mathbf{x}'$ belong to the object region V, and point $\mathbf{y}$ belongs to V', external to the region V. The second term in equation (1.45) describes the complex amplitude of the light incident on the object field, which in diffraction problems can be regarded as a known function:

$$E_0(\mathbf{x}) = \frac{1}{4\pi} \int_{V'} g(\mathbf{y}) \frac{e^{ik_2 R}}{R} dV. \tag{1.46}$$

Solving the equation (1.45), the diffraction field outside the object (in region V') we find, using the integral transform

$$E_2(\mathbf{x}) = \frac{k_1^2 - k_2^2}{4\pi} \int_V E_1(\mathbf{x}') \frac{e^{ik_2 R}}{R} dV + E_0(\mathbf{x}), \quad \mathbf{x} \in V'. \tag{1.47}$$

Equations (1.45) and (1.47) for the two-dimensional problem ($\partial/\partial z = 0$) have the form:

$$E_1(x,y) = \frac{i(k_1^2 - k_2^2)}{4} \int_V E_1(x',y') H_0^{(2)}(k_2 r) dV + E_0(x,y), \quad (x,y) \in V, \tag{1.48}$$

$$r = \left[(x-x')^2 + (y-y')^2\right]^{1/2},$$

$$E_2(x,y) = \frac{i(k_1^2 - k_2^2)}{4} \int_V E_1(x',y') H_0^{(2)}(k_2 r) dV + E_0(x,y), \quad (x,y) \in V', \tag{1.49}$$

where $H_0^{(2)}(x)$ is the Hankel function of second kind of zeroth order, $G(x, y; x', y') = i/4 H_0^{(2)}(kr)$ is the Green function of a homogeneous space for a two-dimensional Helmholtz equation.

Equations (1.48) and (1.49) solve the problem of diffraction of a cylindrical (two-dimensional) electromagnetic wave with TE-polarization (E_0, E_1 and E_2 are projections on the z-axis of the vectors of strength of the electric field) on a uniform cylindrical dielectric object. Similar formulas for TM-polarization can be found in [3].

1.3.2. Stratton–Chu formula

Green vector formulas can be derived by the same procedure by Green's scalar formulas (1.37) and (1.38). The Gauss–Ostrogradskii equation is used:

$$\int_V \mathrm{div}\mathbf{F}dV = \oint_S \mathbf{Fn}dS. \tag{1.50}$$

If the vector field $\mathbf{F}$ as a vector product $\mathbf{F} = [\mathbf{P}, \text{rot } \mathbf{Q}]$ is substituted into (1.50), we can obtain the vector analogue of Green's second formula:

$$\int_V (\mathbf{Q} \text{ rot rot } \mathbf{P} - \mathbf{P} \text{ rot rot } \mathbf{Q})\, d = \oint_S \{[\mathbf{P}, \text{ rot } \mathbf{Q}] - [\mathbf{Q}, \text{rot } \mathbf{P}]\}\, \mathbf{n}\, dS. \tag{1.51}$$

Given the known vector relations

$$\mathbf{Q} \text{ rot rot } \mathbf{P} - \mathbf{P} \text{ rot rot } \mathbf{Q} = \mathbf{P}\Delta Q - \mathbf{Q}\Delta\mathbf{P} + \mathbf{Q} \text{ grad div } \mathbf{P} - \mathbf{P} \text{ grad di} v\, \mathbf{Q} =$$
$$= \mathbf{P}\Delta\mathbf{Q} - \mathbf{Q}\Delta\mathbf{P} + \text{div}\,(\mathbf{Q} \text{ div } \mathbf{P} - \mathbf{P} \text{ div } \mathbf{Q})$$

equation (1.51) can be rewritten as:

$$\int_V (\mathbf{P}\Delta\mathbf{Q} - \mathbf{Q}\Delta\mathbf{P})\, \mathrm{d}V = \oint_S \{\mathbf{n}[\mathbf{P}, \text{rot } \mathbf{Q}] - \mathbf{n}[\mathbf{Q}, \text{ rot } \mathbf{P}] + \mathbf{nP} \text{ div } \mathbf{Q} - \mathbf{nQ} \text{ div } \mathbf{P}\}\, \mathrm{d}S. \tag{1.52}$$

From (1.51) and (1.52) we can easily obtain the integral relations for the electromagnetic field in space. Suppose $\mathbf{P} = \mathbf{E}$, $\mathbf{Q} = \mathbf{a}G$. $G(\mathbf{x} - \mathbf{x}_0) = \exp(ikR)/R$, $\mathbf{a}$ is the unit vector of arbitrary direction, where $R = |\mathbf{x} - \mathbf{x}_0|$, $\mathbf{x}$ is the radius vector of the observation point, $\mathbf{x}_0$ is a point on the surface S. In this case, the function $\mathbf{Q}$ satisfies the vector Helmholtz equation with a point source:

$$\Delta\mathbf{Q} + k^2\mathbf{Q} = -4\pi\mathbf{a}\delta(\mathbf{x} - \mathbf{x}_0), \tag{1.53}$$

where $\text{rot } \mathbf{Q} = [\text{grad } G, \mathbf{a}]$, $\text{div } \mathbf{Q} = (\mathbf{a}, \text{ grad } G)$, $k^2 = \omega^2\varepsilon\mu/c^2$.

The vector of the strength of the electric and magnetic fields of the monochromatic light wave in a homogeneous and isotropic space satisfy the inhomogeneous Helmholtz equations:

$$\Delta\mathbf{E} + \frac{\omega^2\varepsilon\mu}{c^2}\mathbf{E} = -4\pi\mathbf{J}_2,$$
$$\mathbf{J}_2 = \frac{i\omega\mu}{c^2}\mathbf{j} + \frac{i\,\text{grad div } \mathbf{j}}{\omega\varepsilon}, \tag{1.54}$$
$$\Delta\mathbf{H} + \frac{\omega^2\varepsilon\mu}{c^2}\mathbf{H} = -\text{rot}\,\mathbf{j},$$

where $\mathbf{j}$ the density of secondary electric current, ω is the cyclic oscillation frequency of monochromatic light, ε, μ is the dielectric constant and magnetic permeability of the homogeneous medium, c is the speed of light in vacuum.

Using (1.52)–(1.54) and the formula $(\mathbf{n}, [\mathbf{E}, [\text{grad } G, \mathbf{a}]]) = (\mathbf{a}, [\text{grad } G, [\mathbf{E}, \mathbf{n}]])$, we obtain the expression:

$$\mathbf{E}(x)\mathbf{a} = \mathbf{a}\left\{ \int_V \mathbf{J}_2(\mathbf{x}_0)\, G(\mathbf{x}-\mathbf{x}_0)\, dV + \frac{1}{4\pi}\oint_S \{\mathbf{n}\, G(\mathbf{x}-\mathbf{x}_0)\, \mathrm{div}_0\, \mathbf{E}(\mathbf{x}_0)\} dS + \right.$$
$$+\frac{1}{4\pi}\oint_S \{[\mathrm{grad}_0\, G(\mathbf{x}-\mathbf{x}_0), [\mathbf{n}, \mathbf{E}(\mathbf{x}_0)]] + [\mathrm{rot}_0\, \mathbf{E}(\mathbf{x}_0), \mathbf{n}]\, G(\mathbf{x}-\mathbf{x}_0) -$$
$$\left. -\mathrm{grad}_0\, G(\mathbf{x}-\mathbf{x}_0)(\mathbf{n}, \mathbf{E}(\mathbf{x}_0))\}\, dS \right\}. \tag{1.55}$$

Given that div $\mathbf{E} = 0$, and the arbitrariness of the vector $\mathbf{a}$, we obtain the integral representation

$$\mathbf{E}(\mathbf{x}) = \mathbf{E}_0(\mathbf{x}) +$$
$$+\frac{1}{4\pi}\oint_S \{[\mathrm{grad}_0\, G(\mathbf{x}-\mathbf{x}_0), [\mathbf{n}, \mathbf{E}]] + [\mathrm{rot}_0\, \mathbf{E}, \mathbf{n}]G(\mathbf{x}-\mathbf{x}_0) - \mathrm{grad}_0\, G(\mathbf{x}-\mathbf{x}_0)(\mathbf{nE})\}\, dS, \tag{1.56}$$

where

$$\mathbf{E}_0(\mathbf{x}) = \int_V \mathbf{J}_2(\mathbf{x}_0)\, G(\mathbf{x}-\mathbf{x}_0)\, dV.$$

By analogy with (1.56) we can obtain an integral representation for the magnetic field

$$\mathbf{H}(\mathbf{x}) = \mathbf{H}_0(\mathbf{x}) + \frac{1}{4\pi}\times$$
$$\times\oint_S \{[\mathrm{grad}_0\, G(\mathbf{x}-\mathbf{x}_0), [\mathbf{n}, \mathbf{H}(x_0)]] + [\mathrm{rot}_0\, \mathbf{H}(\mathbf{x}_0), \mathbf{n}]\, G(\mathbf{x}-\mathbf{x}_0) - \mathrm{grad}_0\, G(\mathbf{x}-\mathbf{x}_0)(\mathbf{nH}(\mathbf{x}_0))\}\, dS, \tag{1.57}$$

where $\mathbf{H}_0(\mathbf{x}) = \frac{1}{4\pi}\int_V \mathrm{rot}\ \mathbf{j}\ G(\mathbf{x}-\mathbf{x}_0)\, dV.$

In (1.56) and (1.57) $\mathbf{E}_0(\mathbf{x})$ and $\mathbf{H}_0(\mathbf{x})$ are the strengths of the electric and magnetic fields in the incident wave.

Equations (1.56) and (1.57) are called the Stratton–Chu formulas.

Diffraction on a perfectly reflecting object

For example, consider the solution of a problem of electromagnetic wave diffraction by an ideally reflecting object, which occupies a region of space V, with the surface S.

We introduce the notation for the surface density of electric and magnetic currents: $(4\pi/c)\mathbf{j}_e(\mathbf{x}_0) = [\mathbf{n}, \mathbf{H}(\mathbf{x}_0)]$, $(4\pi/c)\mathbf{j}_m(\mathbf{x}_0) = [\mathbf{n}, \mathbf{E}(\mathbf{x}_0)]$. We take into account that rot $(\Phi\mathbf{F}) = \Phi$ rot $\mathbf{F}$ +[grad Φ, $\mathbf{F}$], rot $j_m(\mathbf{x}_0) = 0$, rot $\mathbf{E} = ik\mathbf{H}$, where $k = \frac{\omega}{c}\sqrt{\varepsilon\mu}$,

$$[\mathrm{grad}_0\, G(\mathbf{x}-\mathbf{x}_0), \mathrm{j}_m(\mathbf{x}_0)] = -[\mathrm{grad}\ G(\mathbf{x}-\mathbf{x}_0), \mathrm{j}_m(\mathbf{x}_0)] =$$
$$= -\mathrm{rot}\left(G(\mathbf{x}-\mathbf{x}_0)\mathrm{j}_m(\mathbf{x}_0)\right) + G(\mathbf{x}-\mathbf{x}_0)\ \mathrm{rot}\ \mathrm{j}_m(\mathbf{x}_0).$$

As a result, instead of (1.56) we obtain the following representation for the electric field

$$\mathbf{E}(\mathbf{x}) = \mathbf{E}_0(\mathbf{x}) - \frac{1}{4\pi}\,\mathrm{rot}\oint_S \mathrm{j}_m(\mathbf{x}_0)\,G(\mathbf{x}-\mathbf{x}_0)\,\mathrm{d}S - \\ -\frac{ik}{4\pi}\oint_S \mathrm{j}_e(\mathbf{x}_0)\,G(\mathbf{x}-\mathbf{x}_0)\,\mathrm{d}S + \frac{1}{4\pi}\,\mathrm{grad}\oint_S G(\mathbf{x}-\mathbf{x}_0)\,(\mathbf{n},\mathbf{E}(\mathbf{x}_0))\,\mathrm{d}S. \quad (1.58)$$

Applying the operation rot to both sides of (1.58) and taking into account that rot (grad Φ) = 0, we obtain the following representation for the magnetic field:

$$\mathbf{H}(\mathbf{x}) = \mathbf{H}_0(\mathbf{x}) - \frac{1}{4\pi ik}\,\mathrm{rot}\,\mathrm{rot}\oint_S \mathbf{j}_m(\mathbf{x}_0)\,G(\mathbf{x}-\mathbf{x}_0)\,dS - \frac{1}{4\pi}\,\mathrm{rot}\oint_S \mathbf{j}_e(\mathbf{x}_0)\,G(\mathbf{x}-\mathbf{x}_0)\,dS. \quad (1.59)$$

From equation (1.57) by analogy we obtain an integral representation for the electric field

$$\mathbf{E}(\mathbf{x}) = \mathbf{E}_0(\mathbf{x}) + \frac{1}{4\pi ik}\,\mathrm{rot}\,\mathrm{rot}\oint_S \mathbf{j}_e(\mathbf{x}_0)\,G(\mathbf{x}-\mathbf{x}_0)\,dS - \frac{1}{4\pi}\,\mathrm{rot}\oint_S \mathbf{j}_m(\mathbf{x}_0)\,G(\mathbf{x}-\mathbf{x}_0)\,dS. \quad (1.60)$$

Given the boundary conditions on the perfectly conducting surface $[\mathbf{n},\mathbf{E}] = 0$, $(\mathbf{n}, \mathbf{H}) = 0$, equations (1.59) and (1.60) can be rewritten as:

$$\mathbf{H}(\mathbf{x}) = \mathbf{H}_0(\mathbf{x}) - \frac{1}{4\pi}\,\mathrm{rot}\oint_S \mathbf{j}_e(\mathbf{x}_0)\,G(\mathbf{x}-\mathbf{x}_0)\,dS, \quad (1.61)$$

$$\mathbf{E}(\mathbf{x}) = \mathbf{E}_0(\mathbf{x}) + \frac{1}{4\pi ik}\,\mathrm{rot}\,\mathrm{rot}\oint_S \mathbf{j}_e(\mathbf{x}_0)\,G(\mathbf{x}-\mathbf{x}_0)\,dS. \quad (1.62)$$

To obtain an integral equation of the first kind for the electric current density on the surface of an ideal conductor, we assume that the vector **x** belongs to the surface. Multiplying (1.61) by the vector of the normal at point **x** and taking into account the boundary condition for a perfect conductor, we obtain the integral equation:

$$\left[\mathbf{E}_0(\mathbf{x}),\mathbf{n}(\mathbf{x})\right] = -\frac{1}{4\pi ik}\,\mathrm{rot}\,\mathrm{rot}\oint_S \left[\mathbf{j}_e(\mathbf{x}_0),\,\mathbf{n}(\mathbf{x})\right]G(\mathbf{x}-\mathbf{x}_0)\,dS. \quad (1.63)$$

Thus, the problem of finding the electromagnetic field is divided into two stages:

1) the solution of the integral equation (1.63) with respect $\mathbf{j}_e(\mathbf{x}_0)$;
2) the calculation of the field components from (1.61) and (1.62).

From equation (1.61) we can similarly obtain the Fredholm integral equation of the first kind for the unknown current density on the surface S in terms of known values of the magnetic field of the incident wave:

$$\mathbf{H}_0(\mathbf{x}) = \frac{1}{4\pi}\,\mathrm{rot}\oint_S \mathbf{j}_e(\mathbf{x}_0)\,G(\mathbf{x}-\mathbf{x}_0)\,dS, \quad \mathbf{x} \in S. \quad (1.64)$$

Diffraction on a transmitting object

Consider the solution of the problem of diffraction of an electromagnetic monochromatic wave on a homogeneous dielectric object. For this we consider Maxwell's equations in a homogeneous area of the object V_1 with the characteristics ε_1 and μ and also in the outer region V_2 with the characteristics of the medium ε_2 and μ:

$$\begin{aligned} \text{rot}\mathbf{H}_1 &= -i\frac{\omega\varepsilon_1}{c}\mathbf{E}_1, \\ \text{rot}\mathbf{E}_1 &= i\frac{\omega\mu}{c}\mathbf{H}_1, \quad \mathbf{x}\in V_1, \end{aligned} \tag{1.65}$$

$$\begin{aligned} \text{rot}\mathbf{H}_2 &= -i\frac{\omega\varepsilon_2}{c}\mathbf{E}_2 + \frac{4\pi}{c}\mathbf{j}, \\ \text{rot}\mathbf{E}_2 &= i\frac{\omega\mu}{c}\mathbf{H}_2, \quad \mathbf{x}\in V_2, \end{aligned} \tag{1.66}$$

with the boundary conditions on the surface S of the interface of the media V_1 and V_2

$$\begin{aligned} [\mathbf{n},\mathbf{E}_1]\,|_S &= [\mathbf{n},\mathbf{E}_2]\,|_S, \\ [\mathbf{n},\mathbf{H}_1]\,|_S &= [\mathbf{n},\mathbf{H}_2]\,|_S, \end{aligned} \tag{1.67}$$

and with the radiation condition at infinity

$$[\mathbf{n},\mathbf{E}_2]+[\mathbf{n}[\mathbf{n},\mathbf{H}_2]] = o\left(\frac{1}{r}\right), \quad r\to\infty \tag{1.68}$$

With the Green's vector formula (1.52) we can obtain a Fredholm integral equation of the second kind for the magnetic field strength

$$\begin{aligned} \mathbf{H}_1(\mathbf{x}) &= \frac{\varepsilon_2}{c\varepsilon_1}\int_{V_2} G(\mathbf{x}-\mathbf{x}_0)\text{rot}\mathbf{j}\,dV + \frac{\omega^2\mu(\varepsilon_2-\varepsilon_1)}{4\pi c^2}\int_{V_1} G(\mathbf{x}-\mathbf{x}_0)\,\mathbf{H}_1(\mathbf{x}_0)dV + \\ &+\frac{\varepsilon_2-\varepsilon_1}{4\pi\,\varepsilon_1}\oint_S \{[\text{grad}G(\mathbf{x}-\mathbf{x}_0)[\mathbf{n},\mathbf{H}_1]] - (\mathbf{n},\mathbf{H}_1)\text{grad}G(\mathbf{x}-\mathbf{x}_0)\}\,dS, \quad \mathbf{x}\in V_1. \end{aligned} \tag{1.69}$$

The first term in equation (1.69) can be regarded as a known field incident on the object

$$\mathbf{H}_0(\mathbf{x}) = \frac{\varepsilon_2}{c\,\varepsilon_1}\int_{V_2} G(\mathbf{x}-\mathbf{x}_0)\text{rot}\mathbf{j}\,dV,$$

and the impulse response function $G(\mathbf{x}-\mathbf{x}_0)$ satisfies the equation (1.53).

The magnetic field in the outer region V_2 after solving (1.69) is determined by the integral transform

$$\begin{aligned} \mathbf{H}_2(\mathbf{x}) &= \frac{\varepsilon_1}{\varepsilon_2}\mathbf{H}_0(\mathbf{x}) + \frac{\omega^2\varepsilon_1\mu(\varepsilon_2-\varepsilon_1)}{4\pi c^2\varepsilon_2}\int_{V_1} G(\mathbf{x}-\mathbf{x}_0)\,\mathbf{H}_1(\mathbf{x}_0)dV + \\ &+\frac{\varepsilon_2-\varepsilon_1}{4\pi\,\varepsilon_2}\oint_S \{[\text{grad}G(\mathbf{x}-\mathbf{x}_0)[\mathbf{n},\mathbf{H}_1]] - (\mathbf{n},\mathbf{H}_1)\text{grad}G(\mathbf{x}-\mathbf{x}_0)\}\,dS, \quad \mathbf{x}\in V_2. \end{aligned} \tag{1.70}$$

The strengths of the electric field $\mathbf{E}_1$ and $\mathbf{E}_2$ are located across the known functions of $\mathbf{H}_1$ and $\mathbf{H}_2$ from the Maxwell equations (1.65) and (1.66).

Instead of (1.69) and (1.70), to find the magnetic field of diffraction we can use Green's vector formula (1.52) to obtain the Fredholm integral equation of the second kind to find the electric vector of the diffraction field

$$\mathbf{E}_1(\mathbf{x}) = \frac{-\varepsilon_2}{4\pi\,\varepsilon_1}\int\limits_{V_2} G(\mathbf{x}-\mathbf{x}_0)\left\{\frac{i\omega\mu}{c^2}\mathbf{j} - \frac{i}{\omega\varepsilon_1}\operatorname{grad}\operatorname{div}\mathbf{j}\right\}dV -$$
$$-\frac{\omega^2\mu(\varepsilon_2-\varepsilon_1)}{4\pi\,c^2}\int\limits_{V_1} G(\mathbf{x}-\mathbf{x}_0)\,\mathbf{E}_1(\mathbf{x}_0)dV + \tag{1.71}$$
$$+\frac{\varepsilon_2-\varepsilon_1}{4\pi\,\varepsilon_1}\oint\limits_{S}\{[\operatorname{grad}G(\mathbf{x}-\mathbf{x}_0)[\mathbf{n},\mathbf{E}_1]]-(\mathbf{n},\mathbf{E}_1)\operatorname{grad}G(\mathbf{x}-\mathbf{x}_0)\}dS,\quad \mathbf{x}\in V_1,$$

where the known vector of the strength of the electric field of the incident wave is expressed in terms of current density in the outer region:

$$\mathbf{E}_0(\mathbf{x}) = \frac{-\varepsilon_2}{4\pi\varepsilon_1}\int\limits_{V_2}\left\{\frac{i\omega\mu}{c^2}\mathbf{j} - \frac{i}{\omega\varepsilon_1}\operatorname{grad}\,\operatorname{div}\mathbf{j}\right\}G(\mathbf{x}-\mathbf{x}_0)dV. \tag{1.72}$$

The vector of the strength electric diffraction field in the outer region V_2 is found by solving (1.71) and the integral transformation

$$\mathbf{E}_2(\mathbf{x}) = \frac{\varepsilon_1}{\varepsilon_2}\mathbf{E}_0(\mathbf{x}) - \frac{\omega^2\varepsilon_1\,\mu(\varepsilon_2-\varepsilon_1)}{4\pi\,c^2\varepsilon_2}\int\limits_{V_1} G(\mathbf{x}-\mathbf{x}_0)\,\mathbf{E}_1(\mathbf{x}_0)dV + \tag{1.73}$$
$$+\frac{\varepsilon_2-\varepsilon_1}{4\pi\,\varepsilon_2}\oint\limits_{S}\{[\operatorname{grad}G(\mathbf{x}-\mathbf{x}_0)[\mathbf{n},\mathbf{E}_1]]-(\mathbf{n},\mathbf{E}_1)\operatorname{grad}G(\mathbf{x}-\mathbf{x}_0)\}dS,\quad \mathbf{x}\in V_2.$$

1.4. Integral transformations in optics

In the framework of the scalar theory of diffraction monochromatic light is described by the complex amplitude function $F(\mathbf{x}) = F(x, y, z)$, which satisfies the Helmholtz equation (1.33):

$$(\Delta + k^2)F(x,y,z) = 0, \tag{1.74}$$

where k is the wave number of the light. In a homogeneous and isotropic space without charges and currents the complex amplitude $F(\mathbf{x})$ can be represented by any projection of the vectors of the strength of electric $\mathbf{E}(\mathbf{x})$ and magnetic $\mathbf{H}(\mathbf{x})$ fields of the light wave.

Solving equation (1.74) by using the complex amplitude through the two-dimensional Fourier transform

$$F(x,y,z) = \iint\limits_{-\infty}^{\infty} A(\alpha,\beta,z)\exp[-ik(x\alpha+y\beta)]d\alpha\,d\beta, \tag{1.75}$$

where $A(\alpha, \beta, z)$ is the amplitude of the spatial spectrum of plane waves, we can obtain the decomposition of the complex amplitude with respect to plane waves

$$F(x,y,z)=\iint_{-\infty}^{\infty} A_0(\alpha,\beta)\exp[-ik(x\alpha+y\beta\pm z\sqrt{1-\alpha^2-\beta^2})]d\alpha\, d\beta, \quad (1.76)$$

where $A_0(\alpha, \beta) = A(\alpha, \beta, z = 0)$ is also the amplitude of the spatial spectrum of plane waves at $z = 0$. If we know the direction of light propagation, in the exponent in equation (1.76) we can leave only one sign (when the wave propagates along the z axis we select the plus sign).

We represent the function $A_0(\alpha, \beta)$ via the inverse Fourier transform

$$A_0(\alpha,\beta)=\frac{k^2}{2\pi}\iint_{-\infty}^{\infty} F_0(x,y)\exp[ik(x\alpha+y\beta)dx\, dy. \quad (1.77)$$

From equations (1.76) and (1.77) follows the integral transformation of the complex amplitude of the light field [4]

$$F(x,y,z)=k^2\iint_{-\infty}^{\infty} F_0(x',y')H(x-x',y-y',z)dx'dy', \quad (1.78)$$

where

$$H(x,y,z)=\frac{1}{2\pi}\iint_{-\infty}^{\infty} \exp[-ik(x\alpha+y\beta\pm z\sqrt{1-\alpha^2-\beta^2})]d\alpha\, d\beta. \quad (1.79)$$

$H(x, y, z)$ is the pulse response function of the homogeneous space, $F_0(x,y) = F(x, y, z = 0)$ is the complex amplitude of light at $z = 0$. If $\alpha^2 + \beta^2 > 1$, then the integral exponential factor $\exp(-kz\sqrt{\alpha^2+\beta^2-1})$ appears to be described by the inhomogeneous surface waves that propagate in the plane $z = 0$ and at $z \gg \lambda$ not contributing to the light field. Therefore, if $z \gg \lambda$ the integral in (1.79) can be calculated not in infinite limits, and at $\alpha^2 + \beta^2 < 1$.

1.4.1. Kirchhoff integral

Using the known expansion of the amplitude of a spherical wave in plane waves

$$\frac{e^{ikR}}{R}=ik\iint_{-\infty}^{\infty}\frac{\exp\left[-ik(x\alpha+y\beta-z\sqrt{1-\alpha^2-\beta^2})\right]d\alpha\, d\beta}{\sqrt{1-\alpha^2-\beta^2}}. \quad (1.80)$$

where $R = (x^2 + y^2 + z^2)^{1/2}$ – the pulse response function, defined by equation (1.79) can be written as:

$$H(x,y,z)=-\frac{1}{2\pi k^2}\frac{\partial}{\partial z}\left(\frac{e^{ikR}}{R}\right)=\frac{e^{ikR}}{2\pi k^2 R}\frac{\partial R}{\partial z}\left(R^{-1}-ik\right). \quad (1.81)$$

If we assume that the distance from the plane $z = 0$ to the plane of observation z is much greater than the wavelength of $R \gg \lambda$, $z \gg \lambda$, then instead of (1.81) we can approximately assume that the following equality is satisfied

$$H(x,y,z) = -\frac{i}{2\pi k}\frac{e^{ikR}}{R}\frac{z}{R}. \tag{1.82}$$

Then, instead of the integral transform (1.79) we obtain the Kirchhoff integral

$$F(x,y,z) = \frac{-ik}{2\pi}\int\int_{-\infty}^{\infty} F_0(x',y')\frac{e^{ikR}}{R}\frac{z}{R}dx'dy', \tag{1.83}$$

where $R = [(x-x')^2 + (y-y')^2 + z^2]^{1/2}$. Sometimes, given the fact that $R \approx z$ instead of (1.83) the Kirchhoff integral is written as:

$$F(x,y,z) = \frac{-ik}{2\pi}\int\int_{-\infty}^{\infty} F_0(x',y')\frac{e^{ikR}}{R}dx'dy'. \tag{1.84}$$

The physical meaning of the Kirchhoff integral (1.84) is associated with the Huygens–Fresnel wave principle and consists in the fact that the Kirchhoff integral is an expansion of the complex amplitude of the light field in spherical waves.

1.4.2. Fresnel transform

Integral transforms (1.78) and (1.83) describe the propagation of non-paraxial optical fields in a homogeneous space along the axis z. To describe the propagation of paraxial optical fields that propagate in a small solid angle, we use the Fresnel integral transform.

The complex amplitude of the paraxial light field is represented as:

$$U(x,y,z) = e^{ikz}F(x,y,z). \tag{1.85}$$

and the slowly varying complex amplitude $F(x, y, z)$ satisfies the Fock–Leontovich parabolic equation (1.34) [2]

$$\left(2ik\frac{\partial}{\partial z} + \nabla^2_{xy}\right)F(x,y,z) = 0, \tag{1.86}$$

where $\nabla^2_{xy} = \frac{\partial^2}{\partial x^2} + \frac{\partial^2}{\partial y^2}$ is the transverse Laplacian. Any solution of equation (1.86) can be written in integral form:

$$F(x,y,z) = \frac{-ik}{2\pi z}\int\int_{-\infty}^{\infty} F_0(\xi,\eta)\exp\left\{\frac{ik}{2z}\left[(x-\xi)^2 + (y-\eta)^2\right]\right\}d\xi\, d\eta, \tag{1.87}$$

where $F_0(x, y, z) = F(x, y, z = 0)$.

The Fresnel transform (1.87) is the expansion of the paraxial light field on the parabolic waves, and it is easily obtained from the Kirchhoff integral (1.84), using a Taylor series expansion to the second term of the distance R in the exponent:

$$R = \left[(x-\xi)^2 + (y-\eta)^2 + z^2\right]^{1/2} \approx z + \frac{1}{2z}\left[(x-\xi)^2 + (y-\eta)^2\right].$$

The transition from the Kirchhoff integral (1.84) for the Fresnel integral (1.87) is possible under the condition:

$$\frac{kr^4}{8z^3} << \pi,$$

where r is the effective radius of the light field.

At a considerable distance from the initial plane $z = 0$, when the conditions of the far zone of diffraction (or Fraunhofer diffraction zone)

$$\frac{kr^2}{2z} << \pi, \tag{1.88}$$

instead of the Fresnel integral transform (1.87) we can use the Fourier transform of the parabolic wave multiplier in front of the integral:

$$F(x,y,z) = \frac{-ik}{2\pi z} \exp[\frac{ik}{2z}(x^2 + y^2)] \int_{-\infty}^{\infty}\int F_0(\xi,\eta) \exp[\frac{-ik}{z}(x\xi + y\eta)] d\xi\, d\eta. \tag{1.89}$$

The Fourier transform (1.89) is the expansion of paraxial optical fields on plane waves. The integral on the right-hand side of (1.89), written with the help of the spatial frequencies $u = k\xi/z$, $v = k\eta/z$, has the form of the normal Fourier integral:

$$F(x,y) = \int_{-\infty}^{\infty}\int F_0(u,v) \exp[-i(xu + yv)] du dv. \tag{1.90}$$

The Fourier transforms (1.89) and (1.90) also describe the complex amplitude of the light field in the plane of spatial frequencies of a thin spherical lens.

Conclusion

This chapter introduces the basic differential and integral equations, which are necessary for solving problems of the diffraction of electromagnetic waves. Based on the general system of differential equations for the vectors of electric and magnetic fields of the electromagnetic wave, the wave equation, the Helmholtz equation for monochromatic light, the Fock–Leontovich equation for the paraxial optical fields, as well as the eikonal equation describing the propagation of rays in geometrical optics were derive. Similarly, using the scalar and Green vector theorems, we derived the basic integral relations for the monochromatic electromagnetic field: the Stratton–Chu and Kirchhoff–Helmholtz formulas. We presented the basic Fredholm integral equations of the first and second kind for solving problems of the diffraction of a monochromatic electromagnetic wave by perfectly reflecting and homogeneous dielectric (transmitting) objects. For the scalar complex amplitude, which can be regarded as any of the projections of the vectors of the strength of the electric and magnetic fields, we discussed the widely used integral representations: field expansion in plane waves, the expansion in spherical waves (the Kirchhoff integral), the expansion of the parabolic waves (Fresnel transform).

Many of the relationships in this chapter are used in subsequent chapters for solving direct and inverse problems of diffractive nanophotonics. Chapter 2 presents difference methods for solving Maxwell's equations (variants of the

FDTD method) and finite difference methods for solving the wave equation (BPM-method). Chapter 3 discusses the solution of the Helmholtz equation based on the Galerkin finite element method and the solution of the integral Fredholm equation of the second type, which describes the diffraction of light by dielectric objects. Chapter 4 deals with the solution of the Helmholtz equation on the basis of the Fourier modal method, or the expansion of plane waves for periodic objects (RCWA method).Chapter 5 uses the solution of Maxwell's equations based on the difference FDTD-method in cylindrical coordinates. In Chapter 6 the Helmholtz equation is solved by the method of matched sinusoidal modes. Chapter 7 deals with the paraxial equation of propagation and uses Fresnel and Fourier transform to describe the propagation of laser beams. In Chapter 8, to calculate the force of light pressure on the microparticle, we use an iterative solution of the integral equation of diffraction obtained on the basis of Green's theorem.

References

1. Joannopoulos J.D., Johnson S.G., Winn J.N., Photonic crystal: Molding the Flow of Light, Princeton Univ. Press, Second edition (2008).
2. Prasad P.N., Nanophotonics, Wiley & Son (2004).
3. Kawata S., Ohtsu M., Irie M., Nano-Optics, Springer Verlag (2002).
4. Klimov V.V., Nanoplasmonics, Moscow, Fizmatlit (2009).
5. Novotny L., Hecht B., Principles of Nano-Optics, Cambridge Univ. Press (2006).
6. Born M., Wolf E., Principles of optics, Nauka, Moscow (1973).
7. Solimeno S., Krozinyani B., Di Porto P., Diffraction and waveguide propagation of optical radiation, Moscow, Mir (1989).
8. Ilyinsky A., Kravtsov, V., Sveshnikov A.G., Mathematical models of electrodynamics, Moscow, Vysshaya shkola (1991).
9. Zverev V.A., Radio-optics, Moscow, Sov. Radio (1975).

Chapter 2

Numerical methods for diffraction theory

The FD-TD method actively used at present to meet the challenges of nanophotonics [1, 2] has a long history. Appearing in the middle of the last century, [3] (G. Cron, 1944), the numerical method for solving Maxwell's equations has gone through several stages of development. Previously, only (S.K. Yee, 1966) [4] published explicit difference equations of a high order of approximation of the initial differential problem in time and space. Implicit finite-difference approximations, characterized by absolute stability, were presented in 1997 [5] by D.L. Golovashkin, A.A. Degtyarev and V.I. Soifer, in 1998 [6] the same authors increased the order approximation in time for the implicit approximation, and in 2000 [7] also in space (Zheng, Chen, Zhang).

In 1994, J.–P. Berenger [8] satisfactorily solved the problem of numerical description of the absorption of radiation leaving the boundary of the computational domain.

The problem of modelling of the operation of the source of an incident wave, set by Yee in [4], has been solved with varying degrees of accuracy in many studies to date. The first way to specify the incident wave, which allows to limit the computational domain by the object under study and its immediate neighborhood, was formulated in the work [9] (A. Taflove, M. Brodwin, 1975). A more accurate method was published in 1980 [10] (A. Taflove) using the TF / SF technique (Total–Field/Scattering–Field technique). The increase in the accuracy of this approach in the region enclosed in a shell of a homogeneous medium, is described in a related work in 1999 [11] (D.W. Prather and S. Shi), in which the authors chose to define the emitting conditions numerically, rather than in the analytical form, as previously suggested in [10]. When finding an optical element in the shell of an inhomogeneous medium it is appropriate to apply the methodology of defining the incident wave, described further in section 2.1.4.

The computational complexity of the FD-TD method is reduced by the imposition of a mobile grid area, as proposed in [12] (B. Fidel, E. Heyman, R. Kastner and R.W. Ziokłowski). The specified method is well established in the study of short pulse propagation in a homogeneous medium. In section 2.1.5 a method is proposed for decomposing the grid domain, which reduces the computational complexity when modelling the propagation of monochromatic radiation.

The beam propagation method (BPM) was proposed in the 70s of XX century by Feit and Fleck [13] (M.D. Feit, 1978) and was designed for simulation and analysis of light propagation in gradient refractive optical fibres. Although originally the method was formulated in the field of the scalar theory of light, and it aimed to the gradient media, the basic ideas and principles of the method are more fundamental and remain valid until now. This is confirmed, in particular, by a large number of studies of the method that appeared in the last three decades, as well as studies in which the method is applied to solve research and applied problems.

Interesting also is the fact that a similar approach was independently proposed (and developed independently for a long period of time) in different fields of physics, namely in acoustics. There the method is called the parabolic equation method and is of considerable importance, for example, for the problems of hydroacoustics. At present, these methods are closely related, offer similar mathematical tools and, in fact, are almos identical. However, in some sense, the beam propagation method is more general, as it studies not only scalar cases.

The central idea of the beam propagation method is to reduce the order of differentiation with respect to the selected coordinate in the Helmholtz equation and subsequently solve the problem in the evolutionary form with respect to this coordinate. Feit and Fleck offered a fairly simple way to impose a number of very serious limitations on the scope of the method, but this solution has shown the principal possibility of such an approach.

Later on studies of the method aimed at overcoming its limitations. Thus, the beam propagation method, based on the method of lines, has made possible to perform simulations of light propagation in media with a more contrast refractive index profile [14] (J. Gerdes, 1991).

Further development of computer technology has made possible the effective use of finite-difference methods for solving the consequences of the Helmholtz equation which led to the emergence of a new family of the beam propagation methods (finite-difference BPM), removing a number of requirements for the field distribution in the propagating beam [15] (W. Huang, 1992). However, these methods are, in turn, limited as regards beam propagation: their application is incorrect if most of the energy is distributed at a considerable angle to the axis, considered as the main direction of propagation.

In turn, the application of the finite element approach and the more accurate approximations of differential operators allows us to relax the last restriction, which led to the emergence of yet another family of methods: methods of beam propagation for significant deviations in propagation (wide-angle BPM) [16, 17] (S.L. Chui, 2004 and Kh.Q. Le, 2009).

It should be noted that the studies, which set out the foundations of modern methods of beam propagation were mostly published in the early 90s of the twentieth century. Further development of the method occurred predominantly in the direction of improving the performance of the method through the use of more sophisticated mathematical tools and computational methods.

At present, the beam propagation method is a rather large family of methods with different characteristics which determine which method is used for a specific case. Together, these methods allow to solve a wide range of problems dealing with the propagation of radiation in dielectric media. The main limitations of the method and its mathematical foundations will be discussed in section 2.2.

2.1. The finite-difference time-domain method for solving Maxwell's equations

2.1.1. Explicit difference approximation for Maxwell's equations

The mathematical basis of the finite-difference time-domain method (FD-TD method) are difference expressions for the Maxwell equations and the grid approximations of the boundary and initial conditions corresponding to the boundary value problems for the first and second kind and cyclic. Classical approximations by Yee [4] (from which the FD-TD method is derived) allow the expression of each grid function via values of the functions values at the previous time layers explicitly. The main feature of these approximations is separate location of nodes of the grid area for each projection of the field strength. As shown in [7], this technique raises the order of approximation of the difference scheme of the initial boundary- value problem.

2.1.1.1. One-dimensional case

In the one-dimensional case with the Dirichlet boundary conditions on the area of computer simulation D^1 $(0 < t \leqslant T, 0 \leqslant z \leqslant L_z)$ we traditionally [4] D_h^1 superimpose a grid area in the nodes of which $\{(t_m, z_k)$: $t_m = mh_t$, $m = 0.1,\ldots, M = T/h_t$, $z_k = kh_z$, $k = 0\ldots, K = L_z/h_z\}$ we define the grid projection of the electric field on the axis X– $E_{x_k}^m$. The grid projection of the magnetic field on the axis Y– $H_{y_{k+0.5}}^{m+0.5}$ is defined at the nodes $\{(t_m +0.5, z_{k+0.5})$: $t_{m+0.5} = (m +0.5)\, h_t$, $m = 0.\ 1,\ \ldots, M - 1$, $z_{k+0.5} = (k + 0.5)\, h_z$, $k = 0,\ldots, K - 1\}$. The index k varies in the range D_h^1 indicating the nodes in space, m – in time. Distances between nodes are given by the spatial (h_z) and time (h_t) grid steps. The grid value of of the dielectric constant (ε_k) characterizes the optical element being studied. Figure 2.1 presents the location of nodes D_h^1 in space, without taking into account the time coordinate.

Then Maxwell's equations in the one-dimensional case are usually written as the following difference analogue [4]:

$$\mu_0 \frac{H_{y_{k+0.5}}^{m+0.5} - H_{y_{k+0.5}}^{m-0.5}}{h_t} = -\frac{E_{x_{k+1}}^m - E_{x_k}^m}{h_z}; \tag{2.1}$$

$$\varepsilon_0 \varepsilon_k \frac{E_{x_k}^{m+1} - E_{x_k}^m}{h_t} = -\frac{H_{y_{k+0.5}}^{m+0.5} - H_{y_{k-0.5}}^{m+0.5}}{h_z}. \tag{2.2}$$

Figure 2.2 shows the differential pattern corresponding to (2.1) , (2.2).
By defining the Dirichlet boundary conditions for D_h^1, we set

$$E_{x_0}^m = 0 \text{ and } E_{x_K}^m = 0 \text{ at } \%\, 0 \le m \le M. \tag{2.3}$$

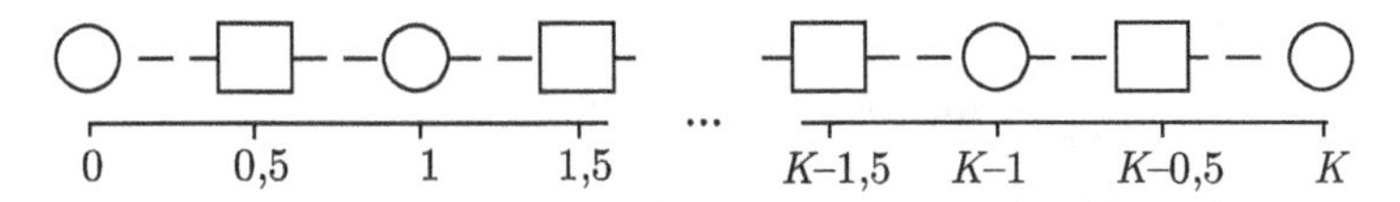

Fig. 2.1. Grid area D^1_h without sampling over time. Circles correspond to the projection $E^m_{x_k}$ and squares to $H^{m+0,5}_{y_{k+0,5}}$.

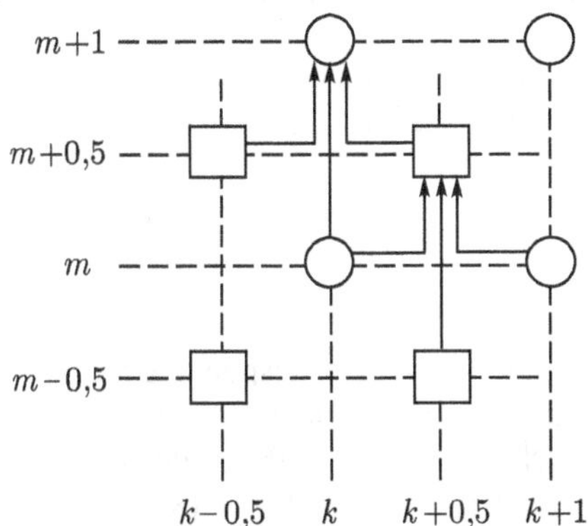

Fig. 2.2. The differential pattern for the construction of (2.1), (2.2). Circles correspond to the projection of the electric field $E^m_{x_k}$, the square – the magnetic field $H^{m+0.5}_{y_{k+0.5}}$.

The initial condition for D^1_h written in

$$E^0_{x_k} = \Phi_1(kh_z)(1 \le k \le K-1). \tag{2.4}$$

The grid projection of the magnetic field is not involved in the formation of the boundary and initial conditions in the field of view of the structure D^1_h, which does not provide for the location of nodes (t_m +0.5, z_k +0.5) at any of its border.

When setting the Neumann boundary conditions we must be impose on D^1 the grid area $\bar{D}^1_h$ [18], in the nodes of which $\{(t_m, z_{k+0.5}): t_m = mh_t, m = 0. 1, ..., M = T/h_t, z_k + 0.5 = (k+0.5) h_z, k = 1, ..., K = L_z / h_z\}$ we define the grid projection of the electric field on the axis $X - E^m_{x_k+0.5}$. The grid projection of the magnetic field on the axis $Y - H^{m+0.5}_{y_K}$ is defined at the nodes $\{(t_{m+0.5}, z_k): t_{m+0.5} - (m+0.5) h_t, m - 0. 1, .., M - 1, z_k - kh_z, k - 0. .., K\}$. Figure 2.3 shows the location of nodes in space, without taking into account the time coordinate.

Redefining the grid area is associated with the imposition on the boundaries $z = 0$ and $z = L_z$ of the nodes of the magnetic field

$$H^{m+0.5}_{y_0} = 0 \text{ and } H^{m+0.5}_{y_K} = 0 \; (0 \le m \le M-1), \tag{2.5}$$

in contrast to the condition (2.3), which would entail the imposition on the boundaries of the nodes of the electric field. The initial condition for $\bar{D}^1_h$ is written in

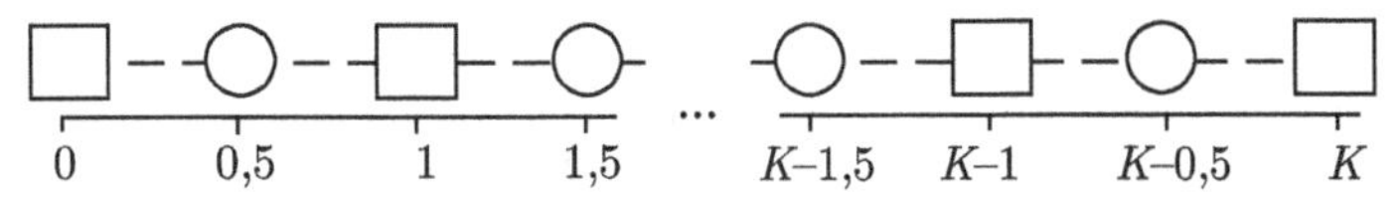

Fig. 2.3. Grid area D^1_h without sampling over time. Circles correspond to the projection $E^m_{x_{k+0.5}}$, squares to $H^{m+0.5}_{y_k}$.

$$E^0_{x_{k+0.5}} = \Phi_1\left(\left(k+0.5\right)h_z\right) \ \left(0 \le k \le K-1\right) \tag{2.6}$$

In this case, when writing the boundary conditions we do not use grid electric field strength, and write the initial condition without the magnetic field.

Then the difference analogue for the Maxwell's equations in the one-dimensional form is:

$$\mu_0 \frac{H^{m+0.5}_{y_k} - H^{m-0.5}_{y_k}}{h_t} = -\frac{E^m_{x_{k+0.5}} - E^m_{x_{k-0.5}}}{h_z}; \tag{2.7}$$

$$\varepsilon_0 \varepsilon_{k+0.5} \frac{E^{m+1}_{x_{k+0.5}} - E^m_{x_{k+0.5}}}{h_t} = -\frac{H^{m+0.5}_{y_{k+1}} - H^{m+0.5}_{y_k}}{h_z}. \tag{2.8}$$

Figure 2.4 shows a differential pattern corresponding to (2.7) and (2.8).

Implementation of the cyclic boundary condition requires the imposition on D^1 of the grid area $\bar{D}^1_h$ (Fig. 2.5), in the nodes of which $\{(t_m, z_k)$: $t_m = mh_t$, $m = 0,..., M = T / h_t$, $z_k = kh_z$, $k = 0,..., K-1$ $(K = L_z / h_z)\}$ we define the grid projection of the electric field on the axis $X - E^m_{x_k}$. The grid projection of the magnetic field on the axis $Y - H^{m+0.5}_{y_k+0.5}$ is defined at the nodes $\{(t_m +0.5, z_{k+0.5})$: $t_{m+0.5} = (m + 0.5)\,h_t$, $m = 0.\ 1,..., M-1$, $z_{k+0.5} = (k + 0.5)\,h_z$, $k = 0,..., K-1\}$.

In contrast to D^1_h, the region $\tilde{D}^1_h$ does not contain a node for $E^m_{x_k}$ because of its redundancy, since the cyclic condition implies the equality of the field strengths at $z = 0$ and $z = L_z$. The difference equations in solving Maxwell's equations coincide with the previously submitted equations, except for the node $k = 0$ (for the definition of the electric field) and $k = K-1$ (for the definition of the magnetic field). For them is true:

$$\mu_0 \frac{H^{m+0.5}_{y_{K-0.5}} - H^{m-0.5}_{y_{K-0,5}}}{h_t} = -\frac{E^m_{x_0} - E^m_{x_{K-1}}}{h_z}; \tag{2.9}$$

$$\varepsilon_0 \varepsilon_k \frac{E^{m+1}_{x_0} - E^m_{x_0}}{h_t} = -\frac{H^{m+0.5}_{y_{0.5}} - H^{m+0.5}_{y_{K-0.5}}}{h_z}. \tag{2.10}$$

The initial condition $\tilde{D}^1_h$ is written as $E^0_{x_k} = \Phi_1(kh_z)$ $(0 \le k \le K-1)$.

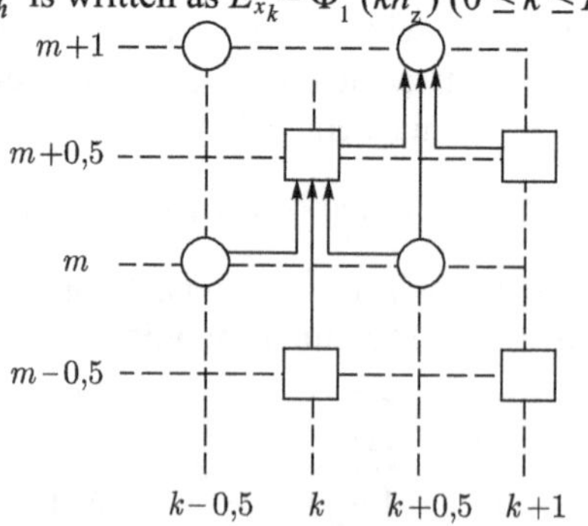

Fig. 2.4. The differential pattern for the construction of (2.7), (2.8). Circles correspond to the projection of the electric field $E^m_{x_{k+0.5}}$, the square – the magnetic field $H^{m+0.5}_{y_k}$.

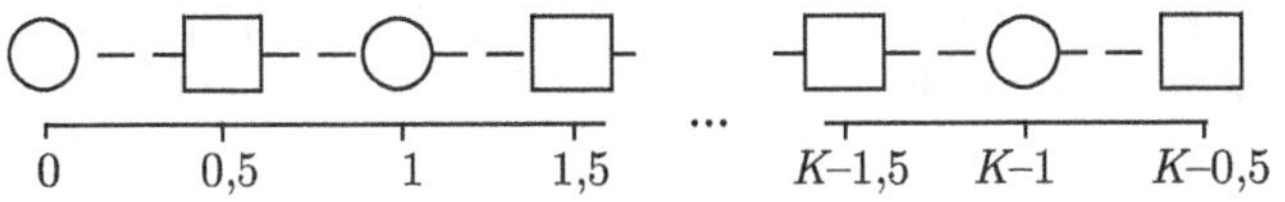

0 0,5 1 1,5 ... K–1,5 K–1 K–0,5

Fig. 2.5. Grid area without sampling over time. Circles correspond to the projection $E^{m}_{x_k}$, squares – $H^{m+0.5}_{y_{k+0.5}}$.

Computational procedures, associated with the proposed schemes, are based on the alternation of time layers: the grid function at the next time step can be expressed solely in terms of functions in the previous two layers (the property of explicit difference approximations). Thus, equations (2.1) and (2.2) are conveniently solved in the form of

$$H^{m+0.5}_{y_{K-0.5}} = H^{m-0.5}_{y_{K-0.5}} - \frac{h_t}{\mu_0 h_z}\left(E^m_{x_0} - E^m_{x_{K-1}}\right); \tag{2.11}$$

$$E^{m+1}_{x_k} = E^m_{x_k} - \frac{h_t}{\varepsilon_0 \varepsilon_k h_z}\left(H^{m+0.5}_{y_{k+0.5}} - H^{m+0.5}_{y_{k-0.5}}\right), \tag{2.12}$$

and (2.7) – (2.8) is rewritten as

$$H^{m+0.5}_{y_k} = H^{m-0.5}_{y_k} - \frac{h_t}{\mu_0 h_z}\left(E^m_{x_{k+0.5}} - E^m_{x_{k-0.5}}\right); \tag{2.13}$$

$$E^{m+1}_{x_{k+0.5}} = E^m_{x_{k+0.5}} - \frac{h_t}{\varepsilon_0 \varepsilon_{k+0.5} h_z}\left(H^{m+0.5}_{y_{k+1}} - H^{m+0.5}_{y_k}\right). \tag{2.14}$$

Accordingly, (2.9) and (2.10) take the form

$$H^{m+0.5}_{y_{K-0.5}} = H^{m-0.5}_{y_{K-0.5}} - \frac{h_t}{\mu_0 h_z}\left(E^m_{x_0} - E^m_{x_{K-1}}\right); \tag{2.15}$$

$$E^{m+1}_{x_0} = E^m_{x_0} - \frac{h_t}{\varepsilon_0 \varepsilon_k h_z}\left(H^{m+0.5}_{y_{0.5}} - H^{m+0.5}_{y_{K-0.5}}\right). \tag{2.16}$$

The advantage of algorithms for solving (2.11)–(2.14) is the possibility of vectorization. Calculations for a time step of (2.11) – (2.13) can be expressed through a single operation of vector addition of the electric field and a single operation saxpy [19] (triad [20]) with a scalar $-\frac{h_t}{\mu_0 h_z}$. For (2.12), (2.14) after the addition of vectors of the magnetic field and in front of saxpy with $-\frac{h_t}{\varepsilon_0 h_z}$ we add a component-wise operation dividing the result of addition by the vector of values of ε_k where $1 \le k \le K-1$ for (2.14) or $\varepsilon_{k+0.5}$, where $0 \le k \le K-1$ for (2.14).

It is known that the difference scheme (2.1)– (2.4 approximates the initial differential problem with the order $O(h^2_t, h^2_z)$ and stable [21]provided $\frac{h_t}{h_z} \le \frac{1}{c} O\left(h_t^2, h_z^2\right)$ [9] (c is the speed of light in the medium). It is obvious that the other two schemes: (2.5)–(2.8) and (2.1), (2.2), (2.4), (2.9) and (2.10) are characterized by the same order of approximation and the stability condition, as derived from (2.1)–(2.4) modifications of the grid area. Shifting in the pattern in Fig. 2.4 the writing below the vertical dashed lines to the right by one position, we obtain the pattern shown in Fig. 2.2.

2.1.1.2. The two-dimensional case

On the two–dimensional area of computer simulation D^2 ($0 < t \le T$, $0 \le y \le L_y$, $0 \le z \le L_z$) we traditionally [4] superimpose the grid area in which the nodes D_h^2, $\{(t_m, y_j, z_k)$: $t_m = mh_t$, $m = 0.\ 1, .., M = T / h_t, y_j = jh_y, j = 0, .., J = L_y / h_y$, $z_k = kh_z$, $k = 0,..., K = L_z / h_z\}$ define the grid projection of the electric field on the axis X $E^m_{x_{j,k}}$. The grid projection of the magnetic field on the axis Z – $H^{m+0.5}_{y_j+0.5,k}$ is define at the nodes $\{(t_{m+0.5}, y_{j+0.5}, z_k)$: $t_{m+0.5} = (m+0.5)\ h_t$, $m = 0.\ 1,..., M–1$, $y_{j+0.5} = (j+0.5)\ h_y$, $j = 0,..., J–1$, $z_k = kh_z$, $k = 1, .., K–1\}$ and the projection of the magnetic field at Y – $H^{m+0.5}_{y_{j,k+0.5}}$. at the nodes $\{(t_{m+0.5}, y_j, z_{k+0.5})$: $t_{m+0.5} = (m+0.5)\ h_t$, $m = 0.\ 1,..., M-1$, $y_j = jh_y$, $j = 1,..., J–1$, $z_{k+0.5} = (k+0.5)\ h_z$, $k = 0,..., K–1\}$. Figure 2.6 presents the location of nodes D_h^2 in space, without taking into account the time coordinate.

In the proposed area the indices j, k denote the nodes in space (directions Y and Z), m – is time. Distances between nodes are given by the spatial (h_y and h_z) and time (h_t) grid steps. The grid value of thedielectric constant ($\varepsilon_{j,k}$) characterizes the optical element being studied.

The system (2.1), (2.4) in the two-dimensional case for the TE-wave is then usually written in the following difference analogue [4] :

$$\mu_0 \frac{H^{m+0.5}_{y_{j,k+0.5}} - H^{m-0.5}_{y_{j,k+0.5}}}{h_t} = -\frac{E^m_{x_{j,k+1}} - E^m_{x_{j,k}}}{h_z}; \tag{2.17}$$

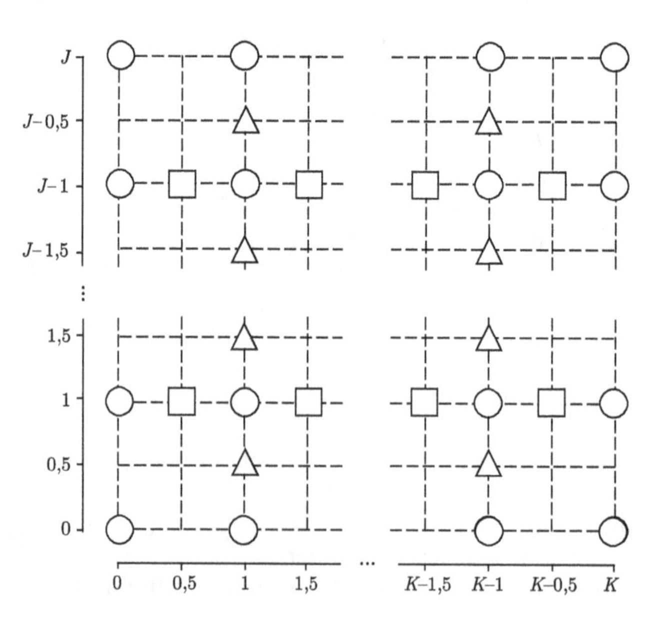

Fig. 2.6. Grid area D_h^2 without sampling over time. Circles correspond to the projections $E^m_{x_{j,k}}$ triangles – $H_{z_j+0.5\ k}^{m+0.5}$, squares – $H^{m+0.5}_{y_{j,k+0.5}}$.

$$\mu_0 \frac{H^{m+0.5}_{z_{j+0.5,k}} - H^{m-0.5}_{z_{j+0.5,k}}}{h_t} = \frac{E^m_{x_{j+1,k}} - E^m_{x_{j,k}}}{h_y}; \tag{2.18}$$

$$\varepsilon_0 \varepsilon_{j,k} \frac{E^{m+1}_{x_{j,k}} - E^m_{x_{j,k}}}{h_t} = \frac{H^{m+0.5}_{z_{j+0.5,k}} - H^{m+0.5}_{z_{j-0.5,k}}}{h_y} - \frac{H^{m+0.5}_{y_{j,k+0.5}} - H^{m+0.5}_{y_{j,k-0.5}}}{h_z}. \tag{2.19}$$

By defining D_h^2 for the Dirichlet boundary conditions, we set

$$\begin{aligned} &E^m_{x_{0,k}} = 0 \quad E^m_{x_{J,k}} = 0 \qquad 0 \le \mathrm{m} \le M \quad 0 \le \mathrm{k} \le K; \\ &E^m_{x_{j,0}} = 0 \quad E^m_{x_{j,K}} = 0 \qquad 0 \le \mathrm{m} \le M \quad 0 \le \mathrm{j} \le J. \end{aligned} \tag{2.20}$$

The initial condition for D_h^2 is written as

$$D_h^2 = \Phi_2(\mathrm{jhy}, \mathrm{khz}) \ (1 \le \mathrm{j} \le \mathrm{J}-1,\ 1 \le \mathrm{k} \le \mathrm{K}-1). \tag{2.21}$$

Grid projections of the magnetic field do not participate in the formation of boundary and initial conditions because of the structure of the region D_h^2, and do not lead to the location of nodes $(t_{m+0.5}, y_j, z_{k+0.5})$ and $(t_{m+0.5}, y_{j+0.5}, z_k)$ at any of its boundaries(Fig. 2.6).

When setting the Neumann boundary conditions we must imposed on grid D^2 the domain $\bar{D}_h^1$ (Fig. 2.7) [18], in the nodes of which $\{(t_m, y_{j+0.5}, z_{k+0.5})$: $t_m = mh_t$, $m = 0.\ 1, .., M = T / h_t$, $y_{j+0.5} = (j + 0.5)\ hy$, $j = 0. .., J-1$ $(J = L_y / h_y)$, $z_{k+0.5} = (k + 0.5)h_z$, $k = 0. .., K-1$ $(K = L_z / h_z)\}$ defined the grid projection of the electric field on the axis $X - E^m_{x_{j+0.5,k+0.5}}$. The grid projection of the magnetic field on the axis $Z - H^{m+0.5}_{z_{j,k+0.5}}$ is defined at the nodes $\{(t_{m+0.5}, y_j, z_{k+0.5})$: $t_{m+0.5} = (m + 0.5)\ h_t$, $m = 0.\ 1,.., M-1$, $y_j = jh_y$, $j = 0. .., J$, $z_{k+0.5} = (k+0.5)\ h_z$, $k = 0. .., K-1\}$ and the projection of the magnetic field on the $Y - H^{m+0.5}_{y_{j+0.5,k}}$ at the nodes $\{(t_{m+0.5}, y_{j+0.5}, z_k)$: $t_{m+0.5} = (m + 0.5)\ h_t$, $m = 0.\ 1, .., M-1$, $y_{j+0.5} = (j + 0.5)\ h_y$, $j = 0. .., J-1$, $z_k = kh_z$, $k = 0. .., K\}$.

Redefining the grid area associated with the imposition of limits on the boundaries $z = 0$ and $z = L_z$, the nodes for $H^{m+0.5}_{y_{j+0.5,k}}$, and at $y = 0$ and $y = L_y$ the nodes for $H^{m+0.5}_{z_{j,k+0.5}}$:

$$\begin{aligned} &H^{m+0.5}_{y_{j+0.5,\,0}} = 0 \text{ and } H^{m+0.5}_{y_{j+0.5,\,K}} = 0 \text{ at } 0 \le m \le M-1 \text{ and } 0 \le j \le J-1; \\ &H^{m+0.5}_{z_{0,\,k+0.5}} = 0 \text{ and } H^{m+0.5}_{z_{J,k+0.5}} = 0 \text{ at } 0 \le m \le M-1 \text{ and } 0 \le k \le K-1 \end{aligned} \tag{2.22}$$

in contrast to the condition (2.20), which would entail the imposition of the nodes of the electric field on the boundaries. The initial condition $\bar{D}_h^1$ is written in

$$E^0_{x_{j+0.5,k+0.5}} = \Phi_2\big((j+0.5)h_y, (k+0.5)h_z\big)(0 \le j \le J-1, 0 \le k \le K-1). \tag{2.23}$$

In this case, when writing the boundary conditions we do not use the grid strength of electric field, and in writing the initial condition the magnetic field is not considered.

Then the system (2.7) (2.8) in the case study of the TE-wave is usually written as the following difference analogue [18]:

$$\mu_0 \frac{H^{m+0.5}_{y_{j+0.5,\,k}} - H^{m-0.5}_{y_{j+0.5,\,k}}}{h_t} = -\frac{E^m_{x_{j+0.5,\,k+0.5}} - E^m_{x_{j+0.5,\,k-0.5}}}{h_z}; \tag{2.24}$$

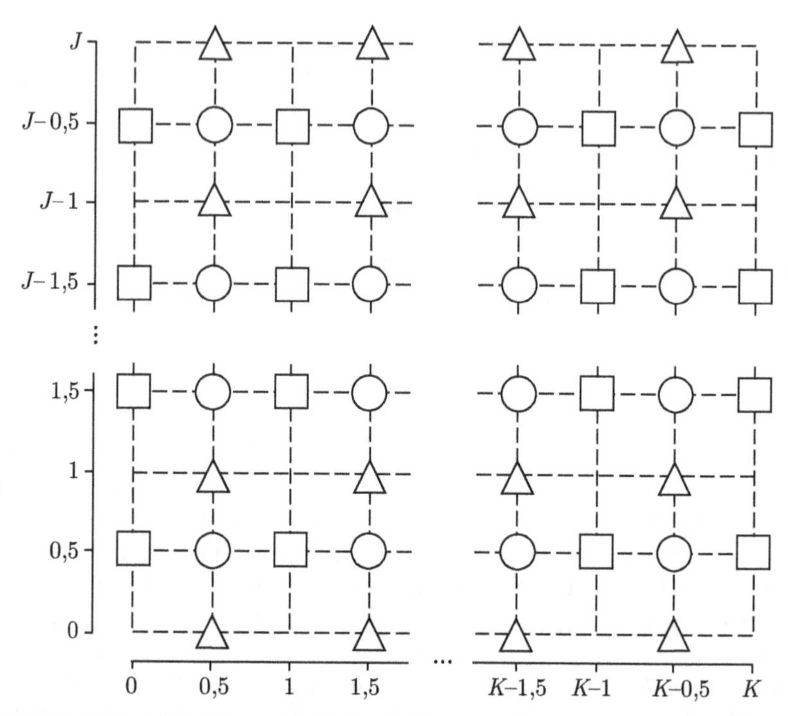

Fig. 2.7. Grid area $\bar{D}_h^2$ without sampling over time. Circles correspond to the projections, triangles – $H_{z_{j,k+0.5}}^{m+0.5}$, squares – $H_{y_{j+0.5,k}}^{m+0.5}$.

$$\mu_0 \frac{H_{z_{j,k+0.5}}^{m+0.5} - H_{z_{j,k+0.5}}^{m-0.5}}{h_t} = \frac{E_{x_{j+0.5,k+0.5}}^{m} - E_{x_{j-0.5,k+0.5}}^{m}}{h_y}; \tag{2.25}$$

$$\varepsilon_0 \varepsilon_{j+0.5,k+0.5} \frac{E_{x_{j+0.5,k+0.5}}^{m+1} - E_{x_{j+0.5,k+0.5}}^{m}}{h_t} = \frac{H_{z_{j+1,k+0.5}}^{m+0.5} - H_{z_{j,k+0.5}}^{m+0.5}}{h_y} - \frac{H_{y_{j+0.5,k+1}}^{m+0,5} - H_{y_{j+0.5,k}}^{m+0,5}}{h_z}. \tag{2.26}$$

Implementation of the cyclic boundary condition requires the imposition on D^2 of the grid domain $\tilde{D}_h^2$ (Fig. 2.8), in the nodes of which $\{(t_m, y_j, z_k)$: $t_m = mh_t$, $m = 0.\ 1, ..., M = T/h_t$, $y_j = jh_y$, $j = 0. ..., J-1$ $(J = L_y/h_y)$, $z_k = kh_z$, $k = 0. ..., K-1$ $(K = L_z/h_z)\}$ we define the grid projection of the electric field on the axis $X - E_{x_{j,k}}^m$. The grid projection of the magnetic field on axis $Z - H_{z_{j+0.5,k}}^{m+0.5}$ is defined at the nodes $\{(t_{m+0.5}, y_{j+0.5}, z_k)$: $t_{m+0.5} = (m+0.5)\,h_t$, $m = 0.\ 1, .., M-1$, $y_{j+0.5} = (j+0.5)\,h_y$, $j = 0. ..., J-1$, $z_k = kh_z$, $k = 0. .., K-1\}$ and the projection of the magnetic field on $Y - H_{y_{j,k+0.5}}^{m+0.5}$ at the nodes $\{(t_{m+0.5}, y_j, z_{k+0.5})$: $t_{m+0.5} = (m+0.5)\,h_t$, $m = 0.\ 1, .., M-1$, $y_j = jh_y$, $j = 0. ..., J-1$, $z_{k+0.5} = (k+0.5)\,h_z$, $k = 0. .., K-1\}$.

In contrast to D_h^2, the region $\tilde{D}_h^2$ does not contain nodes for $E_{x_{j,K}}^m$ $(0 \le j \le J)$ and $E_{x_{J,k}}^m$ $(0 \le k \le K)$, due to their redundancy, because the cyclic condition implies the equality of the field strengths at the opposite boundaries. In addition, on $\tilde{D}_h^2$ there

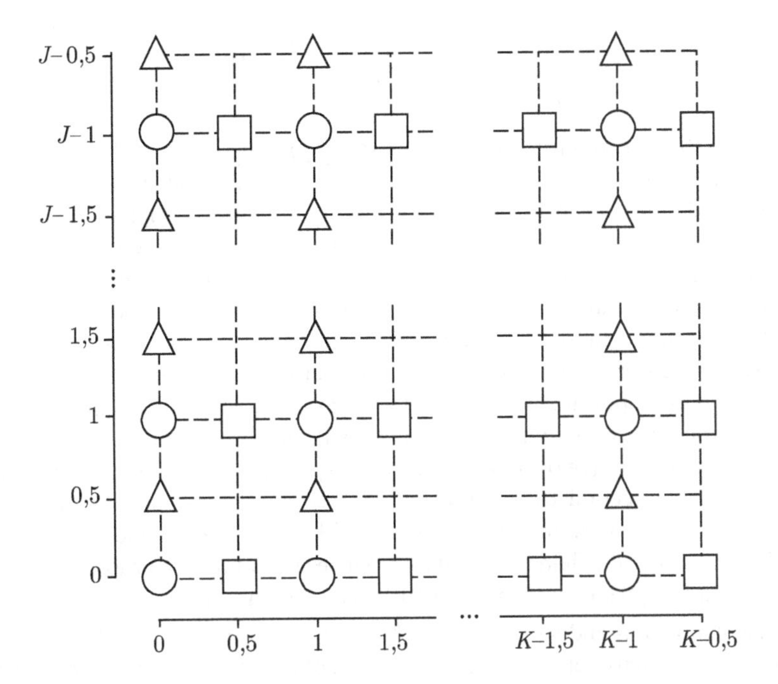

Fig. 2.8 Grid area $\tilde{D}_h^1$ without sampling over time. Circles correspond to the projections, triangles – $H_{z_{j+0.5,k}}^{m+0.5}$, squares – $H_{y_{j,k+0.5}}^{m+0.5}$.

were additional sites for $(0 \le j \le J-1)$ and $(0 \le k \le K-1)$ used in the definition of the electric field which previously defined the electrical wall on $\tilde{D}_h^2$.

Recording of difference equations with cyclic boundary conditions coincides with (2.17)–(2.19) with the following exceptions. Instead of (2.17) for $0 \le j \le J-1$, we have

$$\mu_0 \frac{H_{y_{j,K-0.5}}^{m+0.5} - H_{y_{j,K-0.5}}^{m-0.5}}{h_t} = -\frac{E_{x_{j,0}}^m - E_{x_{j,K-1}}^m}{h_z}. \tag{2.27}$$

Further, (2.18) at $0 \le k \le K-1$ takes the form

$$\mu_0 \frac{H_{z_{J-0.5,k}}^{m+0.5} - H_{z_{J-0.5,k}}^{m-0.5}}{h_t} = \frac{E_{x_{0,k}}^m - E_{x_{J-1,k}}^m}{h_y}. \tag{2.28}$$

For (2.19) at $1 \le j \le J-1$, we have

$$\varepsilon_0 \varepsilon_{j,0} \frac{E_{x_{j,0}}^{m+1} - E_{x_{j,0}}^m}{h_t} = \frac{H_{z_{j+0.5,0}}^{m+0.5} - H_{z_{j-0.5,0}}^{m+0.5}}{h_y} - \frac{H_{y_{j,0.5}}^{m+0.5} - H_{y_{j,K-0.5}}^{m+0.5}}{h_z}, \tag{2.29}$$

for $1 \le k \le K-1$

$$\varepsilon_0 \varepsilon_{0,k} \frac{E_{x_{0,k}}^{m+1} - E_{x_{0,k}}^m}{h_t} = \frac{H_{z_{0.5,k}}^{m+0.5} - H_{z_{J-0.5,k}}^{m+0.5}}{h_y} - \frac{H_{y_{j,k+0.5}}^{m+0.5} - H_{y_{j,k-0.5}}^{m+0.5}}{h_z}. \tag{2.30}$$

and for the node $j = 0$, $k = 0$

$$\varepsilon_0\varepsilon_{0,0}\frac{E^{m+1}_{x_{0.0}}-E^{m}_{x_{0.0}}}{h_t}=\frac{H^{m+0.5}_{z_{0.5,0}}-H^{m+0.5}_{z_{J-0.5,0}}}{h_y}-\frac{H^{m+0.5}_{y_{0,0.5}}-H^{m+0.5}_{0_{0,J-0.5}}}{h_z}. \quad (2.31)$$

The initial condition for $\tilde{D}^2_h$ is $E^0_{x_{j,k}} = \Phi_2(jh_y, kh_z)$ $(0 \le j \le J - 0 \le k \le K-1)$.

Shifting to the right side of (2.17)–(2.19) (2.24)–(2.26) and (2.27)–(2.31), all grid functions defined on the previous time layers, we obtain the computational procedure for calculating the fields shown on the three finite-difference approximations.

Vectorization of such procedures is associated with recording of row- or column-oriented algorithms [22] (depending on the method of storing the matrix in the computer memory) that shortens the duration searching the computer memory [19]. Recording of the vector algorithms in two-dimensional case is very important because of the high computational complexity of the given procedures. We write down the fields in the computer memory in the form of matrices (two-dimensional arrays), the first index of which points to a line, the second to a column. For definiteness, we set the first index j (corresponding to the direction Y in the grid areas), the second k (direction Z).

When storing the fields in the above manner by columns (for example, in using Fortran language) computations by (2.17) and (2.24) are accompanied by operations of calculation of K and K–1 vectors $H^{m+0.5}_{y_{j,k+0.5}}$ $(1 \le j \le J-1)$ and $H^{m+0.5}_{y_{j+0.5,k}}$ $(0 \le j \le J-1)$, respectively. Each operation in the calculation of the values of this vector consists of vector addition and subsequent saxpy. The scalar in saxpy is equal $-h_t/h_z\mu_0$, and the length of the resulting vectors is $J-1$ in the calculation by (2.17) and J by (2.24).

Calculations using (2.18) and (2.25) involve performing operations with the calculation of the K–1 and K vectors $H^{m+0.5}_{z_{j+0.5,k}}$ $(0 \le j \le J-1)$ and $H^{m+0.5}_{z_{j,k+0.5}}$ $(1 \le j \le J-1)$, respectively. As in the previous case, each operation of the calculation of the values of this vector consists of vector addition and subsequent saxpy. The scalar in saxpy equals $-h_t/h_y\mu_0$, and the length of the resulting vectors is J in the computation by (2.18) and J–1 by (2.25).

Similarly, referring to the calculation using (2.19) and (2.26), one should bear in mind the calculation of the K–1 and K vectors $E^{m+1}_{x_{j,k}}$ $(1 \le j \le J-1)$ and $E^{m+1}_{x_{j+0.5,k+0.5}}$ $(0 \le j \le J-1)$ the length J–1 and J. At the same time, to form each vector, we require two vector additions of the grid components of the magnetic field H_y and H_z (the vectors of the same projection are added), two component-wise divisions of the results of these additions to the value of the vector $\varepsilon_{j,k}$ $(1 \le j \le J-1)$ for (2.19) and $\varepsilon_{k+0.5,j+0.5}$ $(0 \le j \le J-1)$ for (2.26), two multiplications of the resulting vectors by scalars $ht/\varepsilon_0 h_y$ and $-ht/\varepsilon_0 h_z$. Then the resulting vectors are added with each other and with the vector $E^m_{x_{j,k}}$ $(1 \le j \le J-1)$ in the calculation by (2.19) and $E^m_{x_{j+0.5,k+0.5}}$ $(0 \le j \le J-1)$ in the calculation by (2.26).

Where $J < K$, the row-oriented algorithms [19] are preferable to column-oriented ones as they permit operations with vectors of greater length, providing a better loading of the conveyor.

When storing the fields in two-dimensional arrays of lines (for example, in using the language C) calculations using (2.17) and (2.24) are accompanied by operations of the calculation of J–1 and J vectors $H^{m+0.5}_{y_{j,k+0.5}}$ $(0 \le k \le K-1)$ and $H^{m+0.5}_{y_{j+0.5,k}}$ $(1 \le k \le K-1)$, respectively. Each operation of the calculation of the values of this vector consists of vector addition and subsequent saxpy. The scalar in saxpy is equal to $-h_t/h_z\mu_0$, and the length of the resulting vector is K in the calculation of (2.17) and K–1 in the case of (2.24).

Implementation of calculations by (2.18) and (2.25) involves performing operations of the calculation of J and J–1 vectors $H^{m+0.5}_{z_{j+0.5,k}}$ $(1 \leq k \leq K-1)$ and $H^{m+0.5}_{z_{j,k+0.5}}$ $(0 \leq k \leq K-1)$, respectively. As in the previous case, each operation of the calculation of the values of this vector consists of vector addition and subsequent saxpy. The scalar in saxpy is equal to $h_t/h_z\mu_0$, and the length of the resulting vectors is K–1 in the calculation of (2.18) and K in (2.25). Similarly, referring to the alculation of (2.19) and (2.26), one should bear in mind the calculation of the J–1 and J vectors $E^{m+1}_{x_{j,k}}$ $(1 \leq j \leq J-1)$ and $E^{m+1}_{x_{j+0.5,\,k+0.5}}$ $(0 \leq j \leq J-1)$ with length K–1 and K. At the same time, to form each vector, we require two vector addition of the grid component of the magnetic field H_y and H_z (the vectors of one projection are added), two component-wise division of the results of these additions by the value of the vector $\varepsilon_{j,k}$ $(1 \leq j \leq J-1)$ for (2.19) and $\varepsilon_{k+0.5,j+0.5}$ $(0 \leq j \leq J-1)$ for (2.26), and two multiplications of the resulting vectors by scalars $\frac{h_t}{\varepsilon_0 h_y}$ and $-\frac{h_t}{\varepsilon_0 h_z}$. The resulting vectors are then added with each other and with the vector $E^{m}_{x_{j,k}}$ $(1 \leq j \leq J-1)$ in the calculation by (2.19) and $E^{m}_{x_{j+0.5,\,k+0.5}}$ $(0 \leq j \leq J-1)$ in the calculation by (2.26).

Do not assume that the shape of the investigated optical element determines imperatively the choice of the algorithmic programming language. This means that in the study of radiation passing through the DOE, elongated along the axis Y, it is appropriate to implement exclusively column-oriented algorithms. In contrast, in the study of optical elements, extending along the axis Z, is reasonable to use only row-oriented methods. In fact, the researcher is free to use any form of writing (row or column-oriented), changing the direction of the axes and rewriting if necessary the difference equations in the new coordinate system.

Of separate interest are block algorithms – the most efficient way to operate with the cache memory of the computer [19]. Their implementation is associated with the storage of fields in two-dimensional arrays of blocks, which requires the development of algorithms for writing (and reading) the values of network functions in the computer memory which from the standard row- or column-oriented functions.

It is known that the difference approximation (2.17)–(2.21) approximates the initial differential problem with the order $O(h_t^2, h_y^2, h_z^2)$ and is stable under the condition $h_t\sqrt{\frac{1}{h_y^2}+\frac{1}{h_z^2}} \leq \frac{1}{c}$ [9]. It is obvious that the other two approximations (2.22) – (2.26) and (2.17)–(2.19) (2.27)–(2.31) (2.22) are characterized by the same order of approximation and the stability condition, as those derived from (2.17) – (2.19) by modifications of the grid area.

2.1.2. Transition from time domain to frequency domain

By studying the propagation of monochromatic light through diffractive optical elements, the researcher usually expects the simulation results in the form of the complex amplitude of the electric field. Strictly speaking, the computational experiment does not generate a monochromatic wave, since prior to the experiment radiation may be absent in the field, but at the selected time T the field in the relevant region can be accurately considered as monochromatic. In computational practice there are several ways of transition from the time domain to the frequency domain in implementing the difference method for the solution of Maxwell's equations.

In [9], one of the first on the issue, the intensity of the resultant electric field was determined by the addition of the intensities at different layers of the grid area in the time chosen for a certain period, followed by averaging. The specified method is similar to the principle of integral intensity sensors used in the formulation of field optical experiments, and is characterized by simplicity of implementation. However, this approach can not provide information about the phase of the complex electric field amplitude and is associated with a large number of additional arithmetic operations. The researcher needs to define the averaging of the intensities in the time interval equal to at least one period of oscillation of the electric field in the steady mode.

The use of Fourier transforms to switch to the frequency domain [23] provides information not only on the modulus of the complex amplitude of the electric field, but also the phase. It is necessary to consider the value of the grid function of strength for all time sections relating to the period of oscillations of the field in the steady mode. Thus, the second way of transformation to the frequency domain also requires large computational costs, differing from the first one by the necessity of applying a Fourier transform.This feature is a major obstacle to constructing the effective vector and parallel procedures for implementing the finite difference solution because of the difficulties in vectorization and parallelization of the fast Fourier transform.

The idea of a third way of transformation to the frequency domain has been known for a long time [24], but it has not been implemented as an algorithm in the literature, available to the authors. This is implemented in this monograph.

Following [24, 25], we represent the field in the form of the complex amplitude $\dot{E}_x = \dot{E}_{x_{re}} + i\dot{E}_{x_{im}}$, assuming

$$E_x = \mathrm{Re}\ \left\{\dot{E}_x \exp(-i\omega t)\right\}, \tag{2.32}$$

where ω is cyclic frequency. Then

$$\mathrm{E_x} = \dot{E}_{x_{re}} \mathrm{Cos\omega t} + \dot{E}_{x_{im}} \mathrm{Sin\omega t}.$$

The intensity of the field can be determined by means of two measurements of E_x at different times, by solving the equation:

$$\mathrm{E_{x1}} = \dot{E}_{x_{re}} \mathrm{Cos\omega t_1} + \dot{E}_{x_{im}} \mathrm{Sin\omega t_1}, \tag{2.33}$$

$$\mathrm{E_{x2}} = \dot{E}_{x_{re}} \mathrm{Cos\omega t_2} + \dot{E}_{x_{im}} \mathrm{Sin\omega t_2}. \tag{2.34}$$

We agree to take t_1 and $t_2 = T$ such that

$$\omega t_2 = \omega t_1 + \pi / 2.$$

Given that here cos $\omega t_1 = \sin \omega t_2$, and $\sin \omega t_1 = -\cos \omega t_2$, instead of (2.33) and (2.34) we get

$$E_{x1} = \dot{E}_{x_{re}} \mathrm{Sin}\omega t_2 - \dot{E}_{x_{im}} \mathrm{Cos}\omega t_2,$$

$$E_{x2} = \dot{E}_{x_{re}} \mathrm{Cos}\omega t_2 + \dot{E}_{x_{im}} \mathrm{Sin}\omega t_2.$$

Squaring the last two equations, we add them together:

$$(E_{x1})^2 + (E_{x2})^2 = \left(\dot{E}_{x_{re}}\right)^2 + \left(\dot{E}_{x_{im}}\right)^2 = \mathrm{I},$$

where I is the unknown quantity which is proportional to light intensity.

For a monochromatic wave, the origin from which t_2 is plotted can be chosen arbitrarily. It is convenient to take $\omega t_2 = \pi / 2 + 2\pi l$, where $l \in N_0$ at which $\cos \omega t_2 = 0$, $\sin \omega t_2 = 1$. Then

$$\dot{E}_{x_{re}} = E_{x1},\ \dot{E}_{x_{im}} = E_{x2}.$$

The foregoing describes the transition in the frequency domain, from E_x to I, which can restrict the choice of two time layers E_{x1} and E_{x2}, not using the FFT in the last layers of time, with further averaging the result. This approach significantly simplifies the implementation of the algorithm and reduces the duration of the calculations on it.

The development of the proposed method of transition [26] associated with the replacement (2.32) by

$$E_x = \mathrm{Re}\left\{\dot{E}_x \exp(-i(\omega t - \pi / 2))\right\}, \tag{2.35}$$

that allows the use of (2.35) for the job of the incident wave, matching a job with the most common form of initial conditions – the lack of fields in D before the computer simulation. Indeed, putting (2.35) $t = 0$. we obtain $E_x = 0$ for $\dot{E}_x = 1$. Instead of (2.33) (2.34) we write

$$E_x(t_1) = \dot{E}_{x_{re}} \mathrm{Sin}\omega t_1 - \dot{E}_{x_{im}} \mathrm{Cos}\omega t_1, \tag{2.36}$$

$$E_x(t_2) = \dot{E}_{x_{re}} \mathrm{Sin}\omega t_2 - \dot{E}_{x_{im}} \mathrm{Cos}\omega t_2. \tag{2.37}$$

Assume further t_1 and $t_2 = T$ such that $\omega t_2 = \omega t_1 + \pi / 2$. Substituting this expression into (2.36) (2.37) we get:

$$E_x(t_1) = -\dot{E}_{x_{re}} \mathrm{Cos}\omega T - \dot{E}_{x_{im}} \mathrm{Sin}\omega T,$$

$$E_x(t_2) = \dot{E}_{x_{re}} \mathrm{Sin}\omega T - \dot{E}_{x_{im}} \mathrm{Cos}\omega T.$$

Solving this system with respect to $\dot{E}_x$, we find:

$$\dot{E}_{x_{re}} = E_x(t_2)\mathrm{Sin}\omega T - E_x(t_1)\mathrm{Cos}\omega T,$$

$$\dot{E}_{x_{re}} = -E_x(t_1)\mathrm{Sin}\omega T - E_x(t_2)\mathrm{Cos}\omega T.$$

In contrast to the method described in [25], this approach allows to take into account the phase of the complex amplitude of the incident wave, if such a wave is determined by the representation (2.35). Such consideration is especially important in decomposing the computational domain, impossible in variants of the transition to the frequency domain.

2.1.3. Application of absorbing layers

The task of limiting the computational domain has an important place in difference solutions of Maxwell's equations. In most cases, the researcher must submit either an optical element, surrounded by a homogeneous medium (e.g. free space), or located at the interface between two semi-infinite media. This is connected with the general tendency of physics to reductionism, when it is attempted to distinguish the phenomena from the surrounding world and consider them separately from external influences. This approach seems most appropriate in the majority of cases.

The researcher is forced to image the computational domain as an infinite homogeneous space extending in any given direction, or as the interface between two such spaces. Otherwise, the processes occurring behind the region have an impact on the processes inside. For example, the wave leaving the region is reflected from an external object and comes back.

However, carrying out simulation for computer engineering, characterized by a given speed, the selected area of memory, and having a limited supply of time, the researcher can not solve the difference problem in infinite space.

Fortunately, this is not required if the area of interest (in which the field distribution is taken as the solution of the problem) is finite, and the processes in this region occur at a given time interval.

In this case it is sufficient to trace the distribution of the scattered field in a homogeneous infinite space outside the region of interest only in places where it has time to spread during the experiment.

Clearly the desire of the researcher is not to study the fate of the radiation leaving the vicinity of the optical element being studied. Is it possible to impose the boundary conditions or carry out the appropriate structuring of the subregion adjacent to the border, allowing outgoing radiation D not to come back?

None of the above boundary conditions provides such an effect. Moreover, none of these conditions allow the scattered radiation to leave the experiment region, which leads to the inevitable distortion of the result. Next, we consider approaches to avoid such a distortion.

2.1.3.1. Formulation of absorbing boundary conditions and the imposition of absorbing layers

The first effective approach to solving the problem [27] is based on the factorization of the wave operator. In the two-dimensional case, writing the d'Alembert operator in the form

$$G \equiv \frac{\partial^2}{\partial x^2} + \frac{\partial^2}{\partial z^2} - \frac{1}{c^2}\frac{\partial^2}{\partial t^2} \equiv D_x^2 + D_z^2 - \frac{1}{c^2}D_t^2,$$

looking for his performance as

$$G \equiv G^+ G^-,$$

where

$$G^+ \equiv D_z + \frac{D_t}{c}\sqrt{1-S^2}$$

corresponds to the wave propagation inside the two-dimensional computational domain from the left border, and

$$G^- \equiv D_z - \frac{D_t}{c}\sqrt{1-S^2}. \tag{2.38}$$

corresponds to the propagation outside the region. At the same time $S \equiv \frac{D_y}{D_t / c}$.

The authors of [27] have shown that the solution at the selected boundary of the equation $G^- U = 0$, where function U characterizes the electromagnetic field equivalent to the boundary condition, which is absorbing the waves and tends to leave D through the left border. At the same time, all plane waves incident on the boundary at any angle are absorbed. For the right edge of the computational domain the absorbing condition is the equation $G^+ U = 0$. Similarly, we seek a factorization of the d'Alembert operator in the formulation of the absorbing conditions at the upper and lower boundaries.

Practical implementation of the approach is associated with the decomposition of the radical from (2.38) into a series. For example, taking

$$\sqrt{1-S^2} \cong 1 - \frac{1}{2}S^2. \tag{2.39}$$

we write

$$G^- \cong D_z - \frac{D_t}{c} + \frac{cD_y^2}{2D_t}.$$

Then, the absorbing boundary conditions take the form:

$$\frac{\partial^2 U}{\partial z \partial t} - \frac{1}{c}\frac{\partial^2 U}{\partial t^2} + \frac{c}{2}\frac{\partial^2 U}{\partial y^2} = 0 \text{ on the left boundary,} \tag{2.40}$$

$$\frac{\partial^2 U}{\partial z \partial t} + \frac{1}{c}\frac{\partial^2 U}{\partial t^2} - \frac{c}{2}\frac{\partial^2 U}{\partial y^2} = 0 \text{ on the right boundary,} \tag{2.41}$$

$$\frac{\partial^2 U}{\partial y \partial t} - \frac{1}{c}\frac{\partial^2 U}{\partial t^2} + \frac{c}{2}\frac{\partial^2 U}{\partial z^2} = 0 \text{ at the upper boundary,} \tag{2.42}$$

$$\frac{\partial^2 U}{\partial y \partial t} + \frac{1}{c}\frac{\partial^2 U}{\partial t^2} - \frac{c}{2}\frac{\partial^2 U}{\partial z^2} = 0 \text{ at the lower boundary.} \tag{2.43}$$

The difference approximation (2.40)–(2.43) for the solution of Maxwell's equations associated with the name Mur [28], and the absorbing boundary conditions are also often referred to as the Mur and the whole approach as a whole.

The 'bottleneck' of the method is the decomposition (2.39) which has been repeatedly improved [29.7] and still remains a source of error.

A better approach does not involve formulation of the boundary conditions other than those listed in section 2.1.1. The absorption of the field leaving the computational domain is achieved by arranging a specific evironment at the borders Γ which does not transmit electromagnetic radiation and does not reflect it.

One of the methods of constructing such an environment is associated with the representation of Maxwell's equations in the form of [8.30]:

$$\text{rot}\,\mathbf{H} = \frac{\partial \mathbf{D}}{\partial t} + \mathbf{j},\ \text{rot}\,\mathbf{E} = -\frac{\partial \mathbf{B}}{\partial t} - \mathbf{j}^*, \tag{2.44}$$

where j^* is the density of magnetic current, which is equal to $\sigma^* H$, σ^* is the magnetic conductivity of the medium. Subject to the conditions

$$\sigma / \varepsilon_0 = \sigma^* / \mu_0 \tag{2.45}$$

the wave impedance in such an environment for $\varepsilon = \mu = 1$ is the wave impedance in vacuum, therefore, there is no reflection from the absorbing layer (in the incidence on it of a plane wave at an arbitrary angle) [7]. In the layer the wave energy spill over into the energy of currents and the field is damped.

In the one-dimensional case, the equations (2.44) can be written as

$$\varepsilon_0 \varepsilon \frac{\partial E_x}{\partial t} + \sigma E_x = -\frac{\partial H_y}{\partial z},\ \mu_0 \frac{\partial H_y}{\partial t} + \sigma^* H_y = -\frac{\partial E_x}{\partial z}. \tag{2.46}$$

The two-dimensional version looks like:

$$\mu_0 \frac{\partial H_y}{\partial t} + \sigma^* H_y = -\frac{\partial E_x}{\partial z},\ \mu_0 \frac{\partial H_z}{\partial t} + \sigma^* H_z = \frac{\partial E_x}{\partial y},$$

$$\varepsilon_0 \varepsilon \frac{\partial E_x}{\partial t} + \sigma E_x = \frac{\partial H_z}{\partial y} - \frac{\partial H_y}{\partial z}.$$

Introducing the propagation and attenuation of the field along different directions by separate equations leads to the splitting of the electric component and conductivities, recording [8,7]:

$$\mu_0 \frac{\partial H_y}{\partial t} + \sigma_z^* H_y = -\frac{\partial \left(E_{xy} + E_{xz}\right)}{\partial z}, \tag{2.47}$$

and

$$\frac{\partial H_z}{\partial t} + \sigma_y^* H_z = \frac{\partial \left(E_{xy} + E_{xz}\right)}{\partial y}, \tag{2.48}$$

$$\varepsilon_0 \varepsilon \frac{\partial E_{xy}}{\partial t} + \sigma_y E_{xy} = \frac{\partial H_z}{\partial y}, \tag{2.49}$$

$$\frac{\partial E_{xz}}{\partial t} + \sigma_z E_{xz} = -\frac{\partial H_y}{\partial z}, \tag{2.50}$$

where $E_x = E_{xy} + E_{xz}$, and condition (2.45) should be observed for the corresponding projections of the conductivities.

Then the attenuation in the direction Z will provide the non-zero conductivity, σ_z^*, σ_z in the solution of (2.47), (2.50). In the direction of $Y - \sigma_z^*, \sigma_y$, in the solution of (2.48), (2.49).

The location of the absorbing layers in the one case (Fig. 2.9) corresponds to the placement of the domain in the shell.

In computational practice [8,7] the conductivities are defined by determining their value in the absorbing layers using the rule

$$\sigma = \sigma_{\max}\left(\frac{L - L_z + z}{L}\right)^q \text{ To the right of the layer } (L_z - L \le z \le L_z),$$

$$\sigma = \sigma_{\max}\left(\frac{L - z}{L}\right)^q \text{ To the left of the layer } (0 \le z \le L),$$

where $q \in R$. Thus, the conductivity of the layer increases towards the boundary and reaches its maximum value there. The magnetic conductivity σ^* is defined in terms of σ (2.45).

The two-dimensional case corresponds to the location of the absorbing layers, shown in Fig. 2.10 [8.7].

In the subareas 2 and 6 (Fig. 2.10) there are non-zero components σ_y and σ_y^*. In subareas 4 and 8 components σ_z and σ_z^*, in 1, 3, 5, 7 both projections σ and σ^* different from zero (provided the attenuation of a wave propagating in any direction).

Then

$$\sigma_y = \sigma_{\max}\left(\frac{L - y}{L}\right)^q, \text{ at } 0 \le y \le L \text{ in subregions 1, 2, 3;}$$

$$\sigma_y = \sigma_{\max}\left(\frac{L - L_y + y}{L}\right)^q, \text{ at } Ly\text{–}L \le y \le Ly \text{ in subregions 5, 6, 7;}$$

$$\sigma_z = \sigma_{\max}\left(\frac{L - z}{L}\right)^q, \text{ at } 0 \le z \le L \text{ in subregions 1, 8, 7;}$$

$$\sigma_z = \sigma_{\max}\left(\frac{L - L_z + z}{L}\right)^q, \text{ at } L_z\text{–}L \le z \le L_z \text{ in subregions 3, 4, 5.}$$

Any projection of the magnetic conductivity is determined by the corresponding projection of the electrical conductivity of (2.45).

2.1.3.2. The difference approximation of Maxwell's equations in absorbing layers

The location of the absorbing layers at the borders of the computational domain allows any boundary conditions to be set. It is traditionally accepted [8,7] to set the electric wall.

Then in the one-dimensional solution (2.46) on D_h^1 we write the Yee explicit difference scheme in the absorption region:

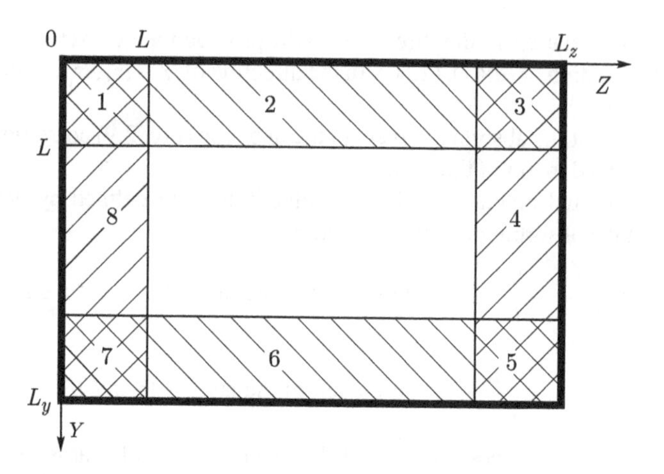

Fig. 2.10. Position of absorbing lauers in the two-dimensional case. The layers are crosshatched, L_z is the length of the computational domain in direction Z; L_y is the length of the computational domain in direction Y; L is the thickness of the absorbing layers.

$$\begin{aligned}
&\mu_0 \frac{H_{y_{k+0.5}}^{m+0.5} - H_{y_{k+0.5}}^{m-0.5}}{h_t} + \sigma_{k+0.5}^{*} H_{y_{k+0.5}}^{m-0.5} = -\frac{E_{x_{k+1}}^{m} - E_{x_k}^{m}}{h_z}; \\
&\varepsilon_0 \varepsilon_k \frac{E_{x_k}^{m+1} - E_{x_k}^{m}}{h_t} + \sigma_k E_{x_k}^{m} = -\frac{H_{y_{k+0.5}}^{m+0.5} - H_{y_{k-0.5}}^{m+0.5}}{h_z},
\end{aligned} \tag{2.51}$$

where the subscripts $1 \le k \le L / h_z$ correspond to left absorbing layer for the electric field and $0 \le k \le L/h_z - 1$ for magnetic, and $K - L / h_z \le k \le K - 1$ to the right layer for the electric field and $K - L / h_z \le k \le K-1$ for the magnetic field (Fig. 2.9).

In the two-dimensional solution for (2.47)–(2.5) on D_h^1 we wrire the Yee explicit difference scheme in the absorption region:

$$\begin{aligned}
&\mu_0 \frac{H_{y_{j,k+0.5}}^{m+0.5} - H_{y_{j,k+0.5}}^{m-0.5}}{h_t} + \sigma_{z_{j,k+0.5}}^{*} H_{y_{j,k+0.5}}^{m-0.5} = -\frac{E_{xy_{j,k+1}}^{m} + E_{xz_{j,k+1}}^{m} - E_{xy_{j,k}}^{m} - E_{xz_{j,k}}^{m}}{h_z}; \\
&\mu_0 \frac{H_{z_{j+0.5,k}}^{m+0.5} - H_{z_{j+0.5,k}}^{m-0.5}}{h_t} + \sigma_{z_{j+0.5,k}}^{*} H_{z_{j+0.5,k}}^{m-0.5} = \frac{E_{xy_{j+1,k}}^{m} + E_{xy_{j+1,k}}^{m} - E_{xz_{j,k}}^{m} - E_{xz_{j,k}}^{m}}{h_y}; \\
&\varepsilon_0 \varepsilon_{j,k} \frac{E_{xy_{j,k}}^{m+1} - E_{xy_{j,k}}^{m}}{h_t} + \sigma_{y_{j,k}} E_{xy_{j,k}}^{m} = \frac{H_{z_{j+0.5,k}}^{m+0.5} - H_{z_{j-0.5,k}}^{m+0.5}}{h_y};
\end{aligned} \tag{2.52}$$

$$\varepsilon_0 \varepsilon_{j,k} \frac{E_{xz_{j,k}}^{m+1} - E_{xz_{j,k}}^{m}}{h_t} + \sigma_{z_{j,k}} E_{xz_{j,k}}^{m} = -\frac{H_{y_{j,k+0.5}}^{m+0.5} - H_{y_{j,k-0.5}}^{m+0.5}}{h_z}.$$

We restrict the grid subdomain 1 (Fig. 2.10) the indices $1 \leq j \leq L / h_y$ and $1 \leq k \leqslant L / h_z$ – for E_x (both components of the cleavage), $1 \leqslant j \leq L / h_y$ and $0 \leq k \leq L/h_z - 1$ – to H_y, $0 \leq j \leq L / h_y -1$ and $1 \leq k \leq L / h_z$ – for H_z; subdomain 2 indices $1 \leq j \leq L / h_y$ and $L / h_z +1 \leq k \leq K–L/h_z–1$ – For E_x, $1 \leq j \leq L / h_y$ and $L / h_z \leq k \leq K–L/h_z–1$ – for H_y, $0 \leq j \leq L/h_y –1$ and $L / h_z +1 \leq k \leq K–L/h_z–1$ – for H_z; subdomain three indices $1 \leq j \leq L / h_y$ and $K–L / h_z \leq k \leq K–1$ – for the E_x, $1 \leq j \leq L / h_y$ and $K–L / h_z \leq k \leq K–1$ – for H_y, $0 \leq j \leq L/h_y –1$ and $K–L / h_z \leq k \leq K–1$ – for H_z; subdomain four indices $L / h_y +1 \leq j \leq J–L/h_y–1$ and $K–L / h_z \leq k \leq K–1$ – for E_x, $L / h_y +1 \leq j \leq J–L/h_y–1$ and $K–L / h_z \leq k \leq K–1$ – for h_y, $L / h_y \leq j \leq J–L/h_y–1$, $K–L / h_z \leq k \leq K–1$ – for H_z; 5 subdomain indexes $J–L / h_y \leq j \leq J–1$ and $K–L / h_z \leq k \leq K–1$ – for the E_x, $J–L / h_y \leq j \leq J–1$ and $K–L / h_z \leq k \leq K–1$ – for H_y, $J–L / h_y \leq j \leq J–1$, $K–L / h_z \leq k \leq K–1$ – for H_z; subdomain six indices $J–L / h_y \leq j \leq J–1$ and $L / h_z +1 \leq k \leq K–L / h_z –1$ – for E_x, $J–L / h_y \leq j \leq J–1$ and $L / h_z \leq k \leq K–L/h_z–1$ – for H_y, $J–L / h_y \leq j \leq J–1$ and $L / h_z +1 \leq k \leq K–L / h_z–1$ – for H_z; subdomain index 7 $J–L / h_y \leq j \leq J–1$ and $1 \leq k \leq L / h_z$ – for E_x, $J–L / h_y \leq j \leq J–1$'s and $0 \leq k \leq L/h_z – 1$ – to H_y, $J–L / h_y \leq j \leq J–1$ and $1 \leq k \leq L / h_z$ – for H_z; subdomain index 8 $L / h_y +1 \leq j \leq J–L/h_y–1$ and $1 \leq k \leq L / h_z$ – For E_x, $L / h_y +1 \leq j \leq J–L/h_y–1$ and $0 \leq k \leq L/h_z–1$ – for H_y, $L / h_y \leq j \leq J–L/h_y–1$ and $1 \leq k \leq L / h_z$ – for H_z. Then the grid subdomain without absorption located in the range of $L / h_y +1 \leq j \leq J–L/h_y–1$ and $L / h_z +1 \leq k \leq K–L/h_z–1$ for E_x, $L / h_y +1 \leq j \leq J–L/h_y– 1$ and $L / h_z \leq k \leq K–L/h_z–1$ – for H_y, $L / h_y \leq j \leq J–L/h_y–1$ and $L / h_z +1 \leq k \leq K–L/h_z–1$ – for H_z. The thickness of the absorbing layers L is selected in multiple steps of discretization (so that the results of all divisions are integers).

2.1.3.3. Association of absorbing layers in vectorization of calculations

The vector algorithms discussed in 2.1, will also be used in the solution (2.51), (2.52) with the following additions.

In the one-dimensional case, the algorithm for solving Maxwell's equations operated with the vector of the grid function of the strength of the electric field of length K–1 and the vector of the grid function of the magnetic field strength of length K. The imposition of absorbing layers reduces the length of the vectors, and their numbers increases due to the need for separate solutions of equations (2.46) in the absorbing layers and (2.1), (2.2) in the subregion without being absorbed. Namely, putting $w = L / h_z$ for the number of nodes in the grid region, coinciding with the absorbing layers, we will continue to operate with two vectors of length w and a single vector of length of K–1–2w in the calculation of the electric field and the two vectors of length w and a single vecto of length K–2w in calculation of the magnetic field.

Reducing the length of vectors as a rule leads to an increase in the duration of the calculations [19.20] (even at a fixed computational complexity of the algorithm), and complicates writing and adjusting the programs.

In an effort to preserve the old length of the vectors, we can write equation (2.46), not only in absorbing layers, but on the whole grid area D_h^1, using in a non-absorbing medium values of the conductivity equal to zero. However, this method is characterized by the need to allocate additional memory to store the input values of the conductivities (even though they are equal to zero), and by an increase in the number of arithmetic operations (even if these are operations with zeros).

Another approach [18], although it does not not reduce the number of vectors to one for each component of the field, it can reduce it from three to two. We write the cyclic boundary condition for for the computational domain instead of setting an electric wall, and solve the equation (2.51) for $\tilde{D}_h^1$ instead of D^1_{h}.

This technique can be used due to the property of absorbing layers not to transmit the electromagnetic radiation. Radiation does not propagate to the edges of the field, therefore, for the calculation accuracy it is not important which of the three boundary conditions is located there.

In addition, this change allows us to look at the computational domain somewhat differently: for the point $z = 0$ we can choose any node $\tilde{D}_h^1$, since the circular area has no edges. Shifting the origin on L to the right (Fig. 2.9) we come to the area in Fig. 2.11, which also corresponds to $\tilde{D}_h^1$.

As a result, two absorbing layers of length L are merged into one layer with length $2L$, for which

$$\sigma = \sigma_{\max}\left(\frac{2L - L_z + z}{L}\right)^q \quad \text{To the left side of the layer } (L_z - 2L \le z \le L_z - L);$$

$$\sigma = \sigma_{\max}\left(\frac{L_z - z}{L}\right)^q \quad \text{To the right side of the layer } (L_z - L \le z \le Lz).$$

Then, in the solution of (2.1), (2.2) in the non–absorbing subregion and (2.46) in the combined absorbing layer, we will operate with two vectors of the grid function of the strength of the electric field of length K–1–2w, 2w, and two vectors of the grid function of magnetic field strength of length K–2w, 2w.

Turning to the two-dimensional case, we note that after the imposition of absorbing layers, the electric field vector and both components of the magnetic field split into three subvectors, regardless of the method of storage: a row- or column-orinted. In the row-oriented method decomposition is performed along the direction Z, in column one – along Y.

Unlike the one-dimensional case the increase in the allocated memory when organizing the calculations will be associated not only with the need to store the values of conductivity (which in the scheme (2.17) – (2.19) was not the case), but also with the placement in the memory the electric field split into two components (in absorbing layers). Therefore, leaving the old length of the vectors we have to place in the memory permeability values and split the components of the electric field in a non-absorbing subregion.

In search of a compromise variant, we can write (2.52) in $\tilde{D}_h^2$ in place of D_h^2 [18], replacing the electric wall at the boundary by the cyclic conditions. In

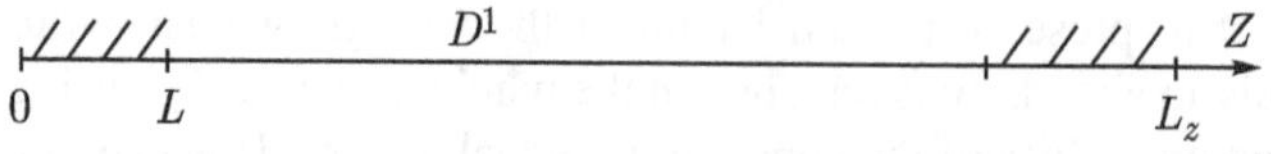

Fig. 2.11. The location of the absorbing layers in the one-dimensional case in the formulation of the cyclic boundary conditions and shift of the origin on the coordinates, $2L$ is the thickness of the combined absorbing layers.

the one-dimensional case, such a change corresponds to the transition from a segment to a ring, in the two-dimensional case from a rectangle to a torus.

Shifting the origin on L down and to the right (Fig. 2.10) we come to the area in Fig. 2.12, which also corresponds to $\tilde{D}_h^2$.

In layer A (Fig. 2.12) the layers 8 and 4 merge (Fig. 2.10); in layer B the layers 1, 3, 5, 7 merge, and in C – 2, 6. The following equalities hold for the projection of the electrical conductivity on the Y-axis:

$$\sigma_y = \sigma_{\max}\left(\frac{2L - L_y + y}{L}\right)^q \text{ at the top of the layers C, B } (L_y - 2L \le y \le L_y - L);$$

$$\sigma_y = \sigma_{\max}\left(\frac{L_y - y}{L}\right)^q \text{ at the bottom of the layers, C, B } (L_y - L \le y \le L_y).$$

The projection of the electrical conductivity on Z satisfies the expression:

$$\sigma_z = \sigma_{\max}\left(\frac{2L - L_z + z}{L}\right)^q \text{ the left side of the layers A, B } (L_z - 2L \le z \le L_z - L);$$

$$\sigma_z = \sigma_{\max}\left(\frac{L_z - z}{L}\right)^q \text{ the right side of the layers A, B } (L_z - L \le z \le L_z).$$

Then, in the solution of (2.18) – (2.19) in a non-absorbing subregion and (2.52) in the absorbing layers, we will operate with the two vectors when setting the values of network functions of the two projections of the magnetic field. The electric field in a non-absorbing subregion is specified by one vector and two vectors of the split components in the absorbing layers.

The proposed layout of the absorbing layers in the network domains $\tilde{D}_h^1$ and $\tilde{D}_h^2$ can not only shorten the calculations, but also greatly simplify the writing and debugging the code. Earlier in the simulation of diffraction in the two[-dimensional region it was necessary to finite difference equations in eight absorbing layers and one non-absorbing subregion, also to match the solutions at all 12 boundaries of the layers and subregions.

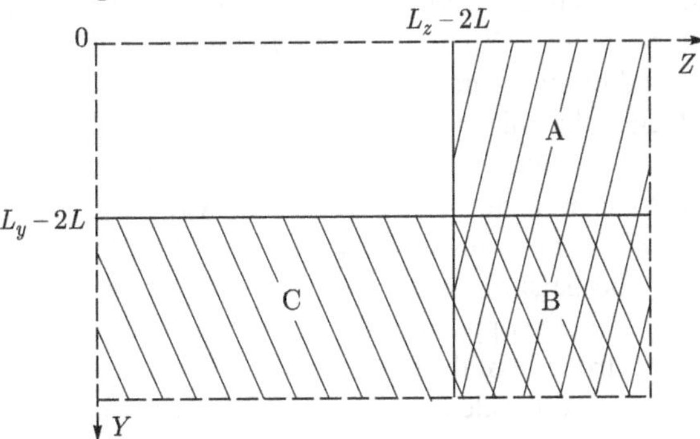

Fig. 2.12. Location of absorbing layers in the two-dimensional case in the formulation of cyclic boundary conditions and the shift of the origin, $2L$ – combined thickness of the absorbing layers.

For example, during the transition from layer 8 (Fig. 2.10) to the central subregion in the direction Z the grid projection of the magnetic field on the Y-axis does not require the determination by the formulas (2.52) for the absorbing layer and not by (2.17)–(2.19) for the non-absorbing subregion, and needs to write a difference equation that contains the electric field of the central subregion and split components from layer 8:

$$\mu_0 \frac{H_{y_{j,k+0.5}}^{m+0.5} - H_{y_{j,k+0.5}}^{m-0.5}}{h_t} = -\frac{E_{x_{j,k+1}}^{m} - E_{xy_{j,k}}^{m} - E_{xz_{j,k}}^{m}}{h_z},$$

where $L/h_y+1 \le j \le J - L/h_y - 1$ and $k = L / h_z$.

After the unification of the layers the solution is obtained in three absorbing layers and a non-absorbing subregion with the coordination of the fields at eight boundaries (Fig. 2.12). The number of subregions declined by more than half and that of the the borders by a third; the code, written in Fortran 90 and Matlab, allowing the organization of vector calculations, was approximately halved

2.1.3.4. Universal grid areas

In addition to simplicity, the presented layout areas $\tilde{D}_h^1$ and $\tilde{D}_h^2$ are characterized by high flexibility. The one-dimensional or two–dimensional difference Yee scheme on overlapping rectangular grid areas with any boundary conditions from section 2.1.1 can be reduced to writing equations on $\tilde{D}_h^1$ or $\tilde{D}_h^2$. Confirming this, we consider several examples, most frequently encountered in modelling the diffraction of laser radiation on microoptics elements. In this case, the mentioned grid areas with the merged absorbing layers will be called universal.

Example 1. A bounded cylindrical element

One of the most common cases is the study of a finite cylindrical optical element fully placed in D^2. Assume that the parameters of absorbing layers (values of the conductivities and the value of the exponent q) have been selected previously for a perfectly conducting shell used normally in solving this problem [7]. Indeed, replacement of electric walls (2.20) by the cyclic condition (2.27) – (2.31) will lead to some (albeit small) drop in accuracy.

In practice, reducing the thickness of the absorbing layers, their parameters are selected in such a way that the weakened scattered wave reaches the electric wall, reflects from it and is finally absorbed on the way back through the layer. Replacement of the electrical wall by the cyclic condition leads to a change in the phase of the back propagating wave which does not reflected from a perfect conductor, and comes from the opposite edge of the computational domain (special features of the toroidal structure of the grid area $\tilde{D}_h^2$). The wave with the new phase will be absorbed in the layer, calculated for a different field, to a lesser extent.

If the new selection of parameters of the absorbing layers for achieve the former accuracy is tedious for the researcher, or he prefers to use the standard values, the following small addition to the calculation algorithm is made to improve the situation.

Instead of the boundaries of the region (the cyclic conditions are already defined there), the electric wall is placed in the absorbing layers. To do this, imagine that the transformation to a torus with the union of absorbing layers takes place in region D^2_h and not $\tilde{D}^2_h$. We merge the electrical walls at the boundaries $z = 0$, $z = L_z$ $(0 \leq y \leq L_z)$ which gives a cylindrical surface; then we combine the electrical walls at the boundaries $y = 0$, $y = L_z$ $(0 \leq z \leq L_z)$ which gives a torus; the origin of the coordinates is then shifted by L to the right and down (Fig. 2.12). This gives the universal grid region (Fig. 2.12) with the electrical walls located on the segments $y = L_y - L$, $0 \leq z < L$ and $z = L_z - L$, $0 \leq y \leq L_y$ (Fig. 2.13)

Performing calculations on such a field, it is not necessary to use the electric and magnetic walls as boundary conditions, writing in their vicinity the difference equations different from (2.52); this will increase the duration of the calculation, offsetting the gains made at the expense of more successful vectorization.

It is reasonable after the transition to the next temporal layer to reset the values of the network functions of the electric field at the specified intervals. It is enough to multiply the two corresponding vectors by a scalar (elementary vector operation is done in hardware), which is equal to zero. The accuracy of the simulation results on the universal grid area after the proposed modification of the algorithm exactly coincides with the simulation results on D^2_h.

Example 2. An infinite two-dimensional diffraction grating

We assume that the investigated element is periodic along the axis Y. It is then sufficient to assumed in the universal grid area that σ_y and σ^*_y are equal to zero to ensure that layer C (Fig. 2.12) does not absorbed the radiation propagating in the directions Y and $-Y$, and layer B absorbs only in the directions Z and $-Z$. In this case C actually enters into a non-absorbing subarea, the layers A and B can be taken as a single absorbing layer. Similarly, modification of the universal domain in the study of an infinite element, periodic along the axis Z, is performed in the same manner

Coordinating solutions on the traditional grid region [23], which combines electric walls at the boundaries $z = 0$, $0 \leq y \leq L_z$ and $z = L_z$, $0 \leq y \leq L_y$ with the cyclic boundary conditions at $y = 0$, $0 \leq z \leq L_z$ and $y = L_y$, $0 \leq z \leq$

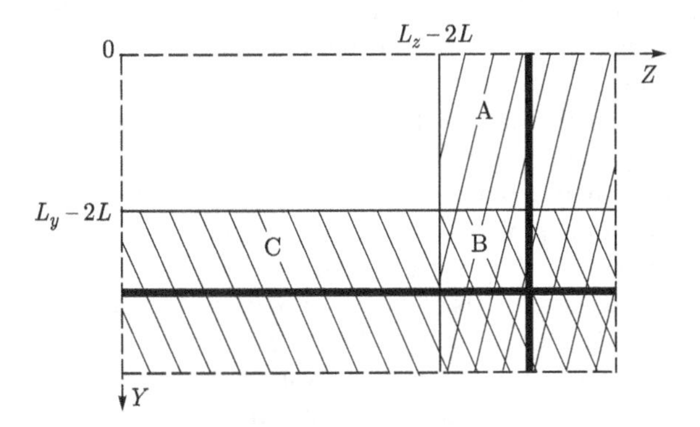

Fig. 2.13. The location of the electrical wall on a universal grid area in the study of an isolated cylindrical optical element. The electric wall is marked by thick bars.

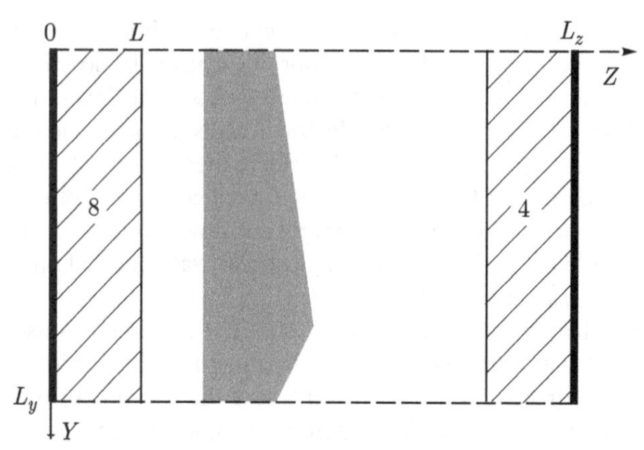

Fig. 2.14. The traditional layout of the computational domain for simulating infinite periodic optical element (a combination of electric walls and cyclic conditions).

L_z (Fig. 2.14), and the universal grid area, in the combined absorbing layer A we place the electrical wall on the interval $z = L_z - L$, $0 \le y \le L_y$ (Fig. 2.15). Then the two solutions on these grids coincide.

Example 3. A bounded symmetric cylindrical element
In the study of the propagation of radiation through a bounded symmetrical cylindrical element the magnetic wall is placed on the axis of symmetry. The grid area $\bar{D}_h^2$ encompasses half of the element and in the variant without combining absorbing layers is shown in Fig. 2.16.

After the combination of layers 4 and 8 (Fig. 2.16) and the transition to $\bar{D}_h^2$, the area of computer simulation will take the form shown in Fig. 2.17.

When the layer 2 in Fig. 2.16a becomes the layer C in Fig. 2.17a, layers 1 and 3 after the merger go to B, sections 4, 8 – to A. Layer 6 in Fig. 2.16b changes to layer C in Fig. 2.17b, layers 5 and 7 after the merger go to B, sections 4, 8 – to A.

Four magnetic walls at the boundaries $\bar{D}_h^2$ change into two walls in the area, located on the segments $y = L$, $0 \le z \le L_z$ and $z = L_z - L$, $0 \le y \le L_y$ (for the case shown in Fig. 2.16a) and at intervals of $y = Ly - L$, $0 \le z \le L_z$ and $z = L_z - L$, $0 \le y \le L_y$ (for the case shown in Fig. 2.17b)

As in *Example 1*, in the universal area one should not regard the magnetic walls as the boundary conditions by writing in their neighborhood new difference equations. It is enough before moving on to the next layer in time to reset the grid components of the vector of projection of the magnetic field projection on the *Z*-axis at $y = L$, $0 \le z \le L_z$ and $z = L_z - L$, $0 \le y \le L_y$ (Fig. 2.17a) or $y = L_y - L$, $0 \le z \le L_z$; also we should reset the grid components of the vector of projection of the magnetic field on the *Y* axis at $z = L_z - L$, $0 \le y \le L_y$ (Fig. 2.17a) or $z = L_z - L$, $0 \le y \le L_y$ (Fig. 2.17b).

It is not compulsory to place the magnetic walls in layers A and B (Fig. 2.17). Surrounded on both sides by an absorbing medium, it will not have a decisive influence on the result of computer simulation. And yet, its absence would cause a slight (usually in the third place for a sufficiently dense mesh) mismatch in the values

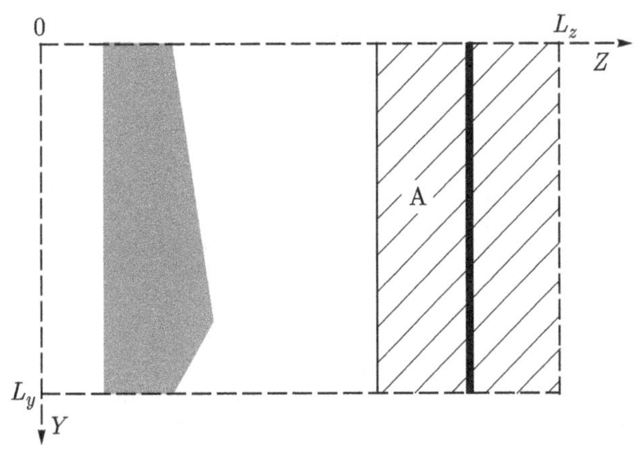

Fig. 2.15. The location of the electrical wall on a universal grid area in the study of an isolated cylindrical optical element.

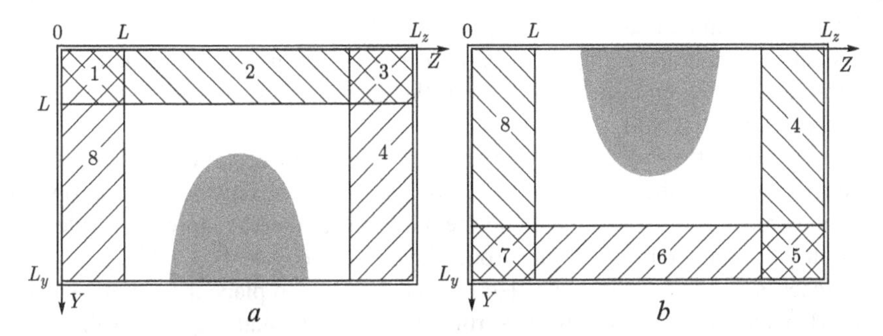

Fig. 2.16. Location of absorbing layers in the field $\bar{D}_h^2$ without their union. Case a) corresponds to the placement into the upper half of the symmetrical elements, case b) – the lower one.

of field strengths in solution on $\bar{D}_h^2$ and $\tilde{D}_h^2$. The reason for this, as in Example 1, is setting the parameters of an absorbing layer on the interaction of the scattered wave with the magnetic wall. Without this absorption at these parameters will be less effective. One should either install a wall, or pick up new parameters of the layers (absorbing the radiation propagating in the direction of Z), to ensure an acceptable accuracy. The latter option requires different settings of the absorption in the *Y*-(where the magnetic wall remains) and *Z*-directions (where the wall is removed).

Example 4. An infinite periodic symmetric element

When considering in Section 2.1 the boundary conditions applied to modelling the propagation of electromagnetic waves through an infinite periodic symmetric element, it was concluded that it is adequate to install the magnetic walls on all boundaries of the computational domain. Taking the axis *Y* as the periodicity direction, the absorbing layers are positioned on $\bar{D}_h^2$ as shown in Fig. 2.18.

The transition to the universal net area is associated with the adoption of conductivity σ_y and σ_y^* equal to zero (as in *Example 2*) and the placement

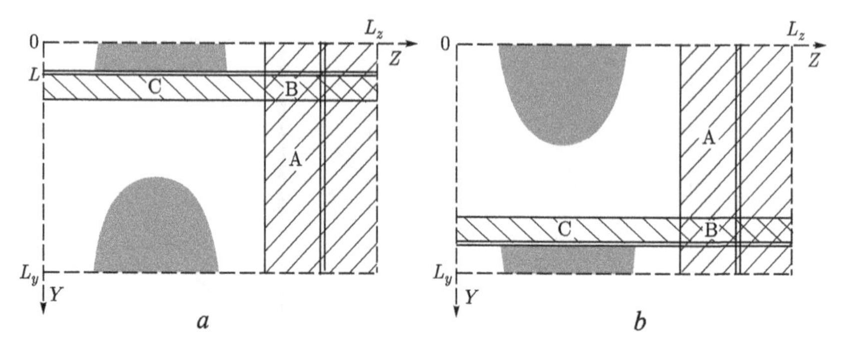

Fig. 2.17. The location of the absorbing layers and magnetic walls (double line) on a universal grid domain in simulation of symmetric elements. Case a) corresponds to the placement in the area in the upper half of the symmetric elements, case b) – the lower one.

of magnetic walls at intervals of $y = L$, $0 \leq z \leq L_z$ and $z = L_z - L$, $0 \leq y \leq L_y$. Then the computational domain takes the form shown in Fig. 2.19.

As in the previous examples, it is not necessary to place the magnetic wall in layer A (Fig. 2.19), and the parameters of the absorbing layers can be selected for the case without the wall.

As seen from the four examples, varying the values of vector components of the conductivities σ, σ^* and placing the electric or magnetic walls inside $\tilde{D}_h^2$ the proposed optical elements can be studied using the universal grid area. Moreover, selecting the optimal parameters of the absorbing layers in areas A, B, C (Fig. 2.12), the researcher is free not to place the electric wall on $\tilde{D}_h^2$ and place the magnetic walls only when they are not inside the absorbing layers (horizontal magnetic walls in Fig s. 2.16, 2.19). This optimization eliminates the need to reset before switching to a new temporary layer of the corresponding components of the electromagnetic field

2.1.4. Incident wave source conditions

Modelling the propagation of radiation through the optical element, in addition to imposing the grid region and write on it difference equations, it is also necsaary to define the field coming from the outside and incident on the element.

Indeed, the result will depend not only on the geometry of the investigated optical element and the material from which it is made, but also on the type of incident electromagnetic wave – the distribution of complex amplitudes of the projections of its vectors in space and time.

A separate task should be matching of the techniques of setting the incident field, boundary conditions, the method of imposing absorbing layers and topography of the investigated element. Some methods of forming the incident wave are used in the study of elements, working on in transmission and reflection, periodical and non-periodical deposited on a substrate or without it, located in free space or formed at the end of the optical waveguide.

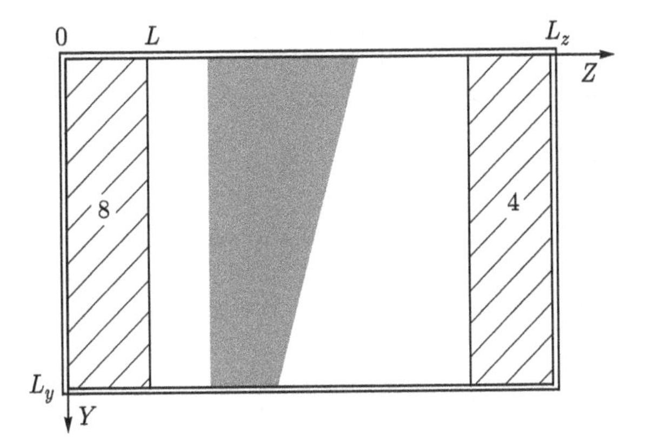

Fig. 2.18. The location of the absorbing layers in the region without combining them in the study of the periodic symmetric element.

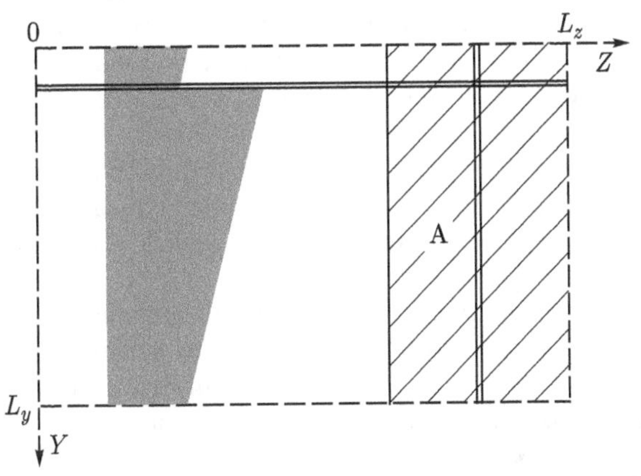

Fig. 2.19. The layout of the universal grid area in the study of the periodic symmetric element.

The choice of the method of forming the incident wave will determine the features of the implementation of algorithms for the difference approximations. The type of method determines both the accuracy and duration of computer simulation. Vector and parallel algorithms are written in different ways for different technology tasks of the incident field.

By limiting the scope of the subject area of optics, we exclude from consideration the methods of excitation of the electromagnetic field by the currents, characteristic of electromagnetic problems in general and in particular the theory of antennas [7]. Moreover, we ignore for a time the physical nature of the radiation source and associated methods for defining the field.

Leaving aside the problem of pulse propagation, we will not deal with the problem of their formation and entry into the computational domain, even though all of the following methods can be easily adapted to solve this problem.

In modern literature there are three main approaches to the task of defining the harmonic incident field in the study of diffraction on the optical structures by the finite difference solution of Maxwell's equations. Let us examine them in the development, with particular emphasis on modifications applied to the study of microelements, in particular, diffractive optical elements.

2.1.4.1. Hard source conditions

The first paper on the difference solution of Maxwell equations in differential form, authored by Yee [4], published in 1966, contained a description of the input method of the radiation in the computational domain, later named [7] using 'the hard source' method.

The method consists of defining for the selected area of space of the vectors of the electromagnetic field through an analytical representation of the form

$$G(\mathbf{x}_0) = \mathrm{Re}[A(\mathbf{x}_0)\exp(-i\omega t + \phi_0))], \tag{2.53}$$

where $\mathbf{x}_0$ are the coordinates of the point from the selected area; $G(\mathbf{x}_0)$ is the formed value of the selected projection of the strength of the component of the electromagnetic field in x_0 before proceeding to the next the temporary layer; $A(x_0)$ is the given complex amplitude of the incident field for the given projection, the power in the exponent determines the phase of the wave (not the phase of the complex amplitude) with an initial phase ϕ_0. and the cofactor with the exponent defines the harmony of the incident monochromatic field; ω is angular frequency; t is time; Re [..] is the operation of selection of the real part of the expression in square brackets.

In the one-dimensional case, organizing Yee computation expressions, for G it is sufficient to accept to either the electric or magnetic field. Let us consider the first option, defining the 'hard' source at node k of the universal grid area with absorbent layers $\tilde{D}_h^1$ with combined absorbing layers of the difference approximation of expression (2.35):

$$E_{x_k}^m = \mathrm{Re}\left[\exp\left(-i\left(\frac{2\pi h_t c}{\lambda}m - \frac{\pi}{2}\right)\right)\right]. \tag{2.54}$$

Furthermore, the geometrical dimensions of the computational domain and objects in the domain are measured in wavelengths (unless otherwise specified), putting (2.54) $\lambda = 1$. The parameters of discretization of the grid region are given in the form of pairs of numbers (Q, Q_t). The first number corresponds to the number of nodes in the grid area in space coinciding with the wavelength, the second – the number of nodes over time during which the plane wave front in a vacuum travels a distance of one wavelength.

Organizing the formulation of a computational experiment in free space [18], we take $L_z = 4\lambda$, $L = \lambda$, and the source is located in the leftmost node $\tilde{D}_h^1$ ($k = 0$). During the time $T = 20\,\lambda/c$ the field in the studied area is stabilized and can be considered monochromatic.

Choosing in $\tilde{D}_h^1$ discretization (Q, Q_t) = (10, 20), we consider the distribution of the module of the complex amplitude of the electric field in the area of computer simulation (Fig. 2.20).

In the subdomain of arrangement of the absorbing layers ($2 \le z \le 4$) the fieldd decaus, and in free space ($0 \le z \le 2$) a plane homogeneous *T*-wave propagates. Complete decay to the zero modulus of the complex amplitude in Fig. 2.20 does not occur, since the layers absorbs over its entire length, and the scattered radiation penetrates the layer from both sides.

By studying the dependence of the error of the difference solution on the discretization parameters (Table 2.1), we note the convergence of the difference method for solving Maxwell's equations to the analytical solution for the chosen parameters of the computational experiment.

Table 2.1 presents the values of the uniform error

$$\varepsilon = \max_k \left| \frac{|B_k| - |A_k|}{|B_k|} \right|, \tag{2.55}$$

where ($1 \le k \le 2Q$), which characterizes the maximum deviation from the analytical solution. In (2.55) value $|B_k|$ is the modulus of the complex amplitude of the electric field in the analytic solution ($|B_k| = 1$ V/m for (2.55)).

In the two-dimensional case, to obtain a homogeneous plane wave in the universal grid region we define a 'hard' a source in the interval $k = 0$; $0 \le j \le J{-}1$

$$E^m_{x_{j,k}} = \mathrm{Re}\left[\exp\left(-i\left(\frac{2\pi h_t c}{\lambda} m - \frac{\pi}{2}\right)\right)\right]. \tag{2.56}$$

Keeping the settings from the previous experiment, we give a square shape to D_2 ($L_y = L_z$).

The results of numerical experiments coincide with the results for the one-dimensional field, as the wave propagates along the direction *Z*.

Note that in the D_h^2 region the assignment of a plane homogeneous wave through the 'hard' source is impossible because of the discontinuity of the strength of the electric field at the edges D_h^2.

Figure 2.21 shows the distribution of the field from the point 'hard' source (in (2.56) $j = k = Q$) in free space, defining a cylindrical wave front wave with discretization parameters $(Q, Q_t) = (20, 40)$.

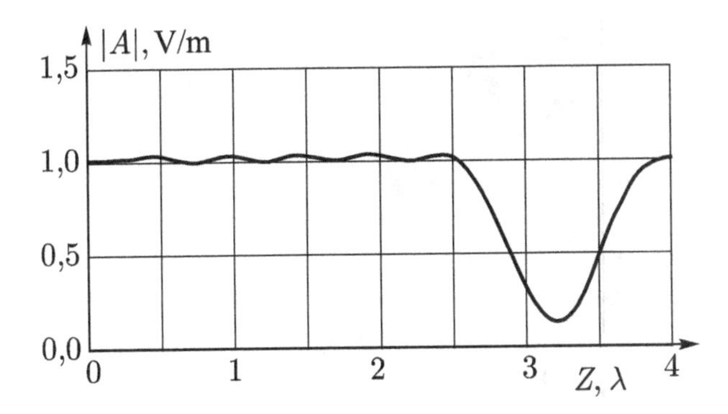

Fig. 2.20. The distribution of the modulus of the complex amplitude of electric field $|A|$ on $\tilde{D}_h^1$ from the 'hard' source.

Table 2.1. Dependence of the errors of numerical experiments for the vacuum on discretization of the grid area (Q, Q_t) and the parameters of absorbing layers σ_{max} and q for the 'hard" source'

(Q,Q_t)	Parameters of absorbing layers		Error values
	σ_{max}, cm/m	Q	ε, %
(10, 20)	0.018	1.5	1,3948
(20, 40)	0.024	2	0.10228
(50, 100)	0.032	2.5	0.03688
(100, 200)	0.037	2.7	0.010405

At the locations of the absorbing layer on a universal grid area ($0 \le y \le 2$, $2 \le z \le 4$ – layer A; $2 \le y \le 4$, $2 \le z \le 4$ – layer B and $2 \le y \le 4$, $0 \le z \le 2$ – layer C in Fig. 2.12) the field attenuates.

Characterized by the simplicity of definition and high accuracy, the 'hard' source is of limited use in computational practice. Definition of the incident wave by the equation (2.53) does not allow the wave reflected from the optical object being studied to pass through the 'hard' source and reach the absorbing layer.

Confirming this, we set up a computer experiment, which differs from the previous experiment by the first position of the radiation source (now $k = Q / 2$, $z = 0.5$) and by discretization of the grid area.

Figure 2.22 shows the result of simulation, when installed the electric wall is situated in a node $3Q / 2$ ($z = 1.5$) and reflects the incident wave back toward the source.

Considering the field in the subdomain $0 \le z \le 0.5$, we note the presence of the homogeneous wave radiated in the direction –Z by the 'hard' source that radiates in both directions. The mentioned wave attanueates in the absorbing layer $2 \le z \le 4$ and fades almost completely. Directly behind the electric wall $z = 1.5$ the field is absent. The field reflected from the wall does not pass over the source, and reflecting from it and the wall interferes with the field incident in the direction Z.

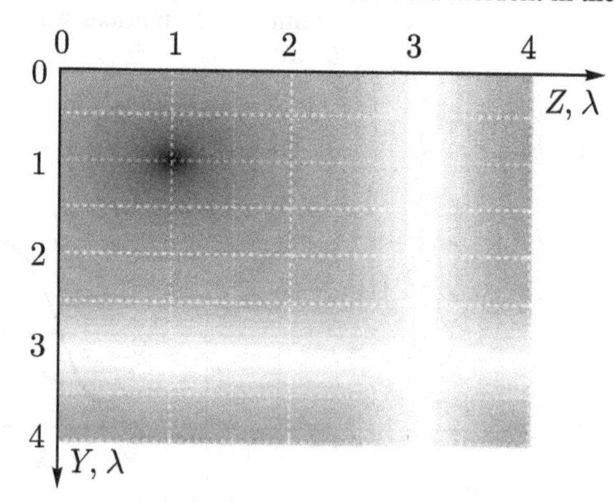

Fig. 2.21. The distribution of the modulus of the complex amplitude of the electric field |A| on $\tilde{D}_h^2$ from a point source.

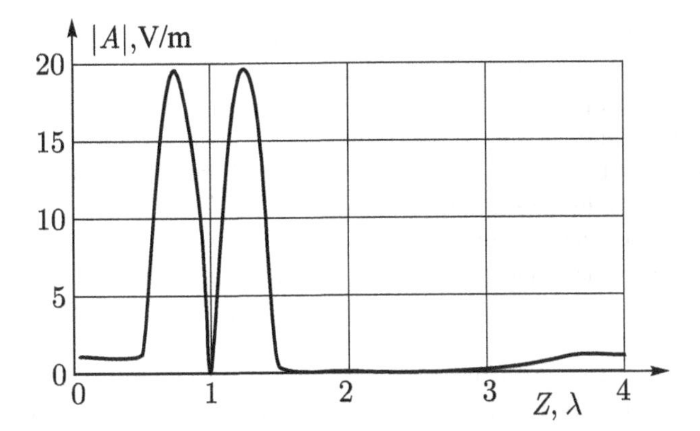

Fig. 2.22. The distribution of the modulus of the complex amplitude of the electric field |A| on $\tilde{D}_h^1$ in wave reflections in between the 'hard' power source and the electric wall.

Consequently, when using the 'hard' source the source must be sufficiently distant from the area of registration of the resulting field so that the wave reflecting from it does not return to the optical element, distorting the diffraction pattern. Using such a method involves a multiple increase in the computational domain and the duration of the experiment which in some cases makes numerical simulation impossible. The 'hard' source is used to solve the auxiliary problems (testing the model, building a 'transparent' source.)

2.1.4.2. The total field formulation method

Obviously, to account for the waves reflected from the object under study, we must somehow find it. In [9, 31] it is suggested to take over the reflected field at the node location of the source result of the calculation of the difference approximation to specify the incident wave before transition to the next grid step on the time coordinate. Then, adding the reflected field with the incident one, we thus define the resulting field in the location of the source. Calculations by this algorithm can be summarized as follows.

Step 1. Calculation of the field by the difference approximation for the entire region of the layer m.

Step 2. Determination of the reflected field as a result of the calculations in step 1 in the location of the source node.

Step 3. The calculation of the resultant field at the node of location of the source by adding the reflected field and the analytically calculated incident field with transition to the next time layer.

Later, in [10] the proposed algorithm was termed the total field formulation method.

Repeating the last experiment with the new source, we obtain the complex amplitude distribution shown in Fig. 2.22.

The wave wave from the electrical wall interferes with the incident wave in the direction Z (on $0.5 \leq z \leq 1.5$), passes through the source and is absorbed on

$2 \le z \le 4$ (Fig. 2.23). In the subregion $0 \le z \le 0.5$ we observe the superimposition of the wave emitted by the source in the direction Z and the wave reflected from the electrical wall passing through the source and retreating in the direction Z. In the analytic solution there are no oscillations in the specified interval, and in Fig. 2.23 the oscillations of the complex amplitude for $0 \le z \le 0.5$ are due to an error introduced into the solution by the source.

Studying the accuracy of the method of the resultant field, let us consider a homogeneous plane wave propagation in a free environment by repeating the experiments with the new source for 'hard' power, the results are presented in Table. 2.1.

Comparison of experimental results for free space with the 'hard' source (Table 2.1) and the source given by the total field formulation (see Table 2.2) in not in favour of the latter. The error of results has increased by an order for all discretizations.

The reason for this [18.32] is an error when setting the reflected wave. Developing the method of the general field, it is necessary to determine the reflected wave not as a grid field after the transition to the next time layer, but by setting it by the difference of such a field and a field in another grid areas, devoid of the optical element and therefore free of the reflected wave. Such a source in [18.32] is named 'transparent'.

We formulate the algorithm for defining the 'transparent' source.

Step 1. Field calculation by the difference approximation for the layer m in area with an optical element.

Step 2. Field calculation by the difference approximation for the layer m in the area without the optical element.

Step 3. Determination of the reflected field in the subregion of definition of the source as the difference of the fields between the subdomains found in the first two steps.

Step 4. Setting the resulting wave in the subregion of the source as the sum of analytically given incident field and the reflected field determined in the previous step. Transition a temporary layer $m+1$.

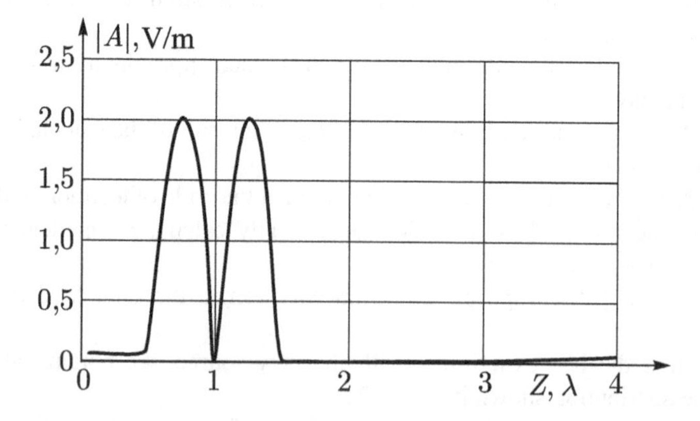

Fig. 2.23. The distribution of the modulus of the complex amplitude of the electric field |A| on $\tilde{D}_h^1$ when using the total field formulation method to specify the incident wave.

Table 2.2. Dependence of the errors of numerical experiments for vacuum on the discretization of the grid area (Q, Q_t) when using the general field method

(Q,Q_t)	(10, 20)	(20, 40)	(50, 100)	(100, 200)
ε, %	8.4609	2.2177	0.37411	0.10187

The calculation results in free space using a ‘transparent’ source coincided with the data from Table 2.1: the new source does not introduce any additional error in the difference solution.

By studying the field in the experiment with an electric wall, we see interference extinguishing of the wave in the subdomain $0 \leq z \leq 0.5$ (Fig. 2.24).

The wave reflected from the wall returned to the source in antiphase. Such an exact match with the analytical solutions indicates the absence of the error introduced by the radiation source in the difference solution.

Attention will be given to the formation of the field in the auxiliary problem. It is made in full accordance with 2.1.4.1, when the ‘hard’ source was used to set a uniform plane wave in free space. In forming the ‘transparent’ source it is important to choose the grid areas of the auxiliary and main tasks to be fully identical. Differences between the two tasks should be found only in the distribution of the refractive index. The main task has an optical element from which the reflected wave arrives to the source; the auxiliary task does not contain such element and not reflected wave forms in it. If we remove the scatterer also from the main task, the complex amplitude of the reflected wave, determined in setting the ‘transparent’ source, vanishes by virtue of the identity problems. This accounts for a full match (in all signs) of the calculation results in free space with the use of ‘hard’ and ‘transparent’ sources.

In the study of diffractive optical elements the source should be placed in the substrate element (not in vacuum), close to the microrelief in order to reduce the area (in three dimensions – volume) of the computational domain and the duration of the simulation. In this case, the auxiliary problem must contain a homogeneous medium with a refractive index of the substrate, that is the medium of the main task in which the source is located.

The above method of setting ‘transparent’ source is also true in the case of the two-dimensional computational domain. If in such a field we must set as a uniform a plane wave as an incident wave, it is sufficient for this purpose to process a one-dimensional auxiliary problem, which describes a wave. Only the main task, which contains a two-dimensional optical element, will be two-dimensional.

It is important to perform similar calculations on a universal grid area, because for the region with electric walls at the boundaries we can not form a homogeneous plane incident wave through the ‘transparent’ source. The boundary conditions, given by the electric walls, do not correspond to a uniform plane wave front. For the same reason, it is inappropriate to install the horizontal electric wall in the subareas B and C of the universal grid area (Fig. 2.12). That, however, in no way limits the researcher in choosing the optical elements for modelling, as noted in previous sections. Magnetic walls are compatible with the spread of a homogeneous plane incident wave; hence, symmetric optical elements can be explored through a ‘transparent’ source.

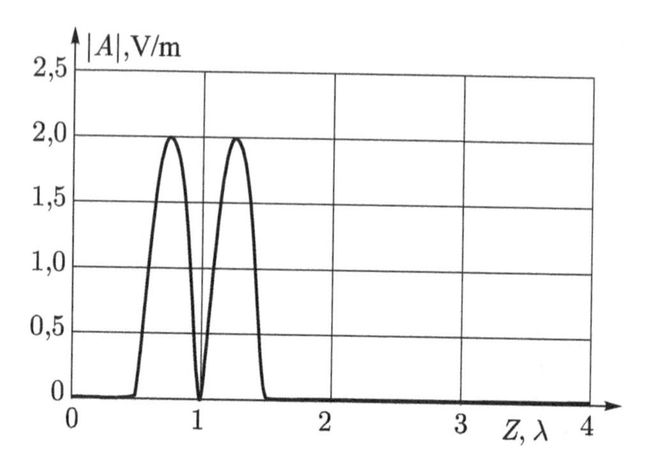

Fig. 2.24. The distribution of the modulus of the complex amplitude of the electric field $|A|$ on $\tilde{D}_h^1$ when using the 'transparent' source to specify the incident wave.

When specifying other types of incident waves it is necessary to use the two-dimensional auxiliary problem, which is modelled by diffraction of a wave in a homogeneous space through a 'hard' two-dimensional source. An arbitrary incident wave is produced by the appropriate choice of the function of the complex amplitude of the incident wave in (2.53) and by varying the form of the subdomain of definition of the source ('hard' in the auxiliary problem and 'transparent' to the main problem).

The method of forming the incident wave from a limited source permits the sharing of the 'transparent' source and arbitrarily oriented electric walls at the boundary of the grid area.

2.1.4.3. The method of separation of the field

The idea of defining the incident wave through the separation of the total and scattered (TF/SF) fields was created after [24] the first publication devoted to the total field formulation [31]; it proved to be more productive and popular [7, 33, 34] up to the development of the 'transparent' source method [32].

The method consists of limiting the subregion of the resulting field (incident and scattered) by the shell different from the computational domain boundaries and located within such a boundary. The shell is not part of the absorbing layers, located between the layers and the optical element. Behind the shell only the scattered field propagates. The expressions used to separate fields contain terms with the analytically defined incident field; thus, the incident wave is introduced into the subarea of the resulting field. The rest of the subdomain of the computational domain does not contain the incident wave.

A one-dimensional case. For the one-dimensional grid region division is performed in nodes and k_L and k_R (Fig. 2.25 in [7]).

Grid functions at $k_L \le k \le k_R$ refer to the resultant field, the rest – to the scattered one. The difference equations of the Yee expression [4] in the specified nodes have the form [7]:

$$\varepsilon_0 \varepsilon_{k_L} \frac{E_{x_{k_L}}^{m+1} - E_{x_{k_L}}^{m}}{h_t} = \frac{H_{y_{k_L+0.5}}^{m+0.5} - \left(H_{y_{k_L-0.5}}^{m+0.5} + \tilde{H}_{y_{k_L-0.5}}^{m+0.5}\right)}{h_x}, \tag{2.57}$$

$$\mu_0 \frac{H_{y_{k_L-0.5}}^{m+0.5} - H_{y_{k_L-0.5}}^{m-0.5}}{h_t} = \frac{\left(E_{y_{k_L}}^{m} - \tilde{E}_{y_{k_L}}^{m}\right) - E_{y_{k_L-1}}^{m}}{h_x}, \tag{2.58}$$

$$\varepsilon_0 \varepsilon_{k_R} \frac{E_{x_{k_R}}^{m+1} - E_{x_{k_R}}^{m+1}}{h_t} = \frac{\left(H_{y_{k_R+0.5}}^{m+0.5} + \tilde{H}_{y_{k_R+0.5}}^{m+0.5}\right) - H_{y_{k_R-0.5}}^{m+0.5}}{h_x}, \tag{2.59}$$

$$\mu_0 \frac{H_{y_{k_R+0.5}}^{m+0.5} - H_{y_{k_R+0.5}}^{m-0.5}}{h_t} = \frac{E_{x_{k_R+1}}^{m} - \left(E_{x_{k_R}}^{m} - \tilde{E}_{x_{k_R}}^{m}\right)}{h_x}, \tag{2.60}$$

where the grid functions under the tilde correspond analytically to the given incident field.

The expression (2.57) defines the electrical component in the left node of the resulting field (Fig. 2.25). To do this, we must subtract from each other the strengths of the magnetic field in the neighbouring sites. It is obvious that the subtraction should be carried out for the field strengths of the same nature, in this case the resultant field strengths, as in (2.57) we calculated precisely the resultant field. But in the difference solution in the node $k_L - 0.5$, only the scattered field is calculated. Therefore, in the right-hand side of (2.57) to the grid function $H_{y_{k_L-0.5}}^{m+0.5}$ we add the value $\tilde{H}_{y_{k_L-0.5}}^{m+0.5}$ of the incident field, which is traditionally given analytically [7].

Similarly, in (2.58) the scattered magnetic field in the node $k_{L-0.5}$ is calculated using the incident electric field $\tilde{E}_{y_{k_L}}^{m}$ in k_L; the resulting electric field in $k_{R-0.5}$ is calculated from (2.59) through the incident magnetic field in the adjancent right node, and the scattered magnetic field in $k_{R+0.5}$ is determined through the incident electric field $\tilde{E}_{x_{k_R}}^{m}$ in (2.60).

Further, setting the parameters of numerical experiments, we take $k_L = 1$, $k_R = 2Q$ for the universal one-dimensional grid domain, calculating the error of propagation of a flat homogeneous electromagnetic wave in a vacuum between the two nodes (Table 2.3).

Comparison of the results from Tables 2.1, 2.2 and 2.3 shows that the analytical task of the incident field in the TF/SF method is preferable to the first version of the general field method, but is inferior in accuracy to the variant of the 'transparent' source.

By studying the separated field method, we temporarily abandon the use of absorbing layers, setting the length of the computational domain to 40λ and placing a subdomain of the general field in the middle, leaving the other settings from the previous experiment the same. The result of calculations for the discretization (100.200) is shown in Fig. 2.26.

The absence of absorbing layers made it possible [35] to observe the surge of the values of the modulus of complex strength of the electric field at the right end of the grid area (Fig. 2.26). This effect is explained by a time delay required for the wave emitted in the node k_L to reach the node k_R. During this delay the wave emitted *in* k_R propagates in the direction of Z. The waves from these nodes then interfere and mutually cancel each other to the right of k_R. The observed surge leads to the

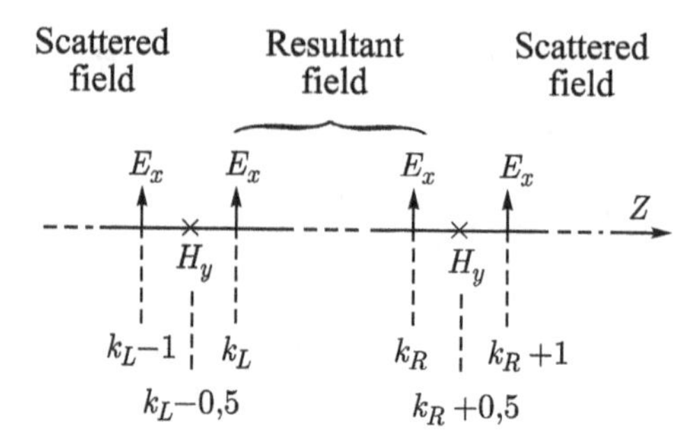

Fig. 2.25. Detail of the one-dimensional grid region [7] with the nodes of of division of resultant and scattered fields.

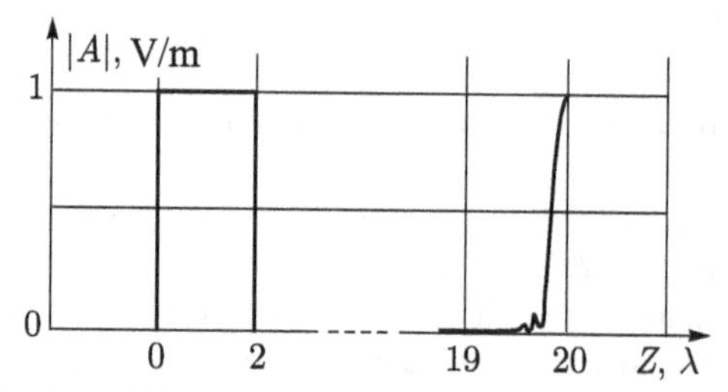

Fig. 2.26. The distribution of the modulus of the complex amplitude of the electric field in the analytic definition of the incident field by the TF/SF method without the imposition of absorbing layers.

conclusion of non-compliance of the expressions (2.57) – (2.60) with the TF / SF method at the initial stage of the calculation in this case.

The numerical definition of the incident field in (2.57) – (2.60) not only improves the accuracy of calculations [11] and reduces their duration [7] in the two- and three-dimensional cases, but also makes it possible to avoid the above-mentioned effect.

Similar to the procedure used in writing expressions for the 'transparent' source, we consider two problems: the primary and secondary, differing in the methods of defining the incident wave (the 'hard' source for the seconday problem). The values of the electric and magnetic fields found in the secondary problem at the nodes $k_L - 0.5$, k_L, k_R and $k_{L+0.5}$ are substituted into equations (2.57) – (2.60) of the first problem as the incident wave. As noted in [11], this leads to the automatic compensation of numerical errors. The errors in defining the phase shift between the components of the incident field are compensated [35] in the adjacent nodes (k_L–0.5 and k_L; k_R and k_L+0.5), because the analytical definition of the amplitude (2.54) in experiments with a 'hard' source (Table 2.1) is associated with smaller errors.

Indeed, numerical experiments confirm the high accuracy of the difference solution obtained on the basis of the approach proposed in [11]. The resulting error for the selected parameters coincided with the sampling results from Table 2.1. Consequently, the numerical definition of the incident wave in (2.57) – (2.60) did not introduce any additional distortions in the difference solution.

Table 2.3. Dependence on the errors of numerical experiments for vacuum on discretization of the grid area (Q, Q_t) with the formulation of the incident wave source conditions by the analytical procedure TF/SF in one dimensional case

(Q, Q_t)	(10, 20)	(20, 40)	(50, 100)	(100, 200)
ε, %	2.2289	0.74774	0.20736	0.066696

Moreover, because of the disappearance of the time interval between the onset of emission of the wave at the node k_R and the arrival there of waves from k_L, no surges were observed in the right side of the domain (Fig. 2.26).

The above technique is successful when the domain is included in the shell of a homogeneous material. In the simulation of the operation of the diffractive optical element (DOE), this condition is not satisfied. We assume that starting at the node $(k_R-k_L)/2$ the left side of the domain is occupied by a homogeneous dielectric half-space (substrate of the DOE) with a refractive index $n = 1.5$. The error of the transmitted wave will be evaluated on a segment of the boundary between the media to k_R (in the area of the resulting field). How do we form the incident wave in equations (2.57) – (2.60)?

The analytical definition of the incident wave in these expressions leads to a significant increase in errors [35] (Table 2.4, column a), compared with experiments in free space (Table 2.3) also as a result of the of non-compliance with the initial condition for the new phase difference of the incident wave between k_L and k_R. By moving the node k_R to the left to the disappearance of this phase difference we can improve the accuracy to a certain extent [35]. However, the study of the optical element does not yield information on the phase shift – it is part of the solution of the problem of diffraction at DOE.

Combining the numerical task of the incident wave (at the nodes $k_L - 0.5$ and k_L; the auxiliary problem is solved for the medium – a homogeneous dielectric) with the analytical task (at the nodes k_R and $k_R+0.5$) does not lead to a steady reduction of errors due to the influence of the analytical task in the right nodes for the whole computational domain [35].

The phase difference between the nodes k_L and k_R can be considered by assuming that the incident wave is not represented by the values of the domains of the second problem is k_R and $k_R+0.5$, and instead it is represented by the nodes that are separated from the data to the appropriate distance to the left (the wave velocity in a vacuum is higher than in a dielectric).However, this view does not account for the difference in phase shift in vacuum (main problem) and the dielectric (auxiliary problem) between i_R and $i_R+0.5$, which leads to an even greater decrease in accuracy. Making amendments to the analytical phase shift corresponds to the analytical task of the incident wave in the right nodes (i_R and $i_R+0.5$), discussed above.

Offering a solution to this problem, in this chapter we study the propagation of the field in a vacuum with the source (2.54), and this third problem is solved simultaneously with the two problems given in [35]. Thus, the value of the incident wave for the main problem (drop at the interface between the dielectric / vacuum) is substituted into equation (2.57), (2.58) of the second problem (which describes

Table 2.4. Dependences of the uniform errors of computational experiments for the domain with the insulator ($n = 15$) / vacuum interface on discretization of the grid area (Q, Q_t) with the formulation of the incident wave source conditions by various methods TF/SF (a – analytical, b – numerical) in the one-dimensional case

(Q,Q_t)	Experimental series	
	a	*b*
(10, 20)	7.9612	4.387
(20, 40)	3.9482	0.9745
(50, 100)	3.7263	0.1515
(100, 200)	1.5870	0.0383

the propagation in a homogeneous dielectric), and in equations (2.59), (2.60) – for the third problem (propagation in free space). At the same time, to account for the distance travelled by the wave in the first problem to the boundary between the two media, the values of the incident field are selected from the third problem not at the nodes k_R and k_R+0.5, but with a corresponding shift to the right.

As a result, we take into account numerically (not analytically) the phase difference of the waves between k_L and k_R, k_L−0.5, and k_L, k_R and k_R+0.5, which improves the accuracy of numerical experiments (Table 2.4, column b).

The proposed method is characterized not only by solutions free from the spike of the values of the modulus of the complex amplitude in the right side of the computational domain, but does not limit the researcher in choosing the location of the site k_R. The initial condition for the numerical formulation of the radiating condition is always satisfied by virtue of its compliance with the two auxiliary problems (second and third).

Disadvantages of this approach includes an increase in computational complexity as a result of adding an additional one-dimensional problem. In the study of diffraction of a plane electromagnetic wave on two-dimensional and three-dimensional objects the indicated increase will not have any significant effect on the duration of the computation because of the dimensionaliyu of the added problem.

The two-dimensional case is characterized by the appearance of two new boundaries of separation parallel to the axis Z (Fig. 2.27 in [7]).

The grid subdomain of the resulting field is enclosed in the space between sections $j_T \leq j \leq j_B$, $k \leq k_L$ (left boundary), $j_T \leq j \leq j_B$, $k \leq k_R$ (right boundary), $j \leq j_T$, $k_L \leq k \leq k_R$ (upper boundary), $j \leq j_B$, $k_L \leq k \leq k_R$ (lower boundary), forming its boundaries.

For the separation of the resulting and scattered fields after the transition to the next temporal layer (for example, by (2.17) – (2.19)) we must carry out calculations:

$$H_{y_{j,k_L-0.5}}^{m+0.5} = H_{y_{j,k_L-0.5}}^{m+0.5} + \frac{h_t}{\mu_0 h_z} \tilde{E}_{x_{j,k_L}}^{m}, \tag{2.61}$$

$$H_{y_{j,k_R+0.5}}^{m+0.5} = H_{y_{j,k_R+0.5}}^{m+0.5} - \frac{h_t}{\mu_0 h_z} \tilde{E}_{x_{j,k_R}}^{m}, \tag{2.62}$$

$$E_{x_{j,k_L}}^{m+1} = E_{y_{j,k_L}}^{m+1} + \frac{h_t}{\varepsilon_0 \varepsilon_{j,k_L} h_z} \tilde{H}_{y_{j,k_L-0.5}}^{m+0.5}, \quad (2.63)$$

$$E_{x_{j,k_R}}^{m+1} = E_{y_{j,k_R}}^{m+1} - \frac{h_t}{\varepsilon_0 \varepsilon_{j,k_R} h_z} \tilde{H}_{y_{j,k_R+0.5}}^{m+0.5}, \quad (2.64)$$

$$H_{z_{j_T-0.5,k}}^{m+0.5} = H_{z_{j_T-0.5,k}}^{m+0.5} - \frac{h_t}{\mu_0 h_y} \tilde{E}_{x_{j_T,k}}^{m}, \quad (2.65)$$

$$H_{z_{j_B+0.5,k}}^{m+0.5} = H_{z_{j_B+0.5,k}}^{m+0.5} + \frac{h_t}{\mu_0 h_y} \tilde{E}_{x_{j_B,k}}^{m}, \quad (2.66)$$

$$E_{x_{j_T,k}}^{m+1} = E_{y_{j_T,k}}^{m+1} - \frac{h_t}{\varepsilon_0 \varepsilon_{j_T,k} h_z} \tilde{H}_{y_{j_T-0.5,k}}^{m+0.5}, \quad (2.67)$$

$$E_{x_{j_B,k}}^{m+1} = E_{y_{j_B,k}}^{m+1} + \frac{h_t}{\varepsilon_0 \varepsilon_{j_B,k} h_z} \tilde{H}_{y_{j_B+0.5,k}}^{m+0.5}, \quad (2.68)$$

where the values under the tilde are the electric and magnetic strength of the incident field which traditionally [24] are given analytically; for the equations (2.61) – (2.64) $j_T \le j \le j_B$, for (2.65) – (2.68) – $k_L \le k \le k_R$. Note that the expression (2.57) – (2.60) permit a similar representation for the one-dimensional case.

Exploring the difference solution with (2.61) – (2.68), let us first of all consider the problem of propagation of a plane homogeneous wave in a vacuum, setting $L_y = L_z = 4\lambda$ and taking in (2.61) – (2.68) $j_T = k_L = 1, j_B = k_R = 2Q$.

Comparison of the results in Tables 2.3 and 2.5 shows the growth of the error of the analytical tasks of the incident wave in vacuum by the TF/SF method in the transition from one-dimensional to two-dimensional case, due to the introduction of additional distortions in the solution from the new boundaries of the division of the resulting and the scattered fields.

As in the one-dimensional case, the difference solution is characterized by the presence of the incident field in the scattered field for some time from the start o propagation. The reason for this is, as in the one-dimensional case, a temporary delay required by the wave emitted at the left boundary of the fields to reach the right boundary. During this delay, the wave emitted by the right boundary propagates in the direction Z. The waves from the boundaries then interfere and mutually cancel each other to the right of the interval $k = k_R, j_T \le j \le j_B$. Consequently, the classical formulation (2.61)–(2.68) from [24] does not correspond to the TF/SF method for which these equations are written.

The solution of this problem can not be considered as the task of artificial definition of the time delay. If the subregion of the resulting field contains an optical element, an analytical calculation of the time delay is difficult. In fact, the result of this calculation is a part of solving the problem of diffraction by the lens.

By improving the method of formation of the incident wave, the authors of [11] turn to the numerical specification of such a wave. Along with the two-dimensional task, the one-dimensional task, characterized as 'hard source conditions' is solved. After the calculation using the Yee algorithm, the numerically determined, in the auxiliary one-dimensional component, components of the incident electromagnetic field (not the analytically calculated) are substituted into (2.61)–(2.68). As in the one-dimensional case this leads to compensation of numerical errors.

Moreover, in the one-dimensional problem radiation appears exclusively in the regfion from the site k_L–1, reaching k_R after a certain time interval. For this reason, on the right of k_R only the scatter filed will form in the two-dimensional problem, as provided by the TF/SF methodology.

The error of modelling, with a numerical assignment of the incident wave, in the two-dimensional case coincides exactly with the results of the one-dimensional case and the results for the 'hard' source conditions (Table 2.1).

Of particular interest is the study of the optical element formed at the interface. In this case, the inclusion of a subdomain of the resultant field in a homogeneous shell is not possible. Consequently, the numerical assignment of the incident wave from [11] is meaningless, suggesting the presence of a homogeneous environment around the scatterer. In [35] for the one-dimensional case it was efficient to define the incident wave through the two auxiliary one-dimensional problems: the first in a homogeneous space filled with the environment, situated in front of an optical element, and second, also in a homogeneous space, but filled with the medium behind the optical element.

Extending the proposed method for the two-dimensional case, we will discuss a basic two-dimensional problem and two auxiliary one-dimensional problems. First, for the medium in front of the optical element (albeit with a refractive index $n = 1.5$) with the 'hard source conditions' at the node k_L–1 and the second for the environment behind the element (for example, free space), with the source in k_L–1–λ (n–1), if the

Table 2.5. Dependence of the errors of numerical experiments for vacuum on discretization of the grid area (Q, Q_t) with the formation of the incident wave on analytical methodology TF/SF in the two-dimensional case

(Q,Q_t)	(10, 20)	(20, 40)	(50, 100)	(100, 200)
ε, %	5.1930	1.5156	1.3372	1.3246

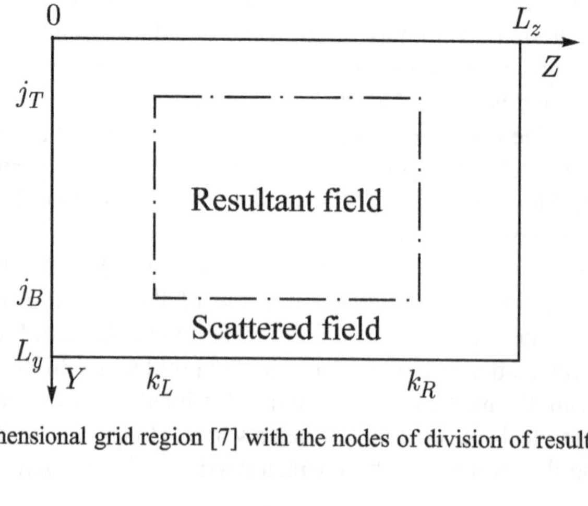

Fig. 2.27. Two-dimensional grid region [7] with the nodes of division of resultant and scattered fields.

boundary between the media divides D in two. The term λ $(n-1)$ provides the phase shift at the interface, taking into account the fact that the distance from the left of the interface of the fields to the boundary between the media is set equal to λ. The value numerically found in the first task for the field strengths of the incident wave are substituted into the formulas (2.61) and (2.63); in the second problem – into the formulas (2.62) and (2.64). The second term on the right side of (2.65)–(2.68) is formed by the two auxiliary problems: to the boundary between the two media by the first task, then by the second one.

Comparing the results of modelling of propagation of the homogeneous plane wave through the medium (n = 1.5) / vacuum boundary with the analytical task of the incident wave (Table 2.6, column a) and by numerical definition (Table 2.6, column b) by the method described in [35], we can the benefits of the proposed method of the formulation of the incident wave source conditions.

However, the two series of experiments are characterized by high values of the errors compared to the one-dimensional case (see Table 2.4). This is explained by the analytical calculation of the phase difference at the interface (the term λ $(n-1)$), which makes the method proposed above not entirely numerical. There is no automatic compensation for the error in determining the phase difference between the nodes of the grid area at the interface.

Improving the definition of the incident wave, it is advisable to apply the method of TF/RF (total field / reflected field), previously used to study the waveguides. The use of TF/RF is based on the separation of the reflected and resultant fields by a a single plane and in contrast to the TF/SF does not provide for the inclusion of an optical element in the shell. In the representation (2.61)–(2.68) the expressions (2.61) and (2.63) are retained and in them $0 \leq j \leq j-1$; the remaining transfomations are superfluous. The dividing plane is perpendicular to OYZ and passes through the segment connecting the nodes k_L, $0 \leq j \leq j-1$ of the universal grid region. The resulting field will be to the right of this section (including the segment), reflected – on the left. Consequently, one auxiliary one-dimensional problem is allowed for the numerical definition of the incident wave in the TF/RF methodology.

In the traditional formulation of TF/RF [7] the strength of the components of the incident electromagnetic field on the upper and lower boundary D with $k = k_L$ (channeled waveguide modes) is assumed to be zero so that this corresponds to the boundary condition for the grid function D_h^2. However, transferring to the general case, when the subjects of research are not only the waveguides but also incidence of a flat homogeneous wave on the optical element, one must refer to the universal grid area.

Then in the incidence of the TEM-wave on a flat boundary the experimental results with numerical assignment of the incident wave method on the basis of the TF/RF method surpass the results of Table 2.6, and coincide with the data for the one-dimensional case of Table 2.4, column b. The accuracy of computational experiments falls three times in the coarse grid and two orders of magnitude for the densest grid region.

The TF/RF method in the study of optical elements on a universal grid area, enclosed in a uniform envelope, yields results that are not inferior in accuracy to the

Table 2.6. Dependence of uniform errors of computational experiments for the domain with the insulator ($n = 1.5$) / vacuum interface on discretisation of the grid area (Q, Q_t) with the formulation of the incident wave source conditions by TF/SF procedures (a – analytical, b – numerical from [35]) in the two–dimensional case

(Q, Q_t)	Experimental series	
	a	b
(10, 20)	13.9267	12.3299
(20, 40)	4.3193	3.8751
(50, 100)	1.4352	1.3202
(100, 200)	1.2614	1.1947

TF/SF methodology, and surpass it in the case of inhomogeneous membranes. In contrast to the TF/SF method, the use of the TF/RF does not define the incident field (numerically or analytically) on the border between the media and, therefore, is not characterized by errors associated with this task.

When specifying other types of incident waves different from the plane homogeneous wave, their intensity is calculated analytically, or by interpolating the results of calculations of the one-dimensional auxiliary problem [7]. Both options lead to a decrease of calculation accuracy when compared with the case of study of the optical diffraction element by means of a numerical definition of incident plane homogeneous wave in the TF/RF method.

Of separate interest is the case of incidence on the optical element of a wave restricted in space, waves, when the analytic definition of the incident wave across the boundary between the fields by the TF / SF method requires the solution of the problem of diffraction on a slit in the mathematical theory of diffraction. There is no need to specify the 'transparent' source or method of implementation the TF/RF procedure.

2.1.4.4. Comparison of methods for the formation of the incident wave

Application of the methods of the resulting and separate fields using their best versions ('transparent' source and TF/RF method) leads to results that differ slightly in terms of accuracy or they match [18,32,35].

Indeed, the formulation of both the 'transparent' source and numerical implementation of the TF/RF methodology require the same grid areas and an additional one-dimensional problem with the 'hard source conditions'. Synchronization of calculations with a general one-dimensional auxiliary problem leads to results with the same accuracy.

However, the implementation of these two approaches and their various modifications is characterized by the following differences.

1. The method of the resulting field does not lead to the separation of the fields on the resulting and scattered, as this is characteristic of the method of the separated field. In the monograph [7] such a separation in the case of the TF / SF method has

an advantage that reduces the thickness of the absorbing layers (and shortens the duration of the calculations) because of the lower intensity of the scattered field, reaching the subregion of location of the absorbing layers.

The above opinion is not justified. As mentioned previously, the analytical definition of the incident wave in the implementation of TF/SF does not comply with this concept, even in studying the propagation of plane waves in a vacuum. The wave propagating from the right side of the border separating the fields within a certain time is emitted as an incident rather than scattered wave.

The transition to the numerical specification of the incident wave brings the TF/SF method in line with the concept of separation of the fields only for the case of the propagation of a plane homogeneous wave in free space.

The TF/RF method is not consistent with the rule of separation of in the region to the right of the segment of introduction of the incident field, but then this area contains absorbing layers.

Moreover, returning to the claim of the authors of [7] of the low intensity of the scattered field in comparison with the resulting field, we note that the intensity of the scattered field (even split correctly) is not always less than the intensity of the resultin field. A suitable example is the field of a metallic reflective optical element that focuses the reflected wave.

Thus, no theoretical or experimental confirmation of the superiority of the method of the separated field over the method of the resulting field is found.

2. The general auxiliary problem is used in the two compared approaches for somewhat different though related purposes. In the method of the total field the auxiliary problem is always used to determine the reflected field, that is, the field scattered by an optical element in a direction opposite to the spread of the incident field. At the same time, to form a 'transparent' source, it is sufficient to have one grid function of the electrical component in the node of definition of the 'hard source conditions' of the auxiliary problem.

The method of the separated field in the TF/SF variant with the numerical specification of the incident wave requires four (one-dimensional case) or $k_R - k_L$ (two-dimensional case) electric and magnetic field components of the auxiliary problem. In the numerical definition of the incident wave by the TF / RF method it is sufficient to have only two components that are selected in the nodes adjacent to the location of the 'hard source conditions' [7]. The further the position of such nodes from the source, the less accurate the solution. The one-dimensional auxiliary problem also is characterized by an error increasing with the distance from the 'hard' source.

3. The use of different methods to define the emission conditions and grid areas not only leads to variations in the accuracy of numerical experiments, but also to changes of their duration.

Suppose that in solving the problem of synthesis of optical elements [36] (e.g., through stochastic optimization), the researcher sets 1000 computational experiments on the two-dimensional region with parameters $L_y = L_z = 4\lambda$, $L = \lambda$, $T = 20\lambda/c$ and sampling (100, 200). Their total duration was 30.33 min and 27.48 min for the experiments with the analytically defined incident wave in the

TF/SF methodology and with the 'transparent' source, respectively (in the grid region D_h^2 without combining absorbing layers). On the universal grid region the durations were found to be 25.56 and 20.54 min for the experiments with TF/SF (analytical definition of the incident wave) and a 'transparent' source. Calculations were performed with a AMD Opteron 244 processor and a program written in the MatLab 7.0 language.

Application of the TF/SF technique involves the formation of the incident wave along the perimeter of the registration subdomain, while the use of a 'transparent' source can restrict considerations to the segment of the area. Because of this, despite the presence of an additional one-dimensional problem, the use of the 'transparent' source reduced the duration of the calculations by 9.37% and 19.64% compared to the classical concept of TF/SF, with different layouts of the absorbing layers. Especially significant differences will be on the regions extending along Z, when most of the calculations in setting the incident wave will be in the upper and lower sections of the shell, dividing the net and the scattered field. When setting the 'transparent' source there is no need for such a shell.

As noted in section 2.1.3, the union of absorbing layers should lead to a reduction in the duration of computations in the vectorization of the algorithm, followed by its implementation in languages that allow vector computing, e.g., MatLab, Fortran 90. or CUDA). Indeed, the above association in the implementation of the TF/SF methodology (with the analytical task of the incident wave) leads to a reduction in the duration of the calculations by 15.73% and by 25.26% when using the 'transparent' source.

Computational complexities of realization of the 'transparent' source and the TF / RF procedure with a numerical specification of the incident wave are distinguished as follows. Both methods involve the calculation of the two tasks. The auxiliary problem can be general (calculations here are identical and consist of four gaxpy operations [22]).

For the TF/RF method in the main problem gaxpy is used in calculations by (2.61) and (2.63) (and also two operations gaxpy). Setting a transparent source is somewhat simpler. Combining the calculation of the resulting field and the field reflected in a single expression, it is possible to confine to a single scalar operation (subtraction of the analytically given incident field and a single grid function of the one-dimensional problem) and one vector – gaxpy (formation of the resulting field as the sum of fields of the main problems in the source region and the result of the previous scalar operation, multiplied by the unit vector).

Neglecting the scalar operations, and considering all the vector equivalent, we recognize that the 'transparent' source is defined by a simpler procedure; its formation requires only 5 operations, against 6 for the TF/RF with a numerical specification of the incident wave. However, against the overall computational complexity of the difference solution such differences are unimportant.

2.1.5. Decomposition of the grid region

High demands on the speed and RAM of the computer system are the charges for the simplicity and universality of the model based on Maxwell's equations. Therefore, despite the fact that the interest solving Maxwell's equations grid methods appeared in the middle of the 20th century [37] and did not weaken further, the practical application of this model was only recorded at the end of the century [7] with the development of computer technology. The task of reducing the computational cost becomes more acute with the spread of the model to the diffraction and refraction optics; especially in the optics of waveguides (where the sizes of the computational domain are characterized not by ones but by tens and hundreds of wavelengths).

Traditionally, the solution of this problem is found in the organisation of parallel computing [38], favoured by the simplicity and efficiency of parallel algorithms for the explicit Yee algorithm. Recognizing the value of this approach, we note that its use in computational practice is important in cases where other methods of reducing the duration of the calculations have been exhausted.

Another common method is based on the imposition of a non-uniform grid on the object under study [39]. This approach is particularly effective when dealing with diffractive optical elements, characterized by a stepped topography. Leaving the same grid area along the contour of the optical element and decimate it in homogeneous areas, we can achieve a significant reduction in computing costs while maintaining accuracy.

Solution of difference equations on a moving grid region in the study of electromagnetic pulse [12] allows the calculation only at the site of localization of the pulse. For diffraction on the interface required definition of two movable grid areas: for transmitted and reflected parts of the pulse.

Note that the three of these approaches are not mutually exclusive but complement each other and can be used together.

This section is devoted to the development of second and third methods in the case of diffraction of monochromatic waves on the optical object.

In the study of multiple reflections inside the optical element (or a system of elements) the use of the mobile grid area is difficult because of numerous reflected waves (following the work [12], we would have to define a separate grid area for each of these waves). Besides, considering a monochromatic wave propagation in a homogeneous space, it seems reasonable to use an expansion to plane waves instead of the difference solution, even in the rarefied region of the grid, as proposed in [39]. Therefore, the appropriate grid domain decomposition is the division into subregions, each of which has the difference solution with subsequent approval at the borders of the subregions [40]. This method allows us to exclude from consideration the homogeneous parts of the optical element (the grid area in not superposed on them) and to study the diffraction process separately in each subregions. This delivers a significant reduction in the duration of the calculations without the use of parallel computing systems.

As an optical element we consider a transparent dielectric diffractive grating. This choice, in the first place, is associated with widespread use of such gratings in optics, and secondly, with the increased interest in the layered diffractive elements [41] whose work is modelled naturally by the proposed concept of decomposition.

2.1.5.1. Decomposition of the one-dimensional grid region

We illustrate the basic techniques of decomposition on the example of the one-dimensional grid area (Fig. 2.1) [26]. For clarity, the absorbing layers will not be merged. This domain may be superimposed on the layered medium in the problem of normal incidence of the TEM-wave. Next, referring to the decomposition of the grid area, we refer to the area in which the electric field is defined. The arguments for the field relevant to the magnetic field are similar. Radiation is applied in the region by means of the 'transparent' source.

Decomposition into two subdomains in free space

Proceeding to analyze the decomposition [40] of the grid domain into two subdomains, we consider the case of the propagation of a homogeneous plane wave in free space, as in the example which reveals the characteristic features of the method.

It is obvious that it is not necessary calculate the field in those parts of the grid area where radiation has already been applied or where it has not yet happened. These considerations are used as a basis of the decomposition in this and following paragraphs.

In decomposition, we isolate from the grid domain $Q = K+1-2R$ central nodes not relatingt to non-absorbing layers, where R is the number of nodes of the area placed under such a layer. Let the first subdomain contains the first $[Q/2]$ nodes ([..] – rounding up to the nearest integer), the second – $Q-[Q/2]+1$ (to the latter $Q-[Q/2]$ nodes of the source area we added one node on the left for overlapping of subdomains). To each of the resulting subdomain we add on the left and right nodes relating to the absorbing layer (Fig. 2.28).

Let the incident waves forms in the first node of the selected main fragment of the original grid area and the transmitted wave is recorded in the last node (Fig. 2.28a). Then, in each resulting subdomain the formation of the incident wave and registration of the transmitted wave is related to to the first and last nodes of the central fragments (Fig. 2.28b, c). At the same time, the strength of the electrical component of the field in the registration node determined by computer simulation is used as the incident wave in the second subdomain. This rule specifies the order of computing experiments: after calculating $M/2$ time layers in the first subregion we tranfer to the second region, where the number of layers in time also equals half of this parameter for the source grid region. For such a time interval the field in both subregions can be regarded as settled, if it were settled at time T in the experiment on the original grid area. The duration T of the wave process is determined prior to computer simulation by physical considerations. To date, it has been selected [42] for the most common types of scattering objects in problems of electrodynamics and microoptics.

Performing the test experiments, we choose as parameters: the length of the incident wave $\lambda = 1$ µm, the length of the source area $L_z = 4\lambda$ (if for both subregions $L_z' = 3\lambda$ with the width of the absorbing layer $L = \lambda$ taken into account); discretization in space of 100 nodes per wavelength, the time step such that for $\tau = 200$ samples (in time), the front of the plane wave in vacuum covered a distance of one wavelength; in 20 wavelengths are introduced into the region. Absorption at the periphery areas here and later is by provided layers with the maximum specific electrical conductivity $\sigma_z^m = 0.0033\ S/\mu m$ at the edge of the area, at the dependence $\sigma_z = \sigma_z^{\max}\left(1-\frac{q}{L}\right)$ for the left layer and $\sigma_z = \sigma_z^{\max}(q/L)^2$ for the right layer, where $0 < q < L$. Then, at the relative error of the value of the strength of the electric field in the registration site in the source area of 0.002%, a similar value in the registration node of the second subregion will be 0.004%. Consequently, we can talk about the decomposition of the grid area as the procedure which does not materially alter the physical nature of the problem.

The gain from the use of decomposition has two aspects associated with a decrease in computing time: saving the memory of a computer system and reducing the number of arithmetic operations. In the study of extended regions ($L_z >> L$), where the capacity of the 'fast' memory of a computer system is insufficient to store the grid functions (whether it is the 'cache' memory compared with the RAM or thge RAM compared to the disk memory), the decomposition can not use 'slow' working memory. In addition, the number of arithmetic operations is reduced from $C \times K \times M$ (where C is the number of operations in the calculation of the two grid functions $E_{x_k}^{m+1}$ and $H_{k+0.5}^{m+0.5}$ in the two nodes according to equations (2.1) and (2.2)) in the case of no decomposition up to $2C \times (2R + Q / 2) \times M / 2$ in the case of decomposition into two subdomains. So, if $L_z >> L$ we can talk about reducing the number of arithmetic operations by half. Decomposition into D subregions (in the case of free space) is associated with the product of $D \times C \times (2R + Q / D) \times M / D$ operations and at $L_z >> L$ reduces the number of arithmetic operations D times.

Note that the above approach can be successfully applied to study the diffraction not only in free space, but also in environments that do not cause the appearance of any significant reflected wave: defect-free waveguides, photorefractive crystals, etc.

Decomposition into two subdomains in the case of an inhomogeneous dielectric medium

Special interested is attracted by the case of an inhomogeneous medium, when the function $\varepsilon(z)$ characterizing the medium has discontinuities. The mathematical model of the propagation of radiation in such an environment should take into account numerous reflections from the boundaries of dielectric layers. For this purpose, the algorithm from the first paragraph 2.1.5.1 is supplemented by the cyclic part, in which this aspect is taken into account, and the number of iterations G will match the number of reflections from the second subregion to the first one and back again.

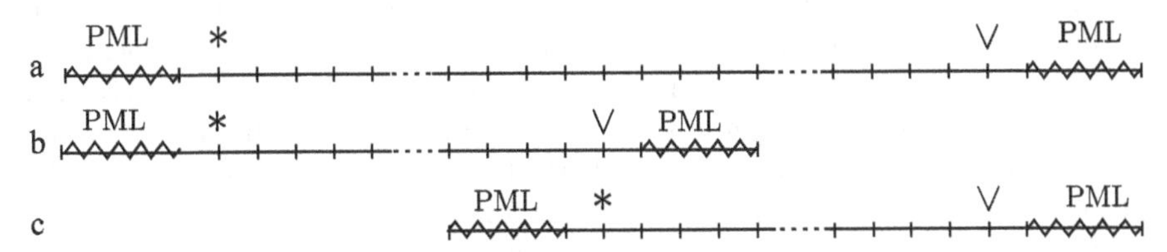

Fig. 2.28. Decomposition of the grid area (s) into two subregions (b and c) and the corresponding steps of the algorithm for the propagation of radiation in free space. The asterisk denotes the location of the radiating condition, the 'upper' tick – the node at the transmitted radiation is recorded, PML (Perfectly Matched Layer) – an absorbing layer.

The possibility of decomposition will be justified by the example of two optical elements (A and B), arranged one after the other (B to A) in the direction of propagation of the incident wave. Evolving over time, the propagation of the incident wave reaches first element A and then B. At a certain stage of this process the wave reflected from B travels back to A and is diffracted by it. Part of this wave is re-reflected to B, but does not yet reach B during this time period. At the same time diffraction on B continues, and the wave reflected from A has not as yet reached B. Note that at this stage, the unknown complex amplitude of the field behind B does not change. Therefore, the calculation of the field diffracted on B, from the onset of appearance of the wave reflected from B to its reflections from A back to B, is superfluous. It can be avoided by the decomposition of the field.

In the first step of the sought algorithm (Fig. 2.29a) the radiation is applied to the first subdomain; the source is located at the left edge, registration is carried out at the right edge (similar to Fig. 2.28b). The second step is characterized by the work with the second subdomain (Fig. 2.29b), the left side of which radiates a wave recorded in the previous step. In contrast to the similar step of the second paragraph of section 2.1.5.1 (Fig. 2.28c) registration is carried out not only the right side of the subdomain (transmitted radiation), but also from the left side (reflected wave) – in the same site of the subdomain in which the emitting condition is defined. The combination of radiation and registration areas in the same site is made possible by the production of the 'transparent' source; the traditional TF/SF method does not allow it. Note that registration of the reflected light from the left edge of the first subdomain is also possible (the first step of the algorithm), but such a wave does not return to the area under investigation and will have no impact on the field behind the transmitting optical element.

The next two steps (Fig. 2.29c, d) are performed in cycle G once and are intended to account for the effect of reflected waves on the field, emerging from the right edge of the second subdomain – at the output of the optical element (required field).Return to the first subdomain of the wave, reflected back from the second subdomain, is the third step of the algorithm (Fig. 2.29c), when the wave recorded at the left edge of the second subdoimain in the previous step is supplied from the right edge of the first subdomain. Registration at the third step is performed for the wave reflected in the first subdomain (right edge), with this wave affecting the desired field, and the wave passed through the subdomain leaves the experiment field in the direction $-Z$ and is not registered. In the fourth step (Fig. 2.29d) the wave previously passed through the optical system is added up with a new wave of re-reflected earlier from the second subdomain (step 2) to the first (step 3) which travelledin the current step to the right edge of the second subdomain. This addition is carried out by the arithmetic addition of the appropriate complex amplitudes according to the principle of superposition [43]. Part of the incident wave reflected in the second subdomain will return to the first subdomain in the next iteration (step 3).

We inspect the adequacy of the model on an example the passage of the T-wave through a plane-parallel plate. At the plate thickness of 1 mm and the refractive index of $n = 2$, the modulus of the complex amplitude of the electric field intensity of the transmitted wave must be equal to the modulus of the complex amplitude of the strength of the corresponding component of the incident wave [43]. Indeed, having placed the investigated plate in the centre of the grid area (Fig. 2.28a) and taking the other parameters of the experiment from the first paragraph 2.1.5.1 without changes, at the value of the modulus of the complex amplitude of the incident wave of 1 V/m (we further examine the relative importance of the module as the result of dividing the complex amplitude of the incident wave by the modulus), we obtain a result corresponding with high accuracy to the theoretically expected result (see Table 2.7).

Note that the value ($|A|$), obtained for $G = 0$, corresponds to the value 0, (8), calculated from the Fresnel formulas without taking reflections into account.

If in the previous experiment (paragraph 2.1.5.1), the correctness of the transfer of the phase of the complex amplitude in transition from the first to the second subdomain was not required, in this case the transmission property of the plate is based on this correctness (otherwise, the phase shift inside the plate would not be the desired value). Consequently, the results from Table. 2.7 confirm the adequacy of the method of transition to the frequency domain, given in section 2.1.2.

In the decomposition of subdomains D into an arbitrary number the volume of the occupied memory of the computer system does not depend on the refractive index of the medium and will remain the same as in paragraph 2.1.5.1. The number of arithmetic operations of the algorithm with dividion into two subdomains is

$$C\times\left(2R+\frac{Q}{2}\right)\frac{M}{2}(2+2G)=C\times\left(2R+\frac{Q}{2}\right)M(1+G) \tag{2.69}$$

with the assumption that in any subdomain the field 'can settle' in $M/2$ steps. Thus, if in (2.69) we set $G = 1$ (one re-reflection), the computational complexity of the algorithm with decomposition will be equal to the complexity of the algorithm without it (if $L_z \gg$ L). In this case, the gain in the computation speed will be achieved only by saving the memory. With the increase in G the value of the expression (2.69) increases, which leads to loss of the advantages of the algorithm with decomposition.

The use of decomposition requires to organize the modelling of reflections and to ensure that the field in each subregion in the transition to the other subregions is settled. Guaranteeing the latter requirement, we set the number of time segments in each subdomain equal to $M/2$. This is a fairly stringent condition and is

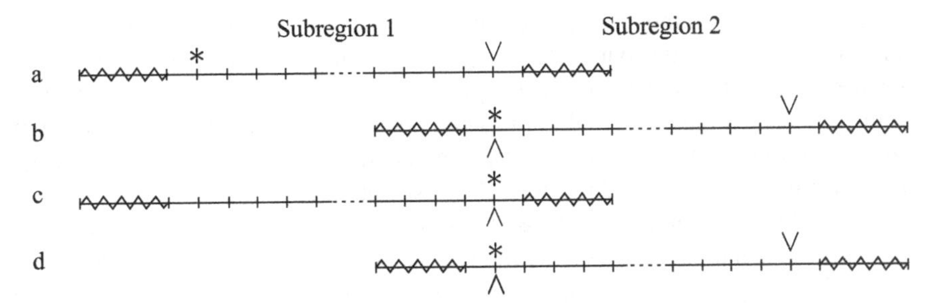

Fig. 2.29. Calculation algorithm of decomposition into two subdomains in the case of an inhomogeneous medium, a, b, c, d – first, second, third and fourth steps of the algorithm, respectively. 'Lower' tick marks the site in which the reflected radiation is registered.

Table 2.7. The dependence of the modulus of the complex amplitude of the electric field (|A|) on the number of reflections in the case of decomposition into two subdomains in the study of TM-wave transmission through a plane-parallel plate

G	$\|A\|$
0	0.8883
1	0.9853
2	0.9955

often redundant. Indeed, the value of M in the initial area is chosen, taking into account all the reflections, and $M/2$ is taken for a single pass of the wave on a subdomain. However, the choice of the number of time layers for a subdomain on the basis of other considerations pose a risk to get an unstable field in the subdomain, containing the entire optical element of a complex configuration.

Decomposition to an arbitrary number of subdomains in the case of an inhomogeneous dielectric medium

The case of the partition of the grid area on D parts has its own peculiarities. Actions by the required algorithm in the first and last subdomains are manufactured in accordance with the rules of the preceding paragraph. For the central subdomains forming of the transmitted and reflected waves takes place by a different mechanism. Each re-reflection in these subdomains is related to the two phases of propagation of the wave, first from right to left – from the penultimate submain to the first, and then from left to right – from the second to the last subdomain. Moreover, according to the rule stated below, in each central subdomain we take into account the value of the reflected wave obtained when passing in the opposite direction.

Let the wave incident in a subdomain of the wave propagates from right to left. After the calculations by the scheme (2.1), (2.2) the amplitude of the reflected wave recorded in the same node in propagation from left to right (in the previous stage of the algorithm) is added to the complex amplitude of the transmitted wave (the left edge of subdomain). The reflected wave (the right edge of the subdomain) will be added up in the next stage of the algorithm (propagation from left to right) with the then registered transmitted wave.

In turn, propagating in the opposite direction, from left to right, after the calculations by the approximation (2.1), (2.2), to the complex amplitude of the recorded transmitted wave (the right edge of the subdomain) we add the amplitude of the reflected wave registered in the same node in propagation from right to left (in the previous step of the algorithm). The resultant reflected wave (the left edge of the subdomain) will be added up in the next stage of the algorithm (propagation from right to left) with the then registered transmitted wave.

Confirming the adequacy of the model, we consider the passage of the TEM-waves through a layered medium consisting of czk (n = 1.5) and flint (n = 1.7) glasses. We set the thickness of each layer as 1 μm. The value of the complex amplitude of the electric field, calculated using the transfer matrix method [43] and equal to 0.8841, is assumed to get an exact solution. Table 2.8 shows the results for the decomposition into three subdomains where each subdomain is characterized by a length of 1 μm (less absorbing layers). In the middle of the first subdomain there is the interface between vacuum and czk glass, in the middle of the second – czk glass/flint glass and in the third – flint glass/vacuum.

The computational complexity of the algorithm with the partition into D subdomains is

$$C\times\left(2R+\frac{Q}{2}\right)\frac{M}{D}\left(D+\left(2+2\left(D-2\right)\right)G\right)=C\times\left(2R+\frac{Q}{D}\right)\frac{M}{D}\left(D+2\left(D-1\right)G\right) \tag{2.70}$$

assuming that on any subdomain the field will stand in M / D steps. By analogy with the theory of parallel computing we evaluate the benefits of such an algorithm by the magnitude of its acceleration [19, 20] – ratio of the number of arithmetic operations of the algorithm without decomposition to the expression (2.70). Then, assuming $L_z \gg L$, the acceleration of the algorithm with decomposition to D subdomains will be

$$S=\frac{D^2}{D+2\left(D-1\right)G}. \tag{2.71}$$

This value will differ from the acceleration the computational process S', generated by the algorithm, and will be equal to the ratio of the duration of the process without regard to the duration of decomposition. The value of S' will be affected not only by the parameters in (2.71), but also by the specific characteristics of the computer system, which explains the frequent mismatch between the acceleration of the recorded algorithm and the calculation process implemented using this algorithm. An example of such a discrepancy will be discussed further in Section 2.1.5.2.

A separate task in grid domain decomposition is the choice of the numnber reflections. On the one hand, the higher the number, the higher the accuracy of determination of the resultant field. However, the increase in G causes a drop in the speed of the algorithm (2.71) for a fixed D. Therefore, G must be choosen earlier from physical considerations, and this must be followed by determining the value D, for which the acceleration can be theoretically achieved. Neglecting the dependence of the number of reflections on the configuration of an optical element (which requires separate consideration in each case), and $G = 0$ in the study of photorefractive crystals without defects and waveguides (no reflected wave), $G = 1$ for elements of optical glass (up to 20% of the energy of the incident wave is reflected, [25]) and $G = 2$ for denser optical media with $n > 2$ (e.g. diamond films). The dependences of the acceleration on the number of subdomains in these common cases are given in Tables 2.9, 2.10.

The obtained high values of acceleration suggests that the decomposition of the grid area is a measure that significantly reduces the duration of computations in the finite difference solution of Maxwell's equations.

Table 2.8. The dependence of the values |A| and error of computer simulation of the number of reflections ε in the case of decomposition into three subdomains in the study of TEM-wave transmission through a vacuum / czk glass / flint glass / vacuum layered medium

G	$\lvert A\rvert$	ε, %
0	0.9447	6.85
1	0.8734	1.21
2	0.8781	0.71

2.1.5.2. Decomposition of two-dimensional grid region

If the one-dimensional case is convenient by its clarity, the two-dimensional (cylindrical optical elements) has also great practical importance. The diffraction grating, cylindrical microlenses and focusators are used in holography, telecommunications and many other areas [36].

One-dimensional decomposition for two-dimensional diffraction gratings
The increase in dimension causes a further projection of the magnetic field (in the case of the *H*-wave), where instead of two difference equations it is necessary to solve three. Formulation of the boundary and initial conditions and the imposition of absorbing layers for such a difference approximation are similar to those discussed in the second paragraph 2.1.5.1 of the case of the grid domain D_h^2 and the standard method of deposition of layers.

Choosing a way to specify the incident wave, it is preferred to stay on the 'transparent' source (as compared with the TF/SF technology [18]). The use of the 'transparent' source does not require additional calculations on the perimeter of the grid subdomain (as in TF/SF) and releases from the decomposition of the incident field into plane waves (necessary when setting TF/SF) in the transition from one subdomain to another.

One-dimensional decomposition of the two-dimensional grid domain does not change the algorithm for the one-dimensional field, as in the new coordinate axes decomposition into subdomains is not performed and their acceleration is the same. Looking at Figs. 2.28 and 2.29, we assume that in this case transverse bars on the axis do not denote a single node of the grid domain (as in the one-dimensional case) and they denote columns of these nodes located along the new axis.

By testing the approach developed here, we give an example of decomposition of the grid domain superimposed on a diffraction grating (n = 1.5), with partition into two subdomains. The lattice spacing is 2.5 mm, the thickness of the substrate 3 μm. A microrelief in the form of convex cylindrical lenses with a radius of curvature of 1.28 μm (so that the height of the microrelief was one wavelength) was deposited on both sides of the substrate. Then the length of the computational domain was 5 μm (without absorbing layers), the length of subdomains 2.5 μm. In sampling in the space of 50 nodes per wavelength, assuming that τ = 100 and entering in the region

Table 2.9. The dependence of the speed of the algorithm S on the number of subdomains D when $G = 1$

D	2	3	4	5	6	7	8	16	32	64	128
S	1	1.29	1.6	1.92	2.25	2.58	2.91	5.57	10.89	21.56	42.67

Table 2.10. The dependence of the speed of the algorithm S on the number of subdomains D when $G = 2$

D	2	3	4	5	6	7	8	16	32	64	128
S	0.67	0.82	1	1.19	1.39	1.58	1.78	3.37	6.48	12.88	25.68

of a train of 200 wavelengths (in the subdomain of 100 wavelengths), we obtain the results shown in Table 2.11.

The closeness of the results obtained with the use of decomposition and in its absence confirms the adequacy of the mathematical model with one-dimensional decomposition of the of the two-dimensional grid domain in space and the adequacy of consideration of a single re-reflection, if the material of the element is optical glass.

The second numerical experiment was formulated in a domain 256 μm long in order to study the dependence of the accelaration of the computational process on the number of subdomains (Table 2.12).

The lattice spacing and sampling parameters did not change compared to the previous experiment. It was assumed that for a satisfactory simulation of any structure of optical glass it is sufficient to consider a single re-reflection. Computational experiments were performed on a Pentium 4 2400 MHz processor, using the Matlab 5.2 software.

At $2 \le D \le 64$ the acceleration of computational processes is greater than that of the algorithms (Tables 2.9, 2.12), due to the reduction of the occupied memory of the computer system in implementing the decomposition. Thus, even for two subdomains the acceleration of the computational process reached values of 1.45, although there was no gain in the number of arithmetic operations. Increasing D increases the effect of absorbing layers. For $D = 128$ the total length of layers $2L$ is already half the total length of the subdomain and, therefore the acceleration of the process becomes less than the acceleration of the algorithm (Tables 2.9, 2.12).

Decomposition with homogeneous dielectric inclusions

Of particular interest is the case of dielectric inclusions, when the diffractive microrelief, deposited on both sides of the optical element, is divided by a homogeneous dielectric layer (substrate). The thickness of the layer is usually many times greater than the height of the diffractive microrelief.

It seems reasonable not to calculate the field inside the substrate by solving the differential equation (as suggested in [39]) and use it as a strict model strict but with less computational complexity. In the case of a diffraction grating such a model will be an expansion in plane waves. The difference solution at the same time is used for finding the field only in the areas of heterogeneities (microrelief of the optical element).

Table 2.11. Intensity of passed orders for computational experiments with a two-dimensional diffraction grating.

Intensity of passed orders	Without decomposition	With decomposition to two subdomains		With decomposition to two non-overlapping subdomains and decomposition into plane waves between subdomains	
		$G = 0$	$G = 1$	$G = 0$	$G = 1$
I_0	0.6861	0.6769	0.6806	0.657	0.6781
$I_1 = I_{-1}$	0.07138	0.0875	0.07165	0.08837	0.07239
$I_2 = I_{-2}$	0.03896	0.02848	0.03861	0.02835	0.03833

Table 2.12. The dependence of the acceleration of the computational process S' on the number of subdomains D when $G = 1$ in the second computational experiment in the first paragraph 2.1.5.2

D	2	4	8	16	32	64	128
S'	1.45	2.34	5.59	10.1	18.06	28.64	39.87

Tables 2.3 and 2.11 show the results of computer simulation for a grating of a test case of the first paragraph 2.1.5.2. The length of the subdomains (without absorbing layers) was 1 μm – the height of the microrelief. The subdomains themselves separated by a distance of 3 mm (thickness of the substrate) at which the wave propagation from left to right and from right to left is modelled by the expansion to plane waves.

It should be noted that in the expansion to plane waves in the substrate the diffraction pattern within the optical element and at a short distance in front of and behind it is different from that without the expansion. This is due to the loss of information about standing waves, neglected in the expansion used. However, the such waves do not carry energy and do not represent much interest [44].

In the case of an optical system consisting of P diffractive surfaces (e.g., layered diffractive elements [41]), it makes sense to decompose into P non-overlapping subdomains [40], which contain only diffraction reliefs separated by homogeneous dielectric inclusions. Inside subdomains there is a difference solution, and the field between the subdomains is defined by the expansion to plane waves. Reducing the computational complexity of the algorithm in this approach evidently exceeds the acceleration (2.71) in the decomposition to contiguous subdomains.

The proposed method of organizing calculations in the different solution of the Maxwell equations can significantly reduce the requirement on the operating memory of the computer system and reduce the number of arithmetic operations by taking into account the locally well-established field and using the superposition principle, based on the linearity of the equations solved. The application of the submitted decomposition of the grid domain can be used not only in solving problems of diffraction optics, but also in other similar applications.

2.1.6. Simulation of the effect of the etching wedge on the focusing of radiation of cylindrical microlenses with a high numerical aperture

Choosing an illustration of the above finite-difference method for solving Maxwell's equations, let us study the propagation of the electromagnetic wave through diffraction microlens with technology manufacture errors (etching wedge).

2.1.6.1. Selection of parameters of computational experiments

As the grid equations we select finite-difference equations in section 2.1.1.2 for the TE-wave; on the optical elements and their surroundings we impose a universal grid area with combined absorbing layers (paragraph 2.1.3.3); a plane homogeneous incident wave is given by the method of the resultant field (section 2.1.4.2).

Attention will be given to diffractive optical elements (DOE), calculated by the quantization of the phase function of the cylindrical refractive microlens with aperture 16 λ (where λ is the wavelength of incident radiation), the numerical aperture sin $\lambda / 4$, the radius of curvature 10 λ and the refractive index of the material $n = 2$ (e.g. for silver chloride at $\lambda = 1$ μm). According to geometrical optics, the focus area of such microlenses is located at a distance of $f = 8\lambda$ from the right pole. In constructing the DOE the aperture of the refractive lenses is divided into seven Fresnel zones with the quantization of the phase function into two (binary microlens) and four (four-level microlens) steps, in accordance with the accepted rules. The thickness (the distance between the poles) of the refractive lens is 4 λ, diffraction lense – λ.

The technology of forming a stepped microrelief by chemical and plasma-chemical etching distorts the calculated profile of the DOE and these distortions are traditionally called the etching wedge. Chemical etching is characterized by an 'internal' wedge (Fig. 2.30b), plasma-chemical – 'external' (Fig. 2.30c). The charac-teristic parameter for these technologies is the wedge parameter $\alpha = \pi / 8$.

Parameters of computer simulation (Fig. 2.31) are defined as follows: $L_y = 18\lambda$ (16λ at the aperture and 2λ for the layer in absorbing in direction Y); $L_z = 18\lambda$ (the incident wave with the intensity 1 W/m^2 is determined in the interval $z = 0$; $0 \leq y \leq L_y$, left pole of the lens is located at around $z\lambda = 1.5\lambda$, the layer absorbing in direction of Z has a thickness of 2 λ). The lens are facing by the reliefs to the left.

The grid area is imposed at the calculated rate of 50 nodes per wavelength in space; 100 nodes in time on the interval in which a plane electromagnetic wave in a vacuum travels distance λ (velocity of propagation c); propagation time $T = 50\ \lambda / c$ was considered sufficient for the assumption of monochromaticity of the field in the computational experiment.

2.1.6.2. Simulation of radiation through a microlens with an etching wedge

In the first series of numerical experiments we investigate the refraction, binary and four-level microlenses with the optical surface, devoid of the etching wedge

(Fig. 2.32, 2.33, 2.34). From the calculated diffraction pattern we determined the effectiveness of DOE and the distance from the right pole of the lens to the position of the maximum of intensity on the main optical axis $-f$. The DOE efficiency is the value of γ, equal to the ratio of the maximum intensity in the main optical axis of the element to the same parameter characterizing the refractive microlens.

Analyzing the results of the first series (Table 2.13, Fig. 2.35), it is appropriate to talk about a significant reduction in efficiency during the transition to the quantized DOE profile.

The focal length of the refractive lens was less than the theoretically calculated value by 1.54λ, the binary microlens by 0.02λ, the four-level microlens by 3.22λ.

The second series of numerical experiments (Fig. 2.36) was carried out to study binary diffractive microlenses with microrelief irregularities inherent in chemical (Table 2.14, the second column) and plasma chemical (Table 2.14,the third column) etching.

In contrast to the results of the previous series of experiments (Table 2.13, third column) the effectiveness of the binary microlenses with technological errors in manufacturing grew by 5% in the case of the 'internal' etching wedge, and 4% for the case with the 'outside' etching wedge. The focus area was not shifted.

The final series of experiments was carried out for four-level microlens (Fig. 2.37, Table 2.15).

In this case, the efficiency of the four-level microlenses with technological errors in manufacturing grew by 11% (for the 'internal' etching wedge) and 1% (for the 'external' etching wedge). The focus area significantly shifted (by 0.56λ) only for lenses with the 'internal' etching wedge.

Describing the results of computational experiments on the whole, it is necessary to note the growth of the efficiency of the DOE (by 11%) with regard to the etching wedge. Obviously, the technological heterogeneities 'smooth' the profile of the DOE, bringing it to the profile of the grayscale lens, thereby reducing the effect of the defect associated with the quantization of the phase function. In this case, the focus area of the diffractive microlenses is in most cases shifted only slightly with the formation of the etching wedge.

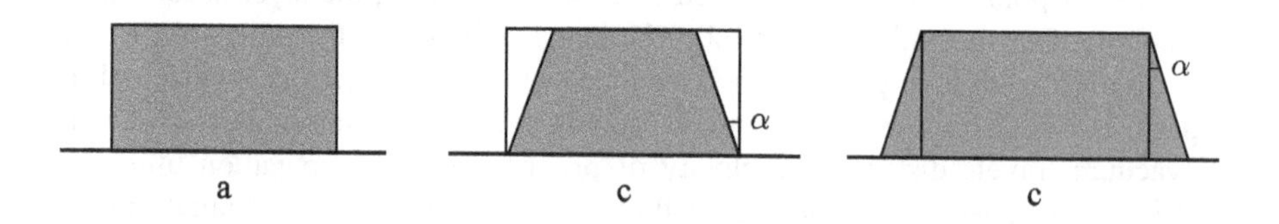

Fig. 2.30. Step of binary lens: a) without the etching wedge, and b) with the 'inner' etching wedge characteristic of liquid etching, c) with the 'outside' etching etching wedge characteristic of plasma chemical etching.

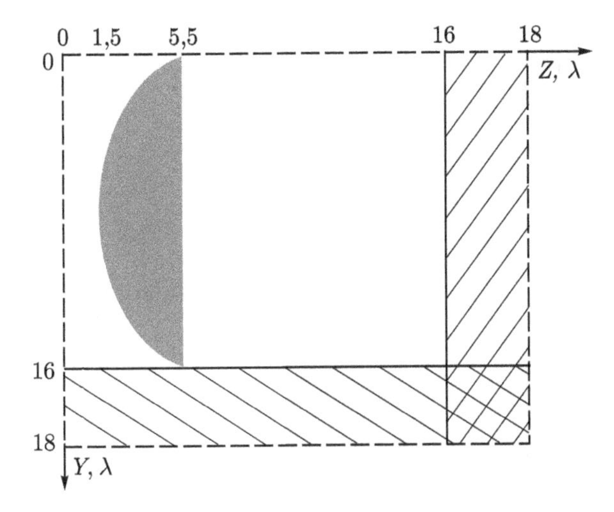

Fig. 2.31. The area of computer simulation. Shaded are the absorbing layers.

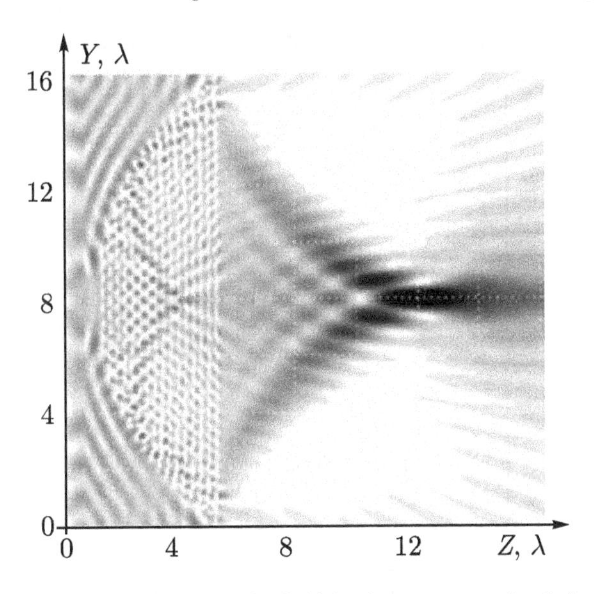

Fig. 2.32. The distribution of the electric field in the computer simulation of refractive microlens.

2.2. Numerical solution of the Helmholtz equations (BPM–approach)

2.2.1. The beam propagation method and its variants

Along with the finite-difference and finite-element methods, the Maxwell equations are widely solved using the methods based on the numerical solution

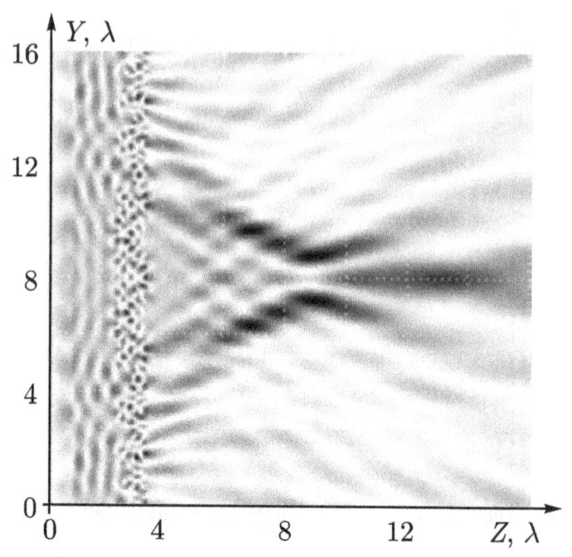

Fig. 2.33. The distribution of the electric field in the computer simulation of binary microlens.

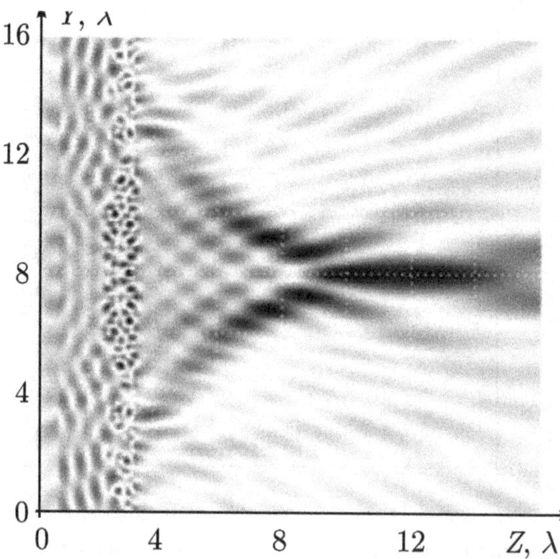

Fig. 2.34. The distribution of the electric field in the computer simulation of a four-level. microlens.

of the Helmholtz equations. The area of possible applications of these methods is significantly narrower, but a significant number of optical problems do not contradict the requirements imposed in the derivation of the Helmholtz equations from the system of Maxwell's equations. In particular, this approach to the calculation of the characteristics of radiation propagating in a medium is widely used in the calculation of fibre and integrated optics.

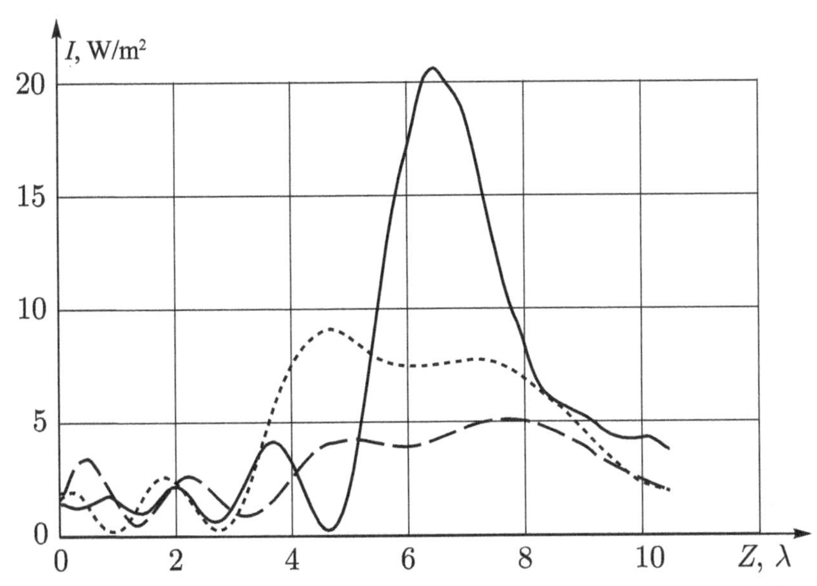

Fig. 2.35. The distribution of the electric field intensity on the main optical axis of refraction (solid line), four-level (dashed curve) and binary (dashed line) diffraction microlens without technological inhomogeneities of the profile. The origin coincides with the right pole of the refractive lens.

Table 2.13. The results of the first series of numerical experiments

Characteristics	Microlenses		
	Refractive	Binary	Four-level
F (λ)	6.46	7.98	4.6
λ	1.0	0.24	0.44

A suitable representative of the methods used to numerically solve the Helmholtz equation is the so-called beam propagation method (BPM) [45]. This section discusses the mathematical foundations of the method and illustrates its capabilities.

If the reader unfamiliar with the method will want to find the descriptions of BPM and its application to solve specific problems, the search result might be a surprise: in various works the BPM means significantly different methods which often differ even from the computational point of view. The fact is that the word 'method' in the title of BPM is not quite true: now BPM is a general name of a family of methods that are significantly different from each other by the mathematical model of propagation in the medium and by numerical methods. Two factors combine all these methods.

First, the BPM methods are based on solving the consequences of the Helmholtz equation. In solution, various constraints are imposed on the environment and propagating beams, which allows to derive the form of equations other than the original one.

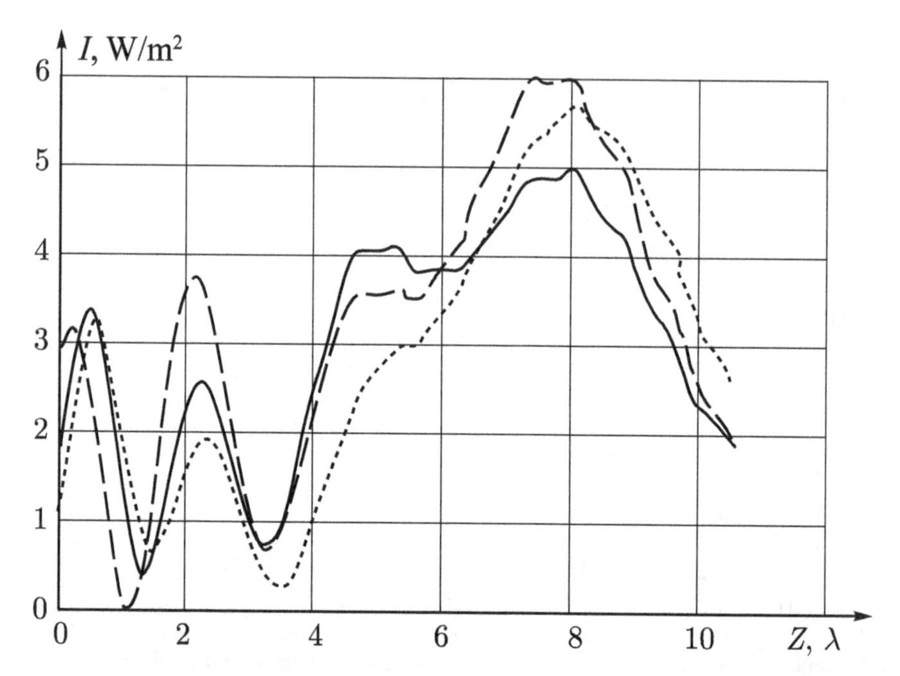

Fig. 2.36. The distribution of the electric field on the main optical axis of the binary microlens without the etching wedge (solid curve), with the 'internal' etching wedge (dash–dotted curve) and with the 'outside' etching wedge (dashed line).

Table 2.14. The results of the second series of numerical experiments

Characteristics	Binary microlenses made by	
	Chemical etching	Plasma chemical etching
$f(\lambda)$	7.92	8.04
γ	0.29	0.28

Table 2.15. The results of the third series of numerical experiments

Characteristics	Four-level microlens made by	
	Chemical etching	Plasma chemical etching
$f(\lambda)$	5.16	4.78
γ	0.55	0.45

Secondly, all these methods solve the problem of beam propagation in space as an evolutionary problem, i.e. as a problem of the initial conditions. It is this feature which gave the name to the first methods of this kind, and later the name was fixed also for new methods for having this property.

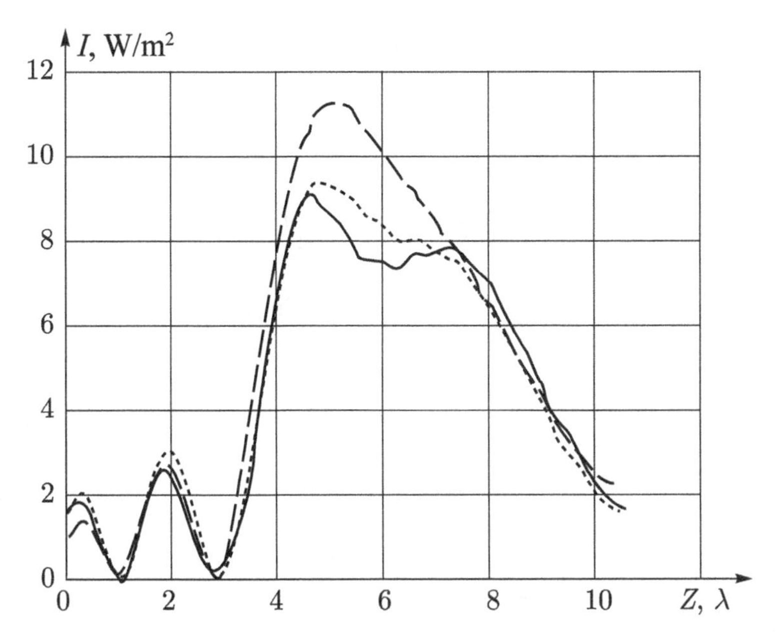

Fig. 2.37. The distribution of the electric field on the main optical axis of the four-level microlens without the etching wedge (solid curve), with the 'internal' etching wedge (dash–dotted curve) and with the 'outside' etching wedge (dashed line).

To explain the differences between the methods of the BPM family, it is necessary to study Helmholtz equations [36, 43], which are the basis for these methods. For simplicity, we use the matrix form of the vector equations derived in the transition from the general vector form to a specific coordinate system. For the Cartesian coordinate system the equations take the form:

$$\frac{\partial^2}{\partial z^2}\mathbf{E}+\mathbf{AE}=0,\ \frac{\partial^2}{\partial z^2}\mathbf{H}+\mathbf{BH}=0, \tag{2.72}$$

where $\mathbf{E}=(E_x, E_y, E_z)^T$ and $\mathbf{H}=(H_x, H_y, H_z)^T$ – the vectors of complex amplitudes of the components of the electric and magnetic fields, respectively, and the matrices $\mathbf{A}$ and $\mathbf{B}$ are matrix differential operators and are as follows:

$$\mathbf{A}=\begin{pmatrix} A_{xx} & A_{xy} & A_{xz} \\ A_{yx} & A_{yy} & A_{yz} \\ A_{zx} & A_{zy} & A_{zz} \end{pmatrix},\ \mathbf{B}=\begin{pmatrix} B_{xx} & B_{xy} & B_{xz} \\ B_{yx} & B_{yy} & B_{yz} \\ B_{zx} & B_{zy} & B_{zz} \end{pmatrix}. \tag{2.73}$$

The components of these operators describe the interaction of the components of the fields:

$$A_{xx}=\frac{\partial^2}{\partial x^2}+\frac{\partial^2}{\partial y^2}+\frac{\partial \ln n^2}{\partial x}\frac{\partial}{\partial x}+\frac{\partial^2 \ln n^2}{\partial x^2}+k_0^2 n^2, \tag{2.74}$$

$$A_{xy} = \frac{\partial \ln n^2}{\partial y}\frac{\partial}{\partial x} + \frac{\partial^2 \ln n^2}{\partial x \partial y}, \tag{2.75}$$

$$A_{xz} = \frac{\partial \ln n^2}{\partial z}\frac{\partial}{\partial x} + \frac{\partial^2 \ln n^2}{\partial x \partial z}, \tag{2.76}$$

$$A_{yx} = \frac{\partial \ln n^2}{\partial x}\frac{\partial}{\partial y} + \frac{\partial^2 \ln n^2}{\partial y \partial x}, \tag{2.77}$$

$$A_{yy} = \frac{\partial^2}{\partial x^2} + \frac{\partial^2}{\partial y^2} + \frac{\partial \ln n^2}{\partial y}\frac{\partial}{\partial y} + \frac{\partial^2 \ln n^2}{\partial y^2} + k_0^2 n^2, \tag{2.78}$$

$$A_{yz} = \frac{\partial \ln n^2}{\partial z}\frac{\partial}{\partial y} + \frac{\partial^2 \ln n^2}{\partial y \partial z}, \tag{2.79}$$

$$A_{zx} = \frac{\partial \ln n^2}{\partial x}\frac{\partial}{\partial z} + \frac{\partial^2 \ln n^2}{\partial z \partial x}, \tag{2.80}$$

$$A_{zy} = \frac{\partial \ln n^2}{\partial y}\frac{\partial}{\partial z} + \frac{\partial^2 \ln n^2}{\partial z \partial y}, \tag{2.81}$$

$$A_{zz} = \frac{\partial^2}{\partial x^2} + \frac{\partial^2}{\partial y^2} + \frac{\partial \ln n^2}{\partial z}\frac{\partial}{\partial z} + \frac{\partial^2 \ln n^2}{\partial z^2} + k_0^2 n^2, \tag{2.82}$$

$$B_{xx} = \frac{\partial^2}{\partial x^2} + \frac{\partial^2}{\partial y^2} + \frac{1}{n^2}\frac{\partial n^2}{\partial x}\frac{\partial}{\partial x} + k_0^2 n^2 - \frac{\nabla_{y,z}^2 n^2}{n^2}, \tag{2.83}$$

$$B_{xy} = \frac{1}{n^2}\left(\frac{\partial n^2}{\partial y}\frac{\partial}{\partial x} + \frac{\partial^2 n^2}{\partial x \partial y}\right), \tag{2.84}$$

$$B_{xz} = \frac{1}{n^2}\left(\frac{\partial n^2}{\partial z}\frac{\partial}{\partial x} + \frac{\partial^2 n^2}{\partial x \partial z}\right), \tag{2.85}$$

$$B_{yx} = \frac{1}{n^2}\left(\frac{\partial n^2}{\partial x}\frac{\partial}{\partial y} + \frac{\partial^2 n^2}{\partial y \partial x}\right), \tag{2.86}$$

$$B_{yy} = \frac{\partial^2}{\partial x^2} + \frac{\partial^2}{\partial y^2} + \frac{1}{n^2}\frac{\partial n^2}{\partial y}\frac{\partial}{\partial y} + k_0^2 n^2 - \frac{\nabla_{x,z}^2 n^2}{n^2}, \tag{2.87}$$

$$B_{yz} = \frac{1}{n^2}\left(\frac{\partial n^2}{\partial z}\frac{\partial}{\partial y} + \frac{\partial^2 n^2}{\partial y \partial z}\right), \tag{2.88}$$

$$B_{zx} = \frac{1}{n^2}\left(\frac{\partial n^2}{\partial x}\frac{\partial}{\partial z} + \frac{\partial^2 n^2}{\partial z \partial x}\right), \tag{2.89}$$

$$B_{zy} = \frac{1}{n^2}\left(\frac{\partial n^2}{\partial y}\frac{\partial}{\partial z} + \frac{\partial^2 n^2}{\partial z \partial y}\right), \tag{2.90}$$

$$B_{zz} = \frac{\partial^2}{\partial x^2} + \frac{\partial^2}{\partial y^2} + \frac{1}{n^2}\frac{\partial n^2}{\partial z}\frac{\partial}{\partial z} + k_0^2 n^2 - \frac{\nabla_{x,y}^2 n^2}{n^2}, \tag{2.91}$$

where operator $\nabla_{\alpha,\beta}$ denotes differentiation with respect to coordinates α and β only.

First, the methods can be divided by the dimensions of the space in which the propagation of radiation is studied. In two-dimensional (2D) cases we usually consider the plane xz, where the z axis is the main direction of propagation. In the differential operators **A** and **B** we omit the derivatives with respect to the coordinate y, which also corresponds to the cylindrical case. In three-dimensional (3D) cases we consider the full form of the operators, and the z axis is also the main direction of propagation. In this section, the methods are described primarily by the two-dimensional case.

Secondly, the methods used are classified by their approximation to the electromagnetic field as scalar and vector methods. For the scalar methods, the equations (2.72) take the form:

$$\frac{\partial^2}{\partial z^2}U + \left(\frac{\partial^2}{\partial x^2} + \frac{\partial^2}{\partial y^2} + k_0^2 n^2\right)U = 0, \tag{2.92}$$

where $U(x,y,z)$ is a complex quantity, the modulus and phase of which characterize the amplitude and phase of the field, k_0 is the wave number in vacuum, and $n(x,y,z)$ the refractive index of the medium. In some cases, the value U can be understood as one of the components of the vector **E** in the case of TE-polarized light, or as one of the components of the vector **H** in the case of TM-polarized radiation. In the vector case the equations (2.72) retain their form and their solution requires the consideration of the nature of the electromagnetic field. This section will address mainly the scalar methods (except for section 2.2.7).

Thirdly, it is easy to see that the equations (2.72) and (2.92) are hyperbolic, which makes it difficult to directly solve them in the evolution relative to z, as required by the characteristic feature of the BPM. The method of producing an evolutionary equation form specifies one of four main varieties types of the methods of the BPM family:

- Fast Fourier Transform Beam Propagation Method (FFT BPM): methods based on the representation of the environment as a set of thin lenses and the calculation with transition to the spectral domain and back based on the Fourier transform [13];
- Method of Lines Beam Propagation Method (MoL BPM): methods based on separation of the lines along the main axis of propagation and transformation of the space of differential operators [14];
- Finite-difference Beam Propagation Method (FD BPM): methods based on finite-difference solution of the Helmholtz equation in the Fresnel form (so-called slowly varying envelope approximation) [15];
- Finite Elements Beam Propagation Method (FE BPM): methods based on the finite element solution of the so-called one-way Helmholtz equations [45].

Mathematical foundations as well as advantages and disadvantages of these approaches will be discussed in subsequent paragraphs of this section. The variety MoL BPM is not considered in this book due to the fact that in general it is inferior in performance to the FD BPM and FE BPM variants.

Fourth, when deriving the equations of evolution type in all varieties of methods additional constraints are imposed on the possibility of the deviation of the basic direction of propagation of radiation from the axis z. This makes it difficult to use the methods in cases such as calculating the propagation in waveguides with bends. To some extent this problem can be solved within the field of a specific version of the method (usually by limiting the quantity of bends and changes in the refractive index of the profile), but there are approaches which solve such problems more accurately:

- Coordinate Transform Method Beam Propagation Method (CTM-BPM): based on the transformation of the original problem to a problem in curvilinear coordinates in which the waveguide under consideration is straight and its shape is compensated by local variations in the refractive index [46, 47];
- Wide-angle Beam Propagation Method (WA BPM): based on more accurate approximations of the operators describing the propagation of radiation (characteristic of the FD BPM and FE BPM) [16, 17].

CTM BPM is not considered in this work, the WA BPM basics will be discussed in section 2.2.4, dedicated to FE BPM.

Fifth, as the evolutionary form of the problem does not allow to take into account the reflection (and re-reflections) on the boundaries of media, these methods without additional modification can not be used for the correct modelling of propagation in the environments with abrupt changes in the refractive index of the profile. However, there are approaches that also use BPM to solve such problems [48, 49]. They were named the Bidirectional Beam Propagation Method (Bidirectional BPM). Some of them, for example, are based on the separation of the layers in which there is an abrupt change in the refractive index of the profile. In these layers transmitted and reflected beams are calculated, and in parts of the environment without abrupt changes in the refractive index of the profile iterative calculations are carried out for the forward and back-propagating beams using one of the BPM methods. The subsequent addition of the complex amplitudes obtained in the calculation of the distributions gives the solution to the problem of propagation of re-reflections. This modification of the method is not considered here.

Sixth, each version of the method can also have its modifications determined by the applied methods and computational algorithms [50–53]. For example, FD BPM requires the choice of the finite-difference approximation for solving the equations, which leads to additional types of the method.

Thus, a particular method of the BPM family belongs to one of the groups in each of the six categories examined and can be further modified. For this reason, the description of all possible methods is a very tedious task so in this section we restrict ourselves to the basic modifications of the method with indication possible areas of modification.

Of special interest is a group of methods Time Domain Beam Propagation Method (TD-BPM) [54,55] developed in recent years: though the title clearly refers to this method as belonging to the beam propagation methods, it consideres a different model of radiation. At the core of this group of methods is not the Helmholtz equation and its consequences, but the wave equation. At the same time, the techniques used to bring the equation to the evolutionary form with respect to time, are similar to techniques used in traditional BPM methods to bring the equation to an evolutionary form with respect to spatial coordinates. Apparently, this similarity was the reason to classify these methods in the BPM family, although in this case we are talking about 'propagation of the beam in time' rather than in space. Consideration of this group is beyond the scope of this monograph.

The main advantage of the BPM family of techniques is the ability to calculate the propagation of radiation over large distances (relative to the wavelength) for large cross-sectional areas of the domain. Given the limitations imposed by this method is not always applicable in the calculation of the propagation of radiation in diffractive elements of nanophotonics, but it is a very advantageous method for analyzing the distribution of the beams formed by the diffractive elements (their field distributions can be obtained by other methods) in optical media with small changes in the profile refractive index. In this case it will be possible to investigate processes occurring in the beam propagating in a medium, even if the cause of these processes is beyond consideration in the simulation in the BPM area.

2.2.2. Solution on the basis of expansion into thin optical elements (FFT BPM)

This method was originally proposed by Feit and Fleck [13] and was intended for the analysis of propagation of radiation in gradient waveguides and other media with an inhomogeneous distribution of the refractive index. The material presented in this section is based on the works [13, 56].

Consider the equation (2.92) for the two-dimensional case. We also assume that the solution permits separation of variablcd and can be represented in the form

$$U(x,z) = \tilde{U}(x) \cdot \exp(-i\beta(z)\, z), \tag{2.93}$$

where i is complex imaginary unit, and $\beta(z)$ is the propagation constant, which is a function of z. However, if the change in the refractive index along the z axis occurs at distances much larger than the wavelength, we can assume that the value $\beta(z)$ is locally constant and does not change in the distance Δz which will be regarded as the sampling step in the direction of propagation.

Then it is easy to show that

$$\frac{\partial^2}{\partial z^2} U(x,z) = -\beta^2(z) U(x,z). \tag{2.94}$$

In view of (2.94), we substitute (2.93) into (2.92) and obtain:

$$\left(\frac{\partial^2}{\partial x^2}-\beta^2(z)+k_0^2n^2(x,z)\right)U(x,z)=0. \tag{2.95}$$

Next, we introduce a formal notation for the operator of differentiation with respect to the transverse coordinates:

$$\frac{\partial^2}{\partial x^2}=\nabla_\tau^2. \tag{2.96}$$

From (2.95) it follows formally that

$$\beta(z)=\left(\nabla_\tau^2+k_0^2n^2(x,z)\right)^{1/2}, \tag{2.97}$$

which allows, starting from (2.93), to write the following relationship for the values of the complex field amplitude at the point $z+\Delta z$:

$$U(x,z+\Delta z)=\exp\left(-i\Delta z\left(\nabla_\tau^2+k_0^2n^2(x,z)\right)^{1/2}\right)U(x,z). \tag{2.98}$$

The exponential expression in the exponent can also be represented as follows:

$$\left(\nabla_\tau^2+k_0^2n^2(x,z)\right)^{1/2}=\frac{\nabla_\tau^2}{\left(\nabla_\tau^2+k_0^2n^2(x,z)\right)^{1/2}+k_0n(x,z)}+k_0n(x,z). \tag{2.99}$$

Up to this point during derivation we did not make any assumptions about the nature of the change in the refractive index. However, it is clear that in the general of its dependence on the coordinates, transformation of (2.98) to the evolution form permitting an efficient solution is associated with difficulties. Furthermore, we assume that the refractive index is formed of two components:

$$n(x,z)=n_*+\delta n(x,z), \tag{2.100}$$

where n_* is a constant (usually chosen equal to the refractive index of the fibre cladding), and $\delta n(x, z)$ describes the deviation from this constant. We also assume that the deviation is small compared with the main component:

$$\frac{\delta n(x,z)}{n_*}\ll 1, \tag{2.101}$$

Then the expression (2.99) can be approximated as follows:

$$\left(\nabla_\tau^2+k_0^2n^2(x,z)\right)^{1/2}\approx\frac{\nabla_\tau^2}{\left(\nabla_\tau^2+k_0^2n_*^2\right)^{1/2}+k_0n_*}+k_0n_*+k_0\delta n(x,z). \tag{2.102}$$

Furthermore, we assume that the complex amplitude satisfies the following condition:

$$U(x,z)=V(x,z)\cdot\exp(-ik_0n_*z), \tag{2.103}$$

where $V(x, z)$ is also a complex amplitude, and the exponent describes the propagation of a constant rate, described by n_*. Substituting (2.103) into (2.98) and using the approximation (2.102), we obtain:

$$V(x,z+\Delta z)=\exp\left(-i\Delta z\left[\frac{\nabla_\tau^2}{\left(\nabla_\tau^2+k_0^2n_*^2\right)^{1/2}+k_0n_*}+\xi(x,z)\right]\right)V(x,z), \quad (2.104)$$

where

$$\xi(x,z)=k_0\,\delta n(x,z). \quad (2.105)$$

Then, formally dividing the exponent into three components, we can write:

$$V(x,z+\Delta z)\approx\exp\left(-i\frac{\Delta z}{2}\frac{\nabla_\tau^2}{\left(\nabla_\tau^2+k_0^2n_*^2\right)^{1/2}+k_0n_*}\right)\times$$

$$\times\exp\left(-i\Delta z\xi(x,z)\right)\exp\left(-i\frac{\Delta z}{2}\frac{\nabla_\tau^2}{\left(\nabla_\tau^2+k_0^2n_*^2\right)^{1/2}+k_0n_*}\right)V(x,z). \quad (2.106)$$

The approximate equality sign is due to the fact that the exponent in the original equation was not an algebraic function but the differential operator, whose division into parts and rearrangement of these parts, generally speaking, lead to some error. Equation (2.106) is the basic equation of the two-dimensional method FFT BPM, and equation (2.103) allows us to recover the unknown complex amplitude of the field.

Besides the fact that equation (2.106) has the desired evolution form, it has a clear physical meaning. Indeed, the first of three exponents describes the distribution in a medium with a homogeneous refractive index n_* at the distance $0.5\Delta z$ (indeed, it suffices to compare (2.106) and (2.98)). Thus, the calculation of propagation of the beam at a distance in an inhomogeneous medium with a refractive index reduces to the calculation of propagation in a homogeneous medium over distance Δz, passing through a thin lens with a phase shift of $-i\Delta z\xi(x, z)$, and then again propagation in a homogeneous medium over distance $0.5\Delta z$. These three steps can be repeated many times so that it is possible to calculate the beam propagation over long distances.

Although the expression (2.106) reflects the essence of the method, from the computational point of view it is difficult to use, so the actual calculations use a slightly different form of it. To obtain this form, it is assumed that the field $V(x,z)$ can be expanded into a Fourier series with a finite number of terms:

$$V(x,z)=\sum_{l=-N/2+1}^{N/2}V_l(z)\exp(ik_{xl}x), \quad (2.107)$$

where the values k_{xl} represent discrete transverse wave numbers and can be calculated as follows:

$$k_{xl}=\frac{2\pi}{L}l, \quad (2.108)$$

where L is the width of the considered computational domain.

Then, for the l-th component of the Fourier expansion V_l the propagation through the section with length with $0.5\Delta z$ with the uniform distribution of the refractive index is described by the following expression:

$$V_l\left(z+\frac{\Delta z}{2}\right)=\exp\left(-i\cdot\frac{\Delta z}{2}\cdot\frac{-k_{xl}^2}{\left(-k_{xl}^2+k_0^2n_*^2\right)^{1/2}+k_0n_*}\right)V_l(z). \qquad (2.109)$$

Taking (2.109) into account, the calculation of propagation over distance Δz is carried out in three stages:

1. transition to the spectral domain using the fast Fourier transform, the calculation of changes in the coefficients using formula (2.109), going back to the original values of the function;
2. the calculation of passage through a thin lens with a phase shift $-i\Delta z\xi(x, z)$;
3. again the transition to the spectral domain, the shift $0.5\Delta z$ to the reverse transition.

This sequence of actions, on the one hand, can be repeated until the required distance of propagation is reached, but on the other hand, at a large number of iterations the computational error begins to accumulate.

The resulting approach and the computational algorithm can be easily generalized to three-dimensional cases. The basic equation in this case takes the form:

$$U(x,y,z+\Delta z)=\exp(-ik_0n_*\Delta z)\times$$

$$\times\exp\left(-i\frac{\Delta z}{2}\frac{\nabla_\tau^2}{\left(\nabla_\tau^2+k_0^2n_*^2\right)^{1/2}+k_0n_*}\right)\exp\left(-i\int_z^{z+\Delta z}\delta n(x,y,z)\Delta z'\right)\times$$

$$\times\exp\left(-i\frac{\Delta z}{2}\frac{\nabla_\tau^2}{\left(\nabla_\tau^2+k_0^2n_*^2\right)^{1/2}}+k_0n_*\right)U(x,y,z)+O\left(\Delta z^3\right), \qquad (2.110)$$

where $\nabla_\tau^2=\frac{\partial^2}{\partial x^2}+\frac{\partial^2}{\partial y^2}$, and the integral in the exponent, responsible for propagation through a thin lens, on the assumption of smallness of the change in the refractive index in this volume, is usually replaced by the value $\delta n\Delta z$.

Consider now the restrictions imposed during the derivation of basic equations and in the preparation of the calculation algorithm which enable us to proceed to the problem of evolutionary type.

1. Requirement of small changes in refractive index (2.101), both in longitudinal and transverse directions.
2. The need to introduce the so-called reference refractive index n_*, the choice of which affects the accuracy of the method.
3. The requirement of the possibility of expanding into a Fourier series with a finite number of terms (2.107), which implies that the accuracy of the method also depends on the chosen number of terms of the series and the nature of distribution of

the field of the propagating beam. Separately, we note that this condition is required to ensure that the formal calculation of the root of the differential operator can be carried out.

Thus, FFT BPM is applicable only in cases of small refractive index changes in the entire considered volume. This feature makes it impossible to apply the method to cases of media with a complex refractive index profile, in particular for the calculation of diffraction photonics devices.

However, this method is convenient and provides sufficient accuracy, for example, in calculating the field in the far zone after passing through the diffraction element.

2.2.3. Solution on the basis of the finite difference method (FD BPM)

The solution described above was based on the reduction of the order of differentiation with respect to the coordinate z by providing opportunities to calculate the square root of the differential operator: indeed, in the FFT BPM we use the Fourier expansion for this purpose (note that in the MoL BPM the reduction to the quadratic form is used for this). Another way to reduce the order of differentiation is the so-called slowly varying envelope approximation (SVEA) [15].

Suppose we have the following expression:

$$U(x, y, z) = \Psi(x, y, z) \cdot \exp(-ik_0 n_* z), \tag{2.111}$$

i.e. it is possible to extract the field changes associated with propagation along the axis z, and the value $\Psi(x, y, z)$ is also a complex amplitude and describes the envelope of the field, while reference refractive index n_* characterizes the accuracy of this envelope. It is assumed that generally the propagation of radiation is similar to the propagation in a homogeneous medium with the refractive index n_*, and the envelope Ψ describes the deviation of the change of the field from the case of propagation along the z axis of a plane wave in a homogeneous medium.

Substitution of (2.111) into (2.92) gives the equation for Ψ:

$$\left(\frac{\partial^2\Psi}{\partial z^2} + 2\frac{\partial\Psi}{\partial z}\left(-ik_0 n_*\right) + \Psi\left(-ik_0 n_*\right)^2 + \left(\frac{\partial^2}{\partial x^2} + \frac{\partial^2}{\partial y^2} + k_0^2 n^2\right)\Psi\right) \times$$
$$\times \exp\left(-ik_0 n_* z\right) = 0, \tag{2.112}$$

and simplifying this equation we obtain:

$$\frac{\partial^2\Psi}{\partial z^2} - 2ik_0 n_* \frac{\partial\Psi}{\partial z} + \left(\frac{\partial^2}{\partial x^2} + \frac{\partial^2}{\partial y^2} + k_0^2 n^2 - k_0^2 n_*^2\right)\Psi = 0. \tag{2.113}$$

Next, we make the assumption of the 'slowly varying envelope' by limiting the value of its second derivative with respect to the z axis:

$$\left|\frac{\partial^2\Psi}{\partial z^2}\right| << 2k_0 n_* \left|\frac{\partial\Psi}{\partial z}\right|. \tag{2.114}$$

This assumption allows us to neglect the value of the second derivative with respect to the first. Equation (2.113) then takes the form:

$$\frac{\partial \Psi}{\partial z}=\frac{-i}{2k_0 n_*}\left(\frac{\partial^2}{\partial x^2}+\frac{\partial^2}{\partial y^2}+k_0^2 n^2-k_0^2 n_*^2\right)\Psi. \tag{2.115}$$

This equation is a generalized form of the Fock–Leontovich equation.

Equation (2.115) can then be solved by finite difference methods. Traditionally, this is carried out using the generalized Crank–Nicolson scheme (although there are other approaches based, for example, on the Douglas scheme [57] and the Galerkin method [58]). Consider its construction for a two-dimensional case, while we assume that the refractive index profile is constant, i.e. $n = n(x)$.

We introduce the following grid in this region:

$$(x_i, z_k)=(x_0+h_x j, z_0+h_z k), \tag{2.116}$$

where $h_x = W/J$ is the sampling step along the axis x; W is the width of the studied region, J is the number of intervals of partition along the axis x; $h_z = L/K$ is the sampling step along the axis z; L is the length of the studied region, K is the number of intervals of patition along the axis z.

We introduce the notation:

$$\Psi_i^k=\Psi\left(x_0+jh_x, z_0+kh_z\right), \tag{2.117}$$

where (x_0, y_0) is the starting corner point of the studied domain.

We approximate the derivatives by difference relations:

$$\left(\frac{\partial}{\partial z}\Psi\right)_j^k=\frac{\Psi_j^{k+1}-\Psi_j^k}{h_z}, j=\overline{0,J},\ k=\overline{0,K-1}, \tag{2.118}$$

$$\left(\frac{\partial^2}{\partial x^2}\Psi\right)_j^k=\frac{\Psi_{j+1}^k-2\Psi_j^k+\Psi_{j-1}^k}{h_x^2}, j=\overline{0,J},\ k=\overline{0,K-1}. \tag{2.119}$$

The initial conditions take the form:

$$\Psi_j^0=\psi_j=\psi\left(x_j\right), \tag{2.120}$$

where $\psi(x)$ is the function describing the distribution of the complex amplitude of the field in the initial plane.

Writing the simplest explicit and implicit schemes for the layers $k+\theta$ and $k+1$ (here θ is a real number that characterizes the displacement of the intermediate point of the scheme and the ratio of explicit and implicit schemes in the scheme under construction $0<\theta<1$):

$$\frac{\Psi_j^{k+\theta}-\Psi_j^k}{\theta h_z}=-\frac{i}{2k_0 n_*}\left(\frac{\Psi_{j+1}^k-2\Psi_j^k+\Psi_{j-1}^k}{h_x^2}+k_0^2 n_j{}^2\Psi_j^k-k_0^2 n_*^2\Psi_i^k\right), \tag{2.121}$$

$$\frac{\Psi_j^{k+1}-\Psi_j^{k+\theta}}{(1-\theta)h_z}=-\frac{i}{2k_0^2n_*^2}\left(\frac{\Psi_{j+1}^{k+1}-2\Psi_j^{k+1}+\Psi_{j-1}^{k+1}}{h_x^2}+k_0^2n_j{}^2\Psi_j^{k+1}-k_0^2n_*^2\Psi_j^{k+1}\right), \quad (2.122)$$

multiplying these expressions by θ and $\theta-1$ and adding up the results, we obtain the finite-difference equation for the generalized Crank–Nicolson scheme:

$$\frac{\Psi_j^{k+1}-\Psi_j^k}{h_z}=-\frac{i\theta}{2k_0n_*}\left(\frac{\Psi_{j+1}^k-2\Psi_j^k+\Psi_{j-1}^k}{h_x^2}+\left(k_0n_j\right)^2\Psi_j^k-k_0^2n_*^2\Psi_j^k\right)-$$

$$-\frac{i(1-\theta)}{2k_0n_*}\left(\frac{\Psi_{j-1}^{k+1}-2\Psi_j^{k+1}+\Psi_{j+1}^{k+1}}{h_x^2}+\left(k_0n_j\right)^2\Psi_j^{k+1}-k_0^2n_*^2\Psi_j^{k+1}\right). \quad (2.123)$$

To obtain a closed system, it is also necessary to specify the boundary conditions that describe the values Ψ^k_j for $j = 0$ and $j = J$. Consider the following boundary conditions.

1. Reflecting boundary conditions [15]. In this case, it is considered that the computational domain is surrounded by a conducting shell, and therefore, $\Psi^k_0 = \Psi^k_J = 0$. In simulation using this type of boundary conditions the boundary of the computational domain is the area where the emerging radiation is reflected back into the region under consideration. This kind of conditions can be successfully applied if it is known in advanced that radiation will not reach the boundaries of the region (for example, when calculating the propagation modes in the waveguide), but otherwise the result is not entirely correct from a physical point of view.

2. Absorbing boundary conditions [7, 58]. By absorbing conditions we usually mean the so-called perfectly matched layers (PML). This type of boundary conditions is widespread for the methods of the FD-TD group, but also applies to the BPM methods. The main idea of these boundary conditions from the computational point of view is to increase the area under consideration by additional sites in which the basic equations have a slightly different form and provide a smooth decay of the field in these areas. Technically, this is realized by considering the complex spatial coordinates in these sites. For example, for a two-dimensional case it is necessary to extend the computational domain with respect to x by the additional width $\tilde{W}$, and the coordinate x within this area is replaced by its complex extension:

$$x\to x+\frac{i}{\omega}\int_0^x\sigma(\xi)d\xi, \quad (2.124)$$

where ω is the cyclic frequency and $\sigma(x)$ is a function describing the profile of PML. At the same time, changing in the basic equations of the method are carried out by the following formal substitutions:

$$\frac{\partial}{\partial x}\to\frac{1}{1+i\sigma(x)\omega^{-1}}\frac{\partial}{\partial x}. \quad (2.125)$$

This approach ensures damping of the field in the additional site, with no reflection of energy back into the computational domain. The quality of the chosen boundary conditions is determined by the width $\tilde{W}$ of the additional sites and by the function $\sigma(x)$ used which must be determined by the specific application.

3. Transparent boundary conditions [60]. This type of boundary conditions is based on the assumption that the field near the boundary of the computational domain can be described by an exponential function. In this case, the values of the complex amplitude at the boundary are calculated as follows:

$$\begin{cases} \dfrac{\Psi_{j-1}^{k}}{\Psi_{j}^{k}} = \exp(-i\alpha h_x), & j \in \{1, J\}, \quad \mathrm{Re}\,\alpha > 0, \\ \dfrac{\Psi_{j-1}^{k}}{\Psi_{j}^{k}} = 1, & j \in \{1, J\}, \quad \mathrm{Re}\,\alpha \le 0, \end{cases} \tag{2.126}$$

where

$$\alpha = -\frac{1}{ih_x} \ln\left(\frac{\Psi_{j}^{k-1}}{\Psi_{j+1}^{k-1}}\right). \tag{2.127}$$

The real part of the number α characterizes the direction of energy propagation in a calculated point (point outside the boundary of the studied region). With the real part of number α being negative, the wave enters the region under consideration, or it otherwise leaves the region. This kind of the conditions (2.126) ensures that there is no flow of energy from outside of the computational domain. However, if the given form of field implies the presence of a field outside the computational domain and propagation into the region (e.g. in the propagation of a plane uniform beam at an angle to the z axis), we should use only the first expression of (2.126) with no additional constraints.

The most popular are the transparent boundary conditions which are simple to implement and describe with sufficient accuracy the majority of cases.

Combining equation (2.123), initial and boundary conditions, we finally obtain the following scheme:

$$\begin{cases} \dfrac{\Psi_j^{k+1} - \Psi_j^k}{h_z} = -\dfrac{i\theta}{2k_0 n_*}\left(\dfrac{\Psi_{j+1}^k - 2\Psi_j^k + \Psi_{j-1}^k}{h_x^2} + \left(k_0 n_j\right)^2 \Psi_j^k - k_0^2 n_*^2 \Psi_j^k\right) - & k = \overline{0, K-1}, \\ -\dfrac{i(1-\theta)}{2k_0 n_*}\left(\dfrac{\Psi_{j-1}^{k+1} - 2\Psi_j^{k+1} + \Psi_{j+1}^{k+1}}{h_x^2} + \left(k_0 n_j\right)^2 \Psi_j^{k+1} - k_0^2 n_*^2 \Psi_j^{k+1}\right), & j = \overline{1, J-1}, \\ \Psi_j^0 = \psi_j, & j = \overline{0, J}, \\ \Psi_0^k = \Psi_1^k \exp(-i\alpha h_x), & k = \overline{1, K}, \\ \Psi_J^k = \Psi_{J-1}^k \exp(-i\alpha h_x), & k = \overline{1, K}. \end{cases}$$

(2.128)

The scheme has a first order approximation with respect to h_x (due to the boundary conditions, the scheme itself is of the second order of approximation) and relative to h_z. Also, it is locally stable under the condition $\theta \geq 0.5$ [15]. Thus, the scheme has a local convergence under the same condition.

Parameter θ together with n_* determine the nature of dispersion and the energy loss in calculations using this scheme [15]

Application of the scheme (2.128) allows the calculation of the incremental value of the complex amplitude Ψ, with each layer requiring the solution of the tridiagonal systems of linear equations (this is usually the sweep method).

If we want to examine only the field amplitude distribution, it is sufficient to consider directly the value Ψ since $|\Psi| \equiv |U|$. In turn, the phase of function Ψ also carries information about the phase of function U, but without taking into account the phase shift occurring in propagation in a homogeneous medium. The reverse transition to the complex amplitude U, obviously, can be accomplished with the aid of equationh (2.111).

Thus, in comparison with the FFT BPM methods, the FD BPM methods impose fewer restrictions. The main advantage is the ability to model propagation in media with an inhomogeneous refractive index profile, but a major role is played by the correct selection of the reference refractive index. However, FD BPM also imposes a restriction on the basic direction of propagation: it is believed that it can not deviate from the z axis by more than 15°.

However, this version of the method may already be effectively used for the calculation and simulation of integrated optics devices and photonics.

2.2.4. Solution on the basis of the finite element method (FE BPM)

A major shortcoming of the already considered versions of BPM is their requirement of a small deviation from the main direction of propagation from the axis z. This requirement can be considerably weakened by the use of one-way Helmholtz equations. These equations can be obtained from (2.72), formally expanding the left side to the product:

$$\begin{gathered}\left(\frac{\partial}{\partial z}-i\sqrt{\mathbf{A}}\right)\left(\frac{\partial}{\partial z}+i\sqrt{\mathbf{A}}\right)\mathbf{E}=0,\\ \left(\frac{\partial}{\partial z}-i\sqrt{\mathbf{B}}\right)\left(\frac{\partial}{\partial z}+i\sqrt{\mathbf{B}}\right)\mathbf{H}=0,\end{gathered} \tag{2.129}$$

where the root of the matrix function is defined on the basis of the eigenvalues of this function [19].

Brackets in (2.129) determine the wave propagation in the forward and reverse directions. If we consider only forward-propagating waves, we obtain the one-way Helmholtz equation:

$$\left(\frac{\partial}{\partial z}+i\sqrt{\mathbf{A}}\right)\mathbf{E}=0,\quad \left(\frac{\partial}{\partial z}+i\sqrt{\mathbf{B}}\right)\mathbf{H}=0, \tag{2.130}$$

whence we can easily obtain expressions describing the explicit dependence of the field components on their current values:

$$\frac{\partial}{\partial z}\mathbf{E} = -i\sqrt{\mathbf{A}}\ \mathbf{E},\ \frac{\partial}{\partial z}\mathbf{H} = -i\sqrt{\mathbf{B}}\ \mathbf{H}. \tag{2.131}$$

Obviously, for the scalar case, the equation takes the form:

$$\frac{\partial}{\partial z}\Psi = -i\sqrt{A}\Psi, \tag{2.132}$$

where

$$A = \frac{\partial^2}{\partial x^2} + \frac{\partial^2}{\partial y^2} + k_0^2 n^2. \tag{2.133}$$

Equation (2.132) already contains only the first derivative with respect to z and has the sought evolutionary form, but its direct solution is difficult because of the need to compute in the general form the power function of the differential operator.

It should be noted that the previously discussed methods actually solve the same problem, but by different means: in the FFT BPM by this shift to the spectral domain, and FD BPM actually used approximation of the power function (this will be demonstrated later).

In the general case it is possible to replace the power function in (2.132) by its approximation [61]. The most frequently used method of calculating power functions, similar to those in (2.132), is the expansion into a series of a specific type. Consider the expansion used for the calculation of the square root.

One of the simplest and traditionally used approximations is the expansion to a Maclaurin series:

$$\sqrt{1+\alpha} \approx 1 + \frac{1}{2}\alpha - \frac{1}{8}\alpha^2 - \ldots, \tag{2.134}$$

however, the requirement of correctness of this approach is that the value α must be close to 0. In the case of (2.132) this condition is clearly not satisfied, since in general the action of the operators of the radicands is not close to the unit operators, as in the case of large values of the parameter α.

Nevertheless, a series of simple steps allows to use the approximation (2.134). Let $\chi \in R$ be a non-negative value, then

$$\sqrt{\alpha} = \sqrt{\frac{\chi^2}{\chi^2}\alpha} = \chi\sqrt{\frac{1}{\chi^2}\alpha} = \chi\sqrt{1 + \left(\frac{1}{\chi^2}\alpha - 1\right)} = \chi\sqrt{1 + \frac{\alpha - \chi^2}{\chi^2}}. \tag{2.135}$$

Although the parameter χ plays absolutely no role in equation (2.135), if χ^2 is selected close to the value of α, then the approximation (2.134) can be used as follows:

$$\sqrt{\alpha} = \chi\sqrt{1+\tilde{\alpha}} \approx \chi\left(1 + \frac{1}{2}\tilde{\alpha} - \frac{1}{8}\tilde{\alpha}^2 - \ldots\right), \tag{2.136}$$

where $\tilde{\alpha} = (\alpha - \chi^2)/\chi^2$.

If the number of terms in the expansion can be taken arbitrarily, the accuracy of the choice of parameter χ does not play a role. However, given the complexity of the operator A, it would be reasonable to choose the parameter value so that the smallest number of members of the expansion series provides a high accuracy.

Traditionally, the value χ is represented as:

$$\chi = k_0 n_*, \tag{2.137}$$

where n_* is the reference refractive index. This kind of parameter χ is easy to obtain, for example, by solving the problem of propagation in a homogeneous medium (in this case, the series will accurately describe the meaning of the root).

In general, the meaning of this normalization is similar to the substitution of (2.111) in the slowly varying envelope approximation, i.e., reduces the problem to propagation in a homogeneous medium in the presence of deviations. The permissible deviation from the propagation in the direction of the z axis is at the same time characterized by the complexity the approximation of the power function and the choice of parameter n_*.

It is easy to verify that the application of the Maclaurin expansion with the accuracy up to the first term gives a result similar to the relations (2.115) obtained in the approximation of the slowly varying envelope. Indeed, in this case

$$\tilde{\alpha} = \frac{A - k_0^2 n_*^2}{k_0^2 n_*^2}, \tag{2.138}$$

and, consequently,

$$\sqrt{A} \approx k_0^2 n_*^2 \left(1 + \frac{1}{2} \cdot \frac{A - k_0^2 n_*^2}{k_0^2 n_*^2}\right) = \frac{1}{2}\left(A + k_0^2 n_*^2\right), \tag{2.139}$$

from this, considering (2.111), equation (2.132) exactly gives (2.115).

However, the use of more sophisticated approximations directly for equation (2.132) is rather difficult from the computational point of view. Instead, the equation itself (2.132) is transformed to the finite-element form.

It is easy to show that equation (2.132) can be written in the form [16]:

$$\frac{\partial}{\partial z} U = i k_0 n_* \sqrt{A} U, \tag{2.140}$$

and from the introduction of sampling and examination of a single step $z_{j+1} = z_j + \Delta z$ we obtain:

$$U_{j+1} = P U_j, \tag{2.141}$$

where U_j approximates the desired solution U at the point z_j, and

$$P = \exp\left(i k_0 n_* \Delta z \sqrt{A}\right), \tag{2.142}$$

where the operator A is evaluated at the point $z_j + \theta \Delta z$ (the parameter θ has the same meaning as in the generalized Crank–Nicholson scheme).

There are two approaches to approximation of the operator (2.142). The first of these (so-called Hoekstra scheme [62]) implies first the approximation to the root of the differential operator, then an approximation of the exponential function with the substitution to it of the approximation of the root. However, a more general approach involves the approximation of the entire operator (2.142) as a whole [16].

A more accurate approximation is achieved by using the Padé approximation [63] instead of the Maclaurin series. In the general case for the studied approximation function we have:

$$P \approx \sum_{l=0}^{p} p_l A^l \Big/ \sum_{l=0}^{q} q_l A^l, \tag{2.143}$$

where $[p, q]$ is the order of the Padé approximation, and $pl \in R$ and $q_l \in R$ – coefficients of the approximation, determined from the equality of the constructed degradable fraction to the expansion of original function into a Taylor series up to the term $p + q$. The approximation of the orders [2.2] [3.2] and [3.3] is used most frequently.

After approximation of the operator by some series we obtain differential equations of the first order with respect to z whose solution for each layer of the calculation is reduced to solving multidiagonal systems of linear equations by the sweep method.

There are also modified methods that combine different Padé approximation (just as in the Crank–Nicolson scheme where explicit and implicit schemes are combined), as well as using other computer techniques (such as the analogue of the scheme of alternating directions) [45].

In general, it should be noted that the application of FE BPM leads to a substantial increase in computational complexity, however, these methods allow us to consider the propagation of radiation at considerable angles to the z axis. However, if radiation in large sections of the simulated regions propagates primarily at small angles to the z axis, the use of methods FE BPM is irrational from the viewpoint of computational complexity. In this case, more effective is the combination of FD BPM and FE BPM at different sites. As regards the accuracy of calculations (but not the speed), this pair of methods is better than both FFT BPM and MoL BPM.

Thus, the group of the FE BPM methods considerably exceeds the earlier versions of the methods in the field of possible applications, but also requires the introduction of the concept of the reference refractive index. But in general, these methods can be used to model more complex processes in the propagation of radiation in the environment which may arise in the case of consideration of complex nanophotonic elements.

2.2.5. Approaches to solving the Helmholtz vector equation

The previous sections discussed the main approaches to reduce the order of differentiation with respect to z and the methods of the BPM familty based on these approaches, but they have been formulated for the scalar case. In general, however, it is often necessary to take into account the vector nature of the electromagnetic field, namely, that in this approximation of the Helmholtz equations the field is described by two vectors of complex amplitudes (for the electric and magnetic fields). Most of these methods can be generalized to the vector case, but before studying them it is important to make some clarifications.

If the profile of the refractive index varies significantly at distances comparable with the wavelength, it is necessary to examine all components of the electromagnetic field (or at least three of them, which in accordance with the consequences of Maxwell's equations can be used to determine other components) and the solution must hold for the general form of the Helmholtz equations (2.72). However, in many cases a change in the refractive index profile is minor, which can significantly simplify the calculation method.

We now assume that the refractive index is practically independent of z. In this case, the field can be 'split' into two components by transferring to the consideration of TE- and TM-polarizations [15, 45].

For the case of TE-polarization (in this case $E_z \equiv 0$), with the independence of the refractive index of z the expression (2.72) for the electric component becomes:

$$\frac{\partial^2}{\partial z^2}\mathbf{E}_\tau + \mathbf{A}_\tau \mathbf{E}_\tau = 0, \tag{2.144}$$

where $\mathbf{E}_\tau = (E_x, E_y)^T$, $\mathbf{A}_\tau$ is the square matrix containing only the elements responsible for the interaction of the tangential components of the electric field. Similarly, when considering only the TM-polarization (in this case $H_z \equiv 0$), the expression (2.72) for the magnetic component takes the form:

$$\frac{\partial^2}{\partial z^2}\mathbf{H}_\tau + \mathbf{B}_\tau \mathbf{H}_\tau = 0, \tag{2.145}$$

where $\mathbf{H}_\tau = (H_x, H_y)^T$, $\mathbf{B}_\tau$ is the square matrix containing only the elements responsible for the interaction of the tangential components of the magnetic field.

In turn, the one-way equations take the following form:

$$\frac{\partial}{\partial z}\mathbf{E}_\tau = -i\sqrt{\mathbf{A}_\tau}\mathbf{E}_\tau, \quad \frac{\partial}{\partial z}\mathbf{H}_\tau = -i\sqrt{\mathbf{B}_\tau}\mathbf{H}_\tau, \tag{2.146}$$

each of which contains a dependence on only z *in* the left side, and the number of equations matches the number of unknowns. Both of these features are due to the transition to the consideration of polarization and exclusion from consideration of the longitudinal components of the field components.

Also, in considering the polarizations it is convenient to consider the field strictly oriented along one axis (e.g., $E_x \equiv 0$ and $H_x \equiv 0$), and then the expressions (2.146) degenerate into the equations for only one field component, i.e. actually we considered the scalar case. Then, using the expression (2.144) and (2.145) and one of the previously described scalar methods, we can solve independently two problems and then, using the resultant distributions E_y and H_y, we can express all other components of the field, using the consequences of Maxwell's equations. It should be noted that the form of the calculated equations may differ from those discussed earlier for the scalar case, since in general it is necessary to take into account the dependence of the refractive index on the transverse coordinates, but the overall logic of inference and methods of calculation in this case remain.

If the condition imposed on the orientation of the field is for whatever reasons not possible (for example, requires modelling of the propagation of the TE-wave waves oriented in a certain way), then it is necessary to consider both transverse field components.

The first approximation to the solution in this case is the use of the so-called semi-vector methods (SV BPM). In this case, we consider both transverse components, however, the relationship between them is ignored (i.e. $A_{xy} = A_{yx} = 0$ or $B_{xy} = B_{yx} = 0$). This allows the solutions for different transverse components to be separated and the problem can be reduced to two scalar cases. In this case, the equations may differ from those previously considered because of the dependence of the refractive index on the transverse coordinates.

A more accurate solution can be obtained by taking into account all components of the matrices $\mathbf{A}_\tau$ and $\mathbf{B}_\tau$. This kind of method is called full vector (full vector, FV BPM). The logic of the derivation of equations is also preserved, but the form of systems of linear equations that must be solved in the course of the calculation is much more complicated: while in the scalar case they had a diagonal structure, in the full vector form they become block-diagonal.

If the refractive index profile depends on the coordinate z, then in general examination of the polarizations is incorrect: in this case, even if the initial field has a polarization, it will be disrupted in the course of propagation. In this case, we need to consider longitudinal components of the field too.

One of the approaches to dealing with this problem is the generalization of the one-way Helmholtz equations [64]. In the first approximation, they link the derivatives of the transverse components of the field with the values of the longitudinal components:

$$\begin{aligned} \frac{\partial}{\partial z}\mathbf{E}_\tau &= -i\left[\mathbf{A}_\tau^{1/2}\mathbf{E}_\tau + \mathbf{A}_\tau^{-1/2}\begin{pmatrix} A_{xz}E_z \\ A_{yz}E_z \end{pmatrix}\right], \\ \frac{\partial}{\partial z}\mathbf{H}_\tau &= -i\left[\mathbf{B}_\tau^{1/2}\mathbf{H}_\tau + \mathbf{B}_\tau^{-1/2}\begin{pmatrix} B_{xz}H_z \\ B_{yz}H_z \end{pmatrix}\right]. \end{aligned} \tag{2.147}$$

However, this notation does not allow a direct solution by the methods discussed earlier because the number of unknowns in this case exceeds the number of equations. If we additionally use the expressions of the longitudinal component through the transverse ones, derived from Maxwell's equations (they are given in matrix form):

$$\begin{aligned} E_z &= \frac{\mu i}{n^2 k_0}\sqrt{\frac{\mu_0}{\varepsilon_0}}\begin{pmatrix} -\dfrac{\partial}{\partial y} & \dfrac{\partial}{\partial x} \end{pmatrix}\mathbf{H}_\tau, \\ H_z &= \frac{-i}{k_0 \mu}\sqrt{\frac{\varepsilon_0}{\mu_0}}\begin{pmatrix} -\dfrac{\partial}{\partial y} & \dfrac{\partial}{\partial x} \end{pmatrix}\mathbf{E}_\tau, \end{aligned} \tag{2.148}$$

we can write the equations in the form of unidirectional equations cross-linking all the components of the field:

$$\begin{aligned} \frac{\partial}{\partial z}\mathbf{E}_\tau &= -i\left[\mathbf{A}_\tau^{1/2}\mathbf{E}_\tau + \mathbf{A}_\tau^{-1/2}\mathbf{C}\mathbf{H}_\tau\right], \\ \frac{\partial}{\partial z}\mathbf{H}_\tau &= -i\left[\mathbf{B}_\tau^{1/2}\mathbf{H}_\tau + \mathbf{B}_\tau^{-1/2}\mathbf{D}\mathbf{E}_\tau\right], \end{aligned} \tag{2.149}$$

where

$$\mathbf{C}=\begin{pmatrix}A_{xz}\\A_{yz}\end{pmatrix}\frac{\mu i}{n^2k_0}\sqrt{\frac{\mu_0}{\varepsilon_0}}\begin{pmatrix}-\frac{\partial}{\partial y} & \frac{\partial}{\partial x}\end{pmatrix},$$
$$\mathbf{D}=\begin{pmatrix}B_{xz}\\B_{yz}\end{pmatrix}\frac{-i}{k_0\mu}\sqrt{\frac{\varepsilon_0}{\mu_0}}\begin{pmatrix}-\frac{\partial}{\partial y} & \frac{\partial}{\partial x}\end{pmatrix}. \quad (2.150)$$

The solution of (2.149) also requires the approximation of the power functions of the matrix differential operator, similar to that considered in section 2.2.5. For example, the Maclaurin series approximation to the first term gives the following approximation:

$$\frac{\partial}{\partial z}\mathbf{E}_\tau=-\frac{i}{2}\left[\left(\frac{1}{k_0n_*}\mathbf{A}_\tau+k_0n_*\mathbf{I}\right)\mathbf{E}_\tau+\left(-\frac{1}{k_0^3n_*^3}\mathbf{A}_\tau+\frac{3}{k_0n_*}\mathbf{I}\right)\mathbf{CH}_\tau\right],$$
$$\frac{\partial}{\partial z}\mathbf{H}_\tau=-\frac{i}{2}\left[\left(\frac{1}{k_0n_*}\mathbf{B}_\tau+k_0n_*\mathbf{I}\right)\mathbf{H}_\tau+\left(-\frac{1}{k_0^3n_*^3}\mathbf{B}_\tau+\frac{3}{k_0n_*}\mathbf{I}\right)\mathbf{DE}_\tau\right], \quad (2.151)$$

where **I** is the square unit matrix.

The transition from the matrix form gives four equations (one for each of the tangential components) in the partial derivatives with the 1st order with respect to z. At the same time, the first two derivatives on the tangential coordinates have the 2nd order for the components of the electric field and the 4th order for the components for the magnetic field, while the latter two, on the contrary, have the 2nd order for the components of the magnetic field and the 4th order for the component for the electric field. In this case the right-hand sides of (2.151) do not contain derivatives of the complex amplitudes with respect to z. This form of equations, describing the propagation in the medium, allows the finite-difference methods to be used for solving them. The numerical solutions are the values of the tangential field components which, in turn, can be used to determine also the longitudinal components. This requires the use of finite-difference analogues of the equations (2.150).

Thus, depending on the conditions imposed on the refractive index profile, it is often possible to reduce the vector problem to a set of independent scalar problems, and in some cases this requires more complicate basic equations and the application of more sophisticated computational methods. However, in most cases various modifications of the BPM make it possible to model of the propagation of radiation in large volumes, taking into account the vector nature of the electromagnetic field.

2.2.6. Examples of application of BPM

To illustrate the capabilities of the method, let us consider a few examples. For simplicity and clarity of the results in the calculations we use the scalar finite-difference version of the method (FD BPM).

Let us start with a simple example from the field of integrated optics: a Y-splitter element for planar waveguides. Figure 2.38a shows the distribution of the refractive index for such an element (the refractive index of the cladding 1.45, the refractive index of the core 1.458). The input receives a beam corresponding to the fundamental mode of the fibre (the wavelength of monochromatic radiation 1.55 μm), and in the propagation through an element the beam splits, after which two beams of lower power propagate through two 'sleeves'. In general, the beam power is determined by the configuration of the waveguides in the element. In this case the parameters of the waveguides were assumed to be the same, so the output power of the beams must also be equal.

Figure 2.38b shows the distribution of the amplitude of the beam in the modelled area and it can be seen that the beam actually diverges into two directions. The amplitude distribution in the input and output planes (Fig. 2.38c and 2.38d, respectively) shows that the beams at the outputs are symmetric, and the power of each beam is less than the initial beam power.

Thus, the BPM can be used to simulate the operation fibre optics devices and to investigate the distribution of radiation power.

Let us now consider a more complex problem associated with the excitation of modes in an optical fibre. One way of excitation of modes other than the fundamental mode of the fibre is the use of diffractive optical elements [65]. In particular, it is possible to implement these elements in the form of micro- and nanorelief on the end of the fibre [66]. Consider a planar optical fibre with a 'step' (binary microrelief) at its end for the excitation of the mode of the 1st order. The distribution of the refractive index of the fibre tip is shown in Fig. 2.39a. A Gaussian beam was fed to the input (see Fig. 2.39b), and the wavelength of the monochromatic radiation was considered equal to 0.63 μm. The height of the relief for the given values of the refractive index and wavelength was approximately 0.68 μm. The intensity here and after is given in relative units $I \sim n|U|^2$.

Generally speaking, this problem does not satisfy the condition of the slow variation of the refractive index imposed in the derivation of the BPM equations. So, we compare the simulation results obtained using the BPM and a more rigorous FD-TD method, discussed earlier in this chapter. To allow comparison, we assume that the field obtained in the framework of the scalar solution corresponds to the component E_y, and the field intensity in the case of a solution by the FD-TD method will be calculated by the method described in section 2.1.2. Figure 2.39c and 2.39d, respectively, show the intensity distribution in the case of simulation using the FD-TD and BMP methods, and Fig. 2.39e and 2.39f show for comparison the intensity for a particular plane.

In Fig. 2.39c in the region with the refractive index $n = 1$ we observed the interference of the incident wave, the wave reflected from the boundary between the media, and also the wave formed at the border of the 'steps' of the microrelief. The front with a complex profile propagates in the fibre region. In simulation using the BPM no interference pattern is found in the region outside the fibre (because the method does not account for the reflection at the boundary between media and the propagation in the opposite direction). However, the wave front propagating in the fibre has the same character as in the simulation by the FD-TD method.

Comparison of Fig. 2.39e and Fig. 2.39f shows that the beams obtained by simulation using the FD-TD and FD-BPM methods are similar. For a numerical comparison of the method we use the following value:

$$\tilde{I} = \frac{\int\limits_{-16.6\times10^{-6}}^{16.6\times10^{-6}} I(x)\,dx}{\int\limits_{-16.6\times10^{-6}}^{16.6\times10^{-6}} I_{\text{í à÷àëü í à}}(x)\,dx}, \tag{2.152}$$

where $I(x)$ is the function of the intensity in relative units, and $I_{\text{initial}}(x)$ is the function of the intensity of the illuminating beam. We also introduce the function characterizing the difference of the solutions obtained in the FD-TD and FD BPM method:

$$I_{\Delta}(x) = |I_{\text{FD-TD}}(x) - I_{\text{FD-BPM}}(x)|, \tag{2.153}$$

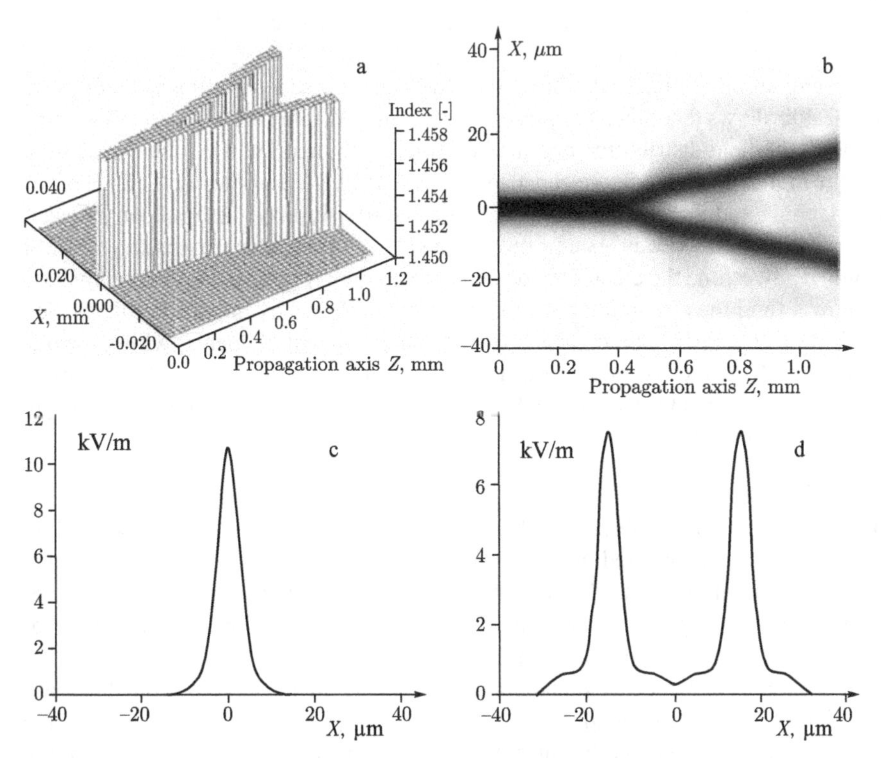

Fig. 2.38. Simulation of the passage of a beam through a Y-splitter element in a planar waveguide: a) the distribution of the refractive index, b) the radiation amplitude distribution, c) the distribution of the amplitude of the input beam, d) distribution of the amplitude of the output beam.

where $I_{FD-TD}(x)$ is the function in Fig. 2.31e and $I_{FD-BPM}(x)$ is the function in Fig. 2.31f.

Numerical calculation of the value (2.152) for the functions $I_{FD-TD}(x)$, $I_{FD-BPM}(x)$ and $I\Delta(x)$ gives the following results:

$$\tilde{I}_{FD-TD} \approx 0.9812,\ \tilde{I}_{FD-BPM} \approx 0.9993,\ \tilde{I}_{\Delta} \approx 0.0484. \qquad (2.154)$$

The total decrease in intensity in the case of simulation using the FD-TD method is explained by the reflection of part of the incident beam from the waveguide surface. In simulation by BPM reflection is not considered, therefore, $\tilde{I}_{FD-BPM}$ is close to 1. The smallness of $\tilde{I}_{\Delta}$ and the fact that it has the same order as the quantity $\tilde{I}_{FD-BPM} - \tilde{I}_{FD-TD}$, suggests that the result of simulation by the BPM is a good approximation to the result of simulations using the FD-TD method.

Thus, there may be cases when BPM methods are not formally applicable, but they do give a reasonably good estimate.

This example will not be complete if we do not also consider the distribution of the beam formed by the relief in the fibre, because the propagation distance is considered insufficient for the formation of the desired mode. Figure 2.40 shows the results of modelling of propagation of the previously considered the beam to a depth of 2 mm in the fibre.

Examination of Fig. 2.40a shows that this leads to a superposition of modes, and Fig. 2.40b shows that this takes place approximately 1 mm from the end of the fibre.

With the field distribution obtained by simulation, and knowing the distribution for the modes of the fibre, we can also calculate the values of the coefficients of the overlapping and thereby determine the mode composition of the beam. In this case the power of the 'steady' beam, propagating in a fibre, is approximately 75% of original power, and the excited mode of the first order represents approximately 93% of power (and the remaining power is mainly transferred by the third order mode).

Thus, the BPM method can be used to study optical elements, which include elements of diffractive photonics. The method is particularly useful in cases in which it is required to investigate the nature of propagation of the formed beams in waveguide environments.

Conclusion

The explicit finite difference schemes for Maxwell's equations, described in this chapter, can be used for simulation of electromagnetic radiation through arbitrary optical elements. Moreover, these schemes are characterized by ease of implementation in the organization of vector computing and by the high order of approximation of the differential problem.

The proposed method of transition for two time segments from the non-stationary solution of Maxwell's equations to the complex amplitude is characterized by low computational complexity, because it is not accompanied by the calculation of the two-dimensional Fourier transform. The universal grid area allows modelling of different types of diffractive optical elements with an infinite aperture, periodic, with a limited aperture and symmetrical. In this way the layout of absorbing layers

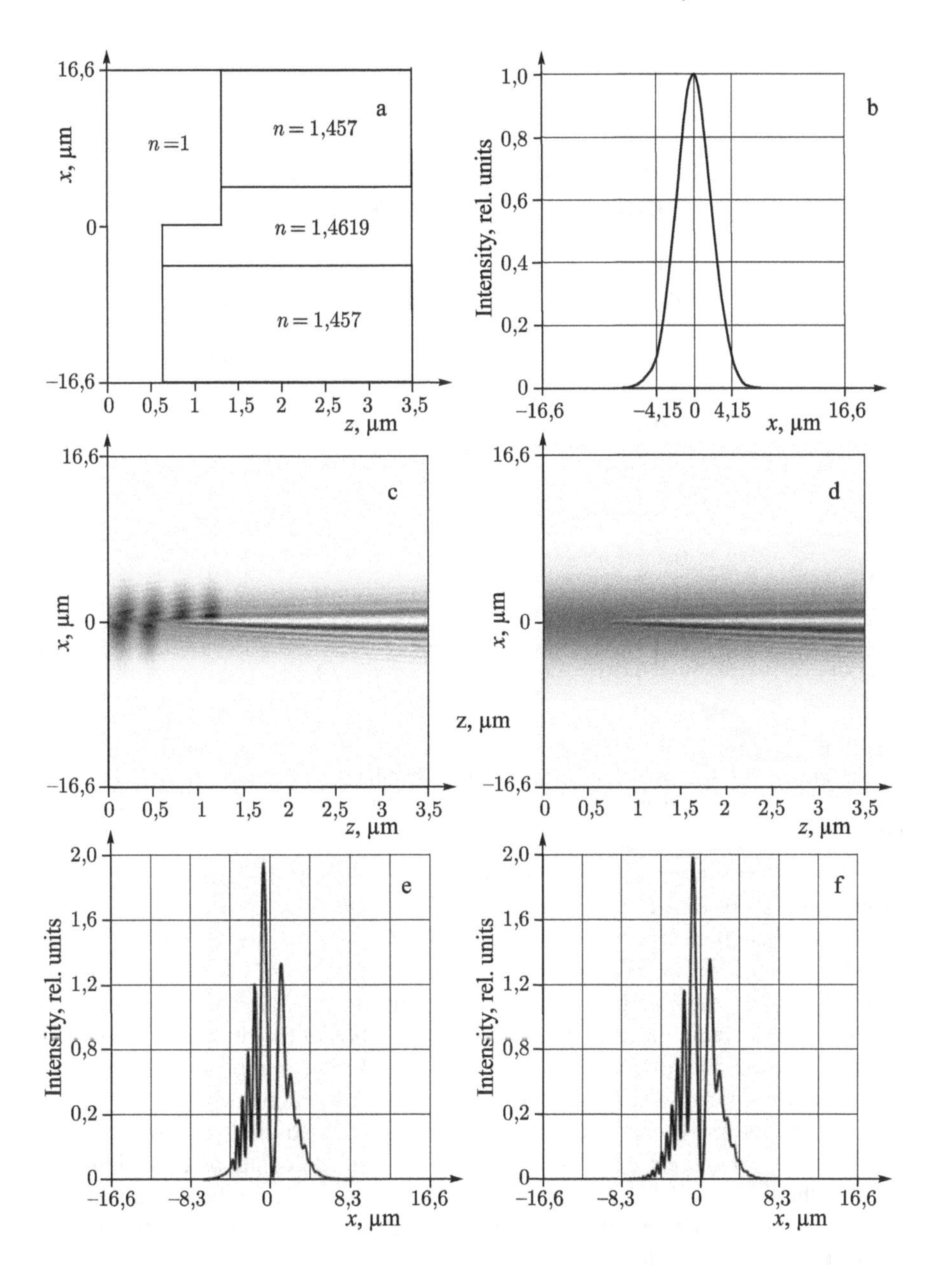

Fig. 2.39. The passage of a Gaussian beam through a microrelief on the face of a planar waveguide with a stepped refractive index profile: a) the distribution of the refractive index, b) the intensity of the illuminating beam, c) the intensity distribution in simulation using the FD-TD method; d) the intensity distribution in simulation using FD BPM; d) the intensity in the plane $z = 3.5$ μm (FD BPM).

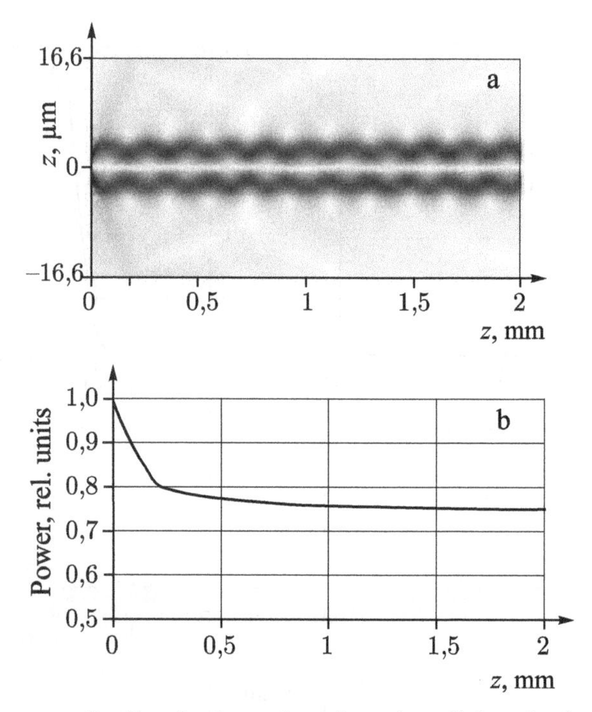

Fig. 2.40. The passage of a Gaussian beam through a microrelief on the face of a planar waveguide with a step–index: a) the amplitude distribution of the radiation in a planar waveguide, and b) the power of the propagating radiation depending on the distance of propagation.

improves the vectorization of the algorithm for calculation and reduces the duration of computing: by up to 25% in the calculation of the characteristic fragment of the subwavelength diffractive microrelief with a linear size of several wavelengths.

The methods of specifying the incident wave formulated on the basis of the methods of the resulting field and the split field in the study of propagation of electromagnetic radiation through the diffractive optical elements are characterized by small errors of the computational experiments and low computational complexity as a result of solving an auxiliary one-dimensional diffraction problem.

Decomposition of the grid area in the difference solution of the Maxwell equations can significantly reduce the duration of the calculations by taking into account the locally established field fragments. For example, in the decomposition of the layered diffractive optical element into 128 subdomains the duration of the calculations is reduced 40 times.

The above review of the existing approaches to the numerical solution of the Helmholtz equation and their classification within the family of the BPM (beam propagation method) methods allows the reader to orient in this field of nanophotonics, understanding the fundamental differences between the variants of the method. The guide discussed solutions to the scalar Helmholtz equation for the cases of paraxial (in the approximation of the smooth envelope) and non-paraxial (based on the approximation of a differential operator) propagation. The finite-difference schemes for solving equations and some variants of the boundary conditions were presented. Also discussed were the approaches to solving the Helmholtz equation for the vector case, including for media with a non-uniform distribution of the refractive index.

References

1. Tsukerman I. Computational methods for nanoscale applications. Particles, plasmons, and waves, NewYork: Springer Science, 2008. 530.
2. Klimov V.V., Nanoplasmonics, Moscow, Fizmatlit, 2009. 480.
3. Cron G. Equivalent circuit of the field equations of Maxwell, I Proc. IRE, 1944. 32. 289–299.
4. Yee K.S. Numerical solution of initial boundary value problems involving Maxwell's equations in isotropic media, IEEE Trans. Antennas Propag., 1966. AP–14. 302–307.
5. Golovashkin D.L., Degtyarev A.A., Soifer V.A., Modeling of waveguide propagation of optical radiation in the electromagnetic theory. Komp. Optika, 1997. 17. 5–9.
6. Golovashkin D.L., Degtyarev A.A., The algorithm of the second order accuracy in time for the solution of Maxwell's equations, Komp. Optika, 1998. 18. 39–41.
7. Taflove A., Hagness S., Computational Electrodynamics: The Finite–Difference Time–Domain Method: 3nd. ed. Boston: Arthech House Publishers, 2005. 852.
8. Berenger Jean-Pierre, A perfectly matched layer for the absorption of electromagnetic waves, Journal of computational physics, 1994. 114. 185–200.
9. Taflove A., Brodwin M. Numerical solution of steady-state electromagnetic scattering problems using the time-dependent Maxwell's equations. IEEE Transactions of Microwave Theory and Techniques, 1975. 23, 8. 623–630.
10. Taflove A., Application of the finite–difference time-domain method to sinusoidal steady-state electromagnetic penetration problems. IEEE Transactions of Microwave Theory and Techniques, 1980. 22, 2. 191–202.
11. Prather D.W. and Shi S., Formulation and application of the finite-difference time–domain method for the analysis of axially symmetric diffractive optical elements, J. Opt. Soc. Am. A., 1999. 16, 5. 1131–1142.
12. Fidel B., Heyman E., Kastner R. and Zioklowski R.W., Hybrid ray–FD-TD moving window approach to pulse propagation, Journal of Computational Physics, 1997. 138, 2. 480–500.
13. Feit M.D. and Fleck J.A., Light Propagation in Graded Index Optical Fibres, Applied Optics. 1978. 17, 24. 3990–3998.
14. Gerdes J., Beam propagation algorithm based on the method of lines, Journal of the Optical Society of America B. 1991. 8, 2. 389–394.
15. Huang W. et al., The finite–difference vector beam propagation method: Analysis and assessment, Journal of Lightwave Technology. 1992. 10, 2. 295–305.
16. Chui S.L. and Lu Y.Y., Wide–angle full vector beam propagation method based on alternating direction implicit preconditioner, Journal of the Optical Society of America A. 2004. 21, 2. 420–425.
17. Le Kh.Q. and Bientsman P., Wide-angle beam propagation method without using slowly varying envelope approximation, Journal of the Optical Society of America B. 2009. 26, 2. 353–356.
18. Golovashkin D.L., Statement of radiating conditions in the simulation of cylindrical diffractive optical elements using finite difference solution of Maxwell's equations. Matem. Modelirovanie, 2007. 19, 2. 3–14.
19. Golub, J., Matrix computations. Springer–Verlag, 1999. 548.
20. Ortega J.M., et, Introduction to parallel and vector methods for solving linear systems: the translation from English, Springer–Verlag, 1991. 364.

21. Samarskii A.AA. and Gulin A.V., Stability of difference schemes. Moscow, Nauka, 1972. 430.
22. Golub M.A., Kazan N.L., Soifer V.A., A mathematical model of laser radiation focusing elements of computer optics, Naycvh. Priborostr., 1992. 3, 1. 8–28.
23. Hiroyuki I., Electromagnetic analysis of diffraction gratings by the finite-difference time-domain method J. Opt. Soc. Am, 1998. 15, 1. 152–157.
24. Umashankar K. and Taflove A.. A novel method to analyze electromagnetic scattering of complex objects, IEEE Trans. Electromagn. Compat. EMC–24, 1982. 397–405.
25. Golovashkin D.L., Diffraction of waves on a two–dimensional dielectric grating by finite difference solution of Maxwell's equations, Matem. Modelirovanie, 2004. 16, 9. 83–91.
26. Golovashkin D.L., Kazan N.L., Decomposition of the grid area at FD-TD, Matem. Modelirovanie, 2007. 19, 2. 48–58.
27. Engquist B., Majda A. Absorbing boundary conditions for the numerical simulation of waves, Mathematics of Computation, 1977. 31. 629–651.
28. Mur G., Absorbing boundary conditions for the finite–difference approximation of the time-domain electromagnetic field equations, IEEE Trans. Electromagnetic Compability, 1981. 22. 377–382.
29. Moore T.G., Blaschak J.G., Taflove A., Kriegsmann G., A.Theory and application of radiation boundary operators, IEEE Trans. Antennas Propagat., 1988. 36, 12. 1797–1812.
30. Katz D.S., Thiele E.T., Taflove A. Validation and extension to three dimensions of the Berenger PML absorbing boundary condition for FD-TD meshes, IEEE Microwave Guided Wave Lett., 1994. 4, 8. 268–270.
31. Taflove A., Brodwin M. Computation of the electromagnetic fields and induced temperatures with a model of the microwave irradiated human eye, IEEE Transactions of Microwave Theory and Techniques, 1975. mtt–23, 11. 888–896.
32. Golovashkin D.L, Diffraction of waves on a two–dimensional perfectly conducting grating by finite difference solution of Maxwell's equations, Matem. Modelirovanie, 2005. 17, 4. 53–61.
33. Anantha V., Taflove A., Efficient modeling of infinite scatterers using a generalized total-field/scattered-field FD-TD boundary partially embedded within PML, IEEE Trans. Antennas Propagat., 2002. 50, 10. 1337–1349.
34. Chang J.H., Taflove A., Three–dimensional diffraction by infinite conducting and dielectric wedges using a generalized total–field/scattered–field FD-TD formulation, IEEE Trans. Antennas Propagat., 2005. 53, 4. 1444–1454.
35. Golovashkin D.L., Kazan, N.L., The method of forming the incident wave at FD-TD (one-dimensional case) Avtometriya, 2006. 42, 6. 78–85.
36. Methods of Computer Optics (second edition, revised), ed. Soifer V.A. Moscow, Fizmatlit, 2002. 688s.
37. Cron G., Equivalent circuit of the field equations of Maxwell, I Proc. IRE, 1944. 32. 289–299.
38. Perlik A.T., Taflove A., Opsahl T., Predicting scattering of electromagnetic fields using FD–TD on a connection machine, IEEE Transactions on Magnetics, 1989. 25, 4. 2910–2912.
39. Shi S., Tao X., Yang L., Prather D.W., Analysis of diffractive optical elements using a nonuniform finite–difference time–domain method, Optical Engineering, 2001. 40, 4. 503–510.
40. Samarskii A.A., Vabishchevich P.N., Computational Heat Transfer. Moscow: URSS, 2002. 784.
41. Borgsmuller S., Noehte S., Dietrich C., Kresse T., Manner R.Computer–generated stratified diffractive optical elements, Applied Optics, 2002. 42, 26. 5274–5282.; a) Taflove A., Umashankar K.R., Review of FD–TD numerical modeling of electromagnetic wave scattering and radar cross section Proc. IEEE, 1989. 77, 5. 682–699.
42. Butikov E.I., Optics, Textbook, Spb, Nevskii Dialekt, 2002.
43. Petit R., Electromagnetic Theory on Gratings, Springer, New York, 1980.

44. Lu, Y.Y., Some Techniques for Computing Wave Propagation in Optical Waveguides, Communications in Computational Physics. 2006. 1, 6. 1056–1075.
45. Heiblum M., Harris J., Analysis of curved optical waveguides by conformal transformation. IEEE Journal of Quantum Electronics. 1975. 11, 2. 75–82.
46. Boertjes D.W., Mullin J.N., Novel wide–angle 3–D BPM implemented on a path-based grid, Microwave and Optical Technology Letters. 1999. 20, 5. 287–290.
47. Rao H., et al., A bidirectional beam propagation method for multiple dielectric interfaces, IEEE Photonics Technology Letters. 1999. 11, 7. 830–832.
48. Zhang H. et al., Improved Bidirectional Beam–Propagation Method by a Fourth–Order Finite–Difference Scheme, Journal of Lightwave Technology. 2007. 25, 9. 2807–2812.
49. Yioltsis T.V., et al., Explicit finite-difference vector beam propagation method based on the iterated Crank-Nicolson scheme, Journal of Optical Society of America A. 2009. 26, 10. 2183–2191.
50. Bekker E.V. et al., Wide-Angle Alternating-Direction Implicit Finite–Difference beam Propagation Method, Journal of Lightwave Technology. 2009. 27, 14. 2595–2604.
51. Cheng H., et al., Nonparaxial Split–Step method With Local One–Dimension Scheme for Three–Dimensional Wide–Angle beam Propagation, Jian–Guo Tian Journal of Lightwave Technology. 2009. 27, 14. 2717–2723.
52. Bhattacharya, D. and Sharma A., Three-dimensional finite difference split–step nonparaxial beam propagation method: new method for splitting of operators, Applied Optics. 2009. 48, 10. 1878–1885.
53. Shibayama J., et al., A Finite–Difference Time–Domain Beam–Propagation Method for TE- and TM Wave Analyses, Journal of Lightwave Technology. 2002. 21, 7. 1709–1715.
54. Masoudi H.M., A Novel Nonparaxial Time–Domain Beam–Propagation Method for Modeling Ultrashort Pulses in Optical Structures. Journal of Lightwave Technology. 2007. 25, 10. 3175–3184.
55. Okoshi T., Analysis methods for electromagnetic wave problems, Editor E. Yamashita. Norwood: Artech House, 1990. Ch.10. 341–369.
56. Yamauchi, J., et al., Wide–angle propagating beam analysis based on the generalized Douglas scheme for variable coefficients, Optics Letters. 1995. 20, 1. 7–9.
57. Fan K., et al., A full vectorial generalized discontinuous Galerkin beam propagation method (GDG BPM) for non-smooth electromagnetic fields in waveguides, Journal of Computational Physics. 2008. 7178–7191.
58. Vassalo C. and Collino F., Highly Efficient Adsorbing Boundary Conditions for the Beam Propagation Method, Journal of Lightwave Technology. 1996. 14, 6. 1570–1577.
59. Hadley, G.R., Transparent Boundary Condition for the Beam Propagation Method, IEEE Journal of Quantum Electronics. 1992. 28, 1. 363–370.
60. Zhang H. et al., Assessment of Rational Approximations for Square Root Operator in Bidirectional Beam Propagation Method, Journal of Lightwave Technology. 2008. 26, 5. 600–607.
61. Stoffer R., et al., New true fourth-order accurate scalar beam propagation methods for both TE ans TM-polarizations, Optical and Quantum Electronics. 1999. 31, 9–10. 705–720.
62. Baker G.A., Graves-Morris P., Padé Approximants, Cambridge: Cambridge University Press, 1996.
63. Gavrilov A.V., A modified beam propagation method and its application to the calculation of propagation in waveguides with a varying refractive index profile, Computer Optics. 2008. T. 32, 1. 15–22.
64. Methods for computer design of diffractive optical elements ed. V.A. Soifer. New York: John Wiley & Sons, Inc., 2002. 765.
65. Pavelyev V.S., et al., Design of On–fibre Diffractive Microrelief for Efficient Graded–index Fibre Mode Excitation Optical Memory & Neural Networks (Information Optics). 2007. 16, 3. 159–166.

Chapter 3

Diffraction on cylindrical inhomogeneities comparable to the wavelength

In opto- and microelectronics, as well as in nanophotonics use is made of sophisticated optical devices with the dimensions of the order of the wavelength of incident light, whose work is described by the non-trivial physical effects, such as multiple scattering on periodic structures (Bragg diffractive gratings), the scattering and diffraction on aperiodic structures (diffractive optical elements (DOE)), the dispersion and non-linear transformation of laser pulses. The effect of these elements can not be predicted on the basis of geometrical optics or scalar diffraction theory, and it is essential to study the propagation of light waves through them using the vector model of diffraction. All this creates a greater need for efficient numerical approaches for modelling the wave propagation of light, if possible taking into account dispersion, scattering, complex interference effects, etc. Also, the use of the vector diffraction model is required when the relevant calculation area is located near or within the optical element. Although analytical solutions of the vector diffraction problem can be obtained for selected objects (sphere, half-plane, cylinder) [1, 2] the boundary conditions on the electromagnetic field for other dielectric structures makes the analytic solution impossible.

Thus, for the problems of modelling the diffraction of light on the elements with dimensions on the order of the wavelength the light must be regarded as electromagnetic radiation, which allows us to transfer many of the developed methods of electromagnetic simulation of microwave and radio waves to the area of optical modelling.

Most numerical methods (the method of moments, the finite difference method, the boundary element method, the finite element method, etc.) came into optics from other areas of science and so far no universal approach has been developed which would cover a large number of optical (and electromagnetic) problems. In general, it is usually necessaty to combine two or three methods to calculate a wide range of problems that lead to the development of various integrated modelling techniques [3–5].

The bulk of the numerical diffraction simulation methods can be classified as differential [32], difference [33, 34], integral [35–38], variational [39–41], the discrete sources [42], and the propagation of rays [43].

In this section, we discuss finite element methods for solving the two-dimensional Helmholtz equation and for the solution of the two-dimensional diffraction integral equation.

3.1 Analysis of diffraction on inhomogeneities by the combined finite element and boundary element method

3.1.1. Analysis of the diffraction of light on non-periodic irregularities

In the problems of modelling the diffraction of light on the optical elements in a homogeneous space with dimensions on the order of the wavelength the light must be regarded as electromagnetic radiation.

To determine the electromagnetic fields at a point of space, integral methods combine the contributions to this point from the field of sources in the volume or on the surface. The popularity of the integral methods is based on their ability to solve unlimited field problems, as the Sommerfeld radiation condition is certainly satisfied in the formulation of the problem. Moreover, the integral methods require knowledge of the field only on the surface of the diffraction element, and not of the total field in a space and this minimizes the number of unknowns. In [21, 22] the authors presented a method based on the combined boundary element method. In [23] a method was developed for calculating the diffraction on a plane-parallel plate with the heterogeneity on the basis of integral equations associated with the numerical solution of the Green tensor. The disadvantage of these methods is that they lead to fully populated matrices and, therefore, require more computer memory and long calculation time. Also, the limit of the application of methods should be chosen at the physical boundary of the object. If the inhomogeneity has a nontrivial form, it leads to an increase in the number of unknowns.

The difference solution of differential Maxwell equations was considered in [7, 24–27]. In [28] the difference solution of the wave equation was described. The disadvantages of this approach are the inability to use the radiation conditions and restrictions on the steps of the grid. To simulate the steady-state problems of passage of radiation by difference schemes, it is necessary to use a finite number of wavelengths of the incident pulse which distorts the wave spectrum. The use of the absorbing boundary conditions [13, 29] as boundary conditions for unbounded diffraction problems allows us to solve approximately the Maxwell's equations by difference schemes and the accuracy of the solution depends on the number of layers on the artificial boundary and the degree of its closure.

In contrast to the methods of finite difference solution of Maxwell's equations, the integral and variational methods do not require the construction of complex absorbing boundary conditions [13, 18].

Variational methods in problems with a limited range of tasks determine solutions of the Helmholtz equation by minimizing the functional relation. In [13], the Helmholtz equation was solved by the Galerkin finite element method using the boundary conditions of the complex type that depend on the unknown parameter which required the use of the border of a certain form. In addition, this method also does not include the Sommerfeld radiation conditions.

In [3] the authors presented a hybrid method based on the finite element method, formulated by the Ritz method, and the boundary element method. In this hybrid method the finite element method is used to solve the Helmholtz equation in the inner part of the inhomogeneous dielectric element of micro-optics and the integral method and the boundary element method are used for the region external to the element where the radiation condition must be satisfied. Both methods are joined at the boundary between inner and outer parts, with the satisfaction of the conditions of continuity of the field. Using the finite element method to determine the field inside the object leads to a tridiagonal matrix, which requires less computer memory and shorter computing time than the methods of the volume integrals [30]. The result of using the boundary element method for determining the field at the boundary is a more accurate solution than using the finite element method with absorbing boundary conditions. But the application of the Ritz method to solve the Helmholtz equation is incorrect because it imposes a requirement of the positivity of the operator of the equation being solved. No conclusion can be made on the definite sign of the operator of the Helmholtz equation.

Description of the method of calculation

In this problem, a source in space illuminates a cylindrical structure. In the absence of the structure the source of the incident field. In the presence of the structure this source creates another field, called the total field. The scattered field is defined as the difference between the total field and the incident field. The purpose of the task is to determine the total or scattered field, characterizing the structure.

Any two-dimensional field can be decomposed into E_z-polarized and H_z-polari zed fields. In the diffraction domain the field is described by a system of differe- ntial equations for various cases of TE- and TM-polarizations. For TE-polarizat- ion ($\mathbf{E}(x,y) = (0,0, E_z(x, y))$ the complex amplitude $u(x, y)$ denotes the total electric field $E_z(x, y)$, which is directed along the axis z (along the generatrix of a cylindrical optical element), the coordinates (x,y) lie in the plane of the normal section. For the TM-polarization ($\mathbf{H}(x,y) = (0,0, H_z(x, y))$ the complex amplitude $u(x, y)$ denotes the total magnetic field $H_z(x, y)$.

The total field $u_\Omega(x, y)$ in the Ω region must satisfy the equation

$$\nabla \cdot \left[\frac{1}{p(x,y)} \nabla u_\Omega(x,y) \right] + k_0^2 q(x,y) u_\Omega(x,y) = f_\Omega, \tag{3.1}$$

where $f_\Omega = jk_0 Z_0 J_z$, $p(x,y) = \mu_r$, $q(x, y) = \varepsilon_r$ for TE-polarization, and

$$f_\Omega = -\frac{\partial}{\partial x}\left(\frac{1}{\varepsilon_r} J_y\right) + \frac{\partial}{\partial y}\left(\frac{1}{\varepsilon_r} J_x\right), \; p(x,y) = \mu_r, \; q(x,y) = \varepsilon_r \quad \text{for TM-polarization.}$$

The values μ_r and ε_r are the ratios of permeability and permittivity to the magnetic and dielectric constants of the medium to the same performance space, i.e. $\mu_r = \mu/$

μ_0 and $\varepsilon_r = \varepsilon/\varepsilon_0$, k_0 is the wave number of waves in free space

$$k_0 = \omega\left(\mu_0\varepsilon_0\right)^{1/2} = \frac{\omega}{c} = \frac{2\pi}{\lambda_0}, \tag{3.2}$$

$Z_0 = \sqrt{\mu_0 / \varepsilon_0}$ is the impedance of free space, J is the density of the electric current source.

In this problem the calculation domain is infinite. However, as is known, the finite element method (FEM) is applicable only to a finite or limited area. Thus, to solve the equation (3.1), an infinite domain Ψ, external to the scatterer, should be limited by the introduction of the artificial boundary Γ. Correspondingly, for the only solution of the problem, boundary conditions must be imposed at this artificial boundary. Such conditions should make the border transparent as possible for the scattered field or, in other words, minimize the non-physical reflections from the boundary. One of the classes of the boundary conditions, designed for this purpose, can be obtained from the boundary integral equations applied to the outer region. These boundary conditions are global in nature, i.e. they relate to the field at a boundary node with the field across the boundary. These boundary conditions prevent reflection at the boundary for all angles of incidence of the waves and lead to the exact solution.

Thus, it is necessary to define the total field $u(x, y)$ in the domains Ω (internal) and Ψ (external) satisfying the above conditions.

Galerkin's solution of equation (1) is based on solving the relations of the form:

$$\iint_{\Omega}\left(-\frac{1}{p}\Delta u_{\Omega}\gamma - qk^2 u_{\Omega}\gamma - f_{\Omega}\gamma\right)d\Omega = 0, \tag{3.3}$$

where γ is an arbitrary function from the domain of equation (1).

Using the first Green's formula:

$$\iint_{\Omega} P\Delta Q \, d\Omega = \int_{\Gamma} P\frac{dQ}{d\mathbf{n}}dl - \iint_{\Omega} \nabla P \nabla Q \, d\Omega,$$

for the functions P and Q, where Ω is the domain of the plane x, y; Γ is its boundary, required anti-clockwise; $dQ/d\mathbf{n}$ is the derivative in the direction of the outward normal to the curve F, we obtain:

$$\iint_{\Omega}\left(\frac{1}{p}\nabla u_{\Omega}(x,y)\nabla\gamma - qk^2 u_{\Omega}(x,y)\gamma\right)d\Omega - \int_{\Gamma}\frac{\gamma}{p}\frac{du_{\Omega}(x,y)}{d\mathbf{n}}d\Gamma = \iint_{\Omega} f_{\Omega}\gamma d\Omega. \tag{3.4}$$

The system of basis functions for Ω is denoted $\left\{\omega_{k,l}^{\Omega}(x,y)\right\}_{k,l=0}^{N_x,N_y}$ and the system of basis functions for Γ $\{\omega_m^{\Gamma}(x,y)\}_{m=1}^{M}$, where N_x, N_y is the number of nodes of the grid covering a rectangular area Ω on the x and y axis, respectively, M is the number of nodes of the grid covering the boundary Γ.

Replacing in (3.4) the arbitrary function γ by the system of basis functions for Galerkin's method, we can write the system of linear equations:

$$\mathbf{A}u + \mathbf{B}v = \mathbf{C}f, \tag{3.5}$$

where $u = \left(u_1, \ldots, u_{NxNy}\right)^T$ is the vector consisting of coefficients

$\{u_{Ny(k)+l} = u_{k,l}\}_{k,l=0}^{N_x,N_y}$ of the expansion:

$$u^{\Omega}(x,y) = \sum_{k,l=0}^{N_x,N_y} u_{k,l}\omega_{k,l}^{\Omega}(x,y). \tag{3.6}$$

The vector $f = \left(f_1, \ldots, f_{NxNy}\right)^T$ is the vector consisting of the coefficients of the expansion:

$$f^{\Omega}(x,y) = \sum_{k,l=0}^{N_x,N_y} f_{k,l}\omega_{k,l}^{\Omega}(x,y). \tag{3.7}$$

Although (3.6) and (3.7) are valid for all points (x, y) in the domain Ω, it is necessary to process separately the values of the field and its partial derivatives on the boundary Γ from the values in the inner region. The decomposition, similar to (3.6) and (3.7), for the field and its partial derivatives at the boundary has the form:

$$u^{\Gamma}(x,y) = \sum_{m=1}^{M} u_m \omega_m^{\Gamma}(x,y), \tag{3.8}$$

$$\mathrm{v}^{\Gamma}(\mathrm{x},\mathrm{y}) = \sum_{m=1}^{M} v_m \omega_m^{\Gamma}(x,y), \tag{3.9}$$

$$f^{\Gamma}(x,y) = \sum_{m=1}^{M} f_m \omega_m^{\Gamma}(x,y), \tag{3.10}$$

where $(x, y) \in \Gamma$, $v = (v_1, \ldots, v_M)^T$ is the vector consisting of the expansion coefficients $v_k = \partial u_k / \partial \mathbf{n}$.

The elements of the matrix **A** are calculated from the equations:

$$a_{N_y k+l, N_y i+j} = \iint_{\Omega_{k,l}} \left(\frac{1}{p(x,y)} \left[\frac{\partial \omega_{k,l}^{\Omega}(x,y)}{\partial x} \frac{\partial \omega_{i,j}^{\Omega}(x,y)}{\partial x} + \frac{\partial \omega_{k,l}^{\Omega}(x,y)}{\partial y} \frac{\partial \omega_{i,j}^{\Omega}(x,y)}{\partial y} \right] - k_0^2 q(x,y) \omega_{k,l}^{\Omega}(x,y) \omega_{i,j}^{\Omega}(x,y) \right) d\Omega,$$

$$k, i = [1, N_x], \; l, j = [1, N_y], \tag{3.11}$$

where $\Omega_{k,j}$ is the domain of decomposition of domain Ω, consisting of nodes k and j.

The elements of the matrix **B** are given by:

$$b_{m,s} = -\oint_{\Gamma_{m,s}} \omega_m^{\Gamma}\omega_s^{\Gamma}\, dl, \qquad m, s = [1, M], \tag{3.12}$$

where $\Gamma_{m,s}$ is the linear region of the boundary Γ, which includes the boundary nodes m and s.

The elements of the matrix **C** are given by:

$$c_{N_y k+l, N_y i+j} = \iint_{\Omega_{k,l}} \omega_{k,l}^{\Omega}(x,y)\omega_{i,j}^{\Omega}(x,y)\, d\Omega, \qquad k,j = [1, N_x],\ l,j = [1, N_y], \tag{3.13}$$

where $\Omega_{k,j}$ is the domain of decomposition of Ω, consisting of nodes k and j.

As a piece-wise linear basis we determine the function of the form:

$$\omega^{\Omega}{}_{k,l}(x,y) = \begin{cases} 1-\dfrac{x_k-x}{h}-\dfrac{y_l-y}{h}, & x,y \in \Omega_{k,l,1}^h \\ 1-\dfrac{x_k-x}{h}, & x,y \in \Omega_{k,l,2}^h \\ 1+\dfrac{y_l-y}{h}, & x,y \in \Omega_{k,l,3}^h \\ 1+\dfrac{x_k-x}{h}+\dfrac{y_l-y}{h}, & x,y \in \Omega_{k,l,4}^h \\ 1+\dfrac{x_k-x}{h}, & x,y \in \Omega_{k,l,5}^h \\ 1-\dfrac{y_l-y}{h}, & x,y \in \Omega_{k,l,6}^h \end{cases}, \tag{3.14}$$

where h is the distance between the adjacent grid points.

Elements ($a_{k,l}^{i,j}$) of the matrix **A**, the elements ($b_{m,s}$) of the matrix **B** and elements ($c_{k,l}^{i,j}$) of the matrix **C** are calculated from equations (3.11) (3.12) and (3.13), respectively. Then, the system of equations (3.5) can be written as:

$$\begin{bmatrix} [\mathbf{A}_{\Omega,\Omega}] & [\mathbf{A}_{\Gamma,\Omega}] & 0 \\ [\mathbf{A}_{\Omega,\Gamma}] & [\mathbf{A}_{\Gamma,\Gamma}] & [\mathbf{B}] \end{bmatrix} \begin{bmatrix} \mathbf{u}_{\Omega} \\ \mathbf{u}_{\Gamma} \\ \mathbf{v}_{\Gamma} \end{bmatrix} = \begin{bmatrix} [\mathbf{C}_{\Omega,\Omega}] & [\mathbf{C}_{\Gamma,\Omega}] \\ [\mathbf{C}_{\Omega,\Gamma}] & [\mathbf{C}_{\Gamma,\Gamma}] \end{bmatrix} \begin{bmatrix} \mathbf{f}_{\Omega} \\ \mathbf{f}_{\Gamma} \end{bmatrix}. \tag{3.15}$$

The system of equations (3.15) has no unique solution, since it consists of N equalities with $N + M$ unknowns: $N = N_x N_y$ is the total number of nodes of the field $u_{k,l}(x, y)$ in the domain Ω and M and derivatives along the normal to the boundary nodes $v_{k,l}(x, y)$.

We define the field in free space Ψ (outside the domain Ω with its boundary Γ). Since this is a homogeneous space, then the field can be formulated in terms of boundary integrals with the appropriate Green's function. The total field $u_\Psi(x, y)$ in domain Ψ must satisfy the following equation:

$$\nabla\cdot\left[\frac{1}{p}\nabla u_\Psi(\xi)\right]+k_0^2 q u_\Psi(\xi)=f_\Psi,\ \xi\in\Psi, \tag{3.16}$$

where $f_\Psi = jk_0 Z_0 J_z^\Psi, p(x,y) = \mu_r, q(x,y) = \varepsilon_r$ forTE-polarization and

$f_\Psi = -\frac{\partial}{\partial x}\left(\frac{1}{\varepsilon_r}J_y^\Psi\right)+\frac{\partial}{\partial y}\left(\frac{1}{\varepsilon_r}J_x^\Psi\right), p(x, y) = \varepsilon_r, q(x, y) = \mu_r$ for TM-polarization,

and J^Ψ is the density of electric current in free space.

To construct the boundary integral equation for the field and its normal derivative, we introduce the Green function u^* which satisfies the Sommerfeld radiation condition and is the fundamental solution of the Helmholtz equation:

$$\nabla^2 u^*(\xi,\eta)+k^2 u^*(\xi,\eta)=-\delta(\xi,\eta),\ \eta\in\Psi. \tag{3.17}$$

The fundamental solution for the Helmholtz equation in the two-dimensional homogeneous space is well known and has the form

$$u^* = (i/4)H_0^{(1)}(kr), \tag{3.18}$$

where

$$r=\sqrt{\left[x_1(\eta)-x_1(\xi)\right]^2+\left[x_2(\eta)-x_2(\xi)\right]^2}$$

$H_0^{(1)}(kr) = J_0(kr)+iY_0(kr)$ is the Hankel function of the first kind and zeroth order, where J_0 is the Bessel function of zeroth order, Y_0 is the Neumann function of zeroth order.

To construct the boundary integral equation for the scattered field and its normal derivative, we use Green's theorem as follows:

$$\begin{gathered}\int_\Psi\left[\nabla^2 u^*(\xi,\eta)+k^2 u^*(\xi,\eta)\right]u(\eta)d\Psi= \\ =-\int_\Gamma q(\eta)u^*(\xi,\eta)d\Gamma+\int_\Gamma u(\eta)q^*(\xi,\eta)d\Gamma+\int_\Psi f_\Psi(\eta)u^*(\xi,\eta)d\Psi.\end{gathered} \tag{3.19}$$

The functions in both integrals on the right side of equation (3.19) $q(\eta) = \partial u(\eta)/\partial\mathbf{n}'$ are the normal derivatives of the field amplitude.

Substituting (3.17) into (3.19) and passing to the limit of the observation point ξ from the inner point to the boundary, we find

$$c(\xi)u(\xi) = -\int_{\Psi} f_{\Psi}(\eta)u^*(\xi,\eta)d\Psi + \int_{\Gamma} q(\eta)u^*(\xi,\eta)d\Gamma - \int_{\Gamma} u(\eta)q^*(\xi,\eta)d\Gamma. \quad (3.20)$$

This equation provides a functional link between the functions u and their normal derivative q at the boundary Γ. The function c in (3.20) is equal to:

$$c(\xi) = 1 - \frac{1}{2\pi}\phi, \quad (3.21)$$

where ϕ is the interior angle of the piecewise-near boundary at point ξ. The first term on the right side is the field produced by a source f_{Ψ} in free space, and may be designated as the incident field u^{in}_{Ψ}.

Thus, equation (3.20) is written as follows:

$$c(\xi)u(\xi) = \int_{\Gamma} q(\eta)u^*(\xi,\eta)d\Gamma - \int_{\Gamma} u(\eta)q^*(\xi,\eta)d\Gamma + u^{in}_{\Psi}(\xi). \quad (3.22)$$

Substituting the function of the complex amplitude in equation (3.22) at the boundary by its approximation by basic piecewise-linear functions at $\xi\in\Gamma$ (3.8) and (3.9), we obtain

$$c_m u_m = \sum_{s=1}^{M} h\left\{ v_s \begin{bmatrix} \int_0^1 \omega_s^{\Gamma}(\rho_s + [\rho_{s+1}-\rho_s]\gamma)u^*(\rho_m,\rho_s+[\rho_{s+1}-\rho_s]\gamma)d\gamma + \\ + \int_0^1 \omega_s^{\Gamma}(\rho_{s-1}+[\rho_s-\rho_{s-1}]\gamma)u^*(\rho_m,\rho_{s-1}+[\rho_s-\rho_{s-1}]\gamma)d\gamma \end{bmatrix} - \right.$$

$$\left. -u_s \begin{bmatrix} \int_0^1 \omega_s^{\Gamma}(\rho_s+[\rho_{s+1}-\rho_s]\gamma)\frac{\partial u^*(\rho_m,\rho_s+[\rho_{s+1}-\rho_s]\gamma)}{\partial \mathbf{n}'}d\gamma + \\ + \int_0^1 \omega_s^{\Gamma}(\rho_{s-1}+[\rho_s-\rho_{s-1}]\gamma)\frac{\partial u^*(\rho_m,\rho_{s-1}+[\rho_s-\rho_{s-1}]\gamma)}{\partial \mathbf{n}'}d\gamma \end{bmatrix}\right\} + u_m^{in},$$

$$m = [1, M], \quad (3.23)$$

which can be represented in the matrix form

$$[\mathrm{D}]\,\bar{\mathbf{u}}_{\Gamma} + [\mathrm{G}]\,\bar{\mathbf{v}}_{\Gamma} = \bar{\mathbf{u}}_{\Gamma}^{in}. \quad (3.24)$$

with elements of the matrices $[D]$ and $[G]$ in the form

$$d_{m,s} = -h\begin{bmatrix} \int_0^1 \omega_s^{\Gamma}(\rho_s+[\rho_{s+1}-\rho_s]\gamma)\frac{\partial u^*(\rho_m,\rho_s+[\rho_{s+1}-\rho_s]\gamma)}{\partial \mathbf{n}'}d\gamma + \\ + \int_0^1 \omega_s^{\Gamma}(\rho_{s-1}+[\rho_s-\rho_{s-1}]\gamma)\frac{\partial u^*(\rho_m,\rho_{s-1}+[\rho_s-\rho_{s-1}]\xi)}{\partial \mathbf{n}'}d\gamma \end{bmatrix} - c_m\delta_{ms}, \quad (3.25)$$

$$g_{m,s} = h \left[\begin{array}{l} \int_0^1 \omega_s^\Gamma \left(\rho_s + [\rho_{s+1} - \rho_s]\gamma \right) u^*(\rho_m, \rho_s + [\rho_{s+1} - \rho_s]\gamma) d\gamma + \\ + \int_0^1 \omega_s^\Gamma \left(\rho_{s-1} + [\rho_s - \rho_{s-1}]\gamma \right) u^*(\rho_m, \rho_{s-1} + [\rho_s - \rho_{s-1}]\gamma) d\gamma \end{array} \right], \quad (3.26)$$

$$m, s = [1, M].$$

The integrals in (3.25) and (3.26) can be evaluated numerically. Combining equations (3.15) and (3.24), we obtain a closed system of linear algebraic equations for solving the problem of diffraction of a plane wave by a cylindrical micro-object

$$\begin{bmatrix} [\mathbf{A}_{\Omega,\Omega}] & [\mathbf{A}_{\Gamma,\Omega}] & \mathbf{0} \\ [\mathbf{A}_{\Omega,\Gamma}] & [\mathbf{A}_{\Gamma,\Gamma}] & [\mathbf{B}] \\ \mathbf{0} & [\mathbf{D}] & [\mathbf{G}] \end{bmatrix} \begin{bmatrix} \mathbf{u}_\Omega \\ \mathbf{u}_\Gamma \\ \mathbf{v}_\Gamma \end{bmatrix} = \begin{bmatrix} [\mathbf{C}_{\Omega,\Omega}] & [\mathbf{C}_{\Gamma,\Omega}] & \mathbf{0} \\ [\mathbf{C}_{\Omega,\Gamma}] & [\mathbf{C}_{\Gamma,\Gamma}] & \mathbf{0} \\ \mathbf{0} & \mathbf{0} & [\mathbf{E}] \end{bmatrix} \begin{bmatrix} \mathbf{f}_\Omega \\ \mathbf{f}_\Gamma \\ \mathbf{u}_\Gamma^{in} \end{bmatrix}, \quad (3.27)$$

where the submatrix $\mathbf{A}_{\Omega,\Omega}$ with the dimension $(N - M) \times (N - M)$ includes the ratio of the field in the internal nodes of the partition grid, submatrix $\mathbf{A}_{\Omega,\Gamma}$ and $\mathbf{A}_{\Gamma\Omega}$, with dimension $(N - M) \times M$ and $M \times (N - M)$, respectively, include the coupling coefficients of the field at the boundary and interior nodes, submatrix $\mathbf{A}_{\Gamma,\Gamma}$ with the size $M \times M$ involves the coupling coefficients of the field at the boundary nodes, the submatrix $\mathbf{B}$ with the size $M \times M$ includes the ratios between the derivatives of the field at the boundary of the field and in the internal nodes of the partition grid, submatrix $\mathbf{D}$ with the size $M \times M$ involves the coupling coefficients of the field of free space at the boundary nodes, the $\mathbf{G}$ with the size $M \times M$ includes the ratio between derivatives of the field at the boundary and the field of free space in the internal nodes of the partition grid. Submatrix $\mathbf{C}_{\Gamma,\Gamma}$ with the dimension $(N - M) \times (N - M)$ includes the ratio of field sources in the internal nodes of the partition grid, submatrices $\mathbf{C}_{\Omega,\Gamma}$ and $\mathbf{C}_{\Gamma,\Omega}$ with the dimension $(N - M) \times M$ and $M \times (N - M)$, respectively, include the ratio of field sources in the boundary and internal nodes, the submatrix $\mathbf{C}_{\Gamma,\Gamma}$ with the size $M \times M$ includes the ratio of the field sources at the boundary nodes. Submatrix $\mathbf{E}$ is the identity matrix with the size $M \times M$. Vectors $\mathbf{u}_\Omega$ and $\mathbf{u}_\Gamma$ are the vectors of the strength of the field in the interior and boundary nodes of the grid, $\mathbf{v}_\Gamma$ is the vector of normal derivatives of the field at the boundary nodes. Vectors $\mathbf{f}_\Omega$ and $\mathbf{f}_\Gamma$ are the vectors of the field sources in the interior and boundary nodes of the grid, $\mathbf{u}_\Gamma^{\text{in}}$ is the vector of the strength of the external incident field at the boundary nodes of the grid. Thus, the dimension of the system of equations (3.27) is $(N + M) \times (N + M)$.

After determining the values of the field and its derivatives on the boundary Γ the field at any point of doimain Ψ is defined by (3.22), where $c(\xi) = 1$.

Examples of calculation

Consider the diffraction of a plane wave (TE- and TM-polarizations) on dielectric and conducting homogeneous cylinders with a circular cross-section for the experimental investigation of convergence of the combined method. The convergence of the algorithm depends on the length of the segment h, the wavelength of the source λ, μ_r and ε_r of the medium. Since the magnetic and dielectric constants of the medium are the variables in the problem, we consider the dependence of solutions on the parameter λ/h that determines the number of grid segments on the same wavelength. Let the plane wave falls on the cylinder, wavelength $\lambda_0 = 0.5$ μm. The radius of the cylinder is 0.25 μm. The relative dielectric permittivity of the conducting cylinder made of is aluminum $\varepsilon = -4.4 + i11.9$. The relative dielectric permittivity of the dielectric cylinder $\varepsilon = 2.25$. The parameters of the homogeneous space surrounding the cylinder are $\varepsilon = \mu = 1$. The calculation were made by the combined finite element and boundary element method. Domain Ω was represented by a square region, the contour Γ was the perimeter of domain Ω. No sources were situated inside Ω. Domain Ω was covered with a square grid consisting of 105 × 105 nodes. The calculation time was 10 min on a PC Pentium 4.

Figure 3.1 shows the results of numerical modelling of diffraction of TE- and TM-waves on a dielectric cylinder. Figure 3.2 shows the simulation results of diffraction on the conducting cylinder.

To evaluate the diffractive processes, we use the directional diagram of scattering, which depends on angular coordinate φ, as defined at points at infinity ($\rho \to \infty$) as

$$\sigma(\varphi) = \lim_{\rho \to \infty} 2\pi\rho \frac{\left|u^{sc}\right|^2}{\left|u^{in}\right|^2}. \tag{3.28}$$

In particular, the directional diagram is determined in the forward direction ($\varphi = 0$), in the opposite direction ($\varphi = \pi$) and in the transverse direction ($\varphi = \pi/2$). The following shows the dependence of σ/λ on parameter λ/h. The results are presented in dB ($\sigma_{\text{DB}} = 10 \log_{10}\sigma$). Figure 3.3 presents an assessment of the diffraction of TE- and TM-waves on a dielectric cylinder.

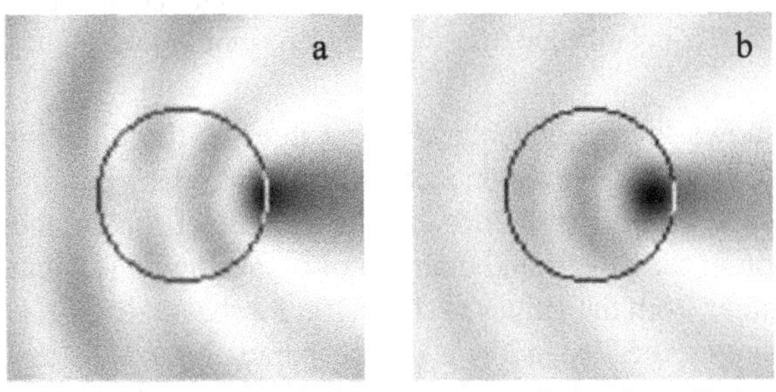

Fig. 3.1. Field intensity distribution of diffraction on a dielectric cylinder (inverted): TE-polarization (a), TM-polarization (b).

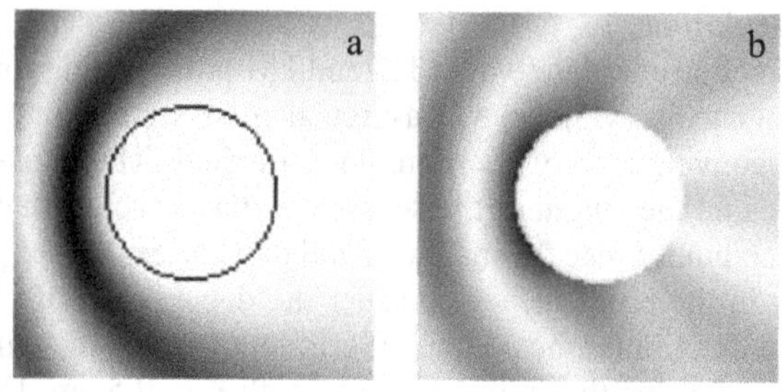

Fig. 3.2. Field intensity distribution of diffraction on a conducting cylinder (inverted): TE-polarization (a), TM-polarization (b).

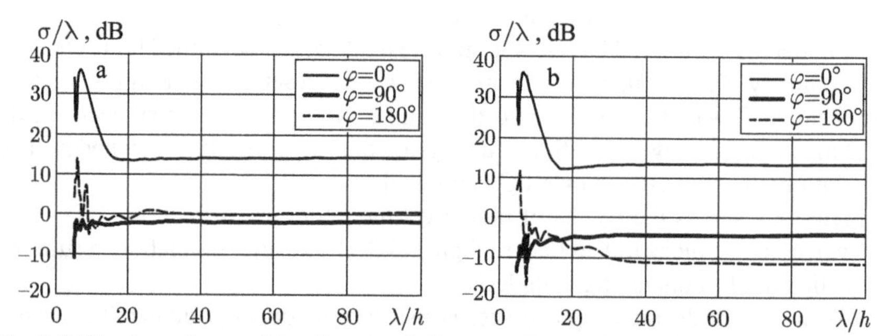

Fig. 3.3. The dependence of the directional diagram of scattering on a dielectric cylinder on paraneter λ/h for the TE-polarization (a) and TM-polarization (b).

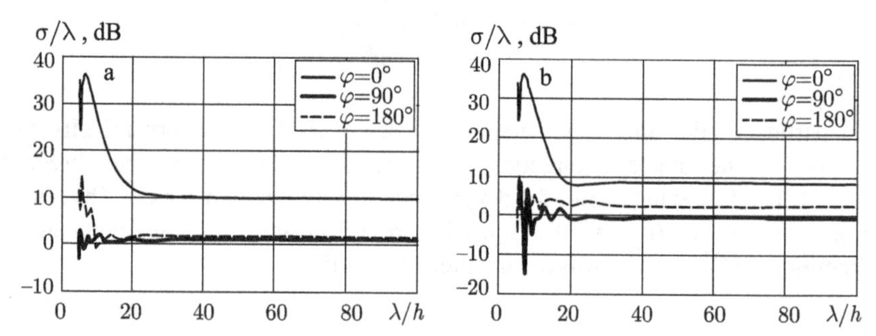

Fig. 3.4. The dependence of the directional diagram of scattering on a conducting cylinder on paraneter λ/h for TE-polarization (a) and TM-polarization (b).

Figure 3.4 shows the dependence of the values of the directional diagram of diffraction of TE- and TM-waves on a conducting cylinder.

Figure 3.5 shows the dependence on the parameter λ/h of the values of the relative deviation of the directional diagram of scattering of TE- and TM-waves on a dielectric cylinder from the values at $\lambda/h = 100$ for the angles φ equal to 0, $\pi/2$, π.

Figure 3.6 shows the dependence on the parameter λ/h of the values of the relative deviation of the directional diagram of scattering of TE- and TM-waves on a conducting cylinder.

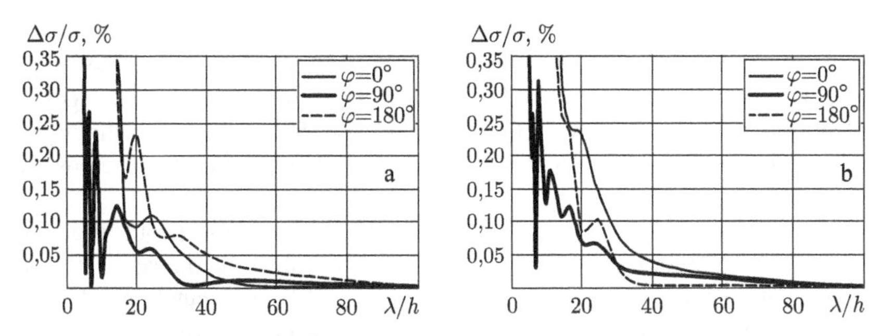

Fig. 3.5. Dependence of the relative deviation of the directional diagram of scattering on a dielectric cylinder on parameter λ/h for TE-polarization (a) and TM-polarization (b).

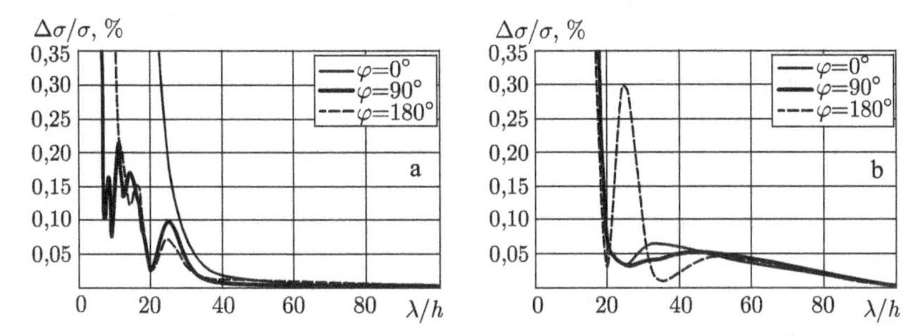

Fig. 3.6. Dependence of the relative deviation of the directional diagram of scattering on a conducting cylinder on parameter λ/h for TE-polarization (a) and TM-polarization (b).

A series of experiments with a dielectric cylinder showed that the relative deviation of the values of the directional diagram of scattering is less than 5% at and $\lambda/h > 40$ and less than 1% at $\lambda/h > 80$ for both polarizations.

Experiments with the conducting cylinder showed that the relative deviation of the values of the directional diagram of scattering is less than 5% at $\lambda/h > 30$ for TE-polarization at $\lambda/h > 50$ for TM-polarization and less than 1% at $\lambda/h > 50$ for TE-polarization and at $\lambda/h > 80$ for TM-polarization.

Thus, the polarization state does not affect the results of modelling dielectric structures by the proposed method, but it must be taken into account when choosing the length of the segment of the coverage grid for the calculation of conductive structures with the corresponding relative error.

3.1.2. Analysis of the diffraction of light on periodic inhomogeneities

The theory of scattering on periodic structures, commonly referred to as diffraction gratings, has many applications in optics, such as electromagnetic and optical communications, visualization tools, determination of the properties of objects and surfaces, electronic and optical components, photonic crystals, diffraction gratings [32]. Numerical methods were developed to simulate the diffraction of light on diffraction gratings. These methods include differential and integral methods, methods based on the propagation of Rayleigh waves and eigenmodes, the variational and finite-difference methods: the method of coupled waves (rigorous coupled wave analysis, RCWA) [33], C-method [34], finite element methods [35–37], the integral methods [21], finite-difference-time domain (FDTD) methods [38, 39].

The variational methods are most effective for inhomogeneous problems with complex geometries. These methods require solving a linear system of equations with tenuous matrices. To reduce the size of the computational domain, the calculation of the field away from the computing domain can be performed using the integral relation. The material of a periodic structure can be dielectric, conducting, superconducting, the size of the inhomogeneities can be arbitrarily small. The corners of the profile of the geometry of the structure can be considered in calculations by the appropriate choice of the sampling grid.

As a special case of the variational methods we can consider the finite element method (FEM) applied to the elliptical Helmholtz equation in the calculation domain. It includes a choice of the discretization scheme, the construction and minimization of the functional relationships. The resulting ratio is converted to a system of linear equations which is incomplete without the use of boundary conditions.

For the boundary-value problem, satisfying the Sommerfeld radiation conditions, we can use methods of integral equations, respectively, the standard method of boundary elements can also be used for periodic tasks. Both methods are joined at the boundary between inner and outer parts, satisfying the conditions of continuity of the field. Using the finite element method to determine the field inside the object leads to a tridiagonal matrix which requires less computer memory and shorter computing time than the methods of the volume integrals. The result of using the boundary element method for determining the field at the boundary is a more accurate solution than using the finite element method with absorbing conditions of the boundary due to the strong dependence on the angle of incidence of the field on the boundary.

In this book, we describe the formulation of a combined method for problems of scattering of light by periodic objects based on the finite element method and the boundary element method (PFEM-BE). The developed PFEM-BE method and the RCWA method [40] were used for a comparative simulation of light diffraction on a dielectric one-dimensional diffraction grating. Comparison of the simulation results is presented for the TE- and TM-polarized waves.

Description of the calculation method

Consider the diffraction of a plane wave with wave vector $\mathbf{k} = k(\sin(\theta), -\cos(\theta), 0)$, $k = k_0\sqrt{\varepsilon}$ for the periodic structure with period d, k_0 is the wave number of the wave in free space $k_0 = 2\pi/\lambda_0$, where λ_0 is the wavelength in free space, ε is the dielectric constant of the medium.

The light, diffracting on the structure, creates a scattered field. In addition to a decaying part, the diffracted light is split into a finite number of reflected and

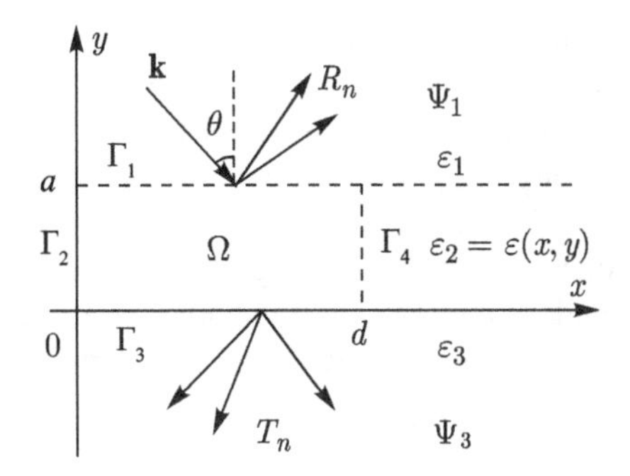

Fig. 3.7. Geometry of the diffraction problem on the periodic structure.

transmitted polarized plane waves whose propagation direction does not depend on the geometry and material of the periodic structure, but depends only on the grating period. The total field is defined as the sum of the incident and scattered fields. The purpose of the task is to determine the amplitude and phase of reflected, transmitted and decaying orders.

The geometry of the problem is shown in Fig. 3.7, where R_n and T_n are the reflection and transmission coefficients of the diffraction orders. For the given geometry of the problem we define three zones with different dielectric constants: the area above the structure at $y > a(\Psi_1)$, where a is the maximum height of the structure, with a dielectric permittivity constant $\varepsilon = \varepsilon_1$, the region of the structure $0 < y < a$ with dielectric permittivity $\varepsilon = \varepsilon(x, y)$, and the $y < 0$ (Ψ_3) with a constant dielectric permittivity $\varepsilon = \varepsilon_3$.

The diffraction of a plane wave on a one-dimensional periodic structure is reduced to two independent problems: the problem of diffraction of a plane wave with TE-polarization ($E_z \neq 0, H_z = 0$) and the problem of diffraction of a plane wave with TM-polarization ($H_z \neq 0, E_z = 0$) [41].

The total field $u_\Omega (x, y)$ in the region Ω ($0 < x < d$, $0 < y < a$) must satisfy the following equation [42]:

$$\nabla \cdot \left[\frac{1}{p(x,y)} \nabla u_\Omega(x,y) \right] + k_0^2 q(x,y) u_\Omega(x,y) = f_\Omega, \tag{3.29}$$

where $f_\Omega = jk_0 Z_0 J_z$, $p(x,y) = \mu_r$, $q(x,y) = \varepsilon_r$ for TE-polarization, and $f_\Omega = -\left[\nabla \times \left(\frac{J^\Omega}{\varepsilon_\Omega} \right) \right] \cdot \mathbf{z}$,

$p(x, y) = \varepsilon_r$, $q(x,y) = \mu_r$ for TM-polarization. Constants μ_r and ε_r are the ratio of magnetic and dielectric constants of the medium to the same parameters of free space, i.e. $\mu_r = \mu / \mu_0$ and $\varepsilon_r = \varepsilon/\varepsilon_0$, $Z_0 = \sqrt{\mu_0 / \varepsilon_0}$ is the impedance of free space, $J^\Omega = (J_x, J_y, J_z)$ is the vector of the electric current density of the source in region Ω. For TE-polarization, the complex amplitude $u(x, y)$ denotes the total electric field $E_z(x, y)$, which is directed along the axis z (along the generatrix of a cylindrical optical element), the coordinates (x, y) lie in the plane of the normal section. For TM-polarization the complex amplitude $u(x,y)$ denotes the total magnetic field $H_z (x, y)$.

To solve (3.29), the computing domain Ω should be limited by the introduction of the artificial boundary $\Gamma = \Gamma_1 + \Gamma_2 + \Gamma_3 + \Gamma_4$ (see Fig. 3.7). Γ_1 and Γ_3 are the fictitious boundaries, infinitely extending parallel to the x-axis of the coordinates $y = a$ and $y = 0$. Accordingly, for the unique solution of the problem boundary conditions must be imposed at the given artificial boundary.

Since the space in zones Ψ_1 and Ψ_3 is homogeneous, the field in these zone can be defined in terms of boundary integrals with the appropriate Green's function. The total field $u_\Psi(x, y)$ in these zones must satisfy the following equation:

$$\nabla \cdot \left[\frac{1}{p} \nabla u_\Psi(\xi) \right] + k_0^2 q u_\Psi(\xi) = f_\Psi, \ \xi \in \Psi, \tag{3.30}$$

where $f_\Psi = jk_0 Z_0 J_z^\Psi$, $p = \mu_\Psi$, $q = \varepsilon_\Psi$ for TM-polarization and $f_\Psi = -\left[\nabla \times \left(\frac{J^\Psi}{\varepsilon_\Psi} \right) \right] \cdot \mathbf{z}$ $p = \varepsilon_\Psi$, $q = \mu_\Psi$ for TE-polarization. J_Ψ is the vector of electric current density of the source in the region Ψ, the region $\Psi = \Psi_1, \Psi_3$.

Application of the Galerkin method to the solution of (3.30) in the region Ψ of the periodic structure is similar to the approach described in §3.1.1, to obtain the system of equations (3.15).

We define the boundary conditions for the field and its derivatives at the boundaries of Γ_1 and Γ_3 to complement the system (3.15). To construct the boundary integral equation for the field and its normal derivative, we introduce the Green function u^* which satisfies the Sommerfeld radiation conditions and is the fundamental solution of the Helmholtz equation in semi-infinite domains Ψ_1 and Ψ_3:

$$\nabla^2 u^*(\xi,\eta) + k^2 u^*(\xi,\eta) = -\delta(\xi,\eta), \xi,\eta \in \Psi_1, \ \Psi_3. \tag{3.31}$$

The fundamental solution for the Helmholtz equation in two-dimensional homogeneous space is well known and equal

$$u^*(\xi,\eta) = (i/4) H_0^{(1)}(kr), \tag{3.32}$$

where $r = \sqrt{[x(\eta) - x(\xi)]^2 + [y(\eta) - y(\xi)]^2}$, $H_0^{(1)}(kr) = J_0(kr) + iY_0(kr)$ is the Hankel function of the first kind and zero order, where J_0 is the Bessel function of zero order, Y_0 is the Neumann function of zero order.

To construct the boundary integral equation for the scattered field and its normal derivative in zones Ψ_1 and Ψ_3, we use Green's theorem as follows:

$$\begin{aligned} \int_\Psi \left[\nabla^2 u^*(\xi,\eta) + k^2 u^*(\xi,\eta) \right] u(\eta) d\Psi = -\int_{\Gamma'} v(\eta) u^*(\xi,\eta) d\Gamma + \\ + \int_{\Gamma'} u(\eta) \frac{\partial u^*(\xi,\eta)}{\partial \mathbf{n}'} d\Gamma + \int_\Psi f_\Psi(\eta) u^*(\xi,\eta) d\Psi, \ \Psi = \Psi_1, \Psi_3, \end{aligned} \tag{3.33}$$

where $v(\eta) = \partial u(\eta)/\mathbf{n}'$ are the normal derivatives of the field, and Γ' represents the infinite boundaries $y = a$ and $y = 0$ for zones Ψ_1 and Ψ_3, respectively.

Substituting (3.31) into equation (3.33), passing to the limit of the observation point ξ from the inner to the boundary and using (3.21), we obtain

$$\frac{1}{2}u(\xi) = -\int_{\Psi} f_{\Psi}(\eta)u^*(\xi,\eta)d\Psi + \int_{\Gamma'} v(\eta)u^*(\xi,\eta)d\Gamma - \int_{\Gamma'} u(\eta)\frac{\partial u^*(\xi,\eta)}{\partial \mathbf{n}'}d\Gamma,\ \Psi = \Psi_1, \Psi_3. \tag{3.34}$$

This equation provides a functional link between the functions u and its normal derivative v on the boundary Γ'. The first term on the right side of (3.34) is the field produced by the source f_{Ψ} in free space, and may be designated as the incident field u_{Ψ}^{in}.

Thus, equation (3.34) is written as follows:

$$\frac{1}{2}u(\xi) = \int_{\Gamma'} v(\eta)u^*(\xi,\eta)d\Gamma - \int_{\Gamma'} u(\eta)\frac{\partial u^*(\xi,\eta)}{\partial \mathbf{n}'}d\Gamma + u_{\Psi}^{in}(\xi). \tag{3.35}$$

Field u and its derivatives v in the case of the diffractive grating are quasi-periodic functions [41, 43–45]:

$$\begin{aligned} u(x,y) &= \tilde{u}(x,y)e^{ik_0\alpha_0 x}, \\ v(x,y) &= \tilde{v}(x,y)e^{ik_0\alpha_0 x}, \end{aligned} \tag{3.36}$$

where $\tilde{u}(x,y)$ and $\tilde{v}(x,y)$ are periodic functions with respect to x with period d.

The integration over the infinite boundary Γ' in (3.35) can be replaced by integration along the boundaries Γ_1 and Γ_3:

$$\begin{aligned} \frac{1}{2}u(x_0,y) &= \int_{\Gamma} \sum_{n=-\infty}^{\infty} \tilde{v}(x,y)e^{ik_0\alpha_0(x+nd)}u^*(x+nd-x_0)\,dx - \\ &-\int_{\Gamma} \sum_{n=-\infty}^{\infty} \tilde{u}(x,y)e^{ik_0\alpha_0(x+nd)}\frac{\partial u^*(x+nd-x_0)}{\partial \mathbf{n}'}dx + u^{in}(x_0,y),\quad x\in[0,d), \end{aligned} \tag{3.37}$$

where $\Gamma = \Gamma_1, \Gamma_3$, $y = a$ and $y = 0$ on the boundaries Γ_1 and Γ_3, respectively.

Substituting the function of the complex amplitude in equation (3.37) on the boundary of its basic approximation of piecewise linear functions (3.8) and (3.9), we obtain:

$$\begin{aligned} \frac{1}{2}u_{\Gamma_i}(x_s) = \sum_{m=1}^{M} h\Bigg\{ & v_{\Gamma_i}(x_m)\sum_{n=-\infty}^{\infty} e^{ik_0\alpha_0(x_m+nd)}\left[\int_{-1}^{1}\omega^{\Gamma_i}(x_m+h\eta)u^*_{\Gamma_i}(x_m+nd-x_s+h\eta)d\eta\right] - \\ & -u_{\Gamma_i}(x_m)\sum_{n=-\infty}^{\infty} e^{ik_0\alpha_0(x_m+nd)}\left[\int_{-1}^{1}\omega^{\Gamma_i}(x_m+h\eta)\frac{\partial u^*_{\Gamma_i}(x_m+nd-x_s+h\eta)}{\partial \mathbf{n}'}d\eta\right]\Bigg\} + u^{in}_{\Gamma_i}(x_s), \end{aligned} \tag{3.38}$$

where $s = [1, N_{x-1}]$, $i = 1, 3$.

The boundary conditions on the field and its derivatives at the boundaries Γ_2 and Γ_4 are periodic of the form:

$$u_{\Gamma_4} = u_{\Gamma_2} e^{ik_0\alpha_0 d}, \quad v_{\Gamma_4} = v_{\Gamma_2} e^{ik_0\alpha_0 d}. \tag{3.39}$$

The expansions (3.8) and (3.9) and boundary conditions (3.39) can be written for the coefficients of the field and its derivatives at the boundaries Γ_2 and Γ_4:

$$u_{\Gamma_4}(x_s) = u_{\Gamma_2}(x_s) e^{ik_0\alpha_0 d}, \quad v_{\Gamma_4}(x_s) = v_{\Gamma_2}(x_s) e^{ik_0\alpha_0 d}, \tag{3.40}$$

where $s = [1, N_y]$. Elements of the matrices **D** and **G**, corresponding to the boundaries Γ_2 and Γ_4, can be written as:

$$d_2^s = 1,\ d_4^s = d_2^s e^{ik_0\alpha_0 d},\ g_2^s = 1,\ g_4^s = g_2^s e^{ik_0\alpha_0 d},\ e_2^s = 0,\ e_4^s = 0,\ s = [1, N_y]. \tag{3.41}$$

Elements of the matrices **D** and **G**, corresponding to the boundaries Γ_1 and Γ_3, can be written as:

$$d_i^{s,m} = -h\left\{\sum_{n=-\infty}^{\infty} e^{ik_0\alpha_0(x_m+nd)}\left[\int_{-1}^{1} \omega^{\Gamma_i}(x_m + h\eta)\frac{\partial u_{\Gamma_i}^{*}(x_m + nd - x_s + h\eta)}{\partial \mathbf{n}'} d\eta\right]\right\},\ e_i^s = \frac{1}{2}\delta_{s,m}, \tag{3.42}$$

$$g_i^{s,m} = h\left\{\sum_{n=-\infty}^{\infty} e^{ik_0\alpha_0(x_m+nd)}\left[\int_{-1}^{1} \omega^{\Gamma_i}(x_m + h\eta) u_{\Gamma_i}^{*}(x_m + nd - x_s + h\eta) d\eta\right]\right\}, \tag{3.43}$$

$$k\sqrt{\varepsilon_i}\tau_i,\quad m, s = [1, N_{x-1}],\ i = 1, 3.$$

The relations (3.38) can be represented in matrix form:

$$[\mathbf{D}]\boldsymbol{u}_\Gamma + [\mathbf{G}]\mathbf{v}_\Gamma = [\mathbf{E}]u_\Gamma^{in} \tag{3.44}$$

with elements of the matrices **D, G** and **E** of the form (3.41) (3.42) and (3.43). Infinite series in (3.42) and (3.43) are approximated by finite sums, integrals can be evaluated numerically. Combining equations (3.15) and (3.44), we obtain a closed system of linear algebraic equations for solving the problem of diffraction of a plane wave on a periodic two-dimensional micro-object:

$$\begin{bmatrix} [\mathbf{A}_{\Omega,\Omega}] & [\mathbf{A}_{\Gamma,\Omega}] & \mathbf{0} \\ [\mathbf{A}_{\Omega,\Gamma}] & [\mathbf{A}_{\Gamma,\Gamma}] & [\mathbf{B}] \\ \mathbf{0} & [\mathbf{D}] & [\mathbf{G}] \end{bmatrix}\begin{bmatrix} \mathbf{u}_\Omega \\ \mathbf{u}_\Gamma \\ \mathbf{v}_\Gamma \end{bmatrix} = \begin{bmatrix} [\mathbf{C}_{\Omega,\Omega}] & [\mathbf{C}_{\Gamma,\Omega}] & \mathbf{0} \\ [\mathbf{C}_{\Omega,\Gamma}] & [\mathbf{C}_{\Gamma,\Gamma}] & \mathbf{0} \\ \mathbf{0} & \mathbf{0} & [\mathbf{E}] \end{bmatrix}\begin{bmatrix} \mathbf{f}_\Omega \\ \mathbf{f}_\Gamma \\ \mathbf{u}_\Gamma^{in} \end{bmatrix}, \tag{3.45}$$

where the submatrix $\mathbf{A}_{\Omega,\Omega}$ with the dimension $(N - M) \times (N - M)$ includes the ratio of the field in the internal nodes of the partition grid, submatrices $\mathbf{A}_{\Omega,\Gamma}$ and

$\mathbf{A}_{\Gamma,\Omega}$ with the dimension $(N-M) \times M$ and $M \times (N-M)$, respectively, include the coupling coefficients of the field at the boundary nodes, the $\mathbf{A}_{\Gamma,\Gamma}$ submatrix with the size $M \times M$ involves the coupling coefficients of the field at the boundary nodes, the submatrix **B** with the size $M \times M$ includes the ratio between the derivatives of the field at the boundary and the field in the internal nodes of the partition grid, submatrix **D** with the size $M \times M$ involves the coupling coefficients of the field of free space at the boundary nodes, the submatrix **G** with the size $M \times M$ includes the ratio between the derivatives of the field at the boundary and the field of free space in the internal nodes of the partition grid. Submatrix $\mathbf{C}_{\Omega,\Omega}$ with the dimension $(N-M) \times (N-M)$ includes the ratio of field sources in the internal nodes of the partition grid, submatrix $\mathbf{C}_{\Omega,\Gamma}$ and $\mathbf{C}_{\Gamma,\Omega}$ with the dimension $(N-M) \times M$ and $M \times (N-M)$, respectively, include the ratio of field sources in the boundary and internal nodes, the submatrix $\mathbf{C}_{\Gamma,\Gamma}$ the size of $M \times M$ includes the ratio of the field sources at the boundary nodes. Submatrix **E** has size $M \times M$. Vector $\mathbf{u}_\Omega$ and $\mathbf{u}_\Gamma$ are the field vectors in the interior and boundary nodes, $\mathbf{v}_\Gamma$ is the vector of normal derivatives of the field at the boundary nodes of the grid. Vectors $\mathbf{f}_\Omega$ and $\mathbf{f}_\Gamma$ are the vectors of the field sources in the interior and boundary nodes of the grid, $\mathbf{u}_\Gamma^{\text{in}}$ is the vector of the external incident field at the boundary nodes of the grid. Thus, the system (3.45) has $N+M$ equations and the same number of unknowns.

The field in the areas of Ψ_1 and Ψ_3 can be represented by a Rayleigh expansion (superposition of plane waves). In the area Ψ_1 the z-components of the fields are as follows:

$$u(x,y) = \exp\left(ik_0\left(\alpha_0 x - \beta_0 y\right)\right) + \sum_{n=-\infty}^{\infty} R_n \exp\left(ik_0\left(\alpha_n x + \beta_n y\right)\right), \tag{3.46}$$

where

$$\alpha_n = \sqrt{\varepsilon_1}\sin(\theta) + n\frac{\lambda_0}{d}, \ \beta_n = \sqrt{\varepsilon_1 - \alpha_n^2}. \tag{3.47}$$

The function $u(x, y)$ is the component $E_z(x, y)$ of the electric field for the case of TE-polarization and the component $H_z(x, y)$ of the magnetic field for TM-polarization.

In zone Ψ_3 the z-component are as follows:

$$u(x,y) = \sum_{n=-\infty}^{\infty} T_n \exp\left(ik_0\left(\alpha_n x - \tilde{\beta}_n y\right)\right), \tag{3.48}$$

where

$$\tilde{\beta}_n = \sqrt{\varepsilon_3 - \alpha_n^2}. \tag{3.49}$$

Rayleigh expansions (3.46) and (3.48) are solutions of the Helmholtz equation,

$$\Delta u + k^2 u = 0, \tag{3.50}$$

at $k^2 = k_0^2 \varepsilon_1$ and $k^2 = k_0^2 \varepsilon_3$ respectively.

After determining the values of the field in the region Ω and its derivatives at the boundaries Γ_1 and Γ_3 the normalized intensities of the reflected and transmitted orders are calculated as follows [32, 36, 37]:

$$I_n^R = |R_n|^2 \frac{\beta_n}{\beta_0},\ I_n^T = \sqrt{\frac{\varepsilon_3}{\varepsilon_1}} |T_n|^2 \frac{\tilde{\beta}_n}{\beta_0},\ \left(\sum_{n \in U_1} I_n^R + \sum_{n \in U_3} I_n^T = 1 \right), \tag{3.51}$$

$$I_n^R = |R_n|^2 \frac{\beta_n}{\beta_0},\ I_n^T = \sqrt{\frac{\varepsilon_1}{\varepsilon_3}} |T_n|^2 \frac{\tilde{\beta}_n}{\beta_0},\ \left(\sum_{n \in U_1} I_n^R + \sum_{n \in U_3} I_n^T = 1 \right) \tag{3.52}$$

for TE- and TM-polarized waves, respectively. Expressions (3.51) and (3.52) determine what portion of the energy of the incident wave will move to the n-th order of diffraction. Note that the intensities of the propagation distribution are for non-absorbing materials, i.e. for $\mathrm{Im}\sqrt{\varepsilon_i} = 0$. The sets U_1 and U_2 in (3.51) and (3.52) are the sets of indices corresponding to the reflected and transmitted diffraction orders:

$$U_1 = \begin{cases} \left\{ n \in Z : \dfrac{\alpha_n^2}{\varepsilon_1} < 1 \right\}, & \mathrm{Im}\sqrt{\varepsilon_1} = 0 \\ 0, & \mathrm{Im}\sqrt{\varepsilon_1} > 0 \end{cases},\quad U_3 = \begin{cases} \left\{ n \in Z : \dfrac{\alpha_n^2}{\varepsilon_3} < 1 \right\}, & \mathrm{Im}\sqrt{\varepsilon_3} = 0 \\ 0, & \mathrm{Im}\sqrt{\varepsilon_3} > 0 \end{cases}. \tag{3.53}$$

To determine the coefficients of reflection R_n and transmission T_n, we use the discrete Fourier transform:

$$R_n = \frac{1}{N} \sum_{k=0}^{N-1} \left[u(x_k, a) - u^{in}(x_k, a) \right] \exp(-ik_0 \alpha_n x_k). \tag{3.54}$$

$$T_n = \frac{1}{N} \sum_{k=0}^{N-1} u(x_k, 0) \exp(-ik_0 \alpha_n x_k). \tag{3.55}$$

These coefficients describe the amplitude and phase shift of propagating plane waves. More precisely, the modules $|R_n|$ and $|T_n|$ are the amplitudes of the n-th reflected and transmitted orders and arg $[R_n/|R_n|]$ and arg $[T_n/|T_n|]$ are the phase shifts. The coefficients with $n \notin U_1$, $n \notin U_3$ describe the decaying waves.

Examples of calculation

For example, consider the diffraction of a plane wave with wavelength $\lambda_0 = 0.6$ μm on a binary dielectric diffraction grating with a fill factor of 0.25 and with a thickness of 0.24 μm. The grating period was varied from 0.2 μm to 2 μm. Accordingly, in simulation by PFEM-BE the length of the segment of the partition grid varied from $\lambda_0/300$ to $\lambda_0/30$.

Consider an example. A plane wave is incident along the normal from the air ($\varepsilon_1 = 1$) on the grating ($\varepsilon_3 = 2.25$). Figures 3.8 and 3.9 show the dependence of the efficiencies I^R_0 and I^T_0 orders of diffraction of TE- and TM-polarized waves, respectively, on the grating period.

Dependences of deviations of intensities ΔI^R_0 and ΔI^T_0 ($\Delta I = \left|I_i^{\text{RCWA}} - I_i^{\text{PFEM-BE}}\right|$) of zero-order diffraction, calculated by RCWA and PFEM-BE methods, of the TE- and TM-polarized waves on the length of the segment of the partition grid h in the PFEM-BE method, are shown in Fig. 3.10. It is difficult to note the explicit dependence of the deviation for $h < \lambda_0/30$. The uniform rate of deviation of the intensities is $4{\cdot}10^{-3}$ and $2{\cdot}10^{-3}$ for the TE- and TM-polarizations, respectively.

In the following example, a plane wave is incident along the normal from the substrate of the grating ($\varepsilon_3 = 2.25$) in air ($\varepsilon_3 = 1$). Figures 3.11 and 3.12 show the dependence of the intensities I^R_0 and I^T_0 of the diffraction orders of TE- and TM-polarized waves on the grating period.

Depending on the deviations of the intensities ΔI^R_0 and I^T_0 of diffraction orders of TE- and TM-polarized waves, the length of the segment of the partition grid h are

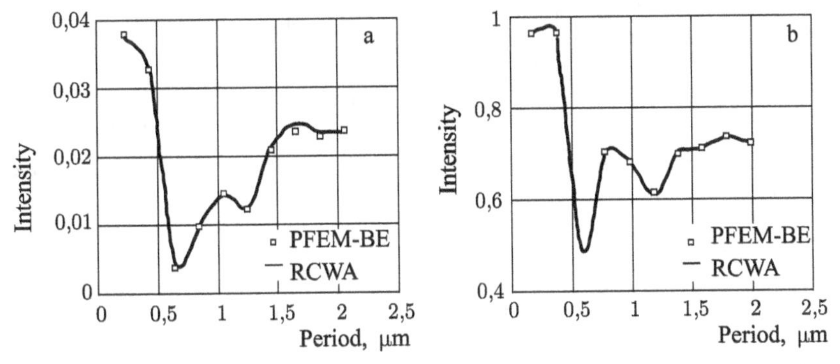

Fig. 3.8. Distribution of the effectiveness of zero-order diffraction of TE-polarized wave: a) I^R_0; b) I^T_0.

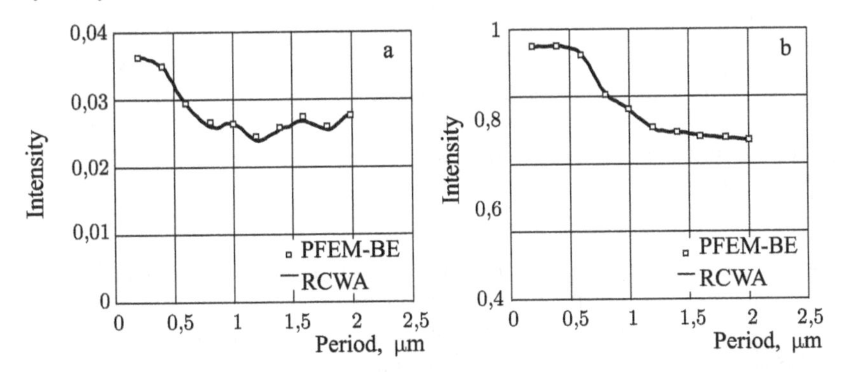

Fig. 3.9. Distribution of the effectiveness of zero order diffraction of TM-polarized wave: a) I^R_0; b) I^T_0.

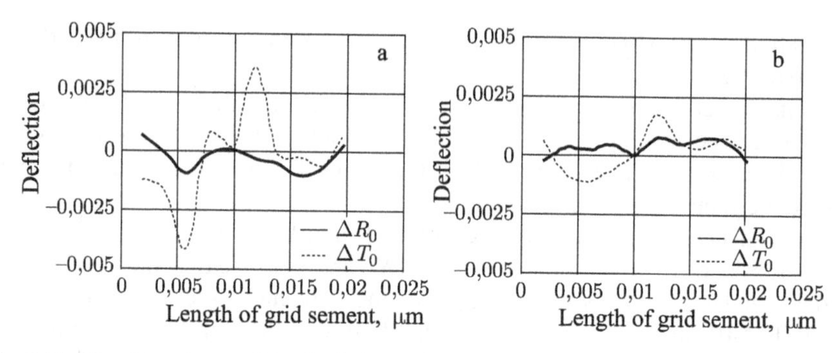

Fig. 3.10. The dependence of the deviations of the efficiency of zero orders on parameter h: a) TE-waves and b) TM waves.

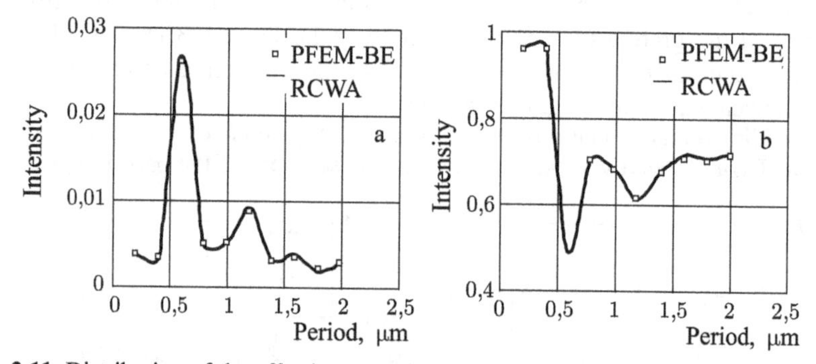

Fig. 3.11. Distribution of the effectiveness of zero-order diffraction of TE-polarized waves: a) I_0^R ; b) I_0^T.

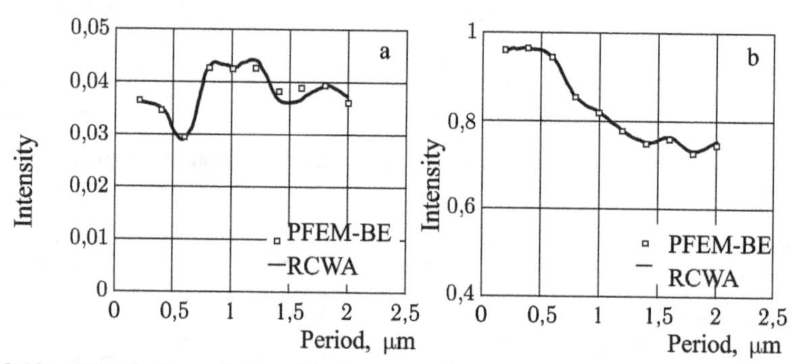

Fig. 3.12. Distribution of the effectiveness of zero-order diffraction of TM-polarized waves: a) I_0^R ; b) I_0^T.

shown in Fig. 3.13. The uniform rate of deviation of the intensities is $7{\cdot}10^{-3}$ and $8{\cdot}10^{-3}$ for the TE- and TM-polarizations, respectively.

Thus, the comparison of results obtained by the PFEM-BE and RCWA methods shows that they are in good agreement.

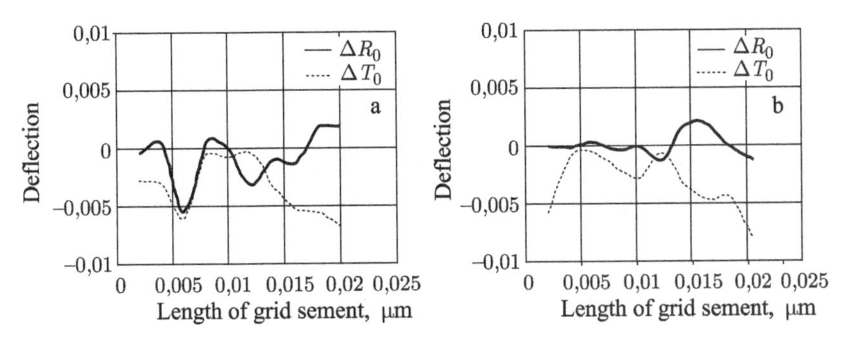

Fig. 3.13. Dependence of the deviations of the efficiency of zero orders on parameter h; a) TE-waves; b) TM-waves.

3.2. Finite element method for solving the two-dimensional integral diffraction equation

3.2.1. TE-polarization

In [1, 30, 42] the scalar problem of diffraction on a transparent body with an inhomogeneous refractive index is reduced to the Fredholm integral equation of the second kind. This chapter discusses the 2D vector diffraction problem for objects with the heterogeneous and, in general, complex refractive index. The resulting integral equation for the cases of TE- and TM-polarization of the incident electromagnetic wave is solved by the finite element method (FEM) [46].

We define the geometry of the problem, as shown in Fig. 3.14.

A cylindrical object has infinite length along the axis z, and its cross-section lies in the plane (x, y). The plane of incidence of th wave coincides with the plane (x, y).

Figure 3.14 gives the following notation: Ω_1 – area of the transparent body bounded by contour S with the function of the dielectric permittivity $\varepsilon_1(x, y)$, magnetic μ_1, $k_2 = 2\pi / \lambda\sqrt{\varepsilon_2\mu_2}$, Ω_2 – the outside homogeneous region with constant properties ε_2 and μ_2. Furthermore, we assume that $\mu_1 = \mu_2 = 1$.

From Maxwell's equations [47]:

$$\text{rot}\,\mathbf{E} + \frac{1}{c}\dot{\mathbf{B}} = 0 \tag{3.56}$$

and the material equation for an isotropic medium

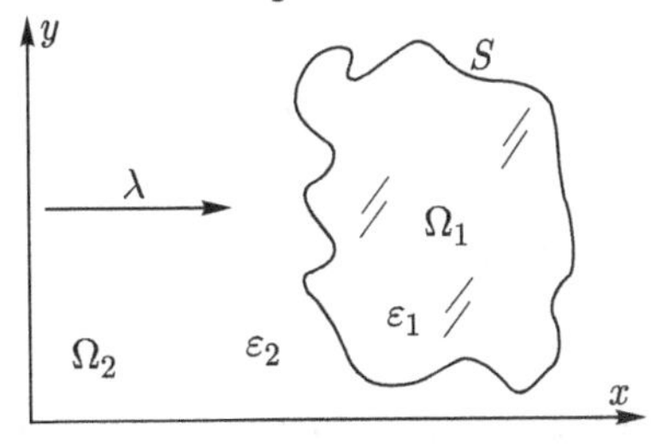

Fig. 3.14. Diffraction of an electromagnetic wave on an inhomogeneous transparent body.

$$\mathbf{B} = \mu \mathbf{H} \tag{3.57}$$

given the fact that

$$\operatorname{rot} \mathbf{E} = \mathbf{i}\left(\frac{\partial E_z}{\partial y} - \frac{\partial E_y}{\partial z}\right) + \mathbf{j}\left(\frac{\partial E_x}{\partial z} - \frac{\partial E_z}{\partial x}\right) + \mathbf{k}\left(\frac{\partial E_y}{\partial x} - \frac{\partial E_x}{\partial y}\right),$$

$$\dot{\mathbf{B}} = \mu \dot{\mathbf{H}} = \mu\left(\mathbf{i}\frac{\partial H_x}{\partial t} + \mathbf{j}\frac{\partial H_y}{\partial t} + \mathbf{k}\frac{\partial H_z}{\partial t}\right),$$

we get

$$\begin{aligned}
&\frac{\partial E_z}{\partial y} - \frac{\partial E_y}{\partial z} + \frac{1}{c}\mu\frac{\partial H_x}{\partial t} = 0, \\
&\frac{\partial E_x}{\partial z} - \frac{\partial E_z}{\partial x} + \frac{1}{c}\mu\frac{\partial H_y}{\partial t} = 0, \\
&\frac{\partial E_y}{\partial x} - \frac{\partial E_x}{\partial y} + \frac{1}{c}\mu\frac{\partial H_z}{\partial t} = 0.
\end{aligned} \tag{3.58}$$

For monochromatic radiation ($E\,e^{-i\omega t}$), system (3.58) takes the form:

$$\begin{aligned}
&\frac{\partial E_z}{\partial y} - \frac{\partial E_y}{\partial z} - \frac{i\omega}{c}\mu H_x = 0, \\
&\frac{\partial E_x}{\partial z} - \frac{\partial E_z}{\partial x} - \frac{i\omega}{c}\mu H_y = 0, \\
&\frac{\partial E_y}{\partial x} - \frac{\partial E_x}{\partial y} - \frac{i\omega}{c}\mu H_z = 0.
\end{aligned} \tag{3.59}$$

Here ω is the angular frequency of oscillations. We denote $k_0=\omega/c=2\pi/\lambda$, where λ is the wavelength of light. In the case of 2D problems, system (3.59) takes the form:

$$\frac{\partial E_z}{\partial y} - ik_0\mu H_x = 0, \tag{3.60}$$

$$-\frac{\partial E_z}{\partial x} - ik_0\mu H_y = 0, \tag{3.61}$$

$$\frac{\partial E_y}{\partial x} - \frac{\partial E_x}{\partial y} - ik_0\mu H_z = 0. \tag{3.62}$$

According to the above reasoning of Maxwell's equations

$$\text{rot}\,\mathbf{H}-\frac{1}{c}\dot{\mathbf{D}}=\frac{4\pi}{c}\mathbf{j} \tag{3.63}$$

and constitutive equation

$$\mathbf{D}=\varepsilon\mathbf{E} \tag{3.64}$$

in the absence of external currents, we obtain

$$\frac{\partial H_z}{\partial y}+ik_0\varepsilon E_x=0, \tag{3.65}$$

$$-\frac{\partial H_z}{\partial x}+ik_0\varepsilon E_y=0, \tag{3.66}$$

$$\frac{\partial H_y}{\partial x}-\frac{\partial H_x}{\partial y}+ik_0\varepsilon E_z=0. \tag{3.67}$$

Using equations (3.60), (3.61) and (3.67), we get a scalar Helmholtz equation

$$\frac{\partial}{\partial x}\left(\frac{\partial E_z}{\partial x}\right)+\frac{\partial}{\partial y}\left(\frac{\partial E_z}{\partial y}\right)+k_0^2\varepsilon\mu E_z=0. \tag{3.68}$$

We denote the field E_z in Ω_1 by E_z^{in}, and in Ω_2 by E_z^{ext}. Then the problem of diffraction on the object Ω_1 is reduced to solving the system of differential Helmholtz equations [30]:

$$\left(\Delta+k_1^2(x,y)\right)E_z^{in}=0,\ (x,y)\in\Omega_1, \tag{3.69}$$

$$\left(\Delta+k_2^2\right)E_z^{ext}=-g_2,\ (x,y)\in\Omega_2, \tag{3.70}$$

where g_2 is the function describing the source in the external field Ω_2, $k_1(x,y)=k_0\sqrt{\varepsilon_1(x,y)\mu}$ is the wave number for the field Ω_1 with an inhomogeneous refractive index, $k_2=k_0\sqrt{\varepsilon_2\mu_2}$ is the wave number for Ω_2. Here $\Delta=\frac{\partial^2}{\partial x^2}+\frac{\partial^2}{\partial y^2}$

For TE-polarization, the boundary conditions (BC) follow from the continuity at the interface between two media of the tangential components of electric and magnetic fields [47]:

$$\begin{gathered}\left[\mathbf{n}_1,\mathbf{E}_1\right]-\left[\mathbf{n}_1,\mathbf{E}_2\right]=0,\\ \left[\mathbf{n}_1,\mathbf{H}_1\right]-\left[\mathbf{n}_1,\mathbf{H}_2\right]=\mathbf{j}_{\text{surf}}.\end{gathered} \tag{3.71}$$

Here $\mathbf{j}_{\text{surf}}=0$, $\mathbf{n}_1$ is the vector the outward normal to the field Ω_1.

Equations (3.71) give the BCs for the fields E_z^{in} and E_z^{ext}.

$$E_z^{in}\big|_S = E_z^{ext}\big|_S,\ (x,y)\in S, \tag{3.72}$$

$$\left.\frac{\partial E_z^{in}}{\partial \mathrm{n}_1}\right|_S = -\left.\frac{\partial E_z^{ext}}{\partial \mathrm{n}_2}\right|_S,\ (x,y)\in S. \tag{3.73}$$

Here $\mathbf{n}_2$ is the vector of the normal (external to the region Ω_2) to the contour S.

The external field E_z^{ext} satisfies the Sommerfeld radiation condition:

$$\frac{\partial E_z^{ext}}{\partial r} - ik_2 E_z^{ext} = o\left(\frac{1}{r}\right) \text{ at } r\to\infty \tag{3.74}$$

For function E_z^{in} and Green's functions G_2 in the region Ω_2 we have the scalar Green's formula [48]:

$$\iint_{\Omega_1}\left(E_z^{\text{in}}\Delta G_2 - G_2\Delta E_z^{\text{in}}\right)dx\,dy = \oint_S\left(E_z^{\text{in}}\frac{\partial G_2}{\partial \mathbf{n}_1} - G_2\frac{\partial E_z^{\text{in}}}{\partial \mathbf{n}_1}\right)dS. \tag{3.75}$$

From equations (3.69) and (3.70) it follows

$$\begin{aligned}\Delta E_z^{in} &= -k_1^2(x,y)E_z^{in},\\ \Delta E_z^{ext} &= -k_2^2E_z^{ext} - g_2.\end{aligned} \tag{3.76}$$

The following equality holds for function G_2

$$\Delta G_2 + k_2^2 G_2 = -\delta(M,M_0), \tag{3.77}$$

where M is the current point at which the integration is carried out, M_0 is the point of observation, i.e. $\delta(M,M_0)=\delta(x',y';x,y)$ – δ-function.

Substituting equations (3.76) and (3.77) into equation (3.75), we obtain

$$\begin{gathered}\oint_S\left(G_2\frac{\partial E_z^{\text{in}}}{\partial \mathbf{n}_1} - E_z^{\text{in}}\frac{\partial G_2}{\partial \mathbf{n}_1}\right)dS + \iint_{\Omega_1}\left(k_1^2-k_2^2\right)E_z^{\text{in}}G_2\,dx\,dy\\ -\iint_{\Omega_1}E_z^{\text{in}}(x',y')\delta(x',y';x,y)\,dx'dy' = 0.\end{gathered} \tag{3.78}$$

Using the filtering properties of the δ-function, we reduce equation (3.78) to the form

$$\oint_S\left(G_2\frac{\partial E_z^{\text{in}}}{\partial \mathbf{n}_1} - E_z^{\text{in}}\frac{\partial G_2}{\partial \mathbf{n}_1}\right)dS + \iint_{\Omega_1}\left(k_1^2-k_2^2\right)E_z^{\text{in}}G_2\,dx\,dy = \begin{cases}E_z^{\text{in}},(x,y)\in\Omega_1\\ 0\ ,(x,y)\in\Omega_2\end{cases}. \tag{3.79}$$

Similarly, applying Green's formula for functions E_z^{ext} and G_2 by using equations (3.76) and (3.77), we obtain

$$\oint_S \left(G_2 \frac{\partial E_z^{\text{ext}}}{\partial \mathbf{n}_2} - E_z^{\text{ext}} \frac{\partial G_2}{\partial \mathbf{n}_2} \right) dl + \iint_{\Omega_2} g_2 G_2 dx\, dy = \begin{cases} 0\ ,(x,y) \in \Omega_1 \\ E_z^{\text{ext}},(x,y) \in \Omega_2 \end{cases}. \tag{3.80}$$

Adding the equations (3.79) and (3.80) with the boundary conditions (3.72) and (3.73), we obtain

$$\iint_{\Omega_1} \left(k_1^2 - k_2^2\right) E_z^{\text{in}} G_2 dx\ dy + E_{0z} = \begin{cases} E_z^{\text{in}},(x,y) \in \Omega_1 \\ E_z^{\text{ext}},(x,y) \in \Omega_2 \end{cases}, \tag{3.81}$$

where
$$E_{0z}(x,y) = \iint_{\Omega_2} g_2 G_2 dx dy$$

is the field in the region Ω_1 or Ω_2 created by the sources with the function $g_2(x, y)$. According to the condition of the problem, the field E_0 is known.

If $(x, y) \in \Omega_1$, the first of equations (3.79) is a Fredholm integral equation of second kind with respect to E_z^{in} and at $E_{0z} \neq 0$ has a unique non-trivial solution [30].

Further assume that the point source is far away from the region Ω_1 and $E_{0z}(x, y)$ can be regarded as a plane wave. Consider the case where a plane wave is incident along the x axis from left to right in the chosen coordinate system (Fig. 3.14):

$$E_{0z} = \exp\left(ik_2 x\right). \tag{3.82}$$

The Green's function in region Ω_2 for 2D optical fields, satisfying the radiation condition, is [49]:

$$G_2(\xi) = \frac{i}{4} H_0^{(1)}(\xi), \tag{3.83}$$

where $\xi = k_2 \sqrt{(x-x')^2 + (y-y')^2}$, $H_0^{(1)}(\xi)$ is the Hankel function of first kind of zero order [50].

The field E_z, determined by solving the system (3.81), is substituted into equation (3.60) (3.61), from which the components H_x, H_y of the magnetic field intensity are determined. Components E_z, H_x, H_y determine the electromagnetic field obtained as a result of diffraction of an electromagnetic wave of TE-polarization on the micro-objects. The existence and uniqueness of the solutions of the 2D problem of diffraction of the TE-polarized electromagnetic wave in an inhomogeneous micro-object is solved using the same procedure as that described in [30], so in this work it is not given.

3.2.2. TM-polarization

Using the equations (3.62) (3.65) and (3.66), we obtain the Helmholtz equation for the projection on the z-axis of the magnetic field strength vector

$$\frac{\partial}{\partial x}\left(\frac{1}{\varepsilon}\frac{\partial H_z}{\partial x}\right)+\frac{\partial}{\partial y}\left(\frac{1}{\varepsilon}\frac{\partial H_z}{\partial y}\right)+k_0^2\mu H_z=0. \tag{3.84}$$

Assuming from the conditions of the problem $\varepsilon_1=\varepsilon_1(x,y)$, $\varepsilon_2=\text{const}$, $\mu_1=\mu_2=1$ and applying equation (3.84) to regions Ω_1, Ω_2 we obtain a system of equations:

$$\begin{cases}\left(\Delta+k_1^2\right)H_z^{\text{in}}-\dfrac{1}{\varepsilon_1}\left(\dfrac{\partial\varepsilon_1}{\partial x}\dfrac{\partial H_z^{\text{in}}}{\partial x}+\dfrac{\partial\varepsilon_1}{\partial y}\dfrac{\partial H_z^{\text{in}}}{\partial y}\right)=0, & (x,y)\in\Omega_1\\ \left(\Delta+k_2^2\right)H_z^{\text{ext}}=-g_2, & (x,y)\in\Omega_2\end{cases}, \tag{3.85}$$

where $k_1^2=k_0^2\varepsilon_1(x,y)$, $k_2^2=k_0^2\varepsilon_2$, g_2 is a function describing the external sources.

The system (3.85) describes the 2D problem of diffraction of the TM-polarized electromagnetic wave on an object with an inhomogeneous refractive index.

From the first equation (3.71) in view of (3.65) and (3.66) we obtain the boundary condition:

$$\left.\frac{1}{\varepsilon_1}\frac{\partial H_z^{\text{in}}}{\partial \mathbf{n}_1}\right|_S=-\left.\frac{1}{\varepsilon_2}\frac{\partial H_z^{\text{ext}}}{\partial \mathbf{n}_2}\right|_S. \tag{3.86}$$

From the second equation of (4.71) we obtain

$$\left.H_z^{\text{in}}\right|_S=\left.H_z^{\text{ext}}\right|_S. \tag{3.87}$$

After a number of auxiliary calculations to convert the first equation of (3.85):

$$\frac{1}{\varepsilon_1}\Delta H_z^{\text{in}}-\frac{1}{\varepsilon_1^2}\left(\frac{\partial\varepsilon_1}{\partial x}\frac{\partial H_z^{\text{in}}}{\partial x}+\frac{\partial\varepsilon_1}{\partial y}\frac{\partial H_z^{\text{in}}}{\partial y}\right)=$$

$$=\frac{1}{\varepsilon_1}\Delta H_z^{\text{in}}+\left(\frac{\partial}{\partial x}\left(\frac{1}{\varepsilon_1}\right)\frac{\partial H_z^{\text{in}}}{\partial x}+\frac{\partial}{\partial y}\left(\frac{1}{\varepsilon_1}\right)\frac{\partial H_z^{\text{in}}}{\partial y}\right)=\frac{\partial}{\partial x}\left(\frac{1}{\varepsilon_1}\frac{\partial H_z^{\text{in}}}{\partial x}\right)+$$

$$+\frac{\partial}{\partial y}\left(\frac{1}{\varepsilon_1}\frac{\partial H_z^{\text{in}}}{\partial y}\right)=\operatorname{div}\left(\frac{1}{\varepsilon_1}\left(\mathbf{i}\frac{\partial H_z^{\text{in}}}{\partial x}+\mathbf{j}\frac{\partial H_z^{\text{in}}}{\partial y}\right)\right)=\operatorname{div}\left(\frac{1}{\varepsilon_1}\operatorname{grad}H_z^{\text{in}}\right). \tag{3.88}$$

In view of the expression (3.88) the first equation of the system (3.85) becomes

$$\operatorname{div}\left(\frac{1}{\varepsilon_1}\operatorname{grad}H_z^{\text{in}}\right)+\frac{k_1^2}{\varepsilon_1}H_z^{\text{in}}=0. \tag{3.89}$$

For the operator $\operatorname{div}\left(\frac{1}{\varepsilon_1}\operatorname{grad}H_z^{\text{in}}\right)$ we have the following Green's integral formula [51]:

$$\iint_{\Omega_1}\left\{G_2\operatorname{div}\left(\frac{1}{\varepsilon_1}\operatorname{grad} H_z^{\text{in}}\right)-H_z^{\text{in}}\operatorname{div}\left(\frac{1}{\varepsilon_1}\operatorname{grad} G_2\right)\right\}dx\,dy=$$

$$=\oint_S\frac{1}{\varepsilon_1}\left(G_2\frac{\partial H_z^{\text{in}}}{\partial n_1}-H_z^{\text{in}}\frac{\partial G_2}{\partial n_1}\right)dl, \tag{3.90}$$

where G_2 is the Green's function for 2D light fields, satisfying the equation

$$\Delta G_2=-k_2^2G_2-\delta\left(M,M_0\right). \tag{3.91}$$

Then for the operator $\operatorname{div}\left(\frac{1}{\varepsilon_1}\operatorname{grad} G_2\right)$ taking into account equations (3.88) and (3.91) we have

$$\operatorname{div}\left(\frac{1}{\varepsilon_1}\operatorname{grad} G_2\right)=\frac{1}{\varepsilon_1}\Delta G_2-\frac{1}{\varepsilon_1^2}\nabla\varepsilon_1\nabla G_2=\frac{-k_2^2G_2-\delta\left(M,M_0\right)}{\varepsilon_1}-\frac{1}{\varepsilon_1^2}\nabla\varepsilon_1\nabla G_2. \tag{3.92}$$

From equation (3.90) with (3.89) and (3.92) we obtain

$$\iint_{\Omega_1}\left\{-\frac{k_1^2}{\varepsilon_1}H_z^{\text{in}}G_2+\frac{k_2^2}{\varepsilon_1}H_z^{\text{in}}G_2+\frac{H_z^{\text{in}}\delta\left(M,M_0\right)}{\varepsilon_1}-H_z^{\text{in}}\nabla\left(\frac{1}{\varepsilon_1}\right)\nabla G_2\right\}dx\,dy=$$

$$=\oint_S\frac{1}{\varepsilon_1}\left(G_2\frac{\partial H_z^{\text{in}}}{\partial n_1}-H_z^{\text{in}}\frac{\partial G_2}{\partial n_1}\right)dl, \tag{3.93}$$

which implies

$$\oint_S\frac{1}{\varepsilon_1}\left(G_2\frac{\partial H_z^{\text{in}}}{\partial n_1}-H_z^{\text{in}}\frac{\partial G_2}{\partial n_1}\right)dl+\iint_{\Omega_1}\frac{\left(k_1^2-k_2^2\right)}{\varepsilon_1}H_z^{\text{in}}G_2\,dx\,dy+$$

$$+\iint_{\Omega_1}H_z^{\text{in}}\nabla\left(\frac{1}{\varepsilon_1}\right)\nabla G_2\,dx\,dy=\begin{cases}\dfrac{H_z^{\text{in}}}{\varepsilon_1}, & (x,y)\subset\Omega_1\\ 0, & (x,y)\in\Omega_2\end{cases}. \tag{3.94}$$

We apply Green's formula (3.90) for functions and taking into account the second of equations (3.85) and equation (3.91) and obtain

$$\frac{1}{\varepsilon_2}\iint_{\Omega_2}G_2g_2\,dx\,dy+\oint_S\left(\frac{G_2}{\varepsilon_2}\frac{\partial H_z^{\text{ext}}}{\partial n_2}-\frac{H_z^{\text{ext}}}{\varepsilon_2}\frac{\partial G_2}{\partial n_2}\right)dl=\begin{cases}0, & (x,y)\in\Omega_1\\ \dfrac{H_z^{\text{ext}}}{\varepsilon_2}, & (x,y)\in\Omega_2\end{cases}. \tag{3.95}$$

Combining (3.94) and (3.95) taking into account the boundary conditions (3.86) and (3.87), we obtain a system of equations

$$\iint_{\Omega_1}\left(k_1^2-k_2^2\right)\frac{H_z^{\text{in}}G_2}{\varepsilon_1}dx\,dy+H_{0z}+\iint_{\Omega_1}H_z^{\text{in}}\nabla\left(\frac{1}{\varepsilon_1}\right)\nabla G_2dx\,dy+$$

$$+\oint_S\left(\frac{\varepsilon_1-\varepsilon_2}{\varepsilon_1\varepsilon_2}\right)H_z^{\text{in}}\frac{\partial G_2}{\partial n_1}dl=\begin{cases}\dfrac{H_z^{\text{in}}}{\varepsilon_1}, & (x,y)\in\Omega_1\\ \dfrac{H_z^{\text{ext}}}{\varepsilon_2}, & (x,y)\in\Omega_2\end{cases}, \quad (3.96)$$

where $H_{0z}=\iint_{\Omega_2}G_2g_2dxdy$ is the given field produced by external sources.

Thus, the problem of diffraction of electromagnetic waves of TM-polarization is reduced to solving a Fredholm integral equation of the second kind with respect to the function $H_z^{\text{in}}(x, y)$. The existence and uniqueness of solutions is proved in [30] so in this work are not given. The Green function has the form (3.83).

By definition of the gradient $\text{grad } G_2=\left(\dfrac{\partial G_2}{\partial x},\dfrac{\partial G_2}{\partial y}\right)$. Since $\xi=\xi(x, y)$, then

$$\frac{\partial G_2}{\partial x}=\frac{\partial G_2}{\partial \xi}\frac{\partial \xi}{\partial x},$$
$$\frac{\partial G_2}{\partial y}=\frac{\partial G_2}{\partial \xi}\frac{\partial \xi}{\partial y}. \quad (3.97)$$

For the Hankel function the following relation holds

$$\frac{dH_0^{(1)}}{d\xi}=-H_1^{(1)}(\xi). \quad (3.98)$$

Given that $\dfrac{\partial \xi}{\partial x}=\dfrac{k_2^2\left(x-x'\right)}{\xi}, \dfrac{\partial \xi}{\partial y}=\dfrac{k_2^2\left(y-y'\right)}{\xi}$, from (3.83) (3.97) (3.98) we have

$$\text{grad}\,G_2=-\frac{k_2^2 i}{4}\left(H_1^{(1)}(\xi)\frac{x-x'}{\xi}, H_1^{(1)}(\xi)\frac{y-y'}{\xi}\right). \quad (3.99)$$

Directional derivative in (3.96) with (3.99) can be calculated by the formula

$$\frac{\partial G_2}{\partial n_1}=n_1\cdot\text{grad }G_2. \quad (3.100)$$

Field H_z, which is determined from the system (3.96), is substituted into equations (3.65) and (3.66) from which the components E_x, E_y of the vector of electric strength are determined. The components H_z, E_x, E_y determine by the electromagnetic field, resulting from diffraction of electromagnetic wave of TM-polarization on a micro-object.

3.2.3. Application of finite element method for solving integral equation

To solve the integral equations of the systems (3.81) and (3.96) we use the finite element method in which by expanding the required fields with respect the basis of interpolating functions these equations were reduced to a system of linear algebraic equations. The system of interpolating functions was represented by linear functions inside cells obtained in discretization of region Ω_1. The linear interpolation functions within the discretization grid have the form [13]

$$\psi_m(x,y)=\begin{cases}1-\dfrac{x_m-x}{\Delta}-\dfrac{y_m-y}{\Delta}, & (x,y)\in\Omega_1(\Delta_1)\\ 1-\dfrac{x_m-x}{\Delta}, & (x,y)\in\Omega_1(\Delta_2)\\ 1+\dfrac{y_m-y}{\Delta}, & (x,y)\in\Omega_1(\Delta_3)\\ 1+\dfrac{x_m-x}{\Delta}+\dfrac{y_m-y}{\Delta}, & (x,y)\in\Omega_1(\Delta_4)\\ 1+\dfrac{x_m-x}{\Delta}, & (x,y)\in\Omega_1(\Delta_5)\\ 1-\dfrac{y_m-y}{\Delta}, & (x,y)\in\Omega_1(\Delta_6)\end{cases}, \tag{3.101}$$

where Δ is the step of the grid, as shown in Fig. 3.15. Δ_i, $i=\overline{1,6}$ are the triangular cells of the neighborhood of the current point m. In Fig. 3.15 $(p+1)$ is the number of discretization nodes along the x axis.

For the case of TE-polarization of the incident wave, expansion in the basis (3.101) takes the form

$$E_z^{in}(x,y)=\sum_{m=1}^{N}C_m\psi_m(x,y), \tag{3.102}$$

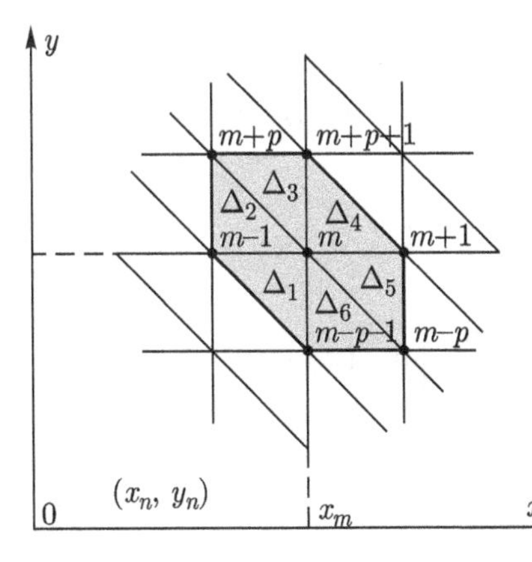

Fig. 3.15. A fragment of triangulation of region Ω_1.

where C_m are unknown coefficients.

Substituting (3.102) in the integral equation of (3.81), we obtain the system of linear algebraic equations for the unknown C_m

$$\sum_{m=1}^{N} C_m D_{mn} = E_{0n}, \tag{3.103}$$

where

$$D_{mn} = \psi_m(x_n, y_n) - \iint_{\Omega_1} \left(k_1^2 - k_2^2\right) \psi_m(x', y') \cdot G_2(x_n, y_n; x', y') dx' dy',$$

$$E_{0n} = E_{0z}(x_n, y_n). \tag{3.104}$$

For the case of TM-polarized incident wave expansion in the basis (3.101) takes the form

$$H_z^{in}(x, y) = \sum_{m=1}^{N} C_m \psi_m(x, y), \tag{3.105}$$

where C_m are unknown coefficients.

Substituting (3.105) in the integral equation of (4.96), we obtain the system of linear algebraic equations for the unknown C_m

$$\sum_{m=1}^{N} C_m D_{mn} = H_{0n}, \tag{3.106}$$

where

$$D_{mn} = \begin{cases} \dfrac{\psi_m(x_n, y_n)}{\varepsilon_1(x_n, y_n)} - \displaystyle\iint_{\Omega_1} \left(k_1^2(x', y') - k_2^2\right) \dfrac{\Psi_m(x', y')}{\varepsilon_1(x', y')} G_2(x', y'; x_n, y_n) dx' dy' - \\ -\displaystyle\iint_{\Omega_1} \psi_m(x', y') \nabla \left(\dfrac{1}{\varepsilon_1(x', y')} \right) \nabla G_2(x', y'; x_n, y_n) dx' dy', \quad (x, y) \in \Omega_1 \setminus S \\ \dfrac{\Psi_m(x_n, y_n)}{\varepsilon_1(x_n, y_n)} - \displaystyle\iint_{\Omega_1} \left(k_1^2(x', y') - k_2^2\right) \dfrac{\Psi_m(x', y')}{\varepsilon_1(x', y')} G_2(x', y'; x_n, y_n) dx' dy' -, \\ -\displaystyle\iint_{\Omega_1} \psi_m(x', y') \nabla \left(\dfrac{1}{\varepsilon_1(x', y')} \right) \nabla G_2(x', y'; x_n, y_n) dx' dy' - \\ -\displaystyle\oint_S \mathbf{n} \nabla G_2(x', y'; x_n, y_n) \psi_m(x', y') \left(\dfrac{\varepsilon_1(x', y') - \varepsilon_2}{\varepsilon_1(x', y') \varepsilon_2} \right) dl, \quad (x, y) \in S \end{cases}$$

$$H_{0n} = \frac{H_{0z}(x_n, y_n)}{\varepsilon_2}. \tag{3.107}$$

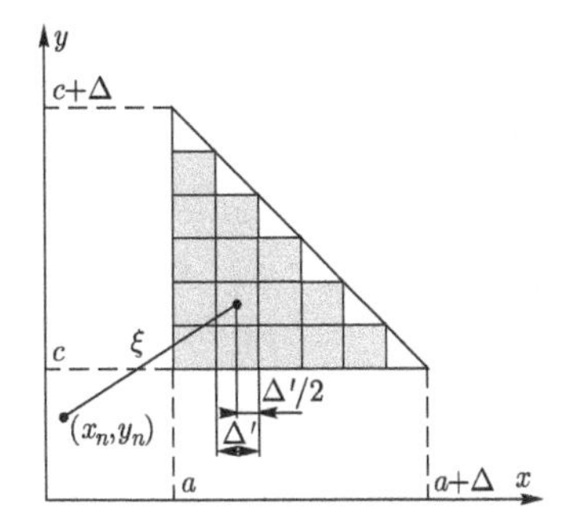

Fig. 3.16. Discretization of the triangular element of the fine grid (Fig. 3.15) for numerical integration.

Since the integrands of the integrals in (3.104) and (3.107) have a complicated form, then their integrals are numerically implemented for each of the six triangles of the current node m. In the case where $m = n$ the Neumann function $Y_0(x_n, y_n; x_m, y_m)$, which is part of the component of the function G_2 has a singularity, i.e. tends to $-\infty$. To calculate the function Y_0 in the neighborhood of zero, we need more detailed discretization of the triangular element formed by the fine grid, shown in Fig. 3.16.

Here (a,c) are the coordinates of the point from which integration starts, and Δ' is the step of the inner grid, shown in Fig. 3.16. The element is divided into squares, and triangles, as shown in Fig. 3.16. The integration in (3.104) is carried out as follows: function $(k_1^2 - k_2^2)\ \psi_m$ is integrated analytically for each square (in each square k_1 and k_2 are assumed constant), and the Green's function G_2 which is a function of distance ξ, is assumed to be constant for each square. Integrals for the area in (3.107) are calculated in a similar way, and to calculate the integral along the contour sections with the same step Δ' are used instead of squares.

Matrices D_{mn} of the systems (3.103), (3.106) are symmetric, fully fitted with the dominant main diagonal. Symmetry D_{mn} is due to the fact that the Green function G_2 is an even function of the distance between the observation point n and the current point m. The prevalence of the main diagonal is due to the fact that the Neumann function Y_0 has a singularity at the origin. Indeed, this feature occurs when $n = m$, i.e. when calculating the diagonal elements. For a single observation point n enumeration happens through all the points of region Ω_1, that is in the same row of the matrix D_{mn} there are N elements (N is the number of points in the region Ω_1). Thus, the resultant system of linear algebraic equations are of order N. To solve this system we used the Gauss method for complex numbers. The number of arithmetic operations performed in the solution of the system of algebraic linear equations can be estimated as $\approx(2/3)N^3$ [52].

For the case of TE-polarization by solving a system of equations (3.103) we obtain the complex coefficients C_m, $m=\overline{1,N}$ which are then substituted into the second equation (3.81) to determine the field in the outer region Ω_2:

$$E_z^{\text{ext}}(x_n,y_n) = E_{0z}(x_n,y_n) + \sum_{m=1}^{N} C_m \cdot \tag{3.108}$$
$$\cdot \iint_{\Omega_1} \left(k_1^2 - k_2^2\right)\psi_m(x',y')\,G_2(x_n,y_n;x',y')\,dx'dy'.$$

Solving the system of equations (3.106) for the case of TM-polarization, we obtain the complex coefficients C_m, $m=\overline{1,N}$ which are then substituted into the second equation (3.96) to determine the field in the outer region:

$$\frac{H_z^{\text{ext}}(x_n,y_n)}{\varepsilon_2}=H_{0z}(x_n,y_n)+\sum_{m=1}^{N}C_m\cdot$$
$$\cdot\left[\iint_{\Omega_1}\left(k_1^2-k_2^2\right)\frac{\Psi_m(x',y')G_2(x_n,y_n;x',y')}{\varepsilon_1(x',y')}dx'dy'+\right.$$
$$+\iint_{\Omega_1}\psi_m(x',y')\nabla\left(\frac{1}{\varepsilon_1(x',y')}\right)\nabla G_2(x_n,y_n;x',y')dx'dy'+ \tag{3.109}$$
$$\left.+\oint_S\left(\frac{\varepsilon_1(x',y')-\varepsilon_2}{\varepsilon_1(x',y')\varepsilon_2}\right)\Psi_m(x',y')\frac{\partial G_2(x_n,y_n;x',y')}{\partial n_1}dl\right].$$

The described method has several advantages and disadvantages compared with other methods. In contrast to the FEM, in Gallagher's formulation [13] the integral equation method does not require specification of boundary conditions and can operate with objects having an arbitrary boundary. However, the use of Green's function considerably complicates the numerical implementation. This method also does not require the calculation of normal and tangential derivatives along the contour of the object from the light field, which distinguishes it from the finite element method [13] and the hybrid finite element method [53]. At the same time, the hybrid method computes several times faster (for the given parameters of the problem and discretization) than the method of integral equations, as in the hybrid method the matrix of the system of linear algebraic equations is tridiagonal, and in this method it is completely filled out (which requires a significant amount of memory in numerical experiments). The method of the integral equation allows the calculation of the diffraction on both homogeneous and heterogeneous objects, and also, importantly, on the combination of several objects, with no need to introduce an artificial field, covering all scattering objects.

3.2.4. Convergence of the approximate solution

The numerical experiment showed that the method has a convergence. To do this, the test object was a homogeneous dielectric cylinder with permittivity $\varepsilon_1=4$ and the square cross-section of the size equal to the wavelength of the incident wave. The cylinder was illuminated by a TE-polarized plane electromagnetic wave with the wavelength $\lambda=1$ μm. The external environment was vacuum with the permittivity $\varepsilon_2=1$. The diffraction pattern with the size of 5×5 μm is shown in Fig. 3.17:

The main discretization grid was 100×100 nodes. Depending on the number of nodes of discretization of the fine grid the value of the maximum in the intensity

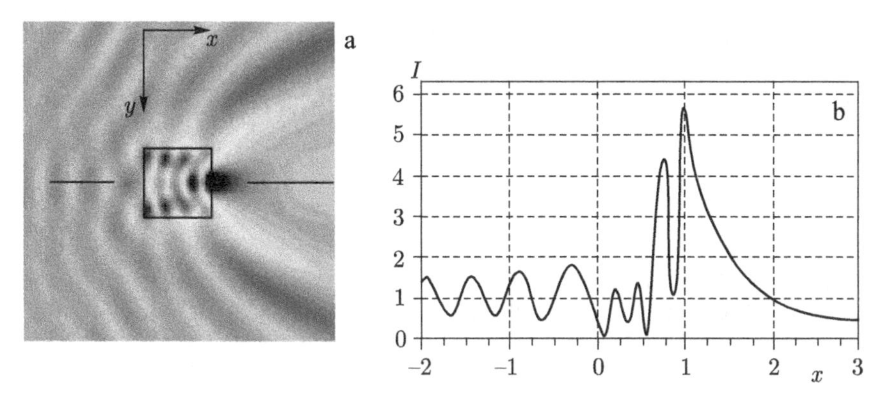

Fig. 3.17. Diffraction of a plane TE-wave by a dielectric cylinder with a square cross-section: a) the intensity distribution in the plane XY; b) cross-section of the intensity of X

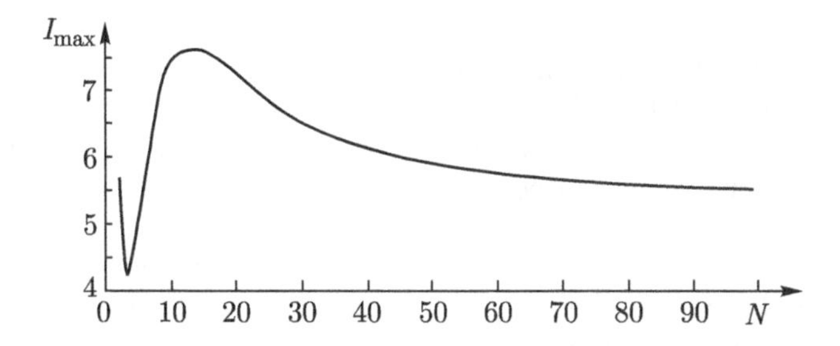

Fig. 3.18. Dependence of the maximum intensity in the 2D pattern of diffraction of a plane TE-wave on a square with the side λ on the number of nodes N of discretization of a 'fine' grid between two adjacent nodes of the main grid.

distribution in the plane XY changes. As the number of nodes of the fine grid increases the maximum value asymptotically approaches a constant value (Fig. 3.18): Fig. 3.17 corresponds to the number of nodes of the fine grid $N = 100$.

3.2.5. The diffraction of light by cylindrical microlenses

The diffraction of light on microlens was analyzed using a collecting lens with a radius of curvature $R = 2.5$ μm (3 μm aperture of the lens), the refractive index of $n = 2$ and a thickness of 0.5 μm. A plane wave with the wavelength $\lambda = 1$ μm impacted on the lens. The diffraction pattern had the dimensions 5×5 μm with discretization grid of 100 × 100 nodes. The fine grid – 50 nodes on the distance between two nodes of the main grid. The aim of light diffraction study on the microlens was to find its focus and compare the results with the geometrical optics approximation of a thin lens. A study was conducted of the impact of its orientation to the incident radiation of flat and convex sides on the focus position.

The modelling results are shown in Figs. 3.19 and 3.20.

As can be seen from Figs. 3.19 and 3.20, the total diffraction pattern at different orientations of the lens has a different intensity distribution in the plane XY. However, from Fig. 3.19c and Fig.3.20c show in that the the intensity at the focus is almost identical and differs by no more than 5%.

From the geometric optics approximation for a thin lens it follows that its focal length is equal to

$$f = \frac{R_1 R_2}{(n-1)(R_1 + R_2)}, \tag{3.110}$$

where R_1 and R_2 are the radii of curvature of the surfaces of the lens, n is the refractive index of the lens. Given that $R_1 = \infty$, the formula (3.110) becomes

$$f = \frac{R_2}{(n-1)}. \tag{3.111}$$

It follows from (3.111), taking into account the lens parameters, that the focal length is $f = 2.5$ μm.

From Fig. 3.19b $f \approx 3.05$ μm, which is different from the theoretical result by 22%, and from Fig. 3.20b $f \approx 3.25$ μm, which is different from the theoretical result by 30%. We conclude that the diffraction of light on the microlens causes shift of the focus of the lens, which depends on the orientation of the lens in relation to the incident plane wave.

Figues 3.19b and Fig. 3.20b also show that the type of diffraction pattern inside and in front of the lenses is significantly different: at the reflection of the plane face of the lens (Fig. 3.19b) there is a local maximum of intensity whose value is 1.3 times larger than the focus intensity.

Microlenses with a continuous profile represent a significant challenge to make, so the following question is of considerable importance: how accurately can these lenses can be approximated by binary lenses?

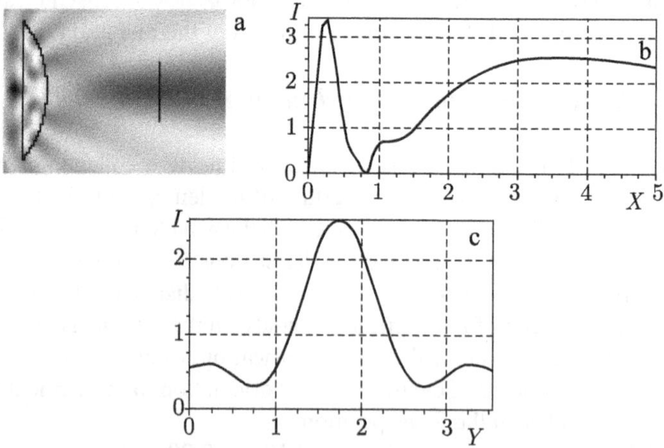

Fig. 3.19. Diffraction of a plane TE-wave on a microlens oriented with the flat edge to the incident wave: a) the intensity distribution in the plane XY; b) cross-section of the intensity with respect to X, c) cross-section of the intensity with respect to Y in focus.

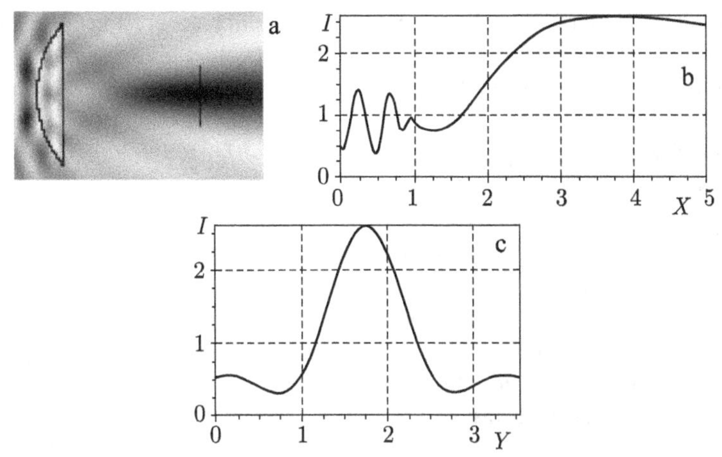

Fig. 3.20. Diffraction of a plane TE-wave by a microlens oriented with the convex surface to the incident wave: a) the intensity distribution in the plane *XY*; b) cross-section of intensity with respect to *X*, c) cross-section of intensity with respect to *Y* in focus.

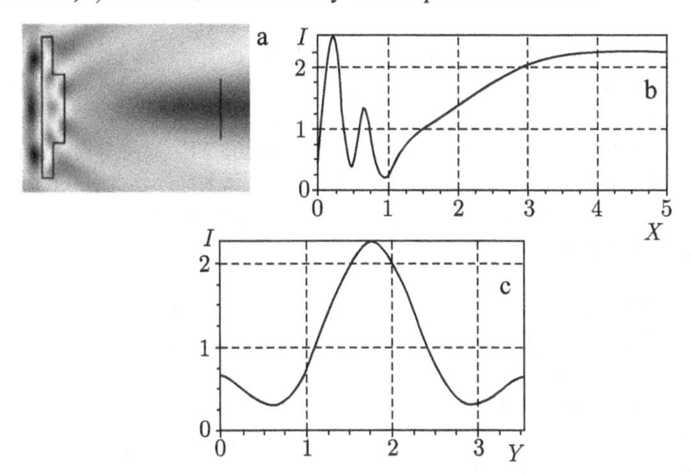

Fig. 3.21. Diffraction of a plane TE wave by a binary microlens: a) the intensity distribution in the plane *XY*; b) cross-section of intensity with respect to *X*, c) cross-section of intensity with respect to *Y* in focuse.

Figure 3.21 shows the diffraction of a plane wave on a binary microlens. Its parameters are the same as in the previous example.

Figures 3.21a and 3.21b show that the focal length of a binary lens is 18–26% greater that the focal length of the ordinary lens. A large part of the energy is reflected back and the focal length increased to about 3.85 μm. The width of the intensity maximum in the cross section for the *Y*-axis of the focus increased compared with the continuous lens by ≈2.3 μm against 2 μm for the normal lens (Fig. 3.19c); the maximum light intensity at the focus of a binary microlens was ≈88% of the maximum intensity of the continuous microlens (Fig. 3.19c).

Note that the numerical results presented in this section differ from those obtained in the hybrid method in [53] also by the fact that the aperture of the lens was 3λ, while in [53] the aperture was 8λ.

3.2.6. Diffraction of light on microscopic objects with a piecewise-uniform refractive index

A feature of the modification the finite element method for solving the Fredholm integral equation of the second kind, discussed in this chapter, is that it allows the calculation of diffraction not only on homogeneous objects, but also on objects with a piecewise-uniform refractive index. In other words, if the piecewise-homogeneous region Ω_1 can be divided into a finite number N of homogeneous subdomains Ω_{1i}, then for the case of TE-polarization the system (3.81) takes the form

$$\sum_{i=1}^{N}\iint_{\Omega_{1i}}\left(k_{1i}^2-k_2^2\right)E_z^{\text{in}}G_2 dx\,dy+E_{0z}=\begin{cases}E_z^{\text{in}},(x,y)\in\Omega_1\\ E_z^{\text{ext}},(x,y)\in\Omega_2\end{cases}. \tag{3.112}$$

To test this assertion, the numerical results of the diffraction problem of a plane TE-polarized electromagnetic wave, obtained by the considered method and the analytical method described in [54, 55, 56], were compared.

Figure 3.22 shows the diffraction pattern of a plane TE-wave with a wavelength $\lambda = 1$ μm for a two-layer dielectric circular cylinder with the characteristics $r_1 = 0.25$ μm, $r_2 = 0.5$ μm, $\varepsilon_1 = 2.25$; $\varepsilon_2 = 4$. Outer space – a vacuum. The dimensions of the diffraction pattern were 3.33×3.33 μm. The sampling grid had 200×200 nodes.

In Fig. 3.22b the intensity curve, displayed by the solid line, corresponds to the analytical solution [56], while the dashed line corresponds to the investigated method (3.112). For the above mentioned parameters of the problem the compared results differ by 4–5%. This allows us to conclude that the method is suitable for calculating the diffraction problems for piecewise-homogeneous (in the general case of inhomogeneous) micro-objects.

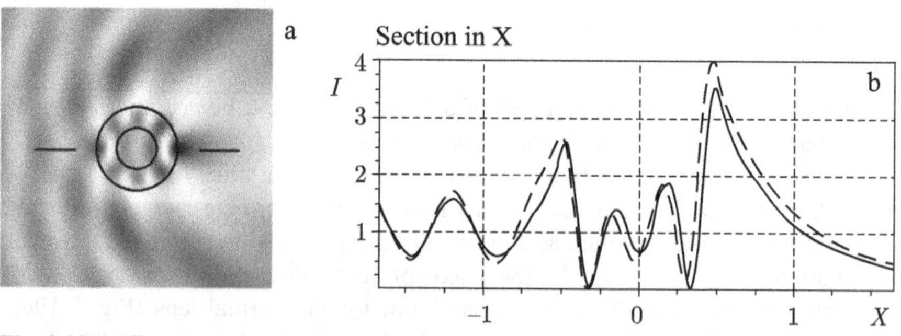

Fig. 3.22. Diffraction of a plane TE-wave on a two-layer microlenses: a) the intensity distribution in the plane *XY*; b) cross-section of intensity with respect to X.

Consider now one of the simplest examples of diffraction of light on piecewise-homogeneous microscopic objects – the diffraction of light on layered films.

The investigated object was a plate with the following parameters (Fig. 3.23):

- dimensions: 0.5 × 3 μm;
- the refractive index:
- left layer with thickness 0.25 μm $n_{1l} = 2$;
- right layer with thickness 0.25 μm $n_{1r} = 1.5$;
- the external environment $n_2 = 1$;
- the wavelength of incident radiation $\lambda = 1$ μm,
- the number of counts in the sampling grid 200 × 200;
- the size of the outer region 10 × 10 μm.

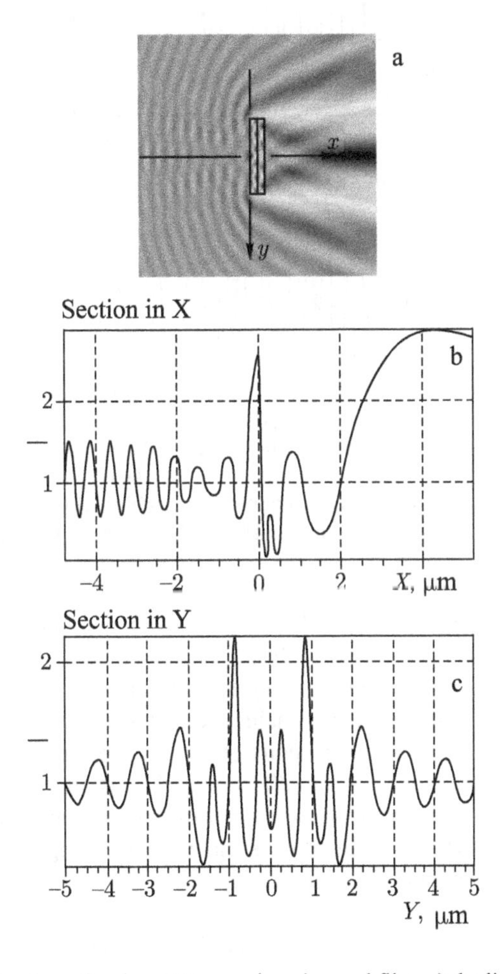

Fig. 3.23. Diffraction of a plane TE-wave by a layered film: a) the light intensity distribution in the plane *XY*; b) cross-section of intensity in *X*, c) cross-section of intensity in *Y*.

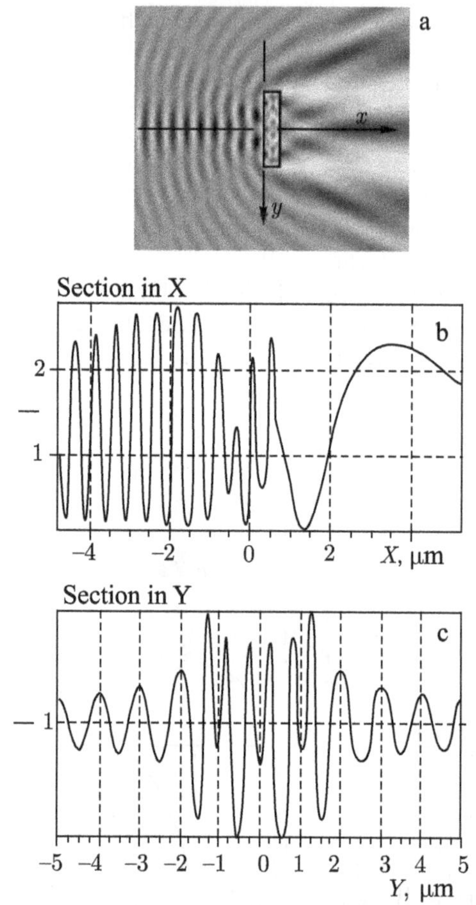

Fig. 3.24. Diffraction of a plane TE-wave on a homogeneous film: a) the light intensity distribution in the plane *XY*; b) cross-section of intensity in *X*, c) cross-section of intensity in *Y*.

Next, a layered film was replaced with a uniform refractive index with the same dimensions as the previous model (Fig. 3.24).

Figure 3.23b and 3.24b show that the layered film has a higher transmittance than the homogeneous film, and the bleaching effect was observed there. This suggests that by replacing the homogeneous object by a piecewise–uniform one we can achieve the required value of the reflection and transmission, which is very important for the design of micro-lenses and other micro-optics objects.

It is interesting to note that both models have the focusing properties, such as microlenses, and the magnitude of the maximum intensity at the focus can be controlled by the choice of layers with different refractive indices. In Fig. 3.23b the maximum intensity at the focus is 3 and in Fig. 3.24b it is 2.4.

3.3. Diffraction of light on inhomogeneous dielectric cylinders

Among the many tasks of light scattering on microscopic objects special attention is paid to the solution of axisymmetric problems of diffraction of electromagnetic waves on bodies of revolution [57–61]. For example, in [62] to solve the scattering problem on 3D axisymmetric particles, the authors suggested the method of separation of electromagnetic fields into two parts: axisymmetric, independent of the azimuthal angle, and asymmetric, whose average over this angle is zero. The scattering problem is considered separately for each of these parts. At the same time, special selection is made of scalar potentials associated with the azimuthal components of the electromagnetic fields used for the axisymmetric part of these fields. For the asymmetric part we used the superposition of Debye potentials and vertical components of the Hertz vector. The formulation of the problem is reduced to solving an integral equation, which requires large computational costs.

In [63] the analytic solution of near-field diffraction on homogeneous metallic and dielectric circular cylinders in the vicinity of the dielectric surface is studied. In [64] the problem of a more general form, where a homogeneous circular cylinder is immersed in a layered medium, is solved.

In [65], the modification of the method of discrete sources was proposed for the two-dimensional problem of diffraction of a plane TE-polarized electromagnetic waves on a two-layer circular dielectric cylinder or a metal cylinder with a dielectric coating.

In [66] the electromagnetic scattering by a multilayer gyrotropic bianisotropic circular cylinder for TE-/TM-polarized incident plane waves was investigated using the method of eigenfunction expansion. Numerical results are presented for a three-layer cylinder.

In [67–70] in the framework of geometrical optics the authors obtained analytical expressions for the dependence of the refractive index on the radial coordinate of gradient optical elements with spherical and transverse cylindrical symmetry (when an infinitely long lateral surface is perpendicular to the direction of incidence of the electromagnetic wave). Note that the Luneberg lens [69] is also used as a lens antenna for centimetric band radio waves [71]. The Luneberg lens focuses the beam of parallel rays to a point on the surface. The inner Luneberg lens [68] focuses a beam of parallel rays in a given internal point lying on a diameter parallel to the incident rays between the centre and the far surface of the lens. The generalized Luneberg lens [67] focuses the incident beam of parallel rays to a point behind the lens, which lies on the continuation of the diameter parallel to the incident rays. In this case, the dependence of the refractive index on the radial variable no longer has an explicit analytical dependence, and is expressed in the form of integral relations. The Eaton-Lippmann lens [70] is a dielectric gradient optical element having a spherical or transverse cylindrical symmetry, which reflects back all the rays falling on it. An explicit analytical dependence of the refractive index on the coordinate for the Eaton-Lippmann lens has a singularity at the origin (in the centre of the lens), which provides back reflection of rays incident at the centre of the lens.

Traces of light rays in all the lenses have been studied well enough. This chapter discusses the passage of the electromagnetic wave through these gradient optical elements, in the case where the radius of the lens is the same (or similar) with the wavelength. In this resonance case, the beam description of diffraction of light is no longer valid and the question arises about the extent of change in the focusing and reflection properties of the given gradient elements.

The analysis of electromagnetic wave diffraction on the gradient cylindrical optical elements, the refractive index of which has a transverse cylindrical symmetry, can be carried out using the method of integral equations described in section 3.2. The numerical solution of Fredholm integral equations of the second kind is generally conducted using methods for solving systems of linear algebraic equations (one of these methods – the finite element method – was also discussed in section 3.1). However, to obtain sufficient accuracy of the system we should consider high-order equations with completely filled matrices, which requires a considerable amount of computing time and a large amount of computer memory. In this connection there was a need to develop a method that would solve the problem of electromagnetic wave diffraction on a transparent body in a short time frame and without significant computational costs.

In this section, the method of separation of variables is used to develop the recurrent analytical method for calculating the diffraction field with TE- and TM-polarizations, in the event of the incidence of the electromagnetic wave on an inhomogeneous dielectric infinite circular cylinder whose generating line extends along the axis z, while the plane (x, y) is the plane of incidence. The heterogeneity of the cylinder is approximated by a piecewise-constant function, and the circular section of the cylinder at the same time will have N concentric rings with constant values of the refractive index within each ring (Fig. 3.25). The method is based on the decomposition of the projection on the z-axis of the vectors of the electric (for TE-polarization) or magnetic (for TM-polarization) fields within each homogeneous ring into a series of cylindrical functions with unknown coefficients. The coefficients themselves are determined from the boundary conditions imposed on the field and their radial derivatives on the lines of discontinuities of the refractive index.

3.3.1. Solution of the problem of diffraction of an arbitrary wave on a cylindrical multilayer dielectric cylinder by separation of variables

Figure 3.25 schematically shows the section of an N-layered circular cylinder in each layer of which the refractive index is constant. The generator of the infinite cylinder is elongated along the axis z, and the plane of incidence of a plane monochromatic electromagnetic wave coincides with the plane (x, y). In this case, the system of six Maxwell equations splits into two independent systems of three equations: for TE-polarization the system of three equations includes the projections of the vectors of the strength of electric and magnetic fields (E_x, H_x, H_y), for TM-polarization the system is formed by the projections of the vectors (H_z, E_x, E_y). For TE-polarization

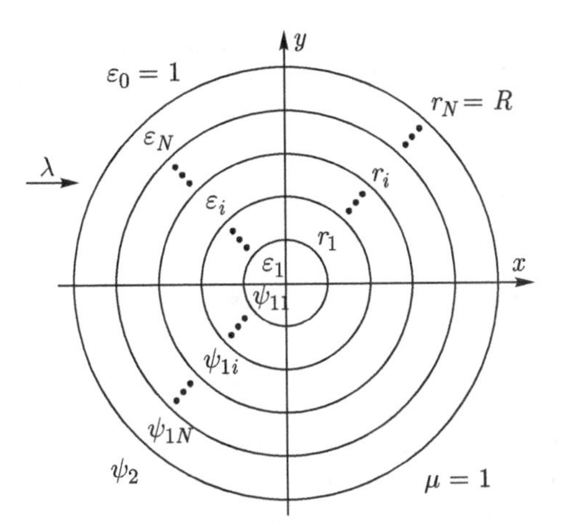

Fig. 3.25. Multilayer dielectric cylinder.

projection E_z satisfies the Helmholtz equation, and the projections H_x and H_y are expressed through E_z, and for TM-polarization, projection H_z satisfies the Helmholtz equation, and the projections E_x and Ey are expressed by H_z.

Thus, to solve the problem we need to solve the Helmholtz equation for the projections E_z and H_z. If we place the centre of the coordinate system (x, y) at the centre of the circular cylinder, then the problem can be solved in cylindrical coordinates $(r,\varphi): x = r\cos\varphi, y = r\sin\varphi$. It is known that partial solutions of the Helmholtz equation in cylindrical coordinates are cylindrical functions, so any solution of the Helmholtz equation in the variables (r, φ), where the refractive index is constant, can be represented as a linear combination of independent cylindrical functions.

TE-polarization

In this section we introduce the notation $E_z = \psi$.

The field amplitude in the inner circle $(0 \le r \le r_1)$ is represented as a series of Bessel functions (Rayleigh series):

$$\psi_{11} = \sum_{m=-\infty}^{+\infty} C_{1m} J_m\left(k\sqrt{\varepsilon_1} r\right)\cos m\varphi. \tag{3.113}$$

The field inside the j-th ring of the dielectric is represented as a series of Bessel and Neumann functions:

$$\psi_{1j} = \sum_{m=-\infty}^{+\infty} \left[C_{(2j-2)m} J_m\left(k\sqrt{\varepsilon_j} r\right) + C_{(2j-1)m} Y_m\left(k\sqrt{\varepsilon_j} r\right)\right]\cos m\varphi, \tag{3.114}$$

where $r_{j-1} < r \le r_j, j = \overline{2,N}$. Here $r_N = R$.

The amplitude of the field outside the dielectric is represented as a series of Hankel functions of the second kind, since they satisfy the Sommerfeld radiation condition:

$$\psi_2 = \psi_0 + \sum_{m=-\infty}^{+\infty} C_{2Nm} H_m^{(2)}(kr)\cos m\varphi, \tag{3.115}$$

where $r > R$. Here we assume that in the freespace the permittivity is equal to unity $\varepsilon_2 = 1$.

In equation (3.115) $\psi_0 = \exp(-ikx) = \exp(-ikr\cos\varphi)$ is the amplitude of the incident plane wave of unit intensity.

To find the unknown coefficients in the series (3.113)–(3.115) we use the boundary conditions. Equating the fields themselves and their radial derivatives at the radii of the jumps of the refractive index r_j, we obtain a system of equations:

$$\begin{cases} \psi_{1j}\big|_{r_j} = \psi_{1(j+1)}\big|_{r_j} \\ \left.\dfrac{\partial \psi_{1j}}{\partial r}\right|_{r_j} = \left.\dfrac{\partial \psi_{1(j+1)}}{\partial r}\right|_{r_j} \end{cases}, j = \overline{1, N-1}, \tag{3.116}$$

$$\begin{cases} \psi_{1N}\big|_R = \psi_2\big|_R \\ \left.\dfrac{\partial \psi_{1N}}{\partial r}\right|_R = \left.\dfrac{\partial \psi_2}{\partial r}\right|_R \end{cases}.$$

For the expansion of ψ_0 into a series in respect of Bessel functions we use the series connected with the generating function [50]:

$$\cos(z\cos\theta) = J_0(z) + 2\sum_{k=1}^{\infty}(-1)^k J_{2k}(z)\cos(2k\theta),$$
$$\sin(z\cos\theta) = 2\sum_{k=0}^{\infty}(-1)^k J_{2k+1}(z)\cos\left[(2k+1)\theta\right]. \tag{3.117}$$

For a plane wave from the equations (3.117) it follows:

$$\psi_0\big|_{r=R} = \cos(kR\cos\varphi) - i\sin(kR\cos\varphi) =$$
$$= J_0(kR) + 2\sum_{m=1}^{\infty}(-1)^m J_{2m}(kR)\cos(2m\varphi) - \tag{3.118}$$
$$-2i\sum_{m=0}^{\infty}(-1)^m J_{2m+1}(kR)\cos\left[(2m+1)\varphi\right]$$

Given that $(-1)^m = (-i)^{2m}$ and $-i\cdot(-1)^m = (-i)^{2m+1}$ from equation (3.118) we have

$$\psi_0\big|_{r=R} = J_0(kR) + 2\sum_{m=1}^{\infty}(-i)^{2m} J_{2m}(kR)\cos(2m\varphi) +$$
$$+2\sum_{m=0}^{\infty}(-i)^{2m+1} J_{2m+1}(kR)\cos\left[(2m+1)\varphi\right] = \tag{3.119}$$
$$= J_0(kR) + 2\sum_{m=1}^{\infty}(-i)^m J_m(kR)\cos m\varphi$$

We check the parity of the function $(-i)^m J_m(kR)\cos m\varphi$:

$$(-i)^{-m} J_{-m}(kR)\cos(-m\varphi) = \frac{1}{(-i)^m}(-1)^m J_m(kR)\cos m\varphi = (-i)^m J_m(kR)\cos m\phi.$$

Here we use the property $J_{-m}(kR) = (-1)^m J_m(kR)$.

Since the function under the sum in equation (3.119) is even, then the expansion of a plane wave in the series becomes

$$\begin{aligned} \psi_0\big|_{r=R} &= \sum_{m=-\infty}^{+\infty}(-i)^m J_m(kR)\cos m\varphi, \\ \frac{\partial\psi_0}{\partial r}\bigg|_{r=R} &= k\sum_{m=-\infty}^{\infty}(-i)^m J'_m(kR)\cos m\varphi. \end{aligned} \tag{3.120}$$

When j = 1 taking into account (3.113) and (3.114) from (3.116) we have:

$$\begin{cases} \displaystyle\sum_{m=-\infty}^{+\infty} C_{1m} J_m\left(k\sqrt{\varepsilon_1}r_1\right)\cos m\varphi = \\ \displaystyle= \sum_{m=-\infty}^{+\infty}\left[C_{2m} J_m\left(k\sqrt{\varepsilon_2}r_1\right) + C_{3m} Y_m\left(k\sqrt{\varepsilon_2}r_1\right)\right]\cos m\varphi \\ \displaystyle\sqrt{\varepsilon_1}\sum_{m=-\infty}^{+\infty} C_{1m} J'_m\left(k\sqrt{\varepsilon_1}r_1\right)\cos m\varphi = \\ \displaystyle= \sqrt{\varepsilon_2}\sum_{m=-\infty}^{+\infty}\left[C_{2m} J'_m\left(k\sqrt{\varepsilon_2}r_1\right) + C_{3m} Y'_m\left(k\sqrt{\varepsilon_2}r_1\right)\right]\cos m\varphi \end{cases}. \tag{3.121}$$

At $j = \overline{2, N-1}$ taking into account (3.114) from (3.116) we have:

$$\begin{cases} \sum_{m=-\infty}^{+\infty} \left[C_{(2j-2)m} J_m\left(k\sqrt{\varepsilon_j} r_j\right) + C_{(2j-1)m} Y_m\left(k\sqrt{\varepsilon_j} r_j\right) \right] \cos m\varphi = \\ = \sum_{m=-\infty}^{+\infty} \left[C_{2jm} J_m\left(k\sqrt{\varepsilon_{j+1}} r_j\right) + C_{(2j+1)m} Y_m\left(k\sqrt{\varepsilon_{j+1}} r_j\right) \right] \cos m\varphi \\ \sqrt{\varepsilon_j} \sum_{m=-\infty}^{+\infty} \left[C_{(2j-2)m} J'_m\left(k\sqrt{\varepsilon_j} r_j\right) + C_{(2j-1)m} Y'_m\left(k\sqrt{\varepsilon_j} r_j\right) \right] \cos m\varphi = \\ = \sqrt{\varepsilon_{j+1}} \sum_{m=-\infty}^{+\infty} \left[C_{2jm} J'_m\left(k\sqrt{\varepsilon_{j+1}} r_j\right) + C_{(2j+1)m} Y'_m\left(k\sqrt{\varepsilon_{j+1}} r_j\right) \right] \cos m\varphi \end{cases} . \tag{3.122}$$

When $j = N$, taking into account (3.114), (3.115) and (3.120) from (3.116) we have:

$$\begin{cases} \sum_{m=-\infty}^{+\infty} \left[C_{(2N-2)m} J_m\left(k\sqrt{\varepsilon_N} r_N\right) + C_{(2N-1)m} Y_m\left(k\sqrt{\varepsilon_N} r_N\right) \right] \cos m\varphi = \\ = \sum_{m=-\infty}^{+\infty} (-i)^m J_m\left(kr_N\right) \cos m\varphi + \sum_{m=-\infty}^{+\infty} C_{2Nm} H_m^{(2)}\left(kr_N\right) \cos m\varphi \\ \sqrt{\varepsilon_N} \sum_{m=-\infty}^{+\infty} \left[C_{(2N-2)m} J'_m\left(k\sqrt{\varepsilon_N} r_N\right) + C_{(2N-1)m} Y'_m\left(k\sqrt{\varepsilon_N} r_N\right) \right] \cos m\varphi = \\ = \sum_{m=-\infty}^{+\infty} (-i)^m J'_m\left(kr_N\right) \cos m\varphi + \sum_{m=-\infty}^{+\infty} C_{2Nm} H_m^{\prime(2)}\left(kr_N\right) \cos m\varphi \end{cases} . \tag{3.123}$$

If the cylinder is impacted by an arbitrary wave (not by the plane wave) satisfying the Helmholtz equation, the function can be written as [72]:

$$\psi_0(r,\varphi) = \int_{-\pi}^{\pi} h(\theta) \exp\left[ikr\cos(\varphi-\theta)\right] d\theta, \tag{3.124}$$

where $h(\theta)$ is an arbitrary function. We expand the exponent in the integrand of (3.124) in a series of cylindrical functions:

$$\exp\left[ikr\cos(\varphi-\theta)\right] = \sum_{m=-\infty}^{+\infty} (-i)^m J_m(kr) e^{im(\varphi-\theta)}, \tag{3,125}$$

then

$$\psi_0(r,\varphi) = \sum_{m=-\infty}^{+\infty} (-i)^m J_m(kr) e^{im\varphi} \int_{-\pi}^{\pi} e^{-im\theta} h(\theta) d\theta. \tag{3.126}$$

If $h(\theta)$ is an even and real function, then its Fourier transform $A_m = \int_{-\pi}^{\pi} e^{-im\theta} h(\theta) d\theta$

is also an even and real function. Thus, for any even real function we can write:

$$\psi_0(r,\varphi)=\sum_{m=-\infty}^{+\infty}(-i)^m A_m J_m(kr)\cos m\varphi, \tag{3.127}$$

where $A_m=\int_{-\pi}^{\pi}h(\theta)\cos m\theta d\theta$.

A plane wave propagating along the optical axis is described by the function $h(\theta)=\delta(\theta)$. In this case, $A_m=1$ for any m.

In view of (3.127), system (3.123) takes the form:

$$\begin{cases}\sum_{m=-\infty}^{+\infty}\left[C_{(2N-2)m}J_m\left(k\sqrt{\varepsilon_N}r_N\right)+C_{(2N-1)m}Y_m\left(k\sqrt{\varepsilon_N}r_N\right)\right]\cos m\varphi= \\ =\sum_{m=-\infty}^{+\infty}(-i)^m A_m J_m(kr_N)\cos m\varphi+\sum_{m=-\infty}^{+\infty}C_{2Nm}H_m^{(2)}(kr_N)\cos m\varphi \\ \sqrt{\varepsilon_N}\sum_{m=-\infty}^{+\infty}\left[C_{(2N-2)m}J'_m\left(k\sqrt{\varepsilon_N}r_N\right)+C_{(2N-1)m}Y'_m\left(k\sqrt{\varepsilon_N}r_N\right)\right]\cos m\varphi= \\ =\sum_{m=-\infty}^{+\infty}(-i)^m A_m J'_m(kr_N)\cos m\varphi+\sum_{m=-\infty}^{+\infty}C_{2Nm}H_m'^{(2)}(kr_N)\cos m\phi\end{cases} \tag{3.128}$$

Because of the completeness and orthogonality of the functions $\cos m\,\varphi$ the required vector of coefficients $C_m=\{C_{km}\}$, $k=1,2N$, for any m is expressed through a system of linear algebraic equations:

$$A_m C_m=B_m, \tag{3.129}$$

where (see equation (3.130) on the next page)

$$B_m=\begin{pmatrix}0\\0\\\vdots\\0\\(-i)^m A_m J_m(kR)\\(-i)^m A_m J'_m(kR)\end{pmatrix}. \tag{3.131}$$

The system of equations with the size $2N\times 2N$ is solved for all m-th coefficients of the expansion into a series of cylindrical functions (3.113)–(3.115). The resulting coefficients are then substituted into equation (3.113)–(3.115).

TM-polarization

The solution to the problem of diffraction of a plane TM-polarized electromagnetic waves on a multilayer dielectric cylinder is analogous to that in the case of TE-polarization, which was reviewed in the previous section. For TM-polarization we introduce the notation $H_z = \psi$. The boundary conditions take the form:

$$\begin{cases} \left.\psi_{1j}\right|_{r_j} = \left.\psi_{1(j+1)}\right|_{r_j} \\ \left.\dfrac{1}{\varepsilon_j}\dfrac{\partial \psi_{1j}}{\partial r}\right|_{r_j} = \left.\dfrac{1}{\varepsilon_{j+1}}\dfrac{\partial \psi_{1(j+1)}}{\partial r}\right|_{r_j} \end{cases}, j = \overline{1, N-1},$$
$$\begin{cases} \left.\psi_{1N}\right|_R = \left.\psi_2\right|_R \\ \left.\dfrac{1}{\varepsilon_N}\dfrac{\partial \psi_{1N}}{\partial r}\right|_R = \left.\dfrac{\partial \psi_2}{\partial r}\right|_R \end{cases}. \qquad (3.132)$$

Here $r_N = R$.

Guided by the above described manner, it is easy to obtain the matrix of the system of equations (3.113)–(3.115), (3.132) (see equation (3.133) on the next page), similar to the matrix (3.130).

Recurrent relations for the unknown coefficients

Because of sparse matrices (3.130) and (3.133), the Gauss method becomes ineffective. In this section, using the sweep method and taking into account the structure of the matrices (3.130) and (3.133), recurrence formulas are obtained for the unknown coefficients. Thus, the system of algebraic equations of any order to the matrix of the form (3.130) or (3.133) is solved. In general, the structure of the system of equations to which we apply the method can be represented as follows:

$$A_m = \begin{pmatrix} J_m\left(k\sqrt{\varepsilon_1}r_1\right) & -J_m\left(k\sqrt{\varepsilon_2}r_1\right) & -Y_m\left(k\sqrt{\varepsilon_2}r_1\right) & & & \\ J'_m\left(k\sqrt{\varepsilon_1}r_1\right)\sqrt{\varepsilon_1} & -J'_m\left(k\sqrt{\varepsilon_2}r_1\right)\sqrt{\varepsilon_2} & -Y'_m\left(k\sqrt{\varepsilon_2}r_1\right)\sqrt{\varepsilon_2} & & & \\ & J_m\left(k\sqrt{\varepsilon_2}r_2\right) & Y_m\left(k\sqrt{\varepsilon_2}r_2\right) & -J_m\left(k\sqrt{\varepsilon_3}r_2\right) & -Y_m\left(k\sqrt{\varepsilon_3}r_2\right) & \\ & J'_m\left(k\sqrt{\varepsilon_2}r_2\right)\sqrt{\varepsilon_2} & Y'_m\left(k\sqrt{\varepsilon_2}r_2\right)\sqrt{\varepsilon_2} & -J'_m\left(k\sqrt{\varepsilon_3}r_2\right)\sqrt{\varepsilon_3} & -Y'_m\left(k\sqrt{\varepsilon_3}r_2\right)\sqrt{\varepsilon_3} & \\ & \ddots & \ddots & \ddots & \ddots & \\ & J_m\left(k\sqrt{\varepsilon_j}r_j\right) & Y_m\left(k\sqrt{\varepsilon_j}r_j\right) & -J_m\left(k\sqrt{\varepsilon_{j+1}}r_j\right) & -Y_m\left(k\sqrt{\varepsilon_{j+1}}r_j\right) & \\ & J'_m\left(k\sqrt{\varepsilon_j}r_j\right)\sqrt{\varepsilon_j} & Y'_m\left(k\sqrt{\varepsilon_j}r_j\right)\sqrt{\varepsilon_j} & -J'_m\left(k\sqrt{\varepsilon_{j+1}}r_j\right)\sqrt{\varepsilon_{j+1}} & -Y'_m\left(k\sqrt{\varepsilon_{j+1}}r_j\right)\sqrt{\varepsilon_{j+1}} & \\ & \ddots & \ddots & \ddots & \ddots & \\ & J_m\left(k\sqrt{\varepsilon_{N-1}}r_{N-1}\right) & Y_m\left(k\sqrt{\varepsilon_{N-1}}r_{N-1}\right) & -J_m\left(k\sqrt{\varepsilon_N}r_{N-1}\right) & Y_m\left(k\sqrt{\varepsilon_N}r_{N-1}\right) & \\ & J'_m\left(k\sqrt{\varepsilon_{N-1}}r_{N-1}\right)\sqrt{\varepsilon_{N-1}} & Y'_m\left(k\sqrt{\varepsilon_{N-1}}r_{N-1}\right)\sqrt{\varepsilon_{N-1}} & -J'_m\left(k\sqrt{\varepsilon_N}r_{N-1}\right)\sqrt{\varepsilon_N} & -Y'_m\left(k\sqrt{\varepsilon_N}r_{N-1}\right)\sqrt{\varepsilon_N} & \\ & & & J_m\left(k\sqrt{\varepsilon_N}r_N\right) & Y_m\left(k\sqrt{\varepsilon_N}r_N\right) & -H_m^{(2)}\left(kr_N\right) \\ & & & J'_m\left(k\sqrt{\varepsilon_N}r_N\right)\sqrt{\varepsilon_N} & Y'_m\left(k\sqrt{\varepsilon_N}r_N\right)\sqrt{\varepsilon_N} & -H'^{(2)}_m\left(kr_N\right) \end{pmatrix}, \quad (3.130)$$

$$\begin{pmatrix} J_m\left(k\sqrt{\varepsilon_1}r_1\right) & -J_m\left(k\sqrt{\varepsilon_2}r_1\right) & -Y_m\left(k\sqrt{\varepsilon_2}r_1\right) & & & \\ J'_m\left(k\sqrt{\varepsilon_1}r_1\right)\frac{1}{\sqrt{\varepsilon_1}} & -J'_m\left(k\sqrt{\varepsilon_2}r_1\right)\frac{1}{\sqrt{\varepsilon_2}} & -Y'_m\left(k\sqrt{\varepsilon_2}r_1\right)\frac{1}{\sqrt{\varepsilon_2}} & & & \\ & J_m\left(k\sqrt{\varepsilon_2}r_2\right) & Y_m\left(k\sqrt{\varepsilon_2}r_2\right) & -J_m\left(k\sqrt{\varepsilon_3}r_2\right) & -Y_m\left(k\sqrt{\varepsilon_3}r_2\right) & \\ & J'_m\left(k\sqrt{\varepsilon_2}r_2\right)\frac{1}{\sqrt{\varepsilon_2}} & Y'_m\left(k\sqrt{\varepsilon_2}r_2\right)\frac{1}{\sqrt{\varepsilon_2}} & -J'_m\left(k\sqrt{\varepsilon_3}r_2\right)\frac{1}{\sqrt{\varepsilon_3}} & -Y'_m\left(k\sqrt{\varepsilon_3}r_2\right)\frac{1}{\sqrt{\varepsilon_3}} & \\ & \ddots & \ddots & \ddots & \ddots & \\ & J_m\left(k\sqrt{\varepsilon_j}r_j\right) & Y_m\left(k\sqrt{\varepsilon_j}r_j\right) & -J_m\left(k\sqrt{\varepsilon_{j+1}}r_j\right) & -Y_m\left(k\sqrt{\varepsilon_{j+1}}r_j\right) & \\ & J'_m\left(k\sqrt{\varepsilon_j}r_j\right)\frac{1}{\sqrt{\varepsilon_j}} & Y'_m\left(k\sqrt{\varepsilon_j}r_j\right)\frac{1}{\sqrt{\varepsilon_j}} & -J'_m\left(k\sqrt{\varepsilon_{j+1}}r_j\right)\frac{1}{\sqrt{\varepsilon_{j+1}}} & -Y'_m\left(k\sqrt{\varepsilon_{j+1}}r_j\right)\frac{1}{\sqrt{\varepsilon_{j+1}}} & \\ & \ddots & \ddots & \ddots & \ddots & \\ & J_m\left(k\sqrt{\varepsilon_{N-1}}r_{N-1}\right) & Y_m\left(k\sqrt{\varepsilon_{N-1}}r_{N-1}\right) & -J_m\left(k\sqrt{\varepsilon_N}r_{N-1}\right) & Y_m\left(k\sqrt{\varepsilon_N}r_{N-1}\right) & \\ & J'_m\left(k\sqrt{\varepsilon_{N-1}}r_{N-1}\right)\frac{1}{\sqrt{\varepsilon_{N-1}}} & Y'_m\left(k\sqrt{\varepsilon_{N-1}}r_{N-1}\right)\frac{1}{\sqrt{\varepsilon_{N-1}}} & -J'_m\left(k\sqrt{\varepsilon_N}r_{N-1}\right)\frac{1}{\sqrt{\varepsilon_N}} & -Y'_m\left(k\sqrt{\varepsilon_N}r_{N-1}\right)\frac{1}{\sqrt{\varepsilon_N}} & \\ & & & J_m\left(k\sqrt{\varepsilon_N}r_N\right) & Y_m\left(k\sqrt{\varepsilon_N}r_N\right) & -H_m^{(2)}\left(kr_N\right) \\ & & & J'_m\left(k\sqrt{\varepsilon_N}r_N\right)\frac{1}{\sqrt{\varepsilon_N}} & Y'_m\left(k\sqrt{\varepsilon_N}r_N\right)\frac{1}{\sqrt{\varepsilon_N}} & -H'^{(2)}_m\left(kr_N\right) \end{pmatrix} \quad (3.133)$$

$$\begin{pmatrix} a_{11} & a_{12} & a_{13} & & & & & 0 \\ a_{21} & a_{22} & a_{23} & & & & & \\ & a_{32} & a_{33} & a_{34} & a_{35} & & & \\ & a_{42} & a_{43} & a_{44} & a_{45} & & & \\ & & \ddots & \ddots & \ddots & & & \\ & & & a_{2N-3,2N-4} & a_{2N-3,2N-3} & a_{2N-3,2N-2} & a_{2N-3,2N-1} & \\ & & & a_{2N-2,2N-4} & a_{2N-1,2N-3} & a_{2N-2,2N-2} & a_{2N-2,2N-1} & \\ & & & & & a_{2N-1,2N-2} & a_{2N-1,2N-1} & a_{2N-1,N} \\ 0 & & & & & a_{2N,2N-2} & a_{2N,2N-1} & a_{2N,2N} \end{pmatrix} \times$$

$$\times \begin{pmatrix} c_1 \\ c_2 \\ c_3 \\ c_4 \\ \vdots \\ c_{2N-3} \\ c_{2N-2} \\ c_{2N-1} \\ c_{2N} \end{pmatrix} = \begin{pmatrix} 0 \\ 0 \\ 0 \\ 0 \\ \vdots \\ 0 \\ 0 \\ b_{2N-1} \\ b_{2N} \end{pmatrix}, \tag{3.134}$$

where c_i, $i = \overline{1,2N}$ are the unknown expansion coefficients in a series of cylindrical functions.

Direct sweep eliminates items that are below the main diagonal of the matrix. As a direct result the system (3.134) becomes:

$$\begin{pmatrix} a_{11} & a_{12} & a_{13} & & & & & & 0 \\ 0 & \tilde{a}_{22} & \tilde{a}_{23} & & & & & & \\ & 0 & \tilde{a}_{33} & a_{34} & a_{35} & & & & \\ & 0 & 0 & \tilde{a}_{44} & \tilde{a}_{45} & & & & \\ & & \ddots & \ddots & \ddots & & & & \\ & & & 0 & \tilde{a}_{2N-3,2N-3} & a_{2N-3,2N-2} & a_{2N-3,2N-1} & & \\ & & & 0 & 0 & \tilde{a}_{2N-2,2N-2} & \tilde{a}_{2N-2,2N-1} & & \\ & & & & & 0 & \tilde{a}_{2N-1,2N-1} & a_{2N-1,N} & \\ 0 & & & & & 0 & 0 & \tilde{a}_{2N,2N} \end{pmatrix} \times \tag{3.135}$$

$$\times \begin{pmatrix} c_1 \\ c_2 \\ c_3 \\ c_4 \\ \vdots \\ c_{2N-3} \\ c_{2N-2} \\ c_{2N-1} \\ c_{2N} \end{pmatrix} = \begin{pmatrix} 0 \\ 0 \\ 0 \\ 0 \\ \vdots \\ 0 \\ 0 \\ b_{2N-1} \\ \tilde{b}_{2N} \end{pmatrix}.$$

The matrix elements (3.135) are marked with a tilde, are related to the initial elements of the matrix (3.134) by the following relations:

$$\begin{aligned} \tilde{a}_{22} &= a_{22} - a_{21}\frac{a_{12}}{a_{11}}, \\ \tilde{a}_{23} &= a_{23} - a_{21}\frac{a_{13}}{a_{11}}, \\ \tilde{a}_{2i-1,2i-1} &= a_{2i-1,2i-1} - a_{2i-1,2(i-1)}\frac{\tilde{a}_{2(i-1),2i-1}}{\tilde{a}_{2(i-1),2(i-1)}}, \\ \tilde{a}_{2i,2i} &= a_{2i,2i} - \left(a_{2i,2i-1} - a_{2i,2(i-1)}\frac{\tilde{a}_{2(i-1),2i-1}}{\tilde{a}_{2(i-1),2(i-1)}}\right)\frac{a_{2i-1,2i}}{\tilde{a}_{2i-1,2i-1}}, \\ \tilde{a}_{2i,2i+1} &= a_{2i,2i+1} - \left(a_{2i,2i-1} - a_{2i,2(i-1)}\frac{\tilde{a}_{2(i-1),2i-1}}{\tilde{a}_{2(i-1),2(i-1)}}\right)\frac{a_{2i-1,2i+1}}{\tilde{a}_{2i-1,2i-1}}. \end{aligned} \tag{3.136}$$

Here $i = \overline{2,N}$, with the exception of the last of equations (3.136), where $i \neq N$

$$\tilde{b}_{2N} = b_{2N} - \left(a_{2N,2N-1} - a_{2N,2N-2} \frac{\tilde{a}_{2(N-1),2N-1}}{\tilde{a}_{2(N-1),2(N-1)}} \right) \frac{b_{2N-1}}{\tilde{a}_{2N-1,2N-1}}.$$

With the reverse course it is possible to obtain an expression for solving systems of linear equations:

$$c_{2N} = \frac{\tilde{b}_{2N}}{\tilde{a}_{2N,2N}},$$

$$c_{2N-1} = \left(b_{2N-1} - a_{2N-1,2N} c_{2N}\right) \frac{1}{\tilde{a}_{2N-1,2N-1}},$$

$$c_{2(N-i)} = -c_{2(N-i)+1} \frac{\tilde{a}_{2(N-i),2(N-i)+1}}{\tilde{a}_{2(N-i),2(N-i)}},$$

$$c_{2(N-i)-1} = -\left(a_{2(N-i)-1,2(N-i)} c_{2(N-i)} + a_{2(N-i)-1,2(N-i)+1} c_{2(N-i)+1}\right) \frac{1}{\tilde{a}_{2(N-i)-1,2(N-i)-1}}. \tag{3.137}$$

Here $i = \overline{1, N-1}$, and in the latter of equations (3.137) at $i = N - 1$:

$$\tilde{a}_{2(N-i)-1,2(N-i)-1} = a_{11}.$$

Raleigh series in ascending integer indices represent the diffracted field in the form of an infinite set of multipoles, and in this set the number of important terms increases with the increase of the ratio of the transverse dimension of the body to the wavelength. The smaller $k\sqrt{\varepsilon_i} r_i$, the more rapidly series converges. Accordingly, the diffraction field for small $k\sqrt{\varepsilon_i} r_i$ has a relatively simple form and becomes more complicated with the increase of $k\sqrt{\varepsilon_i} r_i$. The rapid decrease of the terms of the series, starting with the numbers $m \sim k\sqrt{\varepsilon_i} r_i$, enables us to ignore the remainder of the series, even for large values of $k\sqrt{\varepsilon_i} r_i$.

However, for very large values of $k\sqrt{\varepsilon_i} r_i$ the summation is difficult because of the very long computing time. The transition to the Watson series is effective [1].

3.3.2. The analytical solution for a two-layer cylinder

Study [1] gives an analytic solution of the two-dimensional problem of diffraction of a plane electromagnetic monochromatic wave of TE-/TM-polarization on a homogeneous dielectric circular cylinder. Using the method proposed in section 3.3.1 we obtain an analytic solution of the problem of diffraction of an arbitrary electromagnetic wave on a two-layer dielectric circular cylinder, where the layers are represented in the form of a rod and a shell (Fig. 3.26).

The system of equations (3.113)–(3.115) reduces to:

$$\psi_{11}=\sum_{m=-\infty}^{+\infty}C_{1m}J_m\left(k\sqrt{\varepsilon_1}r\right)\cos m\varphi,\ 0\le r\le r_1;$$

$$\psi_{12}=\sum_{m=-\infty}^{+\infty}\left[C_{2m}J_m\left(k\sqrt{\varepsilon_2}r\right)+C_{3m}Y_m\left(k\sqrt{\varepsilon_2}r\right)\right]\cos m\varphi,\ r_1<r\le r_2; \quad (3.138)$$

$$\psi_2=\psi_0+\sum_{m=-\infty}^{+\infty}C_{4m}H_m^{(2)}(kr)\cos m\varphi,\ r>r_2.$$

The problem is reduced to finding the unknown coefficients C_{1m}, C_{2m}, C_{3m}, C_{4m}. Further, similarly to the case of the N-layered cylinder, as described in Section 3.3.1, we obtain a system of four linear algebraic equations with four unknowns. It is not difficult to solve analytically such a system, so here are only the results are presented.

TE-polarization

$$C_{1m}=\frac{(-i)^m A_m\sqrt{\varepsilon_2}\left(J_m\left(k\sqrt{\varepsilon_2}r_1\right)Y_m'\left(k\sqrt{\varepsilon_2}r_1\right)-J_m'\left(k\sqrt{\varepsilon_2}r_1\right)Y_m\left(k\sqrt{\varepsilon_2}r_1\right)\right)}{\Delta}\cdot$$
$$\cdot\frac{\left(H_m'^{(2)}(kr_2)J_m(kr_2)-H_m^{(2)}(kr_2)J_m'(kr_2)\right)}{\Delta},$$

$$C_{2m}=\frac{(-i)^m A_m\left(Y_m\left(k\sqrt{\varepsilon_2}r_1\right)J_m'\left(k\sqrt{\varepsilon_1}r_1\right)\sqrt{\varepsilon_1}-Y_m'\left(k\sqrt{\varepsilon_2}r_1\right)\sqrt{\varepsilon_2}J_m\left(k\sqrt{\varepsilon_1}r_1\right)\right)}{\Delta}\cdot$$
$$\cdot\frac{\left(H_m^{(2)}(kr_2)J_m'(kr_2)-H_m'^{(2)}(kr_2)J_m(kr_2)\right)}{\Delta},$$

$$C_{3m}=\frac{(-i)^m A_m\left(J_m'\left(k\sqrt{\varepsilon_2}r_1\right)\sqrt{\varepsilon_2}J_m\left(k\sqrt{\varepsilon_1}r_1\right)-J_m\left(k\sqrt{\varepsilon_2}r_1\right)J_m'\left(k\sqrt{\varepsilon_1}r_1\right)\sqrt{\varepsilon_1}\right)}{\Delta}\cdot$$
$$\cdot\frac{\left(H_m^{(2)}(kr_2)J_m'(kr_2)-H_m'^{(2)}(kr_2)J_m(kr_2)\right)}{\Delta},$$

$$C_{4m}=(-i)^m A_m\Big[\Big(Y_m\big(k\sqrt{\varepsilon_2}r_1\big)J'_m\big(k\sqrt{\varepsilon_1}r_1\big)\sqrt{\varepsilon_1}-Y'_m\big(k\sqrt{\varepsilon_2}r_1\big)\sqrt{\varepsilon_2}J_m\big(k\sqrt{\varepsilon_1}r_1\big)\Big)\cdot$$
$$\cdot\Big(J_m\big(k\sqrt{\varepsilon_2}r_2\big)J'_m(kr_2)-J'_m\big(k\sqrt{\varepsilon_2}r_2\big)\sqrt{\varepsilon_2}J_m(kr_2)\Big)+$$
$$+\Big(J'_m\big(k\sqrt{\varepsilon_2}r_1\big)\sqrt{\varepsilon_2}J_m\big(k\sqrt{\varepsilon_1}r_1\big)-J_m\big(k\sqrt{\varepsilon_2}r_1\big)J'_m\big(k\sqrt{\varepsilon_1}r_1\big)\sqrt{\varepsilon_1}\Big)\Big]\cdot$$
$$\cdot\Big(Y_m\big(k\sqrt{\varepsilon_2}r_2\big)J'_m(kr_2)-Y'_m\big(k\sqrt{\varepsilon_2}r_2\big)\sqrt{\varepsilon_2}J_m(kr_2)\Big)\Big]\Big/\Delta$$

Here

$$\Delta=\Big(Y_m\big(k\sqrt{\varepsilon_2}r_1\big)J'_m\big(k\sqrt{\varepsilon_1}r_1\big)\sqrt{\varepsilon_1}-Y'_m\big(k\sqrt{\varepsilon_2}r_1\big)\sqrt{\varepsilon_2}J_m\big(k\sqrt{\varepsilon_1}r_1\big)\Big)\cdot$$
$$\Big(J'_m\big(k\sqrt{\varepsilon_2}r_2\big)\sqrt{\varepsilon_2}H_m^{(2)}(kr_2)-J_m\big(k\sqrt{\varepsilon_2}r_2\big)H_m^{\prime(2)}(kr_2)\Big)+$$
$$+\Big(J'_m\big(k\sqrt{\varepsilon_2}r_1\big)\sqrt{\varepsilon_2}J_m\big(k\sqrt{\varepsilon_1}r_1\big)-J_m\big(k\sqrt{\varepsilon_2}r_1\big)J'_m\big(k\sqrt{\varepsilon_1}r_1\big)\sqrt{\varepsilon_1}\Big)\cdot$$
$$\cdot\Big(Y'_m\big(k\sqrt{\varepsilon_2}r_2\big)\sqrt{\varepsilon_2}H_m^{(2)}(kr_2)-Y_m\big(k\sqrt{\varepsilon_2}r_2\big)H_m^{\prime(2)}(kr_2)\Big)$$

The calculated coefficients are substituted into (3.138).

TM-polarization

$$C_{1m}=\frac{(-i)^m A_m\frac{1}{\sqrt{\varepsilon_2}}\Big(J_m\big(k\sqrt{\varepsilon_2}r_1\big)Y'_m\big(k\sqrt{\varepsilon_2}r_1\big)-J'_m\big(k\sqrt{\varepsilon_2}r_1\big)Y_m\big(k\sqrt{\varepsilon_2}r_1\big)\Big)}{\Delta}\cdot$$
$$\cdot\frac{\Big(H_m^{\prime(2)}(kr_2)J_m(kr_2)-H_m^{(2)}(kr_2)J'_m(kr_2)\Big)}{\Delta},$$

$$C_{2m}=\frac{(-i)^m A_m\left(Y_m\big(k\sqrt{\varepsilon_2}r_1\big)J'_m\big(k\sqrt{\varepsilon_1}r_1\big)\frac{1}{\sqrt{\varepsilon_1}}-Y'_m\big(k\sqrt{\varepsilon_2}r_1\big)J_m\big(k\sqrt{\varepsilon_1}r_1\big)\frac{1}{\sqrt{\varepsilon_2}}\right)}{\Delta}\cdot$$
$$\cdot\frac{\Big(H_m^{(2)}(kr_2)J'_m(kr_2)-H_m^{\prime(2)}(kr_2)J_m(kr_2)\Big)}{\Delta},$$

$$C_{3m}=\frac{(-i)^m A_m\left(J'_m\big(k\sqrt{\varepsilon_2}r_1\big)J_m\big(k\sqrt{\varepsilon_1}r_1\big)\frac{1}{\sqrt{\varepsilon_2}}-J_m\big(k\sqrt{\varepsilon_2}r_1\big)J'_m\big(k\sqrt{\varepsilon_1}r_1\big)\frac{1}{\sqrt{\varepsilon_1}}\right)}{\Delta}\cdot$$
$$\cdot\frac{\Big(H_m^{(2)}(kr_2)J'_m(kr_2)-H_m^{\prime(2)}(kr_2)J_m(kr_2)\Big)}{\Delta},$$

$$C_{4m} = (-i)^m A_m \left[\left(Y_m\left(k\sqrt{\varepsilon_2} r_1\right) J'_m\left(k\sqrt{\varepsilon_1} r_1\right) \frac{1}{\sqrt{\varepsilon_1}} - Y'_m\left(k\sqrt{\varepsilon_2} r_1\right) J_m\left(k\sqrt{\varepsilon_1} r_1\right) \frac{1}{\sqrt{\varepsilon_2}} \right) \cdot \right.$$

$$\cdot \left(J_m\left(k\sqrt{\varepsilon_2} r_2\right) J'_m\left(kr_2\right) - J'_m\left(k\sqrt{\varepsilon_2} r_2\right) J_m\left(kr_2\right) \frac{1}{\sqrt{\varepsilon_2}} \right) +$$

$$\left. + \left(J'_m\left(k\sqrt{\varepsilon_2} r_1\right) J_m\left(k\sqrt{\varepsilon_1} r_1\right) \frac{1}{\sqrt{\varepsilon_2}} - J_m\left(k\sqrt{\varepsilon_2} r_1\right) J'_m\left(k\sqrt{\varepsilon_1} r_1\right) \frac{1}{\sqrt{\varepsilon_1}} \right) \right] \cdot$$

$$\left. \cdot \left(Y_m\left(k\sqrt{\varepsilon_2} r_2\right) J'_m\left(kr_2\right) - Y'_m\left(k\sqrt{\varepsilon_2} r_2\right) J_m\left(kr_2\right) \frac{1}{\sqrt{\varepsilon_2}} \right) \right] \Big/ \Delta$$

Here

$$\Delta = \left(Y_m\left(k\sqrt{\varepsilon_2} r_1\right) Y'_m\left(k\sqrt{\varepsilon_1} r_1\right) \frac{1}{\sqrt{\varepsilon_1}} - J'_m\left(k\sqrt{\varepsilon_2} r_1\right) J_m\left(k\sqrt{\varepsilon_1} r_1\right) \frac{1}{\sqrt{\varepsilon_2}} \right) \cdot$$

$$\left(J'_m\left(k\sqrt{\varepsilon_2} r_2\right) H_m^{(2)}\left(kr_2\right) \frac{1}{\sqrt{\varepsilon_2}} - J_m\left(k\sqrt{\varepsilon_2} r_2\right) H'^{(2)}_m\left(kr_2\right) \right) +$$

$$+ \left(J'_m\left(k\sqrt{\varepsilon_2} r_1\right) J_m\left(k\sqrt{\varepsilon_1} r_1\right) \frac{1}{\sqrt{\varepsilon_2}} - J_m\left(k\sqrt{\varepsilon_2} r_1\right) J'_m\left(k\sqrt{\varepsilon_1} r_1\right) \frac{1}{\sqrt{\varepsilon_1}} \right) \cdot$$

$$\cdot \left(Y'_m\left(k\sqrt{\varepsilon_2} r_2\right) H_m^{(2)}\left(kr_2\right) \frac{1}{\sqrt{\varepsilon_2}} - Y_m\left(k\sqrt{\varepsilon_2} r_2\right) H'^{(2)}_m\left(kr_2\right) \right)$$

The calculated coefficients are substituted into (3.138).

In the equations for the coefficients $C_{km}, k = \overline{1,4}$, the constants A_m are found from (3.127) and characterize the incident light wave.

3.3.3. Diffraction on a gradient microlens
Diffraction of electromagnetic waves on the internal Luneberg lens

Consider the diffraction of a plane TE-polarized electromagnetic wave with a wavelength $\lambda = 1$ μm on a dielectric cylinder of radius $R = 1$ μm, whose refractive index depends on the radius as follows (the internal Luneberg lens) [68]:

$$n^2(r) = \frac{1 + r_1^2 - r^2}{r_1^2}, \; r_1 \le 1, \tag{3.139}$$

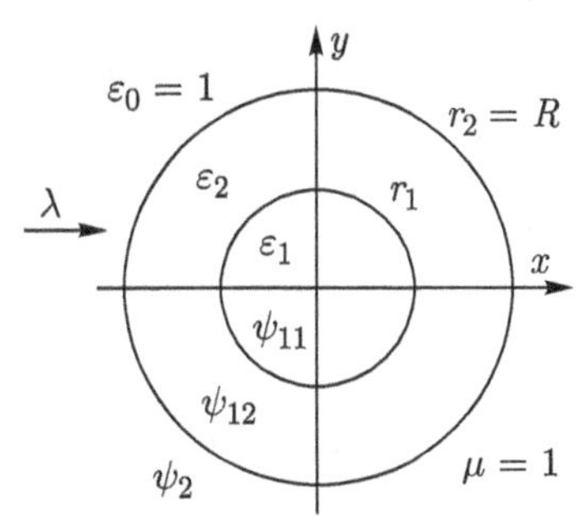

Fig. 3.26. The two-layer dielectric cylinder.

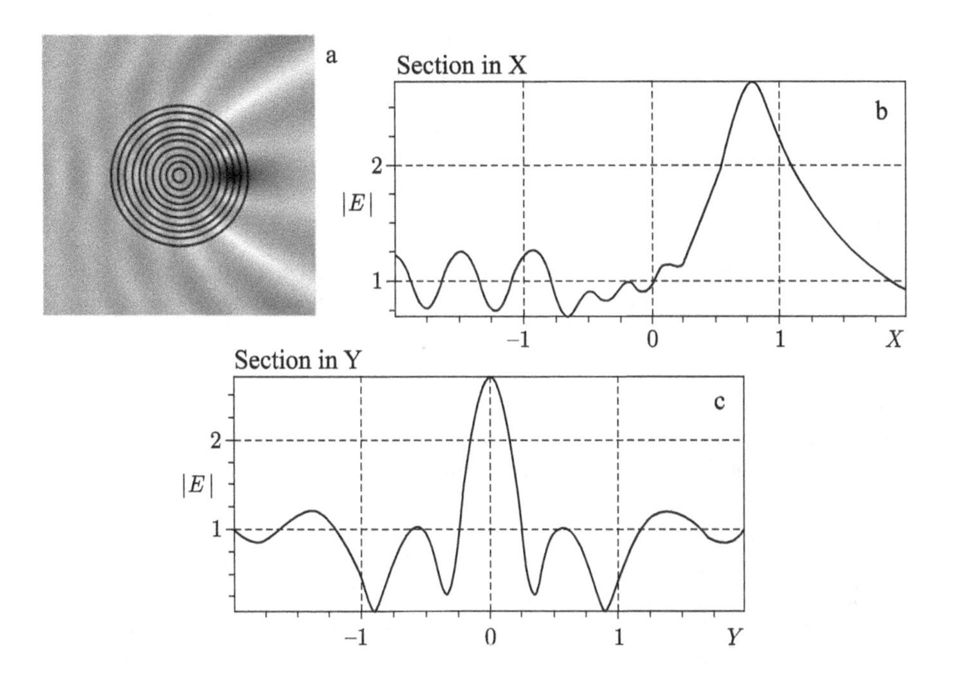

Fig. 3.27. Distribution of electric field amplitude: a) 2D distribution, b) the section on the X axis (horizontal axis) through the focus, c) cross-section on the Y axis (vertical axis) through the focus.

where r_1 is the distance from the centre of the cylinder to the point of the geometrical focus. As an example, consider $r_1 = 0.75$ μm. We select the number of layers of th ecylinder equal to 10 and the maximum order of approximating Bessel functions in the series (3.113)–(3.115) as 20.

The amplitude distribution of the projection of the vector of the strength of electric field $E_z(x, y)$ is shown in Fig. 3.27.

The total size of the diffraction pattern is 4×4 μm (Fig. 3.27a). The number of counts of the sampling grid was 300×300 pixels. The value of the focal length r_1, calculated using the above described method, is $r_1 \approx 0.787$ μm. The relative error

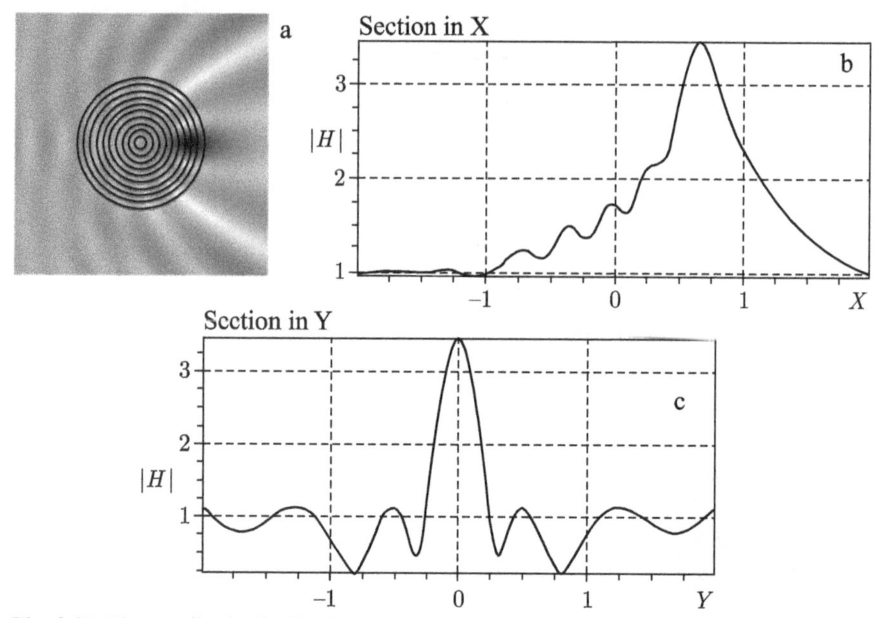

Fig. 3.28. The amplitude distribution of the magnetic field: a) 2D distribution, b) the section on the *X* axis (horizontal axis) through the focus, c) cross-section in *Y* axis (vertical axis) through the focus.

in comparison with the specified value r_1 is ≈4.9%. It should be noted that the calculated value r_1 is compared with the value obtained for ray approximation.

Consider the problem with the same parameters, but for the case of a TM-polarized incident plane wave.

The amplitude distribution of the magnetic field is shown in Fig.3.28.

The value of the focal length r_1 was $r_1 \approx 0.653$ μm. The relative deviation of the calculated focal length from the given value of r_1 is ≈12.9%.

The study of diffraction of light on the Luneberg lens, whose dimensions are comparable with the incident electromagnetic wave, with the developed method was carried out in a series of numerical experiments. The main task was to check how many layers of the lens will be sufficient to obtain a stable value of the focal length, and compare it with the prescribed value, which was used for calculattions by the beam approximation of the refractive index of the lens.

The following parameters of the diffraction pattern were chosen: the size 4×4 μm, number of samples 400×400 pixels, the outer radius of the cylinder 1 μm, the prescribed focal length 0.5 μm.

Based on the results it was concluded that for the wavelengths comparable to the size of the Luneberg lens about 30–40 layers, approximating the lens, are sufficient. Moreover, the deviation of the calculated focal distance from the specified value of the focal length, obtained using the ray approximation, is no more than 10%.

It was also interesting to know the magnitude of the intensity at the focus of the Luneberg lens, and how it changes depending on the number of approximating

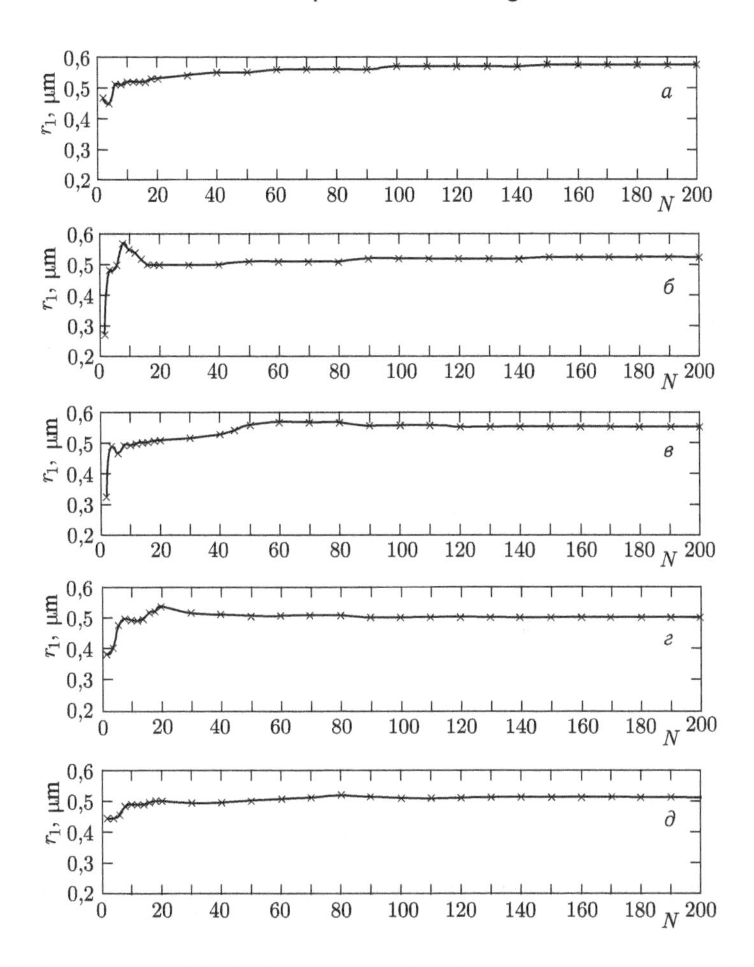

Fig. 3.29. Dependence of the focal length of the Luneberg lens on the number of layers of the cylinder at different wavelengths of the incident wave: a) $\lambda = 1\ \mu m$; b) $\lambda = 0.8\ \mu m$, c) $\lambda = 0.6\ \mu m$, d) $\lambda = 0.4\ \mu m$, e) $\lambda = 0.2\ \mu m$.

segments. The dependences obtained for fixed values of the wavelength of the incident wave are shown in Fig. 3.30.

Based on the numerical results, it is concluded that the developed method provides a stable solution for a given wavelength. It should be noted that with decreasing wavelength the intensity at the focus increases. This is because decreasing wavelength the diffraction effects become weaker, and the light concentrates more and more at the focus. It can be assumed that with a further decrease in wavelength, when the ray approximation holds, the intensity of the focus will tend to 200, and this is the number of pixels placed on the diameter of the circle, and each pixel corresponds to a beam which should theoretically pass through the focus of the Luneberg lens.

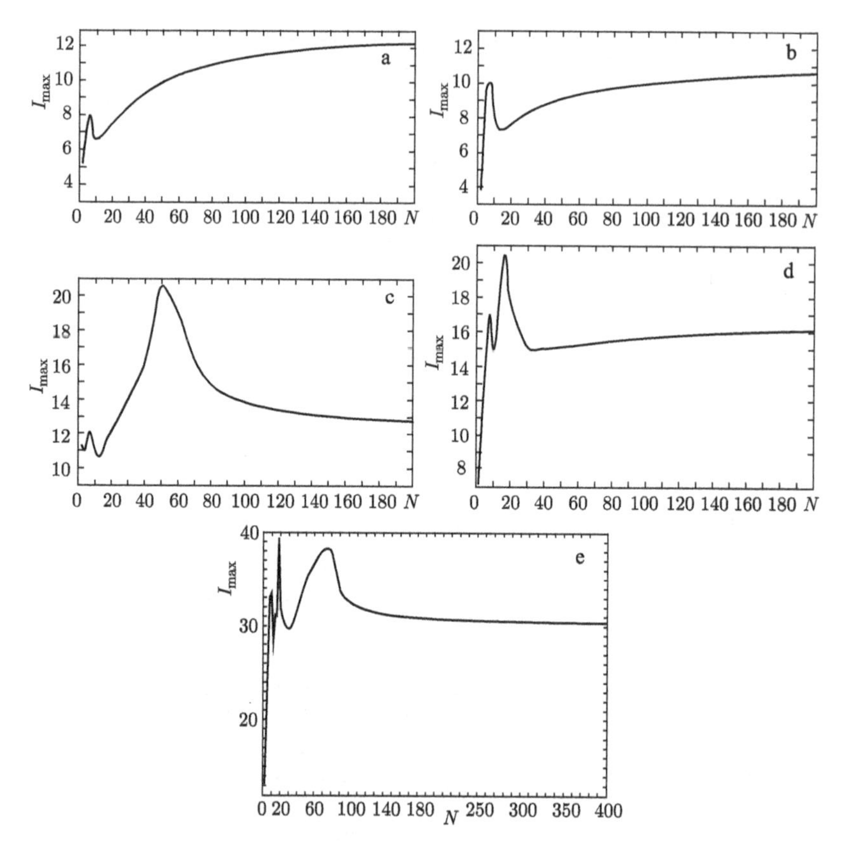

Fig. 3.30. Dependence of the intensity at the focus of the Luneberg lens on the number of layers of the cylinder at different wavelengths of the incident wave: a) $\lambda = 1$ μm; b) $\lambda = 0.8$ μm, c) $\lambda = 0.6$ μm, d) $\lambda = 0.4$ μm, e) $\lambda = 0.2$ μm.

Diffraction of electromagnetic waves on a generalized Luneberg lens

Consider the case where $r_1 > 1$. The refractive index of the generalized Luneberg lens is written as [68]:

$$n(r) = \exp\left[\frac{1}{\pi}\int_{\rho}^{1}\frac{\arcsin(h/r_1)\,dh}{\sqrt{h^2-\rho^2}}\right],\quad \rho = n(r)r,\quad 0<\rho<1. \tag{3.140}$$

The transcendental equation (3.140) was solved numerically with respect to $n(r)$, since the integral in (3.140) is not taken in elementary functions. Let us assume that a plane TE-polarized electromagnetic wave with a wavelength of $\lambda = 0.2$ μm falls on a dielectric cylinder. The focal length is chosen equal to $r_1 = 2.55$ μm, the radius of the lens is $R = 1$ μm. We define the number of layers of the cylinder equal to 100 and the maximum order of the approximating cylinder functions as 35.

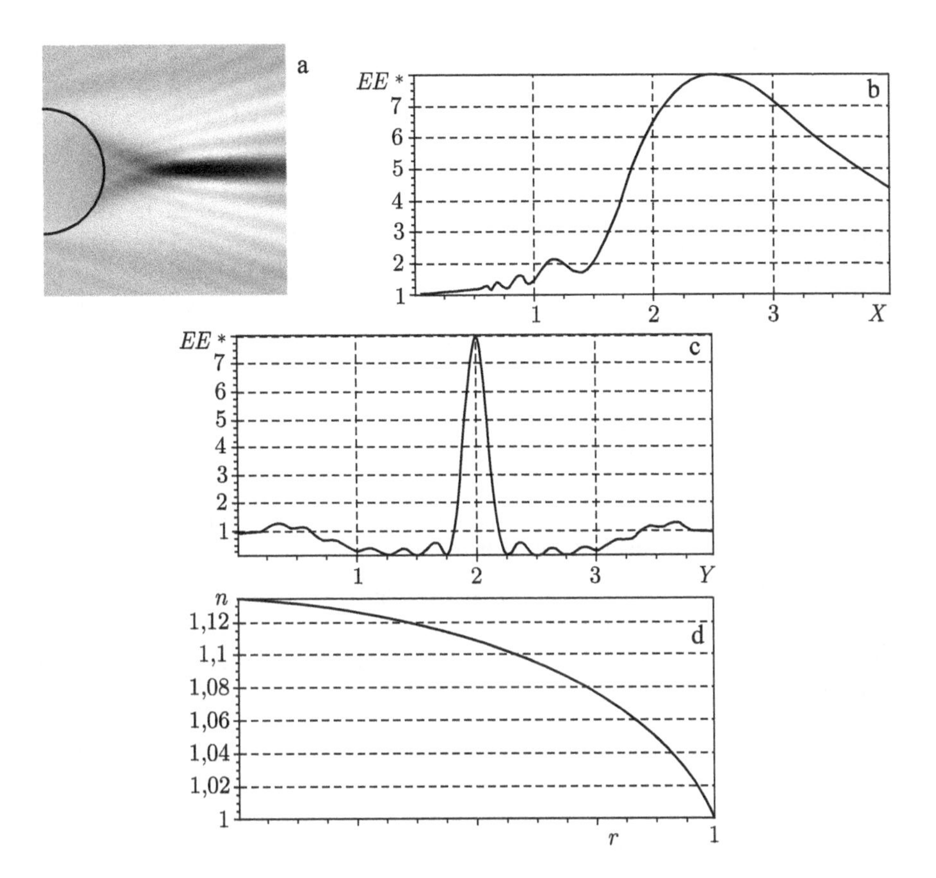

Fig. 3.31. The distribution of the intensity of the electric field: a) 2D distribution, b) the section on the X axis (horizontal axis) through the focus, c) cross-section on the Y axis (vertical axis) through the focus, d) the dependence of the refractive index of the lens on the radial coordinate.

The distribution of the intensity of the electric field is shown in Fig. 3.31:

The total size of the diffraction pattern 4×4 mm. The number of pixels on the sampling grid 400×400. The value of the focal length r_1, calculated using the above described method, was $r_1 \approx 2.48$ μm. The relative error in comparison with the specified value r_1 was equal to ≈3%.

Consider the same case, but for a TM-polarized plane electromagnetic wave.

The intensity distribution of the magnetic field is shown in Fig.3.32.

The total size of the diffraction pattern was 4×4 mm. The number of pixels on the sampling grid 400×400. The value of the focal length r_1, calculated using the above described method, was $r_1 \approx 2.5$ μm. The relative error in comparison with the specified value r_1 was equal to ≈2%.

In all these cases the dependence of the refractive index of the radius decreases monotonically from the centre of the circle to the surface (at a distance of 1 μm from the centre of the refractive index is equal to 1). In the case of the generalized

Luneberg lens (3.140), the refractive index at the centre of the circle is $n \approx 1.134$ and $n \approx 2.236$ for the inner Luneberg lens. Therefore, in Figs. 3.27, 3.28, 3.31 and 3.32 there is almost no Fresnel reflection from the considered gradient Luneberg lenses. Note that for the TM-polarization (Fig. 3.28) back reflection is much less than in the case of TE-polarization (Fig. 3.27).

One may ask: how many layers of a circular cylinder should be used in order to approximate with sufficient accuracy the diffraction field obtained from the generalized Luneberg lens? Below are the results of calculation of the diffraction of a plane TE-polarized electromagnetic wave on circular cylinders approximating the generalized Luneberg lens and consisting of 2, 5 and 10 layers, respectively. Parameters of the problem:

- the size of the diffraction pattern 4×4 μm (400×400 pixels)
- the length of the incident wave is 0.2 μm;
- the outer radius of the cylinder 1 μm;
- the maximum order of the Bessel functions in series of approximations 35;
- focal length, defined on the basis of the ray approximation, 2 μm.

Figure 3.35 shows that even at 10 layers, which approximate the dependence of the refractive index on the radius for the generalized Luneberg lens (3.140), the calculated focus coincides with the geometrical focus. Figure 3.33 shows that although the calculated focal length for the two-layer Luneberg lens is greater than the geometric length, the focusing properties of the lenses are not affected.

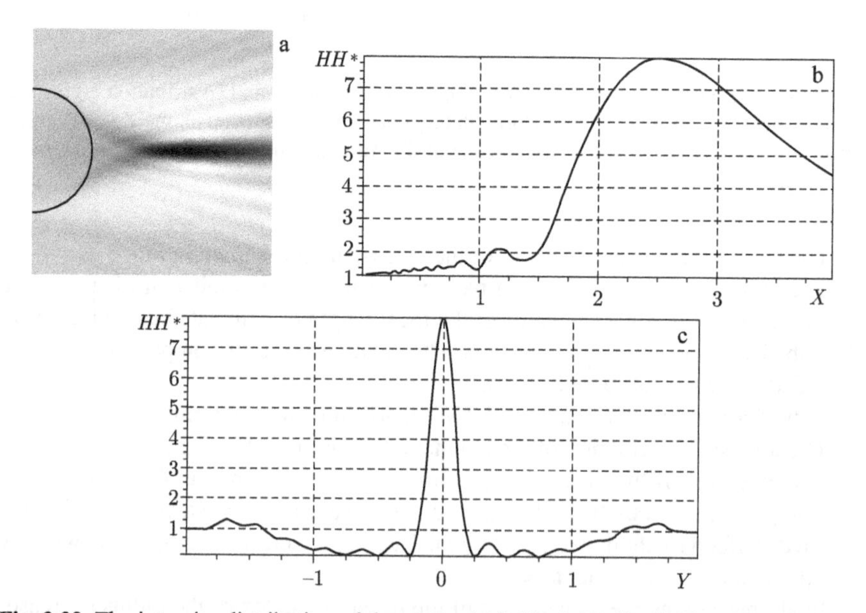

Fig. 3.32. The intensity distribution of the magnetic field: a) 2D distribution, b) the section on the X-axis through the focus, c) cross-section of the Y-axis through the focus.

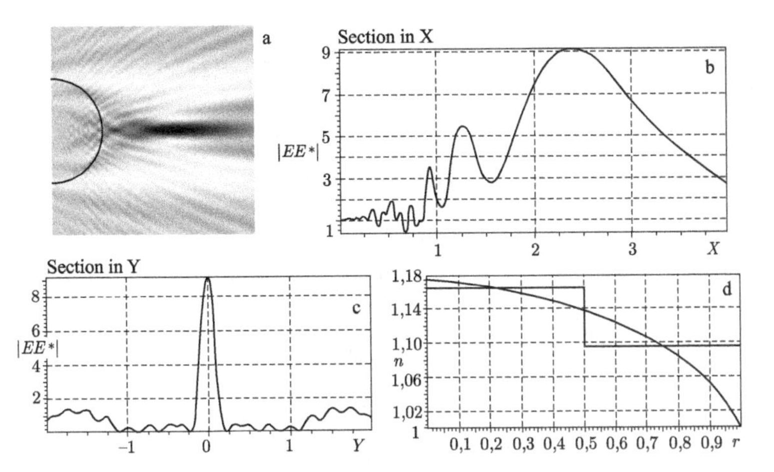

Fig. 3.33. Diffraction on a two-layer cylinder (approximation of generalized Luneberg lens): a) 2D-distribution of the electric field, b) distribution of the electric field along the section on the X-axis, c) the intensity distribution of the electric field along the cross section on the Y-axis; d) the dependence of the refractive index of the lens on the radial coordinate.

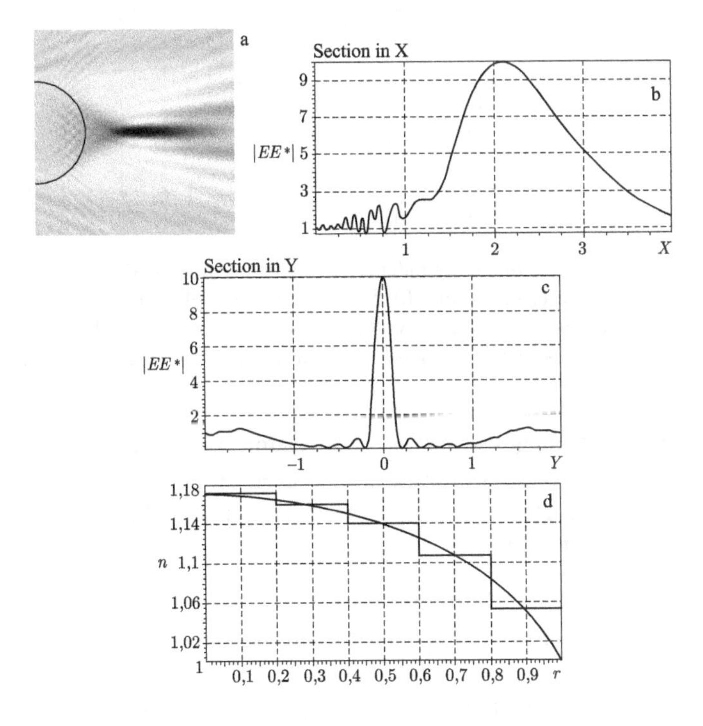

Fig. 3.34. Diffraction on a 5-layer cylinder (approximation of generalized Luneberg lens): a) 2D-distribution of the electric field, b) distribution of the electric field intensity along the section of the X-axis, c) the intensity distribution of the electric field along the cross section on the Y-axis; d) the dependence of the refractive index of the lens on the radial coordinate.

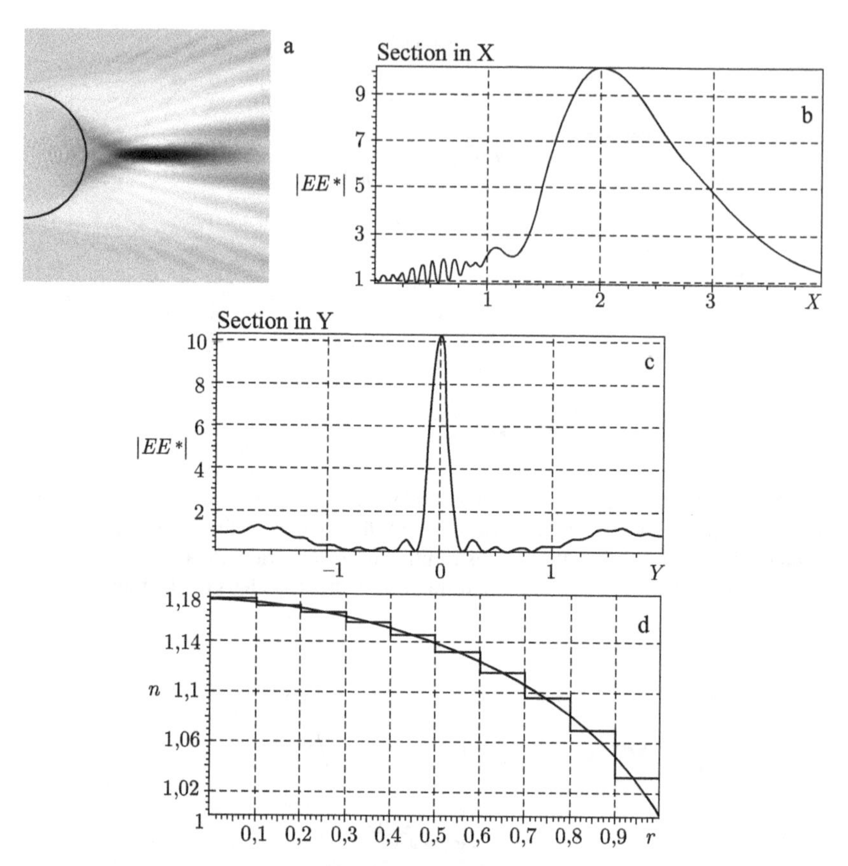

Fig. 3.35. Diffraction of a 10-layer cylinder (approximation of generalized Luneberg lens): a) 2D-distribution of the electric field, b) distribution of the electric field intensity along the section of the *X*-axis, c) the intensity distribution of the electric field intensity along the cross section on the *Y*-axis; d) the dependence of the refractive index of the lens on the radial coordinate.

Diffraction of electromagnetic waves on the Eaton–Lippmann lens

Consider, as another example, diffraction of a plane TE-polarized electromagnetic wave with the wavelength $\lambda = 1$ μm on a dielectric cylinder of unit radius with the refractive index (Eaton–Lippmann lens) [73]:

$$n(r) = n_0 \sqrt{\left(\frac{2R}{r} - 1\right)}. \tag{3.141}$$

The amplitude distribution of the electric field intensity is shown in Fig. 3.36.

Here $n_0 = 2$, $R = 1$ μm. The total size of the diffraction pattern is 5×5 μm. The number of samples of the sampling grid 300×300. The number of layers of the cylinder is 30, and the maximum order of the approximating cylinder functions is 20. Figure 3.36 shows that the lens has a reflective effect, i.e. almost all the light

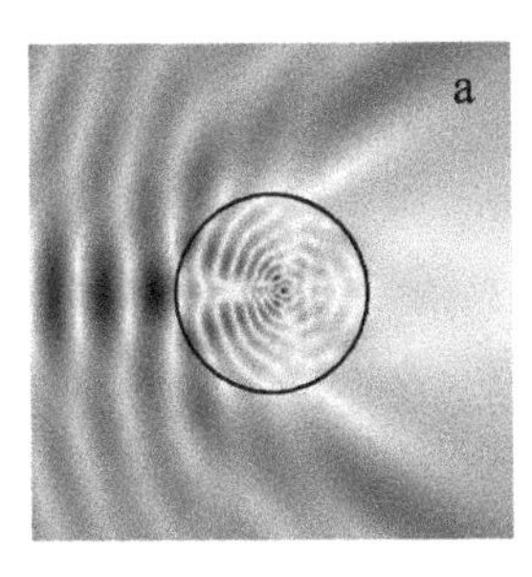

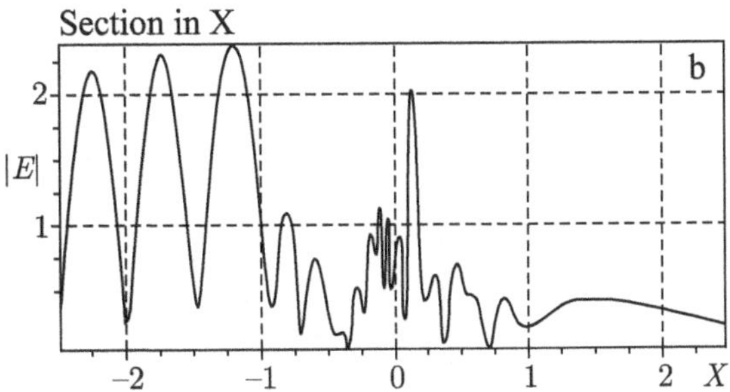

Fig. 3.36. The amplitude distribution of the electric field: a) 2D distribution, b) the section on the X-axis (horizontal axis).

falling on the lens is reflected back. Note that from (3.141) for the refractive index of the Eaton–Lippmann lens it follows that at $r = 0$ that the refractive index tends to infinity. In the example shown in Fig. 3.36 the task parameters are chosen so that in the centre of the lens at $r = 0$ the refractive index is equal to 14.

3.4. Fast iterative method for calculating the diffraction field of a monochromatic electromagnetic wave on a dielectric cylinder

3.4.1. An iterative method for calculating the diffraction of TE-polarized wave

For two-dimensional problem of diffraction of a monochromatic electromagnetic wave with the TE-polarized field components satisfy the Maxwell equations $(\partial/\partial x = 0)$

$$\begin{aligned} &\frac{\partial E_x}{\partial y} - ik_0\mu H_z = 0, \\ &-\frac{\partial E_x}{\partial z} - ik_0\mu H_y = 0, \\ &\frac{\partial H_y}{\partial z} - \frac{\partial H_z}{\partial y} + ik_0\varepsilon E_x = 0, \end{aligned} \tag{3.142}$$

where $k_0 = \omega/c$ is the wave number in vacuum. From equation (3.142) it follows that the Helmholtz equation for the projection of the electric vector:

$$\left(\frac{\partial}{\partial z^2} + \frac{\partial}{\partial y^2} + k^2\right) E_x = 0, \tag{3.143}$$

where $k = \frac{\omega}{c}\sqrt{\varepsilon\mu}$, ε and μ are the dielectric and magnetic permeability of the medium or object.

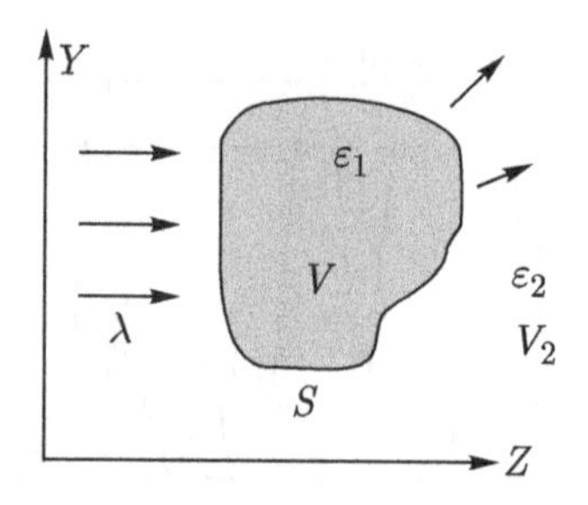

Fig. 3.37. The scheme of two-dimensional diffraction of a monochromatic wave on a cylindrical object with dielectric constant ε_1 in a medium with permittivity ε_2.

We consider the diffraction of TE-polarized electromagnetic waves by a cylinder with cross section V, bounded by contour S.

Figure 3.37 shows a diagram of the problem. A cylindrical object $V(\partial/\partial x = 0)$ with dielectric constant ε_1 ($\mu = 1$ – magnetic permeability), located in a medium V_2 with dielectric constant ε_2 receives a TE-polarized monochromatic electromagnetic wave with the length λ.

Suppose that the field inside the object V is described by function $\phi_1 = E_x$, outside of the object ϕ_2. Then the diffraction problem reduces to solving the differential Helmholtz equations:

$$\begin{aligned} &\left(\Delta + k_1^2(y,z)\right)\phi_1 = 0, \qquad (y,z) \in V, \\ &\left(\Delta + k_2^2(y,z)\right)\phi_2 = -g_2, \ (y,z) \notin V, \end{aligned} \tag{3.144}$$

where $\Delta = \frac{\partial^2}{\partial y^2} + \frac{\partial^2}{\partial z^2}$,

g_2 is the incident field in the medium, k_1 is the wave number in region V with a gradient refractive index; $k_2 = \frac{2\pi}{\lambda}\sqrt{\varepsilon_2 \mu_2}$ is the wave number outside the region V.

For TE-polarization the boundary conditions follow from the continuity of the fields and their derivatives on the boundary between two media:

$$\begin{aligned} &\phi_1\big|_S = \phi_2\big|_S, \qquad (y,z) \in S, \\ &\left.\frac{\partial \phi_1}{\partial \mathbf{n}_1}\right|_S = \left.\frac{\partial \phi_2}{\partial \mathbf{n}_2}\right|_S, \ (y,z) \in S, \end{aligned} \tag{3.145}$$

where $\mathbf{n}_1,\mathbf{n}_2$ are the vectors normal to the contour S, external for the region inside and outside of the object V, respectively.

For functions ϕ_1 and G_2 Green's formula holds:

$$\iint_V \left(\phi_1 \Delta G_2 - G_2 \Delta \phi_1\right) dy\, dz = \oint_S \left(\phi_1 \frac{\partial G_2}{\partial \mathbf{n}} - G_2 \frac{\partial \phi_1}{\partial \mathbf{n}}\right) dl. \tag{3.146}$$

For Green's function we have the following equation:

$$\Delta G_2 + k_2^2 G_2 = -\delta(M, M_0), \tag{4.147}$$

where M is the the current point at which the integral is taken, M_0 is the observation point of observation, so that $\delta(M, M_0) = \delta(y', z', y, z)$ – the delta function.

Substituting (3.144) and (3.147) into (3.146) we get:

$$\oint_S \left(G_2 \frac{\partial \phi_1}{\partial \mathbf{n}} - \phi_1 \frac{\partial G_2}{\partial \mathbf{n}} \right) dl + \iint_V \left(k_1^2(y,z) - k_2^2 \right) \cdot \phi_1 G_2 dy\, dz - \iint_V \phi_1(y', z') \delta(y', z'; y, z) dy' dz' = 0. \tag{3.148}$$

Using the property of the delta function, we rewrite (3.148) in the form of:

$$\oint_S \left(G_2 \frac{\partial \phi_1}{\partial \mathbf{n}} - \phi_1 \frac{\partial G_2}{\partial \mathbf{n}} \right) dl + \iint_V \left(k_1^2(y,z) - k_2^2 \right) \phi_1 G_2 dy dz = \begin{cases} \phi_1, (y,z) \in V, \\ 0, \ (y,z) \notin V. \end{cases} \tag{3.149}$$

Also, applying Green's formula for the function, and on the basis of equations (3.144) and (3.147), we obtain:

$$\oint_S \left(G_2 \frac{\partial \phi_1}{\partial \mathbf{n}} - \phi_1 \frac{\partial G_2}{\partial \mathbf{n}} \right) dl + \iint_{V_2} g_2 G_2 dy\, dz = \begin{cases} 0, \ (y,z) \in V, \\ \varphi_2, (y,z) \notin V. \end{cases} \tag{3.150}$$

Summing equation (3.149) and (3.150) and taking into account the boundary conditions (3.145), we obtain:

$$\iint_{V_2} \left(k_1^2 - k_2^2 \right) \varphi_1 G_2 dy dz + \varphi_0 = \begin{cases} \varphi_1, (y,z) \in V, \\ \varphi_2 . (y,z) \notin V, \end{cases} \tag{3.151}$$

where

$$\phi_0(y,z) = \iint_{V_2} g_2 G_2 dy\, dz. \tag{3.152}$$

Equation (3.151) is a Fredholm integral equation of the second kind and has a unique non-trivial solution for $\phi_0 \neq 0$. It can be rewritten in the form [57, 63] (k_1 and k_2 are constant):

$$\phi(y,z) = \phi_0(y,z) - \frac{ik_0^2(\varepsilon_1 - \varepsilon_2)}{4} \iint_V \varphi\ (\xi, \eta) H_0^{(2)} \left(k_2 \sqrt{(y-\xi)^2 + (z-\eta)^2} \right) d\xi\, d\eta, \tag{3.153}$$

where $f(y, z) = E_z(y, z)$ and $f_0(y, z) = E_{0z}(y, z)$ are the project-ions on the axis X (perpendicular to the plane of Fig. 3.37) of the vectors of the strength of the electric field and of the incident and scattered electromagnetic waves, $k_2 = \frac{2\pi}{\lambda}\sqrt{\varepsilon_2}$

is the wave number of the light outside of the cylinder, $(\xi,\eta)\in V$ are the Cartesian coordinates within the object $H_0^{(2)}(x)$ is the the Hankel function of 2nd kind and zeroth order. If we consider the coordinates (y, z) also belong to only part of the subject $(y,z)\in V$, then the expression (3.153) can be regarded as a Fredholm integral equation of 2nd kind with respect to the amplitude of the diffraction field $\phi(y, z)$, assuming that the function $\phi_0(y, z)$ is given.

The two-dimensional integral in equation (3.153) can be represented as a convolution:

$$\iint_V \phi(y,z)H_0^{(2)}\left(y-\xi,z-\eta\right)d\xi\, d\eta=\left[S(\xi,\eta)\cdot\phi(\xi,\eta)\right]*H_0^{(2)}(\xi,\eta), \quad (3.154)$$

where

$$S(\xi,\eta)=\begin{cases}1,(\xi,\eta)\in V,\\0,(\xi,\eta)\notin V,\end{cases} \quad (3.155)$$

– the functions of the object V (Fig. 3.37). Convolution (3.154) of functions $\phi_S(y,z)=S(\xi,\eta)\cdot\phi(\xi,\eta)$ and $H_0^{(2)}(\xi,\eta)$ can be computed using the three two-dimensional Fourier transforms:

$$\phi_S*H_0^2=\text{Im}^{-1}\left[\text{Im}\left(\phi_S\right)\cdot\text{Im}\left(H_0^2\right)\right], \quad (3.156)$$

where

$$\text{Im}\left[f\right]=\int\int_{-\infty}^{\infty} f(x,y)e^{-i2\pi(x\xi+y\eta)}dx\,dy \quad (3.157)$$

is the direct Fourier transform; Im^{-1} is the inverse Fourier transform.

The iterative algorithm (method of successive approximations) of the solution of (3.153) has the form [74-78]:

$$\phi_{n+1}(y,z)=\gamma\phi_0(y,z)+(1-\gamma)\phi_n(y,z)-\gamma\beta\iint_V \phi_n(\xi,\eta)H_0^{(2}\left(k_2\sqrt{(y-\xi)^2+(z-\eta)^2}\right)d\xi\, d\eta, \quad (3.158)$$

where $(y, z)\in V$, ϕ_{n+1} and ϕ_n is the amplitude of the diffraction field in the region V in (n +1)-th and the n-th iteration steps; γ is the relation constant of the algorithm, which adjusts the speed of its convergence, $\beta = ik^2(\varepsilon_1-\varepsilon_2)/4$. As an initial approximation we can take the incident field:

$$\phi_1(y,z)=\phi_0(y,z). \quad (3.159)$$

3.4.2. An iterative method for calculating the diffraction of TM-polarized wave

For the two-dimensional case of the diffraction of a TM-polarized monochromatic electromagnetic wave on a dielectric object, the projections of the fields H_x, E_y and

E_z satisfy the Maxwell equations:

$$\begin{aligned}&\frac{\partial H_x}{\partial y}+ik_0\varepsilon E_z=0,\\&-\frac{\partial H_x}{\partial z}+ik_0\varepsilon E_y=0,\\&\frac{\partial E_y}{\partial z}-\frac{\partial E_z}{\partial y}-ik_0\mu H_x=0.\end{aligned} \tag{3.160}$$

From equations (3.160) we obtain the Helmholtz equation for the projection of the magnetic field vector:

$$\frac{\partial}{\partial x}\left(\frac{1}{\varepsilon}\frac{\partial H_x}{\partial x}\right)+\frac{\partial}{\partial y}\left(\frac{1}{\varepsilon}\frac{\partial H_x}{\partial y}\right)+k_0^2\mu H_x=0. \tag{3.161}$$

Using equation (3.161) to the problem of diffraction on a dielectric cylinder, we obtain a system of two equations:

$$\begin{cases}\left(\Delta+k_1^2\right)\phi_1-\dfrac{1}{\varepsilon_1}\left(\dfrac{\partial\varepsilon_1}{\partial z}\dfrac{\partial\phi_1}{\partial z}+\dfrac{\partial\varepsilon_1}{\partial y}\dfrac{\partial\phi_1}{\partial y}\right)=0,\ (y,z)\in V,\\ \left(\Delta+k_2^2\right)\phi_2=-g_2, \qquad (y,z)\in V_2,\end{cases} \tag{3.162}$$

where $k_1^2=k_0^2\varepsilon_1$, $\phi=H_x$, g_2 is the external incident field.

The boundary conditions for TM-polarization follow from the equality of tangential components of electric and magnetic fields at the interface and can be expressed:

$$\begin{aligned}&\left.\frac{1}{\varepsilon_1}\frac{\partial\phi_1}{\partial\mathbf{n}}\right|_S=\left.\frac{1}{\varepsilon_2}\frac{\partial\phi_2}{\partial\mathbf{n}}\right|_S,\\&\left.\phi_1\right|_S=\left.\phi_2\right|_S.\end{aligned} \tag{3.163}$$

Assuming that:

$$\begin{aligned}&\frac{1}{\varepsilon_1}\Delta\phi_1-\frac{1}{\varepsilon_1^2}\left(\frac{\partial\varepsilon_1}{\partial z}\frac{\partial\phi_1}{\partial z}+\frac{\partial\varepsilon_1}{\partial y}\frac{\partial\phi_1}{\partial y}\right)=\frac{1}{\varepsilon_1}\Delta\phi_1+\left(\frac{\partial}{\partial z}\left(\frac{1}{\varepsilon_1}\right)\frac{\partial\phi_1}{\partial z}+\frac{\partial}{\partial y}\left(\frac{1}{\varepsilon_1}\right)\frac{\partial\phi_1}{\partial y}\right)=\\&=\frac{\partial}{\partial z}\left(\frac{1}{\varepsilon_1}\frac{\partial\phi_1}{\partial z}\right)+\frac{\partial}{\partial y}\left(\frac{1}{\varepsilon_1}\frac{\partial\phi_1}{\partial y}\right)=\operatorname{div}\left(\frac{1}{\varepsilon_1}\left(\mathbf{i}\frac{\partial\phi_1}{\partial z}+\mathbf{j}\frac{\partial\phi_1}{\partial y}\right)\right)=\operatorname{div}\left(\frac{1}{\varepsilon_1}\operatorname{grad}\phi_1\right),\end{aligned} \tag{3.164}$$

the first equation in (3.162) can be rewritten as:

$$\operatorname{div}\left(\frac{1}{\varepsilon_1}\operatorname{grad}\phi_1\right)+\frac{k_1^2}{\varepsilon_1}\phi_1=0. \tag{3.165}$$

For the $\operatorname{div}\left(\frac{1}{\varepsilon_1}\operatorname{grad}\phi_1\right)$ operator we have the following Green's integral formula:

$$\iint\limits_V \left\{ G_2 \operatorname{div}\left(\frac{1}{\varepsilon_1}\operatorname{grad}\phi_1\right) - \phi_1 \operatorname{div}\left(\frac{1}{\varepsilon_1}\operatorname{grad}G_2\right)\right\} dy\,dz = \oint\limits_S \frac{1}{\varepsilon_1}\left(G_2\frac{\partial \phi_1}{\partial \mathbf{n}} - \phi_1\frac{\partial G_2}{\partial \mathbf{n}}\right) dl. \quad (3.166)$$

The Green function satisfies the following equation:

$$\Delta G_2 = -k_2^2 G_2 - \delta\left(M, M_0\right). \quad (3.167)$$

From equations (3.164) and (3.167) for the operator $div\left(\frac{1}{\varepsilon_1}gradG_2\right)$ we have:

$$\operatorname{div}\left(\frac{1}{\varepsilon_1}\operatorname{grad}G_2\right) = \frac{1}{\varepsilon_1}\Delta G_2 - \frac{1}{\varepsilon_1^2}\nabla\varepsilon_1\nabla G_2 = \frac{-k_2^2 G_2 - \delta\left(M, M_0\right)}{\varepsilon_1} - \frac{1}{\varepsilon_1^2}\nabla\varepsilon_1\nabla G_2. \quad (3.168)$$

From equations (3.165) (3.166) and (3.168) implies the first integral equation:

$$\oint\limits_S \frac{1}{\varepsilon_1}\left(G_2\frac{\partial \phi_1}{\partial \mathbf{n}} - \phi_1\frac{\partial G_2}{\partial \mathbf{n}}\right) dl + \iint\limits_V \frac{k_1^2 - k_2^2}{\varepsilon_1}\phi_1 G_2\,dy\,dz + \iint\limits_{V_2} \phi_1\left(\frac{1}{\varepsilon_1}\right) G_2\,dydz = \begin{cases} \dfrac{\phi_1}{\varepsilon_1}, (y,z) \in V \\ 0, \quad (y,z) \in V_2 \end{cases}. \quad (3.169)$$

Applying the Green theorem for the functions ϕ_2 and G_2, and converting the second equation in (3.162), we obtain the second integral equation:

$$\frac{1}{\varepsilon_2}\iint\limits_{V_2} G_2 g_2\,dy\,dz + \oint\limits_S \left(\frac{G_2}{\varepsilon_2}\frac{\partial \phi_2}{\partial \mathbf{n}_2} - \frac{\phi_2}{\varepsilon_2}\frac{\partial G_2}{\partial \mathbf{n}_2}\right) dl = \begin{cases} 0, \quad (y,z) \in V \\ \dfrac{\phi_2}{\varepsilon_2}, (y,z) \in V_2 \end{cases}. \quad (3.170)$$

Summing equation (3.169) and (3.170) and using the boundary conditions of equations (3.163), we obtain a system of equations:

$$\iint\limits_V \left(k_1^2 - k_2^2\right)\frac{\phi_1 G_2}{\varepsilon_1} dy\,dz + \frac{1}{\varepsilon_2}\iint\limits_{V_2} G_2 g_2\,dydz + \iint\limits_V \phi_1 \nabla\left(\frac{1}{\varepsilon_1}\right)\nabla G_2\,dy\,dz +$$

$$+\oint\limits_S \left(\frac{\varepsilon_1 - \varepsilon_2}{\varepsilon_1\varepsilon_2}\right)\phi_1\frac{\partial G_2}{\partial \mathbf{n}} dl = \begin{cases} \dfrac{\phi_1}{\varepsilon_1}, \quad (y,z) \in V \\ \dfrac{\phi_2}{\varepsilon_2}, \quad (y,z) \in V_2 \end{cases}, \quad (3.171)$$

where $\phi_0(y,z) = \iint\limits_{V_2} g_2 G_2\,dy\,dz$ is the external field incident on an object V, $G_2 = \frac{i}{4}H_0^{(1)}\left(\xi\right)$ is the Green's function, where $\xi = k_2\sqrt{\left(y - y'\right)^2 + \left(z - z'\right)^2}$.

For the Hankel functions have the following relationship:

$$\frac{dH_0^{(1)}}{d\xi} = -H_1^{(1)}(\xi). \tag{3.172}$$

Taking into account that $\frac{\partial \xi}{\partial z} = k_2 \frac{2(z-z')}{\xi}$, $\frac{\partial \xi}{\partial y} = k_2 \frac{2(y-y')}{\xi}$ the gradient of the Green's function occurring in equation (3.171) can be written as:

$$\operatorname{grad} G_2 = \frac{i}{4}\left(k_2 H_1^{(1)}(\xi)\frac{z-z'}{\xi},\ k_2 H_1^{(1)}(\xi)\frac{y-y'}{\xi} \right). \tag{3.173}$$

The system of equations (3.171) takes into account the inhomogeneous dielectric constant inside the object V. We restrict ourselves to the case of a homogeneous cylindrical object and exclude from consideration the term $\iint_V \varphi_1 \nabla\left(\frac{1}{\varepsilon_1}\right)\nabla G_2 dydz$.

If we assume that the object V is homogeneous (ε_1 = const), then equation (3.171) can be rewritten in the form [79]:

$$\phi(y,z) = \phi_0(y,z)\frac{\varepsilon_1}{\varepsilon_2} - \frac{ik_0^2(\varepsilon_1-\varepsilon_2)}{4}\cdot \iint_V \phi(\xi,\eta) H_0^{(2)}\left(k_2\sqrt{(y-\xi)^2+(z-\eta)^2}\right) d\xi\, d\eta +$$
$$+\frac{i(\varepsilon_1-\varepsilon_2)}{4\varepsilon_2}\oint_S \phi(\xi,\eta)\frac{\partial}{\partial \mathbf{n}} H_0^{(2)}\left(k_2\sqrt{(y-\xi)^2+(z-\eta)^2}\right) dl \tag{3.174}$$

inside a cylindrical object with the cross-section V, bounded by contour S, i.e., $(y, z) \in V$ and

$$\varphi(y,z) = \varphi_0(y,z) - \frac{ik_0^2(\varepsilon_1-\varepsilon_2)\varepsilon_2}{4\varepsilon_1}\cdot \iint_V \varphi(\xi,\eta) H_0^{(2)}\left(k_2\sqrt{(y-\xi)^2+(z-\eta)^2}\right) d\xi d\eta +$$
$$+\frac{i(\varepsilon_1-\varepsilon_2)}{4\varepsilon_1}\oint_S \varphi(\xi,\eta)\frac{\partial}{\partial \mathbf{n}} H_0^{(2)}\left(k_2\sqrt{(y-\xi)^2+(z-\eta)^2}\right) dl \tag{3.175}$$

outside the object at $(y, z) \notin V$, where ε_1 is the dielectric constant inside the object, $H_0^{(2)}$ is the the Hankel function of zero order, ϕ_0 is the amplitude of the incident light wave, **n** is the outward normal to the surface of the object.

The integral along the contour in (3.174) and (3.175) can be represented by a Fourier transformation, introducing the function of the contour of the object.

$$G(y,z) = \begin{cases} 1, (y,z) \in S \\ 0, (y,z) \notin S \end{cases}. \tag{3.176}$$

Then

$$\oint_S \phi_m(y,z)\frac{\partial}{\partial \mathbf{n}} H_0^{(2)}\left(k_2\sqrt{(y-\xi)^2+(z-\eta)^2}\right)dl = k_2\,\mathrm{Im}^{-1}\Big[\mathrm{Im}\big(\phi_m(y,z)G(y,z)n_y\big)\times$$

$$\times\mathrm{Im}\left(\frac{\xi}{\sqrt{\xi^2+\eta^2}}H_1^{(2)}\left(k_2\sqrt{\xi^2+\eta^2}\right)\right)+\mathrm{Im}\big(\phi_m(y,z)G(y,z)n_z\big)\times$$

$$\times\mathrm{Im}\left(\frac{\eta}{\sqrt{\xi^2+\eta^2}}H_1^{(2)}\left(k_2\sqrt{\xi^2+\eta^2}\right)\right)\Bigg], \tag{3.177}$$

where $\mathbf{n} = (n_y, n_z)$ is the unit vector of the outward normal to the contour S, $(y, z) \in S$, Im and Im^{-1} are the direct and inverse 2D Fourier transforms. Similarly, the convolution integral over the area of the object V in (3.174) and (3.175) can be rewritten in the Fourier transform using the facility

$$P(\xi,\eta)=\begin{cases}1,(\xi,\eta)\in V\\ 0,(\xi,\eta)\notin V\end{cases}. \tag{3.178}$$

The iterative algorithm (method of successive approximations) of the solution of integral equations (3.174) and (3.175) takes the form [76, 77, 80, 81]:

$$\phi_{n+1}(y,z)=\gamma\phi_0(y,z)\frac{\varepsilon_1}{\varepsilon_2}-\gamma\frac{ik_0^2(\varepsilon_1-\varepsilon_2)}{4}\mathrm{Im}^{-1}\left[\mathrm{Im}\big(\phi_n(\xi,\eta)P(\xi,\eta)\big)\cdot\mathrm{Im}\left(H_0^{(2)}\left(k\sqrt{\xi^2+\eta^2}\right)\right)\right]-$$

$$-\gamma\frac{ik_2(\varepsilon_1-\varepsilon_2)}{4\varepsilon_2}\mathrm{Im}^{-1}\left[\mathrm{Im}\big(\phi_n(\xi,\eta)G(\xi,\eta)n_y\big)\mathrm{Im}\left(\frac{\xi}{\sqrt{\xi^2+\eta^2}}H_1^{(2)}\left(k_2\sqrt{\xi^2+\eta^2}\right)\right)\right]-$$

$$-\gamma\frac{ik_2(\varepsilon_1-\varepsilon_2)}{4\varepsilon_2}\Im^{-1}\Big[\Im\big(\varphi_n(\xi,\eta)G(\xi,\eta)n_z\big)\times$$

$$\times\mathrm{Im}\left(\frac{\eta}{\sqrt{\xi^2+\eta^2}}H_1^{(2)}\left(k_2\sqrt{\xi^2+\eta^2}\right)\right)\Bigg]+(1-\gamma)\phi_n(y,z) \tag{3.179}$$

inside the object $(y, z) \in V$ and

$$\phi_{n+1}(y,z)=\gamma\phi_0(y,z)\varepsilon_2-\gamma\frac{ik_0^2(\varepsilon_1-\varepsilon_2)\varepsilon_2}{4\varepsilon_1}\mathrm{Im}^{-1}\left[\mathrm{Im}\big(\phi_n(\xi,\eta)P(\xi,\eta)\big)\cdot\Im\left(H_0^{(2)}\left(k_2\sqrt{\xi^2+\eta^2}\right)\right)\right]-$$

$$-\gamma\frac{ik_2(\varepsilon_1-\varepsilon_2)}{4\varepsilon_1}\mathrm{Im}^{-1}\left[\mathrm{Im}\big(\phi_n(\xi,\eta)G(\xi,\eta)n_y\big)\mathrm{Im}\left(\frac{\xi}{\sqrt{\xi^2+\eta^2}}H_1^{(2)}\left(k_2\sqrt{\xi^2+\eta^2}\right)\right)\right]-$$

$$-\gamma\frac{ik_2(\varepsilon_1-\varepsilon_2)}{4\varepsilon_1}\mathrm{Im}^{-1}\Big[\mathrm{Im}\big(\varphi_n(\xi,\eta)G(\xi,\eta)n_z\big)\times$$

$$\times\mathrm{Im}\left(\frac{\eta}{\sqrt{\xi^2+\eta^2}}H_1^{(2)}\left(k_2\sqrt{\xi^2+\eta^2}\right)\right)\Bigg]+(1-\gamma)\phi_n(y,z) \tag{3.180}$$

outside the object $(y, z) \notin V$, where ϕ_{n+1} and ϕ_n are the calculated complex field amplitudes at the $(n+1)$-th and (n)-th iterations; γ is the relaxation parameter.

Factors n_z and n_y are the projections of the normal to the surface of the object on the Z- and Y-axis, respectively. If the cylinder is taken as the object with a circular cross section, located in the centre of origin, then $n_z = \frac{z_t}{\sqrt{z_t^2 + y_t^2}}$, $n_y = \frac{y_t}{\sqrt{z_t^2 + y_t^2}}$ where z_t, y_t are the current coordinates of points on the boundary between two media in evaluating the integral along the contour.

In the case of an arbitrary boundary n_z and n_y must be calculated for all points on the boundary for a particular object. For simplicity, we take a smooth border, without breaks. The slope of the tangent at the boundary points is calculated by the methods of mathematical statistics. To calculate the slope of the tangent to the boundary of the object we consider points near the boundary of the study at the radius R_n, where R_n is the number of counts on which the linear regression equation for the Z- and Y-axis is constructed:

$$y = \bar{y} - \frac{R_{\xi\eta}}{\sqrt{D_\xi D_\eta}} \frac{S_y}{S_z}(z - \bar{z}) = \bar{y} - \frac{R_{\xi\eta}}{S_z^2}(z - \bar{z}), \tag{3.181}$$

$$z = \bar{z} - \frac{R_{\xi\eta}}{\sqrt{D_\xi D_\eta}} \frac{S_z}{S_y}(y - \bar{y}) = \bar{z} - \frac{R_{\xi\eta}}{S_y}(y - \bar{y}), \tag{3.182}$$

where $\bar{z} = M_\xi = \frac{1}{N}\sum_{i=1}^{N} z_i$, $\bar{y} = M_\eta = \frac{1}{N}\sum_{i=1}^{N} y_i$ – the expectation of z- and y-coordinates of points taken into account, N is the number of points taken into account, $D_\xi = S_z^2 = \overline{z^2} - \bar{z}^2$, $D_\eta = S_y^2 = \overline{y^2} - \bar{y}^2$ is the variance of the coordinates of points along the Z- and Y-axis, $R_{\xi\eta} = M_{\xi\eta} - M_\xi M_\eta$ is the theoretical correlation. Hence the tangents of the angles of lines of regression to the Z axis and Y is described as follows:

$$tg(\alpha_1) = -\frac{R_{\xi\eta}}{S_z^2}, \tag{3.183}$$

$$tg(\alpha_2) = -\frac{S_z^2}{R_{\xi\eta}} = \frac{1}{tg(\alpha_1)}, \tag{3.184}$$

where α_1 and α_2 are the angles of inclination of the tangent of the regression equations on the Z- and Y-axis. Averaging them, we obtain the slope of the tangent to the surface of the object at the desired point:

$$\alpha = \frac{\alpha_1 + \alpha_2}{2}. \tag{3.185}$$

The required angle of inclination of the surface normal is perpendicular to the tangent to the surface at this point: $\beta = \alpha \pm \pi/2$. Then

$$n_z = \cos(\beta),\ n_y = \sin(\beta). \tag{3.186}$$

3.4.3. Relaxation of the iterative method

For the algorithm (3.158) to have the property of relaxation, the operator $\hat{L}$ in equation (3.158) must be 'contracting' [78]:

$$\|f_{n+1} - g_{n+1}\| = \|\hat{L}f_n - \hat{L}g_n\| \le \|f_n - g_n\|, \tag{3.187}$$

where f and g are arbitrary functions, n is the number of iterations, and the norm function $\|f\|$ is given by:

$$\|f\| = \iint_{\infty} |f(x,y)|^2\, dxdy. \tag{3.188}$$

In view of (3.158), the left-hand side of equation (3.187) can be written as:

$$\|f_{n+1} - g_{n+1}\| = \|(1-\gamma)(f_n - g_n) - \gamma\beta(f_{sn} - g_{sn}) * H_0^{(2)}\|, \tag{3.189}$$

where $f_S = Sf$, $S(\xi, \eta)$ is the function describing the domain of definition of the object (3.155). Using the triangle inequality and Parseval equality we can get an estimate for the norm (3.189):

$$\|f_{n+1} - g_{n+1}\| \le \left\{|1-\gamma| + |\gamma\beta||F|_{\max}\, \pi a^2\right\}\|f_n - g_n\|, \tag{3.190}$$

where a is the radius of the circle in which the object $S(\xi, \eta)$ is inscribed, $|F|_{\max}$ is the maximum absolute value of the function which according to (3.189) is the Fourier transform of the Hankel functions:

$$\mathrm{Im}[f] = \int_{-\infty}^{\infty}\!\!\int H_0^{(2)}\left(k\sqrt{x^2+y^2}\right) e^{-i2\pi(x\xi + y\eta)} dx\, dy. \tag{3.191}$$

The Hankel function, as the Green's function of our problem (equation (3.158)), describes a cylindrical wave and has a logarithmic singularity in the centre at $r = \sqrt{x^2+y^2} = 0$. The Fourier transform of the Hankel function is a function having a singularity at the radius $\rho = \lambda^{-1}$. Indeed, the integral in (3.191) in polar coordinates can be calculated [82]:

$$F(\xi,\eta)=2\pi\int_0^\infty H_0^{(2)}(kr)J_0(2\pi\rho r)rdr=\lambda\delta\left(\rho-\lambda^{-1}\right)+\frac{i}{\pi^2\left(\rho^2-\lambda^{-2}\right)}, \tag{3.192}$$

where $J_0(x)$ is the Bessel function of zero order, $k=2\pi/\lambda$; r and ρ are polar coordinates; $\rho=\sqrt{\xi^2+\eta^2}$, (ξ,η) are the spatial frequencies.

Therefore, strictly speaking, the inequality (3.187) can be satisfied only at $\gamma = 0$, $|F|_{max} = \infty$. But, in practice, the integral in (3.191) is calculated as double the amount of counts in the final grid, and instead of the infinite Hankel function at the origin we select its final value to be close to zero point. Therefore, in practice $|F|_{max} < \infty$ and the inequality (3.190) makes sense.

The inequality (3.187) requires that the expression in curly brackets in (3.190) were less than unity:

$$|1-\gamma|+|\gamma\beta||F|_{max}\,\pi a^2<1. \tag{3.193}$$

If parameter γ is selected in the interval (0,1): $0<\gamma<1$, then (3.193) is satisfied if the following condition is satisfied:

$$|\gamma\beta||F|_{max}\,\pi a^2<1. \tag{3.194}$$

This condition restricts the range of problem parameters under which it can be solved by the integral method (3.158). For example, if we fix the number of counts in the grid and the wavelength λ, then inequality (3.194) gives an upper bound on the dielectric constant which the object can have [74]:

$$\varepsilon_1<\varepsilon_2+\frac{4}{\pi k_0^2|F|_{max}\,a^2}. \tag{3.195}$$

It is preferred to write (3.195) in the form where all the physical parameters of the problem λ, ε and a are in the left-hand side of the equality:

$$\frac{a}{\lambda}\sqrt{\varepsilon_1-\varepsilon_2}<\frac{1}{\sqrt{\pi^3|F|_{max}}}. \tag{3.196}$$

In the following, by numerical experiments, the inequality (3.196) will be refined and it will be shown that the algorithm has relaxaition provided that $\frac{a}{\lambda}\left(\varepsilon_1-\varepsilon_2\right)<0.6$.

If we choose λ from the interval (1, 2): $1<\lambda<2$ (at $\lambda>2$ and $\lambda<0$, the inequality (3.193) is not satisfied in any way), then the inequality (3.193) will be complied under the condition more restrictive than (3.194).

Thus, the analysis of inequality (3.193) shows that the iterative algorithm (3.158) has the property of relaxation (3.187) with the choice of λ from a certain range in the interval (0, 1), provided that the parameters of the problem satisfy the inequality (3.196).

3.4.4. Comparison with the analytical calculation of diffraction of a plane wave

To test the efficiency of algorithms (3.158), (3.179) and (3.180), numerical comparison was carried out of the diffraction fields of a plane TE- (TM-) polarized wave on a dielectric cylinder with a circular cross section, calculated by iterative methods (3.158), (3.179) and (3.180) and using the known analytical formulas [1]. The limits of applicability of analytical formulas are studied in [83–85].

To derive the analytical formulas for the solution of the two-dimensional problem of diffraction on a cylinder with a circular cross section, we use the wave equation (TE-polarization):

$$\Delta E_x + k_2^2 E_x = 0,\ r > R, \tag{3.197}$$

$$\Delta E_x + k_1^2 E_x = 0,\ r < R, \tag{3.198}$$

where $k_1 = \frac{2\pi}{\lambda}\sqrt{\varepsilon_1}$ is the wave number inside the cylinder; $k_2 = \frac{2\pi}{\lambda}\sqrt{\varepsilon_2}$ is the wave number in the medium, R is the radius of the cylinder.

The field outside the cylinder will be sought as an expansion in Hankel functions of the second kind:

$$E_x(\rho,\phi) = E_{0x} + E_0 \sum_{n=-\infty}^{\infty} i^n a_n H_n^{(2)}(k_2 r) e^{in\phi},\ r > R, \tag{3.199}$$

where E_{0x} is the incident wave field, inside the cylinder – in the form of an expansion in Bessel functions:

$$E_x(\rho,\phi) = E_0 \sum_{n=-\infty}^{\infty} i^n b_n J_n(k_1 r) e^{in\phi},\ r < R. \tag{3.200}$$

Given the boundary conditions, we equate the fields and the derivatives at the cylinder boundary:

$$\begin{gathered} E_{0x} + E_0 \sum_{n=-\infty}^{\infty} i^n a_n H_n^{(2)}(k_2 r) e^{in\phi} = E_0 \sum_{n=-\infty}^{\infty} i^n b_n J_n(k_1 r) e^{in\phi}, \\ \left.\frac{\partial E_{0x}}{\partial r}\right|_{r=R} + k_2 E_0 \sum_{n=-\infty}^{\infty} i^n a_n H_n^{(2)'}(k_2 r) e^{in\phi} = k_1 E_0 \sum_{n=-\infty}^{\infty} i^n b_n J_n'(k_1 r) e^{in\phi}. \end{gathered} \tag{3.201}$$

If the incident wave is a plane wave, then

$$E_{0x} = E_0 \exp\left(ik_2 z\right) = E_0 \exp\left(ik_2 r \cos\phi\right) = E_0 \sum_{n=-\infty}^{\infty} i^n J_n\left(k_2 r\right) \exp\left(in\phi\right). \tag{3.202}$$

Then solving the system of equations (3.201), we obtain the desired coefficients:

$$a_n = \left(-i\right)^n \frac{k_1 J_n'(k_1 R) J_n(k_2 R) - k_2 J_n(k_1 R) J_n'(k_2 R)}{k_2 J_n(k_1 R) H_n'^{(2)}(k_2 R) - k_1 J_n'(k_1 R) H_n^{(2)}(k_2 R)} \tag{3.203}$$

$$b_n = \left(-i\right)^n \frac{k_2 J_n(k_2 R) H_n'^{(2)}(k_2 R) - k_2 J_n'(k_2 R) H_n^{(2)}(k_2 R)}{k_2 J_n(k_1 R) H_n'^{(2)}(k_2 R) - k_1 J_n'(k_1 R) H_n^{(2)}(k_2 R)} \tag{3.204}$$

In the case of TM-polarized wave equations will be as follows:

$$\Delta H_x + k_2^2 H_x = 0,\ r > R, \tag{3.205}$$

$$\Delta H_x + k_1^2 H_x = 0,\ r < R. \tag{3.206}$$

The field outside the cylinder will be sought as an expansion in Hankel functions of second kind:

$$H_x\left(\rho,\phi\right) = H_{0x} + H_0 \sum_{n=-\infty}^{\infty} i^n a_n H_n^{(2)}(k_2 r) e^{in\phi},\ r > R, \tag{3.207}$$

where H_{0x} is the incident wave field inside the cylinder – in the form of an expansion in Bessel functions:

$$H_x\left(\rho,\phi\right) = H_0 \sum_{n=-\infty}^{\infty} i^n b_n J_n(k_1 r) e^{in\phi},\ r < R. \tag{3.208}$$

Given the boundary conditions, we equate the fields and the derivatives at the cylinder boundary:

$$\begin{gathered} H_{0x} + H_0 \sum_{n=-\infty}^{\infty} i^n a_n H_n^{(2)}(k_2 r) e^{in\phi} = H_0 \sum_{n=-\infty}^{\infty} i^n b_n J_n(k_1 r) e^{in\phi}, \\ \left.\frac{\partial H_{0x}}{\partial r}\right|_{r=R} + k_2 H_0 \sum_{n=-\infty}^{\infty} i^n a_n H_n^{(2)'}(k_2 r) e^{in\phi} = \frac{k}{\varepsilon_1} H_0 \sum_{n=-\infty}^{\infty} i^n b_n J_n'(k_1 r) e^{in\phi}. \end{gathered} \tag{3.209}$$

If the incident wave is plane wave propagating along the axis Z, then

$$H_{0x} = H_0 \exp(ik_2 z) = H_0 \exp(ik_2 r \cos\phi) = H_0 \sum_{n=-\infty}^{\infty} i^n J_n(k_2 r) \exp(in\phi). \tag{3.210}$$

Then solving the system of equations (3.209), we obtain the desired coefficients:

$$a_n = (-i)^n \frac{k_2 J_n'(k_1 R) J_n(k_2 R) - k_1 J_n(k_1 R) J_n'(k_2 R)}{k_1 J_n(k_1 R) H_n'^{(2)}(k_2 R) - k_2 J_n'(k_1 R) H_n^{(2)}(k_2 R)}, \tag{3.211}$$

$$b_n = (-i)^n \frac{k_1 J_n(k_2 R) H_n'^{(2)}(k_2 R) - k_1 J_n'(k_2 R) H_n^{(2)}(k_2 R)}{k_1 J_n(k_1 R) H_n'^{(2)}(k_2 R) - k_2 J_n'(k_1 R) H_n^{(2)}(k_2 R)}. \tag{3.212}$$

Figures 3.38 and 3.39 show plots of the residuals σ of the amplitude E_x (TE-polarization), H_x (TM-polarization) of the diffraction fields, calculated by (3.158), (3.179) and (3.180) and through a series of Bessel functions (3.199), (3.200), (3.207), (3.208) [1]. The discrepancy was calculated by the formula:

$$\sigma = \sqrt{\frac{\sum_{n,m=1}^{N} \left(\phi_k(n,m) - \phi_A(n,m)\right)^2}{\sum_{n,m=1}^{N} \phi_k^2(n,m)}}, \tag{3.213}$$

where ϕ_A is a field as calculated by an analytic method (3.199), (3.200), (3.207), (3.208), ϕ_k is the field on the kth iteration in an iterative algorithm (3.158), (3.179), (3.180).

Figure 3.38 shows a plot of the dependence of residual value σ on γ for the same number of iterations, N = 100 (TE-polarization) and N = 20 (TM-polarization). Simulation parameters are: diameter of the circular cylinder, D = 1 μm, wavelength $\lambda = 1$ μm, $\varepsilon_1 = 2$ – dielectric permittivity of the cylinder, $\varepsilon_1 = 1$ – the dielectric constant of the medium. The calculated diffraction field 2.5 × 2.5 μm (128×128 samples).

It is evident that in a wide range $0.18 < \gamma < 0.35$ the discrepancy does not exceed 1% for TE-polarization, and does not exceed 6% at $0.35 < \gamma < 0.63$ for the case of TM-polarization.

Figure 3.39 shows plots of the dependence of discrepancy σ of the amplitude $|E_x|$ (TE-polarization), $|H_x|$ (TM-polarization) of the diffraction fields, calculated by the methods (3.158), (3.179), (3.180) and through a series of Bessel functions (3.199), (3.200), (3.207), (3.208) on the number of iterations [86]. The simulation parameters are the same as in Fig. 3.38. The number of terms in the series of the Bessel functions was equal to 15.

The calculation time of 120 iterations of the diffraction field of 128×128 readings on a Celeron 1000 MHz computer is about 13 seconds. The value of the discrepancy σ for 120 iterations reached 0.0064 for TE-polarization and 0.041 for

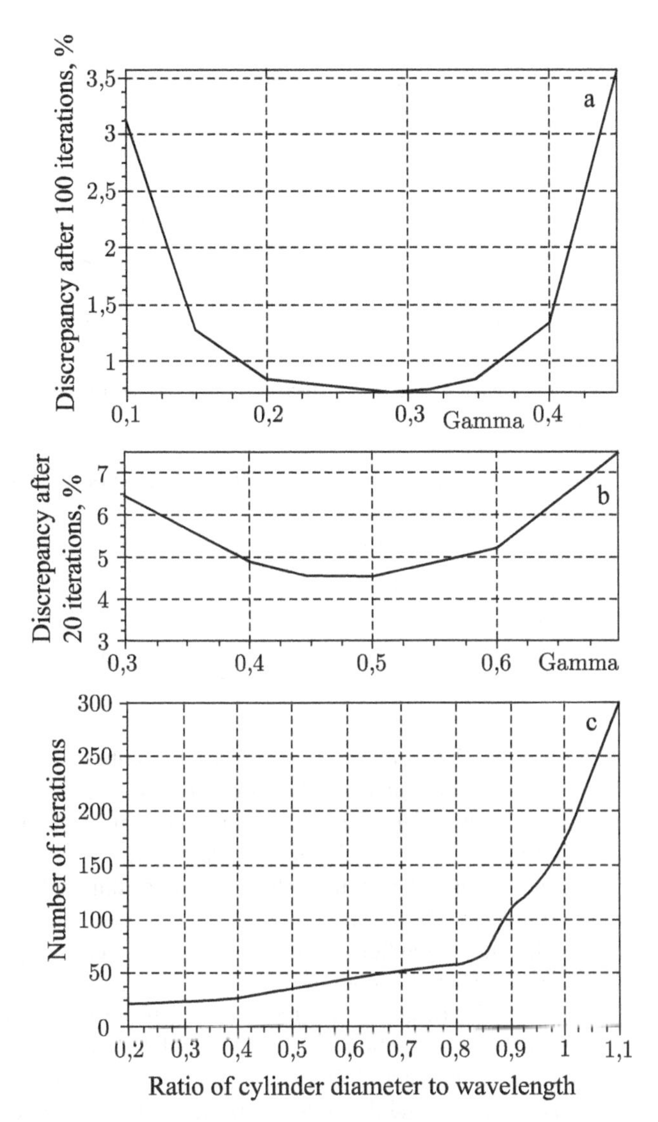

Fig. 3.38. Graphs of the dependence of discrepancy σ on the relaxation parameter γ for the numerical experiment with the parameters $\lambda = 1$ μm, D = 1 μm, $\varepsilon_1 = 2$, $\varepsilon_2 = 1$, $\omega_0 = 0.5$ μm: a) TE-polarization, b) TM-polarization, c) a graph of the dependence of the number of iterations (TE-polarization, $\sigma < 1\%$) on the ratio of the cylinder diameter to the wavelength, at $\varepsilon_1 = 2$, $\gamma = 0$.

TM-polarization. The relaxation constant was chosen as optimum and equal to $\gamma = 0.35$ for TE-polarization.

Consider the dependence of errors in the calculation of the diffraction field by the iterative algorithm on the number of the nodes of the sampling grid.

Table 3.1 shows the discrepancy σ for different numbers of nodes of the sampling grid for the same calculation parameters.

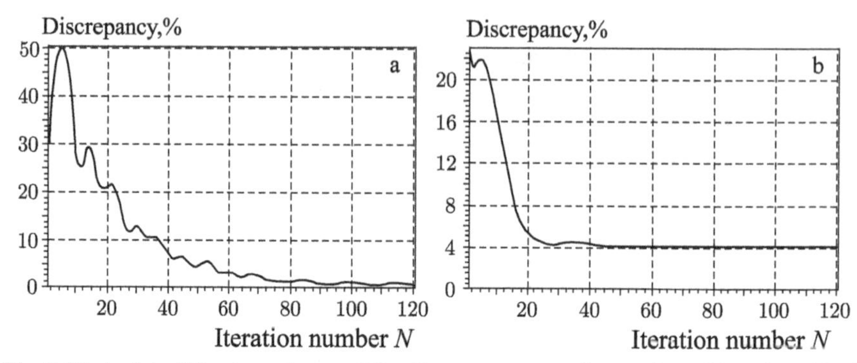

Fig. 3.39. A plot of the dependence of the discrepancy σ on the number of iterations N: a) TE-polarization, b) TM-polarization.

Table 3.1. The dependence of the calculation time and discrepancy σ on the number of samples of the diffraction field

The number of counts, $N \times N$	64×64	128×128	256×256	512×512
σ,%, TE-polarization	2.15	0.6	0.185	0.094
σ,%, TM-polarization	8.91	4.15	3.32	2.83
Calculation time, TE/TM - polarization	4 s/6 s	11 s / 42 s	54 s / 194 s	228 / 860 s

As can be seen from Table 3.1, with an increase in the number of counts the discrepancy gradually diminishes, but the calculation time increases. The time required for the calculation is given for a computer with a Pentium IV Celeron 2400 MHz processor. The discrepancy for TM-polarization is larger with other conditions being equal, due to the additional term in the calculation algorithm – the integral along the contour S. Taking the integral in numerical form is associated with the release of the circuit on the diffraction field which consists of discrete samples, and with the definition of the normals to the contour S.

The convergence of the algorithm for diffraction of a wave on a dielectric cylinder with a circular cross section at a fixed dielectric constant of an object depends on the parameter λ/D, where D is the diameter of the cylinder. Figure 3.38c is a graph of the dependence of the number of iterations (to reach an error σ of less than 1% for TE-polarization) on the ratio of the cylinder diameter to the wavelength. The algorithm converges successfully if $\lambda/D > 0.9$, which is valid for the dielectric constant $\varepsilon_1 < 2$. When $\varepsilon_1 = 2.5$, the graph similar to that shown in Fig. 3.38c can be obtained at $\lambda/D > 1.28$, and $\varepsilon_1 = 3$ if $\lambda/D > 1$. All three groups of the parameters satisfy one inequality: $R/\lambda\,(\varepsilon_1 - 1) < 0.6$, $R = D/2$.

In calculating the diffraction fields for other cylindrical objects, such as a square or a lens, there is no exact analytical solution, and therefore the error σ can not be calculated by the formula (3.213). The process of convergence of the algorithm (3.158) to the solution was monitored with another error σ_2, which is calculated by the formula:

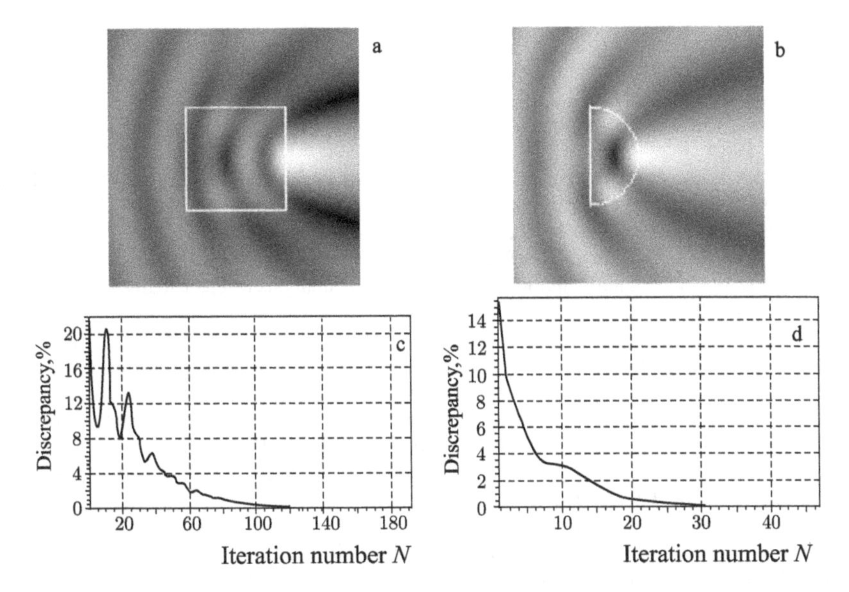

Fig. 3.40. Amplitude distribution $|E_x|$ (in half tones) of diffraction of a plane TE-wave on a dielectric cylinder with a square cross-section (a) and a cylinder with a section in the form of a semi-circle (cylindrical microlens) (b), and the corresponding dependence of the error σ_2 on the number of iterations (c), (d).

$$\sigma_2 = \sqrt{\frac{\sum_{n,m=1}^{N}\left(\phi_{k-1}(n,m) - \phi_k(n,m)\right)^2}{\sum_{n,m=1}^{N}\phi_{k-1}^2(n,m)}}, \tag{3.214}$$

where k is the number of iterations.

Figure 3.40 shows the results of the calculation of diffraction field $|E_x|$ by means of the algorithm (3.158): the calculation of the amplitude of the diffraction of a plane TE-polarized wave on a cylinder with a square cross-section (a) and a cylinder with a semi-circular cross-section (c), and the dependence of the error σ_2 on the number of iterations for the square (b) and a semi-circle (d). For the square the solution was stabilized after 174 iterations ($\sigma_2 = 0.000196$, the calculation time in a Celeron 1000 MHz computer was 17 seconds), and for the semi-circle after 42 iterations ($\sigma_2 =$ 0.000196, computing time 4 seconds). Experimental parameters are the same as for Fig. 3.39: the side of the square (a) and the diameter of the semicircle (c) were equal to a wavelength of 1 μm, the number of samples was equal to 128×128.

References

1. Vaganov R.B. and Katsenelebaum B.Z., Fundamentals of the theory of diffraction, Moscow, Nauka, 1982, 272.
2. Solimeno S., et al., Diffraction and waveguide propagation of optical radiation, Springer-Verlag, 1989.
3. Mirotznik M.A., et al., A hybrid finite elementboundary element method for the analysis of diffractive elements, Journal of Modern Optics, 43. 1996. N. 7. P. 1309–1321.
4. Prather D., Combined scalar-vector method for the analysis of diffractive optical elements, Opt. Eng., 39. 2000. N. 7.P. 1850–1857.
5. Prather D.W.. et al., Field stitching algorithm for the analysis of electrically large diffractive optical elements, Optical Letters, 24. 1999. N. 5. P. 273–275.
6. Montiel F. and Neviere M., Differential theory of gratings: extension to deep gratings of arbitrary profile and permittivity through the R-matrix propagation algorithm, Journal of Optical Society of America, 11. 1994. P. 3241–3250.
7. Golovashkin D.L., et al., Modelling of waveguide propagation of light from the optical radiation in the electromagnetic theory, Komp. Optika, 17. 1997.
8. Brebbia C.A., The boundary Element Method for Engineers, Press, London; Halstead Press, New York, 1978 (Second edition, 1980).
9. Choi M. K., Numerical calculation of light scattering from a layered sphere by the boundary-element method, J. Opt. Soc. Am., 18. 2001. No. 3. P. 577–583.
10. Brebbiya K., et al., Boundary element methods, New York, 1987.
11. Brebbiya K., et al., Application of the boundary element method in engineering, New York, 1982.
12. Davies J.B., Finite element analysis of waveguides and cavities: a review, IEEE Trans. Magn., 29. 1993.P. 15781583.
13. Lichtenberg B., et al., Numerical modelling of diffractive devices using the finite element method, Opt. Eng., 33. 1994. No. 11. P. 3518.
14. Mikhlin S.G., Variational methods in mathematical physics, Moscow, Tekhn. Teoret. Lit., 1957.
15. Syarle F., Finite Element Method for Elliptic Problems, SpringerVerlag, 1980.
16. Blaike R.J.. McNab S.J., Evanescent interferometric lithography, App. Opt., 40. 2001. No. 4. P. 1692–1698.
17. Voznesensky N., Simulation model for light propagation through nanometersized structures, Optical Memory and Neural Networks, 9. 2000. N. 3. P. 175–183.
18. Taflove A., Computational electromagnetics: the finite-difference time-domain method, Artech House, Boston, 1995.
19. Wei X., et al., Finite-element model for three-dimensional optical scattering problems, J. Opt. Soc. Am. A., 24. 2007. No. 3. P. 866–881.
20. Koshiba M. and Saitoh K., Finite-element analysis of birefringence and dispersion properties in actual and idealized holey-fiber structures, App. Opt., 42. 2003. No. 31.
21. Prather D.W. et al., Boundary integral methods applied to the analysis of diffractive optical elements, J. Opt. Soc. Am., 14. 1997. P. 34–43.
22. Tanaka M. and Tanaka K.J., Computer simululation for two-dimensional near-field optics withuseofametal-coateddielectricprobe,Opt. Soc. Am.,2001. Vol. 18. No. 4. P.919–925.
23. Paulus M. and Martin O.J.F., Light propagation and scattering in stratified media: a Green's tensor approach, J. Opt. Soc. Am., 18. 2001. No. 4. P. 854–861.
24. Dou W.B., et al., Diffraction of an electromagnetic beam by an aperture in a conducting screen, J.Opt. Soc. Am., 18. 2001. No. 4. P. 801–806.
25. Lee J.F., et al., Modelling three-dimensional discontinuities in waveguides using non-orthogonal FDTD algorithm, IEEE Trans. Microwave Theory Tech., 40. 1992. P.346–352.
26. Prather D.W. and Shi S., Formulation and application of the finite-difference time-domain method for the analysis of axially symmetric diffractive optical elements, J. Opt. Soc. Am., 16. 1999. No. 5. P. 1131–1142.

27. Shi S., et al., Analysis of diffractive optical elements using a non-uniform finite-difference time-domain method, Opt. Eng., 40. 2001. No. 4. P. 503–510.
28. Gruzdev V. and Gruzdeva A., Finite-difference time-domain modelling of laser beam propagation and scattering in dielectric materials, Proceedings of SPIE, 4436. 2001. P.27–38.
29. Berenger G.P., A perfectly matched layer for the absorption of electromagnetic waves, J. of Comp. Phys., 114. 1994. P. 185–200.
30. Ilyinsky A.S., et al., Mathematical models of electrodynamics, Moscow, Vysshaya shkola, 1991.
31. Kotlyar V.V. and Nesterenko, D.V., A finite element method in the problem of light diffraction by microoptics, Optical Memory and Neural Networks, vol. 9, no. 3. 2000. P.209–219.
32. Electromagnetic Theory of Gratings: Topics in current physics, Ed. by R. Petit, N.Y., Springer-Verlag. 1980.
33. Moharam M.G., et al., Rigorous coupled wave analysis of planar grating diffraction, J. Opt. Soc.Amer. 71. 1981. P. 811–818.
34. Chandezon J., et al., A new theoretical method for diffraction gratings and its applications, J.Opt. (Paris), 11. 1980. P. 235–241.
35. Urbach H.P. Convergence of the Galerkin method for two-dimensional electromagnetic problems, SIAM J. Numer. Anal., 28. 1991. P. 697–710.
36. Bao G., et al., Finite element approximation of time harmonic waves in periodic structures, SIAM J. Numer. Anal., 32. 1995. P. 1155–1169.
37. Elschner J., et al., Finite element solution of conical diffraction problems, Advances in Comp. Math., 16. 2002. P. 139–156.
38. Yee K., Numerical solution of initial boundary value problems involving Maxwell's equations in isotropic media, IEEE Trans. Antennas Propag. AP14. 1966. P. 302–307.
39. Saj W.M., FDTD simulations of 2D plasmon waveguide on silver nanorods in hexagonal lattice, Opt. Express, 13. 2005. P. 4818–4827.
40. GSolver, Rigorous diffraction grating analysis [electronic resource]. http://www.gsolver.com. Grating Solver Development Company.
41. Methods of Computer Optics, Ed. V.A. Soifer, Moscow, Fizmatlit, 2000.
42. Neganov V.A., et al., Linear macroscopic electrodynamics, Moscow, Radio i Svyaz', 2000. Vol.1..
43. Methods For Computer Design of Diffractive Optical Elements, Ed. by V.A. Soifer, N.Y., Wiley Interscience Publication, John Wiley & Sons, Inc. 2002.
44. Moharam M.G., et al., Formulation for Stable and Efficient Implementation of the Rigorous Coupled-Wave Analysis of Binary Gratings, J. Opt. Soc. Amer. 12 (5). 1995. P. 1068–1076.
45. Nesterenko D.V. and Kotlyar V.V., Analysis of the diffraction of light on the elements of the cylindrical microoptics combined finite-element and boundary element, Komp. Optika, 32. N1. 2008. P. 23–28.
46. Kotlyar V.V. and Likhmanov M.A., Analysis of the diffraction of light on the microscopic objects by means of an integral equation on the finite element method, Komp. Optika, 2001. No. 21. 1922.
47. Born M. and Wolf E., Principles of Optics, Moscow, Nauka, 1973.
48. Bronstein I.N. and Semendyaev K.A., Handbook of mathematics for engineers and students of universities, Moscow, Nauka, 1981.
49. Solimeno, S., et al., Diffraction ad waveguide propagation of optical radiation, Springer Verlag, 1989..
50. Handbook of mathematical functions, Ed. M. Abramowitz and I. Stigan. Moscow, Nauka, 1979.
51. Korn G. and Korn T., Handbook of Mathematics for Scientists and Engineers, Moscow, Nauka, 1977.
52. Volkov E.A., Numerical methods, Moscow, Nauka, 1987.

53. Kotlyar V.V. and Nesterenko D.V., Analysis of the problem of light diffraction by micro-hybrid finite-element – boundary elements, Komp. Optika, 2000. No. 20. 10–14.
54. Koltyar V.V. Diffraction of a plane electromagnetic wave by a gradient-index dielectric microcylinder, in: Perspectives in Engineering Optics, Ed. by K. Singh, V.K. Rastogi, Anita Publications, Delhi, 2003. 38–46.
55. Kotlyar V.V. and Likhmanov M.A., Analysis of electromagnetic wave diffraction on a round infinite cylinder with several homogeneous layers, Komp. Optika, 24. 2002. 26–32.
56. Kotlyar V.V. and Likhmanov M.A., Diffraction of a plane electromagnetic wave by a gradient optical element with a transverse cylindrical symmetry, Physics of wave propagation and radio systems, Samara, PGATI, v.5, No. 4, 2002. 37–43.
57. Barton J.P., Electromagnetic field calculations for an irregularly shaped, near-spheroidal particle with arbitrary illumination, Journal of the Optical Society of America, 19. 2002. No.12. 2429–2435.
58. Farafonov V.G., et al., Light scattering by multilayered non-spherical particles: a set of methods, Journal of Quantitative Spectroscopy & Radiative Transfer, 79. 2003. No.80. 599–626.
59. Flores J.R., Gradien-tindex with spherical symmetry, Journal of Modern Optics, 46. 1999. No.11. 1513–1525.
60. Prather D.W. and Shi S., Electromagnetic analysis of axially symmetric diffractive optical elements illuminated by oblique incident plane waves, Journal of the Optical Society of America, 18. 2001. No.11. 2901–2907.
61. Zakharov E.V. and Eremin Yu.A., method for solving axisymmetric problems of diffraction of electromagnetic waves on bodies of revolution, Journal of Computational Mathematics and Mathematical Physics, 19. 1979. No. 5. 1344–1348.
62. Farafonov V.G., et al., A new solution of the light scattering problem for axisymmetric particles, Journal of Quantitative Spectroscopy & Radiative Transfer, 63. 1999. 205–215.
63. Arias-Gonzalez J.R. and Nieto-Vesperinas M., Resonant near-field eigenmodes of nanocylinders on flat surfaces under both homogeneous and inhomogeneous lightwave excitation, Journal of Optical Society of America, 18. 2001. No.3. 657–665.
64. Zakharov E.V., Diffraction of a plane electromagnetic field on a uniform cylindrical body immersed in a layered medium, Izv. AN SSSR, Fiz. Zemli, 1. 1969. 57–62.
65. Anyutin A.P. and Stasevich V.I. Scattering by a multilayer cylindrical structure [electronic resource]. http://at.yorku.ca/cgibin/amca/cacu75.
66. Zhang M., et al., Electromagnetic scattering by a multilayer gyrotropic bianisotropic circular cylinder, Progress In Electromagnetics Research, 40. 2003. 91–111.
67. Flores J.R., Spherically symmetric GRIN amplitude formers, Journal of Modern Optics, 48. 2001. No.7. 1225–1238.
68. Gordon J.M., Spherical gradient-index lenses as perfect imaging and maximum power transfer devices, Applied Optics, 39. 2000. No.22. 3825–3832.
69. Luneberg R.K., Mathematical theory of optics, Brown U. Press, Providence, R.I., 1944.
70. Kotlyar V.V. and Melekin A.S., Abel transformation in problems of synthesis of gradient optical elements, Komp. Optika, 22. 2001. 29–36.
71. Zelkin E.G. and Petrov R.A., Lens antenna. Moscow, Sov. radio, 1971.
72. Miller W., Symmetry and separation of variables. Springer Verlag, 1981.
73. Kotlyar V.V. and Melekhin A.A., Abel transform to calculate the gradient of optical elements with spherically symmetric distribution of the refractive index, Komp. Optika, 24. 2002. 48–52.
74. Kotlyar V.V., et al., A fast method for calculating the electromagnetic wave diffraction on cylindrical dielectric objects, Komp. Optika, 2004. N. 25. 24–28.
75. Kotlyar V.V., et al., A method for fast calculation of the diffraction of TE- and TM--polarized electromagnetic waves by a 2D microscopic objects, Vestnik of Samara State

Aerospace University named after Academician S.P. Korolev, the Second summer school of young scientists in diffractive optics and image processing, 2004. 18–19.
76. Kotlyar V.V., et al., A method for fast calculation of the diffraction of laser radiation at microscopic objects, Opt. Zhurnal, 2005. V. 72. N. 5. 55–61.
77. Kotlyar V.V., et al., Method for rapidly calculating the diffraction of laser radiation at microscopic objects, J. Opt. Technol, 2005. V. 72. 400–405.
78. Vasilenko G.I. and Taratorin A.M., Image restoration, Moscow, Radio i Svyaz', 1986.
79. Kotlyar V.V. and Likhmanov M.A., Analysis of light diffraction by microoptics using finite elements method, Opt.Mem. and Neut. Net., 2001. V. 10. N. 4. 257–265.
80. Kotlyar V.V., et al., Calculation of the Poynting vector and the pressure force of an electromagnetic wave in a homogeneous dielectric cylinder, Izv. SNTs RAN 2005. V. 7. N. 1. 83–91.
81. Kotlyar V.V., et al., Fast calculation of diffraction by microobject, Book of abstracts, Topical meeting on Optoinformatics, 'Optics meets Optika, St. Peterburg, 2004. 52–54.
82. Prudnikov A.P., et al., Integrals and series. Special functions, Moscow, Nauka, 1983.
83. Korsakova S.S., et al., Opt. Zhurnal, 2004. V. 71. N. 7. 65–70.
84. Korsakova S.S., et al., A method of calculating the diffraction and refraction of radiation at a dielectric cylinder, J. Opt. Technol., 2004. V. 71. 472–477.
85. Khonina S.N. et al, Studying diffraction of laser light by a dielectric circular cylinder, Abstracts of the conference Optoinformatics 'Optics meets Optika', St. Peterburg, 2003. 52–54.
86. Kotlyar V.V., et al., Calculation of the Umov–Pointing vector and the electromagnetic wave pressure force on a homogeneous dielectric cylinder, Proceedings of SPIE, Saratov Fall Meeting, Laser Physics and Photonics, Spectroscopy, and Molecular Modelling, 2004. V.5773. 106–118.

Chapter 4

Modelling of periodic diffractive micro- and nanostructures

This chapter describes a numerical method for solving the problem of ligh diffraction by periodic diffractive micro- and nanostructures. The method is used to calculate and study the diffraction structures for a number of modern trends in nanophotonics, including plasmonics, metamaterials, nanometrology.

Section 4.1 deals with the method of rigorous coupled-wave analysis to solve the problem of diffraction of a plane wave on two- and three-dimensional diffraction structures and diffractive gratings. This numerical method for solving Maxwell's equations is best suited for the analysis of micro- and nanostructures described by a periodic function of the dielectric constant. This variant of the method is a modification of the numerically stable modification of the method of rigorous coupled-wave analysis, proposed by M.G. Moharam, D.A. Pommet, E.B. Grann and T.K. Gaylord in 1995 for solving the problem of diffraction on two-dimensional diffractive structures. Attention is given to the current variant of the method, including the case of tensor dielectric and magnetic permittivities and diffraction on three-dimensional structures. The considered variant of the method also includes the application of special rules of expansion into a Fourier series of the product of functions proposed by Lifeng Li in 1996. The application of these rules greatly improves the convergence of the method in solving the diffraction problem on metallic and metallic–dielectric diffractive structures.

Section 4.2 examines surface plasma polaritons (surface plasmon– polaritons), with the study and calculation of the diffraction structures for the formation of interference patterns of surface plasma polaritons. The diffractive structures consist of a dielectric diffraction grating and a metallic layer deposited on the substrate. The calculation and modelling of structures are carried out by rigorous coupled-wave analysis. The parameters of the diffraction structures are calculated from the conditions of of a given set of surface plasma polaritons of different directions at the lower boundary of the metal layer. As a result, a periodic interference pattern of surface plasma polaritons is generated directly under the metallic layer. Periods of of the formed interference patterns are substantially of the subwavelength type. The promising field of application of the considered diffraction structures is nanolithography, based on recording the interference patterns of surface plasma polaritons in a photoresist.

Section 4.3 considers the magneto-optical properties of two-layer metallic-dielectric heterostructures consisting of a metallic diffractive grating and a dielectric magnetized layer. Calculation and study of the magneto-optical properties of structures are based on the method of rigorous coupled-wave analysis. The results show that these structures have resonant magneto-optical effects associated with the rotation of the plane of polarization of the incident wave and with a change in the reflectance (transmittance) of the structure when the magnetization of the layer changes. The above structures can be used as magnetic field sensors, gas sensors, the devices for the intensity modulation of light, controlled by an external magnetic field.

In Section 4.4 we consider the problem of metrology of nanostructures based on the reflectometry method (ellipsometry). Optical reflectometry is a contactless method of measuring the parameters of micro- and nanostructures by measuring the parameters of the light reflected from the structure of the field. The section reviews a number of mathematical methods for determining the geometric parameters of diffraction gratings by measuring the parameters of the wave in the reflected zeroth diffraction order. The reflected field is calculated by rigorous coupled-wave analysis. The effectiveness of the methods is illustrated by the example of determining the parameters of a trapezoidal profile. The error in determining the parameters of a trapezoidal profile is less than 1 nm.

4.1. The method of rigorous coupled-wave analysis for solving the diffraction problem in periodic diffractive structures

This section describes the Fourier modal method (rigorous coupled-wave analysis), designed to address the problem of diffraction of a plane electromagnetic wave on a periodic diffractive structure. This method of solving Maxwell's equations is a variation of the differential method and holds a leading position in terms of functionality and breadth of use. The method is applicable for the analysis of periodic structures with complex geometry, the material of the structure can be anisotropic. Using layers of an anisotropic absorbing material, the method can be used efficiently to solve the problem of diffraction on aperiodic structures [14, 15].

4.1.1. The equation of a plane wave

This subsection is auxiliary. We considered the derivation of the general equation of a plane wave in an isotropic medium used in the following description of the rigorous coupled-wave analysis.

We write the Maxwell equations and constitutive equations in the Gaussian system of units:

$$\begin{cases} \nabla\times\mathbf{E} = -\dfrac{1}{c}\dfrac{\partial}{\partial t}\mathbf{B}, \\ \nabla\times\mathbf{H} = \dfrac{1}{c}\dfrac{\partial}{\partial t}\mathbf{D}, \\ \mathbf{B} = \mu\mathbf{H}, \\ \mathbf{D} = \varepsilon\mathbf{E}. \end{cases} \tag{4.1.1}$$

For a monochromatic field

$$\begin{aligned} \mathbf{E}(x,y,z,t) &= \mathbf{E}(x,y,z)\exp(-i\omega t), \\ \mathbf{H}(x,y,z,t) &= \mathbf{H}(x,y,z)\exp(-i\omega t), \end{aligned} \tag{4.1.2}$$

system (4.1.1) becomes:

$$\begin{cases} \nabla\times\mathbf{E} = ik_0\mu\mathbf{H}, \\ \nabla\times\mathbf{H} = -ik_0\varepsilon\mathbf{E}, \end{cases} \tag{4.1.3}$$

where $k_0 = 2\pi/\lambda$, λ is the wavelength. Expanding the rotor operator, we obtain:

$$\nabla\times\mathbf{E} = \begin{bmatrix} \dfrac{\partial E_z}{\partial y} - \dfrac{\partial E_y}{\partial z} \\ \dfrac{\partial E_x}{\partial z} - \dfrac{\partial E_z}{\partial x} \\ \dfrac{\partial E_y}{\partial x} - \dfrac{\partial E_x}{\partial y} \end{bmatrix} = ik_0\mu\begin{bmatrix} H_x \\ H_y \\ H_z \end{bmatrix}, \quad \nabla\times\mathbf{H} = \begin{bmatrix} \dfrac{\partial H_z}{\partial y} - \dfrac{\partial H_y}{\partial z} \\ \dfrac{\partial H_x}{\partial z} - \dfrac{\partial H_z}{\partial x} \\ \dfrac{\partial H_y}{\partial x} - \dfrac{\partial H_x}{\partial y} \end{bmatrix} = -ik_0\varepsilon\begin{bmatrix} E_x \\ E_y \\ E_z \end{bmatrix}. \tag{4.1.4}$$

To obtain the equation of a plane wave we seek the solution of equations (4.1.4) in the form of

$$\mathbf{\Phi}(x,y,z) = \mathbf{\Phi}\exp(ik_0(\alpha x + \beta y + \gamma z)), \tag{4.1.5}$$

where $\mathbf{\Phi} = \begin{bmatrix} E_x & E_y & E_z & H_x & H_y & H_z \end{bmatrix}^T$ is the column vector of the field components. The values α, β, γ in (4.1.5) determine the direction of propagation of a plane wave. Substituting (4.1.5) into (4.1.4), we obtain:

$$ik_0\begin{bmatrix} \beta E_z - \gamma E_y \\ \gamma E_x - \alpha E_z \\ \alpha E_y - \beta E_x \end{bmatrix} = ik_0\mu\begin{bmatrix} H_x \\ H_y \\ H_z \end{bmatrix}, \quad ik_0\begin{bmatrix} \beta H_z - \gamma H_y \\ \gamma H_x - \alpha H_z \\ \alpha H_y - \beta H_x \end{bmatrix} = -ik_0\varepsilon\begin{bmatrix} E_x \\ E_y \\ E_z \end{bmatrix}. \tag{4.1.6}$$

We represent the equation (4.1.6) in matrix form:

$$\begin{bmatrix} 0 & -\gamma & \beta & -\mu & 0 & 0 \\ \gamma & 0 & -\alpha & 0 & -\mu & 0 \\ -\beta & \alpha & 0 & 0 & 0 & -\mu \\ \varepsilon & 0 & 0 & 0 & -\gamma & \beta \\ 0 & \varepsilon & 0 & \gamma & 0 & -\alpha \\ 0 & 0 & \varepsilon & -\beta & \alpha & 0 \end{bmatrix} \mathbf{\Phi} = \mathbf{A} \cdot \mathbf{\Phi} = \mathbf{0}. \tag{4.1.7}$$

By direct computation we can easily obtain that the determinant of system (4.1.7) has the form:

$$\det \mathbf{A} = \varepsilon\mu\left(\alpha^2 + \beta^2 + \gamma^2 - \varepsilon\mu\right)^2. \tag{4.1.8}$$

The sought non-trivial solution of (4.1.7) exists when the determinant (4.1.8) is zero. In the case $\mu = 1$ the determinant (4.1.8) vanishes if

$$\alpha^2 + \beta^2 + \gamma^2 = \varepsilon. \tag{4.1.9}$$

Let us write the solution of (4.1.6) explicitly. Under the condition (4.1.9), the rank of the system (4.1.7) is equal to four, so to write down the solution we fix the values of the amplitudes E_z and H_z. We introduce the so-called *E*- and *H*-waves. For the *E*-wave $E_z \neq 0$, $Hz \equiv 0$ and for the *H*-wave $H_z \equiv 0$, $E_z \equiv 0$. We represent the desired solution as a superposition of the *E*- and *H*-waves in the form:

$$\mathbf{\Phi} = \frac{1}{\sqrt{\alpha^2+\beta^2}} \begin{bmatrix} -\alpha\gamma \\ -\beta\gamma \\ \alpha^2+\beta^2 \\ \beta\varepsilon \\ -\alpha\varepsilon \\ 0 \end{bmatrix} A'_E + \frac{1}{\sqrt{\alpha^2+\beta^2}} \begin{bmatrix} -\beta \\ \alpha \\ 0 \\ -\alpha\gamma \\ -\beta\gamma \\ \alpha^2+\beta^2 \end{bmatrix} A'_H, \tag{4.1.10}$$

where A'_E and A'_H are the amplitudes of the *E*- and *H*-waves, respectively.

We write the components of the vector $\mathbf{p} = (\alpha, \beta, \gamma)$ representing the direction of the wave, in terms of angles θ and ϕ of the spherical coordinate system:

$$\alpha = \sqrt{\varepsilon}\cos\phi\sin\theta,\ \beta = \sqrt{\varepsilon}\sin\phi\sin\theta,\ \gamma = \sqrt{\varepsilon}\cos\theta, \tag{4.1.11}$$

where θ is the angle between the vector $\mathbf{p}$ and the axis Oz, ϕ is the angle between the plane of incidence and the plane xOz. For these angles, the following relations:

$$\sqrt{\alpha^2+\beta^2} = \sqrt{\varepsilon}\sin\theta,\ \frac{\alpha}{\sqrt{\alpha^2+\beta^2}} = \cos\phi,\ \frac{\beta}{\sqrt{\alpha^2+\beta^2}} = \sin\phi. \tag{4.1.12}$$

We write separately the *y*- and *x*-components of the electric and magnetic fields:

$$\bar{\mathbf{\Phi}} = \begin{bmatrix} E_y \\ E_x \\ H_y \\ H_x \end{bmatrix} = \begin{bmatrix} -\sqrt{\varepsilon}\cos\theta\sin\phi \\ -\sqrt{\varepsilon}\cos\theta\cos\phi \\ -\varepsilon\cos\phi \\ \varepsilon\sin\phi \end{bmatrix} A'_E + \begin{bmatrix} \cos\phi \\ -\sin\phi \\ -\sqrt{\varepsilon}\cos\theta\sin\phi \\ -\sqrt{\varepsilon}\cos\theta\cos\phi \end{bmatrix} A'_H =$$

$$= \begin{bmatrix} \mathbf{R} & 0 \\ 0 & \mathbf{R} \end{bmatrix} \begin{bmatrix} 0 & 1 \\ -\gamma & 0 \\ -\varepsilon & 0 \\ 0 & -\gamma \end{bmatrix} \begin{bmatrix} A'_E \\ A'_H \end{bmatrix}, \quad (4.1.13)$$

where **R** is the rotation matrix of the form:

$$\mathbf{R} = \begin{bmatrix} \cos\phi & \sin\phi \\ -\sin\phi & \cos\phi \end{bmatrix}. \quad (4.1.14)$$

To describe the polarization of the wave, we introduce an angle ψ – the angle between the vector **E** and the plane of incidence. By the plane of incidence we mean the plane containing the vector of direction of the wave **p** and the axis Oz. In this case,

$$A'_H = A\sin\psi,\ A'_E = A\cos\psi. \quad (4.1.15)$$

Note that in solving the problem of diffraction the case $\phi = 0$ corresponds to the so-called flat incidence. In this case, the E-waves are called waves with TM-polarization, and the H-waves – waves with TE-polarization.

To characterize the energy carried by the wave, it is convenient to normalize the coefficients A'_E and A'_H. We compute the z-component of the Poynting vector corresponding to the energy flux through the plane xOy. For the E-wave

$$2S_z = \mathrm{Re}\left|E_x H_y^* - E_y H_x^*\right| = \left|A'_E\right|^2 \mathrm{Re}\left(\varepsilon^* \sqrt{\varepsilon}\right)\cdot\cos\theta = \left|A'_E\right|^2 \left|\varepsilon\right| \mathrm{Re}\sqrt{\varepsilon}\cos\theta, \quad (4.1.16)$$

for the H-wave

$$2S_z = \mathrm{Re}\left(E_x H_y^* - E_y H_x^*\right) = \left|A'_H\right|^2 \mathrm{Re}\left(\sqrt{\varepsilon^*}\right)\cdot\cos = \left|A'_H\right|^2 \mathrm{Re}\sqrt{\varepsilon}\cdot\cos\theta. \quad (4.1.17)$$

We redefine the expressions for the amplitudes in the form of:

$$A'_E = \frac{1}{\left|\varepsilon\right|\sqrt{\mathrm{Re}\sqrt{e}\cdot\cos\theta}} A_E, \quad (4.1.18)$$

$$A'_H = \frac{1}{\sqrt{\mathrm{Re}\sqrt{\varepsilon}\cos\theta}} A_H. \quad (4.1.19)$$

In this case, the values $|A_E|^2$ and $|A_H|^2$ correspond to the energy flux through the plane xOy. Such a normalization provides the fulfillment of the law of conservation of energy while solving the problem of diffraction on diffraction structures of a non-absorbing material.

4.1.2. The method of rigorous coupled-wave analysis in the two-dimensional case

The method of rigorous coupled-wave analysis used to solve the problem of diffraction on periodic diffraction structures consisting of a set of layers [1–13]. In each layer, the dielectric and magnetic permeabilities of materials of the structure are constant in the direction normal to the layer (along the axis Oz). The method is based on the representation of the field in the layers of the structure in the form of segments of Fourier series and on subsequent equating of the tangential field components at the boundaries between the layers. This reduces the solution of the problem to solving a system of linear equations [1–13].

In this section we consider the method for two-dimensional gratings. In the two-dimensional case, the material properties of the structure are constant along the axis Oy. The materials of the structure are given in general terms by the tensors of dielectric permittivity and magnetic permeability.

4.1.2.1. The geometry of the structure and formulation of the problem

Let us consider a grating with a period Λ along the Ox-axis (Fig. 4.1.1). For gratings with a continuous profile (dashed line in Fig. 4.1.1.) the method assumes the approximation of the profile of the grating by a set of binary layers. We assume that the diffractive grating consists of L binary layers (Fig. 4.1.1). The dielectric permittivity ε and magnetic permeability μ of the layers depend only on the variable x. The boundaries of the layers are the lines $z = d_i$, the i-th layer is located in the range $d_i < z < d_{i-1}$.

Above this structure in the range $z > d_0 = 0$ there is a homogeneous dielectric with a refractive index $n_1 = \sqrt{\varepsilon_1}$. Under the structure at $z < d_L$ there is a homogeneous dielectric with a refractive index $n_2 = \sqrt{\varepsilon_2}$.

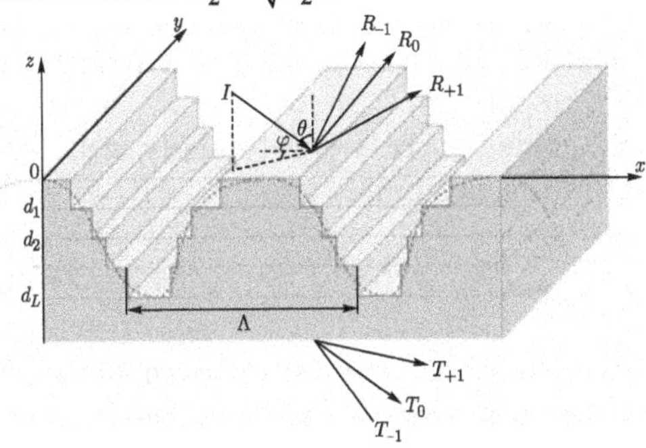

Fig. 4.1.1. Geometry of the problem of diffraction on a two-dimensional grating.

On top of the grating there is a plane monochromatic wave (wavelength λ), whose direction is given by the angles θ and ϕ in a spherical coordinate system (Fig. 4.1.1). The polarization of the wave is determined by the angle ψ between the plane of incidence and the vector **E** (see (4.1.15)).

The solution to the problem of diffraction of light on a periodic structure is to calculate the intensity or complex amplitudes of the diffraction orders. Diffraction orders are the reflected and transmitted plane waves arising from the diffraction of the incident wave on the structure. There are reflected orders with amplitudes R_i, $i = 0, \pm1, \pm2,...$and transmitted orders with amplitudes T_i, $i = 0, \pm1, \pm2,...$ (Fig. 4.1.1). The diffraction orders are also separated into evanescent and propagating. The amplitude of the evanescent orders decreases exponentially with distance from the grating.

4.1.2.2. Presentation of the field above and below the structure

The field above and below the structure is written as a superposition of plane waves (Rayleigh expansion). According to (4.1.5), (4.1.10), we represent the plane wave by the vector of six components:

$$\mathbf{\Phi} = \begin{bmatrix} E_x & E_y & E_z & H_x & H_y & H_z \end{bmatrix}^T. \tag{4.1.20}$$

The incident wave equation has the form:

$$\mathbf{\Phi}^I(x, y, z) = \mathbf{\Phi}^I(\psi)\exp\left(i(k_{x,0}x + k_y y - k_{z,I,0}z)\right), \tag{4.1.21}$$

where $\mathbf{\Phi}^I(\psi)$ indicates the dependence of the incident wave on polarization (see (4.1.15)). Constants $k_{x,0} = k_0 n_1 \sin\theta\cos\phi$, $k_y = k_0 n_1 \sin\theta\sin\phi$, $k_{z,I,0} = \sqrt{(k_0 n_I)^2 - k_{x,0}^2 - k_y^2}$ in (4.1.21) are defined by the angles θ and ϕ representing the directions of the incident wave.

The field above the grating corresponds to a superposition of the incident wave and the reflected diffraction orders:

$$\mathbf{\Phi}^1(x, y, z) = \mathbf{\Phi}^I(x, y, z) + \sum_m \mathbf{\Phi}_m^R(R_m)\exp\left(i(k_{x,m}x + k_y y + k_{z,I,m}z)\right), \tag{4.1.22}$$

where $\mathbf{\Phi}(R) = \mathbf{\Phi0}_E R_E + \mathbf{\Phi}_H R_H$. The expression for $\mathbf{\Phi}(R)$ corresponds to the representation (4.1.10) of the reflected wave as the sum of E- and H-waves with complex amplitudes R_E and R_H, respectively.

The field below the grating can be written similarly, in the form of a superposition of transmitted waves (diffraction orders):

$$\mathbf{\Phi}^2(x, y, z) = \sum_m \mathbf{\Phi}_m^T(T_m)\exp\left(i(k_{x,m}x + k_y y - k_{z,II,m}(z - d_L))\right). \tag{4.1.23}$$

Directions of orders in (4.1.22), (4.1.23) are given by the values $k_{x,i}$, k_y, $k_{z,I,i}$, $k_{z,II,i}$ that are called propagation constants and have the form:

$$k_{x,i} = k_0 \left(n_1 \sin\theta \cos\phi + i\frac{\lambda}{\Lambda} \right),$$
$$k_y = k_0 n_1 \sin\theta \sin\phi, \quad (4.1.24)$$
$$k_{z,l,i} = \sqrt{(k_0 n_l)^2 - k_{x,i}^2 - k_y^2},$$

where the index l is taken to be unity for the field over the grating and 2 – for the field under it. The submitted expression for $k_{x,i}$ follows from the Bloch–Floquet theorem [16, 17]. From a physical point of view the form of $k_{x,i}$ ensures that the so-called quasi-periodicity conditions

$$\mathbf{\Phi}^l(x+\Lambda, y, z) = \mathbf{\Phi}^l(x, y, z)\exp\left(ik_{x,0}\Lambda\right), \; l = 1,2. \quad (4.1.25)$$

is fulfilled.

According to (4.1.25), the amplitude of the field does not change with the shift in the period. The waves with real $k_{z,l,i}$ are propagating, with imaginary – evanescent.

4.1.2.3. The system of differential equations to describe the field inside the layer

Let us now consider the field inside the single l-th layer. For simplicity, we omit the index l.

The field in each layer is described by the basic Maxwell equations for a monochromatic field in the form of:

$$\begin{cases} \nabla \times \mathbf{E} = ik_0 \ddot{\boldsymbol{\mu}} \mathbf{H}, \\ \nabla \times \mathbf{H} = -ik_0 \ddot{\boldsymbol{\varepsilon}} \mathbf{E}, \end{cases} \quad (4.1.26)$$

where $\ddot{\boldsymbol{\varepsilon}}, \ddot{\boldsymbol{\mu}}$ in general are tensors:

$$\ddot{\boldsymbol{\varepsilon}} = \begin{bmatrix} \varepsilon_{1,1} & \varepsilon_{1,2} & \varepsilon_{1,3} \\ \varepsilon_{2,1} & \varepsilon_{2,2} & \varepsilon_{2,3} \\ \varepsilon_{3,1} & \varepsilon_{3,2} & \varepsilon_{3,3} \end{bmatrix}, \quad \ddot{\boldsymbol{\mu}} = \begin{bmatrix} \mu_{1,1} & \mu_{1,2} & \mu_{1,3} \\ \mu_{2,1} & \mu_{2,2} & \mu_{2,3} \\ \mu_{3,1} & \mu_{3,2} & \mu_{3,3} \end{bmatrix}. \quad (4.1.27)$$

The tensor components depend only on the coordinate x. Expanding the operator of the rotor (4.1.26), we obtain:

$$\begin{cases} \dfrac{\partial E_z}{\partial y} - \dfrac{\partial E_y}{\partial z} & = ik_0(\mu_{1,1}H_x + \mu_{1,2}H_y + \mu_{1,3}H_z), \\ \dfrac{\partial E_x}{\partial z} - \dfrac{\partial E_z}{\partial x} & = ik_0(\mu_{2,1}H_x + \mu_{2,2}H_y + \mu_{2,3}H_z), \\ \dfrac{\partial E_y}{\partial x} - \dfrac{\partial E_x}{\partial y} & = ik_0(\mu_{3,1}H_x + \mu_{3,2}H_y + \mu_{3,3}H_z), \\ \dfrac{\partial H_z}{\partial y} - \dfrac{\partial H_y}{\partial z} & = -ik_0(\varepsilon_{1,1}E_x + \varepsilon_{1,2}E_y + \varepsilon_{1,3}E_z), \\ \dfrac{\partial H_x}{\partial z} - \dfrac{\partial H_z}{\partial x} & = -ik_0(\varepsilon_{2,1}E_x + \varepsilon_{2,2}E_y + \varepsilon_{2,3}E_z), \\ \dfrac{\partial H_y}{\partial x} - \dfrac{\partial H_x}{\partial y} & = -ik_0(\varepsilon_{3,1}E_x + \varepsilon_{3,2}E_y + \varepsilon_{3,3}E_z). \end{cases} \tag{4.1.28}$$

We represent the components of the electric and magnetic fields in the form of Fourier series in the variable x:

$$\begin{cases} E_x & = & \sum\limits_j S_{x,j}(z)\exp\left(i(k_{x,j}x + k_y y)\right), \\ E_y & = & \sum\limits_j S_{y,j}(z)\exp\left(i(k_{x,j}x + k_y y)\right), \\ E_z & = & \sum\limits_j S_{z,j}(z)\exp\left(i(k_{x,j}x + k_y y)\right), \\ H_x & = -i & \sum\limits_j U_{x,j}(z)\exp\left(i(k_{x,j}x + k_y y)\right), \\ H_y & = -i & \sum\limits_j U_{y,j}(z)\exp\left(i(k_{x,j}x + k_y y)\right), \\ H_z & = -i & \sum\limits_j U_{z,j}(z)\exp\left(i(k_{x,j}x + k_y y)\right). \end{cases} \tag{4.1.29}$$

Representations (4.1.29) are written taking into account the quasi-periodicity of the field components with respect to x. We confine ourselves to the finite number of terms in the expansions (4.1.29) correspond to $-N \le j \le N$. We form from the quantities $S_{x,j}$, $S_{y,j}$, $S_{z,j}$, $U_{x,j}$, $U_{y,j}$, $U_{z,j}$ column vectors $\mathbf{S}_x$, $\mathbf{S}_y$, $\mathbf{S}_z$, $\mathbf{U}_x$, $\mathbf{U}_y$, $\mathbf{U}_z$ containing $2N$+1 elements.

Consider the Fourier series expansion of the product of two functions $\varepsilon_a(x)$ and $E_b(x,y,z)$:

$$\varepsilon_a(x)E_b(x)=\left(\sum_m e_{a,m}\exp\left(i\frac{2\pi}{\Lambda}mx\right)\right)\cdot\left(\sum_l S_{b,l}(z)\exp\left(i(k_{x,l}x+k_yy)\right)\right)=$$

$$=\sum_{j=l+m}\left(\sum_l e_{a,j-l}S_{b,l}(z)\right)\exp\left(i(k_{x,j}x+k_yy)\right). \tag{4.1.30}$$

We restrict ourselves in (4.1.30) by a finite number of terms in the expansion at $-N\le j\le N$ and in the following expression:

$$[\varepsilon_a(x)E_b(x)]=[\![\varepsilon_a(x)]\!]\mathbf{S}_b=[\![\varepsilon_a(x)]\!][E_b(x)], \tag{4.1.31}$$

where the square brackets denote the vector consisting of the Fourier coefficients of the expansion of the $\varepsilon_a(x)E_b(x)$ and $E_b(x)$ functions, and $[\![\varepsilon_a(x)]\!]$ – the Toeplitz matrix of the Fourier coefficients, which has the following form:

$$\mathbf{Q}=\begin{bmatrix} e_0 & e_1 & e_2 & \cdots & \cdots & e_{2N} \\ e_{-1} & e_0 & e_1 & e_2 & \ddots & e_{2N-1} \\ e_{-2} & e_{-1} & e_0 & e_1 & \ddots & \vdots \\ \vdots & e_{-2} & e_{-1} & \ddots & \ddots & \vdots \\ \vdots & \ddots & \ddots & \ddots & \ddots & e_1 \\ e_{-2N} & e_{1-2N} & \cdots & \cdots & e_{-1} & e_0 \end{bmatrix}. \tag{4.1.32}$$

Substituting (4.1.29) into (4.1.28) and equating the coefficients of equal Fourier harmonics, we obtain:

$$\begin{cases} ik_0\mathbf{K}_y\mathbf{S}_z-\dfrac{d\mathbf{S}_y}{dz} & =k_0(\mathbf{M}_{1,1}\mathbf{U}_x+\mathbf{M}_{1,2}\mathbf{U}_y+\mathbf{M}_{1,3}\mathbf{U}_z), \\ \dfrac{d\mathbf{S}_x}{dz}-ik_0\mathbf{K}_x\mathbf{S}_z & =k_0(\mathbf{M}_{2,1}\mathbf{U}_x+\mathbf{M}_{2,2}\mathbf{U}_y+\mathbf{M}_{2,3}\mathbf{U}_z), \\ ik_0\mathbf{K}_x\mathbf{S}_y-ik_0\mathbf{K}_y\mathbf{S}_x & =k_0(\mathbf{M}_{3,1}\mathbf{U}_x+\mathbf{M}_{3,2}\mathbf{U}_y+\mathbf{M}_{3,3}\mathbf{U}_z), \\ ik_0\mathbf{K}_y\mathbf{U}_z-\dfrac{d\mathbf{U}_y}{dz} & =k_0(\mathbf{E}_{1,1}\mathbf{S}_x+\mathbf{E}_{1,2}\mathbf{S}_y+\mathbf{E}_{1,3}\mathbf{S}_z), \\ \dfrac{d\mathbf{U}_x}{dz}-ik_0\mathbf{K}_x\mathbf{U}_z & =k_0(\mathbf{E}_{2,1}\mathbf{S}_x+\mathbf{E}_{2,2}\mathbf{S}_y+\mathbf{E}_{2,3}\mathbf{S}_z), \\ ik_0\mathbf{K}_x\mathbf{U}_y-ik_0\mathbf{K}_y\mathbf{U}_x & =k_0(\mathbf{E}_{3,1}\mathbf{S}_x+\mathbf{E}_{3,2}\mathbf{S}_y+\mathbf{E}_{3,3}\mathbf{S}_z), \end{cases} \tag{4.1.33}$$

where $\mathbf{K}_x=\operatorname{diag}_j\dfrac{k_{x,j}}{k_0},\mathbf{K}_y=\operatorname{diag}_j\dfrac{k_y}{k_0}$ the diagonal matricess, and $\mathbf{E}_{i,j}=[\![\varepsilon_{i,j}]\!]$ and $\mathbf{M}_{i,j}=[\![\mu_{i,j}]\!]$ is the Toeplitz matrix of the form (4.1.32), consisting of the expansion coefficients in the Fourier series of functions $\varepsilon_{i,i}(x)$ and $\mu_{i,j}(x)$.

We make the change of variables $z'=k_0z$ and transform the system (4.1.33) to the form:

$$\begin{cases} -\dfrac{d\mathbf{S}_y}{dz'} = \mathbf{M}_{1,1}\mathbf{U}_x + \mathbf{M}_{1,2}\mathbf{U}_y + \mathbf{M}_{1,3}\mathbf{U}_z - i\mathbf{K}_y\mathbf{S}_z, \\ -\dfrac{d\mathbf{S}_x}{dz'} = -\mathbf{M}_{2,1}\mathbf{U}_x - \mathbf{M}_{2,2}\mathbf{U}_y - \mathbf{M}_{2,3}\mathbf{U}_z - i\mathbf{K}_x\mathbf{S}_z, \\ -\dfrac{d\mathbf{U}_y}{dz'} = \mathbf{E}_{1,1}\mathbf{S}_x + \mathbf{E}_{1,2}\mathbf{S}_y + \mathbf{E}_{1,3}\mathbf{S}_z - i\mathbf{K}_y\mathbf{U}_z, \\ -\dfrac{d\mathbf{U}_x}{dz'} = -\mathbf{E}_{2,1}\mathbf{S}_x - \mathbf{E}_{2,2}\mathbf{S}_y - \mathbf{E}_{2,3}\mathbf{S}_z - i\mathbf{K}_x\mathbf{U}_z, \\ \mathbf{E}_{3,3}\mathbf{S}_z = i\mathbf{K}_x\mathbf{U}_y - i\mathbf{K}_y\mathbf{U}_x - \mathbf{E}_{3,1}\mathbf{S}_x - \mathbf{E}_{3,2}\mathbf{S}_y, \\ \mathbf{M}_{3,3}\mathbf{U}_z = i\mathbf{K}_x\mathbf{S}_y - i\mathbf{K}_y\mathbf{S}_x - \mathbf{M}_{3,1}\mathbf{U}_x - \mathbf{M}_{3,2}\mathbf{U}_y. \end{cases} \tag{4.1.34}$$

We rewrite the first four equations of the system in matrix form:

$$\frac{d}{dz'}\begin{bmatrix} \mathbf{S}_y \\ \mathbf{S}_x \\ \mathbf{U}_y \\ \mathbf{U}_x \end{bmatrix} = -\begin{bmatrix} \mathbf{0} & \mathbf{0} & \mathbf{M}_{1,2} & \mathbf{M}_{1,1} \\ \mathbf{0} & \mathbf{0} & -\mathbf{M}_{2,2} & -\mathbf{M}_{2,1} \\ \mathbf{E}_{1,2} & \mathbf{E}_{1,1} & \mathbf{0} & \mathbf{0} \\ -\mathbf{E}_{2,2} & -\mathbf{E}_{2,1} & \mathbf{0} & \mathbf{0} \end{bmatrix}\begin{bmatrix} \mathbf{S}_y \\ \mathbf{S}_x \\ \mathbf{U}_y \\ \mathbf{U}_x \end{bmatrix} - \begin{bmatrix} -i\mathbf{K}_y & \mathbf{M}_{1,3} \\ -i\mathbf{K}_x & -\mathbf{M}_{2,3} \\ \mathbf{E}_{1,3} & -i\mathbf{K}_y \\ -\mathbf{E}_{2,3} & -i\mathbf{K}_x \end{bmatrix}\begin{bmatrix} \mathbf{S}_z \\ \mathbf{U}_z \end{bmatrix}. \tag{4.1.35}$$

We express in the last two equations (4.1.34) $\mathbf{S}_z$ and $\mathbf{U}_z$, and substitute these expressions into (4.1.35). The result is a system of linear differential equations with respect to x- and y-Fourier components of the fields:

$$\frac{d}{dz'}\begin{bmatrix} \mathbf{S}_y \\ \mathbf{S}_x \\ \mathbf{U}_y \\ \mathbf{U}_x \end{bmatrix} = \mathbf{A}\begin{bmatrix} \mathbf{S}_y \\ \mathbf{S}_x \\ \mathbf{U}_y \\ \mathbf{U}_x \end{bmatrix}, \tag{4.1.36}$$

where

$$\mathbf{A} = \mathbf{J} - \begin{bmatrix} i\mathbf{K}_y & -\mathbf{M}_{1,3} \\ i\mathbf{K}_x & \mathbf{M}_{2,3} \\ -\mathbf{E}_{1,3} & i\mathbf{K}_y \\ \mathbf{E}_{2,3} & i\mathbf{K}_x \end{bmatrix}\begin{bmatrix} \mathbf{E}_{3,3}^{-1} & \mathbf{0} \\ \mathbf{0} & \mathbf{M}_{3,3}^{-1} \end{bmatrix}\begin{bmatrix} \mathbf{E}_{3,2} & \mathbf{E}_{3,1} & -i\mathbf{K}_x & i\mathbf{K}_y \\ -i\mathbf{K}_x & i\mathbf{K}_y & \mathbf{M}_{3,2} & \mathbf{M}_{3,1} \end{bmatrix}, \tag{4.1.37}$$

where

$$\mathbf{J} = -\begin{bmatrix} \mathbf{0} & \mathbf{0} & \mathbf{M}_{1,2} & \mathbf{M}_{1,1} \\ \mathbf{0} & \mathbf{0} & -\mathbf{M}_{2,2} & -\mathbf{M}_{2,1} \\ \mathbf{E}_{1,2} & \mathbf{E}_{1,1} & \mathbf{0} & \mathbf{0} \\ -\mathbf{E}_{2,2} & -\mathbf{E}_{2,1} & \mathbf{0} & \mathbf{0} \end{bmatrix}. \tag{4.1.38}$$

Thus, we obtain a system of linear first order differential equations for the vectors $\mathbf{S}_x$, $\mathbf{S}_y$, $\mathbf{U}_x$, $\mathbf{U}_y$. Note that the matrix **A** has the dimension $4(2N+1)\times 4(2N+1)$.

Correct rules of Fourier-expansion of the product of functions
Derivation of the system of differential equations (4.1.36) was based on the representation in (4.1.28) of the components of electromagnetic fields and the components of the tensor of dielectric and magnetic permeability in the form of segments of Fourier series. Fourier series representation of products of functions in (1.4.28) is a delicate question. The formulas (1.4.30)–(1.4.32) are of limited use.

Consider two periodic functions

$$f(x)=\sum_m f_m \exp(iKmx),\ g(x)=\sum_m g_m \exp(iKmx),\ K=\frac{2\pi}{\Lambda} \tag{4.1.39}$$

and the Fourier series expansion of their product

$$h(x)=f(x)g(x)=\sum_m h_m \exp(iKmx). \tag{4.1.40}$$

As a product of the Fourier coefficients of the standard values we use

$$h_j=\sum_{m=-N}^{N} f_{j-m}g_m, \tag{4.1.41}$$

obtained by direct multiplication of the series (4.1.39), (4.1.40). Equation (4.1.41) to compute the Fourier coefficients is called the Laurent rule. In matrix notation (4.1.41) can be written as [5]

$$[h]=[[f]][g], \tag{4.1.42}$$

where, as in (4.1.31), the square brackets denote the vectors consisting of Fourier coefficients of expansion of functions, and $[[f]]$ is the Toeplitz matrix of the Fourier coefficients.

As shown in [5, 6], the use of (4.1.42) is correct if there are no values of x for which the functions $f(x)$ and $g(x)$ become discontinuos at the same time. Using the Laurent rules (4.1.42) for the product of functions with coincident points of discontinuity leads to disruption of the convergence of Fourier series at the points of discontinuity.

If the functions $f(x)$ and $g(x)$ are discontinuous at the same time, but the function $h(x)=f(x)g(x)$ is continuous, it is correct to use of the so-called inverse Laurent rule [5]

$$[h]=\left[\left[\frac{1}{f}\right]\right]^{-1}[g]. \tag{4.1.43}$$

In the products $\mu_{i,1}(x)\cdot H_x(x,y,z)$ and $\varepsilon_{i,1}(x)\cdot E_x(x,y,z)$in (4.1.28) both decomposed functions have discontinuities at the same points x corresponding to the vertical boundaries of the media in the layers. Accordingly, the use of the Laurent rules (4.1.30)–(4.1.32) when writing (4.1.33) is erroneous. Errors in the use of the Laurent are large when working with diffraction gratings of conductive materials. In particular, the erroneous use of the Laurent rule leads to slow convergence of solutions for binary metal gratings for the case of TM-polarization [5–10]. Convergence in this case refers to the stabilization of the results of calculation of the amplitudes of the diffraction orders with increasing number of Fourier harmonics N in the representation of the field components (4.1.29).

Continuous components at the vertical boundaries of the media are the tangential components E_y, E_z, H_y, H_z and normal components of electric displacement and magnetic induction D_x, B_x. We express the discontinuous components by the continuous field components

$$\begin{aligned} E_x &= \frac{1}{\varepsilon_{11}} D_x - \frac{\varepsilon_{12}}{\varepsilon_{11}} E_y - \frac{\varepsilon_{13}}{\varepsilon_{11}} E_z, \\ H_x &= \frac{1}{\mu_{11}} B_x - \frac{\mu_{12}}{\mu_{11}} H_y - \frac{\mu_{13}}{\mu_{11}} H_z \end{aligned} \tag{4.1.44}$$

and substitute into Maxwell's equations (4.1.28). As a result, we obtain:

$$\begin{cases} \dfrac{\partial E_z}{\partial y} - \dfrac{\partial E_y}{\partial z} = ik_0 B_x, \\ \dfrac{\partial E_x}{\partial z} - \dfrac{\partial E_z}{\partial x} = ik_0 \left(\dfrac{\mu_{2,1}}{\mu_{1,1}} B_x + \left(\mu_{2,2} - \mu_{2,1} \dfrac{\mu_{1,2}}{\mu_{1,1}} \right) H_y + \left(\mu_{2,3} - \mu_{2,1} \dfrac{\mu_{1,3}}{\mu_{1,1}} \right) H_z \right), \\ \dfrac{\partial E_y}{\partial x} - \dfrac{\partial E_x}{\partial y} = ik_0 \left(\dfrac{\mu_{3,1}}{\mu_{1,1}} B_x + \left(\mu_{3,2} - \mu_{3,1} \dfrac{\mu_{1,2}}{\mu_{1,1}} \right) H_y + \left(\mu_{3,3} - \mu_{3,1} \dfrac{\mu_{1,3}}{\mu_{1,1}} \right) H_z \right), \\ \dfrac{\partial H_z}{\partial y} - \dfrac{\partial H_y}{\partial z} = -ik_0 D_x, \\ \dfrac{\partial H_x}{\partial z} - \dfrac{\partial H_z}{\partial x} = -ik_0 \left(\dfrac{\varepsilon_{2,1}}{\varepsilon_{1,1}} D_x + \left(\varepsilon_{2,2} - \varepsilon_{2,1} \dfrac{\varepsilon_{1,2}}{\varepsilon_{1,1}} \right) E_y + \left(\varepsilon_{2,3} - \varepsilon_{2,1} \dfrac{\varepsilon_{1,3}}{\varepsilon_{1,1}} \right) E_z \right), \\ \dfrac{\partial H_y}{\partial x} - \dfrac{\partial H_x}{\partial y} = -ik_0 \left(\dfrac{\varepsilon_{3,1}}{\varepsilon_{1,1}} D_x + \left(\varepsilon_{3,2} - \varepsilon_{3,1} \dfrac{\varepsilon_{1,2}}{\varepsilon_{1,1}} \right) E_y + \left(\varepsilon_{3,3} - \varepsilon_{3,1} \dfrac{\varepsilon_{1,3}}{\varepsilon_{1,1}} \right) E_z \right). \end{cases} \tag{4.1.45}$$

Equations (1.4.45) do not contain products of functions showing discontinuties at the same time. Accordingly, in the transition to the spatial frequency domain we can use the direct Laurent rule (4.1.31), (4.1.32). According to (4.1.44), the vectors of the Fourier coefficients of the functions D_x, B_x have the form

$$[D_x] = \left[\!\left[\frac{1}{\varepsilon_{11}}\right]\!\right]^{-1}\left([E_x] + \left[\!\left[\frac{\varepsilon_{12}}{\varepsilon_{11}}\right]\!\right][E_y] + \left[\!\left[\frac{\varepsilon_{13}}{\varepsilon_{11}}\right]\!\right][E_z]\right),$$
$$[B_x] = \left[\!\left[\frac{1}{\mu_{11}}\right]\!\right]^{-1}\left([H_x] + \left[\!\left[\frac{\mu_{12}}{\mu_{11}}\right]\!\right][H_y] + \left[\!\left[\frac{\mu_{13}}{\mu_{11}}\right]\!\right][H_z]\right). \tag{4.1.46}$$

Performing in (4.1.45) the transition to the spatial–frequency domain and performing transformations, we obtain a system of differential equations (4.1.36) - (4.1.38), where the matrices $\mathbf{M}_{i,j}$ and $\mathbf{E}_{i,j}$ have the form:

$$\mathbf{M}_{1,1} = \left[\!\left[\frac{1}{\mu_{11}}\right]\!\right]^{-1}, \ \mathbf{M}_{1,2} = \left[\!\left[\frac{1}{\mu_{11}}\right]\!\right]^{-1}\left[\!\left[\frac{\mu_{12}}{\mu_{11}}\right]\!\right], \ \mathbf{M}_{1,3} = \left[\!\left[\frac{1}{\mu_{11}}\right]\!\right]^{-1}\left[\!\left[\frac{\mu_{13}}{\mu_{11}}\right]\!\right],$$

$$\mathbf{M}_{2,1} = \left[\!\left[\frac{\mu_{21}}{\mu_{11}}\right]\!\right]\left[\!\left[\frac{1}{\mu_{11}}\right]\!\right]^{-1}, \ \mathbf{M}_{2,2} = \left[\!\left[\mu_{22} - \frac{\mu_{12}\mu_{21}}{\mu_{11}}\right]\!\right] + \left[\!\left[\frac{\mu_{21}}{\mu_{11}}\right]\!\right]\left[\!\left[\frac{1}{\mu_{11}}\right]\!\right]^{-1}\left[\!\left[\frac{\mu_{12}}{\mu_{11}}\right]\!\right],$$

$$\mathbf{M}_{2,3} = \left[\!\left[\mu_{23} - \frac{\mu_{13}\mu_{21}}{\mu_{11}}\right]\!\right] + \left[\!\left[\frac{\mu_{21}}{\mu_{11}}\right]\!\right]\left[\!\left[\frac{1}{\mu_{11}}\right]\!\right]^{-1}\left[\!\left[\frac{\mu_{13}}{\mu_{11}}\right]\!\right],$$

$$\mathbf{M}_{3,1} = \left[\!\left[\frac{\mu_{31}}{\mu_{11}}\right]\!\right]\left[\!\left[\frac{1}{\mu_{11}}\right]\!\right]^{-1}, \ \mathbf{M}_{3,2} = \left[\!\left[\mu_{32} - \frac{\mu_{31}\mu_{12}}{\mu_{11}}\right]\!\right] + \left[\!\left[\frac{\mu_{31}}{\mu_{11}}\right]\!\right]\left[\!\left[\frac{1}{\mu_{11}}\right]\!\right]^{-1}\left[\!\left[\frac{\mu_{12}}{\mu_{11}}\right]\!\right],$$

$$\mathbf{M}_{3,3} = \left[\!\left[\mu_{33} - \frac{\mu_{31}\mu_{13}}{\mu_{11}}\right]\!\right] + \left[\!\left[\frac{\mu_{31}}{\mu_{11}}\right]\!\right]\left[\!\left[\frac{1}{\mu_{11}}\right]\!\right]^{-1}\left[\!\left[\frac{\mu_{13}}{\mu_{11}}\right]\!\right],$$

$$\mathbf{E}_{1,1} = \left[\!\left[\frac{1}{\varepsilon_{11}}\right]\!\right]^{-1}, \ \mathbf{E}_{1,2} = \left[\!\left[\frac{1}{\varepsilon_{11}}\right]\!\right]^{-1}\left[\!\left[\frac{\varepsilon_{12}}{\varepsilon_{11}}\right]\!\right], \ \mathbf{E}_{1,3} = \left[\!\left[\frac{1}{\varepsilon_{11}}\right]\!\right]^{-1}\left[\!\left[\frac{\varepsilon_{13}}{\varepsilon_{11}}\right]\!\right],$$

$$\mathbf{E}_{2,1} = \left[\!\left[\frac{\varepsilon_{21}}{\varepsilon_{11}}\right]\!\right]\left[\!\left[\frac{1}{\varepsilon_{11}}\right]\!\right]^{-1}, \ \mathbf{E}_{2,2} = \left[\!\left[\varepsilon_{22} - \frac{\varepsilon_{12}\varepsilon_{21}}{\varepsilon_{11}}\right]\!\right] + \left[\!\left[\frac{\varepsilon_{21}}{\varepsilon_{11}}\right]\!\right]\left[\!\left[\frac{1}{\varepsilon_{11}}\right]\!\right]^{-1}\left[\!\left[\frac{\varepsilon_{12}}{\varepsilon_{11}}\right]\!\right],$$

$$\mathbf{E}_{2,3} = \left[\!\left[\varepsilon_{23} - \frac{\varepsilon_{13}\varepsilon_{21}}{\varepsilon_{11}}\right]\!\right] + \left[\!\left[\frac{\varepsilon_{21}}{\varepsilon_{11}}\right]\!\right]\left[\!\left[\frac{1}{\varepsilon_{11}}\right]\!\right]^{-1}\left[\!\left[\frac{\varepsilon_{13}}{\varepsilon_{11}}\right]\!\right],$$

$$\mathbf{E}_{3,1} = \left[\!\left[\frac{\varepsilon_{31}}{\varepsilon_{11}}\right]\!\right]\left[\!\left[\frac{1}{\varepsilon_{11}}\right]\!\right]^{-1}, \ \mathbf{E}_{3,2} = \left[\!\left[\varepsilon_{32} - \frac{\varepsilon_{31}\varepsilon_{12}}{\varepsilon_{11}}\right]\!\right] + \left[\!\left[\frac{\varepsilon_{31}}{\varepsilon_{11}}\right]\!\right]\left[\!\left[\frac{1}{\varepsilon_{11}}\right]\!\right]^{-1}\left[\!\left[\frac{\varepsilon_{12}}{\varepsilon_{11}}\right]\!\right],$$

$$\mathbf{E}_{3,3} = \left[\!\left[\varepsilon_{33} - \frac{\varepsilon_{31}\varepsilon_{13}}{\varepsilon_{11}}\right]\!\right] + \left[\!\left[\frac{\varepsilon_{31}}{\varepsilon_{11}}\right]\!\right]\left[\!\left[\frac{1}{\varepsilon_{11}}\right]\!\right]^{-1}\left[\!\left[\frac{\varepsilon_{13}}{\varepsilon_{11}}\right]\!\right]. \tag{4.1.47}$$

The system of the differential equations (4.1.36) with the matrices $\mathbf{M}_{i,j}$, $\mathbf{E}_{i,j}$ in the form of (4.1.47) is called a system, obtained using the correct rules of the Fourier expansions for the products of functions.

Form of the matrix of the system at different dielectric permittivity tensors
We consider some special types of dielectric tensors and the corresponding matrices of the system of differential equations (4.1.36)–(1.4.38). The systems of differential equations are given for the case of correct rules for Fourier expansions.

Consider the case of an isotropic material. For an isotropic material $\mu = 1$, and ε are scalars. In this case the $\mathbf{E}_{i,j}$, $\mathbf{M}_{i,j}$ matrices take the following form:

$$\begin{aligned}
&\mathbf{E}_{1,1} = \mathbf{E}^* = \left[\!\left[\frac{1}{\varepsilon}\right]\!\right]^{-1}, \mathbf{E}_{2,2} = \mathbf{E}_{3,3} = \mathbf{E} = [\![\varepsilon]\!],\\
&\mathbf{E}_{1,2} = \mathbf{E}_{1,3} = \mathbf{E}_{2,1} = \mathbf{E}_{2,3} = \mathbf{E}_{3,1} = \mathbf{E}_{3,2} = \mathbf{0},\\
&\mathbf{M}_{1,1} = \mathbf{M}_{2,2} = \mathbf{M}_{3,3} = \mathbf{I},\\
&\mathbf{M}_{1,2} = \mathbf{M}_{1,3} = \mathbf{M}_{2,1} = \mathbf{M}_{2,3} = \mathbf{M}_{3,1} = \mathbf{M}_{3,2} = \mathbf{0},
\end{aligned} \tag{4.1.48}$$

where $\mathbf{I}$ is the identity matrix of dimension $(2N+1)\times(2N+1)$. Substituting (4.1.48) into (4.1.37)–(4.1.38), we obtain a matrix of the system of differential equations in the form of:

$$\mathbf{A} = -\begin{bmatrix} \mathbf{0} & \mathbf{0} & \mathbf{K}_y\mathbf{E}^{-1}\mathbf{K}_x & \mathbf{I}-\mathbf{K}_y\mathbf{E}^{-1}\mathbf{K}_y \\ \mathbf{0} & \mathbf{0} & \mathbf{K}_x\mathbf{E}^{-1}\mathbf{K}_x-\mathbf{I} & -\mathbf{K}_x\mathbf{E}^{-1}\mathbf{K}_y \\ \mathbf{K}_y\mathbf{K}_x & \mathbf{E}^*-\mathbf{K}_y^2 & \mathbf{0} & \mathbf{0} \\ \mathbf{K}_x^2-\mathbf{E} & -\mathbf{K}_x\mathbf{K}_y & \mathbf{0} & \mathbf{0} \end{bmatrix}. \tag{4.1.49}$$

Consider the case of planar incidence, where $k_y = 0$ ($\phi = 0$) as in (4.1.24) and the vector of direction of the incident wave vector lies in the plane xOz. In this case $\mathbf{K}_y = 0$ in (4.1.49) and the system of differential equations (4.1.36) splits into two independent systems

$$\begin{aligned}
&\frac{d}{dz'}\begin{bmatrix} \mathbf{U}_y \\ \mathbf{S}_x \end{bmatrix} = \mathbf{A}^{TM}\begin{bmatrix} \mathbf{U}_y \\ \mathbf{S}_x \end{bmatrix}, \ \mathbf{A}^{TM} = -\begin{bmatrix} \mathbf{0} & \mathbf{E}^* \\ \mathbf{K}_x\mathbf{E}^{-1}\mathbf{K}_x-\mathbf{I} & \mathbf{0} \end{bmatrix},\\
&\frac{d}{dz'}\begin{bmatrix} \mathbf{S}_y \\ \mathbf{U}_x \end{bmatrix} = \mathbf{A}^{TE}\begin{bmatrix} \mathbf{S}_y \\ \mathbf{U}_x \end{bmatrix}, \ \mathbf{A}^{TE} = -\begin{bmatrix} \mathbf{0} & \mathbf{I} \\ \mathbf{K}_x^2-\mathbf{E} & \mathbf{0} \end{bmatrix}.
\end{aligned} \tag{4.1.50}$$

In the case of plane incidence this result allows to reduce the solution of the diffraction problem to two independent problems of diffraction of waves with TM- and TE-polarization.

In section 4.3 we study structures containing layers of a magnetized material. For the magnetized materials the permittivity is given by the tensor [18, 19]:

$$\vec{\vec{\boldsymbol{\varepsilon}}} = \begin{bmatrix} \varepsilon & ig\cos\theta_M & -ig\sin\theta_M\sin\phi_M \\ -ig\cos\theta_M & \varepsilon & ig\sin\theta_M\cos\phi_M \\ ig\sin\theta_M\sin\phi_M & -ig\sin\theta_M\cos\phi_M & \varepsilon \end{bmatrix} \tag{4.1.51}$$

where ε is the main dielectric constant of the medium, g is the modulus of the gyration vector of the medium, proportional to the magnetization [18, 19], θ_M and ϕ_M are the spherical coordinates describing the direction of the magnetization vector. In the optical range $\mu = 1$.

We consider three basic cases, corresponding to the direction of the magnetization vector in three coordinate axes.

At vertical magnetization (magnetization vector perpendicular to the plane of the layers of the structure) $\theta_M = 0$ and tensor (4.1.51) takes the form:

$$\vec{\vec{\varepsilon}} = \begin{bmatrix} \varepsilon & ig & 0 \\ -ig & \varepsilon & 0 \\ 0 & 0 & \varepsilon \end{bmatrix}. \tag{4.1.52}$$

In this case,

$$\mathbf{A} = -\begin{bmatrix} \mathbf{0} & \mathbf{0} & \mathbf{K}_y\mathbf{E}^{-1}\mathbf{K}_x & \mathbf{I}-\mathbf{K}_y\mathbf{E}^{-1}\mathbf{K}_y \\ \mathbf{0} & \mathbf{0} & \mathbf{K}_x\mathbf{E}^{-1}\mathbf{K}_x-\mathbf{I} & -\mathbf{K}_x\mathbf{E}^{-1}\mathbf{K}_y \\ i\mathbf{E}^*\mathbf{H}+\mathbf{K}_y\mathbf{K}_x & \mathbf{E}^*-\mathbf{K}_y^2 & \mathbf{0} & \mathbf{0} \\ \mathbf{K}_x^2-\mathbf{E}_2 & i\mathbf{H}\mathbf{E}^*-\mathbf{K}_x\mathbf{K}_y & \mathbf{0} & \mathbf{0} \end{bmatrix}, \tag{4.1.53}$$

where $\mathbf{E}^* = \left[\!\left[\frac{1}{\varepsilon}\right]\!\right]^{-1}$, $\mathbf{E} = [\![\varepsilon]\!]$, $\mathbf{H} = \left[\!\left[\frac{g}{\varepsilon}\right]\!\right]$, $\mathbf{E}_2 = \left[\!\left[\varepsilon - \frac{g^2}{\varepsilon}\right]\!\right] + \mathbf{H}\mathbf{E}^*\mathbf{H}$.

For horizontal magnetization (magnetization vector is parallel to the layer plane and directed along the axis Ox) $\theta_M = \frac{\pi}{2}$, $M = 0$, and tensor (4.1.51) takes the form

$$\vec{\vec{\varepsilon}} = \begin{bmatrix} \varepsilon & 0 & 0 \\ 0 & \varepsilon & ig \\ 0 & -ig & \varepsilon \end{bmatrix}. \tag{4.1.54}$$

In the case of (4.1.54), the matrix of the system takes the form

$$\mathbf{A} = -\begin{bmatrix} \mathbf{K}_y\mathbf{E}^{-1}\mathbf{G} & \mathbf{0} & \mathbf{K}_y\mathbf{E}^{-1}\mathbf{K}_x & \mathbf{I}-\mathbf{K}_y\mathbf{E}^{-1}\mathbf{K}_y \\ \mathbf{K}_x\mathbf{E}^{-1}\mathbf{G} & \mathbf{0} & \mathbf{K}_x\mathbf{E}^{-1}\mathbf{K}_x-\mathbf{I} & -\mathbf{K}_x\mathbf{E}^{-1}\mathbf{K}_y \\ \mathbf{K}_y\mathbf{K}_x & \mathbf{E}^*-\mathbf{K}_y^2 & \mathbf{0} & \mathbf{0} \\ \mathbf{G}\mathbf{E}^{-1}\mathbf{G}+\mathbf{K}_x^2-\mathbf{E} & -\mathbf{K}_x\mathbf{K}_y & \mathbf{G}\mathbf{E}^{-1}K_x & -\mathbf{G}\mathbf{E}^{-1}\mathbf{K}_y \end{bmatrix}, \tag{4.1.55}$$

where $\mathbf{G} = [[g]]$.

At the direction of the magnetization vector along the axis Oy $\theta_M = \varphi_M = \frac{\pi}{2}$. In this case from (4.1.51) we get:

$$\overset{\leftrightarrow}{\varepsilon} = \begin{bmatrix} \varepsilon & 0 & ig \\ 0 & \varepsilon & 0 \\ -ig & 0 & \varepsilon \end{bmatrix}. \tag{4.1.56}$$

For the tensor (4.1.56), the matrix of the system takes the form

$$\mathbf{A} = -\begin{bmatrix} \mathbf{0} & \mathbf{K}_y\mathbf{E}_2^{-1}\mathbf{H}\mathbf{E}^* & \mathbf{K}_y\mathbf{E}_2^{-1}\mathbf{K}_x & \mathbf{I}-\mathbf{K}_y\mathbf{E}_2^{-1}\mathbf{K}_y \\ \mathbf{0} & \mathbf{K}_x\mathbf{E}_2^{-1}\mathbf{H}\mathbf{E}^* & \mathbf{K}_x\mathbf{E}_2^{-1}\mathbf{K}_x-\mathbf{I} & -\mathbf{K}_x\mathbf{E}_2^{-1}\mathbf{K}_y \\ \mathbf{K}_y\mathbf{K}_x & \mathbf{E}^*-\mathbf{K}_y^2-\mathbf{E}^*\mathbf{H}\mathbf{E}_2^{-1}\mathbf{H}\mathbf{E}^* & -\mathbf{E}^*\mathbf{H}\mathbf{E}_2^{-1}\mathbf{K}_x & \mathbf{E}^*\mathbf{H}\mathbf{E}_2^{-1}\mathbf{K}_y \\ \mathbf{K}_x^2-\mathbf{E} & -\mathbf{K}_x\mathbf{K}_y & \mathbf{0} & \mathbf{0} \end{bmatrix}. \tag{4.1.57}$$

4.1.2.4. Representation of the field inside the layer

To directly view the field in the layer, we examine the eigen decomposition of the matrix:

$$\mathbf{A} = \mathbf{W}\mathbf{\Lambda}\mathbf{W}^{-1}, \tag{4.1.58}$$

where $\mathbf{\Lambda} = \underset{i}{\operatorname{diag}}\,\lambda_i$ is the diagonal matrix of the eigenvalues of the matrix $\mathbf{A}$, and $\mathbf{W}$ is the matrix of eigenvectors. Then the solution of systems of differential equations (4.1.36) can be written as:

$$\begin{bmatrix} \mathbf{S}_y \\ \mathbf{S}_x \\ \mathbf{U}_y \\ \mathbf{U}_x \end{bmatrix} = \mathbf{W}\exp(\mathbf{\Lambda}z')\mathbf{C}' = \mathbf{W}\exp(\mathbf{\Lambda}k_0 z)\mathbf{C}' = \sum_i c_i' w_i \exp(\lambda_i k_0 z). \tag{4.1.59}$$

We split the last sum into two, depending on the sign of the real part λ_i and write the expression in matrix notation:

$$\begin{aligned} \begin{bmatrix} \mathbf{S}_y \\ \mathbf{S}_x \\ \mathbf{U}_y \\ \mathbf{U}_x \end{bmatrix} &= \sum_i c_i' w_i \exp(\lambda_i k_0 z) = \\ &= \sum_{i:\mathrm{Re}\lambda_i<0} c_i w_i \exp(\lambda_i k_0 (z-d_l)) + \sum_{i:\mathrm{Re}\lambda_i>0} c_i w_i \exp(\lambda_i k_0 (z-d_{l-1})) = \\ &= \mathbf{W}^{(-)}\exp\left(\mathbf{\Lambda}^{(-)}k_0(z-d_l)\right)\mathbf{C}^{(-)} + \mathbf{W}^{(+)}\exp\left(\mathbf{\Lambda}^{(+)}k_0(z-d_{l-1})\right)\mathbf{C}^{(+)}, \end{aligned} \tag{4.1.60}$$

where $\boldsymbol{\Lambda}^{(+)}$ and $\boldsymbol{\Lambda}^{(-)}$ are the diagonal matrices of eigenvalues whose real parts are positive and negative, respectively, $\mathbf{W}^{(+)}$ and $\mathbf{W}^{(-)}$ are the corresponding matrices of eigenvectors, $\mathbf{C}^{(-)}$, $\mathbf{C}^{(+)}$ are the vectors of arbitrary constants. The representation (4.1.60) is focused on computer calculation. The exponents in (4.1.60) always have a negative real part. This ensures that no overflow occurs.

In some cases, the calculation of the eigenvectors and eigenvalues can be greatly speeded up taking into account the special form of the matrix **A** [4]. In particular, the matrix A in (4.1.49), (4.1.50), (4.1.53) has the following block structure

$$\mathbf{A} = \begin{bmatrix} \mathbf{0} & \mathbf{A}_{12} \\ \mathbf{A}_{21} & \mathbf{0} \end{bmatrix}. \tag{4.1.61}$$

We write for **A** the matrices of eigenvectors and eigenvalues as

$$\mathbf{W} = \begin{bmatrix} \mathbf{W}_{11} & \mathbf{W}_{12} \\ \mathbf{W}_{21} & \mathbf{W}_{22} \end{bmatrix}, \quad \boldsymbol{\Lambda} = \begin{bmatrix} \boldsymbol{\Lambda}_{11} & \mathbf{0} \\ \mathbf{0} & \boldsymbol{\Lambda}_{22} \end{bmatrix}. \tag{4.1.62}$$

Since $\mathbf{A} = \mathbf{W}\,\boldsymbol{\Lambda}\,\mathbf{W}^{-1}$, then

$$\begin{bmatrix} \mathbf{0} & \mathbf{A}_{12} \\ \mathbf{A}_{21} & \mathbf{0} \end{bmatrix} \cdot \begin{bmatrix} \mathbf{W}_{11} & \mathbf{W}_{12} \\ \mathbf{W}_{21} & \mathbf{W}_{22} \end{bmatrix} = \begin{bmatrix} \mathbf{W}_{11} & \mathbf{W}_{12} \\ \mathbf{W}_{21} & \mathbf{W}_{22} \end{bmatrix} \cdot \begin{bmatrix} \boldsymbol{\Lambda}_{11} & \mathbf{0} \\ \mathbf{0} & \boldsymbol{\Lambda}_{22} \end{bmatrix},$$

$$\begin{bmatrix} \mathbf{A}_{12}\mathbf{W}_{21} & \mathbf{A}_{12}\mathbf{W}_{22} \\ \mathbf{A}_{21}\mathbf{W}_{11} & \mathbf{A}_{21}\mathbf{W}_{12} \end{bmatrix} = \begin{bmatrix} \mathbf{W}_{11}\boldsymbol{\Lambda}_{11} & \mathbf{W}_{12}\boldsymbol{\Lambda}_{22} \\ \mathbf{W}_{21}\boldsymbol{\Lambda}_{11} & \mathbf{W}_{22}\boldsymbol{\Lambda}_{22} \end{bmatrix}$$

and we have

$$\begin{aligned} \mathbf{A}_{12}\mathbf{A}_{21}\mathbf{W}_{11} &= \mathbf{A}_{12}\mathbf{W}_{21}\boldsymbol{\Lambda}_{11} = \mathbf{W}_{11}\boldsymbol{\Lambda}_{11}\boldsymbol{\Lambda}_{11}, \\ \mathbf{A}_{12}\mathbf{A}_{21}\mathbf{W}_{12} &= \mathbf{A}_{12}\mathbf{W}_{22}\boldsymbol{\Lambda}_{22} = \mathbf{W}_{12}\boldsymbol{\Lambda}_{22}\boldsymbol{\Lambda}_{22}. \end{aligned} \tag{4.1.63}$$

We introduce the matrix $\mathbf{B} = \mathbf{A}_{12}\,\mathbf{A}_{21}$and represent (4.1.63) as

$$\mathbf{B}\cdot\mathbf{W}_{11} = \mathbf{W}_{11}\boldsymbol{\Lambda}_{11}^2, \qquad \mathbf{B}\cdot\mathbf{W}_{12} = \mathbf{W}_{12}\boldsymbol{\Lambda}_{22}^2. \tag{4.1.64}$$

According to (4.1.64) $\mathbf{W}_{11}$, $\boldsymbol{\Lambda}_{11}^2 = \boldsymbol{\Lambda}$, $\mathbf{W}_{12}$, $\boldsymbol{\Lambda}_{22}^2$ are the matrixces of eigenvectors and diagonal matrices of eigenvalues of the same matrix **B**. So

$$\mathbf{W}_{11} = \mathbf{W}_{12}, \quad \boldsymbol{\Lambda}_{22} = -\boldsymbol{\Lambda}_{11} \text{ and } \mathbf{W}_{21} = \mathbf{A}_{21}\mathbf{W}_{11}\boldsymbol{\Lambda}_{11}^{-1}, \quad \mathbf{W}_{22} = -\mathbf{W}_{21}. \tag{4.1.65}$$

The relations (4.1.65) determine the eigenvalues and eigenvectors of the matrix **A** through the eigenvalues and eigenvectors of the matrix **B** twice smaller in the form of:

$$\mathbf{W} = \begin{bmatrix} \mathbf{W}_{11} & \mathbf{W}_{11} \\ \mathbf{A}_{21}\mathbf{W}_{11}\boldsymbol{\Lambda}_{11}^{-1} & -\mathbf{A}_{21}\mathbf{W}_{11}\boldsymbol{\Lambda}_{11}^{-1} \end{bmatrix}, \quad \boldsymbol{\Lambda} = \begin{bmatrix} \sqrt{\boldsymbol{\Lambda}_{11}} & 0 \\ 0 & -\sqrt{\boldsymbol{\Lambda}_{11}} \end{bmatrix}. \tag{4.1.66}$$

In [4] it is noted that halving the dimension of the eigenvalue problem is equivalent to reducing system (4.1.36) from $4(2N+1)$ first order differential equations to a system $2(2N+1)$ of second order differential equations.

4.1.2.5. 'Stitching' of the electromagnetic field on the layer boundaries

The general representation of the field in the layer was described above. To obtain a solution which satisfies the Maxwell equations, it is necessary to equate the tangential field components at the layer boundaries. Equating the tangential field components is equivalent to equating functions (4.1.60), i.e. the Fourier coefficients for each fixed z. We write the preliminary solutions (4.1.60) on the upper and lower boundaries of all layers. For convenience, we assume that the matrix of eigenvectors $\mathbf{W}$ is given by:

$$\mathbf{W} = \begin{bmatrix} \mathbf{W}^{(-)} & \mathbf{W}^{(+)} \end{bmatrix}. \tag{4.1.67}$$

In addition, we introduce the vector of unknown constants $\mathbf{C}$ as follows:

$$\mathbf{C} = \begin{bmatrix} \mathbf{C}^{(-)} \\ \mathbf{C}^{(+)} \end{bmatrix}. \tag{4.1.68}$$

Given this notation the solution of (4.1.60) on the upper and lower boundaries of the layer has the form

$$\begin{bmatrix} \mathbf{S}_y(d_{l-1}) \\ \mathbf{S}_x(d_{l-1}) \\ \mathbf{U}_y(d_{l-1}) \\ \mathbf{U}_x(d_{l-1}) \end{bmatrix} = \mathbf{W}^{(-)(-)}\mathbf{C}^{(-)} + \mathbf{W}^{(+)}\mathbf{C}^{(+)} = \mathbf{W}\begin{bmatrix} \mathbf{X}^{(-)} & \mathbf{0} \\ \mathbf{0} & \mathbf{I} \end{bmatrix}\mathbf{C} = \mathbf{NC}, \tag{4.1.69}$$

$$\begin{bmatrix} \mathbf{S}_y(d_l) \\ \mathbf{S}_x(d_l) \\ \mathbf{U}_y(d_l) \\ \mathbf{U}_x(d_l) \end{bmatrix} = \mathbf{W}^{(-)}\mathbf{C}^{(-)} + \mathbf{W}^{(+)}\mathbf{X}^{(+)}\mathbf{C}^{(+)} = \mathbf{W}\begin{bmatrix} \mathbf{I} & \mathbf{0} \\ \mathbf{0} & \mathbf{X}^{(+)} \end{bmatrix}\mathbf{C} = \mathbf{MC}, \tag{4.1.70}$$

where

$$\mathbf{X}^{(+)} = \exp\left(\mathbf{\Lambda}^{(+)}k_0\left(d_l - d_{l-1}\right)\right), \ \mathbf{X}^{(-)} = \exp\left(\mathbf{\Lambda}^{(-)}k_0\left(d_{l-1} - d_l\right)\right). \tag{4.1.71}$$

Equating the tangential components at the interface between adjacent layers, we obtain the equation

$$\mathbf{M}_{i-1}\mathbf{C}_{i-1} = \mathbf{N}_i\mathbf{C}_i, i = 2,\ldots,L, \tag{4.1.72}$$

where the index $i = 2$ corresponds to the condition at the lower boundary of the topmost layer, the index $i = L$ – at the upper boundary of the bottommost layer.

The relations (4.1.72) must be supplemented by the conditions of equality of the tangential components of the field above the structure (4.1.22) and at the upper boundary of the 1st layer, as well as of the field under the structure (4.1.23) and at the lower boundary of the L-th layer. Given the fact that in the layers of the field components are represented by segments of Fourier series of dimension $2N+1$ in the representations in the fields above and below the grating (4.1.22)–(4.1.24) it is necessary to take $2N+1$ waves at $-N \leq i \leq N$.

By adding these relations for the upper and lower boundaries of the diffraction structure, we obtain the following system of linear equations:

$$\begin{cases} \mathbf{D} + \mathbf{P}\mathbf{R} = \mathbf{N}_1\mathbf{C}_1, \\ \mathbf{M}_{i-1}\mathbf{C}_{i-1} = \mathbf{N}_i\mathbf{C}_i, \ i = 2,\ldots,L, \\ \mathbf{M}_L\mathbf{C}_L = \mathbf{F}\mathbf{T}, \end{cases} \tag{4.1.73}$$

where $\mathbf{R}$ and $\mathbf{T}$ are the vectors of complex amplitudes of reflected and transmitted orders, respectively. These vectors have the form:

$$\mathbf{R} = \begin{bmatrix} \mathbf{R}_E \\ \mathbf{R}_H \end{bmatrix}, \quad \mathbf{T} = \begin{bmatrix} \mathbf{T}_E \\ \mathbf{T}_H \end{bmatrix}, \tag{4.1.74}$$

$\mathbf{R}_E$ and $\mathbf{T}_E$ are the the vectors of complex amplitudes of E-waves, $\mathbf{R}_H$ and $\mathbf{T}_H$ are the vectors of complex amplitudes of H-waves.

The vector $\mathbf{D}$ in (4.1.73) represents the incident wave and has the form:

$$\mathbf{D} = \begin{bmatrix} \cos\theta\sin\phi\cdot\boldsymbol{\delta}_i \\ \cos\theta\cos\phi\cdot\boldsymbol{\delta}_i \\ -in_1\cos\phi\cdot\boldsymbol{\delta}_i \\ in_1\sin\phi\cdot\boldsymbol{\delta}_i \end{bmatrix}\cos\psi + \begin{bmatrix} \cos\phi\cdot\boldsymbol{\delta}_i \\ -\sin\phi\cdot\boldsymbol{\delta}_i \\ in_1\cos\theta\sin\phi\cdot\boldsymbol{\delta}_i \\ in_1\cos\theta\cos\phi\cdot\boldsymbol{\delta}_i \end{bmatrix}\sin\psi, \tag{4.1.75}$$

where $\boldsymbol{\delta}_i$ is a column vector, in which only one element in the middle is different from zero and unity. The column vector $\boldsymbol{\delta}_i$ has the dimension $2N+1$, vector $\mathbf{D}$ – dimension $4(2N+1)$.

The matrices $\mathbf{P}$ and $\mathbf{F}$ in (4.1.73) correspond to the reflected and transmitted orders, respectively, and have the form

$$\mathbf{P} = \begin{bmatrix} -n_1\cos\boldsymbol{\Theta}^{(1)}\sin\boldsymbol{\Phi} & \cos\boldsymbol{\Phi} \\ -n_1\cos\boldsymbol{\Theta}^{(1)}\cos\boldsymbol{\Phi} & -\sin\boldsymbol{\Phi} \\ -in_1^2\cos\boldsymbol{\Phi} & -in_1\cos\boldsymbol{\Theta}^{(1)}\sin\boldsymbol{\Phi} \\ in_1^2\sin\boldsymbol{\Phi} & -in_1\cos\boldsymbol{\Theta}^{(1)}\cos\boldsymbol{\Phi} \end{bmatrix}, \tag{4.1.76}$$

$$\mathbf{F}=\begin{bmatrix} -n_2\cos\mathbf{\Theta}^{(2)}\sin\mathbf{\Phi} & \cos\mathbf{\Phi} \\ -n_2\cos\mathbf{\Theta}^{(2)}\cos\mathbf{\Phi} & -\sin\mathbf{\Phi} \\ -in_2^2\cos\mathbf{\Phi} & -in_2\cos\mathbf{\Theta}^{(2)}\sin\mathbf{\Phi} \\ in_2^2\sin\mathbf{\Phi} & -in_2\cos\mathbf{\Theta}^{(2)}\cos\mathbf{\Phi} \end{bmatrix}, \tag{4.1.77}$$

where $\mathbf{\Phi}$, $\mathbf{\Theta}^{(1)}$,$\mathbf{\Theta}^{(2)}$ are the diagonal matrices of angles corresponding to diffraction orders. They satisfy the following relations:

$$\sin\mathbf{\Phi}=\operatorname*{diag}_i\frac{k_y}{\sqrt{k_{x,i}^2+k_y^2}},\quad \cos\mathbf{\Phi}=\operatorname*{diag}_i\frac{k_{x,i}}{\sqrt{k_{x,i}^2+k_y^2}}, \tag{4.1.78}$$

$$\cos\mathbf{\Theta}^{(1)}=\operatorname*{diag}_i\frac{-k_{z,I,i}}{k_0n_1},\quad \cos\mathbf{\Theta}^{(2)}=\operatorname*{diag}_i\frac{k_{z,II,i}}{k_0n_2}. \tag{4.1.79}$$

The expressions (1.4.75)–(4.1.77) follow directly from the general formulas (4.1.12), (4.1.13) for a plane wave taking into account the type of propagation constants of the orders (4.1.24).

According to (4.1.73), the solution of the diffraction problem is reduced to solving a system of linear equations. Sequential exclusion of the coefficients $\mathbf{C}_i$ in (4.1.73) reduces the system (4.1.73) to a system of equations for the coefficients **R** and **T**. Indeed, from the last two equations in (4.1.73) we obtain

$$\mathbf{M}_{L-1}\mathbf{C}_{L-1}=\mathbf{N}_L\mathbf{M}_L^{-1}\mathbf{F}\mathbf{T}. \tag{4.1.80}$$

Substituting (1.4.80) into the equation with the index $i = L - 1$ in (4.1.73), we obtain

$$\mathbf{M}_{L-2}\mathbf{C}_{L-2}=\mathbf{N}_{L-1}\mathbf{M}_{L-1}^{-1}\mathbf{N}_L\mathbf{M}_L^{-1}\mathbf{F}\mathbf{T}. \tag{4.1.81}$$

Continuing this process to the equation with the index $i = 2$, we have

$$\mathbf{M}_1\mathbf{C}_1=\left(\prod_{i=2}^{L}\mathbf{N}_i\mathbf{M}_i^{-1}\right)\mathbf{F}\mathbf{T}. \tag{4.1.82}$$

Finally, substituting (4.1.82) in the first equation of (4.1.73), we obtain the desired system of linear equations for the coefficients **R** and **T** in the form of:

$$\mathbf{D}+\mathbf{P}\mathbf{R}=\left(\prod_{i=1}^{L}\mathbf{N}_i\mathbf{M}_i^{-1}\right)\mathbf{F}\mathbf{T}. \tag{4.1.83}$$

4.1.2.6. Numerically stable implementation of the method

The straightforward callculations based on (4.1.83) can lead to a numerical instability of the problem [3]. Problems are associated with the calculation of the matrix

$$\mathbf{M}_i^{-1} = \left(\mathbf{W}_i \begin{bmatrix} \mathbf{I} & \mathbf{0} \\ \mathbf{0} & \mathbf{X}_i^{(+)} \end{bmatrix} \right)^{-1} = \begin{bmatrix} \mathbf{I} & \mathbf{0} \\ \mathbf{0} & \left(\mathbf{X}_i^{(+)}\right)^{-1} \end{bmatrix} \mathbf{W}_i^{-1} \qquad (4.1.84)$$

in (1.4.83). The diagonal matrix

$$\left(\mathbf{X}^{(+)}\right)^{-1} = \exp\left(-\boldsymbol{\Lambda}_i^{(+)} k_0 \left(d_i - d_{i-1}\right)\right) \qquad (4.1.85)$$

contains the exponents of the positive values that can lead to overflow [3]. Consider the variant of writing a system of linear equations, avoiding computing the matrices (4.1.85). We denote

$$\mathbf{F}_L = \mathbf{F},\ \mathbf{T}^{(L)} = \mathbf{T}. \qquad (4.1.86)$$

Given this notation, equation (1.4.83) takes the form

$$\mathbf{D} + \mathbf{P}\mathbf{R} = \left(\prod_{i=1}^{L} \mathbf{N}_i \mathbf{M}_i^{-1} \right) \mathbf{F}_L \mathbf{T}^{(L)}. \qquad (4.1.87)$$

Transform (1.4.87) to the equation

$$\mathbf{D} + \mathbf{P}\mathbf{R} = \left(\prod_{i=1}^{L-1} \mathbf{N}_i \mathbf{M}_i^{-1} \right) \mathbf{F}_{L-1} \mathbf{T}^{(L-1)}, \qquad (4.1.88)$$

where

$$\mathbf{F}_{L-1} \mathbf{T}^{(L-1)} = \mathbf{N}_L \mathbf{M}_L^{-1} \mathbf{F}_L \mathbf{T}^{(L)}. \qquad (4.1.89)$$

Equation (4.1.88) has the same form as (4.1.87), but contains one less factor. Substituting into (4.1.89) the concrete form of matrices $\mathbf{N}_L$ and $\mathbf{M}_L^{-1}$, we get

$$\mathbf{F}_{L-1} \mathbf{T}^{(L-1)} = \mathbf{W}_L \begin{bmatrix} \mathbf{x}_L^{(-)} & \mathbf{0} \\ \mathbf{0} & \mathbf{I} \end{bmatrix} \begin{bmatrix} \mathbf{I} & \mathbf{0} \\ \mathbf{0} & \mathbf{x}_L^{(+)} \end{bmatrix}^{-1} \mathbf{W}_L^{-1} \mathbf{F}_L \mathbf{T}^{(L)}. \qquad (4.1.90)$$

We introduce the notation

$$\mathbf{A}_L = \begin{bmatrix} \mathbf{A}_L^{(-)} \\ \mathbf{A}_L^{(+)} \end{bmatrix} = \mathbf{W}_L^{-1} \mathbf{F}_L \qquad (4.1.91)$$

and rewrite (4.1.90) to the form

$$\mathbf{F}_{L-1} \mathbf{T}^{(L-1)} = \mathbf{W}_L \begin{bmatrix} \mathbf{x}_L^{(-)} & \mathbf{0} \\ \mathbf{0} & \mathbf{I} \end{bmatrix} \begin{bmatrix} \mathbf{I} & \mathbf{0} \\ \mathbf{0} & \mathbf{x}_L^{(+)} \end{bmatrix}^{-1} \mathbf{A}_L \mathbf{T}^{(L)} = \mathbf{W}_L \begin{bmatrix} \mathbf{x}_L^{(-)} \mathbf{A}_L^{(-)} \\ \left(\mathbf{x}_L^{(+)}\right)^{-1} \mathbf{A}_L^{(+)} \end{bmatrix} \mathbf{T}^{(L)}. \qquad (4.1.92)$$

In order to avoid possible overflows in the calculation of the matrix $(\mathbf{X}_L^{(+)})^{-1}$, we draw the following substitution [3]

$$\mathbf{T}^{(L)} = \left(\mathbf{A}_L^{(+)}\right)^{-1} \mathbf{X}_L^{(+)} \mathbf{T}^{(L-1)}. \tag{4.1.93}$$

As a result, we obtain 0

$$\mathbf{F}_{L-1}\mathbf{T}^{(L-1)} = \mathbf{W}_L \begin{bmatrix} \mathbf{X}_L^{(-)}\mathbf{A}_L^{(-)}\left(\mathbf{A}_L^{(+)}\right)^{-1}\mathbf{X}_L^{(+)} \\ \mathbf{I} \end{bmatrix} \mathbf{T}^{(L-1)}. \tag{4.1.94}$$

Thus, the original system (4.1.87) is reduced to the system (1.4.88) of the same species, but containing at least one factor less. The transition to the system (4.1.88) is numerically stable. Numerical stability is achieved by replacing the variables (4.1.93).

By repeating these steps and successively reducing the number of factors in (4.1.88), we obtain the system

$$\mathbf{D} + \mathbf{P}\mathbf{R} = \mathbf{F}_0\mathbf{T}^{(0)}. \tag{4.1.95}$$

The matrix $\mathbf{F}_0$ in (4.1.95) can be calculated by successive application of recurrence relations

$$\mathbf{F}_{i-1} = \mathbf{W}_i \begin{bmatrix} \mathbf{X}_i^{(-)}\mathbf{A}_i^{(-)}\left(\mathbf{A}_i^{(+)}\right)^{-1}\mathbf{X}_i^{(+)} \\ \mathbf{I} \end{bmatrix}, \quad \begin{bmatrix} \mathbf{A}_i^{(-)} \\ \mathbf{A}_i^{(+)} \end{bmatrix} = \mathbf{W}_i^{-1}\mathbf{F}_i \tag{4.1.96}$$

with a decrease of the index i from L down to 1. The solution of systems of linear equations determines the complex amplitudes of the reflected orders $\mathbf{R}$ and vector $\mathbf{T}^{(0)}$. The amplitudes of the transmitted orders $\mathbf{T} = \mathbf{T}^{(L)}$ are determined through the consistent application of the formula (4.1.93):

$$\mathbf{T} = \mathbf{T}^{(L)} = \left(\mathbf{A}_L^{(+)}\right)^{-1} \mathbf{X}_L^{(+)} \cdot \left(\mathbf{A}_{L-1}^{(+)}\right)^{-1} \mathbf{X}_{L-1}^{(+)} \cdots \left(\mathbf{A}_1^{(+)}\right)^{-1} \mathbf{X}_1^{(+)} \mathbf{T}^{(0)}. \tag{4.1.97}$$

Note that the expressions (4.1.93)–(4.1.97) can be obtained in the present form only in the case where the number of eigenvalues of the matrix $\mathbf{A}$ with positive and negative signs of the real parts is equal. Otherwise $\mathbf{A}_L^{(+)}$, (4.1.91) will not be a square and then the inverse matrix $(\mathbf{A}_L^{(+)})^{-}$ cannot be used. In particular, the above condition for the eigenvalues is always satisfied for gratings of a homogeneous material and gratings of a magnetic material with a dielectric tensor in the form (4.1.52). It can be shown that this condition is also satisfied when the diffraction structure contains homogeneous magnetic layers with the dielectric tensor in the form of (4.1.54), (4.1.56).

4.1.2.7. Characteristics of diffraction orders

In the analysis of the field outside the near zone of the grating the researcher is usually interested not in the complex amplitudes (4.1.74) but it the intensity of the reflected and transmitted propagating diffraction orders. The propagating orders are determined by the real values $k_{z,I,i}$ in (4.1.24). The intensity of the diffraction orders is defined as the Poynting vector flux through the plane z = const, normalized to the flux of the incident wave [17]. Given the normalization (4.1.18) (4.1.19), the intensity of the orders can be found from the following expressions:

$$\mathbf{I}^R = \left|\frac{\operatorname{Re}\left(\cos\boldsymbol{\Theta}^{(1)}\right)}{\cos\theta}\right|\left(|n_1\mathbf{R}_E|^2 + |\mathbf{R}_H|^2\right), \tag{4.1.98}$$

$$\mathbf{I}^T = \left|\frac{n_2\operatorname{Re}\left(\cos\boldsymbol{\Theta}^{(2)}\right)}{n_1\cos\theta}\right|\left(|n_2\mathbf{T}_E|^2 + |\mathbf{T}_H|^2\right), \tag{4.1.99}$$

where the diagonal matrices $\cos\boldsymbol{\Theta}^{(1)}$, $\cos\boldsymbol{\Theta}^{(2)}$ are defined in (4.1.79), and the squaring of the vectors $\mathbf{R}_E, \mathbf{R}_H, \mathbf{T}_E, \mathbf{T}_H$ is element-wise. For the evanescent diffraction orders $\operatorname{Re}\left(\cos\boldsymbol{\Theta}^{(l)}\right) = 0,\ l = 1,2$, respectively, their intensities are equal to zero.

In general, the propagating diffraction orders are plane waves with elliptical polarization. Indeed, each diffraction order corresponds to the superposition of the *E*- and *H*-waves. We denote by E_E, E_H the complex amplitudes of electric field vectors in the *E*- and *H*-waves. Note that the electric field vector of the *E*- and *H*-waves are perpendicular to each other and perpendicular to the direction of wave propagation. When adding two perpendicular oscillations an elliptically polarized wave is obtained in a general case. The polarization ellipse is characterized by two parameters: the angle φ of the main axis of the polarization ellipse and the ellipticity parameter χ [20]. The ellipticity parameter is the ratio of the lengths a, b of the axes of the polarization ellipse in the form tg $\chi = a/b$. The parameters are determined by the complex amplitudes E_E, E_H of the form [20]

$$\operatorname{tg}(2\varphi) = \frac{2\operatorname{Re}(E_E/E_H)}{1-|E_E/E_H|^2},$$
$$\sin(2\chi) = \frac{2\operatorname{Im}(E_E/E_H)}{1+|E_E/\mathrm{E_H}|^2}. \tag{4.1.100}$$

In conclusion, let us make some remarks on the choice of the parameter N that determines the length of the segments of the Fourier series approximating the components of the electric and magnetic fields in the zone of the grating. For a given N the number of calculated orders is equal to $2N+1$, from $-N$ to $+N$. The parameter

N must be greater than the number of propagating orders. If the grating consists of dielectrics only (all refractive indices are real numbers), and the parameter N satisfies the above condition, then the energy conservation law in the following form should be satisfied:

$$\sum I^R + \sum I^T = 1. \tag{4.1.101}$$

However, if the grating contains absorbing materials, the sum (4.1.101) should be less than unity. In general, the choice of N is made in the computational experiment from the conditions for stabilization of the values of the intensities of the orders.

4.1.3. Fourier modal method in a three-dimensional case

Consider the described method for the case of three-dimensional periodic diffraction structures. The z axis is directed perpendicular to the plane of the grating. The functions of dielectric and magnetic permeabilities of the grating are assumed to be periodic with respect to variables x, y, and with periods Λ_x and Λ_y, respectively. As in the planar case, we assume that the grating consists of L binary layers, and the dielectric permittivity and magnetic permeability in each layer does not depend on the variable z.

A method of solving the diffraction problem in three-dimensional case is similar to the considered two-dimensional case. Here are the main features of the three-dimensional problem.

In the diffraction of a plane wave in a three-dimensional diffraction grating formed by two-dimensional set of reflected and transmitted diffraction orders. In this case, the field above and below the structure is as follows:

$$\mathbf{\Phi}^1(x,y,z) = \mathbf{\Phi}^I(x,y,z) + \sum_n \sum_m \mathbf{\Phi}^R_{n,m}\left(R_{n,m}\right)\exp\left(i(k_{x,n}x + k_{y,m}y + k_{z,I,n,m}z)\right), \tag{4.1.102}$$

$$\mathbf{\Phi}^2(x,y,z) = \sum_n \sum_m \mathbf{\Phi}^T_{n,m}\left(T_{n,m}\right)\exp\left(i(k_{x,n}x + k_{y,m}y - k_{z,II,n,m}(z - d_L))\right), \tag{4.1.103}$$

where $\mathbf{\Phi}^I(x,y,z)$ is the incident wave. We consider the incident wave given in the form (4.1.21). The propagation constants of the diffraction orders with the index (n, m) are described by the following expressions:

$$\begin{aligned} k_{x,n} &= k_0\left(n_1 \sin\theta\cos\phi + n\frac{\lambda}{\Lambda_x}\right), \\ k_{y,m} &= k_0\left(n_1 \sin\theta\sin\phi + m\frac{\lambda}{\Lambda_y}\right), \\ k_{z,l,n,m} &= \sqrt{(k_0 n_l)^2 - k_{x,n}^2 - k_{y,m}^2}, \end{aligned} \tag{4.1.104}$$

where the index $l = 1$ is for the reflected orders and $l = 2$ for the transmitted ones. The form of the propagation constants ensures fulfilling the two-dimensional quasi-periodicity condition

$$\Phi^{1,2}(x+\Lambda_x, y+\Lambda_y, z) = \Phi^{1,2}(x,y,z)\exp\left(ik_{x,0}\Lambda_x + ik_{y,0}\Lambda_y\right). \tag{4.1.105}$$

According to (4.1.105), the amplitude of the field does not change with the shift to an integer number of periods. The waves with the real $k_{z,l,n,m}$ are propagating, those with imaginary – evanescent.

The distribution of an electromagnetic field in each layer, as in the two-dimensional case, is described the basic Maxwell equations for monochromatic fields in the form (4.1.26)–(1.4.29). We represent the components of the electric and magnetic fields in the form of two-dimensional Fourier series with respect to the variables x, y:

$$\begin{cases} E_x &= \sum\limits_n \sum\limits_m S_{x,n,m}(z)\exp\left(i\left(k_{x,n}x + k_{y,m}y\right)\right), \\ E_y &= \sum\limits_n \sum\limits_m S_{y,n,m}(z)\exp\left(i\left(k_{x,n}x + k_{y,m}y\right)\right), \\ E_z &= \sum\limits_n \sum\limits_m S_{z,n,m}(z)\exp\left(i\left(k_{x,n}x + k_{y,m}y\right)\right), \\ H_x &= -i\sum\limits_n \sum\limits_m U_{x,n,m}(z)\exp\left(i\left(k_{x,n}x + k_{y,m}y\right)\right), \\ H_y &= -i\sum\limits_n \sum\limits_m U_{y,n,m}(z)\exp\left(i\left(k_{x,n}x + k_{y,m}y\right)\right), \\ H_z &= -i\sum\limits_n \sum\limits_m U_{z,n,m}(z)\exp\left(i\left(k_{x,n}x + k_{y,m}y\right)\right). \end{cases} \tag{4.1.106}$$

The equations (4.1.106) are written taking into account the quasi-periodicity of the components of the fields in the variables x, y. We confine ourselves to the finite number of terms in the expansions (4.1.106), corresponding to $-N_x \le n \le N_x$, $-N_y \le m \le N_y$. Substituting (4.1.106) into the system (4.1.28) and equating the coefficients for the same Fourier harmonics, we obtain a system of differential equations in the form

$$\begin{cases} ik_0\mathbf{K}_y\mathbf{S}_z - \dfrac{d\mathbf{S}_y}{dz} &= k_0(\mathbf{M}_{1,1}\mathbf{U}_x + \mathbf{M}_{1,2}\mathbf{U}_y + \mathbf{M}_{1,3}\mathbf{U}_z), \\ \dfrac{d\mathbf{S}_x}{dz} - ik_0\mathbf{K}_x\mathbf{S}_z &= k_0(\mathbf{M}_{2,1}\mathbf{U}_x + \mathbf{M}_{2,2}\mathbf{U}_y + \mathbf{M}_{2,3}\mathbf{U}_z), \\ ik_0\mathbf{K}_x\mathbf{S}_y - ik_0\mathbf{K}_y\mathbf{S}_x &= k_0(\mathbf{M}_{3,1}\mathbf{U}_x + \mathbf{M}_{3,2}\mathbf{U}_y + \mathbf{M}_{3,3}\mathbf{U}_z), \\ ik_0\mathbf{K}_y\mathbf{U}_z - \dfrac{d\mathbf{U}_y}{dz} &= k_0(\mathbf{E}_{1,1}\mathbf{S}_x + \mathbf{E}_{1,2}\mathbf{S}_y + \mathbf{E}_{1,3}\mathbf{S}_z), \\ \dfrac{d\mathbf{U}_x}{dz} - ik_0\mathbf{K}_x\mathbf{U}_z &= k_0(\mathbf{E}_{2,1}\mathbf{S}_x + \mathbf{E}_{2,2}\mathbf{S}_y + \mathbf{E}_{2,3}\mathbf{S}_z), \\ ik_0\mathbf{K}_x\mathbf{U}_y - ik_0\mathbf{K}_y\mathbf{U}_x &= k_0(\mathbf{E}_{3,1}\mathbf{S}_x + \mathbf{E}_{3,2}\mathbf{S}_y + \mathbf{E}_{3,3}\mathbf{S}_z). \end{cases} \tag{4.1.107}$$

The form of the resulting system is identical to that of the system (4.1.33) for the two-dimensional case. The difference lies in the particular representation of vectors and matrices in the system. The vectors $\mathbf{S}_x$, $\mathbf{S}_y$, $\mathbf{S}_z$, $\mathbf{U}_x$, $\mathbf{U}_y$, $\mathbf{U}_z$ in (4.1.107) are obtained by row-by-row reordering the elements of the matrices $S_{x,j,k}$, $S_{y,j,k}$, $S_{z,j,k}$, $U_{x,j,k}$, $U_{y,j,k}$, $U_{z,j,k}$, $-N_x \le j \le N_x$, $-N_y \le k \le N_y$. This means that the element of the vector $\mathbf{S}_x$ with the number

$$l(i,j) = i\left(2N_y + 1\right) + j. \tag{4.1.108}$$

correspond to the value $S_{x,i,j}$. The vectors introduced in this way have the dimension $(2N_x + 1)(2N_y + 1)$ equal to the total number of the calculated diffraction orders. For example, the vector $\mathbf{S}_x$ has the form:

$$\mathbf{S}_x = \left[S_{x,-N_x,-N_y}, S_{x,-N_x,1-N_y}, \ldots, S_{x,-N_x,N_y}, S_{x,1-N_x,-N_y}, S_{x,1-N_x,1-N_y}, \ldots, S_{x,N_x,N_y} \right]^T.$$

The matrices in $\mathbf{K}_x, \mathbf{K}_y, \mathbf{E}_{i,j}, \mathbf{M}_{i,j}$ (4.1.107) have the dimension $(2N_x+1)(2N_y + 1)\times(2N_x + 1)(2N_y + 1)$. The matrices $\mathbf{K}_x$ and $\mathbf{K}_y$ are determined by the following expressions:

$$\begin{aligned} K_{x,l(i,j),l(n,m)} &= k_{x,i}\delta_{i-n}\delta_{j-m}/k_0, \\ K_{y,l(i,j),l(n,m)} &= k_{y,j}\delta_{i-n}\delta_{j-m}/k_0, \end{aligned} \tag{4.1.109}$$

where $-N_x \le i,n \le N_x$, $-N_y \le j,m \le N_y$.

Consider the form of the matrices $\mathbf{E}_{i,j}, \mathbf{M}_{i,j}$ obtained by using the normal Laurent rules for the expansion into Fourier series of the products of functions. The matrices $\mathbf{E}_{i,j}, \mathbf{M}_{i,j}$ are composed of the Fourier coefficients of the tensors of dielectric permittivity and magnetic permeability, the structure of the matrices is the same and has the form

$$T_{l(i,j),l(n,m)} = e_{i-n,j-m}, \tag{4.1.110}$$

where $e_{i,j}$ are Fourier coefficients, $-N_x \le i,n \le N_x$, $-N_y \le j,m \le N_y$,

Since the form of systems of differential equations in two- and three-dimensional cases is identical, then all the subsequent derivations will hold identical. The system of differential equations for the vectors $\mathbf{S}_x$, $\mathbf{S}_y$, $\mathbf{U}_x$, $\mathbf{U}_y$, in (4.1.107) has the form (4.1.36)–(4.1.38). In particular, for the grating of an isotropic material, and when $\varepsilon = (x, y)$ and $\mu = 1$ are scalars, the matrix of the system of differential equations (4.1.36) has the form

$$\mathbf{A} = -\begin{bmatrix} \mathbf{0} & \mathbf{0} & \mathbf{K}_y\mathbf{E}^{-1}\mathbf{K}_x & \mathbf{I} - \mathbf{K}_y\mathbf{E}^{-1}\mathbf{K}_y \\ \mathbf{0} & \mathbf{0} & \mathbf{K}_x\mathbf{E}^{-1}\mathbf{K}_x - \mathbf{I} & -\mathbf{K}_x\mathbf{E}^{-1}\mathbf{K}_y \\ \mathbf{K}_y\mathbf{K}_x & \mathbf{E} - \mathbf{K}_y^2 & \mathbf{0} & \mathbf{0} \\ \mathbf{K}_x^2 - \mathbf{E} & -\mathbf{K}_x\mathbf{K}_y & \mathbf{0} & \mathbf{0} \end{bmatrix}, \tag{4.1.111}$$

where the matrix $\mathbf{E}$ is given by (4.1.109) and is composed of the Fourier coefficients of functions $\varepsilon(x, y)$. Equation (4.1.111) can be obtained from the general expressions (4.1.37), (4.1.38) for

$$\begin{aligned}
&\mathbf{E}_{1,1} = \mathbf{E}_{2,2} = \mathbf{E}_{3,3} = \mathbf{E}, \\
&\mathbf{E}_{1,2} = \mathbf{E}_{1,3} = \mathbf{E}_{2,1} = \mathbf{E}_{2,3} = \mathbf{E}_{3,1} = \mathbf{E}_{3,2} = \mathbf{0}, \\
&\mathbf{M}_{1,1} = \mathbf{M}_{2,2} = \mathbf{M}_{3,3} = \mathbf{I}, \\
&\mathbf{M}_{1,2} = \mathbf{M}_{1,3} = \mathbf{M}_{2,1} = \mathbf{M}_{2,3} = \mathbf{M}_{3,1} = \mathbf{M}_{3,2} = \mathbf{0}.
\end{aligned} \tag{4.1.112}$$

Consider the transition to the spatial frequency representation (4.1.107), using the correct rules of Fourier expansions of the products of functions. Derivation of formulas will be carried out for the case of an isotropic material. The system of Maxwell's equations (4.1.28) contains only the following three products: εE_z, εE_x, εE_y.

The tangential component E_z is continuous, and therefore the product εE_z is expanded into a Fourier series with the Laurent rules (4.1.42). In this case, the corresponding matrix $E_{3,3}$ in (4.1.107) has the form (4.1.110).

We assume that the boundaries of two media with different dielectric constants in each layer are parallel to the coordinate axes [6]. Consider the product $D_x = \varepsilon E_x$. The product D_x is continuous at the points of the interfaces parallel to the axis Oy. Indeed, at these points $D_x = \varepsilon E_x$ is a normal component of electric displacement. At the same time, the component E_x and the function of the dielectric constant ε are discontinuous at these boundaries. Thus, the product $D_x = \varepsilon E_x$ is continuous in x for any fixed y. Accordingly, for the expansion of $D_x = \varepsilon E_x$ into the Fourier series with respect to variable x we used the inverse Laurent rule (4.1.43)

$$d_{x,i}(y,z) = \sum_n \varepsilon_{x\,i,n}(y) S_{x,n}(y,z), \tag{4.1.113}$$

where $d_{x,i}(y,z)$, $S_{x,n}(y,z)$ are the Fourier coefficients of the functions D_x, εE_x, and $\varepsilon_{xi,n}(y)$ are the elements of the Toeplitz matrix $\left[\!\left[1/\varepsilon_x\right]\!\right]^{-1}$ formed from the Fourier coefficients with respect to variable x of the function $1/\varepsilon(x, y)$. In points of the boundaries of the media parallel to the axis Ox, the component E_x is tangential and therefore continuous in y. The Fourier coefficients $S_{x,n}(y,z)$ of function E_x are also continuous. Therefore, for the expansion in (4.1.113) of terms $\varepsilon_{x\,i,n}(y) S_{x,n}(y,z)$ into a Fourier series in y we apply the Laurent rule (4.1.42)

$$d_{x,i,j}(z) = \sum_{n,m} \varepsilon_{x\,i,n,j-m} S_{x,n,m}(z), \tag{4.1.114}$$

where $\varepsilon_{x\,i,n,\,j-m}$ are the Fourier coefficients with the number $(j–m)$ of the function $\varepsilon_{xi,n}(y)$.

Repeating a similar argument for $D_y = E_y$, we obtain

$$d_{y,j}(x,z)=\sum_{m}\varepsilon_{y\,j,m}(x)S_{y,m}(x,z), \tag{4.1.115}$$

$$d_{y,i,j}(z)=\sum_{n,m}\varepsilon_{y\,i-n,j,m}S_{y,n,m}(z), \tag{4.1.116}$$

where $\varepsilon_{yj,\,m}(y)$ are the elements of the Toeplitz matrix $[\![1/\varepsilon_y]\!]^{-1}$ formed from the Fourier coefficients with respect to y of the function $1/\varepsilon(x, y)$, and $\varepsilon_{y\,i-n,j,m}$ is the Fourier coefficient with the number $(i-n)$ of the function $\varepsilon_{y\,j,m}(x)$.

Equations (4.1.114) and (4.1.116) were obtained using the correct rules of expansion into Fourier series of the products $D_x=\varepsilon E_x$, $D_y=\varepsilon E_y$. Accordingly, in the transition from the system (1.4.28) to the spatial frequency domain (4.1.107), the matrices $\mathbf{E}_{1,1}$ and $\mathbf{E}_{2,2}$ will have the form

$$\begin{aligned}E_{1,1\,l(i,n),l(j,m)}&=\varepsilon_{x\,i,n,j-m},\\E_{2,2\,l(i,n),l(j,m)}&=\varepsilon_{y\,i-n,j,m},\end{aligned} \tag{4.1.117}$$

where $l(i, j)$ is defined in (4.1.108), $-N_x\le i,n\le N_x$, $-N_y\le j,m\le N_y$. Matrices in (4.1.117) have dimension $(2N_x+1)\,(2N_y+1)\times(2N_x+1)\,(2N_y+1)$.

As a result, the matrix $\mathbf{A}$ of the system of linear differential equations takes the form:

$$\mathbf{A}=-\begin{bmatrix}\mathbf{0}&\mathbf{0}&\mathbf{K}_y\mathbf{E}^{-1}\mathbf{K}_x&\mathbf{I}-\mathbf{K}_y\mathbf{E}^{-1}\mathbf{K}_y\\\mathbf{0}&\mathbf{0}&\mathbf{K}_x\mathbf{E}^{-1}\mathbf{K}_x-\mathbf{I}&-\mathbf{K}_x\mathbf{E}^{-1}\mathbf{K}_y\\\mathbf{K}_y\mathbf{K}_x&\mathbf{E}_{1,1}-\mathbf{K}_y^2&\mathbf{0}&\mathbf{0}\\\mathbf{K}_x^2-\mathbf{E}_{2,2}&-\mathbf{K}_x\mathbf{K}_y&\mathbf{0}&\mathbf{0}\end{bmatrix}. \tag{4.1.118}$$

Detailed description of the correct rules of expansion into a Fourier series for the general case of the tensors of dielectric permittivity and magnetic permeability can be found in [8, 9].

Subsequent operations on the 'stitching' of solutions at the layer boundaries and numerically stable implementation of the calculation of the matrix of the system of linear equations for the amplitudes of the diffraction orders are also the same. Matrices $\mathbf{E}$, $\mathbf{F}$ in the systems of linear equations that represent the tangential field components (4.1.102), (4.1.103) on the upper and lower boundaries of the lattice also have the form (4.1.76), (4.1.77), where

$$\begin{aligned}&\sin\boldsymbol{\Phi}=\operatorname*{diag}_{l(i,j)}\frac{k_{y,j}}{\sqrt{k_{x,i}^2+k_{y,j}^2}},\quad\cos\boldsymbol{\Phi}=\operatorname*{diag}_{l(i,j)}\frac{k_{x,i}}{\sqrt{k_{x,i}^2+k_{y,j}^2}},\\&\cos\boldsymbol{\Theta}^{(1)}=\operatorname*{diag}_{l(i,j)}\frac{-k_{z,I,i,j}}{k_0n_1},\quad\cos\boldsymbol{\Theta}^{(2)}=\operatorname*{diag}_{l(i,j)}\frac{k_{z,II,i,j}}{k_0n_2},\end{aligned} \tag{4.1.119}$$

here $\operatorname{diag}_{l(i,j)} f(i,j)$ denotes the diagonal matrix composed of elements $f(i,j)$, arranged in the ascending order of magnitude $l(i,j)$.

4.1.4. Examples of calculation of diffraction gratings

The Fourier modal method can be used to calculate a wide class of periodic structures, including beam splitters, polarizers, antireflection structures, etc. In this section we consider several typical examples of these devices.

4.1.4.1. Grating polarizers

Figure 4.1.2 shows the geometry of the grating polarizer (period d), intended for the transmission of the component of the incident wave with TM-polarization and reflection of the component with TE-polarization. Such gratings are of great practical importance in backlight systems of LCD monitors.

The grating in Fig. 4.1.2 has the properties of the polarizer when the projection of the vector of direction of the incident wave on the grating plane is parallel to the axis Oy (when $\phi = \pi/2$ in (4.1.24)). Figure 4.1.2 schematically shows the directions of the incident wave in this geometry for the symmetric range of angles $\theta \in [-\theta_{\max}, \theta_{\max}]$.

Calculation of the geometric parameters of the structure in Fig.4.1.2 was carried out using a gradient optimization procedure [20] at the wavelength $\lambda = 550$ nm and the refractive indices of materials $n_1 = 1.5$, $n_2 = 1.72$. In practice, the structure should have polarizing properties for a certain range of incidence angles $\theta \in [-\theta_{\max}, \theta_{\max}]$. Therefore, as the merit function we select the following integral criterion

$$\varepsilon(h_1, h_2, h_3, d, r) = \int_{-\theta_{\max}}^{\theta_{\max}} \left(T_{\mathrm{TM}}(\theta) - T_{\mathrm{TE}}(\theta)\right) d\theta \to \max, \qquad (4.1.120)$$

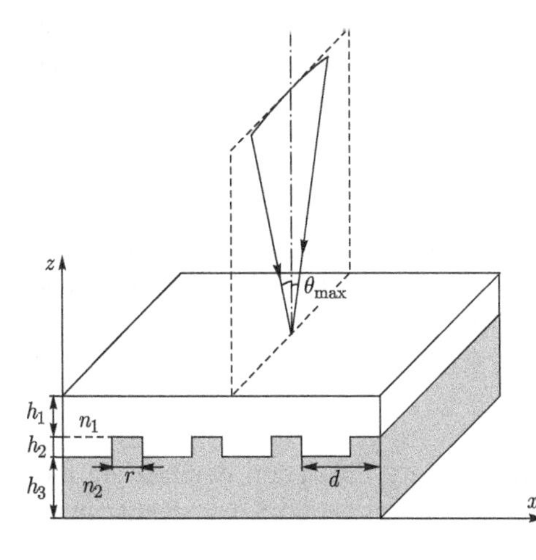

Fig. 4.1.2. Geometry of the grating polarizer geometry and the incident configuration.

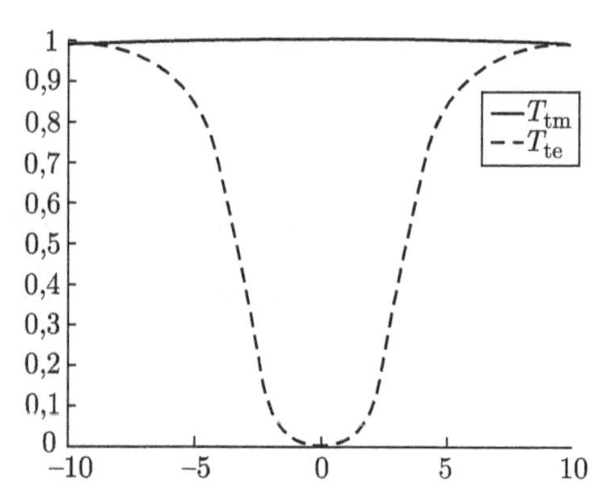

Fig. 4.1.3. The transmission coefficient of the structure vs. the angle of incidence at TM- (solid line) and TE-polarized (dashed line) incident wave.

where $T_{TM}(\theta)$, $T_{TE}(\theta)$ is the intensity of the transmitted zeroth diffraction orders in the case of TM-polarization ($\mathbf{H} = (H_x, 0, 0)$, $\mathbf{E} = (0, E_y, E_z)$ and the TE-polarization ($\mathbf{E} = (E_x, 0, 0)$, $\mathbf{H} = (0, H_y, H_z)$) of the incident wave, respectively. The calculation $T_{TM}(\theta)$, $T_{TE}(\theta)$ in (4.1.120) is carried out by the Fourier modal method. As a result, optimization at $\theta_{max} = 2°$ yielded the following parameters: period d = 482 nm, h_1 = 205 nm, h_2 = 285 nm, h_3 = 998 nm, r = 233 nm. Figure 4.1.3 shows the calculated transmission of the structures, depending on the angle of incidence of the TM- and TE-polarized incident waves. Figure 4.1.3 shows that the calculated structure in the required range of angles completely transmits the radiation with TM-polarization. The transmission for a wave with TE-polarization is zero at normal incidence and is less than 10% at $\theta \in [-2°, 2°]$.When using the grating in Fig. 4.1.2a it is difficult to achieve good polarization properties at $\theta_{max} > 2°$. The polarization properties in a large angular range can be achieved using a sandwich-type structure obtained by repeating the grating in Fig. 4.1.2a along the axis *Oz*. In this case the vertical periods of the sandwich structure are identical and have the form shown in Fig. 4.1.2.

We calculated the structure containing four vertical periods for the interval of angles of incidence $\theta \in [-10°, 10°]$. Optimization of the function (4.1.120) at $\theta_{max} = 10°$ resulted in the following parameters: period d = 439 nm, h_1 = 38 nm, h_2 = 460 nm, r = 287 nm. Calculated transmission spectra for a structure with four vertical periods are shown in Fig. 4.1.4.

Figure 4.1.4 shows that the calculated structure in the angular range $\theta \in [-10°, 10°]$ has the transmission coefficient for a wave with TM-polarization above 97%. In this case the transmission for a wave with TE-polarization does not exceed 10%, and at $\theta \in [-5°, 5°]$ it is less than 1.5%.

4.1.4.2. The beam splitter

Gratings are widely used as beam splitters. Binary diffraction gratings are the easiest to manufacture. The binary grating contains at period K rectangular lines

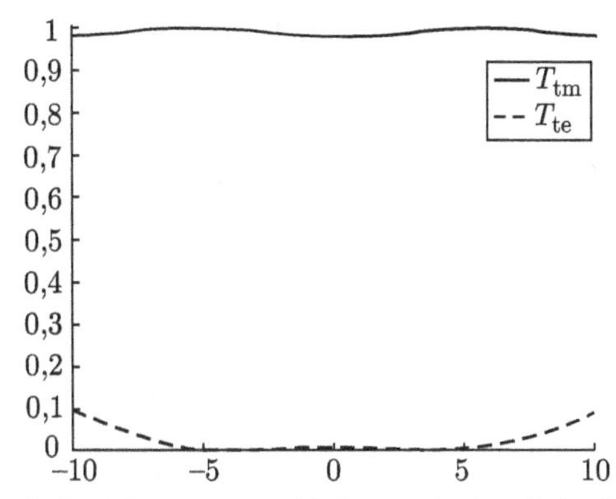

Fig. 4.1.4. The transmission of a structure with four vertical periods, depending on the angle of incidence with the incident wave with TM-polarization (solid line) and TE-polarization (dashed line).

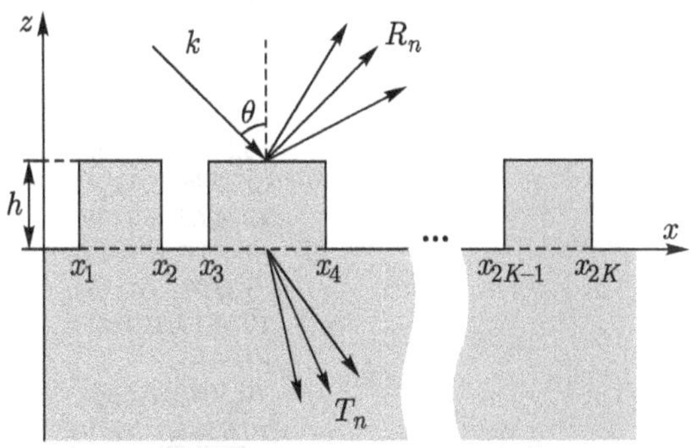

Fig. 4.1.5. Geometry of a binary dielectric grating.

of the same height but different widths (see Fig. 4.1.5). The task of calculating the beam splitter is formulated as the problem of calculating the coordinates of the profile lines $x_1,..., x_{2K}$ and the height of the lines a from the conditions of formation of the given intensities of the diffraction orders. Depending on the type of grating (dielectric transmissive or metallic reflective) the profile is calculated from the condition of the formation of given intensities of reflected or transmitted diffraction orders [20].

The Fourier modal method was used for the calculation of binary dielectric gratings with equal intensity of transmitted $2N + 1$ orders. Order numbers are symmetrical, from $-N$ to N. Calculation of grating parameters $\mathbf{p} = (x_1,..., x_{2K,\ a})$ was performed using the gradient optimization procedure [20] at the period $d = 5.5\ \lambda$, at normal incidence of the wave ($\theta = 0$) and at the refractive indices of the grating material and the substrate $n = 1.5$. The merit function was the quadratic error function

$$\varepsilon(\mathbf{p}) = \sum_{j=-N}^{N} \left(I_j^T(\mathbf{p}) - I\right)^2 \to \min, \tag{4.1.121}$$

Table 4.1.1. The results of calculation of binary dielectric gratings within the framework of electromagnetic theory ($d = 5.5\lambda$, $n = 1.5\theta = 0$)

Number of orders $2N+1$	Number of lines K	Height of lines (a/λ)	Coordinates of the profile	E, (%)	δ (%)
		TM-polarization			
3	1	0.675	(0.283, 0.7719)	80.1	3.1
5	2	0.9	(0.1800, 0.4841), (0.5525, 0.8567)	80.1	3.1
7	2	0.885	(0.2392, 0.4447) (0.5919, 0.7974)	87.7	0.34
9	3	1.7	(0.1000, 0.1924) (0.3842, 0.4777) (0.6217, 0.7154)	94.6	1.6
11	3	1.61	(0.1542, 0.3454), (0.4858, 0.5729), (0.713, 0.9043)	94.4	6.9
		TE-polarization			
3	1	0.65	(0, 0.5)	85	0.005
5	2	0.9	(0.1832, 0.4785), (0.5579, 0.8535)	80.2	0.08
7	2	0.875	(0.2584, 0.4296), (0.6067, 0.7779)	83.8	0.7
9	3	1.0	(0.0117, 0.1945), (0.3198, 0.4838), (0.7819, 0.9545)	89.4	3.6
11	3	1.57	(0.0446, 0.3383), (0.5049, 0.556), (0.8268, 0.8779)	90.7	5.6

where $I^T_j(\mathbf{p})$ is the intensity of transmitted orders. The results of calculations of the gratings are given in Table 4.1.1 for TE- and TM-polarization of the incident wave. In the last two columns of the table are the values of energy efficiency

$$E = \sum_{j=-N}^{N} I_j \tag{4.1.122}$$

and the standard error of formation of the given equal intensity of the orders

$$\delta = \frac{1}{\bar{I}}\left[\frac{1}{2N+1}\sum_{j=-N}^{N}\left(I_j - \bar{I}\right)^2\right]^{\frac{1}{2}}, \tag{4.1.123}$$

where $\bar{I} = E/(2N+1)$ is the average intensity. The coordinates of the profile in Table 4.1.1 are normalized to the value of the period.

The calculation results show the possibility of formation of 5–11 equal orders at the energy efficiency of 80–90% and a low rms error. Note that the methods of the scalar theory can not be used in the calculation of diffraction gratings with the specified period [20].

Table 4.1.2. The intensity of the order of three-dimensional binary grating

Indices of orders (i,j)	−1	0	+1
−1	0.0881	0.0885	0.0879
0	0.0885	0.0875	0.0885
+1	0.0879	0.0885	0.0881

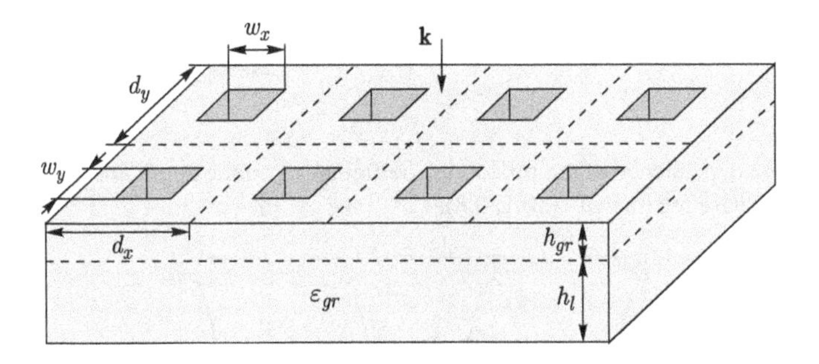

Fig. 4.1.6. Geometry of the dielectric binary grating

Three-dimensional binary diffraction gratings are used to generate the required two-dimensional set of diffraction orders. The Fourier modal method was used to calculate a three-dimensional binary dielectric grating to form nine equal intensity orders with numbers (i,j), $i,j = -1, 0, 1$. In this case, it suffices to use a simple binary grating having one rectangular recess per period (Fig. 4.1.6). Calculation of the parameters of the structure was performed at a wavelength $\lambda = 808$ nm, equal periods $d_x = d_y = d = 2362$ nm, the refraction index of the grating material $n = \sqrt{\varepsilon_{gr}} = 1.54$ for the case of normally incident waves with the angle of polarization $\psi = 45°$. In this case, the electric vector of the incident wave coincides with the bisector of the 1st quadrant. The geometrical parameters of the structure $h_{gr} = 1955$ nm, $w_x = 0.9d$ were determined using the gradient optimization procedure. In the calculations it was assumed that the total thickness of the grating $h_{gr} + h_l$ is 0.6 mm.

The calculated values of intensities of the orders of the grating at the given parameters are given in Table. 4.1.2. Table 4.1.2 shows the high uniformity of intensity distribution in the orders, the mean square error of (4.1.123) is less than 1%. The energy efficiency of the grating is 79.4%.

4.1.4.3. Subwavelength antireflection coatings

The Fourier modal method was used for the calculation of binary subwavelength antireflection gratings. The term 'subwavelength grating' means that the grating has only zero reflected and transmitted propagating orders. The remaining orders correspond to evanescent waves. For the existence of only zero order at normal incidence, the periods of the grating must satisfy the condition $d_x, d_y < \lambda / \sqrt{\varepsilon_{gr}}$.

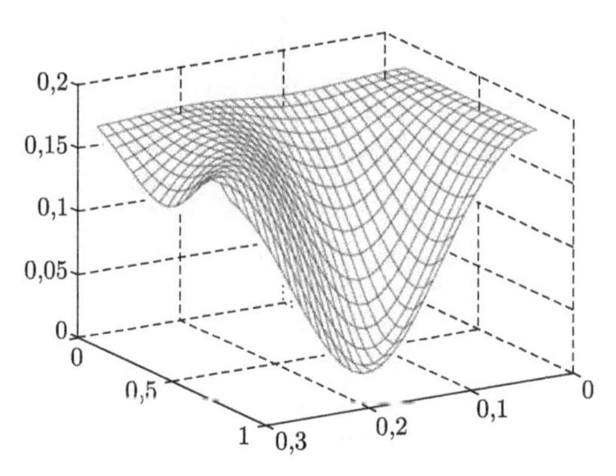

Fig. 4.1.7. The dependence of the reflection coefficient I_{00}^{R} on the depth $h/\lambda \in (0, 0.3)$ and on the side of a square recess $w/d \in (0.1, 0.8)$.

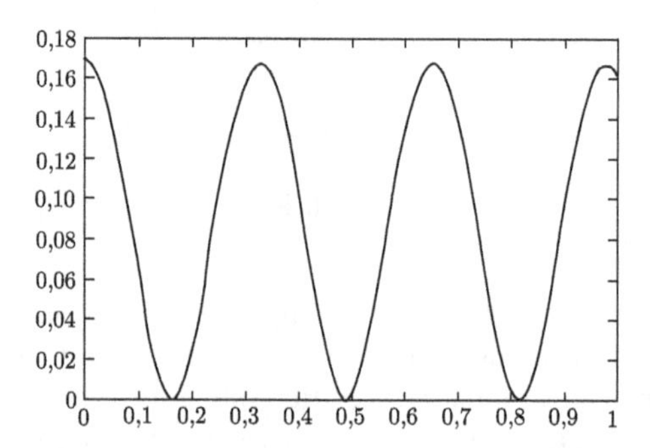

Fig. 4.1.8. The dependence of the reflection coefficient I_{00}^{R} on the depth (h/λ) at the optimum size of the side of the recess $w/d = 0.8$.

The antireflective coating was a simple binary grating with equal periods $d_x = dy = d$, as shown in Fig. 4.1.6. Calculations of the antireflective coating consisted of calculating the intensity of the zero reflected order I_{00}^{R} at different depths of the recess h_{gr} and the size of the recess $w_x = w_y = w$ for a fixed period d. The values h_{gr}, w, ensuring minimum reflection I_{00}^{R} can be considered as optimal grating parameters.

The following parameters were chosen for calculations: $\lambda = 10.6$ μm, $\varepsilon_{gr} = 5.76$, $d = 0.25, \lambda = 2.65$ μm. These values of permittivity ε_s and wavelength correspond to the case of synthesis of the antireflective coating of ZnSe (zinc selenide) for a CO_2 laser. The problem of synthesis of such coatings is especially important for high-power CO_2 lasers. Figure 4.1.7 shows a graph of the function I_{00}^{R} (h, r, d) at $h \in (0, 3.5)$ μm and $w/d \in (0.1, 0.8)$. The calculation was performed for a normal incidence of a plane wave. The incident wave was represented by a superposition of E- and H-waves with equal coefficients. Such a representation simulates the case of unpolarized light.

Figure 4.1.7 shows that at $w/d \approx 0.8$ and $h_{gr} \approx 1.8$ μm there is a pronounced minimum reflection coefficient I_{00}^{R} (h, r, d). Figure 4.1.8 shows a plot of the

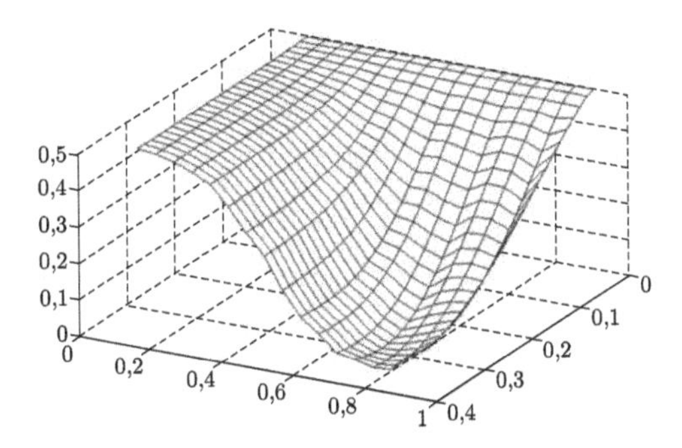

Fig. 4.1.9. The dependence of the reflection coefficient on the depth $h/\lambda \in (0, 0.4)$ and the size of the recess $w/d \in (0, 1.0.9)$.

reflection coefficient I^R_{00} (h, r, d) at the optimal size of the recess $w = 0.8d$ and different heights. From the graph in Fig 4.18 one can see that the reflection coefficient decreases to zero at the depths 0.18λ, 0.49λ and 0.82λ. Thus, on a plane zinc selenide – air interface the reflection coefficient is approximately 17%

Investigations were carried out into the possibility of using the simplest binary gratings (Fig. 4.1.6) as a a reflective coating for tungsten in the visible spectrum at $\lambda = 0.55$ μm. In this case, the dielectric constant is complex and $\lambda = 0.55$ μm $\varepsilon_{gr} = 4.8 + 19.11i$. The graph of the reflection coefficient I^R_{00} (h, r, d) for a tungsten binary grating with the period $d = 0.85\lambda$ is shown in Fig. 4.1.9. Figure 4.1.9 shows that the minimum reflection coefficient is achieved when the size of the recess is $w/d \approx 0.75$ and depth $h_{gr} \approx 0.35$.

The graph of the reflection coefficient at the optimum size of the recess is shown in Fig. 4.1.10. The graph shows that the reflection coefficient for a plane tungsten–air boundary is close to 50%. With increasing depth the reflection coefficient decreases, reaching zero at depth $h_{gr} = 0.33\lambda$. In contrast to dielectric gratings (Fig. 4.1.8), the secondary minima are not very pronounced. Figure 4.1.11 shows a graph for the reflection coefficient at the optimum size of the recess, depending on the wavelength

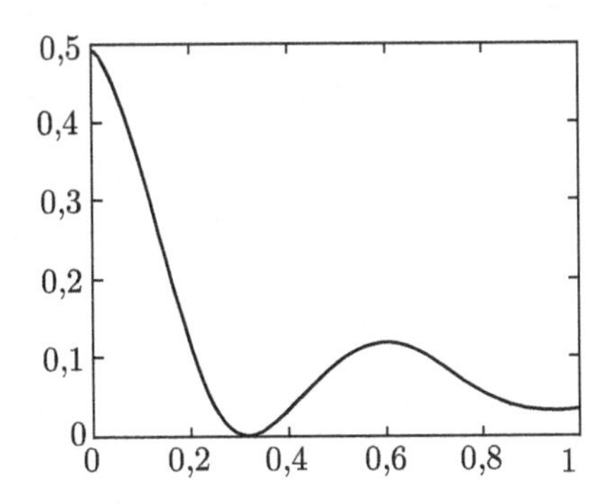

Fig. 4.1.10. Dependence of the reflection coefficient of the depth (h/λ) at the optimal size of the recess $w/d = 0.75$.

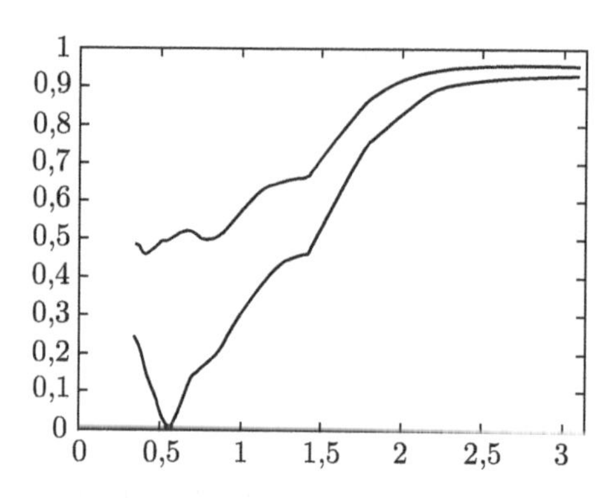

Fig. 4.1.11. Dependence of the reflection coefficient on wavelength $w/d = 0.75$, $h_{gr} = 0.33\lambda$.

range from 0.35 μm to 3.5 μm. The top graph in Fig. 4.1.11 shows the reflection coefficient for a planar tungsten–air interface. Figure 4.1.11 shows the presence of a sharp minimum of the reflection coefficient near the calculated wavelength $\lambda = 0.55$ μm. At the same time across the entire visible range the reflection coefficient for a binary grating is half the coefficient than for the planar interface.

4.2. Formation of high-frequency interference patterns of surface plasma polaritons by diffraction gratings

Due to diffraction, the light can not be focused to an arbitrarily small spot. The minimum diameter of the spot is about half a wavelength. Thus, in the best diffraction-limited microscopy systems the maximum attainable resolution is of the order of hundreds of nanometers. Using the interference patterns of surface plasma polaritons (SPP) we can achieve superresolution of about a tenth of a wavelength of the light used.

In section 2.1 we derive the equations of SPP from Maxwell's equations and examine the characteristics of the SPP. Section 2.1.3 examines the Kretschmann scheme for SPP excitation and the formation of interference patterns in the framework of the scheme.

The paragraphs 2.2 and 2.3 discuss the formation of interference patterns of the SPP with a diffraction structure consisting of a dielectric diffraction grating (one- or two-dimensional) and a metal film placed under the grating in the region of the substrate [21–25]. Diffraction gratings are used for excitation at the lower surface of the metal film at the lower surface of the metal film, which forms an interference pattern. The excitation of SPP and the formation of interference patterns are carried out uisng higher diffraction orders (with numbers $\pm m$, $m > 1$). This allows to generate high-frequency interference patterns with a period several times smaller than the wavelength of incident light with a low-frequency diffraction microrelief with a period several times greater than the wavelength of incident radiation [21–24]. These diffraction structures are used in surface plasmon interference nanolithography). In this case the interference pattern of SPP is recorded in the resist, which is located directly below the metal film and then the appropriate nano- or microstructure is produced [26–30]. When using electron beam lithography for the production of a similar structure with a substantially subwavelength period the required sample screen size (resolution) should not be more than a quarter period of the interference pattern. Using the interference patterns of the SPP we can achieve resolution of a few tens of nanometers (about a tenth of a wavelength).

Section 2.4 describes the integral representation of the electromagnetic field at the interface of two media through the angular spectrum of SPP and also describes the calculations of the diffraction structures for the transformation and focusing of the SPP. The calculation of the diffraction structures is based on the phase modulation of SPP, formed during the passage of a wave through the dielectric block, situated directly on the surface of propagation of SPP. The given phase modulation takes place both as a result of the variation of the height of the block above the surface at the fixed length and as a result of the change of the length of the block at the fixed height. The calculation of the 'lens' of surface plasma polaritons is discussed as an example.

4.2.1. Surface plasma polaritons (SPP)

4.2.1.1. The equation of a surface plasma polariton

Consider the derivation of the equation of the surface plasma polariton (SPP) at the interface between two semi-infinite media from Maxwell's equations. Let the interface be the plane $z = 0$, with the media 1 and 2 corresponding to the regions $z > 0$ and $z < 0$, respectively.

We write a general representation of the field in the media 1 and 2. The index of the number of the medium in the field components and dielectric constants will be introduced later, before applying the boundary conditions at the interface. Since the properties of the medium do not depend on the variables x, y, then the electric and magnetic fields in the media 1 and 2 have the form

$$\begin{aligned} \mathbf{E}(x,y,z) &= \mathbf{E}(z)\exp\left(ik_0\left(\alpha x+\beta y\right)\right), \\ \mathbf{H}(x,y,z) &= \mathbf{H}(z)\exp\left(ik_0\left(\alpha x+\beta y\right)\right), \end{aligned} \tag{4.2.124}$$

where $k_0 = 2\pi/\lambda$, λ is the wavelength in vacuum. Substituting (4.2.1) into Maxwell's equations for a monochromatic field (4.1.3), we obtain:

$$\begin{aligned} &ik_0\beta H_z - \frac{\partial H_y}{\partial z} = -ik_0\varepsilon E_x, && ik_0\beta E_z - \frac{\partial E_y}{\partial z} = ik_0 H_x, \\ &\frac{\partial H_x}{\partial z} - ik_0\alpha H_z = -ik_0\varepsilon E_y, && \frac{\partial E_x}{\partial z} - ik_0\alpha E_z = ik_0 H_y, \\ &ik_0\alpha H_y - ik_0\beta H_x = -ik_0\varepsilon E_z, && ik_0\alpha E_y - ik_0\beta E_x = ik_0 H_z. \end{aligned} \tag{4.2.125}$$

We rewrite equation (4.2.125) in the form of

$$E_x = \frac{1}{ik_0\left(\varepsilon-\beta^2\right)}\left(\frac{\partial H_y}{\partial z} - ik_0\alpha\beta E_y\right),$$
$$H_x = \frac{-1}{ik_0\left(\varepsilon-\beta^2\right)}\left(\varepsilon\frac{\partial E_y}{\partial z} + ik_0\alpha\beta H_y\right), \tag{4.2.126}$$

$$E_z = \frac{-1}{ik_0\left(\varepsilon-\beta^2\right)}\left(\beta\frac{\partial E_y}{\partial z} + ik_0\alpha H_y\right),$$
$$H_z = \frac{1}{ik_0\left(\varepsilon-\beta^2\right)}\left(ik_0\varepsilon\alpha E_y - \beta\frac{\partial H_y}{\partial z}\right), \tag{4.2.127}$$

where the components E_y and H_y satisfy the Helmholtz equation:

$$\frac{\partial^2 E_y}{\partial z^2} + k_0^2\left(\varepsilon-\alpha^2-\beta^2\right)E_y = 0,$$
$$\frac{\partial^2 H_y}{\partial z^2} + k_0^2\left(\varepsilon-\alpha^2-\beta^2\right)H_y = 0. \tag{4.2.128}$$

Solving (4.2.5), we obtain equations for the components E_y and H_y in the media 1 and 2:

$$E_y^{(1)}(x,y,z) = e_1\exp\left(ik_0\left(\alpha x+\beta y\right)\right)\exp\left(-k_0\gamma_1 z\right),$$
$$H_y^{(1)}(x,y,z) = h_1\exp\left(ik_0\left(\alpha x+\beta y\right)\right)\exp\left(-k_0\gamma_1 z\right),$$
$$E_y^{(2)}(x,y,z) = e_2\exp\left(ik_0\left(\alpha x+\beta y\right)\right)\exp\left(k_0\gamma_2 z\right),$$
$$H_y^{(2)}(x,y,z) = h_2\exp\left(ik_0\left(\alpha x+\beta y\right)\right)\exp\left(k_0\gamma_2 z\right), \tag{4.2.129}$$

where

$$\gamma_i^2 = \alpha^2+\beta^2-\varepsilon_i, \tag{4.2.130}$$

index i = 1, 2 indicates the number of the medium, e_i, h_i are arbitrary constants. Representation of the field (4.2.6) corresponds to the SPP because it is exponentially decaying in the direction z, perpendicular to the interface.

Substituting (4.2.6) into (4.2.3) we get a second pair of tangential components E_x and H_x in the form

$$E_x^{(1)}=\frac{-1}{i\left(\varepsilon_1-\beta^2\right)}\left(h_1\gamma_1+e_1 i\alpha\beta\right)\exp\left(ik_0\left(\alpha x+\beta y\right)\right)\exp\left(-k_0\gamma_1 z\right),$$

$$H_x^{(1)}=\frac{-1}{i\left(\varepsilon_1-\beta^2\right)}\left(-e_1\varepsilon_1\gamma_1+h_1 i\alpha\beta\right)\exp\left(ik_0\left(\alpha x+\beta y\right)\right)\exp\left(-k_0\gamma_1 z\right),$$

$$E_x^{(2)}=\frac{1}{i\left(\varepsilon_2-\beta^2\right)}\left(h_2\gamma_2-e_2 i\alpha\beta\right)\exp\left(ik_0\left(\alpha x+\beta y\right)\right)\exp\left(k_0\gamma_2 z\right),$$

$$H_x^{(2)}=\frac{-1}{i\left(\varepsilon_2-\beta^2\right)}\left(e_2\varepsilon_2\gamma_2+h_2 i\alpha\beta\right)\exp\left(ik_0\left(\alpha x+\beta y\right)\right)\exp\left(k_0\gamma_2 z\right).$$

(4.2.131)

We write the conditions of equality of the tangential components of electric and magnetic fields at the interface of the media at $z = 0$:

$$\begin{cases} E_y^{(1)}\left(x,y,0\right)=E_y^{(2)}\left(x,y,0\right), \\ H_y^{(1)}\left(x,y,0\right)=H_y^{(2)}\left(x,y,0\right), \\ E_x^{(1)}\left(x,y,0\right)=E_x^{(2)}\left(x,y,0\right), \\ H_x^{(1)}\left(x,y,0\right)=H_x^{(2)}\left(x,y,0\right). \end{cases} \quad (4.2.132)$$

From the first two equations in (4.2.9) we obtain

$$e_1=e_2,\ h_1=h_2. \quad (4.2.133)$$

Substituting (4.2.131), (4.2.133) into the second two equations of (4.2.9), we have:

$$\begin{cases} e_1\left(\dfrac{\varepsilon_1\gamma_1}{\varepsilon_1-\beta^2}+\dfrac{\varepsilon_2\gamma_2}{\varepsilon_2-\beta^2}\right)+h_1\left(\dfrac{i\alpha\beta}{\varepsilon_2-\beta^2}-\dfrac{i\alpha\beta}{\varepsilon_1-\beta^2}\right)=0, \\ e_1\left(\dfrac{i\alpha\beta}{\varepsilon_1-\beta^2}-\dfrac{i\alpha\beta}{\varepsilon_2-\beta^2}\right)+h_1\left(\dfrac{\gamma_1}{\varepsilon_1-\beta^2}+\dfrac{\gamma_2}{\varepsilon_2-\beta^2}\right)=0. \end{cases} \quad (4.2.134)$$

A non-trivial solution of (4.2.11) will exist when the determinant of system (4.2.11) is zero. Thus, we obtain

$$\left(\frac{\gamma_1}{\varepsilon_1-\beta^2}+\frac{\gamma_2}{\varepsilon_2-\beta^2}\right)\left(\frac{\varepsilon_1\gamma_1}{\varepsilon_1-\beta^2}+\frac{\varepsilon_2\gamma_2}{\varepsilon_2-\beta^2}\right)-\alpha^2\beta^2\left(\frac{1}{\varepsilon_1-\beta^2}-\frac{1}{\varepsilon_2-\beta^2}\right)^2=0.$$

(4.2.135)

A direct substitution shows that equation (4.2.12) becomes an identity when the following condition is fulfilled

$$k_0^2\left(\alpha^2+\beta^2\right)=k_0^2\frac{\varepsilon_1\varepsilon_2}{\varepsilon_1+\varepsilon_2}. \tag{4.2.136}$$

Equation (4.2.136) is the dispersion relation of SPP. Under the condition (4.2.13) component H_z in (4.2.127) is identically zero. Thus, the vector **H** of the SPP is situated in the plane of the interface between two media.

Direct analysis shows that the fields in the equations (4.2.6) and (4.2.8) have the damping form with respect to z under the condition Re $(\varepsilon_1+\varepsilon_2)<0$. This condition can occur at the interface between metal and dielectric. The dielectric constant of metals with high conductivity has a large negative real part and a small imaginary part, which ensures the fulfillment of the above conditions. For convenience, we replace the indices 1 and 2, denoting the number of the medium, by the indices m and d denoting the metal and the dielectric, respectively, and introduce the quantity

$$k_{\text{SPP}}=k_0\sqrt{\frac{\varepsilon_m\varepsilon_d}{\varepsilon_m+\varepsilon_d}}. \tag{4.2.137}$$

Quantity k_{SPP} is called the propagation constant of SPP and determines the projection of the wave vector of the SPP on the plane xOy.

In a particular case $\beta=0$ we obtain the SPP propagating along the axis Ox. In this case, equation (4.2.12) takes the form

$$\left(\frac{\gamma_1}{\varepsilon_1}+\frac{\gamma_2}{\varepsilon_2}\right)(\gamma_1+\gamma_2)=0,$$

which immediately yields the dispersion equation $k_0^2\alpha^2=k_{\text{SPP}}^2$. Note that at $\beta=0$ it follows from (4.2.134) that $e_1=0$. Thus, from (4.2.129) and (4.2.131) we have:

$$\mathbf{E}^{(i)}=\left(E_x^{(i)},0,E_z^{(i)}\right),\ \mathbf{H}^{(i)}=\left(0,H_y^{(i)},0\right),\ i=1,2. \tag{4.2.138}$$

The expressions (4.2.138) show that the SPPs propagating along the axis Ox, are TM-polarized.

It should be noted that the value k_{SPP} is complex, since the dielectric constant ε_m is complex:

$$\varepsilon_m=\varepsilon_m'+i\varepsilon_m''. \tag{4.2.139}$$

Thus, the SPP decays also in the direction of propagation. Note that since $\varepsilon_m'<0$ then the inequality

$$\left|\sqrt{\frac{\varepsilon_m}{\varepsilon_m+\varepsilon_d}}\right|>1. \tag{4.2.140}$$

is fulfilled. According to (4.2.140), we obtain

$$\left|k_{\text{SPP}}\right| > \sqrt{\varepsilon_d}\,k_0. \tag{4.2.141}$$

Inequality (4.2.141) shows that for excitation of the SPP we can not use a plane wave incident on a medium with permittivity ε_d. SPPs are excited using special configurations containing a prism made of a material with higher dielectric constant or diffraction gratings.

4.2.1.2. The properties of surface plasma polaritons

To characterize the SPP, we use quantities such as the wavelength, the propagation length, the depth of penetration into the dielectric and metallic media [31]. For convenience, k_{SPP} is presented in the form

$$k_{\text{SPP}} = k'_{\text{SPP}} + ik''_{\text{SPP}}, \tag{4.2.142}$$

where k'_{SPP} and k''_{SPP} are the real and imaginary parts, respectively.

The sections 2.2 and 2.3 of this chapter discuss the SPP on the interface between silver and a dielectric with $\varepsilon_d = 2.56$. Accordingly, the characteristics of the SPP will be given for this pair of materials. The dependence of the dielectric constant of silver on the wavelength is shown in Fig. 4.2.1. At $\lambda > 368$ nm Re $(\varepsilon_m) + \varepsilon_d < 0$ and hence for such values of wavelength the SPP can exist at the interface.

The length of SPP is determined from the expressions

$$\lambda_{\text{SPP}} = 2\pi / k'_{\text{SPP}} = \lambda / \operatorname{Re}\left(\sqrt{\frac{\varepsilon_m \varepsilon_d}{\varepsilon_m + \varepsilon_d}}\right). \tag{4.2.143}$$

If

$$\left|\varepsilon'_m + \varepsilon_d\right| \gg \left|\varepsilon''_m\right|, \tag{4.2.144}$$

an approximate expression holds

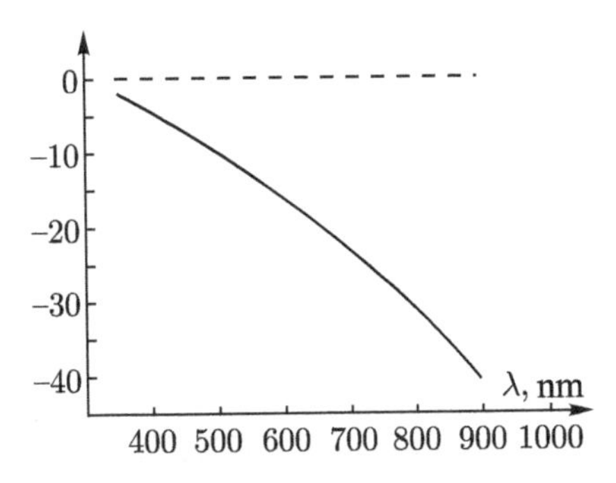

Fig. 4.2.1. The dependence of the dielectric constant of silver on the wavelength (the real part – solid line, imaginary part – dotted line).

$$k'_{\rm SPP} \approx k_0 \sqrt{\frac{\varepsilon'_m \varepsilon_d}{\varepsilon'_m + \varepsilon_d}}. \tag{4.2.145}$$

In view of (4.2.145) we obtain an approximate expression for the wavelength of SPP:

$$\lambda_{\rm SPP} \approx \lambda \sqrt{\frac{\varepsilon'_m + \varepsilon_d}{\varepsilon'_m \varepsilon_d}}. \tag{4.2.146}$$

The normalized wavelength of the SPP is defined by

$$\frac{\lambda_{\rm SPP}}{\lambda} = 1 / \operatorname{Re}\left(\sqrt{\frac{\varepsilon_m \varepsilon_d}{\varepsilon_m + \varepsilon_d}} \right). \tag{4.2.147}$$

When the condition (2.4.21) is fulfilled

$$\frac{\lambda_{\rm SPP}}{\lambda} \approx \sqrt{\frac{\varepsilon'_m + \varepsilon_d}{\varepsilon'_m \varepsilon_d}}. \tag{4.2.148}$$

Since $\varepsilon'_m < 0$ then, according to (4.2.23), we obtain

$$\lambda_{\rm SPP} < \lambda. \tag{4.2.149}$$

This fact is the basis for the use of SPP in photolithography systems in the formation of nanostructures with subwavelength dimensions.

Figure 4.2.2 shows the dependence of the real (solid line) and imaginary (dashed line) parts of the value $k_{\rm SPP}/k_0$ on the wavelength, and Fig. 4.2.3 is the same graph for the normalized wavelength of SPP. It is easy to see that the extrema of these quantities are obtained under the condition

$$\varepsilon'_m = -\varepsilon_d. \tag{4.2.150}$$

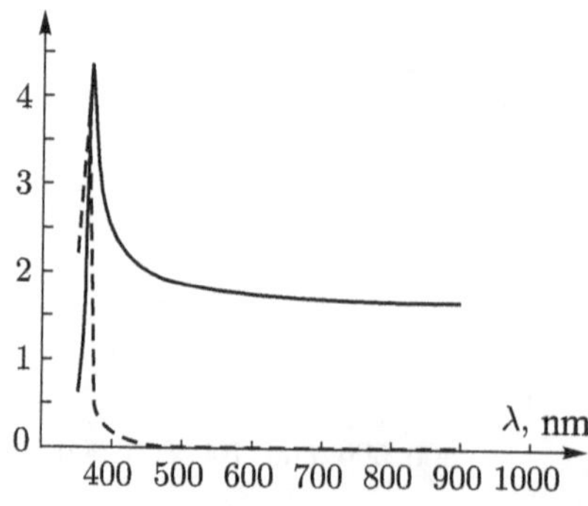

Fig. 4.2.2. The dependence of the normalized propagation constants of the SPP on the wavelength (the real part – solid line, imaginary part – dashed line).

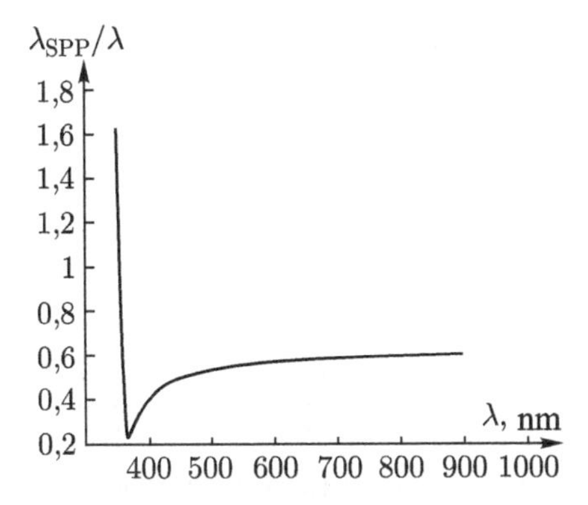

Fig. 4.2.3. The dependence of the normalized SPP length of the free-space wavelength. EW.

Condition (4.2.150) holds for $\lambda \approx 368$ nm. The wavelength at which the condition (4.2.150) is fulfilled is called the resonance wavelength [28, 31]. Note that in the vicinity of the resonance wavelength the imaginary part of magnitude k_{SPP}/k_0 also increases. As shown below, this leads to a decrease in the length of propagation of the SPP.

The SPP propagation length

$$\delta_{\text{SPP}} = \frac{1}{2k''_{\text{SPP}}}. \tag{4.2.151}$$

is defined as the distance at which the intensity of the wave decreases e times. When the condition (4.4.144) is fulfilled, the approximate equality holds

$$k''_{\text{SPP}} \approx k_0 \frac{\varepsilon''_m}{2\left(\varepsilon'_m\right)^2}\left(\frac{\varepsilon'_m \varepsilon_d}{\varepsilon'_m + \varepsilon_d}\right)^{\frac{3}{2}}. \tag{4.2.152}$$

Substituting (4.2.152) into (4.2.151), we obtain an approximate expression for the length of the SPP propagation in the form:

$$\delta_{\text{SPP}} \approx \lambda \frac{\left(\varepsilon'_m\right)^2}{2\pi\varepsilon''_m}\left(\frac{\varepsilon'_m + \varepsilon_d}{\varepsilon'_m \varepsilon_d}\right)^{\frac{3}{2}}. \tag{4.2.153}$$

When the condition

$$\left|\varepsilon'_m\right| \gg \left|\varepsilon_d\right| \tag{4.2.154}$$

is fulfilled, we can write a simple approximate expression for:

$$\delta_{\text{SPP}} \approx \lambda \frac{\left(\varepsilon'_m\right)^2}{2\pi\varepsilon''_m \varepsilon_d^{3/2}}. \tag{4.2.155}$$

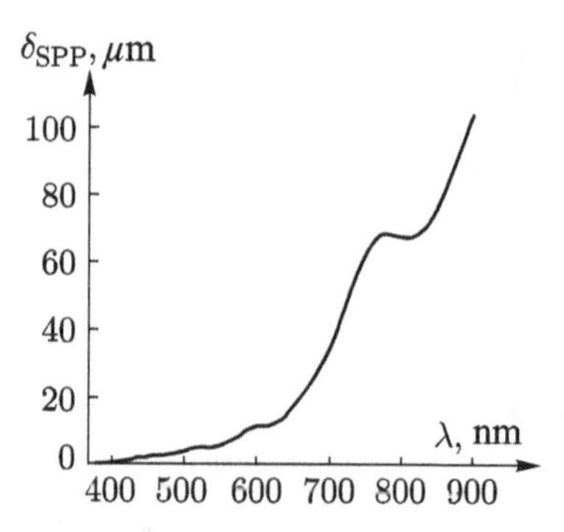

Fig. 4.2.4. Dependence of the propagation length of the SPP on the wavelength.

Equation (4.2.155) shows that the SPP propagation length is directly proportional to the real part and inversely proportional to the imaginary part of the permittivity of the metal. Figure 4.2.4 shows the propagation length of the SPP (4.2.151) on the wavelength of light for the given pair of materials. The graph shows that when the value of the wavelength approaches the resonant value $\lambda > 368$ nm the propagation length tends to zero. The practical use of the SPP is impossible in the vicinity of the resonance wavelength.

The penetration depth of the SPP into the medium is defined as the distance at which the wave amplitude decreases by e times.According to (4.2.129)–(4.2.131), (4.2.136), the damping of the SPP is determined by the quantity

$$k_{z,l} = k_0 \gamma_l = \sqrt{k_{\text{SPP}}^2 - \varepsilon_l k_0^2}, \tag{4.2.156}$$

where the superscript l denotes the metallic (m) or dielectric (d) medium. The penetration depth takes the form

$$\delta_l = 1 \Big/ \left| \operatorname{Re}\left(k_{z,l}\right) \right|. \tag{4.2.157}$$

When the condition (2.4.144) is fulfilled, the following approximate expressions hold:

$$\delta_d \approx \frac{1}{k_0} \left| \frac{\varepsilon_m' + \varepsilon_d}{\varepsilon_d^2} \right|^{\frac{1}{2}}, \tag{4.2.158}$$

$$\delta_m \approx \frac{1}{k_0} \left| \frac{\varepsilon_m' + \varepsilon_d}{\left(\varepsilon_m'\right)^2} \right|^{\frac{1}{2}}. \tag{4.2.159}$$

These formulas for the penetration depths δ_d, δ_m are of practical importance, since they allow to determine the minimum thickness of material required for the

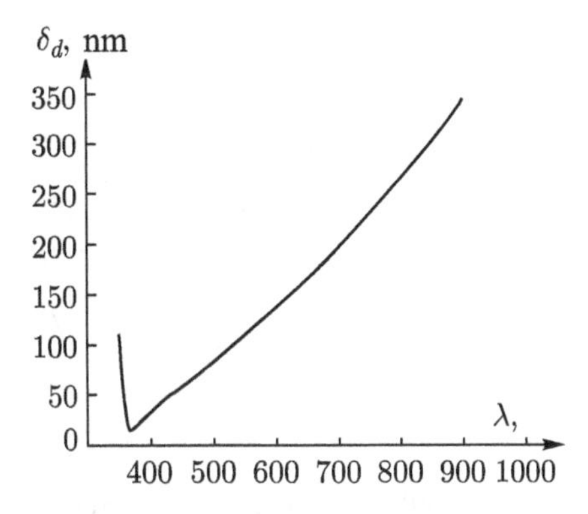

Fig. 4.2.5. The dependence of the penetration depth of SPP in a dielectric medium on the wavelength.

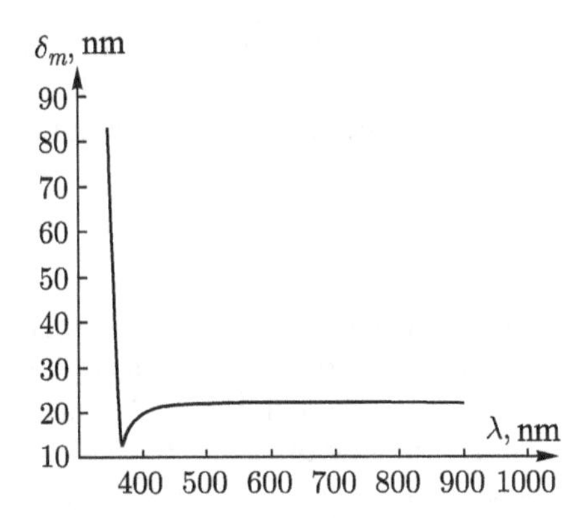

Fig. 4.2.6. The dependence of the penetration depth of SPP in the metal environment on the wavelength.

excitation and existence of SPP. Figures 4.2.5 and 4.2.6 show the dependences of the depth of penetration of the SPP in the dielectric and the metal environment on the wavelength. It is seen that for values of the wavelength far from the resonance value, the penetration depth of SPP in the metal environment is similar to a constant value equal to 22 nm.

4.2.1.3. Excitation of surface plasma polaritons

One of the most common schemes used for the excitation of the SPP is the Kretschmann scheme [32–38]. The Kretschmann scheme includes a glass prism with a metal film at the bottom (Fig. 4.2.7a). The metal film is assumed to be made of good electrical conductors (silver, gold). At a certain angle of incidence of the wave with TM-polarization from the side of the prism SPPs are excited at the lower boundary of the metal film [32–38]. The configuration in Fig. 4.2.7a is described

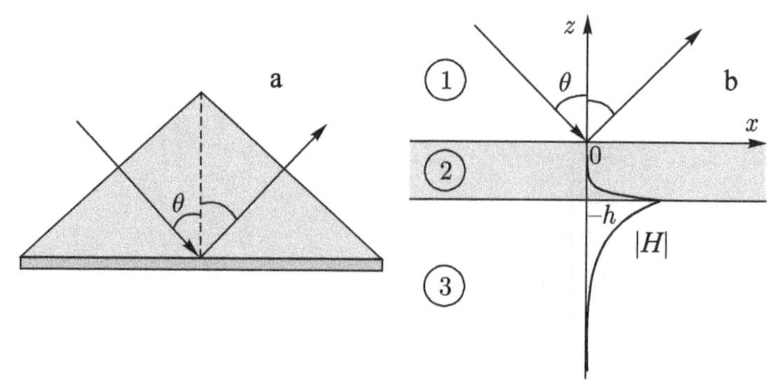

Fig. 4.2.7. Geometry of SPP excitation (a) and the equivalent model (b).

by the model of the three-layer medium in Fig. 4.2.7b. The dielectric constants 1–3 in Fig. 4.2.7b correspond to the materials of the prism (ε_{pr}), the metal layer (ε_m) and the substrate (ε_d).

SPP excitation occurs at an angle of incidence θ, defined by the condition of equality of the projection of the wave vector of the incident wave to the direction of propagation of SPP (axis Ox) to the propagation constant of SPP. This condition has the form

$$k_{x,0} = k_0 \sqrt{\varepsilon_{\text{pr}}} \sin\theta = \text{Re}\left(k_{\text{SPP}}\right). \tag{4.2.160}$$

Condition (4.2.160) is approximate. This is due to the fact that formula (4.2.137) defines the propagation constant of SPP for the boundary of semi-infinite media, and the metal film in Fig. 4.2.7 has a finite thickness. In practice, the film thickness is 40–50 nm. The error of determining the angle θ from the formula (4.2.160) is about 0.1° and in most practical problems is not essential. The exact definition of the angle θ can be found by solving the problem of diffraction of a plane wave with TM-polarization on a homogeneous metal layer. The SPP excitation angle is determined by a sharp minimum that appears in the reflection spectrum. Figure 4.2.8 shows a typical plot of the reflection coefficient $R(\theta)$ calculated with the following parameters: $\lambda = 550$ nm, $h = 50$ nm, $\varepsilon_{pr} = 4$, $\varepsilon_m = -12.922 + 0.447i$ (Ag), $\varepsilon_d = 2.25$.

With these parameters, the surface wave excitation occurs at an angle of incidence $\theta_{SPP} = 63.18°$. Note that the angle obtained from (4.2.160) is in this case $\theta_{SPP} = 63.28°$.

Consider in the model of SPP excitation (Fig. 4.2.7) not one but two symmetrically incident TM-waves:

$$H_{0y}^{(1)}(x,z) = \exp\left(ik_{x,0}x - ik_z^{(1)}z\right) + \exp\left(-ik_{x,0}x - ik_z^{(1)}z\right), \tag{4.2.161}$$

where $(k_z^{(1)})^2 = k_0^2 \varepsilon_{pr} - k_{x,0}^2$. In this case, two SPPs propagating in opposite directions will be excited along the boundary of zones 2 and 3. As a result, the interference pattern of the SPP will form directly under the metal film. From Maxwell's equations we can easily obtain the distribution of the electric field intensity

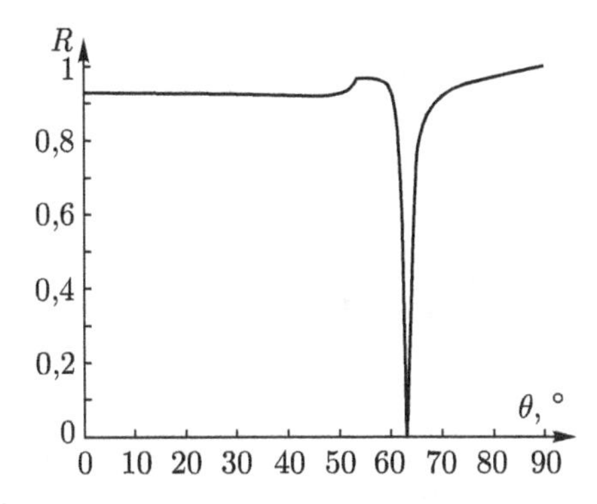

Fig. 4.2.8. The reflection coefficient as a function of incidence angle.

($|\mathbf{E}|^2=|E_x|^2+|E_z|^2$) under the film (when $z \le -h$) in the form

$$\left|\mathbf{E}_{\text{SPP}}(x,z)\right|^2 \sim \left(\left(k_{x,0}^2-\left|k_z^{(3)}\right|^2\right)\sin^2\left(k_{x,0}x\right)+\left|k_z^{(3)}\right|^2\right)\exp\left(2\left|k_z^{(3)}\right|(z+h)\right), \tag{4.2.162}$$

where $(k_z^{(3)})^2 = k_0^2\,\varepsilon_d - k_{x,0}^2$. Equation (4.2.162) describes the interference pattern periodic with respect to the axis Ox and exponentially decaying along the axis Oz. The period of the interference pattern is the same as the function $\sin^2(k_{x,0}x)$, i.e.

$$d_{\text{ip}} = \frac{\pi}{k_{x,0}} = \frac{\pi}{\text{Re}\left(k_{\text{SPP}}\right)} = \frac{\lambda}{2\sqrt{\varepsilon_{\text{pr}}}\sin\theta_{\text{SPP}}}, \tag{4.2.163}$$

where θ_{SPP} is the angle of incidence, defined by the condition (4.2.160).

The contrast of the interference pattern is given by

$$K = \frac{\max\limits_x\left\{\left|\mathbf{E}_{\text{SPP}}(x,-h)\right|^2\right\}-\min\limits_x\left\{\left|\mathbf{E}_{\text{SPP}}(x,-h)\right|^2\right\}}{\max\limits_x\left\{\left|\mathbf{E}_{\text{SPP}}(x,-h)\right|^2\right\}+\min\limits_x\left\{\left|\mathbf{E}_{\text{SPP}}(x,-h)\right|^2\right\}}. \tag{4.2.164}$$

When the condition (4.2.144) is fulfilled, the following simple expression can be obtained for contrast:

$$K = \frac{\left|k_{x,0}\right|^2-\left|k_z^{(3)}\right|^2}{\left|k_{x,0}\right|^2+\left|k_z^{(3)}\right|^2} \approx 1-\frac{2\varepsilon_d}{\left|\varepsilon_m'\right|+\varepsilon_d}. \tag{4.2.165}$$

If a photorecording material is placed in the area under the metal film, the interference pattern of the SPP can be written and used for making a diffraction grating with period d_{ip}. The width of the step is controlled by exposure time.

Practical use of the considered scheme for the formation of interference patterns of the SPP is inconvenient for several reasons. In particular, two coherent beams are required, the scheme is not compact, and the formation of an interference pattern in an optically dense dielectric medium requires a prism made of a material with a high dielectric constant, etc.

4.2.2. Formation of one-dimensional interference patterns of surface plasma polaritons

Consider the formation of one-dimensional interference patterns of SPP with a diffraction structure consisting of a binary dielectric diffraction grating and a metallic film below the grating [21–24]. The geometry of the structure is shown in Fig. 4.2.9. Above and below the structure there are homogeneous dielectric media with refractive indices n_{I} and n_{II}, respectively. The grating has a single ridge with width w and height h_{gr} per period d. The dielectric constant of the material lattice is $\varepsilon_{\mathrm{gr}}$. Between the grating and the metal film there is a homogeneous layer of the material with thickness h_l. The metal film has thickness h_m and the dielectric constant ε_m. The diffraction grating is used for excitation at the lower boundary of the metal film of two counterpropagating SPPs, intended to generate an interference pattern below a metal film.

Consider the normal incidence of a TM-polarized wave on the structure. At normal incidence $\theta = 0$, $\phi = 0$ and the propagation constants (4.1.24) have the form

$$k_{x,m} = \frac{2\pi m}{d},\ k_y = 0,\ k_{z,l,m} = \sqrt{(k_0 n_l)^2 - k_{x,m}^2 - k_y^2}, \tag{4.2.166}$$

where the index l is taken to be unity for the field over the grating and 2 – for the field under it. The condition of SPP excitation by transmitted diffraction orders with the numbers $\pm m$ has the form

$$\sqrt{k_{x,m}^2 + k_y^2} = \mathrm{Re}\left(k_{\mathrm{SPP}}\right), \tag{4.2.167}$$

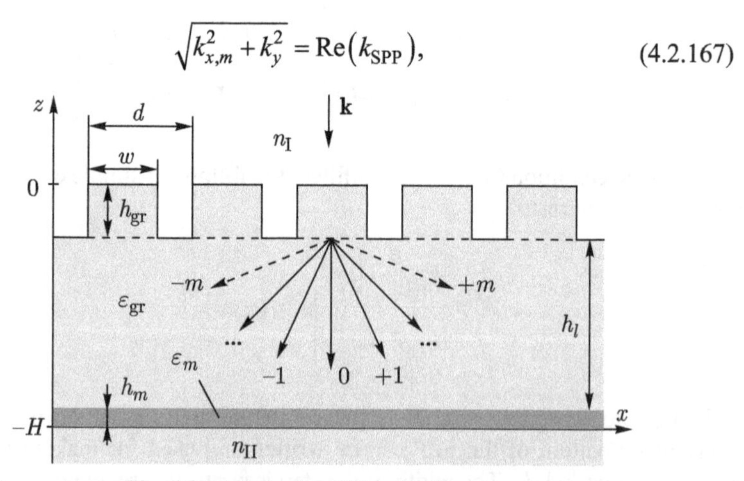

Fig. 4.2.9. Geometry of the structure.

where the propagation constant of SPP k_{SPP} is defined by (4.2.137) at $\varepsilon_d = n_{II}^2$. According to (4.2.166) and (4.2.167) the period of the diffraction grating for the excitation of the SPP by orders with numbers $\pm m$ should be determined from the relation

$$d = 2\pi m / \mathrm{Re}\left(k_{\mathrm{SPP}}\right). \tag{4.2.168}$$

The period of the interference pattern formed by the SPP

$$d_{ip} = \frac{d}{2m} \tag{4.2.169}$$

is $2m$ times smaller than the period of the grating.

It should be noted that the transmitted diffraction orders with numbers that are different from $\pm m$ will distort the interference pattern. Therefore, the calculation and study of interference patterns of the SPPs generated by the diffraction structure in Fig. 4.2.9 should be based on the rigorous solution of the diffraction problem.

Calculation of the interference pattern of the SPP for the structure shown in Fig. 4.2.9 was conducted by the Fourier modal method with the following parameters: $\lambda = -550$ mm, $n_I = 1$, $n_{II} = 1.6$, $\varepsilon_m = -12.922 + 0.447i$. The value of n_{II} corresponds to the photoresist, and ε_m is the dielectric constant of silver for the selected wavelength. The dielectric constant of the material of the grating ε_{gr} was also chosen equal to 2.56. Note that the selected wavelength is far from the resonance value $\lambda \approx 368$ nm. The SPP propagation length, calculated by the formula (4.2.151), in this case equals 5739 nm.The grating period $d = 1539$ nm was calculated from (4.2.168) with $m = 5$. According to (4.2.169), the period of the interference pattern of the SPP $d_{ip} = 154$ nm, that is 10 times smaller than the period of the diffraction grating forming the pattern. The values of other geometrical parameters of structure h_{gr}, h_l, h_m, w were determined from the condition of maximizing the quality of the interference pattern formed on the lower boundary of the metal film at $z = -H$. The merit function was the function

$$F\left(h_{gr}, h_l, h_m, w\right) = \frac{\int_0^d \left| \left|\mathbf{E}_{\mathrm{SPP}}(x,-H)\right|^2 - \left|\mathbf{E}(x,-H)\right|^2 \right| dx}{\max_x \left\{ \left|\mathbf{E}_{\mathrm{SPP}}(x,-H)\right|^2 \right\}} \cdot \frac{1}{\max_x \left\{ \left|\mathbf{E}_{\mathrm{SPP}}(x,-H)\right|^2 \right\}} \to \min_{h_{gr}, h_l, h_m, w}, \tag{4.2.170}$$

The first factor in (4.2.170) is a measure of proximity of the intensity of the calculated interference pattern $|\mathbf{E}(x, -H)|^2$ to the 'ideal' interference pattern $|\mathbf{E}_{SPP}(x, -H)|^2$. The ideal interference pattern is understood to be the intensity of the field, which is formed at the lower boundary of the metal film when only diffraction orders with the numbers $\pm m$ are taken into account. The intensity of the ideal interference pattern has the form

$$\left|\mathbf{E}_{\text{SPP}}\left(x,-H\right)\right|^2 = 4\frac{\left|T_m\right|^2}{k_0 n_{\text{II}}^2}\left[\sin^2\left(k_{x,m}x+\frac{\Delta\phi}{2}\right)\left(\left|k_{x,m}\right|^2-\left|k_{z,\text{II},m}\right|^2\right)+\left|k_{z,\text{II},m}\right|^2\right], \tag{4.2.171}$$

where $|T_m|$ is the modulus of the complex transmission coefficient of the m-th order (because of symmetry $|T_m|=|T_{-m}|$), $\Delta\phi$ is the phase difference of the coefficients of orders $\pm m$, k_{xm}, $k_{\text{II},zm}$ are the constanst of propagation (4.2.166). Equation (4.2.171) is identical in structure to (4.2.162) for the interference pattern formed in the Kretschmann scheme. An estimate of the contrast of the interference pattern is also given by (4.2.165). The second factor in (4.2.170) is responsible for maximizing the field intensity at the maxima of the interference pattern. Minimizing the merit function was carried out numerically using the gradient optimization techniques. The calculation of the intensity $|\mathbf{E}(x, H)|^2$ in (4.2.171) was performed by the Fourier modal method. The optimization resulted in the following values of geometrical parameters: $h_{gr} = 435$ nm, $h_l = 0$, $h_m = 65$ nm, $w = 0.5d$. Figure 4.2.10 shows a plot of the calculated intensity of the field which is formed directly under the metal layer at the above parameters. The graph is normalized to the intensity of the incident wave.

Figure 4.2.10 shows the formation of an interference pattern with the calculated period d_{ip} = 154 nm. Note that the period d_{ip} is not only 10 times smaller than the period of the diffraction grating, but also 3.57 times smaller than the wavelength. The field enhancement factor, showing the value of the intensity at the maxima of the interference relative to the intensity of the incident wave, is greater than 45. The contrast of the resulting interference pattern given by

$$K = \frac{\max\limits_x\left\{\left|\mathbf{E}\left(x,-H\right)\right|^2\right\}-\min\limits_x\left\{\left|\mathbf{E}\left(x,-H\right)\right|^2\right\}}{\max\limits_x\left\{\left|\mathbf{E}\left(x,-H\right)\right|^2\right\}+\min\limits_x\left\{\left|\mathbf{E}\left(x,-H\right)\right|^2\right\}}, \tag{4.2.172}$$

is equal to 0.701. The resulting value of contrast is close to the theoretical estimate 0.67, calculated by the formula (4.2.165). Figure 4.2.11 shows the distribution of the normalized intensity of the electric field in a dielectric substrate below a

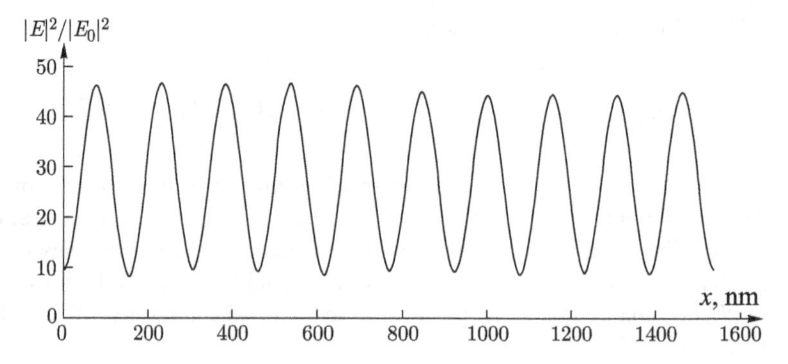

Fig. 4.2.10. The interference pattern within a grating period at λ = 550 nm.

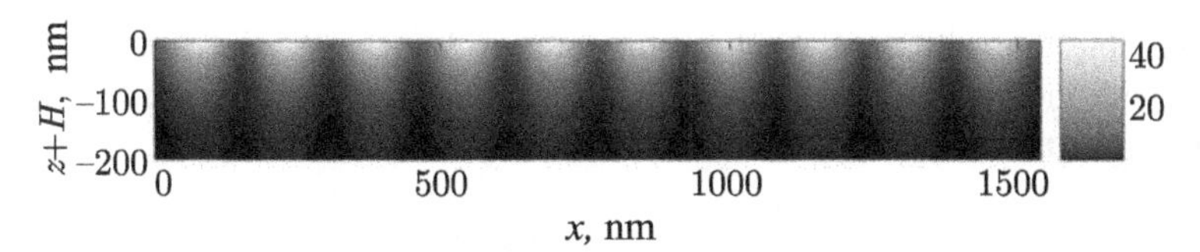

Fig. 4.2.11. Distribution of the normalized intensity of the electric field under the metal film.

metal film. The penetration depth of SPP in the medium below the metal film (n_{II} = 1.6, photoresist) is equal to 111 nm, which agrees with the theoretical value. The theoretical value is defined in (4.2.157) and at the parameters under consideration is 110 nm.

As another example, we calculated the diffraction structure for a wavelength of 436 nm, which corresponds to a InGaN/GaN semiconductor laser. The length of propagation of the SPP is 1521 nm. The period of the grating at $m = 5$ is 1051 nm. Geometrical parameters of structures h_{gr} = 835 nm, $h_l = 0$, h_m = 73 nm, $w = 0.43d$ were found in the optimization function (4.2.170), as in the previous case. The calculated interference pattern is shown in Fig. 4.2.12.

The period of the interference pattern is 105 nm, which is 4.15 times less than the wavelength of incident light. The field enhancement factor at the maxima of the interference pattern is close to 30, the contrast is 0.48. The resulting value of contrast is close to the estimate (2.4.42) which is equal to 0.43.

With further approach of the wavelength to the resonant value $\lambda \approx 368$ nm the SPP propagation length decreases, which leads to deterioration of the quality of the resultant interference patterns. In particular, at a wavelength of 420 nm the propagation length of the SPP is smaller than the period of the diffraction grating, calculated for $m = 5$. In this case, the high-quality interference pattern can no longer be generated.

In the considered scheme (see Fig. 4.2.9) it is possible to control the frequency of the interference pattern by changing the wavelength and the angle of incidence [21, 23]. Consider using different wavelengths for the formation of interference patterns for various periods. Let the grating period d in (4.2.168) be calculated from the condition of excitation of SPPs by orders with numbers $\pm m$ at some wavelength λ. The SPP propagation constant $k_{SPP} = k_{SPP}(\lambda)$ depends on the wavelength. Therefore,

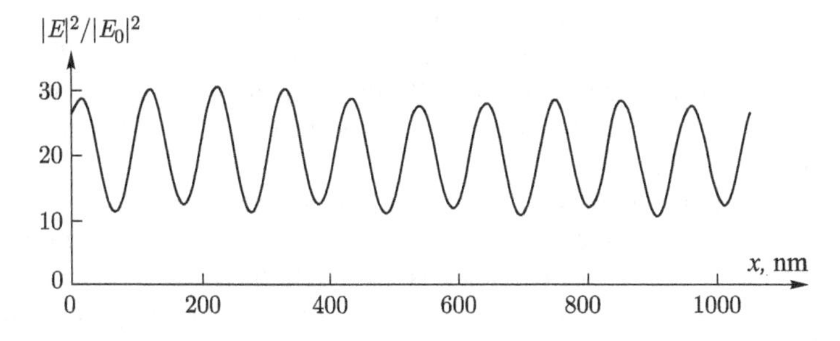

Fig. 4.2.12. The interference pattern within a grating period at λ= 436 nm.

it is possible to excite the SPP at a different wavelength $\lambda' \neq \lambda$ using diffraction orders with the numbers $\pm n$, $n \neq m$. The wavelengths that will excite the SPP by orders with numbers $\pm n$ can be found from the equation

$$\frac{2\pi n}{d} = \mathrm{Re}\left(k_{\mathrm{SPP}}(\lambda')\right). \quad (4.2.173)$$

In particular, for the above case λ = 550 nm, m = 5, SPP are excited by the orders ±4 and orders ±3 at λ' = 852 nm. The corresponding interference patterns will have periods of 192 nm and 257 nm (8 and 6 times smaller than the grating period, respectively). The geometrical parameters of the structure were determined similarly to the previous case, using an optimization procedure. The merit function has the form:

$$F\left(h_{\mathrm{gr}}, h_l, h_m, w\right) = \sum_{n=3}^{5} \left[\frac{\int_0^d \left| \left|\mathbf{E}_{\mathrm{SPP}}(x, -H; \lambda_n)\right|^2 - \left|\mathbf{E}(x, -H; \lambda_n)\right|^2 \right| dx}{\max_x \left\{ \left|\mathbf{E}_{\mathrm{SPP}}(x, -H; \lambda_n)\right|^2 \right\}} \times \right.$$

$$\left. \times \frac{1}{\max_x \left\{ \left|\mathbf{E}_{\mathrm{SPP}}(x, -H; \lambda_n)\right|^2 \right\}} \right] \to \min_{h_{\mathrm{gr}}, h_l, h_m, w}, \quad (4.2.174)$$

where the index n corresponds to the number of diffraction order exciting the SPP, $\left|\mathbf{E}(x, -H; \lambda_n)\right|^2$ is the calculated intensity of the interference pattern at the appropriate wavelength λ_n, $\left|\mathbf{E}_{SPP}(x, -H; \lambda_n)\right|^2$ is the intensity of the ideal interference pattern formed by the orders of a wavelength λ_n. As a result of minimizing (4.2.174) we obtained the following values of the geometric parameters of the structure: $w = 0.37d$, h_{gr} = 1000 nm, $h_l = 0$, h_m = 70 nm. Calculated plots of the normalized intensity of the interference patterns directly under the metal film are shown in Fig. 4.2.13a–c.

Contrast values and the field enhancement factor are (0.71, 25), (0.78, 37) and (0.87, 76) respectively. The obtained values of the contrast of interference patterns are close to the theoretical estimates (4.2.165), constituting 0.67, 0.77 and 0.86, respectively. The penetration depth of SPP in the medium under the metal film (n_{II} = 1.6) was 111 nm, 172 nm and 305 nm for the wavelengths of 550 nm, 659 nm and 852 nm, respectively.

The second way to control the frequency of the interference pattern is to change the angle of incidence at fixed wavelength. Consider the case when the projection of the wave vector of the incident wave is parallel to the grooves of a diffraction grating (Fig. 4.2.14). In this case $\phi = 90°$ in (4.2.14) and the propagation constants have the form

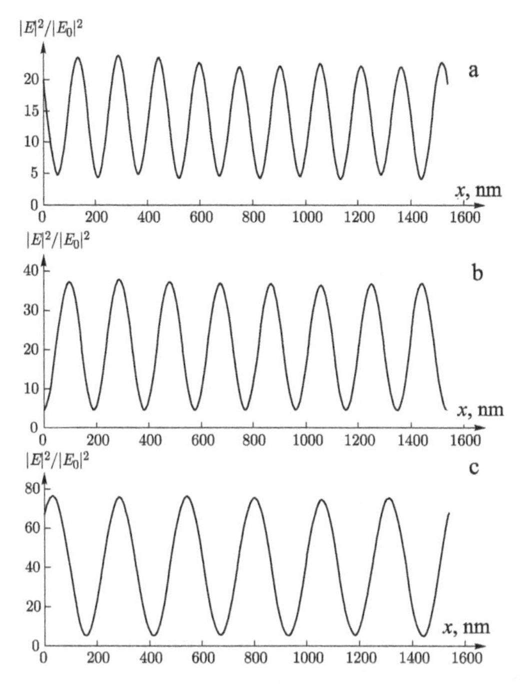

Fig. 4.2.13. Interference pattern formed at wavelengths 550 nm (a), 659 nm (b), 852 nm (c).

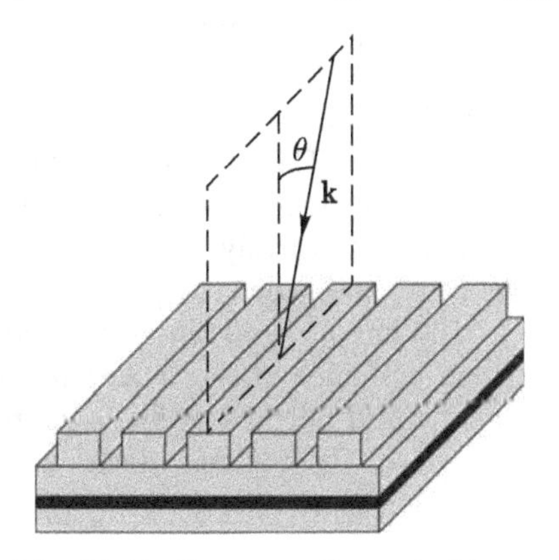

Fig. 4.2.14. Geometry of wave incidence.

$$k_{x,m} = \frac{2\pi m}{d},\ k_y = k_0 n_I \sin\theta, \tag{4.2.175}$$

where θ is the angle of incidence. The condition of SPP excitation has the form

$$\sqrt{k_{x,m}^2 + k_y^2(\theta)} = \mathrm{Re}(k_{SPP}). \tag{4.2.176}$$

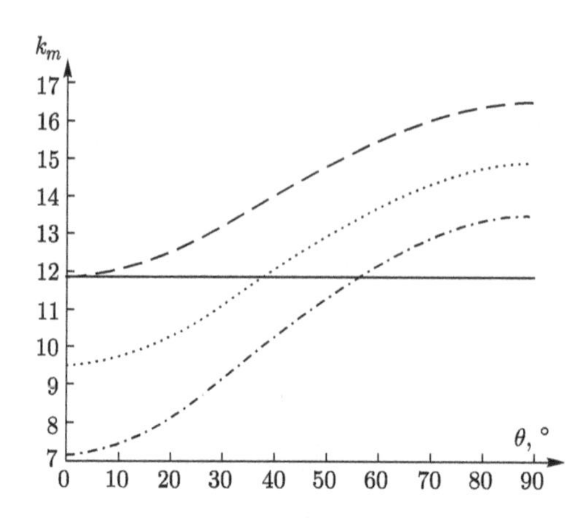

Fig. 4.2.15. Dependence of $k_m(\theta)$ on the angle of incidence (m = 3 – dot-and-dash line, m = 4 – dotted line, m = 5 – dashed line, the value Re (k_{SPP}) – solid line).

Condition (4.2.176) can be satisfied for various combinations of order number m and angle of incidence θ.

Figure 4.2.15 shows the plots of the dependence of the modulus of the projection of the wave vector $k_m(\theta)=\sqrt{k_{x,m}^2+k_y^2(\theta)}$ on the angle of incidence for m = 3 (dot-and-dash line), m = 4 (dotted line) and m = 5 (dashed line). The solid line corresponds to the propagation constant of SPP. The graphs were obtained for d = 2641 nm, the values of other parameters λ, ε_m, n_I, ε_{gr} are the same as in the previous cases. Conditions $k_m(\theta)$ = Re(k_{SPP}) are satisfied for values of the angle θ of 56.4° (m = 3), 38.7° (m = 4) and 0° (m = 5).This means that at these angles the SPPs are excited by ±3, ±4, and ±5 diffraction orders. The periods of the interference patterns are 440 nm, 330 nm and 264 nm, respectively.

Similar to the previous cases, the geometric parameters of the structure w = 0.554d, h_{gr} = 668 nm, h_l = 100 nm, h_m = 40 nm, were found in the optimization of conditions for maximizing the quality of generated interference patterns. The calculated graphs of the normalized intensity at the lower boundary of the metal layer are shown in Fig. 4.2.16a–c. The contrast values and gain factor values are (0.86, 57), (0.92, 27), (0.96, 30), respectively.

4.2.3. Formation of two-dimensional interference patterns of surface plasma polaritons

The scheme shown in Fig. 4.2.9 is obviously extended to the case of forming two-dimensional interference patterns. In this case, SPPs are excited using three-dimensional dielectric gratings with a metal film on the substrate (Fig. 4.2.17). The diffraction grating has periods d_x, d_y. On the period the grating contains a rectangular hole with the size w_x, w_y and depth h_{gr}. The remaining notation is the same as for the previously discussed two-dimensional structure in Fig. 4.2.9.

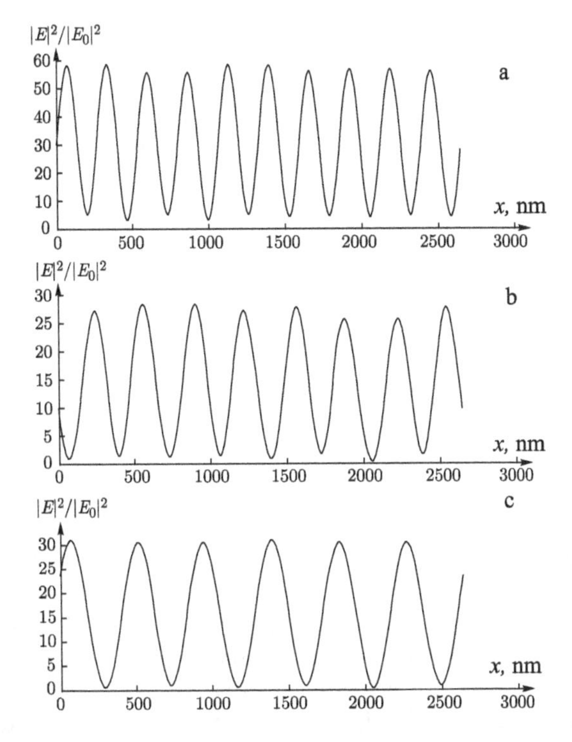

Fig. 4.2.16. Interference pattern formed at angles of incidence 0° (a), 38.7° (b), 56.4° (c).

We consider normal incidence of a plane wave of different polarizations. In this case, $\theta = 0$, $\phi = 0$ in (4.1.24). When determining the period of the grating $d_x = d_y = d$ according to (4.2.168), SPPs are excited by orders with numbers $(\pm m, 0)$, $(0, \pm m)$. In this symmetric configuration it will be assumed that $w_x = w_y = w$.

We define the polarization of the wave by the angle ψ between the direction of the vector **E** and the axis Ox. At normal incidence of waves with $\psi = 0$ ($\mathbf{E} = (E_x, 0, 0)$, $\mathbf{H} = (0, H_y, 0)$), the SPPs are excited by the orders with numbers $(\pm m, 0)$. Similarly, if $\psi = 90°$ ($\mathbf{E} = (0, E_y, 0)$, $\mathbf{H} = (H_x, 0, 0)$) the SPPs are excited with the orders of numbers $(0, \pm m)$. In these cases, one-dimensional interference patterns will be generated at the lower boundary of the metal film analogous to those considered in section 2.2. In order to form two-dimensional interference patterns it is necessary to excite four SPPs by symmetric diffraction orders with numbers $(\pm m, 0)$, $(0, \pm m)$. In this case, the incident wave must contain perpendicular components of the electric field directed along the axes Ox, Oy. This condition is satisfied by the elliptically polarized wave, corresponding to a superposition of normally incident waves with the following vectors of the electric field:

$$\mathbf{E}_{01} = \left(E_{0x}, 0, 0\right)\exp(-ik_0 n_I z),$$
$$\mathbf{E}_{02} = \left(0, E_{0y}, 0\right)\exp(-ik_0 n_I z + i\delta), \tag{4.2.177}$$

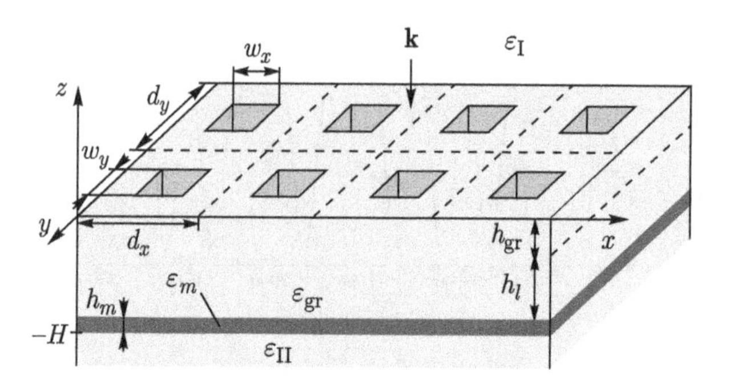

Fig. 4.2.17. Geometry of the structure.

where δ is the phase difference. It will be assumed that E_{0x} = cos ψ, E_{0y} = sin ψ. The case $\delta = n\pi$, n = 0, ± 1, ± 2, ... corresponds to a linearly polarized incident wave. When $\delta = \pm\pi/2 + 2\pi n$, n = 0, ± 1, ± 2,... and $\psi = 45°$ we have a wave with circular polarization.

We analyze the type of interference pattern formed at $\psi = 45°$ and $\delta = 0$. Because of the symmetry of the diffraction structure the complex amplitudes of the diffraction orders ($\pm m$, 0), (0, $\pm m$) will be assumed to be the same. In fact, the phases of diffraction orders may vary, but it only leads to a shift in the interference pattern, and not to change of its form. In subsequent calculations the complex amplitudes of the diffraction orders ($\pm m$, 0), (0, $\pm m$) are omitted.

For SPPs propagating along the axis Ox, $\beta = 0$ in (4.2.124) and $\mathbf{H}_{\text{SPP}}$ = (0, $H_{y,\text{SPP}}$, 0), $\mathbf{E}_{\text{SPP}}$ = ($\text{E}_{x,\text{SPP}}$, $E^{(x)}_{z,\text{SPP}}$). Here and below, the upper index in the z-component of the electric field indicates the direction of propagation of SPP. Non-zero components of the electromagnetic field for the SPP propagating along the Ox-axis and corresponding to the orders ($\pm m$, 0) may be obtained from (4.2.124) and (4.2.125) as:

$$H_{y,\text{SPP}}(x,z) = \cos(k_m x)\exp(-ik_{z,\text{II},m}(z+H)), \tag{4.2.178}$$

$$\begin{aligned} E_{x,\text{SPP}}(x,z) &= -\frac{k_{z,\,\text{II},m}}{k_0 n_{\text{II}}^2}\cos(k_m x)\exp(-ik_{\text{II},zm}(z+H)), \\ E^{(x)}_{z,\text{SPP}}(x,z) &= \frac{k_m}{ik_0 n_{\text{II}}^2}\sin(k_m x)\exp(-ik_{z,\,\text{II},m}(z+H)), \end{aligned} \tag{4.2.179}$$

where $k_m = \dfrac{2\pi m}{d}$, $k_{z,\,\text{II},m} = \sqrt{k_0^2 n_{\text{II}}^2 - k_m^2}$. For SPPs propagating along the axis Oy, $\alpha = 0$ in (4.2.1) and $\mathbf{H}_{\text{SPP}} = (H_{x,\text{SPP}}, 0, 0)$, $\mathbf{E}_{\text{SPP}} = \left(0, E_{y,\text{SPP}}, E^{(y)}_{z,\text{SPP}}\right)$. Similarly, the components of the electromagnetic field for SPPs propagating along the axis Oy and the corresponding orders (0, $\pm m$) are as follows:

$$H_{x,\mathrm{SPP}}(y,z)=\cos(k_m y)\exp\left(-ik_{z,\mathrm{II},m}(z+H)\right), \tag{4.2.180}$$

$$\begin{aligned} E_{y,\mathrm{SPP}}(y,z)&=\frac{k_{z,\mathrm{II},m}}{k_0 n_{\mathrm{II}}^2}\cos(k_m y)\exp\left(-ik_{z,\mathrm{II},m}(z+H)\right),\\ E_{z,\mathrm{SPP}}^{(y)}(y,z)&=-\frac{k_m}{ik_0 n_{\mathrm{II}}^2}\sin(k_m y)\exp\left(-ik_{z,\mathrm{II},m}(z+H)\right). \end{aligned} \tag{4.2.181}$$

From (4.2.179) and (4.2.181) we obtain the electric field components at the lower boundary of the metal film (at $z=-H$) as follows:

$$\begin{aligned} E_{x,\mathrm{SPP}}(x,-H)&=-\frac{k_{z,\mathrm{II},m}}{k_0 n_{\mathrm{II}}^2}\cos(k_m x);\\ E_{y,\mathrm{SPP}}(y,-H)&=\frac{k_{z,\mathrm{II},m}}{k_0 n_{\mathrm{II}}^2}\cos(k_m y);\\ E_{z,\mathrm{SPP}}(x,y,-H)&=\frac{k_m}{ik_0 n_{\mathrm{II}}^2}\left[\sin(k_m x)-\sin(k_m y)\right]. \end{aligned} \tag{4.2.182}$$

In the case where the condition (4.2.144) is satisfied, we have approximate expressions

$$k_m\approx k_0\sqrt{\frac{\varepsilon_d\varepsilon_m'}{\varepsilon_d+\varepsilon_m'}},\quad k_{z,\mathrm{II},m}\approx k_0\frac{\varepsilon_d}{\sqrt{\varepsilon_d+\varepsilon_m'}},\quad \left|\frac{k_m}{k_{z,\mathrm{II},m}}\right|^2\approx\left|\frac{\varepsilon_m'}{\varepsilon_d}\right|^2, \tag{4.2.183}$$

where $\varepsilon_d=n_{II}^2$. Consequently,

$$\frac{\max\limits_{x,y}\left|E_{z,\mathrm{SPP}}(x,y,-H)\right|^2}{\max\limits_{x}\left|E_{x,\mathrm{SPP}}(x,-H)\right|^2}=\frac{\max\limits_{x,y}\left|E_{z,\mathrm{SPP}}(x,y,-H)\right|^2}{\max\limits_{y}\left|E_{y,\mathrm{SPP}}(x,-H)\right|^2}\approx 4\frac{\left|\varepsilon_m'\right|}{\varepsilon_d}. \tag{4.2.184}$$

For the considered wavelengths and materials, the relation $\left|\varepsilon_m'\right|\gg\varepsilon_d$ holds so that from (4.2.184) it follows that

$$\begin{aligned} \max_{x,y}\left|E_{z,\mathrm{SPP}}(x,y,-H)\right|^2&\gg\max_{x}\left|E_{x,\mathrm{SPP}}(x,-H)\right|^2,\\ \max_{x,y}\left|E_{z,\mathrm{SPP}}(x,y,-H)\right|^2&\gg\max_{y}\left|E_{y,\mathrm{SPP}}(y,-H)\right|^2. \end{aligned} \tag{4.2.185}$$

According to (4.2.185), the type of interference pattern is determined by the component $E_{z,\mathrm{SPP}}$. We omit the inessential constant in (4.2.182) and introduce the quantity

$$\tilde{E}_{z,\mathrm{SPP}}(x,y,-H)=\sin(k_m x)-\sin(k_m y). \tag{4.2.186}$$

The structure of the interference pattern is determined by the extrema of the function

$$F(x,y) = \left|\tilde{E}_{z,\mathrm{SPP}}(x,y,-H)\right|^2 = \\ = \sin^2(k_m x) + \sin^2(k_m y) - 2\sin(k_m x)\sin(k_m y). \quad (4.2.187)$$

Extrema of the function (4.2.187) coincide with the extrema of the function $\left|E_{z,SPP}(x,y,-H)\right|^2$. To find the extrema of the function (4.2.187), we equate the partial derivatives to zero and obtain the following system of equations:

$$\begin{cases} \cos(k_m x)\left[\sin(k_m x) - \sin(k_m y)\right] = 0, \\ \cos(k_m y)\left[\sin(k_m y) - \sin(k_m x)\right] = 0. \end{cases} \quad (4.2.188)$$

The system (4.2.188) breaks down into a system of equations

$$\begin{cases} \cos(k_m x) = 0, \\ \cos(k_m y) = 0 \end{cases} \quad (4.2.189)$$

and the equation

$$\sin(k_m x) - \sin(k_m y) = 0. \quad (4.2.190)$$

Assuming $k_m = \frac{2\pi}{d} m$, we get a solution of (4.2.189) as:

$$\begin{cases} x = \frac{dl_x}{2m} + \frac{d}{4m}, \\ y = \frac{dl_y}{2m} + \frac{d}{4m}, \end{cases} \quad (4.2.191)$$

where l_x, l_y are integers. The solutions of (4.2.190) have the form

$$\left[\begin{array}{l} y = x - \frac{dl}{m}, \\ y = -x + \frac{dl}{m} + \frac{d}{2m}. \end{array}\right. \quad (4.2.192)$$

where l is an integer. For the solutions of (4.2.68) and (4.2.69) the sufficient extremum condition should be verified

$$F''_{xx}(x_0,y_0) \cdot F''_{yy}(x_0,y_0) - \left(F''_{xy}(x_0,y_0)\right)^2 > 0. \quad (4.2.193)$$

Moreover, if $F''_{xx}(x_0,y_0) > 0$, then at the point (x_0, y_0) the function has a minimum, if $F''_{xx}(x_0,y_0) < 0$ a maximum. The second derivatives of (4.2.187) have the form

$$F''_{xx} = 2k_m^2\left[\cos(2k_m x)+\sin(k_m x)\sin(k_m y)\right],$$
$$F''_{yy} = 2k_m^2\left[\cos(2k_m y)+\sin(k_m x)\sin(k_m y)\right], \qquad (4.2.194)$$
$$F''_{xy} = -2k_m^2\cos(k_m x)\cos(k_m y).$$

Substituting (4.2.194) into (4.2.193), we obtain a sufficient condition for an extremum in the form of

$$\cos(2k_m x)\cos(2k_m y)+\sin(k_m x)\sin(k_m y)\left[\cos(2k_m x)+\cos(2k_m y)\right]+$$
$$+\sin^2(k_m x)\sin^2(k_m y)-\cos^2(k_m x)\cos^2(k_m y)>0. \qquad (4.2.195)$$

Substituting into (4.2.195) the solutions (4.2.191), we obtain:

$$1-\cos(\pi l_x)\cos(\pi l_y)>0. \qquad (4.2.196)$$

The last inequality holds when l_x and l_y are the numbers of different parity. In this case, the resultant extremes are maxima. It can be shown that the solution (4.2.192), representing a set of straight lines, are not extrema. Maxima of the function $F(x,y)\sim\left|E_{z,\mathrm{SPP}}(x,y,-H)\right|^2$ and the zeros of its first derivatives for m = 3, λ = 550 nm are shown in Fig. 4.2.18. The maxima are shown by dots, the dotted line shows the zeros (4.2.192) that are not extrema. In this case, the grating period is determined from the relation (4.2.168) and is equal to 923 nm. According to Fig. 4.2.18, the interference pattern is rotated relative to the coordinate axes by an angle of 45 °, and its period is $\sqrt{2}d/2m$.

Consider the case when the incident wave is circularly polarized. In this case, taking into account the phase difference $\delta=\pi/2$ in (4.2.177), the components of the electric field can be obtained in the form of

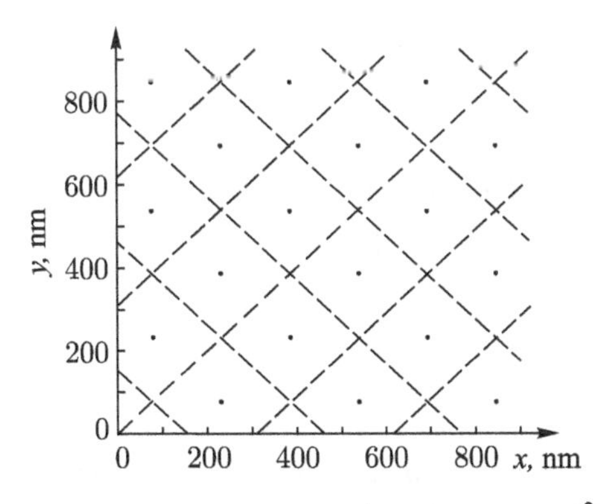

Fig. 4.2.18. Maxima of the function $F(x,y)\sim\left|E_{z,SPP}(x,y,-H)\right|^2$ (dots) and zeros of its first derivatives (dotted lines) in the case of linear polarization of the incident wave at λ = 550 nm, m = 3 (d = 923 nm).

$$E_{x,\mathrm{SPP}}(x,-H)=-i\frac{k_{z,\mathrm{II},m}}{k_0 n_{\mathrm{II}}^2}\cos(k_m x),$$
$$E_{y,\mathrm{SPP}}(y,-H)=\frac{k_{z,\mathrm{II},m}}{k_0 n_{\mathrm{II}}^2}\cos(k_m y), \tag{4.2.197}$$
$$E_{z,\mathrm{SPP}}(x,y,-H)=\frac{k_m}{ik_0 n_{\mathrm{II}}^2}\left[i\sin(k_m x)-\sin(k_m y)\right].$$

When (4.2.144) is fulfilled, the following approximate equality holds

$$\frac{\max\limits_{x,y}\left|E_{z,\mathrm{SPP}}(x,y,-H)\right|^2}{\max\limits_{x}\left|E_{x,\mathrm{SPP}}(x,-H)\right|^2}=\frac{\max\limits_{x,y}\left|E_{z,\mathrm{SPP}}(x,y,-H)\right|^2}{\max\limits_{y}\left|E_{y,\mathrm{SPP}}(x,-H)\right|^2}\approx 2\frac{\left|\varepsilon_m'\right|}{n_{\mathrm{II}}^2}=2\frac{\left|\varepsilon_m'\right|}{\varepsilon_d}. \tag{4.2.198}$$

Note that the maximum value of $\left|E_{z,\mathrm{SPP}}(x,y,-H)\right|^2$ for the circular polarization is two times smaller than the linear one. Accordingly, the factors in front of the ratio $\left|\varepsilon_m'\right|/\varepsilon_d$ in (4.2.184) and (4.2.198) differ two-fold. Since in these cases $|\varepsilon'_m| \gg \varepsilon_{md}$ the form of the interference pattern is determined by the component $E_{z,\mathrm{SPP}}$. As in the case of linear polarization of the incident wave, we introduce the function

$$\tilde{E}_{z,\mathrm{SPP}}(x,y,-H)=i\sin(k_m x)-\sin(k_m y). \tag{4.2.199}$$

The structure of the interference pattern is determined by the extrema of the function

$$F(x,y)=\left|\tilde{E}_{z,\mathrm{SPP}}(x,y,-H)\right|^2=\sin^2(k_m x)+\sin^2(k_m y). \tag{4.2.200}$$

To find the extrema, we equate the partial derivatives of the function (4.2.77) to zero and we obtain the following equations:

$$\begin{cases}\sin(2k_m x)=0,\\ \sin(2k_m y)=0.\end{cases} \tag{4.2.201}$$

Solution of equations (2.4.78) is the set of points

$$\begin{cases}x=\dfrac{dl_x}{4m},\\ y=\dfrac{dl_y}{4m},\end{cases} \tag{4.2.202}$$

where l_x, l_y are integers. At these points we need to check the condition of extremum (4.2.193). We write the second partial derivatives of the function $F(x,y)$:

$$\begin{aligned} F''_{xx} &= 2k_m^2 \cos(2k_x x), \\ F''_{yy} &= 2k_m^2 \cos(2k_y y), \\ F''_{xy} &= 0. \end{aligned} \tag{4.2.203}$$

In view of (4.2.203), the extremum condition becomes

$$\cos(2k_m x)\cos(2k_m y) > 0. \tag{4.2.204}$$

Substituting into (4.2.204) the solution (4.2.202), we obtain

$$\cos(\pi l_x)\cos(\pi l_y) > 0. \tag{4.2.205}$$

Inequality (4.2.82) is satisfied when l_x and l_x have the same parity. For even l_x, l_y function $\left|E_{z,\mathrm{SPP}}(x,y,-H)\right|^2$ has a minimum, for odd l_x, l_y maximum. Extrema of the function $\left|E_{z,\mathrm{SPP}}(x,y,-H)\right|^2$ and the zeros of its first derivatives are shown in Fig. 4.2.19. The crosses show a set of points that are not extrema, but in which the first derivatives are zero, the points – the maxima, the circles the minima. According to Fig. 4.2.19, the interference pattern is oriented parallel to the coordinate axes, and its period is $d/2m$.

Figure 4.2.20 shows the calculated interference pattern formed directly under the metal film in the case of normally incident waves with the angle of polarization $\psi = 0$ ($\mathbf{E} = (E_x, 0, 0)$, $\mathbf{H} = (0, H_y, 0)$. The calculations were performed using the Fourier modal method for the following geometric parameters: $h_{gr} = 260$ nm, $h_l = 0$, $h_m = 70$ nm. The parameters of the materials coincide with the parameters of the two-dimensional structure in section 4.2.2. The period $d = 923$ nm was determined from the relation (4.2.168) with $m = 3$, $\lambda = 550$ nm. In this case, SPPs are excited by orders with the numbers (± 3, 0) and the one-dimensional interference pattern is formed. The period of the interference pattern is 154 nm, and the contrast and

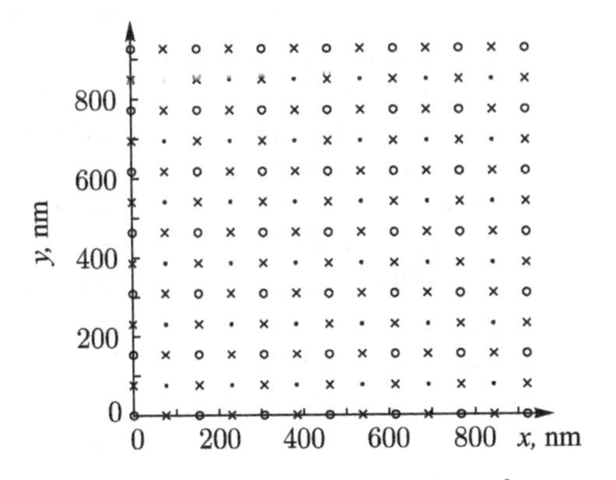

Fig. 4.2.19. Extrema of function $F(x,y) \sim \left|E_{z,\mathrm{SPP}}(x,y,-H)\right|^2$ (points – maxima, circles – minima) and the zeros of its first derivatives (crosses) in the case of circular polarization of the incident wave at $\lambda = 550$ nm, $m = 3$ ($d = 923$ nm).

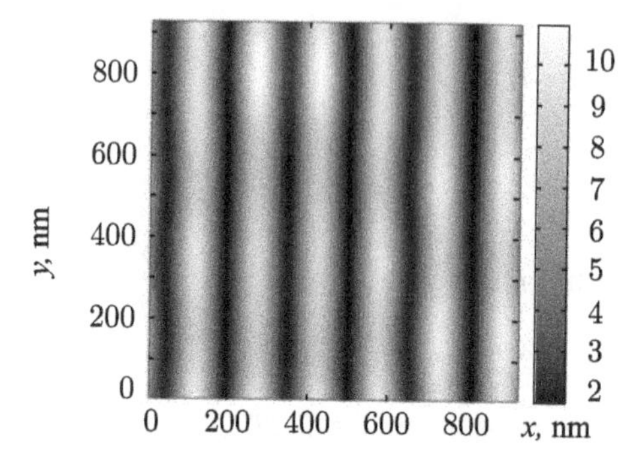

Fig. 4.2.20. Distribution of the electric field intensity directly under the metal film in the case of TM-polarization ($\psi = 0$, $\delta = 0$) at $\lambda = 550$ nm.

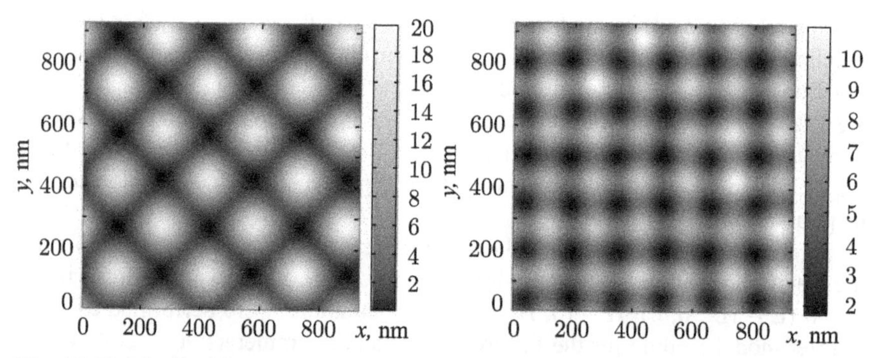

Fig. 4.2.21. Distribution of the electric field intensity directly under the metal film in the case of linear polarization ($\psi = 45°$, $\delta = 0$) at $\lambda = 550$ nm.
Fig. 4.2.22 (right). Distribution of the electric field intensity directly under the metal film in the case of circular polarization ($\psi = 45°$, $\delta = 0$) at $\lambda = 550$ nm.

the gain factor of the electric field at the peaks of the pattern are 0.73 and 10.9, respectively. In this case the interference patterns are given within the period of the three-dimensional structure at $x \in [0, 923]$ nm, $y \in [90, 923]$ nm.

Let us consider now the case $\psi = 45°$. Figures 4.2.21 and 4.2.22 show the calculated interference patterns formed directly under the metal film for the cases of linear and circular polarization of the incident wave. The period of the interference pattern in the case of linear polarization of the incident wave is equal to 218 nm, and in the case of circular polarization to 154 nm. The contrast and the gain factor of the electric field at the maxima of the interference pattern amount to (0.99, 20) and (0.73, 10.9), respectively. The structure of the calculated patterns coincides with the theoretically derived patterns shown in Fig. 4.2.18 and 4.2.19.

In general, the angle ψ may be different from 45°, which corresponds to different amplitudes of the TE- and TM-components. The phase difference δ can have arbitrary values different from 0 (linear polarization) and 90° (circular polarization). Consider two examples. Figure 4.2.23 shows the calculated interference pattern for the case when the incident wave is linearly polarized and the angle $\psi = 20°$. The contrast of

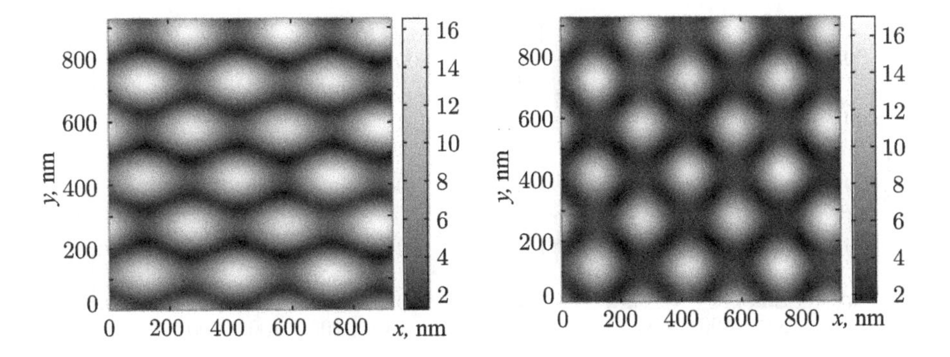

Fig. 4.2.23. Distribution of the electric field intensity directly under the metal film in the case of linear polarization for ($\psi = 20°$, $\delta = 0$) at $\lambda = 550$ nm.
Fig. 4.2.24 (right). Distribution of the electric field intensity directly under the metal film in the case of elliptical polarization ($\psi = 45°$, $\delta = 0$) at $\lambda = 774$ nm.

the formed interference pattern is equal to 0.86, and the gain factor of the electric field at the peaks of the patterns is 16.6. Figure 4.2.23 shows clearly a change of the pattern as compared to Fig. 4.2.21. The interference maxima in Fig. 4.2.23 have a more elongated shape. Figure 4.2.24 shows the calculated interference pattern for the elliptically polarized incident wave. The contrast of the formed interference pattern is equal to 0.83 and the gain factor of the electric field at the peaks of the pattern is 16.5. In contrast to Figs. 4.2.21–4.2.23 the form of the interference maxima in Fig. 4.2.24 is close to the rhombus.

The results of simulations in Figs. 4.2.20–4.2.24 show the formation of two-dimensional interference patterns of SPP of high quality. The gain factor of the electric field at the peaks of the patterns exceeds 10. The form of the patterns can be controlled by changing the polarization of the incident wave.

For the considered three-dimensional structure we also can control the frequency and type of interference pattern by changing the wavelength and the angle of incidence. In particular, as in the two-dimensional case, interference patterns of different frequencies can be produced by using different wavelengths. The wavelengths are determined by (4.2.173). For the above structures the SPPs are excited at 550 nm. According to (4.2.173), the same structure will excite SPPs by the orders (±2, 0), (0, ±2) at a wavelength of 774 nm. The interference patterns, which are formed at a wavelength of 550 nm, are shown above in Figs. 4.2.2 and 4.2.22. Figures 4.2.25 and 4.2.26 show the calculated interference patterns formed by the structure for the cases of linear and circular polarization at a wavelength of 774 nm. The contrast values and the gain factor of the electric field at the maxima of the interference pattern amount to (0.99, 76.6) and (0.85, 38.7), respectively.

This scheme with a diffraction grating can be used to create radially symmetric interference patterns of SPP [24]. In this case, a radial diffraction grating (diffractive axicon) with a metal film on the substrate (Fig. 2.4.27) is used. In [39] an approximation for the field formed by a radiation diffraction grating was obtained. In [39] it was shown that the radial field can be locally approximated

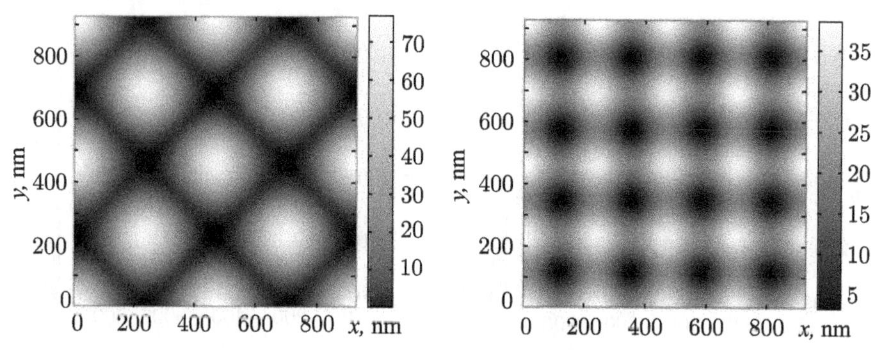

Fig. 4.2.25. Distribution of the electric field intensity directly under the metal film in the case of linear polarization ($\psi = 45°$, $\delta = 0$) at $\lambda = 774$ nm.
Fig 4.2.26 (right). Distribution of the electric field intensity directly under the metal film in the case of circular polarization ($\psi = 45°$, $\delta = 90°$) at $\lambda = 550$ nm.

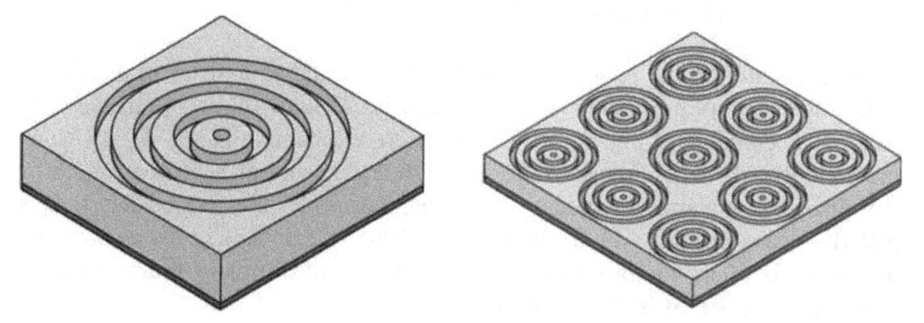

Fig. 4.2.27. Radial binary diffraction grating with metal films to form radially symmetric interference patterns of the SPP.
Fig. 4.2.28 (right). Periodic raster of radial diffraction gratings with metal film.

by a field from a linear (two-dimensional) diffraction grating. This assumption is not valid in the central region of the diffraction pattern [39]. Outside the central region the diffraction structure in Fig. 4.2.27 along the radius can excite SPP at the lower boundary of the metal film as with the two-dimensional diffraction grating in Fig. 4.2.9.

For the excitation of SPPs counterpropagating along the radius in opposite directions, the incident wave must have components with the polarization perpendicular to the grating grooves. In particular, this condition is satisfied by waves with circular and radial polarization. Within the approximation of [39], the period of the radial grating is calculated by the formula (4.2.45), as in the case of a linear grating. In this case the produced radial interference pattern is obtained by 'unfolding' (rotation) of the interference pattern formed by the linear diffraction grating. In particular, when using as the period of the radial structure the linear grating, forming an interference pattern in Fig. 4.2.10, we obtain a radial interference pattern with a period $d_{ip} = 154$ nm. The radial section of the interference pattern within the period of the radial grating will also have the form shown in Fig. 4.2.10.

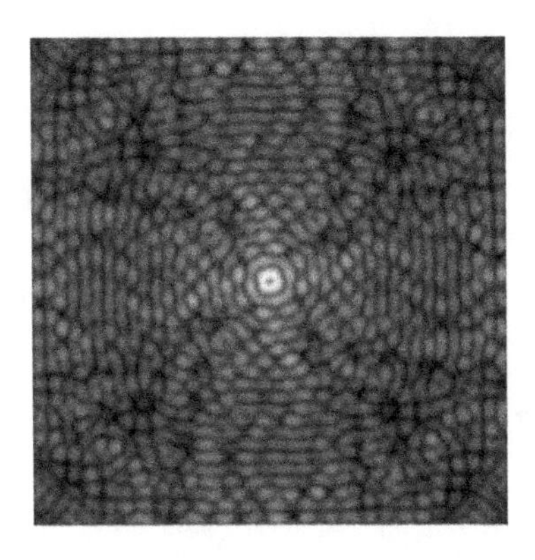

Fig. 4.2.29. Distribution of the electric field intensity under the metal film for the periodic structure shown in Fig. 4.2.28.

For an accurate description of the field in the central part of the interference pattern we should use the Fourier modal method in the three-dimensional case. A rigorous modelling of the radial structure containing a sufficient number of radial periods is the task of high computational complexity. Using personal computers it is possible to model three-dimensional periodic structures with the size of a period not exceeding 15–20 wavelengths. This size of the period is not sufficient for the approximation described in [39], which allows locally to reduce the solution of the three-dimensional diffraction problem to the problem of diffraction on the two-dimensional grating.

The structure shown in Fig. 4.2.27 may provide focusing of the SPP in the centre of the diffraction pattern and act as a lens of the SPP. The Fourier modal method is suitable for simulation of periodic diffraction structures. Therefore, to study the lens properties of the structure shown in Fig. 4.2.27 the periodic raster, shown in Fig. 4.2.28, was modelled. Figure 4.2.29 shows the calculated distribution of the intensity of the electric field under the metal film formed by the periodic structure in Fig. 4.2.28 at a normally incident wave with circular polarization at λ = 550 nm. The intensity distribution is given within the period of the raster cell d = 6200 nm. The radial period of the ring structure in the cell of the raster d_r = 310 nm was determined from the relation (4.2.168) for m = 1, λ = 550 nm. With these parameters, the period of the raster cell contains nine circular steps. The parameters of the materials match those of the one-dimensional structure discussed in section 2.2. The width of the steps of the ring structure w = 171 nm, height h_{gr} = 550 nm, thickness of the dielectric layer h_l = 0, the thickness of the metal layer h_m = 55 nm were selected from the conditions of maximization of the energy density in a one-dimensional interference pattern.

Figure 4.2.29 shows the formation of the peak of the intensity at the centre of the ring-shaped diffraction pattern. The value of the electric field at the circular peak is 19.6 times greater than the intensity of the incident wave. The average radius of the circular peak is 170 nm. The average width of the ring, determined by the level of decline of 0.5, is about 90 nm. The annular shape of the peak with a dip in the centre is due to different signs of the components E_z in the SPP in the centre of the picture from the opposite direction [28, 40]. Figure 4.2.30 shows the profile of the

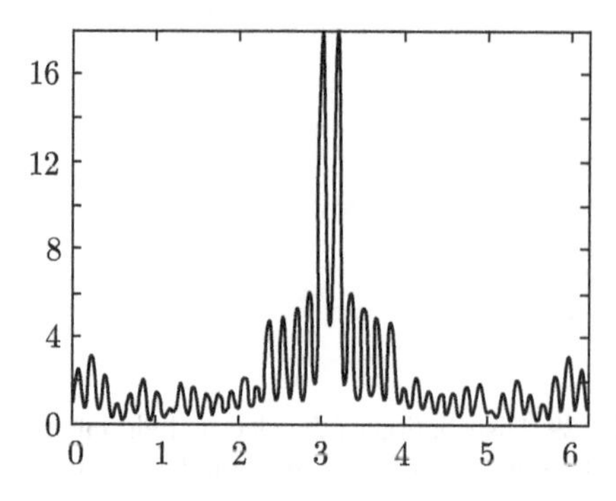

Fig. 4.2.30. Central profile of the diffraction pattern.

diffraction pattern corresponding to the central section along the axis *Ox*. The above calculation results confirm the possibility of using the structure in Fig. 4.2.27 and 4.2.28 to focus the SPP.

4.2.4. Diffractive optical elements for focusing of surface plasma polaritons

Section 2.1.1 deals with an equation of SPP at the interface between metal and dielectric with dielectric permittivities ε_m, ε_d. The interface is the plane $z = 0$, while the media ε_m, ε_d correspond to the regions $z < 0$ and $z > 0$, respectively.

As the propagation axis of the SPP we choose the axis *Ox*. In this case, the field is conveniently described by component H_y which, according to (4.2.129), has the form

$$H_y^{(d)}(x,y,z) = \exp\left(ik_0\left(\alpha x + \beta y\right)\right)\exp\left(-k_0\gamma_d z\right),$$
$$H_y^{(m)}(x,y,z) = \exp\left(ik_0\left(\alpha x + \beta y\right)\right)\exp\left(k_0\gamma_m z\right). \tag{4.2.206}$$

The constants α, β in (4.2.206) must satisfy the dispersion equation (4.2.136). In what follows we use the SPP, in which the constants α, β are defined as

$$\alpha = \sqrt{k_{\text{SPP}}^2 - k_0^2\beta^2},\ \beta \in R, \tag{4.2.207}$$

where R is the set of real numbers. For real values of β the component H_y is limited with respect to the variable y and at the interface $z = 0$ has the form

$$H_y(x,y,0) = \exp\left(ik_0\beta y\right)\exp\left(i\sqrt{k_{\text{SPP}}^2 - k_0^2\beta^2}\,x\right). \tag{4.2.208}$$

Equation (4.2.208) gives integral representations for the function H_y at the boundary $z = 0$ similar to the representation of the field through the angular spectrum of plane waves and the Kirchhoff integral, which are widely used in the scalar diffraction theory [20]. Indeed, we write the general solution at $z = 0$ as a superposition of the SPP

$$H_y(x,y)=\int_{-\infty}^{+\infty} I(\beta)\exp(ik_0\beta y)exp\left(i\sqrt{k_{\text{SPP}}^2-k_0^2\beta^2}\,x\right)d\beta. \quad (4.2.209)$$

The function $I(\beta)$ is defined by the values of the fields at $x = 0$ in the form of

$$I(\beta)=\frac{k}{2\pi}\int_{-\infty}^{+\infty} H_y(0,y)exp(-ik_0\beta y)dy. \quad (4.2.210)$$

Equation (2.4.210) is identical to the integral representation of the field through the angular spectrum of plane waves used in the scalar diffraction theory [20]. Equation (4.2.209) allows us to write the Kirchhoff integral for SPP in the form of

$$H_y(x,y)=\int_{-\infty}^{+\infty} H_y(0,u)G(x,y-u)du, \quad (4.2.211)$$

where

$$G(x,y)=\frac{k}{2\pi}\int_{-\infty}^{+\infty}\exp\left(i\sqrt{k_{\text{SPP}}^2-k_0^2\beta^2}\,x\right)\exp(ik_0\beta y)d\beta = \\ =\frac{ik_{\text{SPP}}x}{2\sqrt{x^2+y^2}}H_1^1\left(k_{\text{SPP}}\sqrt{x^2+y^2}\right), \quad (4.2.212)$$

where $H_1^1(x)$ is the Hankel function of first kind, the first order [41]. Replacing the Hankel function by the asymptotic expression for large arguments [41], we obtain

$$G(x,y)=\sqrt{\frac{ik_{\text{SPP}}}{2\pi\sqrt{x^2+y^2}}}\frac{x}{\sqrt{x^2+y^2}}\exp\left(ik_{\text{SPP}}\sqrt{x^2+y^2}\right). \quad (4.2.213)$$

Equations (2.4.211), (2.4.213) are analogous to the Kirchhoff integral in the two-dimensional case. Expanding $\sqrt{x^2+y^2}\approx x+y^2/2x$ at $y/x\ll 1$ in (4.2.213), we obtain the kernel of the integral transformation for the paraxial Fresnel approximation

$$G(x,y)=\sqrt{\frac{ik_{\text{SPP}}}{2\pi x}}\exp(ik_{\text{SPP}}x)\exp\left(\frac{ik_{\text{SPP}}y^2}{2x}\right). \quad (4.2.214)$$

The Kirchhoff integral and the representation of the field in the form of the angular spectrum of plane waves are widely used in the calculation of diffractive optical elements (DOE) as part of the scalar diffraction theory [12, 13]. The existence of such relations for SPP can directly transfer the methods of calculating the DOE developed as part of the scalar theory, to the calculation of DOE for transformation and focusing of the SPP.

In the scalar diffraction theory the propagation of the incident wave passing through the DOE is described by the phase modulation of the input wave field [12, 13]. The phase shift at each point is calculated by solving the model problem of passage of the incident wave through the dielectric plate. The plate thickness is

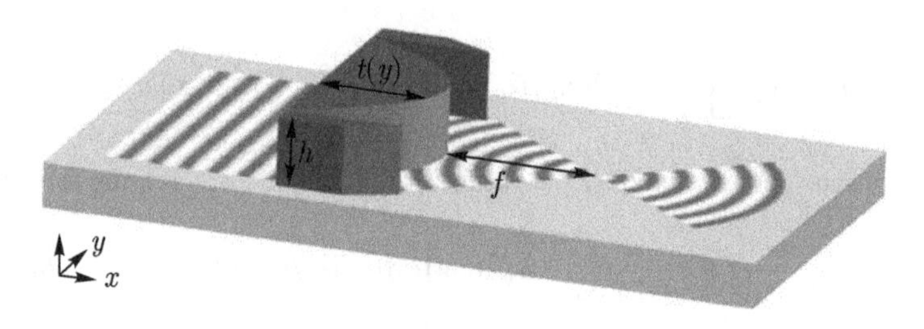

Fig. 4.2.31. Dielectric DOE, located on the surface of propagation of the SPP.

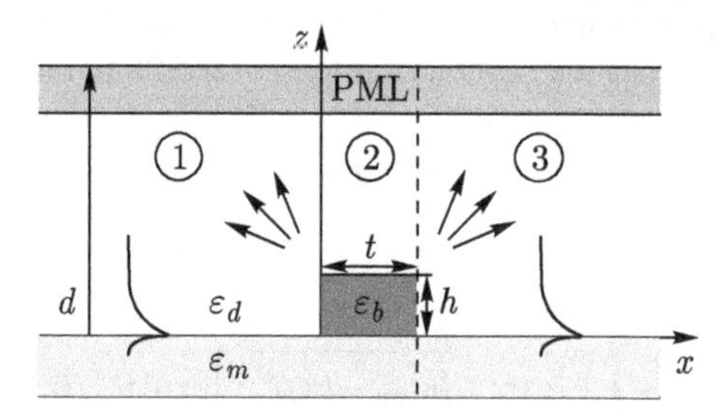

Fig. 4.2.32. Geometry of the model problem.

equal to the thickness of the DOE at the given point. Similarly, the DOE calculation for SPP is based on a phase modulation that occurs when SPP passes through the dielectric rectangular notch located directly on the surface of SPP propagation [42, 43].

Consider the passage of the SPP propagating along the axis *Ox*, through a dielectric DOE, located directly on the surface of wave propagation (Fig. 4.2.31). The DOE is described by the function of 'length' and is made of a material having a dielectric constant ε_b. The height of the DOE in the direction of the *z*-axis will be assumed constant. Calculation of the field behind an optical element at each point *y* is reduced to solving a model problem of passing of SPP through the dielectric rectangular ridge with length *t*(*y*). The geometry of the model problem is shown in Fig. 4.2.32. SPP $H_y^{(1)}(x) = \exp(ik_{\text{SPP}}x)$ in medium 1 is incident on a rectangular step from the left, and on the right the output SPP forms in medium 3

$$H_y^{(3)}(x) = T_0 \cdot \exp\left(ik_{\text{SPP}}(x-t)\right). \tag{4.2.215}$$

In the present work the problem of diffraction of SPP on a dielectric step is solved using the Fourier modal method, discussed in section 1 of the chapter. The Fourier modal method is most suitable for the simulation of periodic diffraction structures and therefore the function of the dielectric constant $\varepsilon(z)$, $|z| < d/2$ was assumed to be periodic with period *d*. The periodicity in this case is introduced artificially [14]. Perfectly matched layers (PML) were used to eliminate the interaction between the periods at the boundaries of the periods [14, 44].

Figure 4.2.33 shows the calculated dependence of the modulus and phase of the transmission coefficient of the SPP on the length *t* and height *h* of the dielectric

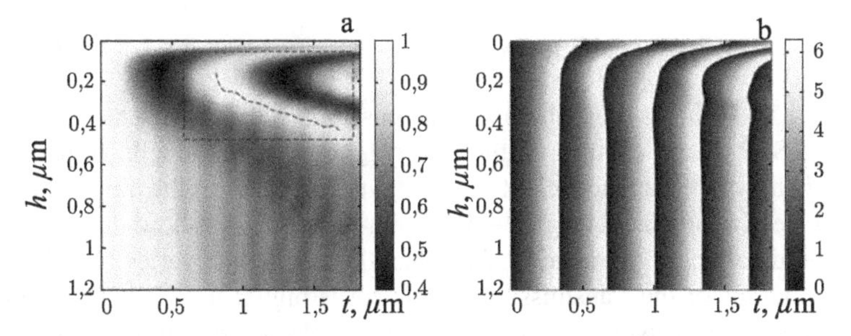

Fig. 4.2.33. Modulus $|T_0(h, t)|$ (a) and phase $\phi(h, t) = \arg(T_0(h, t))$(b) of the transmission coefficient of the dielectric step.

step. The calculation was performed with the following parameters: $\lambda = 550$ nm, $\varepsilon_m = -13.686 + 0.444i$, $\varepsilon_d = 1$, $\varepsilon_b = 2.25$. This value ε_m corresponds to the dielectric constant of silver at this wavelength.

Figure 4.2.33b shows that the dependence of the phase of the step length becomes close to linear with increasing step height. The phase begins to take a linear form at the height $h \approx 0.25$ μm. The depth of penetration of the SPP into the dielectric of the lens at the above parameters is $\delta_d = 0.13$ μm. The penetration depth is defined as the distance at which the wave amplitude decreases by e times, and is calculated by the formulas (4.2.156)–(4.2.158). Thus, the dependence is quasi-linear at $h \approx 2\delta_d$ that corresponds to a decrease in the amplitude of SPP by e^2 times. When $h > 4\delta_d$ the phase is well described by the expression

$$\varphi(t,h) = k_0\sqrt{\frac{\varepsilon_b\varepsilon_m}{\varepsilon_b + \varepsilon_m}}\, t = k_{\text{SPP}}^b t, \tag{4.2.216}$$

where k^b_{SPP} is the SPP wave vector in the step region. Equation (4.2.216) is similar to the formula of geometrical optics used for the phase shift of a plane wave upon trnasmission through a layer with thickness t. The maximum value of the length of the step $t_{\max} = 1.67$ in Fig. 4. 2.33 was chosen from the condition $\Delta\phi(h) = (k^b_{\text{SPP}} - k_{\text{SPP}})$ $h_{\max} = 2\pi$. This condition provides a range of phase difference $[0, 2\pi]$ between the SPP passing through the step with the dielectric constant ε_b and SPP propagating in the initial medium. We introduce a function $\Delta(h,t) = \arg(T_0(h,t)) - k_{\text{SPP}}^b t$ that characterizes the error of formula (4.2.216) for fixed h. The maximum value of error $\Delta(h, t)$ at $h = 1$ μm is less than $\pi/17$. In the calculation of the DOE such an error in the phase function is not essential in most cases. The modulus of the transmission coefficient at $h = 1$ μm is more than 0.7.

The linear relationship of the phase shift with the length of the dielectric step allows to create the given phase distribution by changing the length of the step. Thus, the wavefront transformation and focusing of the SPP can be carried out using dielectric diffractive structures with varying length and a fixed height [42–44]. According to (4.2.216), 'the DOE microrelief length' for the formation of a given phase function $\phi(y)$ has the form

$$t(y) = \varphi(y) / (k_{\text{SPP}}^{b} - k_{\text{SPP}}). \quad (4.2.217)$$

Figure 4.2.33 shows that there is the possibility of phase changes of the transmitted SPP due to changes in the step height at a fixed length. In particular, Fig. 4.2.34 shows the dependence of the modulus and phase of the transmitted SPP on the step height for a fixed length $t = 1055$ nm. Figure 4.2.34 shows the possibility of phase modulation in the range $[0, 2\pi]$ when the step height varies from 0 to 180 nm. The modulus of the transmission coefficient is higher than 0.7. Thus, for the transformation and focusing SPP we can also use dielectric structures with variable height and fixed length [42, 44]. An example of such a structure is shown in Fig. 2.4.35.

For example, consider the calculation of the lens of the SPP using modulation by varying the length and height of the step. According to (4.2.211), (4.2.213), the phase function of diffractive lenses with a focal point f has the form

$$\phi(y) = -\text{mod}_{2\pi}\left(\text{Re}\left(k_{\text{SPP}}\right)\sqrt{y^2 + f^2} + \phi_0\right), \quad (4.2.218)$$

where ϕ_0 is an arbitrary constant. Figure 4.2.36a shows the length of the lens microrelief and transmission amplitude calculated at the focus $f = 8\lambda_{\text{SPP}}$ and the aperture of the lens $2a = 10\lambda_{\text{SPP}}$. The length of the microrelief is normalized to the wavelength of the SPP. The length of the element was calculated using (4.2.217), (4.2.218). The height of the lens is constant at 1 μm. Figure 4.2.36b shows the distribution of values $|H_y(x, y)|$ formed by the lens and calculated by the formulas

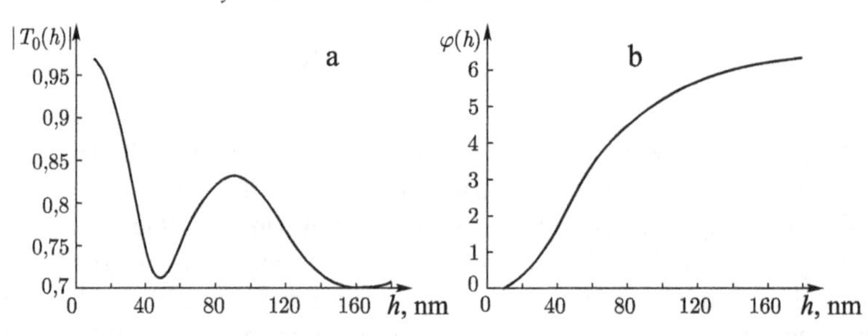

Fig. 4.2.34. Dependences $|T_0(h)|$ (a), $\phi(h)$ (b) at a step length $t = 1055$ nm.

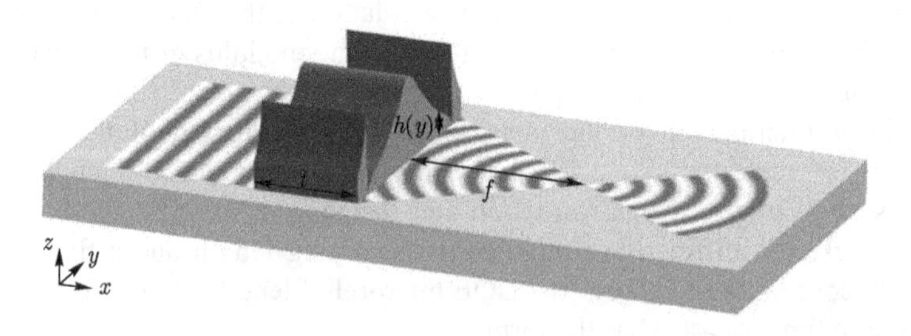

Fig. 4.2.35. Dielectric DOE with variable height and fixed width.

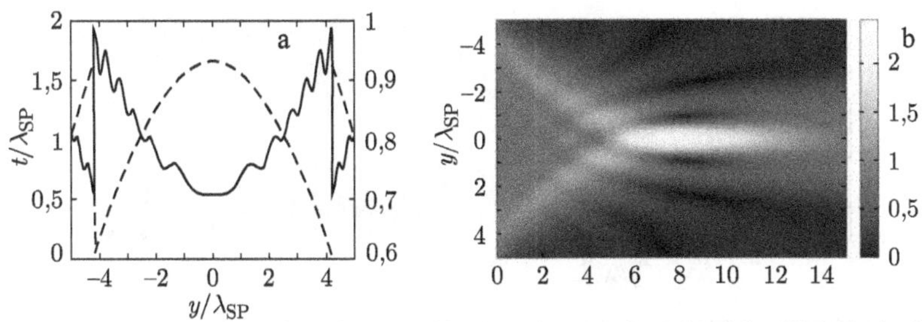

Fig. 4.2.36. Function of the length, normalized to the wavelength of the SPP (dashed line) and the transmission amplitude function (solid line) (a) forming the distribution $|H_y(x, y)|$ (b).

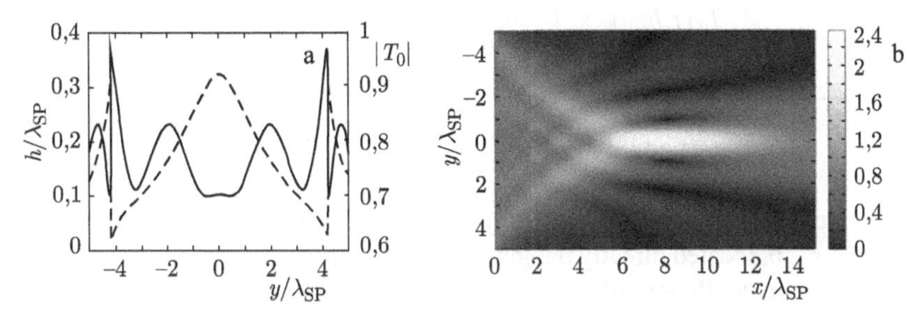

Fig. 4.2.37. The function of height normalized to the wavelength of the SPP (dashed line) and the transmission amplitude function (solid line) (a) forming the distribution $|H_y(x, y)|$ (b).

(4.2.211), (2.4.212). Figure 4.2.36b shows focusing to a point on the line $x = f = 8\lambda_{SPP}$.

Figure 4.2.37a shows the function of the height of the microrelief of the lens and the transmission amplitude for the lens calculate for the fixed length of the step t = 1055 nm (Fig. 4.2.35). The lens parameters coincide with the previous example. The formed distribution of $|H_y(x, y)|$ is shown in Fig.4.2.37b and also shows the focusing to a point. The graphs in Figs. 4.2.36b and 4.2.37b are similar in structure.

The diffraction efficiency of lenses can be estimated by the formula [43]

$$T_e = 100 \times \int_{-\infty}^{+\infty} \frac{|I(\beta)|^2 \operatorname{Re}\left(\sqrt{k_{SPP}^2 - k_0^2 \beta^2}\right)}{\operatorname{Re}(k_{SPP})} d\beta \ (\%). \tag{4.2.219}$$

The diffraction efficiency is 60.5% for the lens shown in Fig. 4.2.31, and 56.5% for the lens shown in Fig. 4.2.35.

The highest efficiency of the lens can be achieved by modulating SPP due to simultaneous changes in the length and height of the lens [44]. Indeed, we assume that the lens is located at $-L \le x \le 0$, where L is the maximum length of the lens. For each value of y the height h and length t of the steps can be determined from the condition of the maximum modulus of the transmission coefficient

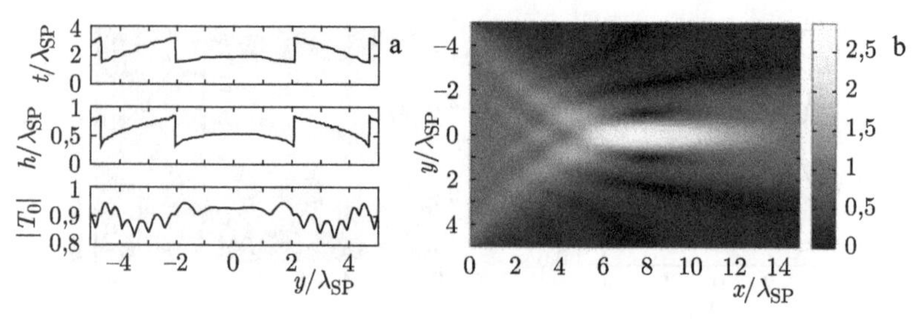

Fig. 4.2.38. Function of length and height, normalized to the wavelength of the SPP, and the function of the transmission amplitude (a) forming the distribution $|H_y(x, y)|$ (b).

$$\left|T_0\left(t,h;y\right)\right|\exp\left(-\mathrm{Im}\left(k_{SPP}\right)\left(L-t\right)\right)\to\max, \tag{4.2.220}$$

where h, t are determined from the formation of a given phase

$$\mathrm{mod}_{2\pi}\left(\phi\left(t,h\right)+k_{\mathrm{SPP}}\left(L-t\right)\right)=\phi\left(y\right). \tag{4.2.221}$$

Equation (4.2.221) implies that the given phase distribution (4.2.218) is formed on the line $x = 0$, located directly behind the lens. The phase consists of the phase of the transmission coefficient of the step $\phi(t, h)$ and the phase shift acquired by SPP during propagation over the distance $(L–t)$ to the line $x = 0$. The exponential factor in (4.2.220) determines the damping of the SPP at a distance $(L–t)$. According to Fig. 4.2.33b, there are many points (t, h) that provide a given phase shift modulo 2π.

Figure 4.2.38 shows the calculated length of the relief, the relief height and transmission amplitude of the lens, obtained from the condition of maximizing the transmittance (4.2.220) under the constraint (4.2.221). Maximization was carried out by exhaustive search of values (t, h) in region D, marked by the dashed rectangle in Fig. 4.2.33a. In this case the optimal values (t, h) are on the dashed curve in Fig. 4.2.33a. This curve passes through the maxima of the modulus of the transmission coefficient. The modulus of the transmission coefficient in Fig. 4.2.38a exceeds 0.83, which is much more than for the lens in Figs. 4.22.36 and 4.2.37. The resultant distribution of $|H_y(x, y)|$ is shown in Fig. 4.2.38b and has higher maximal intensity than in Figs. 4.2.36 and 4.2.37.

The energy efficiency of the lenses is 80.6%. This is more than 20% greater than that of the considered lens, based on changing only one parameter (length or height). Optimization of (t, h) in a wider area increases the efficiency of the lens by another 3–4%, but the functions of the length and the height of the relief have substantially irregular appearance.

Thus, the mechanism of phase modulation of the SPP due to simultaneous changes in length and step height is most effective in achieving high energy efficiency of the DOE.

4.3. Diffractive heterostructures with resonant magneto-optical properties

This section discusses the magneto-optical properties of two-layer metal-dielectric heterostructures consisting of a metallic grating and a dielectric magnetized layer. Calculation and study of the magneto-optical properties are based on the Fourier modal method to solve the problem of diffraction on these periodic structures.

We consider three basic geometries of the magnetization of the layer: polar, equatorial (transverse) and meridional. In the case of polar geometry, the magnetization vector is perpendicular to the layer plane. The structure has resonant magneto-optical effects associated with the rotation of the polarization plane [46–52]. At the meridional and equatorial geometry the magnetization vector lies in the plane of the layer. The direction of the magnetization vector is perpendicular and parallel to the grooves of a diffraction grating. In this case, the structure has magneto-optical effects due to the strong dependence of the transmission and reflection of the structure on the magnetization [53–59].

4.3.1. Magneto-optical effect in the polar geometry

4.3.1.1. The geometry of the structure

The geometry of the studied two-layer structure is shown in Fig. 4.3.1. The top layer is a binary diffraction grating made of gold. The bottom layer is a uniformly magnetized layer and its magnetization vector is directed along the normal to the surface.

The dielectric constant of the lattice is described by a periodic piecewise constant function

$$\varepsilon(x)=\begin{cases}\varepsilon_{\mathrm{gr}}, & x\in[0,d-r),\\ 1, & x\in[d-r,d),\end{cases} \tag{4.3.222}$$

where r is the slit size, d is the grating period, $\varepsilon_{\mathrm{gr}}$ is the permittivity of the material

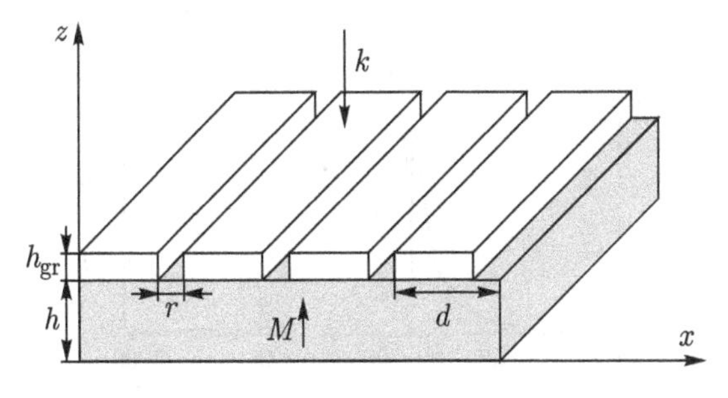

Fig. 4.3.1. Geometry of a two-layer structure consisting of a diffraction grating and a magnetized layer.

of the grating. In the case where the magnetization vector is normal to the surface along the axis Oz (polar geometry), the dielectric permittivity tensor is described by (4.1.52). The magnetic permeability is equal to unity [18, 19]. The appropriate form of the matrix of the system of differential equations, representing the field in the magnetized layer, is shown in (4.1.53). For a uniform magnetic layer in the matrices **E***, **E**, **H**, $\mathbf{E}_2$ in (4.1.53), only the zeroth Fourier coefficients differ from zero.

4.3.1.2. The study of magneto-optical effects

Figure 4.3.2 shows the calculated plots of the transmittance coefficient and the Faraday angle as a function of wavelength. The calculation was performed by the Fourier modal method at normal incidence of waves with TM-polarization (vector **H** is parallel to the grating grooves). The transmittance coefficient refers to the intensity of the 0th transmitted order of diffraction. The Faraday angle corresponds to the angle between the principal axis of the polarization ellipse for the wave in the zeroth transmitted order of diffraction and the axis Ox (see (4.1.100)). The Faraday angle characterizes the rotation of the polarization plane for the transmitted wave. The spectra in Fig. 4.3.2 are calculated for the following geometric parameters of the structure: period $d = 750$ nm, the width of the hole $r = 75$ nm, the thickness of the grating $h_{gr} = 75$ nm, the magnetic layer thickness $h = 537$ nm. The dielectric permittivity of the grating material $\varepsilon_{gr}(\lambda)$ was represented by the reference data for gold [57]. For the dielectric permittivity tensor of the magnetic layer (4.1.52), we used parameters $\varepsilon = 5.5 + 0.0025i$, $g = (1-0.15i)\cdot 10^{-2}$. These parameters correspond to the material Bi:YIG (bismuth-substituted yttrium iron garnet), which is one of the most common materials in magneto-optics [58].

The transmission graph in Figure 4.3.2 has a sharp peak at 43% at a wavelength $\lambda = 884$ nm. The transmission peak coincides with the negative peak of the Faraday angle. The value of the Faraday angle at the peak is 2.25° of which is almost five times larger than that of the homogeneous magnetic plate, placed in an optically matched medium (a medium with the same dielectric constant $\varepsilon = 5.5 + 0.0025i$). Note

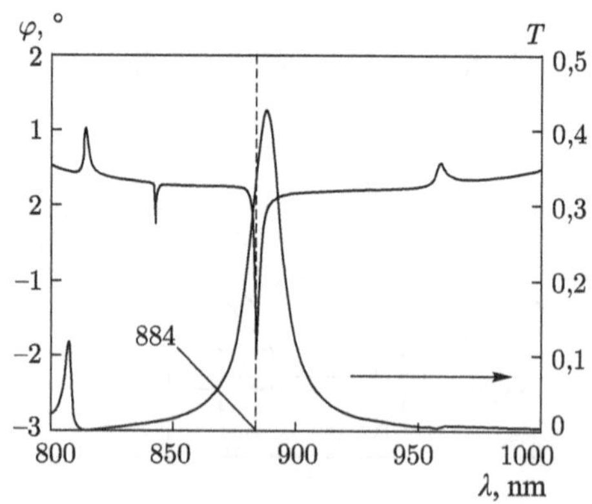

Fig. 4.3.2. The transmission (lower curve, the axis on the right) and the Faraday angle (upper curve, the axis on the left) as a function of wavelength.

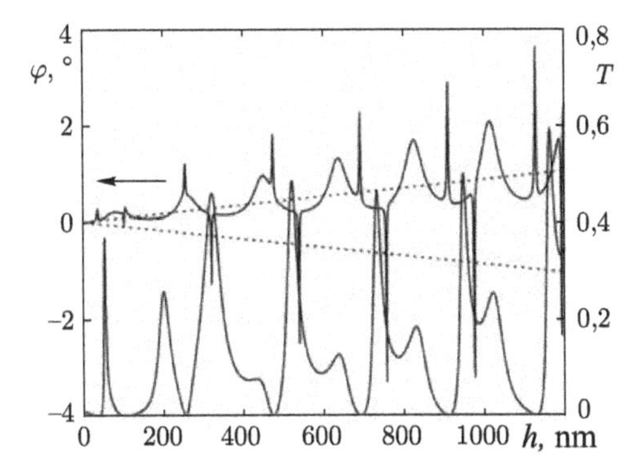

Fig. 4.3.3. The transmission of the structure (lower curve, right axis), the Faraday angle of structures (upper curve, left axis) and the Faraday angle for the plate in the optically matched medium (dashed lines) as a function of wavelength.

that the parameter of ellipticity (see (4.1.100)) is just 0.53°. This means that the transmitted wave is in fact linearly polarized in the rotated plane of polarization.

Figure 4.3.3 shows the calculated transmission graphs and the Faraday angle in relation to the plate thickness at a wavelength $\lambda = 884$ nm. The graphs in Fig. 4.3.3 show a series of transmittance peaks and a series of negative and positive peaks of the Faraday angle. The transmittance peaks are related to the phenomenon of extraordinary transmission characteristic of the grating of good conductors [61–65].

The Faraday angle peaks may be due to an increase in the optical path length of light in a magnetized layer, due to the excitation of modes by the diffraction grating and the waveguide propagation of radiation in the layer [46–52]. To test the hypothesis, we find the estimates of the thickness of the magnetic layer at which modes will be excited in it. For simplicity, we neglect the magnetic component in (4.1.52) and assume $g = 0$ (a more accurate calculation can be performed using the formulas from [64] for the modes of the magnetic waveguide). Possible directions of propagation of modes in the layer will be assumed to coincide with the directions of the propagating diffraction orders of the grating. The angles between the directions of propagation of the modes in the layer and normal to the plane of the layer are determined from the formula of the grating (4.1.24) as

$$\theta_i = \arcsin\frac{m\lambda}{n_f d}, \tag{4.3.223}$$

where m is the number of the diffraction order, n_f is the refractive index of the material of the layer.

Since the size of the holes in this grating is only about 10% of the period, we may use the formulas for a plane-parallel waveguide. Consider the calculation of modes of the plane-parallel waveguide, as shown in Fig. 4.3.4. If $z > h$ we have a material with a dielectric constant $\varepsilon = \varepsilon_1$, and at $0 < z < h$ there is the waveguide layer with $\varepsilon = \varepsilon_2$, and at $z < 0$ there is a material with $\varepsilon = \varepsilon_3$.

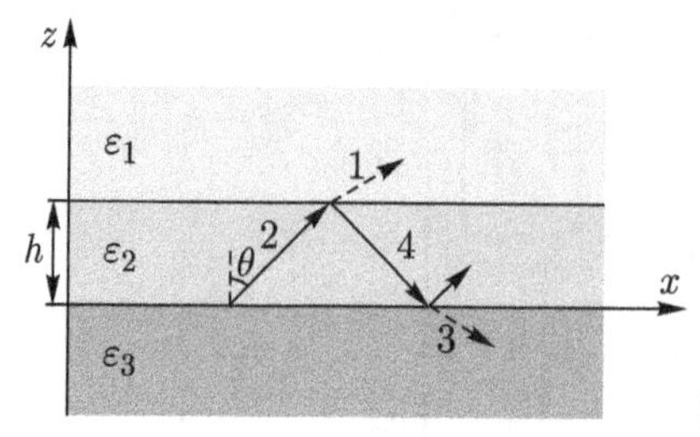

Fig. 4.3.4. Plane-paralell waveguide.

The conditions for the existence of the TE- and TM-modes are given by [67]:

$$\exp\left(-2ik_0\gamma_2h^{\mathrm{TE}}+2\pi im\right)=\frac{(\gamma_2-\gamma_1)(\gamma_2-\gamma_3)}{(\gamma_2+\gamma_1)(\gamma_2+\gamma_3)}, \quad (4.3.224)$$

$$\exp\left(-2ik_0\gamma_2h^{\mathrm{TM}}+2\pi im\right)=\frac{(\gamma_2\varepsilon_1-\gamma_1\varepsilon_2)(\gamma_2\varepsilon_3-\gamma_3\varepsilon_2)}{(\gamma_2\varepsilon_1+\gamma_1\varepsilon_2)(\gamma_2\varepsilon_3+\gamma_3\varepsilon_2)}, \quad (4.3.225)$$

where $k_0 = 2\pi/\lambda$, λ is the wavelength, h^{TE}, h^{TM} is the thickness of the waveguide (the superscript denotes the type of mode), m is an integer specifying the order of the mode,

$$\gamma_i=\sqrt{\varepsilon_i-\varepsilon_2\sin^2\theta},\ i=1,2,3, \quad (4.3.226)$$

where the angle θ determines the direction of the mode in the waveguide (Fig. 4.3.4). From (4.3.224) and (4.3.225) we obtain values of thicknesses of the waveguides in the form of

$$h_m^{\mathrm{TE}}=\frac{1}{k_0\gamma_2}\left(-\frac{1}{2i}\ln\frac{(\gamma_2-\gamma_1)(\gamma_2-\gamma_3)}{(\gamma_2+\gamma_1)(\gamma_2+\gamma_3)}+\pi m\right), \quad (4.3.227)$$

$$h_m^{\mathrm{TE}}=\frac{1}{k_0\gamma_2}\left(-\frac{1}{2i}\ln\frac{(\gamma_2\varepsilon_1-\gamma_1\varepsilon_2)(\gamma_2\varepsilon_3-\gamma_3\varepsilon_2)}{(\gamma_2\varepsilon_1+\gamma_1\varepsilon_2)(\gamma_2\varepsilon_3+\gamma_3\varepsilon_2)}+\pi m\right). \quad (4.3.228)$$

At a given angle of propagation of the mode θ the expressions (4.3.227) and (4.3.228) allow us to calculate the thicknesses at which TE- and TM-modes exist in the structure. At the above parameters of the diffraction gratings the propagating orders, able to excite modes, are ± 1 orders of diffraction. Therefore, as the propagation angle in (4.3.226), we choose the angle $\theta_1 = 30.16°$ corresponding to the direction of the ± 1 orders. The values of the thickness at which modes propagate in the layer calculated by the formulas (4.3.227) and (4.3.228) are shown in Table 4.3.1.

The values obtained are indicated in Fig. 4.3.5 by the vertical dashed lines. Figure 4.3.5 shows that the thicknesses h_m^{TE} agree well with the position of negative peaks of the Faraday angle, and thicknesses h_m^{TM} – with the position of positive

Table 4.3.1. Estimates of the thickness at which modes propagate in the layer

Order of mode m	h_m^{TE} for TE-modes (nm)	h_m^{TM} for TM-modes (nm)
0	111.9	39.9
1	329.8	257.8
2	547.7	475.8
3	765.6	693.7
4	983.5	911.6
5	1201.4	1129.5

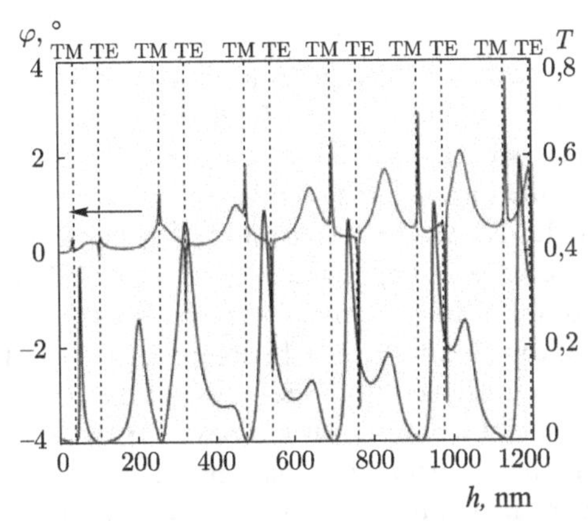

Fig. 4.3.5. The transmittance (lower curve, the axis on the right) and the Faraday angle (upper curve, left axis) as a function of the thickness of the magnetic layer. The dotted line shows the thickness corresponding to the propagation conditions for the TE- and TM-modes.

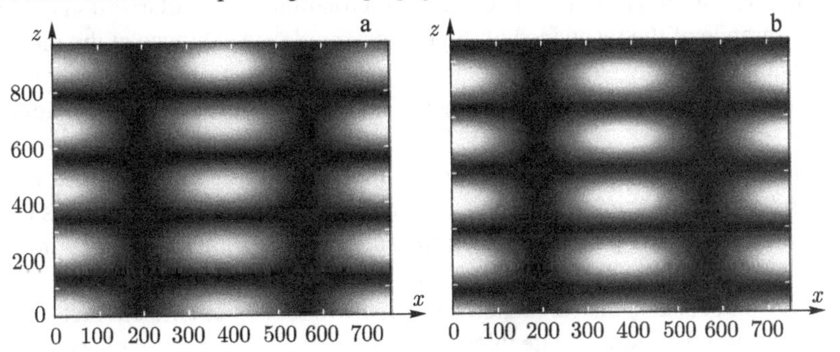

Fig. 4.3.6. Distribution of the field $|E_y(x, z)|$ in a magnetized layer with a thickness of 974 nm (left) and in an ideal dielectric waveguide (right).

peaks. In this case the positive peaks of the Faraday angle also coincide with the transmittance minima.

It is interesting to compare the field distribution in the magnetized layer with the field of the TE-mode corresponding to a dielectric waveguide. For the TE-mode the electric field contains only a component $E_y(x, z)$. Figure 4.3.6 shows the distribution of the field $|E_y(x, z)|$, calculated in the magnetized layer with the thickness of the layer of 974 nm. The field is given within a single period at $x \in [0, 750]$ nm, $z \in [0, 974]$ nm. The specified thickness of the layer corresponds to a negative peak of the

Faraday angle in Fig. 4.3.5 and is close to the estimate $h_4^{\mathrm{TE}} = 984$ nm obtained by the formula (4.3.227).

For comparison, Fig. 4.3.6 on the right shows the field distribution in the waveguide, calculated by the formula [67]

$$\left|E_y(x,z)\right| = C \cdot \left| \left(\cos\left(ik_0\gamma z\right) - \frac{\gamma_c}{\gamma} \sin\left(ik_0\gamma z\right) \right) \cos\left(k_0\beta x\right) \right|, \qquad (4.3.229)$$

where $\beta = \sqrt{\varepsilon}\sin\theta$, $\gamma = \sqrt{\varepsilon_{\mathrm{gr}} - \beta^2}$, $\gamma_c = \sqrt{\varepsilon - \beta^2}$. The field (4.3.229) corresponds to a superposition of two modes propagating in the direction of the +1st and –1st diffraction orders of the grating. Figure 4.3.6 shows that the component of the field $E_y(x, z)$ in a magnetized layer has a pronounced modal nature and is close, with the accuracy to the translation along the axis z, to the estimate (4.3.8) of the field in the waveguide. The field component $E_y(x, z)$ has a similar mode structure at other thicknesses, corresponding to the negative peaks of the Faraday angle. For the positive peaks of the Faraday angle the mode structure has a field component $H_y(x, z)$. In this case the mode structure of the components $E_y(x, z)$, $H_y(x, z)$ disappears quickly with increasing distance from the peaks of the Faraday angle. Thus, the calculations confirm the relationship of the peaks of the Faraday angle with the waveguide propagation of radiation in a magnetized layer [48–50].

The following explanation of the Faraday angle resonance is offered [46–51]. At the incident wave with TM-polarization the rotation of the polarization plane and the appearance of the TE-component are due to the presence of the magnetized layer in this two-layer system. The maximum rotation of the polarization plane and the appearance of the resonance of the Faraday angle are achieved at the maximum conversion of the wave with TM-polarization to the wave with TE-polarization. The specified conversion is maximum when TE-modes are excited in the magnetized layer.

Note that at thicknesses corresponding to excitation of TM-modes in the film the transmittance is minimum. A qualitative explanation of these minima is that the gold grating acts as a polarizer which transmits waves with TM-polarization from the layer surface and reflecting waves with TE-polarization.

Figure 4.3.7 shows the calculated plots of the reflection coefficient factor and the Kerr angle as a function of plate thickness ($\lambda = 884$ nm). The Kerr angle corresponds to the angle between the principal axis of the polarization ellipse for the wave in the zero reflected diffraction order and the axis Ox. The reflection graph in Fig. 4.3.7 has maxima – at thicknesses of excitation of TM-modes, which confirms the above assumption. The graph of the Kerr angle has peaks at thicknesses of excitation of the TE modes. The Kerr angle peaks are explained as follows. Due to imperfection of the waveguide propagation there is a partial passage of a wave with TE-polarization from the layer into the region above the grating. As a result, the wave reflected from the grating acquires the TE-component, which leads to the formation of the Kerr angle peaks.

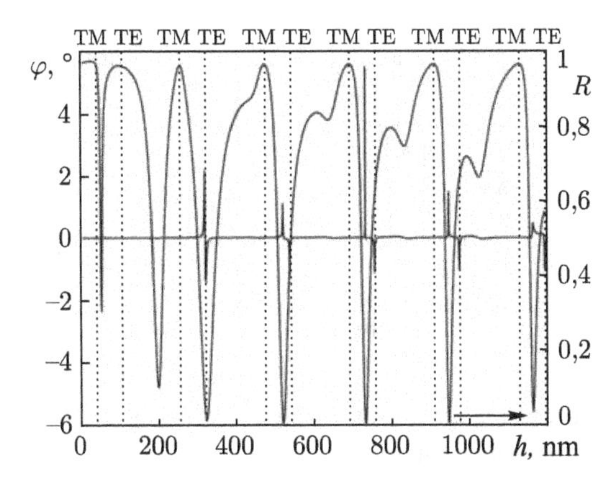

Fig. 4.3.7. Reflection and Kerr angle in dependence on the thickness of the magnetized layer at normal incidence of waves.Vertical lines indicate the thickness of excitation of modes.

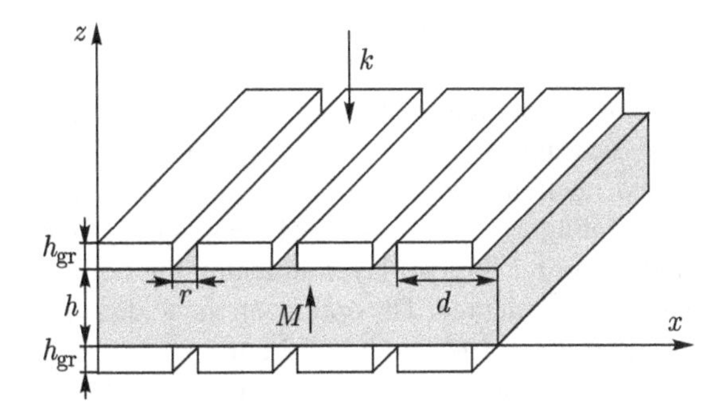

Fig. 4.3.8. Geometry of the three-layer structure.

4.3.1.3. Investigation of three-layer structure

The transmittance peaks in the spectrum of a diffraction grating made of a material with high conductivity are associated with the phenomenon of extraordinary transmittance [61–65]. In [63] it is shown that the three-layer structure containing two subwavelength metal gratings, separated by a dielectric layer, can significantly increase the transmittance as compared with a single diffraction grating. In this connection it is interesting to study the magneto-optical properties of the three-layer structure comprising two identical gold gratings, separated by a magnetized dielectric layer of Bi:YIG (Fig. 4.3.8).

Calculation of the parameters of the three-layer structure was carried out using the gradient optimization procedure [12]. The merit function was selected the ratio of the product of the transmittance coefficient and the Faraday angle for the zero transmitted order to a similar product for the magnetic layer in an optically matched medium:

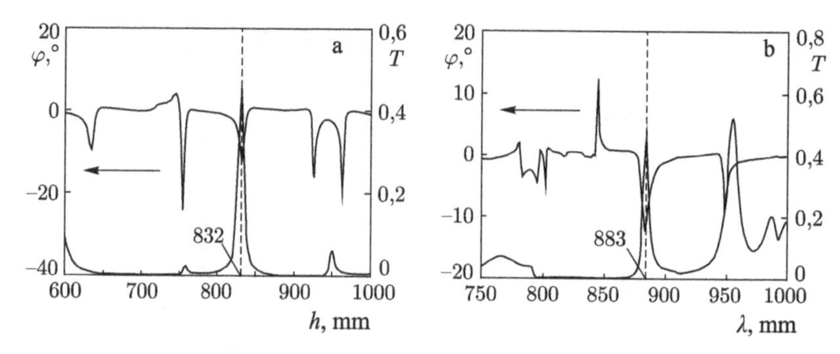

Fig. 4.3.9. The transmittance and Faraday angle at normal incidence as a function of layer thickness (left) and wavelength (right).

$$\frac{T_0 \mid \phi_0 \mid}{\tilde{T}_0 \mid \tilde{\phi}_0 \mid} \to \max, \tag{4.3.230}$$

where T_0 and ϕ_0 are the transmittance and Faraday angle for a three-layer structure, and $\tilde{T}_0$ and $\tilde{\phi}_0$ are the transmittance and Faraday angle for the magnetic layer in an optically matched medium. Optimization was performed on four parameters: the thickness of the grating h_{gr}, the thickness of the magnetic layer h, the period of grating d, slit width r. As for the two-layer structure, the normal incidence of the TM-polarized wave was considered. The optimization yielded the following values: $d = 832$ nm, $r = 362$ nm, $h_{gr} = 194$ nm, $h = 832$ nm.

Figure 4.3.9 shows dependences of the transmittance and the Faraday angle on the thickness of the magnetized layer and the wavelength. For the magnetized layer thickness of 832 nm and 883 nm wavelength, the peak of the Faraday angle greater than 12° at a transmittance of 46% was detected. This value of the Faraday angle is 17 times greater than for the homogeneous magneitc layer in the optically matched environment.

Note that adding additional optimization parameters, such as variable thickness and size of holes in the grating, leads to a further increase in the value of the angle of Faraday only by 6–7% with a 3–5% increase of transmittance. The transition from the two-layer to a three-layer structure more than doubles the best value of the merit function (4.3.230).

We summarize the results of this section. The presented studies show that the two-layer structure, consisting of a binary gold diffraction grating and the magnetized layer, has both the transmission resonances and resonances of the Faraday and Kerr angles. An explanation of the resonances of the Faraday and Kerr angles is offered and their connection with the waveguide propagation of radiation in the magnetic layer described. We considered a three-layer structure consisting of gold diffraction gratings, separated by a magnetized layer. It is shown that the transition to a three-layer structure provides a further increase in the values of the resonance of the Faraday angle by up to 17 times.The presence of two types of resonances (transmittance and Faraday and Kerr angles) makes promising the use of the investigated structures as active element of optical sensors, using the values of the resonance changes with the variation of the physical or geometrical parameters of the structure.

4.3.2. Magneto-optical effect in meridional geometry

4.3.2.1. The geometry of the structure and type of magneto-optical effect

The geometry of the studied two-layer structure is shown in Fig. 4.3.10. The top layer is a binary diffraction grating made of gold. Below the grating there is a uniformly magnetized layer; its magnetization vector lies in the layer plane and is perpendicular to the grating grooves.

As in the previous case, the dielectric constant of the grating is described by the function (4.3.222). In the case where the magnetization vector is directed along the axis Ox (meridional geometry), the dielectric permittivity tensor is described by (4.1.54). The magnetic permeability is equal to unity [18,19]. The appropriate form of the matrix of the system of differential equations representing the field in the magnetized layer is shown in (4.1.55). For the uniform magnetic layer in the matrices E^*, E, G in (4.1.55), only the zeroth Fourier coefficients differ from zero.

The results of numerical studies of the examined structure by the Fourier modal method show that at a specific combination of geometric parameters the transmittance and reflection of the structure strongly depend on the gyration g (see (4.1.54)), which determines the magnetization of the layer. The transmittance and reflection are the intensities of the zero transmitted and reflected diffraction orders. Figure 4.3.11 shows the dependence of the transmittance and reflectance on the wavelength for the magnetized layer (solid lines) and in the absence of magnetization (dashed lines). In the absence of magnetization $g = 0$ in (1.4.54) and the dielectric constant of the layer is a scalar. The calculations were performed at normal incidence of waves with TM-polarization for the following parameters: period $d = 886$ nm, the width of the hole $r = 88$ nm, the thickness of the grating $h_{gr} = 302$ nm, the magnetic layer thickness $h = 883$ nm. As before, the following parameters were used for the dielectric tensor of the magnetic layer: $\varepsilon = 5.5 + 0.0025i$, $g = (1–0.15i)\cdot 10^{-2}$ corresponding to the material Bi:YIG (bismuth-

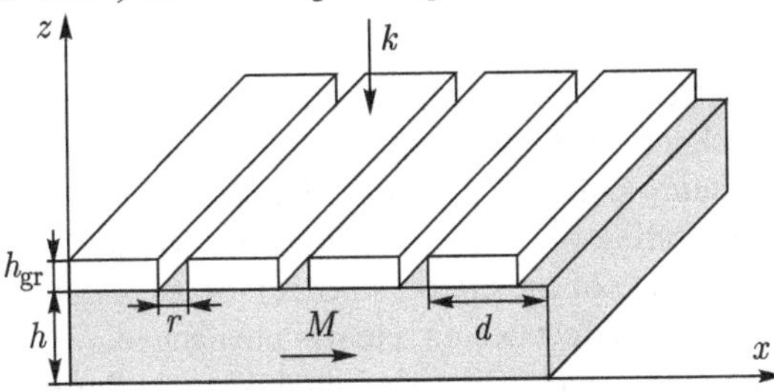

Fig. 4.3.10. Geometry of the two-layer structure containing a diffraction grating and a dielectric layer magnetized in the plane.

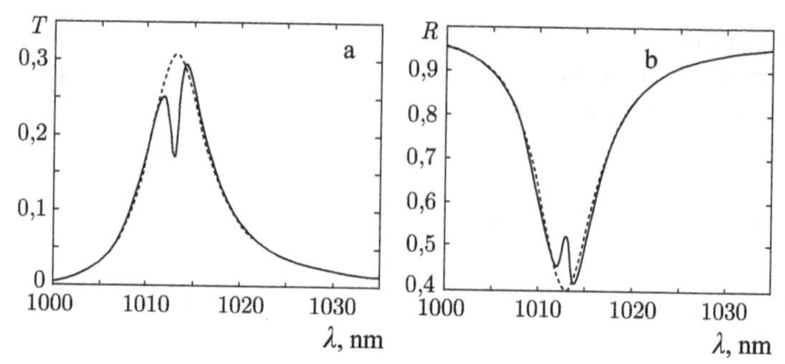

Fig. 4.3.11. Transmittance (left) and reflection (right) in the case of the magnetized layer (solid line) and in the absence of magnetization (dashed line).

substituted yttrium iron garnet). The values of the permittivity of the grating material were represented by the reference data for gold [60]. Refractive indexes above and below the structure were assumed to be unity.

Figure 4.3.11 shows that in the case of the magnetized layer the transmission spectrum of the structure has a sharp minimum at a wavelength $\lambda = 1013$ nm. This effect is also seen as a peak in the reflection spectrum. The narrow spectral width of the observed effect suggests its resonant nature. By analogy with the effect of changes in the reflectance coefficient observed in reflection from the magnetic material [68, 69], we call this effect the intensity effect. Under the magnitude of the intensity effect in transmission (reflection) we the modulus of the difference of the transmittance coefficients (reflectance) in the magnetized layer ($g \neq 0$) and in the absence of magnetization ($g = 0$):

$$I_T(g) = \left|T_0(0) - T_0(g)\right|, \tag{4.3.231}$$

$$I_R(g) = \left|R_0(0) - R_0(g)\right|, \tag{4.3.232}$$

where $T_0(0)$, $T_0(g)$, $R_0(0)$, $R_0(g)$ is the intensity of transmitted and reflected zeroth diffraction orders.

4.3.2.2. Investigation of the magneto-optical effect

To explain the nature of the resonance peaks in Fig. 4.3.11 studies were conducted of the mutual dependence of the structure parameters (height and width of steps, the period, the thickness of the magnetic layer, and layer and grating materials) for which the structure under consideration has a pronounced intensity effect. It was found that the grating parameters, such as the width and height of the step, affect only the magnitude of the effect (4.3.231), (4.3.232). At the same time, the wavelength at which this effect occurs remains unchanged.

When changing parameters of the structures such as the period, the dielectric constant of the magnetized material of the layer and its thickness, the minimum in the transmission spectrum shifts. This indicates the relationship of the observed effect with one of the orders of the grating in the magnetized layer. In the investigated case of normal incidence the diffraction orders are symmetrical. The orders with the numbers $\pm m$ are distributed under the same angles and their effect on the transmission spectrum occurs because of symmetry at the same wavelength. Consider incidence under a small angle of 0.2° (relevant spectra are shown in Fig. 4.3.12). At oblique incidence the moduli of the propagation constants $k_{x,m}$ for the orders with the numbers $-m$ and $+m$ will vary. Consequently, their influence on the spectrum will occur at different wavelengths, which is observed in Fig. 4.3.12 in the form of two local minima.

We equate the propagation constants of the orders

$$k_0\left(\sin\theta+m\frac{\lambda_1}{d}\right)=-k_0\left(\sin\theta-m\frac{\lambda_2}{d}\right), \tag{4.3.233}$$

where θ is the angle of incidence, λ_1 and λ_2 are the wavelengths at which the maxima of the intensity effect are observed. From (4.3.233) we can determine the number of the order on the basis of the difference in the wavelength corresponding to minima in the transmission spectrum in Fig. 4.3.12, as follows:

$$m=\frac{2d\sin\theta}{\lambda_2-\lambda_1}. \tag{4.3.234}$$

The distance between the minima in Fig. 4.3.12 corresponds to the second transmitted diffraction order.

In the considered diffraction structure, there are different types of resonances associated with the excitation of modes in the dielectric layer and surface plasma polaritons (SPP) between the slits of a diffraction grating. To determine what kind of resonance is responsible for the observed effect, it is necessary to study the influence of the main component ε of the dielectric permittivity tensor of the magnetic layer (4.1.54) at the position of the maximum intensity effect in the spectrum. As shown above, the desired resonance should be associated with the

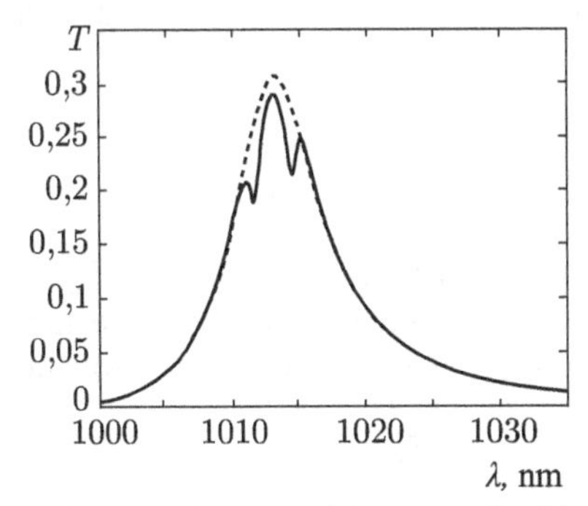

Fig. 4.3.12. Transmission sp[ectra in the case of the magnetized layer (solid line) and in the absence of magnetization (dashed line) at oblique angles of incidence of the wave under the angle 0.2°.

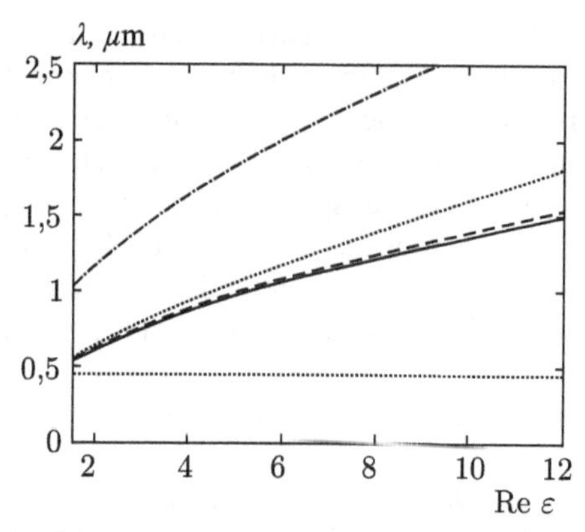

Fig. 4.3.13. Wavelength of the resonance structure depending on the dielectric layer. Intensity effect and mode TE_2 – solid line, excitation of the SPP at the top and bottom borders – dotted lines, the emergence of 2nd order (Rayleigh–Wood anomaly) – dashed; mode TE_1 – dot-and-dash.

second diffraction order. The corresponding dependence for the various resonances of the structure is shown in Fig. 4.3.13.

The solid line in Fig. 4.3.13 describes the studied intensity effect. The intensity effect curve coincides with the condition of existence of the TE-mode with the propagation direction as in the second order diffraction in the layer. Since the size of the slits in this grating is only about 10% of the period, the conditions of the TE-mode were calculated using equation (4.3.224) for a plane-parallel waveguide. The value of the magnetization is not taken into account in the calculation of modes. Strictly speaking, the gyrotropic waveguide modes are not TE- and TM-modes but the right and left elliptically polarized modes. Exact relations for them are given in [68]. Given their complexity, the formulas [68] are inconvenient for analysis. As shown by further calculations, the influence of magnetization on the conditions for the existence of modes can be neglected.

Near the curve of the intensity effect the dotted line shows the Rayleigh–Wood anomaly associated with the emergence of the second propagating diffraction order. The wavelength corresponding to the Rayleigh–Wood anomaly, is defined by the formula

$$\lambda = \frac{d\sqrt{\varepsilon}}{m}, \tag{4.3.235}$$

where m is the number of the order. The dotted lines in Fig. 4.3.13 show the graphs of the resonances associated with excitation of SPP on the upper and lower boundaries of the grating. The graphs were calculated using the equation

$$k_{x,2} = \mathrm{Re}\left(k_{\mathrm{SPP}}\right), \tag{4.3.236}$$

where k_{SPP} is the constant of propagation of SPP, as defined in (4.2.14). The dash-and-dot line in Fig. 4.3.13 shows the condition of existence of the TE-mode with the propagation direction corresponding to the first order of diffraction. The last three

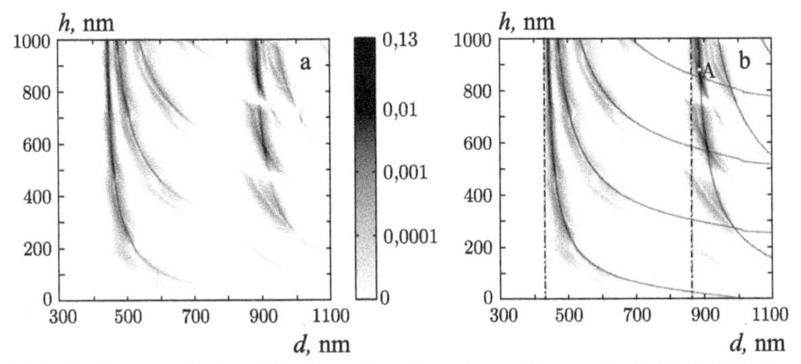

Fig 4.3.14. a) The magnitude of the intensity effect depending on the height of the layer and the grating period. b) The magnitude of the intensity effect superimposed with the curves of the existence of modes (continuous lines) and the terms of the Rayleigh–Wood anomalies (dot-and-dash line).

graphs of the resonances are separated by a large distance from the curve of the intensity effect. Thus, the analysis of Fig. 4.3.13 shows a close relationship of the intensity effect with the excitation of the TE-mode by the second-order diffraction and the Rayleigh–Wood anomaly.

Figure 4.3.14a shows the dependence of the intensity effect (4.3.10) on two structural parameters: the period and height of the layer. Figure 4.3.14b shows the same distribution, but with superimposed curves corresponding to the conditions of existence of the TE-modes with the propagation directions of the 1st and 2nd orders of diffraction. The curves of the modes corresponding to the 1st order are on the left, and to the 2nd order – on the right. Figure 4.3.14 shows the existence of analytic curves of existence of the TE-modes with maxima of the intensity effect. In addition, the maximum value of the intensity effect is found around lines $d = m\lambda/\sqrt{\varepsilon}$, $m = 1,2$ that match the Rayleigh–Wood anomalies for the 1st and 2nd orders of diffraction (dot-and-dash lines in Fig. 4.3.14). Point A indicates the parameters corresponding to the graphs in Fig. 4.3.11. Point A is located on the curve of the TE-modes, corresponding to the 2nd order diffraction near the Rayleigh–Wood anomaly.

Thus, Fig. 4.3.14 shows that the intensity effect occurs at the same wavelength as the waveguide TE-mode excited by one of the diffraction orders in the magnetic layer. In addition, the maximum effect is reached near the Rayleigh–Wood anomalies for the corresponding order. Note that near the Rayleigh–Wood anomalies the corresponding diffraction order propagates almost parallel to the layer plane. Such propagation, as well as mode propagation, gives an increase in the optical path length in the magnetic layer and leads to increased magneto-optical effect.

The relationship of the intensity effect with waveguide TE-modes can give the following explanation for the magneto-optical effect [53–57]. With the incident wave with TM-polarization in the structure with the non-magnetized layer only TM-modes can be excited. In the structure with the magnetized layer the modes of oppositTE-polarization (TE-modes) can be excited. The change of the transmission and reflection coefficients is due to the excitation of the TE-modes in the magnetized layer. Part of the energy of the incident TM-mode is transferred to the given TE-type

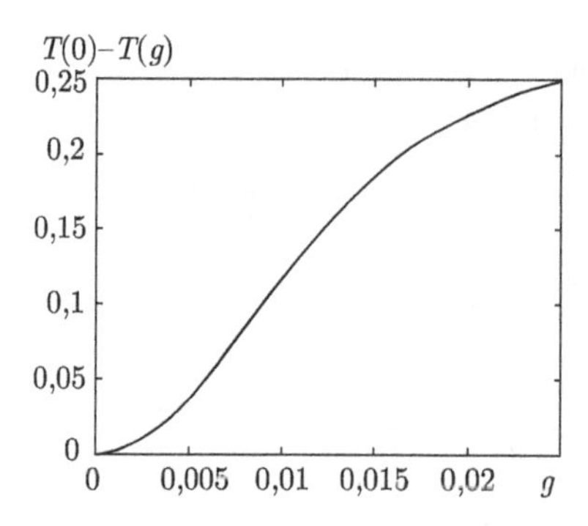

Fig. 4.3.15. Dependence of the intensity effect on g.

mode causing the redistribution of the energy between the diffraction orders (energy is transferred from the zero transmitted order to the zero reflected order). The specified conversion is maximum when the excitation conditions of the TE-mode in the layer are satisfied.

Consider the effect of gyration g on the value of the intensity effect. Changing the sign of g does not change the value of the transmittance ($T(+g) = T(-g)$). Figure 4.3.15 shows the calculated value of the intensity effect depending on g. From Fig. 4.3.15 it can be seen that the effect at small g is quadratic with respect to gyration. Similar effect quadratic with respect to gyration for homogeneous magnetic films have been observed experimentally in [68, 69]. They were called the orientation effects due to their dependence on the relative orientation of magnetization of the film and the polarization. The orientation effects, described in [69], were quite weak and observed in reflected light. The relative change in reflectivity was one percent. This effect is stronger by several orders [53–57].

For the above structure it was assumed that it is in a medium with a refractive index of 1. Such a structure is complicated for practical implementation. Processing operations of depositing layers and the formation of a diffraction grating are performed on a substrate. In this regard, a structure located on the substrate of SiO_2 ($\varepsilon = 2.1$) was studied. The parameters used for the dielectric permittivity tensor of the magnetic layer (4.1.54) were: $\varepsilon = 5.06 + 0.0004i$, $g = (1.53 - 0.003i)\cdot 10^{-2}$ at $\lambda = 1200$ nm. These parameters correspond to the material of bismuth-substituted dysprosium iron garnet.

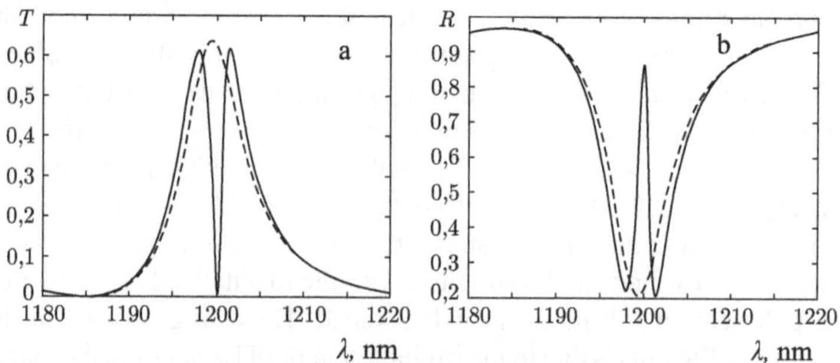

Fig. 4.3.16. Transmission (left) and reflection (right) in the case of the magnetized layer (solid line) and in the absence of magnetization (dashed line).

The transmission and reflectance spectra of the structure-on-the-substrate are shown in Fig. 4.3.16. Calculation of the spectra was carried out at normal incidence of waves with TM-polarization for the following parameters: period $d = 552$ nm, the width of the hole $r = 55$ nm, the thickness of the grating $h_{gr} = 362$ nm, the magnetic layer thickness $h = 1935$ nm.

Figure 4.3.16 shows a large intensity effect near the wavelength of 1200 nm. The reduction of the transmittance at this wavelength is about 60% for the structure with the magnetized layer. The coefficient of reflection from the magnetized layer structure respectively increased by the same amount. The intensity effect of this structure is much stronger than in the previous case. This is due to the fact that the material in question at a wavelength of 1200 nm has practically zero absorption. In this case, the mode propagating in the magnetic layer is decays much more slowly.

To confirm the connection of the intensity effect with excitation of the TE- modes, the dependence of the strength of the effect on the period and height of the layer was calculated. The results are presented in Fig. 4.3.17a, b for the transmission and reflection, respectively. On the distribution in Fig. 4.3.17 we superimposed curves corresponding to the conditions of existence of the TE-modes with the propagation directions of the 1st and 2nd orders of diffraction in a magnetic layer.

As before, the maximum value of the intensity effect is reached in the vicinity of the Rayleigh–Wood anomalies for the 1st and 2nd orders of diffraction indicated by the dot-and-dash lines. At the dimensions of the period of 829 nm and 1067 nm the magneto-optical effect is virtually absent. This is due to a violation of the conditions of total internal reflection at the lower boundary of the dielectric layer for the TE-mode with the direction of the 1st order of diffraction.

We summarize the results of this section. At certain wavelengths, the diffraction structure in Fig. 4.3.10 has the resonant magneto-optical effect expressed in the change of the transmission and reflection coefficients with the change of the value of the gyration of the magnetic layer. As the orientational magneto-optical effects for homogeneous magnetic films, the effect is quadratic in the magnetization, but several orders of magnitude greater than their magnitude. In the case of incidence of TM-polarized light, the spectral position and magnitude of this effect are

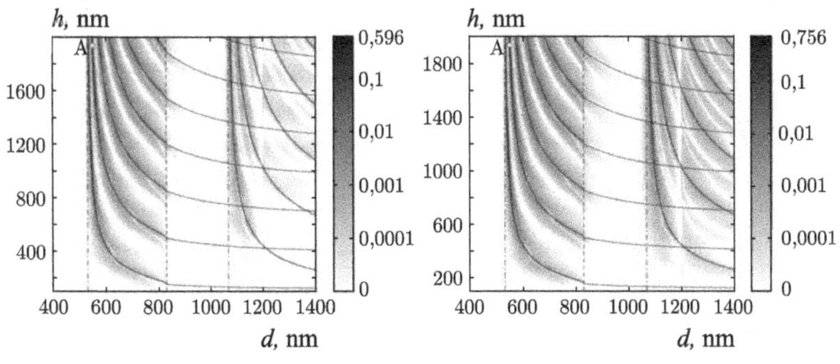

Fig. 4.3.17. The intensity effect in transmission (a) and reflection (b) as a function of layer height and period of the grating with superimposed curves of the existence of modes (continuous lines), the Rayleigh–Wood anomalies (dot-and-dash line) and the boundary of the total internal reflection (dashed line).

determined by the conditions of excitation of TE-modes in the dielectric layer. This effect is related to the conversion of the incident wave energy to the waveguide TE-mode. This causes a dip in transmission and peak in reflection. In the absence of magnetization of the incident wave with TM-polarization can not be converted into the TE-polarized light and the corresponding TE mode is excited. The properties of the observed magneto-optical effect are investigated. It is shown that it is maximal near the Rayleigh–Wood anomalies corresponding to the 'sliding' angle distribution of waveguide modes.

It should be mentioned that in a general case in the excitation of the TE-mode in the magnetised layer, the variation of the transmission and reflection coefficients may differ from that shown in Fig. 4.3.11 and for wind 3.16 where reflection is intensified with a simultaneous attenuation of transmission [57]. In particular, there may be the peaks in reflection accompanied by the minimum of transmission, or there may by simultaneous appearance of minima both in reflection and transmission. In general, the complicated nature of the energy redistribution between transmission to reflection can take place [57].

The considered intensity effect can be applied in practice and can be used in new devices of integrated optics to modulate the light intensity by changing the external magnetic field.

4.3.3. The magneto-optical effect in the equatorial geometry

4.3.3.1. The geometry of the structure and type of magneto-optical effect

Consider a binary grating of gold (Fig. 4.3.18), located on a substrate of a magnetic material. The magnetization vector is directed along the *Oy* axis along the grating grooves (equatorial geometry). The dielectric permittivity tensor of the substrate material has the form (4.1.56). The appropriate form of the matrix of the system of differential equations, representing the field in the layer, is given in (4.1.57).

Figure 4.3.19 shows the calculated dependence of the reflection coefficient on the wavelength for three values of gyration in (4.1.56): $-g$, 0, $+g$. The case in which

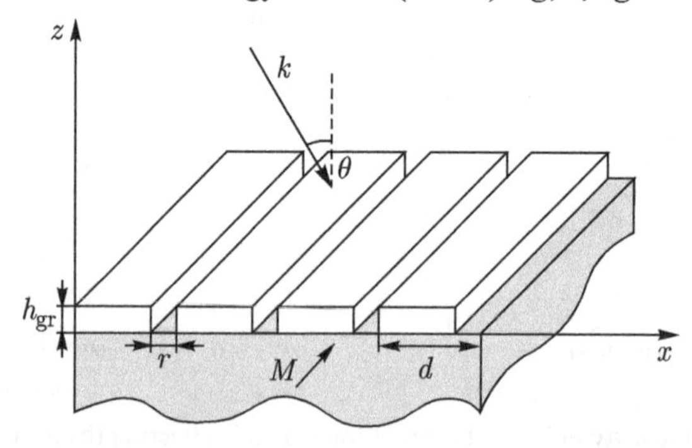

Fig. 4.3.18. Geometry of the diffraction grating on a substrate magnetized in equatorial geometry.

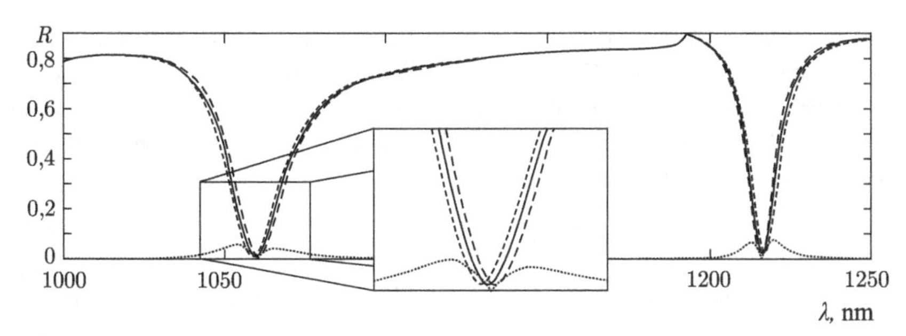

Fig. 4.3.19. The dependence of the intensity of the reflected zeroth order on the wavelength in the absence of magnetization ($g = 0$) and opposite directions of magnetization ('$+g$' – long dashed line, '$-g$' – short dashed line). The magnitude of the intensity effect (4.3.237) – dotted line.

$g = 0$ coresponds to the non-magnetized substrate, the cases with $-g$ and $+g$ correspond to the two opposite directions of the magnetization vector. The calculation was performed for oblique incidence of waves with TM-polarization at an angle $\theta = 12^\circ$ with the following structure parameters: period $d = 485$ nm, the slit width $r = 0.05d = 24.25$ nm, the thickness of the grating $h_{gr} = 163$ nm. For the dielectric permittivity tensor of the magnetic layer (4.1.56), we used the parameters $\varepsilon = 5.06 + 0.0004i$, $g = (1.53–0.003i)\cdot 10^{-2}$ corresponding to the material of bismuth-substituted dysprosium iron garnet.

Figure 4.3.19 shows that, in contrast to the previously considered case, the introduction of magnetization does not lead to a peak in the reflection spectrum but results in a shift of the spectra corresponding to the cases $-g$ and $+g$. In this case, we define the intensity effect in reflection by the formula

$$I_R(g) = |R_0(-g) - R_0(g)|. \tag{4.3.237}$$

The magnitude of the intensity effect is shown in Fig. 4.3.19 by the dotted line. The maximum magnitude of the effect is close to 8%. It should be noted that the strength of the effect (4.3.16) for films of ferromagnetic materials is only about 0.1% [19]. In [69] a similar intensity effect in a multilayer system of magnetic films was studied. It was shown in [69] that the intensity effect increases in the generation of SPP in the system, but the effect is also less than 1% [69]. Thus, the magnitude of the intensity effect in Fig. 4.3.19 is significant when compared with similar effects for homogeneous magnetized films.

4.3.3.2. Explanation of the magneto-optical effect

According to Fig. 4.3.19, the intensity effect (4.3.237) is observed in the vicinity of the wavelengths $\lambda = 1060$ nm and $\lambda = 1220$ nm. The shift in Fig.4.3.19 is not even on g ($R(-g) \neq R(+g)$). Note that in the different minima of the spectrum shift occurs in different directions. For example, in Fig. 4.3.19 at a wavelength $\lambda = 1060$ nm the shift of the reflection minimum at magnetization '$+g$' takes place in the direction of longer wavelengths and a wavelength $\lambda = 1220$ nm – to shorter wavelengths.

The considered effects are associated with the excitation of an SPP at the lower boundary of the grating [56, 58, 59]. Indeed, the wavelengths $\lambda = 1060$ nm and $\lambda = 1220$ nm are in good agreement with the condition of SPP excitation on the lower boundary of the grating by the orders ±1. The condition for excitation of the SPP by the orders $\pm m$ has the form

$$k_{x,\pm m} = \pm \mathrm{Re}\left(k_{\mathrm{SPP}}\right), \tag{4.3.238}$$

where $\pm m$ is the number of orders. From (3.4.17) we obtain the wavelength as

$$\lambda_m = \frac{d}{m}\left(\sin\theta \pm \frac{\mathrm{Re}\, k_{\mathrm{SPP}}}{k_0}\right), \tag{4.3.239}$$

From equation (4.3.239) at $m = \pm 1$ we get $\lambda_{-1} = 1056$ nm and $\lambda_{+1} = 1226$ nm, respectively. Thus, the minima of reflection in Fig. 4.3.19 at $g = 0$ are in good agreement with the condition of SPP excitation. The shift of the spectra for the values of gyration $\pm g$ are associated with the dependence of the conditions of SPP excitation on the value of g. In the next section (see (4.3.258)) we derived an equation for the propagation constants of the SPP at the interface between the metal and the magnetic medium with the dielectric permittivity tensor (4.1.56) as:

$$k_{\mathrm{SPP}}(g) = \pm k_{\mathrm{SPP}}(0) - ik_0 g \frac{\varepsilon_1^2}{(\varepsilon+\varepsilon_1)^{3/2}(\varepsilon-\varepsilon_1)} + o(g), \tag{4.3.240}$$

where $k_{\mathrm{SPP}}(0)$ is the propagation constant of SPP (4.2.14) for the boundary between two isotropic media. In this case, as ε_1 in (4.3.19) we use the values of the dielectric permittivity of the material of the grating (gold), and as ε, g – the components of the permittivity tensor of the substrate material. According to (4.3.240), the dependence of the propagation constants of the SPP on g is close to linear (for small g). This explains the shift of the minima of reflection on the wavelength in Fig. 4.3.19 [56, 58,59]. In addition, from (4.3.240) it follows that the shift of the minima, corresponding to the orders with the numbers +1 and –1, will occur in different directions. This effect was also observed in Fig. 4.3.19.

Equation (4.3.240) allows us to estimate the displacement of the minima of the spectra in Fig. 4.3.19. Let the wavelengths $\lambda_{\pm g,m}$ correspond to the conditions of SPP excitation (3.4.17) for the values $\pm g$, respectively. From (4.3.18) and (4.3.19) it is easy to get the shift magnitude in the form [55]

$$\Delta\lambda_m = |\lambda_{+g,m} - \lambda_{-g,m}| = \left|\mathrm{Re}\left(k_{\mathrm{SPP}}(+g)\right) - \mathrm{Re}\left(k_{\mathrm{SPP}}(-g)\right)\right| \frac{d}{m} \approx$$
$$\approx 2\left|\mathrm{Im}\frac{g\varepsilon_1^2}{(\varepsilon+\varepsilon_1)^{3/2}(\varepsilon-\varepsilon_1)}\right| \frac{d}{m}, \tag{4.3.241}$$

where m is the order number. From (4.3.241) at $m = \pm 1$ we obtain $\Delta\lambda_{-1} = 2.2$ nm, $\Delta_{\lambda+1} = 1.7$ nm. These values are consistent with the distances between the minima of the spectra in Fig. 4.3.19, corresponding to the cases '$-g$' and '$+g$'. When $\lambda =$

1060 nm and $\lambda = 1220$ nm the distance between the minima of the the shifted spectra in Fig. 4.3.19 is 2 nm and 1.4 nm, respectively.

Let us obtain an approximate formula for the magnitude of the effect (4.3.237). Let $R_0(\lambda)$, $R_{\pm g}(\lambda)$ be the reflection spectra for the non-magnetized substrate and the values of $\pm g$, respectively. Figure 4.3.19 shows that in the vicinity of the wavelength (4.3.239) the spectra have approximately the same shape and differ only in the shift. Thus, according to Fig.4.3.19, we have

$$R_0(\lambda) \approx R_{\pm g}(\lambda \pm \Delta\lambda_m / 2). \tag{4.3.242}$$

In this case, the magnitude of the intensity effect can be approximately represented as

$$I_R(q,\lambda) \approx \left| R_0(\lambda - \Delta\lambda_m / 2) - R_0(\lambda + \Delta\lambda_m / 2) \right| \approx R_0'(\lambda)\Delta\lambda_m. \tag{4.3.243}$$

Substituting (4.3.20) into (4.3.22) and taking account of the smallness of Im(g), we obtain

$$I_R(g,\lambda) \approx 2gR_0'(\lambda) \left| \mathrm{Im} \frac{\varepsilon_1^2}{(\varepsilon+\varepsilon_1)^{3/2}(\varepsilon-\varepsilon_1)} \right| \frac{d}{m}. \tag{4.3.244}$$

Formula (4.3.22) describes the linear dependence of the the intensity effect on magnetization. Figure 4.3.20 shows the calculated dependence of the magnitude of the intensity effect (4.3.237) on g at λ = 1065 nm. The values on the abscissa are normalized by the amount Re(g) = 0.0153 used in calculating the spectra in Fig. 4.3.19. Figure 4.3.20 shows that the considered effect at small g is linear with respect to gyration [56, 58].

We summarize the results. The structure consisting of a gold grating on a dielectric substrate, magnetized parallel to the grating grooves, has an odd intensity effect on magnetization. The intensity effect is due to the dependence of the conditions of SPP excitation of waves at the lower boundary of the grating on the magnetization of the substrate material. An analytical estimate of the magnitude of the observed effect was obtained.

4.3.3.3. The equation of a surface plasma polariton at the boundary of a magnetized medium

This item is auxiliary. It considers the derivation of equation (4.3.240) for the propagation constants of the SPP at the boundary between the metal and the

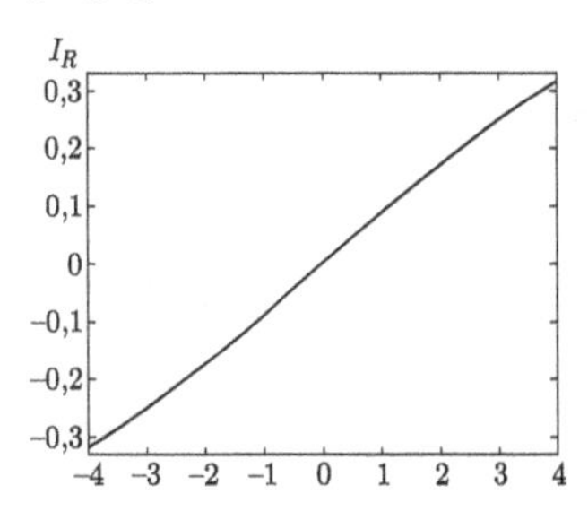

Fig. 4.3.20. Dependence of the intensity effect on g.

magnetized medium. This equation was used in the preceding section to explain the intensity effect.

To derive the equation for the propagation constants of the SPP, it is necessary to obtain the equation of a plane wave propagating in a magnetized medium described by the dielectric permittivity tensor (4.1.56).

Similar to (4.1.5), we seek the equation of a plane wave in the form of

$$\mathbf{\Phi}(x,y,z)=\mathbf{\Phi}\exp\left(ik_0(\alpha x+\beta y+\gamma z)\right). \tag{4.3.245}$$

Substituting (4.3.245) into Maxwell's equations (4.1.26) and taking into account (4.1.56), we obtain:

$$\begin{cases} ik_0\begin{bmatrix}\beta E_z-\gamma E_y\\ \gamma E_x-\alpha E_z\\ \alpha E_y-\beta E_x\end{bmatrix}=ik_0\begin{bmatrix}H_x\\ H_y\\ H_z\end{bmatrix}, \\ ik_0\begin{bmatrix}\beta H_z-\gamma H_y\\ \gamma H_x-\alpha H_z\\ \alpha H_y-\beta H_x\end{bmatrix}=-ik_0\begin{bmatrix}\varepsilon E_x+igE_z\\ \varepsilon E_y\\ \varepsilon E_z-igE_x\end{bmatrix}. \end{cases} \tag{4.3.246}$$

We present (4.3.26) in matrix form:

$$\begin{bmatrix} 0 & -\gamma & \beta & -1 & 0 & 0\\ \gamma & 0 & -\alpha & 0 & -1 & 0\\ -\beta & \alpha & 0 & 0 & 0 & -1\\ \varepsilon & 0 & ig & 0 & -\gamma & \beta\\ 0 & \varepsilon & 0 & \gamma & 0 & -\alpha\\ -ig & 0 & \varepsilon & -\beta & \alpha & 0 \end{bmatrix}\mathbf{\Phi}=\mathbf{A\Phi}=\mathbf{0}. \tag{4.3.247}$$

The non-trivial solutions of (4.3.27) are determined from the condition of zero determinant of the system:

$$\det\mathbf{A}=g^2(\alpha^2+\gamma^2-\varepsilon)+(\alpha^2+\beta^2+\gamma^2-\varepsilon)^2\varepsilon=0. \tag{4.3.248}$$

Consider the particular case $\beta=0$. In this case, the wave propagates in the plane xOz. At $\beta=0$ the matrix of the system (4.3.26) can be written in the block-diagonal form:

$$\begin{bmatrix} \gamma & -\alpha & -1 & 0 & 0 & 0\\ \varepsilon & ig & -\gamma & 0 & 0 & 0\\ -ig & \varepsilon & \alpha & 0 & 0 & 0\\ 0 & 0 & 0 & -\gamma & -1 & 0\\ 0 & 0 & 0 & \alpha & 0 & -1\\ 0 & 0 & 0 & \varepsilon & \gamma & -\alpha \end{bmatrix}\begin{bmatrix}E_x\\ E_z\\ H_y\\ E_y\\ H_x\\ H_z\end{bmatrix}=\mathbf{0}. \tag{4.3.249}$$

According to (4.3.249), in the case of $\beta = 0$ the TE- and TM-waves which propagate independently of each other can be introduced into the magnetized material. In the TE-wave the components E_x, E_z, H_y are equal to zero, while in the TM-wave the components E_y, H_x, H_z are equal to zero. The equation (4.3.248) reduces to two independent equations. For the TE-wave equation (4.3.27) takes the form:

$$\det\begin{bmatrix} -\gamma & -1 & 0 \\ \alpha & 0 & -1 \\ \varepsilon & \gamma & -\alpha \end{bmatrix} = \alpha^2 + \gamma^2 - \varepsilon = 0. \quad (4.3.250)$$

In this case, from (4.3.249) we obtain the following nontrivial solution:

$$\begin{bmatrix} E_y \\ H_x \\ H_z \end{bmatrix} = A_H \begin{bmatrix} 1 \\ -\gamma \\ \alpha \end{bmatrix}. \quad (4.3.251)$$

In the case of the TM-wave equation (4.3.248) takes the form

$$\det\begin{bmatrix} \gamma & -\alpha & -1 \\ \varepsilon & ig & -\gamma \\ -ig & \varepsilon & \alpha \end{bmatrix} = \alpha^2 + \gamma^2 - \varepsilon + \frac{g^2}{\varepsilon} = 0. \quad (4.3.252)$$

In this case the non-trivial solution can be obtained from (4.3.249) as

$$\begin{bmatrix} E_x \\ E_z \\ H_y \end{bmatrix} = A_E \begin{bmatrix} -\alpha\gamma - ig \\ \varepsilon - \gamma^2 \\ -\alpha\varepsilon - ig\gamma \end{bmatrix}. \quad (4.3.253)$$

Consider the derivation of the equation for the SPP for the interface of two media: metal ($z > 0$) and a magnetized material ($z < 0$). As in the case of conventional materials the SPP is represented by two plane waves with TM-polarization – one in each medium. The equation of the TM-wave in the metal has the form (4.1.10) under the conditions $\beta = 0$, $A'_H = 0$. The wave equation in the magnetized material is given in (4.3.252) and (4.3.253).

The SPP propagation constant is determined from the boundary conditions of equality of the tangential components at the interface (at $z = 0$). The values of the constants in the equations of the waves α will be assumed to be equal. This requirement is necessary to satisfy the boundary conditions.

Tangential field components at the wave in the metal at the interface have the form

$$\begin{bmatrix} E_x \\ H_y \end{bmatrix} = A_1 \begin{bmatrix} -\alpha\gamma_1 \\ -\alpha\varepsilon_1 \end{bmatrix}, \quad (4.3.254)$$

where ε_1 is the dielectric constant of the metal, $\gamma_1^2 = \varepsilon_1 - \alpha^2$, A_1 is the normalized amplitude. Similarly, the tangential field components in the wave in the magnetic medium at the interface have the form

$$\begin{bmatrix} E_x \\ H_y \end{bmatrix} = A_2 \begin{bmatrix} -\alpha\gamma - ig \\ -\alpha\varepsilon - ig\gamma \end{bmatrix}, \tag{4.3.255}$$

where $\gamma^2 = \varepsilon - \frac{g^2}{\varepsilon} - \alpha^2$, A_2 is the normalized amplitude. We equate the tangential components and obtain the following system of linear equations

$$\begin{bmatrix} \gamma_1 & \alpha\gamma + ig \\ \varepsilon_1 & \alpha\varepsilon + ig\gamma \end{bmatrix} \begin{bmatrix} A_1 \\ A_2 \end{bmatrix} = 0. \tag{4.3.256}$$

To find non-trivial solutions, we equate the determinant of the system to zero and obtain the following equation

$$F\left(k_{\mathrm{SPP}}(g), g\right) = k_0\alpha(\gamma_1\varepsilon - \gamma\varepsilon_1) + ik_0 g(\gamma_1\gamma - \varepsilon_1) = 0, \tag{4.3.257}$$

where $k_{\mathrm{SPP}}(g) = k_0\alpha(g)$ is the propagation constant propagation of SPP. In general, the solution of equation (3.4.36) has a complicated form. The case of small g is of interest for practices. For small g we expand the function $F(k_{\mathrm{SPP}}(g), g)$ into a Taylor series up to linear terms and equate the coefficients to zero. As a result we obtain the following expression for the propagation constants of the SPP

$$\begin{aligned} k_{\mathrm{SPP}}(g) &= \pm k_0\sqrt{\frac{\varepsilon\varepsilon_1}{\varepsilon + \varepsilon_1}} - ik_0 g \frac{\varepsilon_1^2}{(\varepsilon + \varepsilon_1)^{3/2}(\varepsilon - \varepsilon_1)} + o(g) = \\ &= \pm k_{\mathrm{SPP}}(0) - ik_0 g \frac{\varepsilon_1^2}{(\varepsilon + \varepsilon_1)^{3/2}(\varepsilon - \varepsilon_1)} + o(g). \end{aligned} \tag{4.3.258}$$

The first term in (4.3.258) corresponds to the propagation constant of SPP at the interface between isotropic media, and the second term describes the correction linear with respect to g.

4.4. Metrology of periodic micro- and nanostructures by the reflectometry method

Optical reflectometry is a rapid, non-destructive method of contactless measurement of the parameters of micro- and nanostructures. The method consists of determining the parameters of the structure by measuring the characteristics of the reflected field for various parameters of incident radiation (wavelength, polarization, angle of incidence). The accuracy of determining the geometric parameters of structures may exceed 1 nm [72–78]. In this section we consider the inverse problem of estimating the parameters of a diffraction grating. The solution of the inverse problem consists of determining the geometrical parameters of the grating profile which ensure agreement between the calculated and measured values of the intensity of the reflected zeroth diffraction order. To solve the direct problem that consists in calculating the reflected field with the known parameters of the diffraction structure, we use the Fourier modal method. In this section we describe some methods for solving the given inverse problem. The effectiveness of method is illustrated by the example of determining the parameters of a trapezoidal profile.

4.4.1. Formulation of the problem

Let $\mathbf{g}=(g_1,...,g_N)$ be the vector of the estimated parameters of the two-dimensional profile of a diffraction grating. The task is to determine **g** by measuring the parameters of the reflected zeroth diffraction order at different parameters of the incident wave $\mathbf{v}_i=(\lambda_i,\theta_i,\varphi_i)$, $i=1,...,M$, where λ_i is the wavelength, θ_i is the angle of incidence, ϕ_i is the angle that determines the polarization of the incident wave. ϕ_i is the angle between the electric field vector and the plane of incidence. An elliptically polarized beam forms in the zeroth reflected order. As the measured parameters we use the intensity of the reflected order I_R or the parameters of the polarization ellipse [77, 78]. The polarization ellipse is described by real parameters tg(Ψ), cos (Δ) determined from the equation

$$\tan(\Psi)\exp(i\Delta)=\frac{R_E}{R_H}, \tag{4.4.259}$$

where R_E, R_H are the complex amplitudes of the waves of the *E*- and *H*-type, forming an elliptically polarized beam [79, 80]. The polarization parameters are also represented by the values

$$\alpha=\sqrt{\frac{\tan(\Psi)-1}{\tan(\Psi)+1}},\ \beta=\cos(\Delta)\sqrt{1-\alpha^2}. \tag{4.4.260}$$

4.4.2. Methods for estimating the geometric parameters of the profile of the grating

For simplicity, we assume that the measured parameter is the intensity of the reflected zero order I^R. Estimation of parameters of the profile is based on minimizing the error function

$$\varepsilon(\mathbf{g})=\left\|\mathbf{S}(\mathbf{g})-\mathbf{S}_0\right\|\to\min, \tag{4.4.261}$$

representing the difference between the calculated $\mathbf{S}(\mathbf{g})=\left(I^R(\mathbf{g};\mathbf{v}_1),...,I^R(\mathbf{g};\mathbf{v}_M)\right)$ and measured $\mathbf{S}_0=\left(I^R(\mathbf{v}_1),...,I^R(\mathbf{v}_M)\right)$ values of intensity. Vectors **S**(**g**), $\mathbf{S}_0$ are called signatures. The calculated values of the intensity $I^R(\mathbf{g};\mathbf{v}_i)$ in **S**(**g**) are functions of the determined parameters **g**, and values $\mathbf{v}_i$ are treated as parameters. The calculation of **S**(**g**) in (4.4.261) (solution of the direct problem) is based on a rigorous solution of diffraction using the above Fourier modal method.

The optimization problem (4.4.261) was solved by three different methods. Method 1 is based on direct minimization of the error function (4.4.261) using the principal axis method [79]. This method does not use the knowledge of the gradient of the minimized error function. Its advantages are the quadratic convergence near the minimum and the ability to find the minimum of the ravine functions. The main drawback of the method of direct optimization 1 are significant computing costs associated with repeated solution of the direct diffraction problem in the calculation of **S**(**g**). This limits the scope of the method for parameter estimation problems in real time.

Method 2 is based on a neural network [72, 74, 75, 78]. For the neural network the main computational burden falls on the stage of network training. The calculation of the yield of the trained network is fast, which makes it a convenient tool for estimates of the parameters in real time. We used a network of the 'multilayer perceptron' type. The number of input neurons equals the number of points at $\mathbf{S}_0$, and the number of output neurons – the number of the parameters.

Method 3 is based on constructing polynomial approximations for the intensities of the reflected zero order [74]

$$\tilde{I}^R(\mathbf{g};\mathbf{v}_m)=\sum_{i_1 i_2 \dots i_N,\ \sum i_j \le K_m} a_{i_1 i_2 \dots i_N}(\mathbf{v}_m)\, g_1^{i_1} g_2^{i_2}\cdot\ldots\cdot g_N^{i_N}, \qquad (4.4.262)$$

where K_m is the degree of the polynomial. Typically, the ranges of parameters $a_i < g_i < b_i$, $i = 1,\dots,N$ are known in advance (defined by technology) and equal a few tens of nanometers. The use of polynomial approximations can be considered as a variant of the perturbation theory. To calculate the coefficients of the polynomial $a_{i_1 i_2 \dots i_N}(\mathbf{v}_m)$ we choose a training set $\mathbf{T}=\{\mathbf{g}_i,\ i=1,\dots,L\}$ of 'characteristic' grating parameters. Using the training set of parameters, we calculate the intensity of the reflected zero order. Further, the coefficients in (4.4.262) are calculated by the least squares method by minimizing the square of the difference between $\tilde{I}^R(\mathbf{g};\mathbf{v}_m)$ and $I^R(\mathbf{g};\mathbf{v}_m)$ at $\mathbf{g}\in\mathbf{T}$. The choice of the training set depends on the considered structure. Representation (4.4.262) allows us to efficiently solve the optimization problem (4.4.261). As for the neural network, the main computational load is at the preliminary stage of constructing polynomials (4.4.262). The calculation of **S**(**g**) in (4.4.261) instead of the rigorous solution of diffraction problems is reduced to the computation of polynomials. Representation (4.4.262) can also be used to analytically calculate the derivatives of the error function (4.4.261).

Method 4 is based on constructing a table of signatures **S**(**g**) and the subsequent spline interpolation between the nodes of the table [76, 78]. The table of signatures is built as follows. The range of allowed values of the parameters $a_i < g_i < b_i$, $i=1,\dots,N$ is covered by a uniform grid with a certain step Δ_i = 1,..., N determined for each parameter. In each node of the grid, defined by a set of indices $\mathbf{k}=(i_1,\dots,i_N)$, we calculate the corresponding signature $\mathbf{S}(\mathbf{g}_\mathbf{k})$. Thus, a 'table' of the structure parameters and their signatures is formed. Next, the table is used to construct the N-dimensional spline $\tilde{\mathbf{S}}(\mathbf{g})$. Spline $\tilde{\mathbf{S}}(\mathbf{g})$ is used to solve the optimization problem (4.4.261). Thus, instead of calculating the signatures **S**(**g**) requiring the solution of the diffraction problem, the values of the interpolation spline $\mathbf{S}(\mathbf{g})$ are used.

4.4.3. Determining the parameters of a trapezoidal profile

We present the results of methods 1–4 for a silicon grating (period $d = 140$ nm) with a symmetric trapezoidal profile. Such a profile is described by three parameters, $\mathbf{g} = (h, p, \alpha)$, h – height, p – the length of the lower base, α – the angle of side walls (Fig. 4.4.1). We assume that $h \in [310, 340]$ nm, $p \in [86, 106]$ nm, $\alpha \in [1, 5]^\circ$.

To calculate the signatures, the following parameters of the illuminating beam were used: polarization of the incident wave – constant, TM (E-wave), the angle of incidence α = constant, 71°, wavelength – variable, ranging from 300 to 900 nm. The number of wavelengths in the calculation of signatures in the methods 1, 2, 4, was 20, in the method of 3 it was 50. In the calculations it was found that the specified number of wavelengths allows us to solve the problem of estimating the parameters with high accuracy (greater than 0.1 nm).

The performance of the methods was evaluated by the following numerical experiment. In the above range of parameters (h, p, α) we randomly generated three parameters $\mathbf{g}$ and calculated the vector of reflection coefficients $\mathbf{S}_0(\mathbf{g})$. The signature $\mathbf{S}_0(\mathbf{g})$ is considered as the measured signal. Using $\mathbf{S}_0(\mathbf{g})$ and the methods 1–4 we obtain estimates of the parameters $\tilde{\mathbf{g}}$ that are compared to the original settings. By comparing the generated parameters $\mathbf{g}$ and of the estimates $\tilde{\mathbf{g}}$ we determine the accuracy of estimated parameters $(\Delta h, \Delta p, \Delta\alpha) = \tilde{\mathbf{g}} - \mathbf{g}$.

Figures 4.4.2a–c shows the results of method 1 at 200 random trials. Along the x-axis are the numbers of tests in the numerical experiment, and along the vertical axis – the absolute deviation of the estimated parameters from the generated values.

The results (Fig. 4.4.2) show the high accuracy of the method 1, the error in determining the parameters is negligible. However, at such a high accuracy, this method has a drawback – a relatively long time of searching parameters (5–10 seconds on a standard personal computer). This limits the scope of the method for parameter estimation problems in real time.

To estimate the parameters by method 2, a network of the 'multilayer perceptron' type was used [70, 72, 74, 75, 78]. A schematic view of the network is shown in Fig. 4.4.3. The number of neurons in the hidden layer is chosen experimentally and is equal to 20. The activation function of neurons was a logistic sigmoid. The network training was carried out by the Levenberg–Marquardt method [80, 81]. The training set was formed by 8000 signatures generated for the parameters $\mathbf{g} = (h, p, \alpha)$ on a uniform $20 \times 20 \times 20$ grid in the above specified ranges. The training

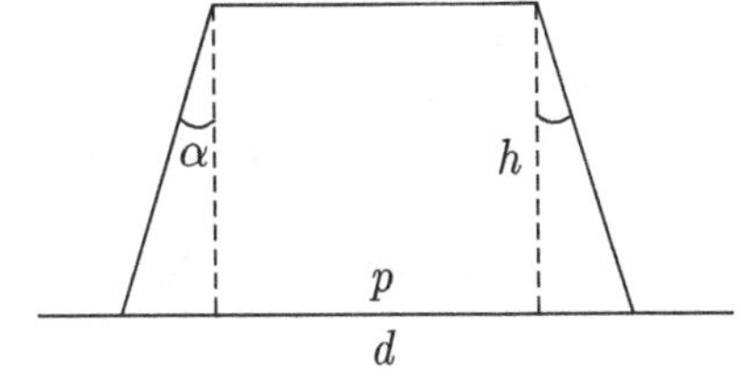

Fig. 4.4.1. The grating parameters of a trapezoidal profile.

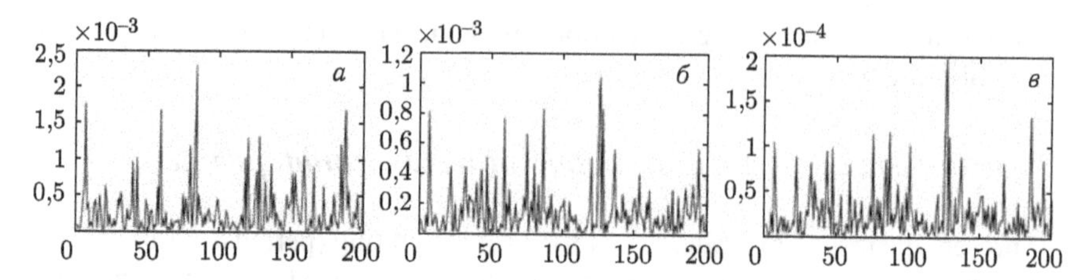

Fig. 4.4.2. Absolute deviations of the generated parameter from the found estimates at 200 trials for method 1: (a) $|\Delta h|$ in nanometers, (b) $|\Delta p|$ in nanometers; (c) $|\Delta\alpha|$ (deg).

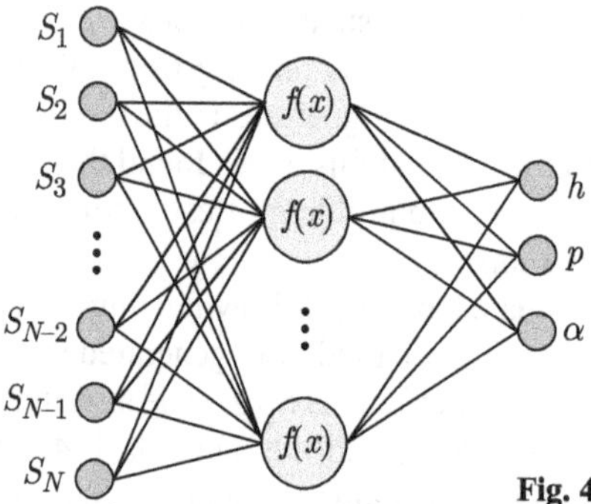

Fig. 4.4.3. The configuration of the used neural network.

time of the network on a standard personal computer is 3–4 hours. Figures 4.4.4a–c demonstrate the results of the trained network on 200 randomly generated triples of parameters (h, p, α).

The results in Fig. 4.4.4 show that the neural network also determines the parameters with high accuracy exceeding 0.1 nm. Such an accuracy exceeds practical requirements. The time required to solve the problem is a few tenths of a second, which is less than that of method 1.

To estimate the parameters by method 3, approximations were constructed for the intensities of the zero reflected order in the form of polynomials of second degree

$$\tilde{R}_m\left(\mathbf{g}=(h,p,\alpha);\lambda_m\right)=\sum_{i_1 i_2 i_3,\sum i_j\le 6} a_{i_1 i_2 i_3}\left(\lambda_m\right)h^{i_1}p^{i_2}\alpha^{i_3}. \qquad (4.4.263)$$

The training set $\mathbf{T}$ for the construction of polynomial approximations (4.4.5) was formed from 2000 signatures corresponding to the set of triples of parameters (h, p, α) randomly generated in the specified above range. The time for constructing approximations (4.4.263) on a standard personal computer is 3–4 hours.

Figures 4.4.5a–c show the results of method 3 at 200 randomly generated triples of parameters. Parameter estimation is accomplished by minimizing the error (4.4.261) using the Levenberg–Marquardt algorithm [80, 81]. In addition, for the calculations of $\mathbf{S}(\mathbf{g})$ in (4.4.261), we used polynomial approximation (4.4.263).

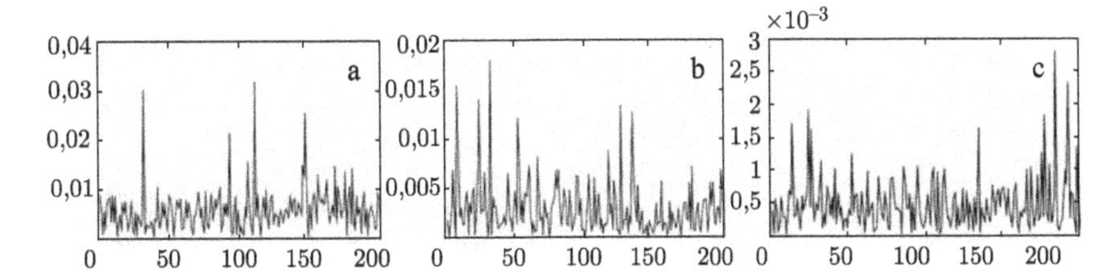

Fig. 4.4.4. Absolute deviations of the generated parameters from the found estimates at 200 trials for method 2: (a) $|\Delta h|$ in nanometers, (b) $|\Delta p|$ in nanometers, (c) $|\Delta\alpha|$ (deg).

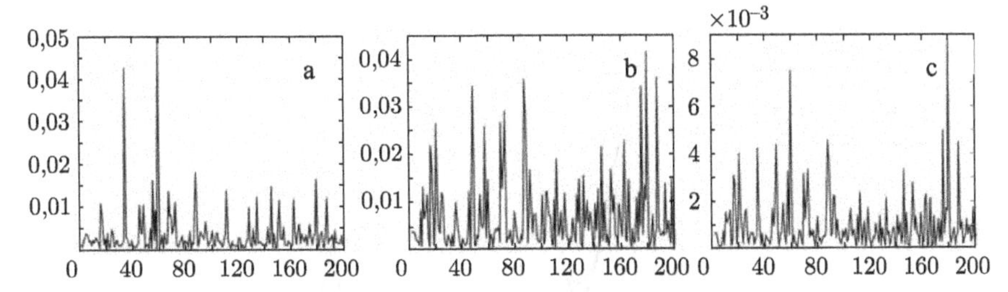

Fig. 4.4.5. Absolute deviations of the generated parameters from the found estimates at 200 trials for method 2: (a) $|\Delta h|$ in nanometers, (b) $|\Delta p|$ in nanometers, (c) $|\Delta\alpha|$ (deg), in degrees.

Figures 4.4.5a–c show that the method 3 also determines the parameters of a trapezoidal profile with a high degree of accuracy greater than 0.1 nm. The time required for solving the problem, as for method 2, is a few tenths of a second.

To estimate the parameters by method 4, a table of signatures was constructed with discretization steps of 1 nm with respect to parameters h and p and 0.4° with respect to the parameter α. The interpolation spline was constructed using the table. The duration of calculation of the table and construction of the spline approximating the tabular data on a standard personal computer is about 2–3 hours.

Figures 4.4.6a–c show the results of the method 4 at 200 randomly generated triples of parameters (h, p, α). Parameter estimation is accomplished by minimizing the error (4.4.261) using the Levenberg–Marquardt algorithm [82, 83]. In addition, **S(g)** in (4.4.261) was calculated by the spline approximation of the table of signatures.

Figures 4.4.6 show that the method 4 also defines the parameters of a trapezoidal profile with a high degree of accuracy greater than 0.1 nm. The time for solving the problem, as for the methods 2 and 3, is tenths of a second.

The methods were used to estimate the parameters of the profile of the diffraction structures based on actual measurements. In Fig. 4.4.7 the dotted lines show the measured spectra $\alpha(\lambda)$, $\beta(\lambda)$ (see (4.4.2)) of a silicon grating with period $d = 140$ nm, made by ellipsometer UV1280SE (TLA TENCOR). It was assumed that the grating profile is trapezoidal (Fig. 4.4.1). The profile parameters were estimated by all four methods considered. The methods gave virtually identical estimates of the parameters: $h = 321.6$ nm, $p = 98.16$ nm, $\alpha = 2.9°$. The calculated spectra of the trapezoidal structure with estimated parameters are shown in Fig. 4.4.7 by solid

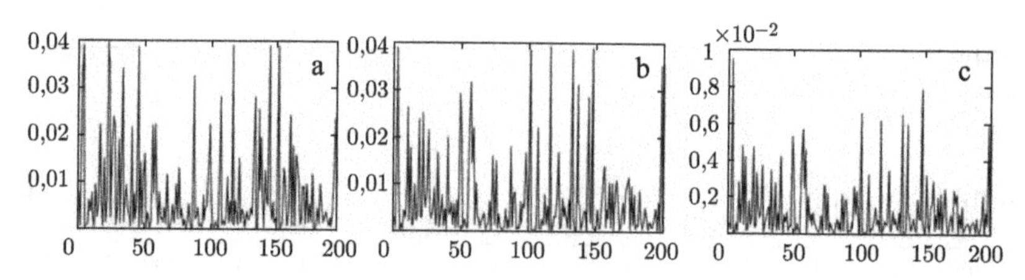

Fig. 4.4.6. Absolute deviations of the generated parameters from the found estimates at 200 trials for method 4: (a) $|\Delta h|$ in nanometers, (b) $|\Delta p|$ in nanometers, (c) $|\Delta\alpha|$ (deg).

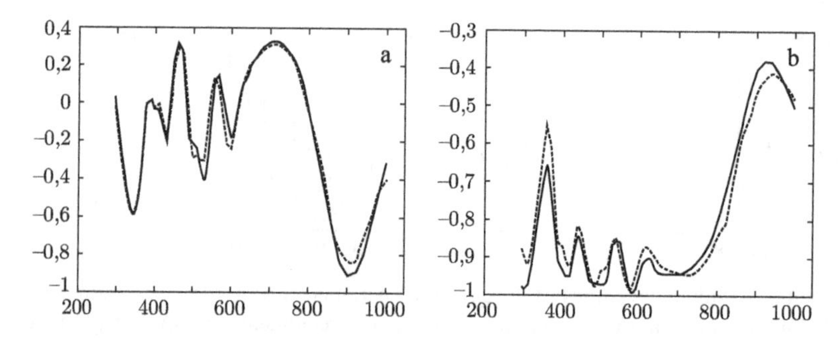

Fig. 4.4.7. Dotted line – measured spectra $\alpha(\lambda)$ (a), $\beta(\lambda)$ (b). The solid lines – calculated spectra of the trapezoidal grating with the estimated parameters.

lines. Comparison of the calculated and measured spectra show a good agreement of the spectra. To assess the degree of coincidence of the spectra in the ellipsometric measurements we used sample correlation coefficient (goodness of fit – GOF). For the case under consideration $GOF_\beta = 0.962$. The difference in the spectra in Fig. 4.4.7 is caused by the inaccuracy of the model. The measured structure has a profile that is different from the ideal trapezoidal profile.

We summarize the main results of the section. Numerical experiments show the high accuracy of the proposed methods of parameter estimation of the diffraction patterns. The mathematical accuracy of the parameters of the trapezoidal profile in the computational experiment is better than 0.1 nm. The results of experimental studies demonstrate the applicability of the method to real ellipsometric measurements.

Conclusion

The Fourier modal method for solving the problems of diffraction of a plane wave on periodic two-dimensional and three-dimensional diffraction structures and the diffraction gratings within the framework of the electromagnetic theory was investigated. In the two-dimensional case, the method is applied for the general case of structures made of anisotropic materials described by the tensors of dielectric and magnetic permittivity. The presented calculation examples show the high efficiency of the method in the problems of calculating the diffraction gratings – polarizers, beam dividers, and subwavelength antireflection coatings.

A method of the formation of one-dimensional and two-dimensional interference patterns of surface plasma polaritons (SPP) using diffraction structures, containing the dielectric diffraction grating (two- or three-dimensional) and a metallic layer is developed. The calculations and simulation of these structures are based on the presented Fourier modal method. The simulation results show that the investigated structures can be used to form interference patterns with a high contrast (greater than 70%) and high intensity. The intensity of the interference maxima is an order of magnitude greater than the intensity of the incident wave. The period of the interference patterns is 2.5–3.5 times smaller than the wavelength. The form of the calculated patterns coincides with the theoretical estimates. The calculation results show that it is possible to produce interference patterns with different structures, period and form of the interference maxima with the variation of the polarisation parameters and the wavelength of incident light. The investigated structures can be used for the production of nanostructures on the basis of recording the interference patterns of the SPP by contact photolithography.

The integral representation of the electromagnetic field at the boundary of the media as the angular spectrum of the surface electromagnetic waves and the corresponding Kirchoff integral was derived from the Maxwell's equations. On the basis of the exact solution of the diffraction problem of surface plasma polaritons on a rectangular dielectric step, situated directly on the propagation surface, it was shown possible to carry out the phase modulation of surface plasma polaritons both by changing the length of the step at a fixed height and by changing the height of the step at a fixed length. A combined method of modulation of the surface plasma polaritons by the simultaneous variation of the length and height of the step has been proposed. The modulation methods were used for calculating the diffraction lenses for the focusing of the surface plasma polaritons. When using the modulation method based on the variation of only the length or only the height of the step, the diffraction efficiency of the lenses is approximately 55%. When using the combined modulation method, the diffraction efficiency of the lenses is greater than 80%. The investigated types of modulation can be used for calculating the diffractive optical elements of a general type for carrying out the specific transformations of the surface plasma polaritons.

The magneto-optical properties of the metal–dielectric heterostructures, consisting of the metallic diffraction grating and the dielectric magnetized layer have been investigated. These properties were studied for three main cases in which the layer is magnetised in the polar, meridional and equatorial geometries. The results of the calculations using the Fourier modal method show that in the case of polar geometry, the heterostructure is characterised by resonance magneto-optical effects associated with the resonance rotation of the polarisation plane of the transmitted and reflected light. In meridional and equatorial geometry, the structure has magneto-optical effects associated with the resonance changes of the transmittance and reflection coefficients with the variation of magnetization. The investigated effects are of both fundamental and applied significance because they can be used in new optical devices for the modulation of the intensity and polarisation of light by changing the external magnetic field.

The methods of solving the inverse problem of evaluation of the parameters of the diffraction grating using the results of measurements of the intensity of the zeroth reflected diffraction order have been presented. The efficiency of the method is illustrated by the example of determination of the parameters of the trapezoidal profile. The results of the numerical experiments indicate the high accuracy of the presented methods. The mathematical accuracy of determination of the linear parameters for the four investigated methods is greater than 0.1 nm. This accuracy is greater than the physically justified requirements. The experimental results show that the methods can be used for actual ellipsometric measurements.

References

1. Moharam M. G., Gaylord T.K. Rigorous coupled-wave analysis of planar-grating diffraction, J. Opt. Soc. Am. 1981. V. 71 (7). P. 811–818.
2. Moharam M.G., Grann E.B., Pommet D.A., Gaylord T.K. Formulation for stable and efficient implementation of the rigorous coupled-wave analysis of binary gratings, J. Opt. Soc. Am. A. 1995. V. 12 (5). P. 1068–1076.
3. Moharam M. G., Pommet D.A., Grann E. B., Gaylord T.K. Stable implementation of the rigorous coupled-wave analysis for surface-relief gratings: enhanced transmittance matrix approach, J. Opt. Soc. Am. A. 1995. V. 12 (5). P. 1077–1086 .
4. Peng S., Morris G.M., Efficient implementation of rigorous coupled-wave analysis for surface-relief gratings, J. Opt. Soc. Am. A. 1995. V. 12 (5). P. 1087–1096 .
5 . Lifeng Li, Use of Fourier series in the analysis of discontinuous periodic structures, J. Opt. Soc. Am. A. 1996. V. 13 (9). P. 1870–1876 .
6. Lifeng Li New formulation of the Fourier modal method for crossed surface-relief gratings, J. Opt. Soc. Am. A. 1997. V. 14 (10). P. 2758–2767 .
7. Popov E., Nevie're M. Grating theory: new equations in Fourier space leading to fast converging results for TM-polarization, J. Opt. Soc. Am. A. 2000. V. 17 (10). P. 1773-1784 .
8. Lifeng Li, Fourier modal method for crossed anisotropic gratings with arbitrary permittivity and permeability tensors, J. Opt. A: Pure Appl. 2003. V. 5 (4). P. 345–355 .
9. Popov E., Nevie're M. Maxwell equations in Fourier space: fast-converging formulation for diffraction by arbitrary shaped, periodic, anisotropic media, J. Opt. Soc. Am. A. 2001. V. 18 (11). P. 2886–2894 .
10. Watanabe K., Petit R., Nevie're M., Differential theory of gratings made of anisotropic-materials, J. Opt. Soc. Am. A. 2002. V. 19 (2). P. 325–334 .
11. Zhou C., Li L., Formulation of the Fourier modal method for symmetric crossed gratings in symmetric mountings, J. Opt. A: Pure Appl. Opt. 2004. V. 6. P. 43–50.
12. Diffractive computer optics, Ed. V.A. Soifer. Fizmatlit, Moscow, 2007.
13. Methods for Computer Design of Diffractive Optical Elements, Ed. by V.A. Soifer. New York: Wiley Interscience Publication; John Wiley & Sons, Inc., 2002. P. 159 .266.
14. Silberstein E., Lalanne P., Hugonin J.-P., Cao Q., Use of grating theories in integrated optics, J. Opt. Soc. Am. A. 2001. V. 18 (11). P. 2865–2875.
15. Cao Q., Lalanne P., Hugonin J.-P., Stable and efficient Bloch-mode computational method for one-dimensional grating waveguides, Opt. Soc. Am. A. 2002. V. 19 (2). P. 335–338.
16. Gans M.J., A General Proof of Floquet's Theorem, IEEE Transactions on Microwave Theory and Techniques. 1965. V. 13 (3). P. 384–385.

17. Electromagnetic Theory of Gratings: Topics in current physics. V. 22, Ed. byR. Petit. N.Y.: Springer-Verlag, 1980.
18. Visovsky S., Postava K., Yamaguchi T., Lopusnık R., Magneto-Optic Ellipsometry in Exchange-Coupled Films, Appl. Opt. 2002. V. 41 (19). P. 3950–3960.
a) Zvezdin K., Kotov V.A. Modern Magneto-Optics and Magneto-Optical Materials. Bristol: IOP, 1997.
19. Born M., Wolf E., Principles of optics. 4th Edition. Pergamon Press, 1968.
20. Bezus E.A., Bykov D.A., Doskolovich L. L., Kadomin I. I. Diffraction gratings for generating varying-period interference patterns of surface plasmons, J. Opt. A: Pure Appl. Opt. 2008. V. 10. P. 095204–095208.
21. Doskolovich L.L., Kadomina E.A., Kadomin I. I. Nanoscale photolithography by means of surface plasmon interference, J. Opt. A: Pure Appl. Opt. 2007. V. 9. P. 854–857.
22. Bezus E.A., et al., Komp. Optika, 2008. V. 32, No. 3. P. 234–237.
23. Doskolovich L.L., et al., Radiotekhnika, 2008. N. 3. 75–79.
24. Jiao X., Wang P., Zhang D., Tang L., Xie J., Ming H. Numerical simulation of nanolithography with the subwavelength metallic grating waveguide structure. Optics Express. 2006. V. 14, No. 11. P. 4850–4860.
25. Luoa X., Ishihara T., Surface plasmon resonant interference nanolithography technique, Appl. Phys. Lett. 2004. V. 84 (23). P. 4780–4783.
26. Srituravanich W., Fang N., Sun C., Luo Q., Zhang X. Plasmonic Nanolithography, Nano Lett. 2004. V. 4 (6). P. 85-88.
27. Srituravanich W., Fang N., Sun C., Luo Q., Zhang X., Surface Plasmon Interference Nanolithography, Nano Lett. 2005. V. 5 (5). P. 957-961.
28. Srituravanich W., Durant S., Lee H., Sun C., Zhanga X., Deep subwavelength nanolithography using localized surface plasmon modes on planar silver mask, J. Vac. Sci. Technol. B. 2006. V. 23 (6). P. 2636-2639.
29. Derouard M., Hazart J., Polarization-sensitive printing of surface plasmon interferences, Opt. Express. 2007. V. 15 (7). P. 4238–4246.
30. Barnes W. L. Surface plasmon-polariton length scales: a route to sub-wavelength optics, J. Opt. A: Pure Appl. Opt. 2006. V. 8. P. 87–93.
31. Kurihara K., Suzuki K. Theoretical understanding of an absorption-based surface plasmon resonance sensor based on Kretchman's theory, Anal. Chem. 2002. V. 74. No. 3. P. 696–701.
32. Depine R. A., Gigli M. L. Resonant excitation of surface modes at a single flat uniaxial-metal interface, J. Opt. Soc. Am. A. 1997. V. 14 , No. 2. P. 510–519.
33. Patskovsky S., Kabashin A., Meunier M., Properties and sensing characteristics of surface-plasmon resonance in infrared light, J. Opt. Soc. Am. A. 2003. V. 20. No. 8.
34 Homola J., Yee S., Gauglitz G. Surface plasmon resonance sensors, Review, Sensors and Actuators B. 1999. V. 54. P. 3–15;
a) Bouhelier, Wiederrecht G. P. Excitation of broadband surfaces plasmon polaritons: Plasmonic continuum spectroscopy, Phys. Rev. B. 2005. V. 71. P. 195406.
35. Barker A.S., Jr. Optical Measurements of Surface Plasmons in Gold, Phys. Rev. B. 1973. V. 8, No. 12. P. 5418–5426.
36. Berger C., Beumer T., Kooyman R., Greve J. Surface plasmon resonance multisensing, Anal. Chem. 1998. V. 70. P. 703–706.
37. Vahimaa P., Kettunen V., Kuittinen M., Turunen J., Friberg A. T. Electromagnetic analysis of non-paraxial Bessel beams generated by diffractive axicons, J. Opt. Soc. Am. A. 1997. V. 14 (8). P. 1817–1824.
38. Liu Z., Steele J.M., Srituravanich W., Pikus Y., Sun C., Zhang X. Focusing Surface Plasmons with a Plasmonic Lens, Nano Lett. 2005. V. 5 (9). P. 1726–1729.
39. Korn G., Korn T. Handbook of mathematics (for scientists and engineers), Moscow Nauka, 1974.

40. Bezus E.A., Komp. Optika, 2009. V. 33 (2). C. 122–128.
41. Kim H., Hahn J., Lee B. Focusing properties of surface plasmon polariton floating dielectric lenses, Opt. Express. 2008. V. 16 (5). P. 3049-3057.
42. Bezus E.A., Doskolovich L.L., Kazanskiy N.L., Soifer V.A., Kharitonov S.I., Design of diffractive lenses for focusing surface plasmons, J. Opt. 2010. V. 12. P. 015001–015007.
43. Berenger J.P., A perfectly matched layer for the absorption of electromagnetic waves, J. of Comput. Physics. 1994. V. 114 (2). P. 185–200.
44. Belotelov V. I., Doskolovich L.L., Zvezdin A.K., Extraordinary magneto-optical effects and transmission through metal-dielectric plasmonic systems, Phys. Rev. Lett. 2007. V. 98. P. 077401 (4).
45. Belotelov V. I., Doskolovich L.L., Kotov V.A., Zvezdin A.K., Magnetooptical Properties of Perforated Metallic films, J. of Magnetism and Magnetic Materials. 2007. V. 310, Issue 2, Part 3. P. e843–e845.
46. Belotelov V. I., Doskolovich L.L., Kotov V.A., Bezus E. A., Bykov D.A., Zvezdin A.K., Magnetooptical effects in the metal-dielectric gratings, Optics Communications. 2007. V. 278. P. 104–109.
47. Belotelov V. I., Doskolovich L.L., Kotov V.A., Bezus E.A., Bykov D.A., Zvezdin A.K. Faraday effect enhancement in metal-dielectric plasmonic systems, Proc. of the SPIE. 2007. V. 6581. P. 65810S.
48. Doskolovich L.L., et al., Komp. Optika, 2007. V. 31, N. 1. P. 4–8.
49. Belotelov V.I., Volkova Z.A., Doskolovich L. L., Zvezdin A.K., Kotov V.A., Magnetooptical Effects in Metal-Dielectric Plasmonic Systems, Bulletin of the Russian Academy of Sciences: Physics. 2007. V. 71, No. 11. P. 1530–1532. Allerton Press, Inc., 2007.
50. Belotelov V.I., et al., Komp. Optika, 2007. V. 31, N. 3. P. 4–9.
51. Belotelov V. I., Bykov D.A., Doskolovich L.L., Kalish A.N., Kotov V.A., Zvezdin A.K. Giant Magnetooptical Orientational Effect in Plasmonic Heterostructures, Optics Lett. 2009. V. 34 (4). P. 398–400.
52. Belotelov V.I., et al., Fiz. Tverd. Tela, 2009. V. 51, N. 8. 1562–1567.
53. Bezus E.A., Bykov D.A., Vestnik CGAU. 2008. V. 2 (15). P. 51–58.
54. Bykov D.A., et al., Izv. Samarsk. Nauch. Tsentra RAN, 2009, V. 11, N. 5, 72–77.
55. Bykov D.A., et al., ibid, 2009. V. 11, N. 5. C. 72–77.
56. Belotelov V.I., Bykov D.A., Doskolovich L.L., Kalish A. N., Zvezdin A.K., Extraordinary transmission and giant magneto-optical transverse Kerr effect in plasmonic nanostructured films, J. Opt. Soc. Am. B. 2009. V. 26 (8). P. 1594–1598.
57. Belotelov V.I., et al., Zh. Eksp. Teor. Fiz., 2010. VC. 137 (4).
58. Palik E.D., Handbook of optical constants of solids, Ed. by Palik, Edward D. Academic-Press Handbook Series. New York, Academic Press, 1985.
59. Ebbesen T.W., Lezec H. J., Ghaemi H. F., Thio T., Wolff P.A. Extraordinary optical transmission through sub-wavelength hole arrays, Nature. 1998. V. 391. P. 667–669.
60. Ghaemi H.F., Thio T., Grupp D.E., Ebbesen T.W., Lezec H.J., Surface plasmon enhance optical transmission through subwavelength holes, Phys. Rev. B. 1998. V. 58. P. 6779–6782.
61. Yong-Hong Ye, Jia-Yu Zhang, Enhanced light transmission through cascaded metal films perforated with periodic hole arrays, Opt. Lett. 2005. V. 30 (12). P. 1521–1523.
62. Barnes W.L., Murray W.A., Dintinger J., Devaux E., Ebbesen T.W., Surface plasmon polaritons and their role in the enhanced transmission of light through periodic arrays of subwavelength holes in a metal film, Phys. Rev. Lett. 2004. V. 92.
63. Porto J.A., Garcia-Vidal F.J., Pendry J.B., Transmission resonances on metallic gratings with very narrow slits, Phys. Rev. Lett. 1999. V. 83. P. 2845–2848.
64. Hlawiczka P.A., A gyrotropic waveguide with dielectric boundaries: The longitudinally magnetised case, J. of Physics D: Appl. Phys. 1978. V. 11 (8). P. 1157–1166.
65. Ginés Lifante. Integrated Photonics: Fundamentals. John Wiley and Sons Ltd., 2003.
66. Krinchik G.S., Ganshina E.A., Quadratic magneto-optical reflection effects in ferromagnetic, Zh. Eksp. Teor. Fiz., 1973. V. 35 (5 (11)). P. 1970–1978.

67. Ganshina E.A., Magneto-optical spectroscopy of ferro- and ferrimagnetics, M.V. Lomonosov University, Moscow, 1994.
68. Matyáš M., Čapek V., Explicit dispersion relation for a gyrotropic waveguide, J. Opt. Soc. Am. A. 1988. V. 5 (11). P. 1901–1904.
69. Bonod N., Reinisch R., Popov E., Nevie're M., Optimization of surface-plasmon enhanced magneto-optical effects, J. Opt. Soc. Am. B. 2004. V. 21, No. 4. P. 791–797.
70. Robert S., Mure-Ravaud A., Lacour D., Characterization of optical diffraction gratings by use of a neural method, JOSA A. 2002. V. 19 (1). P. 24–32.
71. Logofatu P., Apostol D., Damian V., Nascov V., et al., Scatterometry, an optical metrology technique for lithography, Semiconductor Conf. 2004, CAS 2004 Proc. V. 2. P. 517–520.
72. Wei S., Li L., Measurement of photoresist grating profiles based on multiwavelength scatterometry and artificial neural network, Applied Optics. 2008. V. 47, No. 13. P. 2524–2532.
73. Doskolovich L.L., et al., Komp. Optika, 2008. V. 32, N. 1. P. 29–32.
74. Babin S.V., et al., Komp. Optika, 2009. V. 33 (2). P. 156–161.
75. Babin B.S., Doskolovich L.L., Ishibashi Y., Ivanchikov A., Kazanskiy N., Kadomin I., Mikami A., Yamazaki Y. SCATT: software to model scatterometry using the rigorous electromagnetic theory, Proc. SPIE. 2009. V. 7272. Advanced Lithography. 72723X; DOI: 10.1117/12.816904.
76. Babin B.S., Doskolovich L.L., Kadomina E., Kadomin I., Volotovskiy S., Restoring pattern CD and cross-section using scatterometry: various approaches, Proc. SPIE. 2009. V. 7272. Advanced Lithography. 727,243 ; DOI: 10.1117/12.816436.
77. Johs B., French R.H., Kalk F.D., McGahan W.A., Woollam J.A., Optical analysis of complex multilayer structures using multiple data types, Proc. SPIE. V. 2253. Optical Interference coating, Ed. by F.Abeles. 1994. P. 1096–1106.
78. Tompkins H.G., McGahan W.A., Spectroscopic Ellipsometry and Reflectometry. John Wiley & Sons, Inc., 1999.
79. Brent R.P., Algorithms for Minimization Without Derivatives. Englewood Cliffs, NJ: Prentice-Hall, 1973.
80. Marquardt D.W. An algorithm for least-squares estimation of nonlinear parameters, J. Soc. Indust. Appl. Math. 1963. V. 11. P. 431–441.
81. Press W.H., Flannery B.P., Teukolsky S.A., Vetterling W.T., Numerical Recipes in C: The Art of Scientific Computing, Cambridge University Press, 1993.

Chapter 5

Photonic crystals and light focusing

5.1 One- and two-dimensional photonic crystals

Photonic crystals (PCs) [1,2] are the structures with nanoresolution and periodic modulation of the refractive index, which have a photonic band gap. Band gaps define the frequency of electromagnetic radiation, which can not exist in this structure. For optical PCs at a wavelength of 1.3 μm the size of the band gap equals tens of nanometers. Accordingly, total reflection is observed in the fall of electromagnetic radiation on the PC, the frequency of which lies in the band gap. This property determines the prospects of using photonic crystal structures as waveguides, antireflection coatings, frequency filters, metamaterials, photonic crystal lens, working at a given light frequency.

5.1.1. Photonic band gaps

Based on the general theory of light propagation in the PC [3], we consider the solution of Maxwell's equations for a dielectric medium without free charges and currents which simulates the photonic crystal. The system of Maxwell's equations in this case has the following form (in the SI system):

$$\begin{cases} \nabla \mathbf{D} = 0, \\ \nabla \mathbf{B} = 0, \\ \nabla \times \mathbf{E} = -\dfrac{\partial \mathbf{B}}{\partial t}, \\ \nabla \times \mathbf{H} = \dfrac{\partial \mathbf{D}}{\partial t}, \end{cases} \tag{5.1}$$

where $\mathbf{D} = \varepsilon\varepsilon_0\mathbf{E}$, $\mathbf{B} = \mu\mu_0\mathbf{H}$, $\sqrt{(\varepsilon_0\mu_0)^{-1}} = c$,

From (5.1) we obtain the following relation:

$$\varepsilon^{-1}(r)(\nabla\times\mathbf{H})=\varepsilon_0\frac{\partial\mathbf{E}}{\partial t}. \tag{5.2}$$

Applying the 'rot' operation to the expression (5.2) with (5.1) taken into account, we obtain:

$$\nabla\times(\varepsilon^{-1}(r)(\nabla\times\mathbf{H}))=-\frac{\mu}{c^2}\frac{\partial^2}{\partial t^2}\mathbf{H}. \tag{5.3}$$

Hence, for monochromatic waves, we have:

$$\frac{\partial^2}{\partial t^2}\mathbf{H}(r,t)=-w^2\mathbf{H}(r,t). \tag{5.4}$$

In view of (5.4) instead of (5.3) with $\mu = 1$ can be written:

$$\nabla\times(\varepsilon^{-1}(r)(\nabla\times\mathbf{H}))=\left(\frac{w}{c}\right)^2\mathbf{H}. \tag{5.5}$$

Since the value of $\varepsilon(r)$ in this case is real, then equation (5.5) is the task of finding the eigenvalues ω^2/c^2 of a Hermitian operator $L=\nabla\times(\varepsilon^{-1}(r)\nabla\times)$ in the equation

$$L\mathbf{H}=\left(w/c\right)^2\mathbf{H}. \tag{5.6}$$

Consider the one-dimensional structure of the PC with a period $d = a + b$, where a and b are the sizes of sites of the dielectric constants ε_1 and ε_2, respectively. It is known [1, 3] that the eigenfunctions of equation (5.6) in a periodic medium have the Bloch form:

$$\varphi=e^{ikx}u(x), \tag{5.7}$$

where x is the coordinate, $k = 2\pi/\lambda$ is the wave number. Eigenfunctions of the operator L are determined on the basis of their form (5.7) and boundary conditions defined by the function

$$\varepsilon(x)=\begin{cases}\varepsilon_1, nd\le x<a+nd,\\ \varepsilon_2, a+nd\le x<(n+1)d,\end{cases} \tag{5.8}$$

where n is an integer. The eigenfunctions in the regions with dielectric constants ε_1 and ε_2 will be, respectively, form

$$\begin{aligned}\varphi_1(x)&=Ae^{ik_1x}+Be^{-ik_1x},\\ \varphi_2(x)&=Ce^{ik_2x}+De^{-ik_2x},\end{aligned} \tag{5.9}$$

where A, B, C, D are unknown coefficients.

Since at the boundaries of zones with different dielectric constants there must be continuous both the eigenfunctions and their derivatives, we can form a system of equations:

$$\begin{cases} A+B=e^{-iQd}(Ce^{ik_2d}+De^{-ik_2d}), \\ k_2(A-B)=k_2e^{-iQd}(Ce^{ik_2d}-De^{-ik_2d}), \\ Ae^{ik_1a}+Be^{-ik_1a}=Ae^{ik_2a}+Be^{-ik_2a}, \\ k_1(Ae^{ik_1a}-Be^{-ik_1a})=k_2(Ae^{ik_2a}-Be^{-ik_2a}). \end{cases} \quad (5.10)$$

In the matrix form, this system of equations for A, B, C and D can be written as:

$$M(k_1,k_2,Q)V=0, \quad (5.11)$$

where

$$M(k_1,k_2,Q)=\begin{pmatrix} 1 & 1 & -e^{id(k_2-Q)} & -e^{-id(k_2+Q)} \\ k_1 & -k_1 & -k_2e^{id(k_2-Q)} & k_2e^{-id(k_2+Q)} \\ e^{ik_1a} & e^{-ik_1a} & -e^{ik_2a} & -e^{-ik_2a} \\ k_1e^{ik_1a} & -k_1e^{-ik_1a} & -k_2e^{ik_2a} & k_2e^{-ik_1a} \end{pmatrix}, \quad (5.12)$$

$$V=\begin{pmatrix} A \\ B \\ C \\ D \end{pmatrix}.$$

This system of equations has a nontrivial solution if $\det M=0$. Expanding the determinant we can obtain in an implicit form the dispersion relation $\omega(Q)$:

$$\cos(k_1a)\cos(k_2b)-\frac{1}{2}\frac{\varepsilon_1+\varepsilon_2}{\sqrt{\varepsilon_1\varepsilon_2}}\sin(k_1a)\sin(k_2b)=\cos(Qd), \quad (5.13)$$

where $k_i=\sqrt{\varepsilon_i}\frac{w}{c}$, $i=1, 2$, Q is the Bloch wave number. Since $|\cos(Qd)|\le 1$, band gaps form in the spectrum, i.e. the values of k_i for which

$$\left|\cos(k_1a)\cos(k_2b)-\frac{1}{2}\frac{\varepsilon_1+\varepsilon_2}{\sqrt{\varepsilon_1\varepsilon_2}}\sin(k_1a)\sin(k_2b)\right|>1. \quad (5.14)$$

In these areas, the propagation of radiation in a crystal is impossible. Or conversely, given a one-dimensional photonic crystal (a, b, ε_1, ε_2), from the inequality (5.14) we find $w=2\pi\nu$ – the cyclic frequency of light that can pass through a photonic crystal.

Consider the simplest one-dimensional theory. We can now consider some examples of the model. Consider the diffraction of a plane wave on photonic crystals in two dimensions. In all the examples in this section, a wave of unit intensity (E_0 = 1 V/m). Modelling was performed using the FullWAVE 6.0 software and the FDTD method implemented in it.

5.1.2. Plane wave diffraction on photonic crystals without defects

Consider the incidence of a plane electromagnetic wave with TE-polarization on a photonic crystal (PC). PC parameters are taken from [4] to obtain comparable results: $n_1 = 3.25$ – the refractive index of the medium, $n_2 = 1$ – the refractive index of a hole, $r = 0.25$ μm – the radius of the holes, $T_z = 0.6$ μm and $T_x = 1$ μm – distance between the centres of the holes on the optical axis z and x, respectively. The fill factor on the axis z: $\Lambda_z = 0.83$, on the axis x: $\Lambda_x = 0.5$, also taken from [4]. Figure 5.1 shows the band gap for a given crystal for TE-polarization. The length of the electromagnetic wave $\lambda = 1.55$ μm. Figure 5.1 shows that this wavelength lies in the photonic band gap, the reflection coefficient $R \approx 0.89$. Figure 5.2 shows that the light of this wavelength does not enter the crystal further the first three layers of holes. Now consider the same photonic crystal, but change the wavelength at $\lambda = 2.2$ μm. Figure 5.3 shows that at this wavelength PC transmits electromagnetic radiation, the reflection coefficient is small, $R \approx 0.2$.

That is, the band gap in Fig. 5.1 is not an exact function-rectangle (rect) and the light reflection coefficient reaches the maximum value of 1 in connection with the fact that the simulation considers the finite size of PC on the coordinates x and z.

5.1.3. Propagation of light in a photonic crystal waveguide

Consider the same photonic crystal as in the preceding paragraph by removing the three central rows of holes. The wavelength is chosen on the basis of Fig. 5.1 so that the transmission coefficient was minimal.

This condition is satisfied for $\lambda = 1.55$ μm. Figure 5.4 shows the light propagation in a photonic crystal waveguide. Figure 5.4 shows that light travels only in the way that we created in a photonic crystal by removing three adjacent rows of holes.

Now let us reduce the radiation source to the dimensions of the waveguide in order to assess the power loss when light passes through a waveguide (Fig. 5.5). The energy loss at a distance of 5 μm (the length of the waveguide) was approximately 0.51%. That is, if we focus the light at the entrance of the photonic-crystal waveguide about 1 μm wide, the light passes almost completely through it to the exit.

5.1.4. Photonic crystal collimators

Optimization methods [5] for the structure of photonic crystal waveguides to reduce the beam divergence at the exit of the fibre have recently been developed. For conventional optical fibres this problem is solved by means of structuring the output of the fibre tip.

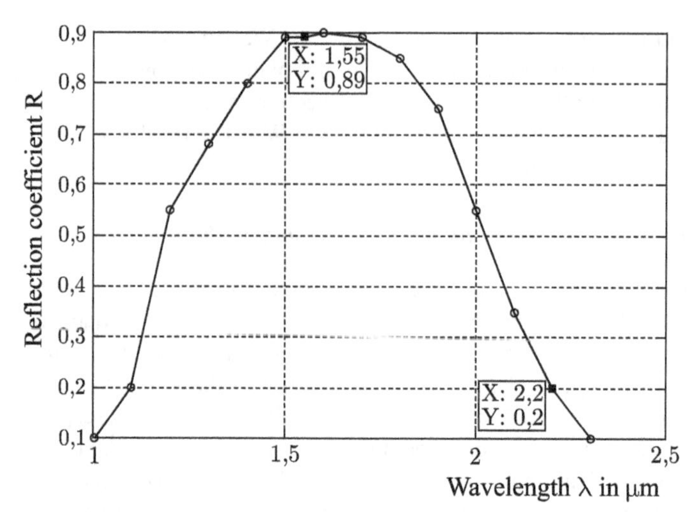

Fig. 5.1. Photonic band gap.

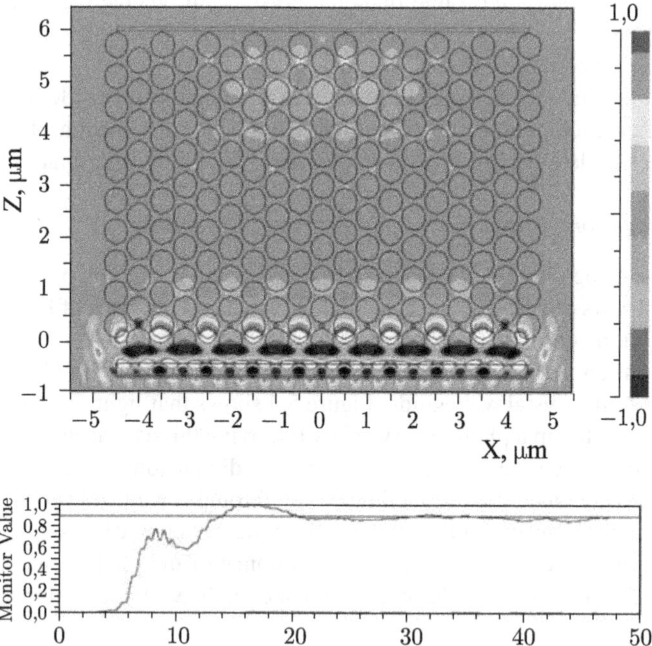

Fig. 5.2. Propagation of light in a photonic crystal in the band gap. Light travels from the bottom up.

Figure 5.6a shows a schematic of a two-dimensional photonic crystal waveguide, the shell of which consists of periodically spaced (period 228 nm) dielectric nanorods ($\varepsilon = 3.38$, silicon) with a diameter of 114 nm. To create a waveguide a row

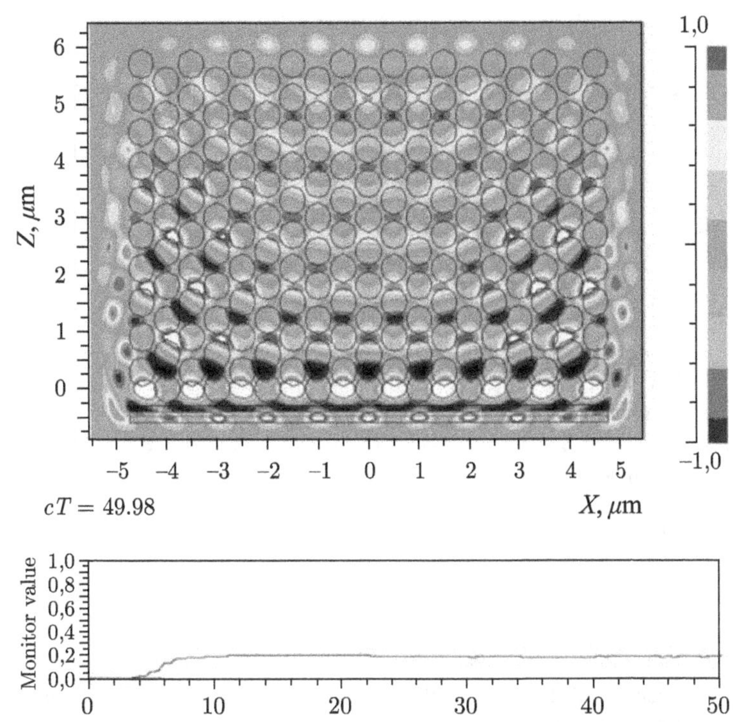

Fig.5.3. Light propagation in a photonic crystal outside the band gap. Light travels from the bottom up.

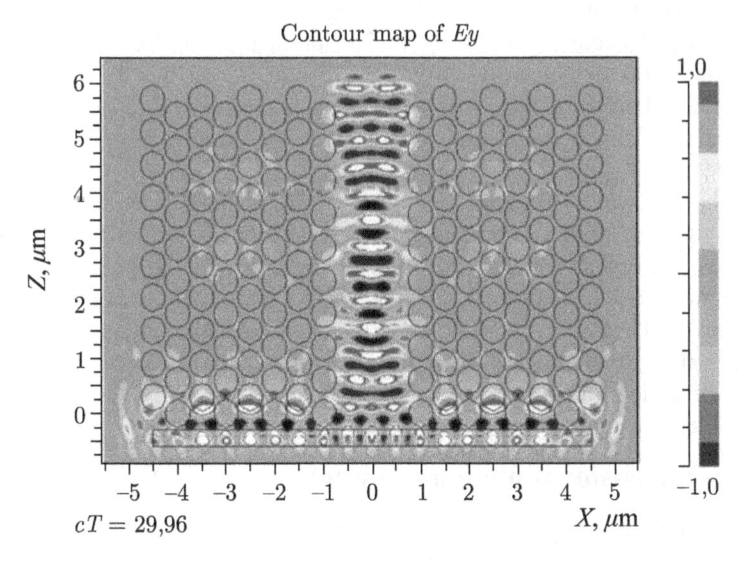

Fig. 5.4. Propagation of light in a photonic crystal waveguide.

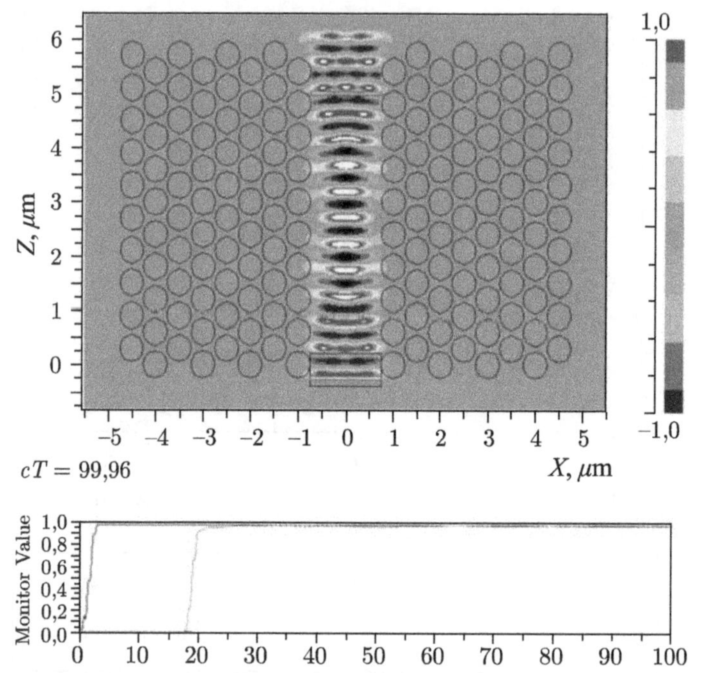

Fig. 5.5. Propagation of light in a photonic crystal waveguide with a narrow light source. Light travels from the bottom up.

of nanorods is removed. The size of this 'defect' in the periodic structure of the nanorods is 342 nm and a half period. The light wavelength 633 nm. Figure 5.6b shows an instantaneous picture of the diffraction of light on this structure, calculated by the FDTD method with FullWAVE software. We see that the light does not go into the shell and propagates inside the waveguide side with a refractive index of 1. At the exit from the waveguide the light wave is strongly divergent, extending at an angle of 140°.

Note that the width of the waveguide (342 nm) is slightly more than half the length of light (633 nm). Some modernization of the structure of the waveguide near the exit can greatly reduce the beam divergence. Thus, Fig. 5.7a shows the PC-waveguide in which two rods in the back row near the output waveguide were removed. This led to the situation in which the radiation after the waveguide diverges at an angle of only 30° (Fig. 5.7b). Note that in the scalar theory of diffraction the full angle of divergence can be estimated as $2\lambda/\pi r = 2.35$, or 130°, r is the radius of the waveguide.

5.2. Two-dimensional photonic crystal gradient Mikaelian lens

Modern technology allows manufacture of the optical micro- and nano-objects with dimensions comparable to the wavelength of light. So the question arises of computer simulation of light diffraction on such objects. To solve this problem, we must solve directly Maxwell's equations. One of the most widely used methods for the

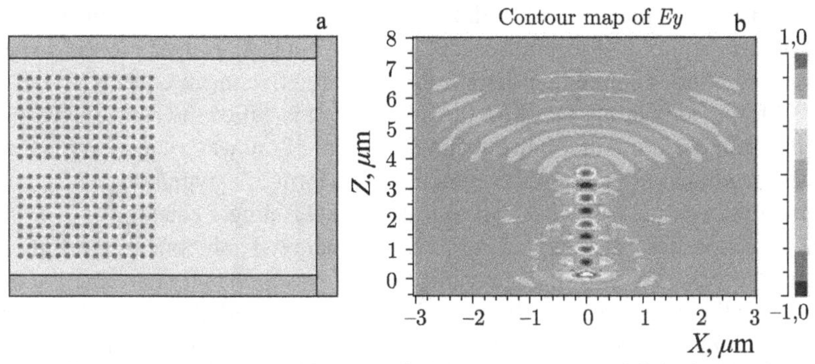

Fig. 5.6. Photonic-crystal waveguide (a) and diffraction pattern of light inside the waveguide and exit (b). Light travels from left to right (a) and bottom up (b).

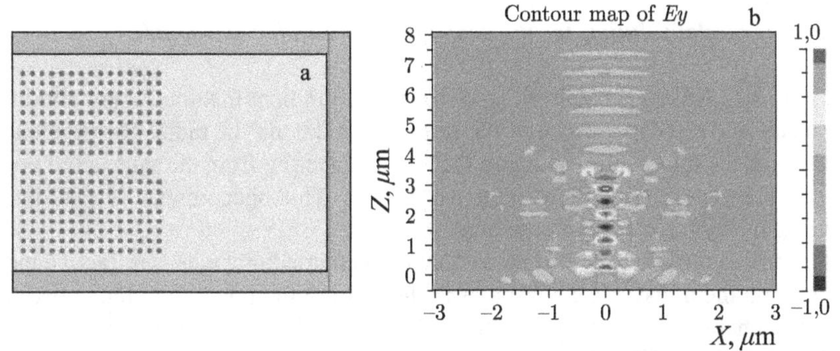

Fig. 5.7. Photonic-crystal waveguide with a collimator (a) and diffraction pattern of light inside the waveguide and at the exit from it (b). Light travels from left to right (a) and bottom up (b).

numerical solution of these equations is the 'finite-difference time-domain method' (FDTD) [6]. This method worked well, thanks to its versatility in solving diffraction problems [7, 8]. It is used in this section.

Photonic crystals, which we briefly discussed in the previous section, are structure with a periodically varying refractive index (this period should be smaller than the wavelength), which allow the manipulation of light at the nanometer scale [9]. Recently, they have attracted increasing attention due to a number of their interesting properties. One of the fundamental properties of these materials is that they do not transmit light at certain wavelengths. This spectral region is called a 'photonic band gap'. Currently, the most interesting are photonic crystals in which the band gap lies in the visible or near infrared regions [10–13].

Through the development of production technology of photonic crystals it is possible to create a photonic crystal lens. The PC lens is a photonic crystal in which the radius of the hole changes according to a specific law, which ensures focusing of light. The period of the crystal lattice remains constant. These lenses, for example, solve the problem of focusing light on the input of the photonic crystal waveguide, being a more compact alternative to microlens and tapered waveguides.

It is known that the gradient medium with the radial dependence of the refractive index in the form of a hyperbolic secant, proposed in [14], is used for self-focusing of laser radiation. The gradient lens with such refractive index collects all the rays parallel to the axis in the focus on the surface and is called the Mikaelian lens. In [15] it is proposed to search for the mode solution of the wave equation for the 2D gradient medium with the refractive index in the form of a hyperbolic secant, similar to finding soliton solutions of the non-linear Schrödinger equation.

This section describes the paraxial and non-paraxial solutions in the form of a hyperbolic secant for two-dimensional gradient waveguide whose refractive index depends on the transverse coordinate in the form of a hyperbolic secant. For a cylindrical gradient lens we find a similar photonic crystal lens which can be produced by photo- or electronic lithography. The FDTD method implemented in the programming language C ++ is used for comparative simulation of a plane wave passing through both microlens.

5.2.1. The modal solution for the gradient secant-index waveguide

In 1951, A.L. Mikaelian showed [14] that in the gradient medium with cylindrical symmetry and the dependence of the refractive index on the radial coordinate as a function of the hyperbolic secant all the rays, emanating from the same axial point, at some distance again gather in the axial focus. This phenomenon is called self-focusing of light in a gradient medium.

It can be shown that a two-dimensional gradient medium, whose refractive index depends on the transverse coordinates in the form of the function of the hyperbolic secant is characterized by spreading of the light field, which retains its structure, showing modal (soliton) properties, and its complex amplitude is proportional to the same function of the hyperbolic secant.

Indeed, suppose that the distribution of the refractive index in the 2D model of the gradient medium depends only on the transverse coordinates in the form of the function of the hyperbolic secant:

$$n(y) = n_0 ch^{-1}\left(\frac{kn_0 y}{\sqrt{2}}\right), \tag{5.15}$$

where n_0 is the maximum refractive index of the medium on the optical axis, k is the wave number of light in vacuum. In the case of TE-polarization the only non-zero projection of the vector of the strength of the electric field of a monochromatic electromagnetic wave $E_x(y, z)$ satisfies the Helmholtz equation:

$$\left[\frac{\partial^2}{\partial z^2} + \frac{\partial^2}{\partial y^2} + \frac{k^2 n_0^2}{ch^2\left(kn_0 y/\sqrt{2}\right)}\right] E_x(y,z) = 0, \tag{5.16}$$

where z is the direction along the optical axis. Then the modal solution of (5.16) in the form of a 'soliton' will look like:

$$E_x(y,z) = E_0 \mathrm{ch}^{-1}\left(\frac{kn_0 y}{\sqrt{2}}\right)\exp\left(\frac{ikn_0 z}{\sqrt{2}}\right), \tag{5.17}$$

where E_0 is a constant. The word 'soliton' is in quotes, as in our case there is no non-linearity, and the solution (5.17) just looks like a soliton solution and is the mode of the given gradient medium. Interestingly, the solution (5.17) holds in the paraxial case. If instead of the gradient medium (5.15) we choose a somewhat different dependence of the refractive index on the transverse coordinates:

$$n_1(y) = n_0\sqrt{1+\mathrm{ch}^{-2}\left(\frac{kn_0 y}{\sqrt{2}}\right)}, \tag{5.18}$$

where $n_1(0)=\sqrt{2}n_0$ is the maximum refractive index, and $n_1(\infty)=n_0$ is the minimum refraction index, the solution of the paraxial equation

$$\left[2ik\frac{\partial}{\partial z}+\frac{\partial^2}{\partial y^2}+\frac{k^2 n_0^2}{\mathrm{ch}^2\left(kn_0 y/\sqrt{2}\right)}\right]E_{1x}(y,z)=0, \tag{5.19}$$

will be similar to the complex amplitude (5.17):

$$E_{1x}(y,z) = E_0\,\mathrm{ch}^{-1}\left(\frac{kn_0 y}{\sqrt{2}}\right)\exp\left(\frac{ikn_0^2 z}{4}\right). \tag{5.20}$$

Note that the solutions (5.17) and (5.20) have finite energy:

$$W=\int_{-\infty}^{\infty}\left|E_x(y,z)\right|^2 dy = \left|E_0\right|^2\int_{-\infty}^{\infty}\mathrm{ch}^{-2}\left(kn_0 y/\sqrt{2}\right)dy = 2\left|E_0\right|^2. \tag{5.21}$$

Modal solutions, similar to (5.17) and (5.20), can be found for the 3D gradient waveguide with a refractive index:

$$n(x,y) = n_0\mathrm{ch}^{-1}\left(bx + y\sqrt{\frac{(kn_0)^2}{2}-b^2}\right), \tag{5.22}$$

where b is an arbitrary parameter. A method of producing such a ch^{-1} solutions can be found in recent papers of I.V. Alimenkov [16,17] in which 3D-soliton solutions were found for non-linear Schrödinger equation with the Kerr non-linearity of the third order, when the refractive index of the non-linear medium is described by:

$$n^2(x,y,z) = n_0^2 + \alpha I(x,y,z), \tag{5.23}$$

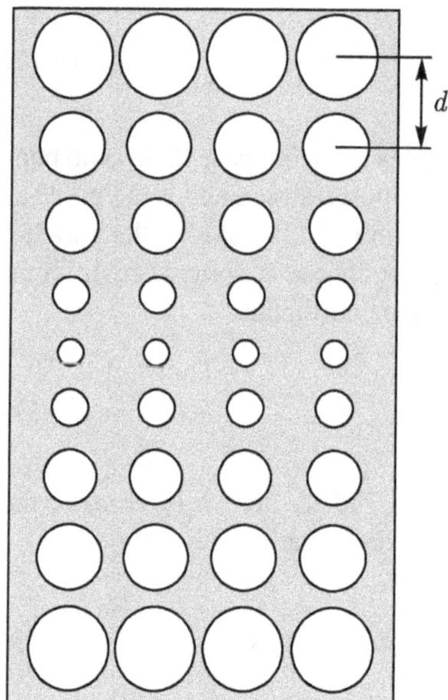

Fig. 5.8. Photonic crystal gradient lens.

where α is a constant, $I(x,y,z) = |E(x,y,z)|^2$ is the intensity of one of the components of the electric field vector of the light wave. The analogy between the soliton ch^{-1} solution of the non-linear medium (5.23) and a linear gradient medium with a refractive index (5.15) or (5.18) was first pointed out by A.W. Snayder [15].

The next section shows how to replace the cylindrical gradient lens (GL) by a 2D photonic crystal lens.

5.2.2. Photonic crystal gradient lens

A two-dimensional photonic crystal gradient lens (PCGL) consists of a photonic crystal in which the radius of holes varies according to a definite law. Like a conventional lens, PCGL can focus a parallel light beam to a point. However, PCGL can be more compact and can be easier to manufacture. Figure 5.8 shows schematically the PCGL.

The cylindrical GL [14] is a gradient lens, whose refractive index varies from centre to edge of the lens according to the law:

$$n(y) = \frac{n_0}{\mathrm{ch}\left(\dfrac{\pi|y|}{2L}\right)}, \tag{5.24}$$

where L is the width of the lens along the axis z, n_0 is the refractive index in the centre.

We choose the equivalent PCGL from a material with a refractive index n and thickness along the optical axis a so that it could be replaced by the GL. For this we require that at discrete points of this lens the optical path length is equal to the optical path length in the GL. The optical path length in the GL will be:

$$\Delta_1 = \frac{Ln_0}{\mathrm{ch}\left(\dfrac{\pi|y|}{2L}\right)}. \tag{5.25}$$

The optical path length in the PCGL is:

$$\Delta_2 = N\left[2r(y)+(d-2r(y))n\right], \tag{5.26}$$

where N is the number of holes in a row, d is a constant of the crystal or the distance between the centres of holes, $r(y)$ is the radius of holes, varying from row to row. Equating the optical lengths (5.25) and (5.26), we obtain the following expression for the radius

$$r(y) = \frac{d}{2(n-1)}\left(n - n_0 \frac{1}{ch\left(\dfrac{\pi|y|}{2L}\right)}\frac{L}{a}\right). \tag{5.27}$$

Suppose that in each column of the lens there are M holes. Then the obtained dependence should be performed at the points $y=\pm dm$ and m varies from 0 to $M/2$. In this case the radius of the hole should also be subjected to certain conditions. First, the radius must be non-negative. It follows from (5.27) that the minimum radius is attained at $y=0$. Applying a non-negativity condition to it, we obtain the following relation for the parameters of the GL and the corresponding PCGL:

$$na \geq n_0 L. \tag{5.28}$$

Secondly, the diameter of the hole obviously must be less than the constant of the crystal. The maximum radius is attained at $y = b/2$, where b is the aperture of the lens. The above condition imposes the following restriction on the aperture of the lens:

$$\mathrm{ch}\frac{\pi b}{4L} < n_0 \frac{L}{a}. \tag{5.29}$$

Third, the period of the lattice, as mentioned above, should be subject to the condition $d < \lambda$. In addition, in the numerical simulation of the photonic crystal lens the discretization step should be chosen so small that the radius varies from row to row. The fact is that it may happen that the change in radius from row to row may be less than the discretization step. In this case, the radius does not change and the desired effect would not be reached.

Simulation of light diffraction on 2D micro-lenses was performed using a finite difference solution of Maxwell's equations by the FDTD-method. The C ++ language in the MS Visual Studio 6.0 was used to implement the Yee algorithm [18] in the two-dimensional case for TE-polarization. Input of radiation in the computational domain is implemented using the 'total field–scattered field' condition [19]. The boundary conditions are represented by perfectly absorbing Berenger layers (J.P. Berenger) [20].

GL has the property of focusing light to a point on the surface. In our numerical experiments we used the GL with the following parameters: wavelength λ = 1.5 μm, L = 3 μm, n_0 = 1.5, b = 4 μm. The distribution of the squared modulus of the complex amplitude of the electric field when light passes through such a lens is shown in Fig. 5.9, and the cross section of intensity in Fig. 5.10.

The graph shows that the focus of such a lens is exactly the same as its front surface.

We now simulate the passage of light through PCGL with parameters $a = L$ = 3.4 μm, $n = n_0$ = 1.5, d = 0.25 4 μm. Figure 5.11 shows the dependence of the radius of holes (Fig. 5.8) in the PCGL on the number of some of these holes.

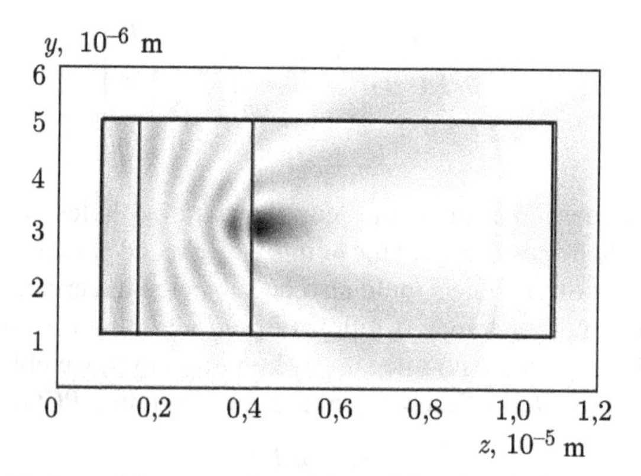

Fig. 5.9. The distribution of the squared modulus of the electric field $|E_x|^2$ (negative) of the GL, the location of the lens is indicated by two vertical lines.

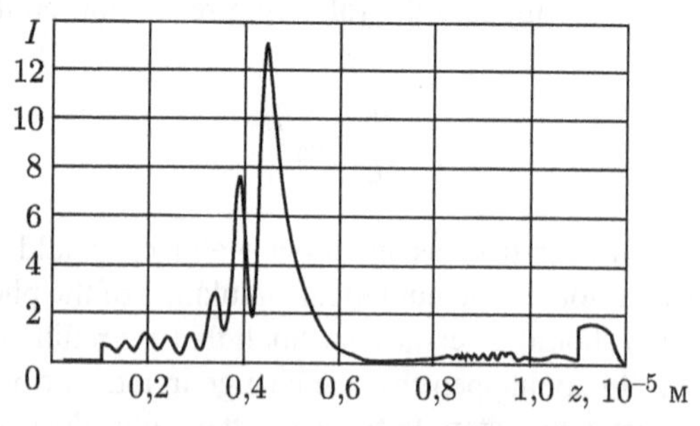

Fig. 5.10. The distribution of the squared modulus of the electric field in the cross section along the optical axis of the GL.

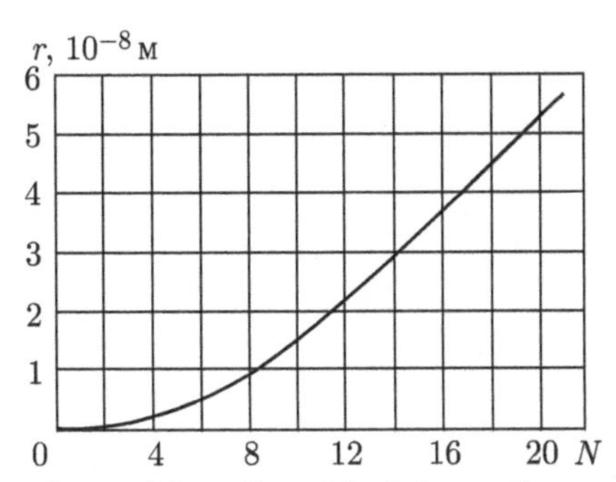

Fig. 5.11. Dependence of the radius of the holes on the number of the row.

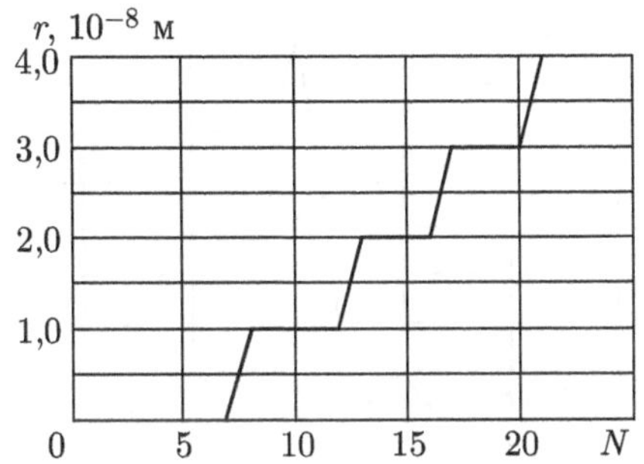

Fig. 5.12. Dependence of the radius of the holes on the number of rows in the sample of 100 samples per wavelength.

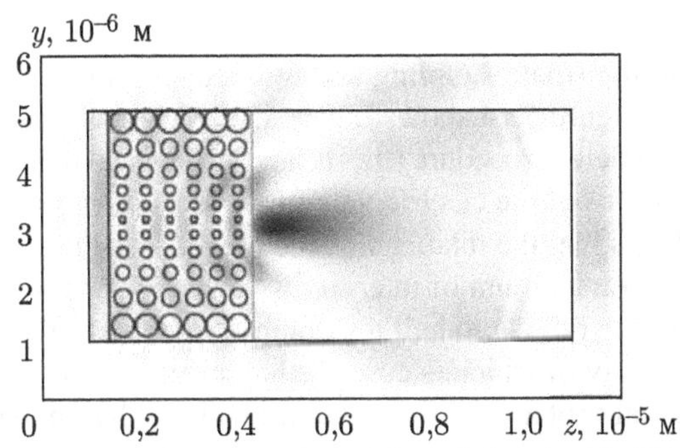

Fig. 5.13. The distribution of the squared modulus of the electric field $|E_x|^2$ (negative) in PCGL.

However, even with large sample $\lambda/h = 100$, it turns out to be quite an inaccurate approximation of the radius. This is shown in Fig. 5.12.

The period of the nanostructure of the holes is 250 nm, the minimum hole diameter 10 nm, maximum 40 nm. Figures 5.13 and 5.14 show the distribution of the squared modulus of the complex amplitude of light passing through the lens.

We see that the focus of this lens is at a distance of $f = 3.3$ μm from the beginning of the lens, i.e. accurately corresponds to the GL with $L = 3$ μm. The intensity in the focus is $I_f = 7.5$, which is less than in the GL (PCGL efficiency is 70% of the efficiency of GL), and the depth of focus of the PCGL on the z-axis is twice that of the

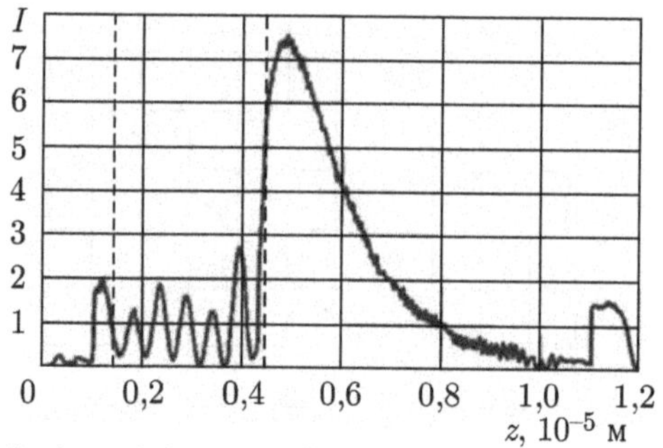

Fig. 5.14. The distribution of the squared modulus of the electric field along the main optical axis of the PCGL.

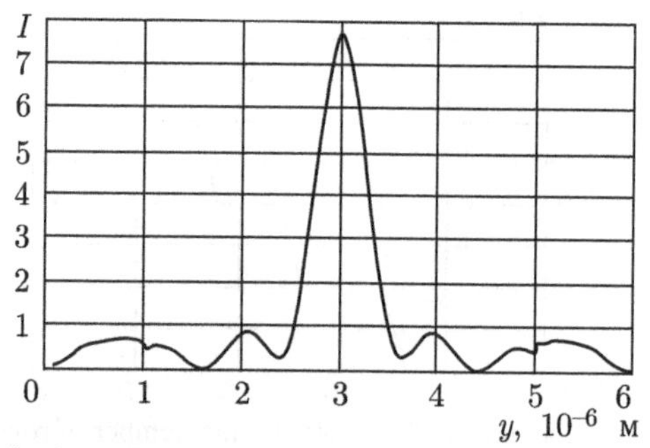

Fig. 5.15. The distribution of the squared modulus of the electric field in the focal plane of the PCGL.

GL. This is due to the small sampling and rough approximation (Fig. 5.12) of the curve in Fig. 5.11. Figure 5.14 shows 'noise' caused by the coupling terms of 'the total field–scattered field' procedure (this is not the physical noise). The distribution of the squared modulus of the electric field in the focal plane is shown in Fig. 5.15.

Figure 5.15 shows that the diameter of the focal spot of the PCGL according to the full width of half maximum of intensity is equal to FWHM = 0.42λ. FWHM is the abbreviation of the full width half maximum. Note that in the scalar case for the focal spot, the intensity of which is described by the sinc-function, it is known that the width of the focal spot according to the full width of half maximum of intensity is 0.44λ/NA, where NA is the numerical aperture of the lens. In the case of PCGL NA = 0.67, so the width of the spots according to the full width of half maximum of intensity is equal to $0.29\ \lambda$ / NA. This is 1.5 times less than in the scalar case.

5.2.3. The photonic crystal lens for coupling two waveguides

In recent years, various micro-and nanophotonic devices have been actively investigated for coupling two waveguides of different types, for example, a conventional single-mode fibre with a wire or a planar waveguide or a planar waveguide with a photonic crystal (PC) waveguide. The following nanophotonic device are available for coupling of two waveguide structures: an adiabatically tapered ridge waveguides for coupling with the PC-waveguides [21–27]; waveguide structures can not only couple with each other by output to input, but can also overlap parallel to each other [28], Bragg gratings in a waveguide [29–32] to extract the radiation from the fibre, the fibre with the Bragg grating can be placed on the surface of a planar waveguide [33], a parabolic micromirror at an angle to enter into a planar waveguide [34], conventional refractive lenses or microlenses [35-38], Veselago superlens with negative refraction: plane [39–46] or with a concave surface [47–49], the interface in the millimeter range: superlens [50, 51] and PC-lens [52]. Work is also being carried out to combine two different PC waveguides [53].

Tapered waveguides can be highly effective coupling devices if the widths of the mode in the ridge waveguide and in the PC-waveguide are comparable with each other. In this case, the effectiveness of coupling (i.e. the ratio of the energy at the output of the device to the energy at the input) can reach 80% [21], 90% [24], 95% [22], and even nearly 100% [26]. If the width of the ridge waveguide (1.6 μm) is several times larger than the width of PC-waveguide (200 nm), the effectiveness of coupling is reduced to 60% [23]. At an even greater difference in the widths of the mating waveguides the size of the adiabatically narrowing (tapered) waveguide is relatively large: in the compression of the mode of the single-waveguide fibre with the core diameter of 4.9 μm to the size of the mode of a planar waveguide with a width of 120 nm, the length of the taper is 40 μm [25], and the waveguide with a cross section of 0.3×0.5 μm narrows to 75 nm in diameter at a distance of 150 μm [27].

Coupling devices, which transfer radiation from a single-mode fibre to planar waveguides or photonic crystal waveguides with the grating on the waveguide also have tapered areas. For example, the tapering of a Gaussian beam with a waist diameter of 14 μm to the size of the waveguide with a width of 1 μm is carried out using a taper only 14 μm long [29, 30]. In this regard, the experimental efficiency of coupling is 35% [10], without a mirror layer on the reverse side of the waveguide, and 57% [29] with a mirror. Input of the Gaussian beam with a wavelength of 1.3 μm into the waveguide was performed with a diffraction grating on the waveguide [30]. A similar device is connected to the grating on a silicon waveguide with a period of 630 nm and a 20–40 μm taper, but for a wavelength of 1.55 μm the experimental efficiency was 33% (with a mirror 54%) [31]. Higher quality has an input device for transferring radiation from a single-mode fibre with a diffraction grating in silicon with a period of 610 nm and a width of 10 μm to a wire waveguide 3 μm wide with an experimental coupling efficiency of 69% [32]. The computed communication efficiency over 90% has the J-coupler, which connects a wide waveguide (10 μm) with an PC-waveguide (420 nm) using a parabolic mirror with the size of 15×20 μm for a wavelength of 1.3 μm [34]. In this case both the waveguide and the parabolic mirror are made of a silicon film (refractive index $n = 3.47$).

Conventional refractive lenses and microlenses have also been successfully applied in coupling problems. For example, a silicon waveguide ($n = 3.092$) 1–2 μm wide has a lens at the end which allows this waveguide to be coupled with a silicon PC-waveguide ($n = 3.342$) with a calculated 90% efficiency [35]. Modelling has shown [37] that the single-mode fibre with a diameter 3.10 μm (wavelength

$\lambda = 1.55$ μm) with a collimating lens of VK7 glass (numerical aperture NA = 0.1) with radius R = 1.77 mm and with a focusing silicon microlens with a radius of 123 μm can be coupled with a PC-waveguide with a cross section of the mode of 0.19×0.27 μm with an efficiency of 80%. At the same time the microlens creates inside the PC-waveguide a focal spot with a diameter FWHM = 0.24λ (numerical aperture of the waveguide NA = 2.2).

A special place among the couplers is occupied by the devices based on 2D superlenses (or Veselago lens), which are based on the phenomenon of negative refraction. A superlens with an effective refractive index close to −1 can be produced using photonic crystals. The superlens is used to image a point source. The first image appears inside the lens and the second image behind the lens at a distance of $2B–A$, where B is the thickness of plane-parallel lenses, A is the distance from the lens to the source [39, 43]. In [41] it is shown that if a 2D point light source is described by the Hankel functions $H_0(kr)$, k is the wave number, r is the distance from the source to the observation point, then the image will be proportional to the Bessel function $J_0(kr)$. That is the spot image formed by the superlens has a diameter FWHM = 0.35λ. In [44] simulation of the 2D photonic-crystal superlens have shown that if the lens consists of two layers of dielectric rods (dielectric constant $\varepsilon = 12.96$) for the wavelength $\lambda = 1.55$ μm with the radius $r = 0.45a$, where a is the period of the grating rods, then at the cyclic frequency $w = 0.293a/\lambda$ the refractive index is equal to $n = -1$, and a point source located at a distance $A = 0.26\lambda$ from the lens is imaged at approximately the same distance on the other side of the lens, and the width of the image spot is FWHM = 0.36λ. In some studies attention was paid to Veselago lenses not in the form of a plane-parallel PC layer but with one surface being concave. For example, in [47] it was shown that PC-lens of a rectangular grating of rods with $\varepsilon = 10$ and the magnetic permeability $\mu = 1.5$ with a period of $a = 0.48$ cm, the radius of rods $r = 0.4a$, has an effective refractive index of $n = -0.634$. Also, if this 2D lens is plane-concave with a radius of curvature R = 3.31 cm, the focus of such a superlens would be located at a distance $f = R/(1-n)$, for TE-polarization $f = 1.69$ cm, while for TM-polarization, $f = 2.38$ cm. The radiation frequency is equal to $w = 0.48a/\lambda$. In [48], the results are presented of modelling of the input of radiation in a PC-waveguide with a superlens with a concave surface. The PC-lens had a thickness of 8.6a and an aperture of 38a, while the PC consisted of a 2D grating of holes with a period $a = 465$ nm in GaAs ($\varepsilon = 12.96$) and a diameter $2r = 372$ nm. In the focus of the lens at a distance of 7.56λ ($\lambda = 1.55$ μm) there formed a focal spot with a radius of radius 0.5λ, if the lens was illuminated with a Gaussian beam with a waist radius of 3λ. Then radiation behind the lens travelled to the 3W PC-waveguide (3W means that the width of the waveguide is equal to three grating periods of the PC) with a width of 3a (approximately λ). Unfortunately, the efficiency of input to such a structure was not given in [48]. In [49] the authors also discussed the results of modelling the input of radiation from a single-mode fibre to a PC waveguide with PC-superlens (plano-concave, $n = -1$). The thickness of the lens was $16a = 4.8\lambda$, aperture 25a, and it consisted of a triangular grating of holes with a period $a = 0.305\lambda$ and radius $r = 0.4a$ in GaAs. The radius of curvature of the concave lens surface was $R = 2.1\lambda$, the focal length $f = 1.05\lambda$.

The effective input in the PC waveguide with $\varepsilon = 12.96$, $r' = 0.2a$, $a' = 0.312\lambda$ was equal to 95%. The width of the waveguide is equal to one period of photonic-crystal lattice a, and angular frequency $w = 0.315a/\lambda$. Unfortunately, the size of the focal spot of this lens was not given.

A different type of PC lens was studied in [54–56]. The grating of holes in such 2D PC lens has a constant period, but the size of the holes is changed according to some function. In the Mikaelian gradient lens [14] all rays parallel to the optical axis and falling perpendicular to its flat surface are collected in a point on the optical axis on the opposite flat surface. Such an axially-symmetric lens has the dependence of the refractive index on the radial coordinate (distance from the optical axis) in the form (6.24). In [43] the authors simulated a 2D Mikaelian lens with an aperture of 12 μm, consisting of 7 columns of holes with a period of 0.81 μm for the wavelength $\lambda = 1.55$ μm. The efficiency of input from a wide waveguide (12 μm) into the PC-waveguide 1.5 μm wide with the effective refractive index $n = 1.73$ was 55%. The PC-waveguide consists of a lattice of holes with a period of 0.63 μm and a diameter of 0.4 μm. In this study the characteristics of the focal spot of the lens are again not given. In [55, 56] a similar PCGL, but with different parameters was simulated. The lens thickness 3 μm, 12 columns of holes, aperture of the lens 4 μm, the refractive index of 1.5, wavelength 1.5 μm. The diameter of the focusing spot was FWHM = 0.42λ, and the focal spot diameter from zero to zero intensity was equal to 0.8λ.

This section examines the ultracompact nanophotonics device enabling the effective coupling of 2D waveguides of different widths using PCGL. The device was manufactured by the 'silicon on silica' technology, the width of the input waveguide was 4.5 μm, the width of the output waveguide 1 μm, the size of the PCGL 3×4 μm. The lens consisted of a matrix of holes 12×17 with the period of the lattice of holes 250 nm, and the diameter of the holes varied from centre to the periphery from 160 to 200 nm. The device operates in the wavelength range 1.5–1.6 μm. The calculated efficiency of coupling ranged from 40% to 80%, depending on the width of the output waveguide. PCGL focuses light into a small focal spot in the air just behind the lens with the diameter equal to FWHM = 0.36λ, which is 1.4 times smaller than the scalar diffraction limit of resolution in the 2D case, which is determined by the width of the sinc-function and is equal to FWHM = 0.44λ.

Modelling of photonic crystal waveguide lens

The photonic crystal gradient lens, which is modelled in the work, consisted of a matrix of 12 × 17 holes in silicon (the effective refractive index for TE-waves is $n = 2.83$), the lattice constant of holes 250 nm, the minimum diameter of the holes on the optical axis 186 nm, the maximum diameter of the holes on the edge of the lens 250 nm. The thickness of the lens along the optical axis 3 μm, the width of the lens (aperture) 5 μm. Wavelength $\lambda = 1.55$ μm.

Modelling was performed using the difference method for solving Maxwell's equations FDTD, implemented in the programming language C++. Figure 5.16a shows a 2D PC lens in silicon, as described above, and Fig. 5.16b shows the two-dimensional halftone diffraction pattern (averaged over time) of a plane wave

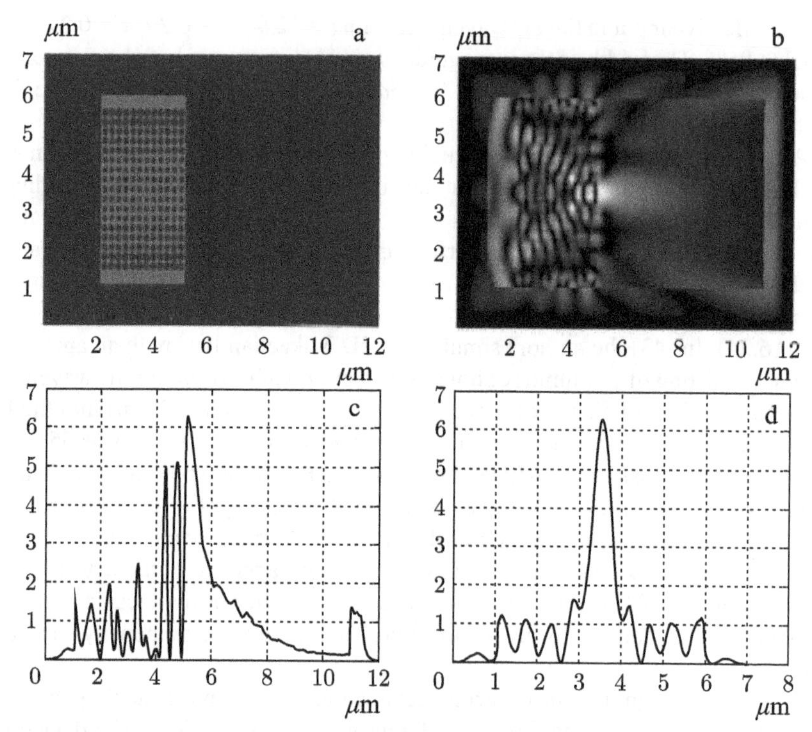

Fig. 5.16. 2D PCGL 12×17 holes in silicon, the size 3×4 mm (a), the field of diffraction of light (a plane TE-wave) or 2D-intensity distribution $|E_x|^2$, y is the vertical axis, z is the horizontal axis (b), the intensity distribution along the optical axis (c) and in the focus plane (d).

of the TE-polarization with an amplitude E_x (x-axis is perpendicular to the plane Fig. 5.16). Figure 5.16c and Fig. 5.16d show the distribution of intensity $|E_x(y, z)|^2$ along the optical z-axis and along the line y perpendicular to the optical axis where the focus is situated. Figure 5.16c and d shows that the size of the focal spot is FWHM = 0.36λ, and the longitudinal size of the focus is FWHM = 0.52λ.

PCGL with the parameters of the previous example (Figure 5.16a) was simulated but PCGL was located at the output of the waveguide in silicon 5 μm wide and 5 μm long (plus the length of the lens 3 μm, total length of the waveguide with a lens along the optical axis 8 μm) (Fig. 5.17a).

The diffraction field (intensity $|E_x(y, z)|^2$), calculated by the FDTD method and averaged over time, is shown in Fig. 5.17b (wavelength 1.45 μm). Figure 5.17c shows the intensity distribution along the optical axis. A comparison of Figs. 5.16c and Fig. 5.17c shows that the intensity of the focus increased, and the amplitude of modulation of the intensity inside the lenses decreased. This is due to the fact that the difference in the refractive indices between the lens and the waveguide (Fig. 5.17c) is much smaller than the difference between the lens and the air (Fig. 5.16c) and, therefore, the amplitude of the wave reflected from the interface is smaller. Figure 5.17d shows the distribution of intensity in the lens focus along a line parallel to the y axis. Figure 5.17d shows that the diameter of

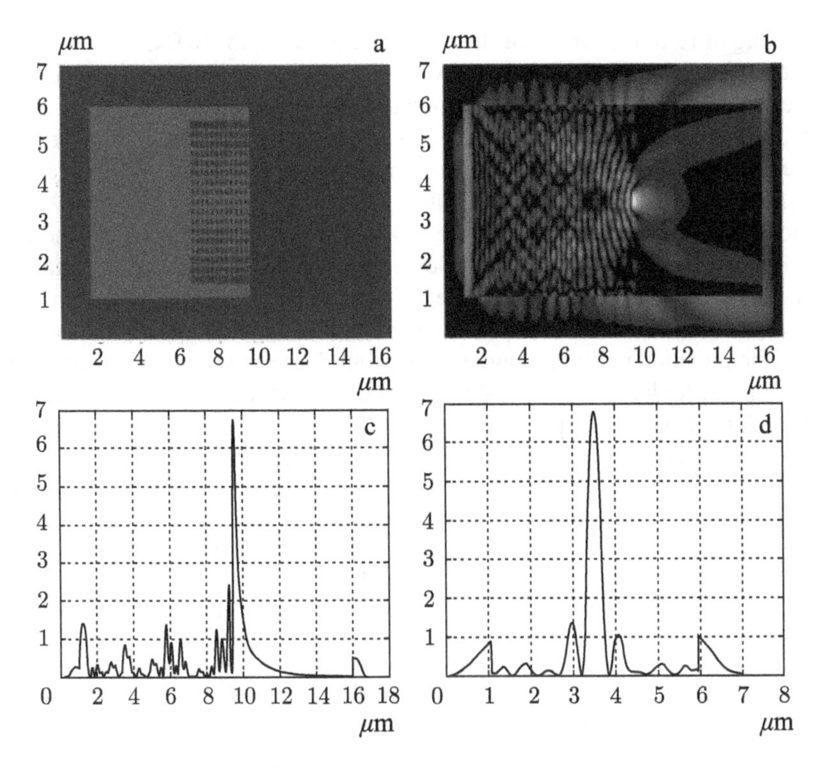

Fig. 5.17. 2D PCGL at the output of the waveguide (a), half-tone diffraction pattern of a plane TE-wave with amplitudes E_x, incident on the input of the waveguide 5 μm long, the output of which contains a lens 3 μm long (b), the intensity distribution $|E_x|^2$ along the optical axis (c), and the focus of the lens (d). Intensity is given in arbitrary units.

the focal spot at half intensity is FWHM = 0.31λ. A comparison of Figs. 5.17d and Fig. 5.16d shows that in addition to decreasing diameter of the focal spot in the case of PCGL in the waveguide, the sidelobes of the diffraction pattern at the focus were also smaller.

Note that the scalar theory in the 2D case describes a diffraction-limited focus by the sinc-function: $E_x(y, z) = \mathrm{sinc}\,(2\pi y/(\lambda NA))$ which is at a maximum numerical aperture $NA = 1$ gives the diffraction limit of the focal spot with the diameter at half intensity FWHM = 0.44λ. For the superlens [41], the limiting value of the focal spot is described by the Bessel function $J_0\,(kr)$ and gives the value of the diameter at half intensity FWHM = 0.35λ. Thus, the lens in Fig. 5.17a focuses light to a spot smaller than the diffraction limit.

The simulation showed that in the wavelength range 1.3–1.6 μm the intensity at the focus has two maximum values for the wavelengths of 1450 nm and 1600 nm (both maxima are about 20 nm wide). At other wavelengths in this range the intensity of the focus is 2–3 times smaller. With increasing wavelength the focus shifts to the lens surface, and at λ = 1.6 μm focus is inside the lens.

Modelling of relations between the two waveguides with PCGL

Figure 5.18a shows the coupling of two 2D waveguides using PCGL. The width of the input waveguide is 5 μm, the output waveguide 0.5 μm. PCGL in silicon (n = 2.83) has a matrix of 12×10 holes with the grating period of 0.25 μm. The diameters of the holes are the same as in previous examples. The wavelength 1.55 μm. Both waveguides are 6 μm long.

Modelling was performed uing the FDTD method, implemented in the FullWAVE 6.0 software (the company RSoft). Figure 5.18b shows the instantaneous diffraction pattern of a TE-wave. The effectiveness of coupling is 45%. Part of the radiation (20%) is reflected from the lens back into the input waveguide, another part of the radiation passes through the lens, but does not fall into a narrow waveguide. Fig. 5.18c shows an enlarged fragment of the diffraction pattern in Fig. 5.18b at the outlet of a narrow output waveguide. Unfortunately, in this program the y axis is not the transverse axis, as shown in Fig. 5.16 and Fig. 5.17, and the axis x is transverse. The intensity distribution $|E_y(x, z)|^2$ along the transverse axis x at the outlet of

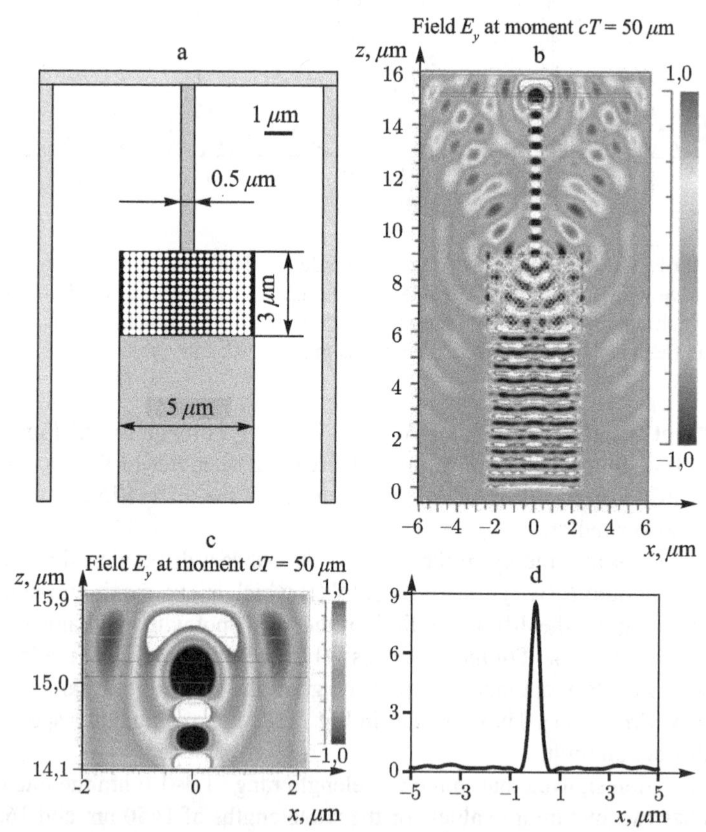

Fig. 5.18. Scheme of the coupling of two planar waveguides using PCGL (a), the instantaneous diffraction pattern of the TE-wave, calculated by FDTD using FullWAVE 6.0 software (b) and the enlarged fragment of the pattern at the output of the waveguide 0.5 μm wide (c), the intensity distribution at the exit of the waveguide (d).

a narrow waveguide is shown in Fig. 5.18d. Figure 5.18d shows that the diameter of the laser spot at the output at half intensity is FWHM = 0.32λ. Note that the focus in the output waveguide 1 μm wide (ceteris paribus) had a smaller diameter FWHM = 0.21λ, where λ is the wavelength in vacuum. This is lower than previously reported in [37] (FWHM = 0.24λ).

Simulation of the gap between the waveguides

Figure 5.19 shows a 2D scheme of coupling of two coaxial waveguides with a gap between them. The width of the input waveguide with PCGL W_1 = 4.6 μm, the output W_2 = 1 μm, the gap between the waveguides Δz = 1 μm. Other parameters are: λ = 1.55 μm, n = 1.46, PC-lens consists of a 12×17 matrix of holes with a period of a = 0.25 μm and the hole diameter from 186 to 250 nm. Figure 5.19a shows in white the waveguide material (n = 1.46), in black and gray the air (n = 1). Figure 5.19b shows the instantaneous pattern of the amplitude $E_y(x, z)$ for the TE-wave, calculated using FullWAVE 6.0 software for the circuit in Fig. 5.19a. Figure 5.19c shows the

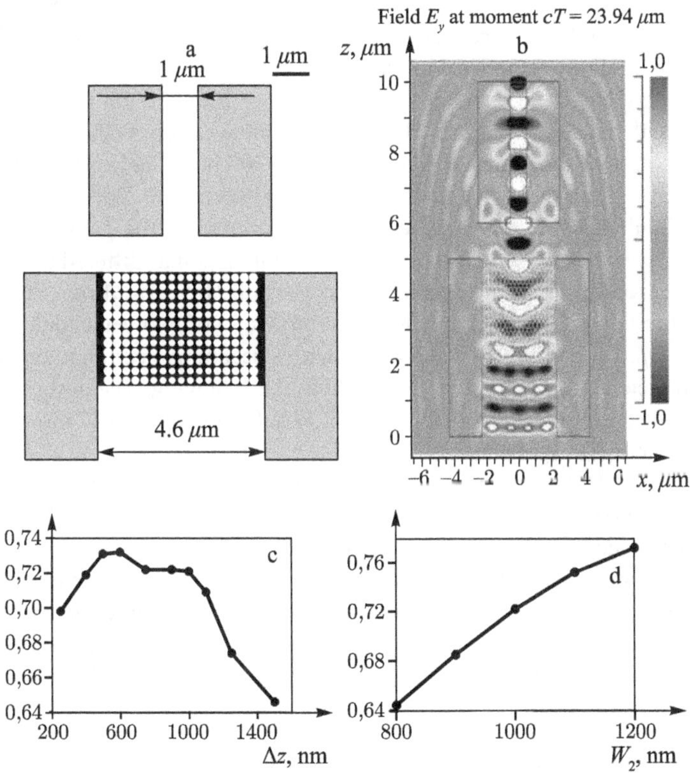

Fig. 5.19. 2D scheme of coupling between the two waveguides with PCGL at the gap Δz = 1 μm between the waveguides (white – material, black color – air) (a); instantaneous amplitude distribution $E_y(x, z)$ of the TE-wave, calculated by FullWAVE software (b); dependence of the effectiveness of coupling on the size the gap between the waveguides Δz (c) and the width of the output waveguide W_2 (*d*).

dependence of the effectiveness of coupling (the ratio of the intensity of light at the exit of the narrow waveguide I to the intensity at the entrance to the wide waveguide I_0) on the distance between the waveguides Δz. Figure 5.19c shows that the most effective coupling of 73% is achieved at the gap between the waveguides equal to 0.6 μm. Note that in the gap between the waveguides there is the waveguide material ($n = 1.46$) and not air.

Figure 5.19d shows the calculated dependence of the efficiency of coupling for the case shown in Fig. 5.19a on the width of the output waveguide W_2 at the gap size of $\Delta z = 1$ μm. Figure 5.19d shows that with increasing width of the output waveguide W_2 the efficiency of coupling almost linearly increases.

Production of two 2D waveguides connected by PCGL

Planar waveguides by the scheme in Fig. 5.19a were recorded on a PMMA resist by direct writing technology with an electron beam at a voltage of 30 kV with a lithograph ZEP520A (University of St Andrews, Scotland). Processing of the resist in order to eliminate sections 'illuminated' by the electron beam was carried out using xylene. After that, the mixture of gases CHF_3 and SF_6 and the technology of reactive ion etching (RIE) were used for further plasma etching of the materials. That is, the pattern of 2D waveguides with PCGL (Fig. 5.19a) was transferred to a silicon film (SOI technology: silicon-on-insulator): a silicon film 220 nm thick on a fused silica layer with thickness of 2 μm. The etching depth was about 300 nm. The diameter of holes in the PCGL ranged from 160 nm to 200 nm. The length of the entire sample (length of the two waveguides) was 5 mm. Several similar structures were produced simultaneusly on the same substrate differing in the gaps between the waveguides $\Delta z = 0$ μm, 1 μm, 3 μm and a few structures that differed in the offset between the axes of the two waveguides is $\Delta x = 0$ μm, ± 0.5 μm, ± 1 μm. Figure 5.20 shows a magnified (7000 times) photograph (top view) of two waveguides with an gap of $\Delta z = 1$ μm and with PCGL produced with a scanning electron microscope.

The parameters of the sample in Fig. 5.20 as the following. The design width of the waveguide $W_1 = 4.5$ μm and $W_2 = 1$ μm, PCGL consists of a 12×17 matrix of holes with a period of 250 nm.

Figure 5.21 shows a part of the relief profile of the two waveguides fabricated in a silicon film on fused silica (a) and a section of the matrix of 6×6 holes of the PCGL (b) obtained with a scanning probe microscope.

Figure 5.22 shows the sections of the output (a) and input (b) waveguides. It is seen that the depth of etching of both waveguides is about the same and equal to 300 nm, and the width of the output waveguide at the tip is 1 μm, and at the base 2 μm (Fig. 5.22a). Similarly, the width of the input waveguide at the tip is 4.5 μm and at the base of the trapezoid 5 μm (Fig. 5.22b).

Characterization of two waveguides with PCGL

Figure 5.23 shows the optical arrangement for the transmission spectrum of two planar waveguides connected by PCGL. The wideband light source (1450–1700 nm), operating on the basis of amplified spontaneous emission, is coupled with an optical fibre. The light at the output from the fibre is collimated and is incident on

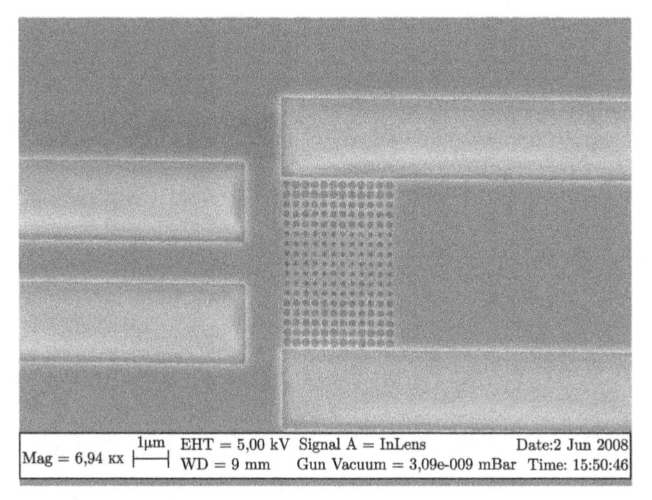

Fig. 5.20. Photograph of planar waveguides made in a silicon film coupled with PCGL and obtained with a scanning electron microscope with a magnification of 7000.

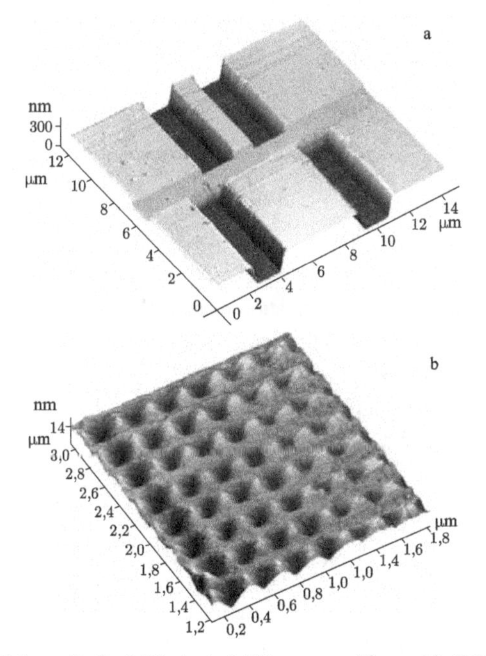

Fig. 5.21. Profile of the relief of fabricated 2D waveguides with PCGL obtained with a scanning probe microscope (atomic force microscope) SolverPro (Zelinograd): profile of waveguides (a), section 6×8 matrix of holes in PCGL (b). On the horizontal axes there are microns, on the vertical axis – nanometers.

a polarizer which separates TE-polarization. Further, using the microscope objective the radiation is focused on the surface of the input waveguide. A small fraction of light energy enters the waveguide and passes through the sample.

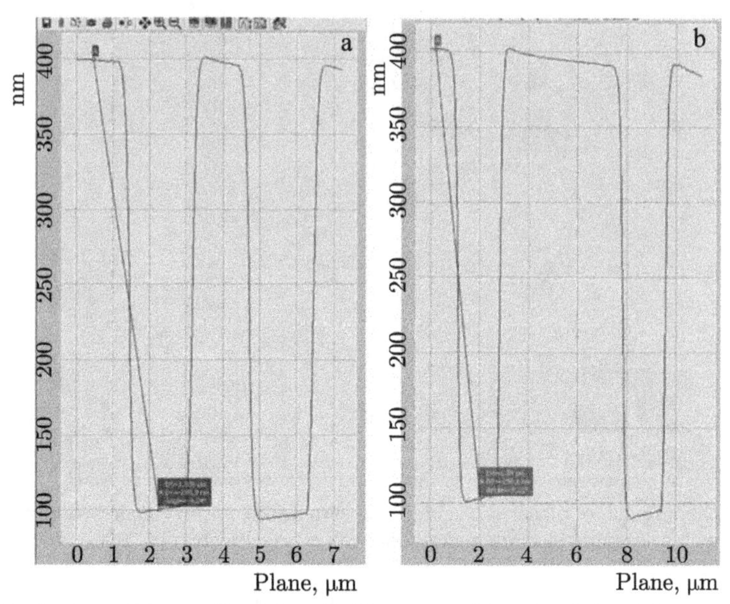

Fig. 5.22. Sections of the profile of the relief for narrow (s) and wide input (b) waveguides.

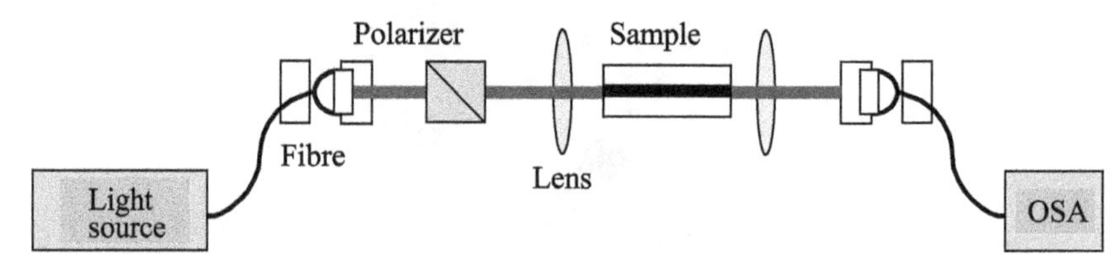

Fig. 5.23. Optical scheme for the investigation of nanophotonic devices, consisting of two waveguides and PCGL.

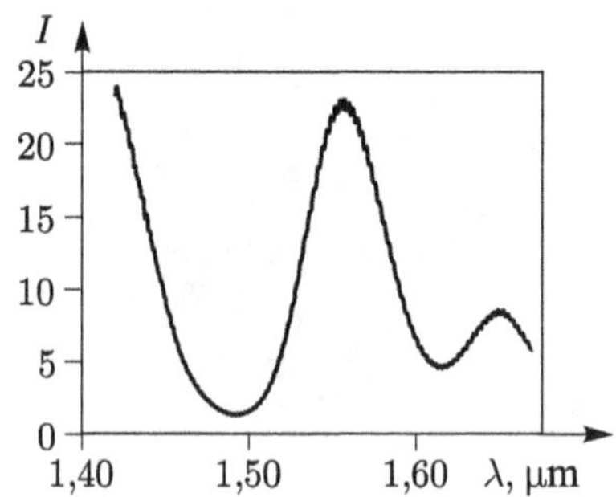

Fig. 5.24. The emission spectrum of the light source used in the optical system shown in Fig. 5.23.

At the exit of the narrow waveguide there is a second microlens, which gathers the light and focuses it on the input end of a multimode optical fibre connected to an optical spectrum analyzer (OSA). Figure 5.24 shows the spectrum of the radiation source, whose maximum is at the wavelength of 1.55 μm. The intensity of radiation is given in arbitrary units.

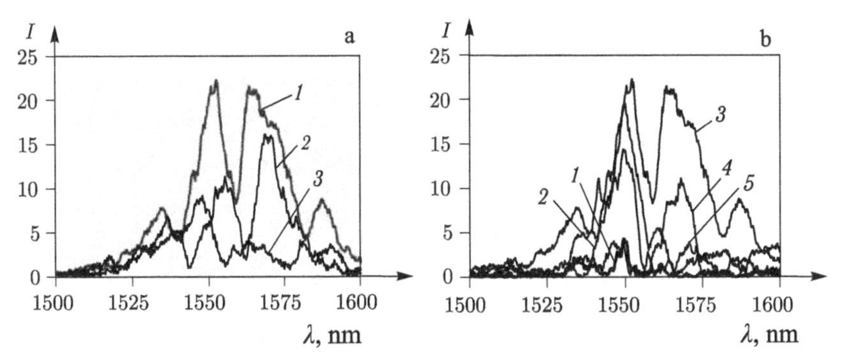

Fig. 5.25. The transmission spectra measured by the optical system shown in Fig. 5.23 for the samples shown in Fig. 5.20 at the following gaps between the waveguides (a): $\Delta z = 0$ μm (curve 1), $\Delta z = 1$ μm (curve 2) and $\Delta z = 3$ μm (curve 3), as well as at the following offsets from the optical axis of the output waveguide (b): $\Delta x = 0$ (curve 1), $\Delta x = -0.5$ μm (curve 2), $\Delta x = 0.5$ μm (curve 3), $\Delta x = -1$ μm (curve 1) and $\Delta x = 1$ μm (curve 5).

Figure 5.25 shows the transmission spectra of the samples in the range of 1.5–1.6 μm at the following gaps Δz between the waveguides on the optical axis (a): 0 μm (curve 1), 1 μm (curve 2) and 3 μm (curve 3), as well as the following displacement Δx from the optical axis of the output waveguide (B): $\Delta x = 0$ (curve 3), $\Delta x = -0.5$ μm (curve 2), $\Delta x = 0.5$ μm (curve 3), $\Delta x = -1$ μm (curve 4) and $\Delta x = +1$ μm (curve 5). Figure 5.25a (curve 1) shows that the transmission spectrum has four local maxima at wavelengths around 1535 nm, 1550 nm, 1565 nm and 1590 nm. Two of these peaks (at wavelengths of 1550 nm and 1565 nm) have the intensity 3 times greater than the other two. This is most likely due to the fact that the intensity of radiation spectrum of the source (Fig. 5.24) is several times smaller at these wavelengths.

With increasing axial distance $\Delta z = 1$ μm between the waveguides (Fig. 5.25a, curve 2) the transmission spectrum on average retains its structure, but the local maxima decrease in magnitude and are shifted to the 'red' region of the spectrum. With further increase of the distance $\Delta z = 3$ μm between the waveguides (Fig. 5.25a, curve 3) the local maxima not only further decrease but also acquire a 'blue' shift. 'Red' shift is about 10 nm, and 'blue' shift is also –10 nm (to a maximum near the central wavelength of 1.55 μm). Figure 5.25b shows that at the displacement of the output waveguide with the optical axis by 1 μm (curves 1 and 5) the intensity of the output is reduced by 8 times (wavelength 1.55 μm). This means that the diameter of the focal spot, formed by PCGL in silicon, is less than 1 μm.

To compare experiment with theory, we compared the transmission spectra. Figure 5.26 shows the smoothed experimental transmission spectrum (a) of the nanophotonic device (Fig. 5.20, but with no gap) and the calculated spectrum (b). Figure 5.26 shows that the two peaks of the spectrum in both cases occur at the same wavelengths (1535 nm and 1550 nm), the third maximum is shifted by 5 nm,

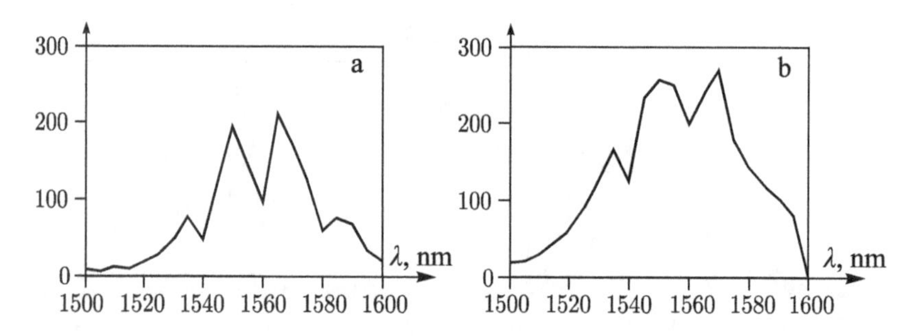

Fig. 5.26. The smoothed experimental (a) and calculated (b) transmission spectra for the two waveguides with PCGL without gaps (Fig. 5.20) in the wavelength range 1.5–1.6 μm. The vertical axis – arbitrary units, horizontal axis – nanometers.

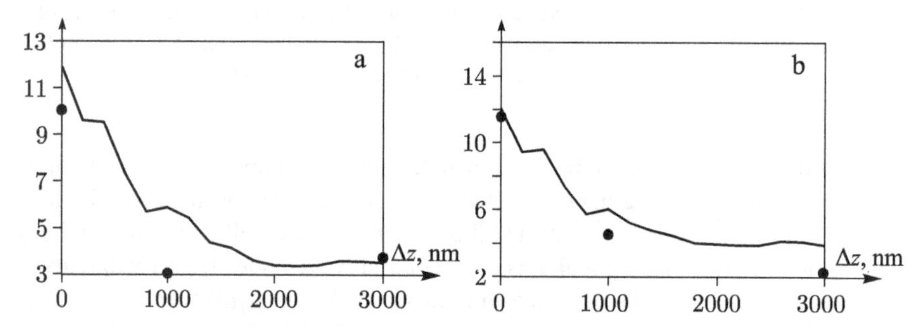

Fig. 5.27. Dependence of the transmission function of the two waveguides with PCGL on the gap between then Δz for the wavelengths 1550 nm (a) and 1565 nm (b): continuous curve – calculation, the individual dots – experiment. The vertical axis – arbitrary units along the abscissa – nanometers.

and 4th maximum does not occur in the calculations. The standard deviation of the two spectra in Fig. 5.26 was 29%.

Another comparison between theory and experiment shown in Fig. 5.27. This figure shows the dependence of the transmission function of the device (Fig. 5.20) in arbitrary units on the distance Δz between two coaxial waveguides for the wavelengths of 1550 nm (a) and 1565 nm (b): the continuous curve shows the calculations and the experiments are indicated by individual dots.

Figure 5.27 shows that on average the calculations and experiments consistently reflect a decrease in the intensity at the output of the waveguide as the distance between the waveguides increases. Note that the experimental points (all except one) lie below the theoretical curve, which is caused by absorption of light in real waveguides with a length of 5 mm. Figure 5.27 also shows that the calculated curve of the output intensity is reduced by half when the gap between the waveguides is $\Delta z = 600$ nm, which agrees with the longitudinal size of the focal spot of the PCGL (Fig. 5.16c).

5.3. Sharp focusing of radially-polarized light

A large number of optical devices use sharp focusing of laser light: optical memory drives, photolithography, confocal microscopy, optical micromanipulation. Studies of the formation of the minimum focal spot with superresolution extends are continuing. The scalar paraxial theory of diffraction shows that the diameter of the focal spot of the Airy disk, the amplitude of which is described by the function $2J_1(x)/(x)$ at half intensity is equal to FWHM = 0.51λ/NA, where λ is the wavelength of light, NA is the numerical aperture of the focusing lens. The area of the Airy disk at half intensity is HMA = $0.204\lambda^2$ (HMA = half maximum area) with NA = 1. This is the area of a two-dimensional region bounded by a closed curve drawn in the focal plane of the diffraction pattern at the points where the light intensity is equal to half the maximum value. If we use a narrow annular aperture and a spherical lens to form a diffraction pattern, which describes the Bessel function $J_0(x)$, then the diameter of the focal spot is smaller FWHM = 0.36λ/NA. The diameter of the focal spot at sharp focused light depends on the type of polarization of the beam and the type of optical element, performing focusing. It has been experimentally shown [57] that with the help of a Leico microlens plan apo 100x with NA = 0.9 in the air the laser beam with radial polarization can be focused to a spot with the area HMA = $0.16\lambda^2$ and the diameter FWHM = 0.451λ. This is carried out using the fundamental mode of a helium–neon laser with a wavelength of 632.8 nm and a ring-shaped amplitude mask, covering the centre part 3 mm in diameter of the incident beam with a diameter 3.6 mm. In [57] it is also indicated that, in theory, for a laser beam with linear polarization of the field in the same conditions we can expect that the focal spot will have a large area of HMA = $0.26\lambda^2$ and larger diameter FWHM = 0.575λ. If the beam is circularly polarized, HMA = $0.22\lambda^2$. Note that if the focal spot is circular, the HMA = $\pi D^2/4$, where FWHM = D.

Most works on modelling of sharp focusing of the laser beam use the Debye vector theory or the analogous Richards–Wolf theory. In these theories, the electromagnetic field in the image of a point source, situated at infinity, by the aplanatic optical system is expressed in integral form as an expansion in plane waves. Some studies use the Rayleigh–Sommerfeld diffraction theory. Thus, in [28] on the basis of the Debye formulas that are true if the focal length is much larger than the wavelength, it is shown that with the aid of a parabolic mirror or a flat diffractive lens with numerical aperture NA = 0.98 the radially polarized hollow Gaussian beam with the amplitude $r\exp(-r^2/w^2)$, where r is the radial coordinate, w is the waist radius of the Gaussian beam, can be focused by an aplanatic lens in the focal spot with the area of HMA = $0.210\lambda^2$ and HMA = $0.157\lambda^2$ respectively. It was also calculated [58] that for a parabolic mirror with a numerical aperture NA = 1 the area of the focal spot is less than HMA = $0.154\lambda^2$. And if the Gaussian beam is restricted by a narrow annular aperture, the area of the focal spot will be even less HMA = $0.101\lambda^2$.

In [59] attention is given to the non-paraxial propagation of spirally polarized Laguerre–Gauss beams (LG). It is shown that such beams are also candidates for sharp focusing. To obtain radially-polarized laser beams, we can use a conventional interferometer with two spiral phase plates, giving a delay at half wavelength and rotated relative to each other by 90° around the optical axis [60]. In [61], the Richards–Wolf (RW) formula is used to simulate focusing of a linearly polarized beam with a planar aplanatic lens with a high numerical aperture with an annular (2 or 3 rings) binary phase plate. The parameters of the mask at which superresolution by 20% along the optical axis is achieved. Using the RW formulas, in [62] it is shown that by focusing radially-polarized TEM_{11} laser mode, which has two light rings in its cross-section, with the help of an aplanatic lens with NA = 1.2 in water (n = 1.33) a dark area appears in the focal region surrounded on all sides by light (optical bottle). The longitudinal dimension of this field is 2λ, and transverse is λ. In [63], using the RW-formulas the authors considered optimal distribution of the optical vortex with circular polarization. It is shown that at the topological charge $n = 1$ and the choice of such a sign that the spiral spin of the phase of an optical vortex compensates the rotation of the polarization in the opposite direction, in the focal plane (NA = 0.9) there is a circular focal spot with the diameter smaller than the wavelength.

With the help of the Rayleigh–Sommerfeld (RS) integral the authors of [64] studied the non-paraxial propagation of the LG modes with radial polarization but no spiral phase component. It is shown that when the non-paraxiality parameter $f = (kw)^{-1}$, where k is the wave number of light, w is the waist radius of the Gaussian beam, is selected equal to 0.5, the light spot diameter at the Fresnel distance from the waist is about 0.4λ for $p = 3$, where p is the order of the Laguerre polynomial. In [65] the Maxwell's equations in the cylindrical coordinates are solved using a series expansion of the non-paraxiality parameter $f = \theta/2$, where θ is the angle of diffraction, with the accuracy up to θ^5. As an example we consider the diffraction of the beam of the axicon–Gauss type. It is shown that at the diffraction angle $\theta = 0.75°$ the waist radius is 0.424λ. On the basis of the RS-integral analytical expressions were obtained in [66] describing the non-paraxial propagation of the elegant LG modes in the cross section of which there is always an annular intensity distribution. In [67] the authors reported on a new form of resist for lithography PMMA-DR1, which has polarization-filtering properties and responds only to the longitudinal component of the electric vector of the electromagnetic wave. It was shown experimentally that a radially-polarized beam of an argon laser $\lambda = 514$ nm, passing an axicon with NA = 0.67, forms a focal spot with a diameter FWHM = 0.89λ, but after writing on the resist a spot with the size of 0.62λ forms.

Using the RW formulas it is shown in [68] that radially polarized higher laser modes R-TEM_{p1} can reduce the diameter of the focal spot. Thus, when NA = 1, and by focusing with aplanatic lenses for numbers of the modes $p = 0, 1, 2, 3$, we obtain focal spots with a diameter of FWHM = 0.582λ, 0.432λ, 0.403λ, 0.378λ. In [69] using the RW formulas attention is given to vector diffraction and focusing by an aplanatic lens of a linearly-polarized beam with elliptical radial symmetry with an eccentricity of 0.87. The numerical aperture was NA = 0.9. In this case an elliptical focal spot area of HMA = $0.56\lambda^2$ formed. In [70] the non-paraxial propagation (5th order corrections) of radially-polarized LG beams R-TEM_{p1} was studied. It is shown that when the angle of diffraction $\theta = 2(kw)^{-1}$ is more than 0.5 the non-paraxial corrections of 5th order are no longer sufficient to describe the mode R-TEM_{21}. In

[71] using the RW formula it is shown that when illuminating the exit pupil of a spherical lens with plane, Gaussian or Bessel–Gaussian beams of radially-polarized light the localized focal spot diameter is equal to FWHM = 0.6λ, 1.2λ, 1.4λ respectively, at NA = 1.4, λ = 632.8 nm, n = 1.5. In [72] also using the RW formulas it is shown that for the incident radially polarized Bessel–Gaussian beam $J_1(2r)\exp(-r^2)$ and a binary phase Fresnel plate the focal spot has a diameter of FWHM = 0.425λ/NA. When a three-zone optimized plate was added to the Fresnel zone plate, the resultant diameter of the focal spot was even smaller, FWHM = 0.378λ/NA.

With the help of a parabolic mirror with a diameter of 19 mm and NA = 0.999, and with a radially polarized laser beam with a wavelength of 632.8 nm the authors of [73] experimentally obtained a focal spot with the least to date area of HMA = $0.134\lambda^2$. Radial polarization was obtained from the linear polarization of the laser beam with four half-wave plates, arranged in four quadrants of the aperture of the beam and rotated by 45° (along the bisector in each quadrant). The incident beam had an Bessel–Gauss amplitude. Modelling was carried out with the aid of the Debye formulas. The distribution of the intensity in the focal plane was measured with a fluorescent bulb with a diameter of 40 nm. For comparison, recall that the radius of the Airy disk in the scalar approximation is equal to 0.61λ at NA = 1, and in [73] a focal spot with a radius from maximum to first minimum was equal to 0.45λ. Recall that the best experimental results for aplanatic lenses is HMA = $0.16\lambda^2$ [57].

In [74] the authors proposed and experimentally tested a method way to convert linear polarization to radial or azimuthal using photonic crystal fibre of length 24 mm. In [75] calculations were carried out to determine the forces acting from the the focused laser beams (NA = 1.25 in water, angle of convergence 140°) on a spherical particle with the refractive index n = 1.59 and a radius equal to $4\lambda/n$. Several types of incident beams were considered: Gaussian, LG mode with radial, azimuthal and circular polarization. It was found that the efficiency of optical trapping is higher for radial polarization of light beams. In [76] the authors investigated numerically and experimentally a radially polarized laser beam, which is focused through a uniaxial crystal at the output of a Nd laser with a hemispherical cavity.

An interesting result was obtained in [77], where with the aid of the Debye formula it was shown that radially-polarized LG modes of even orders LG_p^0 with numerical aperture NA = 0.85 after passage through a special ring amplitude mask are focused in a small focal spot with almost no side lobes, with the area HMA = $0.276\lambda^2$. Using the amplitude mask does not reduce the size of the focal spot, but reduces the level of side lobes in the focal diffraction pattern and also five times the depth of field.

In [78] the scalar version of the RW formulas was used for analytical investigation of the function of the output pupil optimum for obtaining high resolution. In the two-dimensional case, a photonic crystal lens can be used for the radiation propagating in a planar waveguide and to focus this radiation at the output of the waveguide. In [56] it is shown that the 2D photonic crystal lens, which implements a gradient Mikaelian microlens (refractive index decreases according to the function of the hyperbolic secant), allows to focus the laser light in a focal spot smaller than that predicted by the scalar theory. In the paraxial case, the two-dimensional diffraction pattern in focus is described by the sinc-function for which FWHM = 0.44λ/NA.

The photonic crystal lens can also be used to produce a spot with a diameter of FWHM = 0.42λ with numerical aperture NA = 0.67. In [79], the FDTD method is applied to simulate the focusing of linearly polarized microwave radiation with a frequency of 30 GHz (wavelength λ = 10 mm) with binary phase Fresnel lenses made of a material with a dielectric constant ε = 4. Focal spots for different lenses were formed at distances of 2λ, λ and 0.5λ from the flat surface of the Fresnel lens and had the respective diameters: 1.04λ, 0.90λ and 0.80λ (here the diameter is twice the radius of the maximum intensity to the first minimum). The shape of the focal spot was close to a square.

In this section, using the R-FDTD-method, which applies to the radially symmetric case, we review the results of simulation of sharp focusing of a plane electromagnetic wave with linear, azimuthal and radial polarization using the micro-optics element: a biconvex spherical lenses, gradient lenses and conical axicon. It is shown that the smallest focal spot can be achieved by focusing a radially-polarized ring-shaped Gaussian beam on a micro-axicon with a numerical aperture NA = 0.65. The area of the focal spot at half intensity is HMA = 0.096λ^2, and the diameter of FWHM = 0.35λ.

5.3.1. Richards–Wolf vector formulas

According to the Debye vector theory, the vector of the strength of the electric field of the electromagnetic wave in the focal region in cylindrical coordinates (r, ψ, z) is expressed through the amplitude $I(\theta)$ of a converging spherical wave in the coordinates of the exit pupil of the aplanatic optical system in the form of (linear polarization vector is directed along the axis y):

$$\begin{aligned} E_x(r,\psi,z) = \frac{-iA}{2\pi}\int_0^{\alpha} d\theta \int_0^{2\pi} d\phi \sin\theta\sqrt{\cos\theta}\sin 2\phi \times \\ \times(1-\cos\theta)I(\theta)\exp\left[ikz\cos\theta - ikr\sin\theta\cos(\psi-\phi)\right], \end{aligned} \tag{5.30}$$

$$\begin{aligned} E_y(r,\psi,z) = \frac{iA}{2\pi}\int_0^{\alpha} d\theta \int_0^{2\pi} d\varphi \sin\theta\sqrt{\cos\theta} \times \\ \times\left[(1+\cos\theta)+(1-\cos\theta)\cos 2\phi\right]\times \\ \times I(\theta)\exp\left[ikz\cos\theta - ikr\sin\theta\cos(\psi-\phi)\right], \end{aligned} \tag{5.31}$$

$$\begin{aligned} E_z(r,\psi,z) = \frac{iA}{\pi}\int_0^{\alpha} d\theta \int_0^{2\pi} d\phi \sin^2\theta\sqrt{\cos\theta}\cos\phi \times \\ \times I(\theta)\exp\left[ikz\cos\theta - ikr\sin\theta\cos(\psi-\phi)\right], \end{aligned} \tag{5.32}$$

where A is a constant, α = arcsin (NA), k is the wave number of the light. For example, a Gaussian function in the pupil plane will have the form:

$$l(\theta) = \exp\left(\frac{-\rho^2}{w^2}\right) = \exp\left[-\left(\frac{\beta \sin\theta}{\sin\alpha}\right)^2\right], \tag{5.33}$$

where β is a constant.

Richards and Wolf, on the basis of the Debye formulas (5.30)–(5.32), obtained simpler formulas, integrating with respect to the azimuthal angle ϕ for radially-polarized light [80]:

$$E_r(r,z) = A\int_0^\alpha \sin 2\theta \sqrt{\cos\theta}\; l(\theta) \exp\left[ikz\cos\theta\right] J_1(kr\sin\theta) d\theta, \tag{5.34}$$

$$E_z(r,z) = 2iA\int_0^\alpha \sin^2\theta \sqrt{\cos\theta}\; l(\theta) \exp\left[ikz\cos\theta\right] J_0(kr\sin\theta) d\theta, \tag{5.35}$$

where $J_0(x)$ and $J_1(x)$ are Bessel functions. Equations (5.34) and (5.35) show that the radially-polarized wave does not depend on the angle ψ, has only two electrical components E_r and E_z, and it can also be seem that E_r $(r = 0) = 0$ for any z and any function $l(\theta)$. The focus is at the origin of the coordinates $(r, \psi, z) = (0, \psi, 0)$. When replacing an aplanatic lens by a Fresnel zone plate, instead of the factor $(\cos\theta)^{1/2}$ in (5.34) and (5.35) we should use another factor $(\cos\theta)^{-3/2}$ [58].

In the Cartesian coordinates for linearly polarized light (polarization vector is directed along the axis y) the RW formulas take the form:

$$\begin{aligned} E_x(r,\psi,z) &= -iA\sin 2\psi \int_0^\alpha \sin\theta \sqrt{\cos\theta} \times \\ &\times l(\theta)(1-\cos\theta)\exp\left[ikz\cos\theta\right] J_2(kr\sin\theta) d\theta, \end{aligned} \tag{5.36}$$

$$\begin{aligned} E_y(r,\psi,z) = &-iA\cos 2\psi \int_0^\alpha \sin\theta \sqrt{\cos\theta} l(\theta)(1-\cos\theta)\exp\left[ikz\cos\theta\right] J_2(kr\sin\theta) d\theta - \\ &-iA\int_0^\alpha \sin\theta \sqrt{\cos\theta} l(\theta)(1+\cos\theta)\exp\left[ikz\cos\theta\right] J_0(kr\sin\theta) d\theta, \end{aligned} \tag{5.37}$$

$$\begin{aligned} E_z(r,\psi,z) &= -2A\cos\psi \int_0^\alpha \sin^2\theta \sqrt{\cos\theta} \times \\ &\times l(\theta)\exp\left[ikz\cos\theta\right] J_1(kr\sin\theta) d\theta. \end{aligned} \tag{5.38}$$

In the case of azimuthal polarization of light instead of the radial component E_r in (5.34) the azimuthal component of the electric field differs from zero:

$$E_\psi(r,z) = 2A\int_0^\alpha \sin\theta\sqrt{\cos\theta}l(\theta)\exp\left[ikz\cos\theta\right]J_1(kr\sin\theta)d\theta. \quad (5.39)$$

From (5.35) and (5.39) is can be seed that the azimuthally polarized wave does not depend on angle ψ. these formulas (5.30)–(5.39) were used for simulation in [57, 58, 61–63, 66, 68, 69, 71–73, 75, 77], studying the sharp focusing of laser light.

5.3.2. The minimum focal spot: an analytical estimation

In [78] the intensity distribution of light at the focus of a radially-symmetrical optical systems with high numerical aperture is analyzed using the scalar form of the RW formula:

$$U(r,z) = -ikf\int_0^\alpha P(\theta)\exp\left[ikz\cos\theta\right]J_0(kr\sin\theta)d\theta, \quad (5.40)$$

where $U(r, z)$ is the complex amplitude of the light near the focus, $P(\theta)$ is the function of the pupil of the optical system, f is the focal length. If we compare (5.40) with (5.35) for the longitudinal component of the radially-polarized light, we can conclude that the scalar amplitude, describing the non-paraxial focusing of light (5.40), is proportional to the longitudinal component of a converging spherical wave with radial polarization (5.35). Using the reference integrals in [81], we can estimate the minimum possible diameter at the focal spot of the non-paraxial optical system. Using the reference integral

$$\int_0^\pi \begin{Bmatrix}\sin(bx)\\ \cos(bx)\end{Bmatrix} J_\nu(c\sin x)dx = \pi\begin{Bmatrix}\sin(b\pi/2)\\ \cos(b\pi/2)\end{Bmatrix} J_{(\nu-b)/2}(c/2)J_{(\nu+b)/2}(c/2), \quad (5.41)$$

and setting in (5.40) $P(\theta) = \sin\theta$, $\alpha = \pi$, and in (5.41) $\nu = 0$, $b = 1$, $c = kr$, we obtain from equation (5.40) for the uniform pupil in the focal plane $z = 0$ the following complex amplitude:

$$U_1(r,z=0) = -2ikf\sin(kr)/(kr). \quad (5.42)$$

It follows from equation (5.42) that the minimum focal spot diameter (twice the distance from the maximum to first minimum) is

$$D_1 = \lambda, \quad (5.43)$$

and the diameter of the focal spot at half intensity is FWHM = 0.44λ, and the spot area at half intensity is HMA = $0.152\lambda^2$. The latter figure is consistent with the calculations in [58].

The result with the same order of magnitude can be obtained by choosing a uniform pupil function in the form of $P(\theta) = 1$. Then instead of (5.40) with (5.41) we obtain ($v = 0$, $b = 0$):

$$U_2(r, z=0) = -ikf\pi J_0^2(kr/2). \tag{5.44}$$

From (5.44) it follows that in this particular case, the focal spot diameter (twice the distance from the maximum to first minimum of intensity) is equal to

$$D_2 = 1.53\lambda. \tag{5.45}$$

If the focusing lens is illuminated with a narrow annular field with the pupil function $P(\theta) = \delta(\theta - \alpha)$, then from (5.40) we obtain for the amplitude in the focus:

$$U_3(r, z=0) = -2ikfJ_0(krNA). \tag{5.46}$$

From (5.46) it follows that the diameter of the focal spot, similar to (5.43) and (5.45) is equal to (NA = 1)

$$D_3 = 0.76\lambda, \tag{5.47}$$

and the diameter of such a focal spot at half intensity is FWHM = 0.36λ, and the spot area at half intensity is HMA = $0.101\lambda^2$. The latter figure is consistent with the calculations in [58].

The formulas (5.43) (5.45) and (5.47) give only an estimate of the minimum diameter of the focal spot with a scalar formula (5.40), but in sharp focusing it is required to take into account the vector nature of the field, when all three components of the electric field give a comparable contribution to the formation of focal pattern. The value of (5.47) can be considered as the accurate minimum diameter of the focal spot, which can be formed by the focusing optical system illuminated by radially-polarized light. This follows from the fact that the scalar equation (5.40) coincides with the expression for the longitudinal component of the field (5.35), while the radial component (5.34) of radially-polarized light at the optical axis is zero. But the Debye and Richards–Wolf formulas are approximate (they were obtained under the condition that the focal length of the optical system is much larger than the wavelength), therefore we will consider the rigorous solution of the diffraction problem on the basis of the numerical solution of Maxwell's equations. Only at the exact solution of the diffraction problem, and if the focal length is comparable to the wavelength, can we hope to obtain the area of the focal spot smaller than HMA = $0.101\lambda^2$.

5.3.3. Maxwell's equations in cylindrical coordinates

In [82] a method was proposed for calculating diffraction of the electromagnetic wave on a radially symmetric optical element on the basis of the difference solution of Maxwell's equations in cylindrical coordinates. The Maxwell's equations in the cylindrical coordinates (r, φ, z) in SI units are as follows:

$$\frac{1}{r}\frac{\partial H_z}{\partial \phi}-\frac{\partial H_\phi}{\partial z}=\varepsilon\varepsilon_0\frac{\partial E_r}{\partial t}+\sigma E_r, \quad (5.48)$$

$$\frac{\partial H_r}{\partial z}-\frac{\partial H_z}{\partial r}=\varepsilon\varepsilon_0\frac{\partial E_\phi}{\partial t}+\sigma E_\phi, \quad (5.49)$$

$$\frac{1}{r}\frac{\partial\left(rH_\phi\right)}{\partial r}-\frac{1}{r}\frac{\partial H_r}{\partial \phi}=\varepsilon\varepsilon_0\frac{\partial E_z}{\partial t}+\sigma E_z, \quad (5.50)$$

$$\frac{1}{r}\frac{\partial E_z}{\partial \phi}-\frac{\partial E_\phi}{\partial z}=-\mu\mu_0\frac{\partial H_r}{\partial t}, \quad (5.51)$$

$$\frac{\partial E_r}{\partial z}-\frac{\partial E_z}{\partial r}=-\mu\mu_0\frac{\partial H_\phi}{\partial t}, \quad (5.52)$$

$$\frac{1}{r}\frac{\partial\left(rE_\phi\right)}{\partial r}-\frac{1}{r}\frac{\partial E_r}{\partial \phi}=-\mu\mu_0\frac{\partial H_z}{\partial t}, \quad (5.53)$$

where μ and ε are relative magnetic and electric permeability, μ_0 and ε_0 are magnetic and electric permittivity of vacuum, σ is conductivity, E_v and H_v are the amplitude of the electric and magnetic fields, the index v takes the values r, ϕ, z. We expand the components of the electromagnetic field in a Fourier series in the azimuthal angle ϕ:

$$E_\gamma(r,z,\phi,t)=\frac{E_{\gamma 0}(r,z,t)}{2}+ \\ +\sum_{k=1}^{\infty}\left[E_{\gamma,k}^{(1)}(r,z,t)\cos(k\phi)+E_{\gamma,k}^{(2)}(r,z,t)\sin(k\phi)\right], \quad (5.54)$$

$$H_\gamma(r,z,\phi,t)=\frac{H_{\gamma 0}(r,z,t)}{2}+ \\ +\sum_{k=1}^{\infty}\left[H_{\gamma,k}^{(1)}(r,z,t)\cos(k\phi)+H_{\gamma,k}^{(2)}(r,z,t)\sin(k\phi)\right]. \quad (5.55)$$

Substituting (5.54) and (5.55) into (5.48)–(5.53), we can calculate the derivatives of ϕ. Then instead of (5.48)–(5.53) for $k = 0$, we have six of the Maxwell equations in the functions that do not depend on the angle ϕ:

$$-\frac{\partial H_{\phi,0}}{\partial z}=\varepsilon\varepsilon_0\frac{\partial E_{r,0}}{\partial t}+\sigma E_{r,0}, \quad (5.56)$$

$$\frac{\partial H_{r,0}}{\partial z} - \frac{\partial H_{z,0}}{\partial r} = \varepsilon\varepsilon_0 \frac{\partial E_{\phi,0}}{\partial t} + \sigma E_{\phi,0}, \tag{5.57}$$

$$\frac{1}{r}\frac{\partial\left(rH_{\phi,0}\right)}{\partial r} = \varepsilon\varepsilon_0 \frac{\partial E_{z,0}}{\partial t} + \sigma E_{z,0}, \tag{5.58}$$

$$-\frac{\partial E_{\phi,0}}{\partial z} = -\mu\mu_0 \frac{\partial H_{r,0}}{\partial t}, \tag{5.59}$$

$$\frac{\partial E_{r,0}}{\partial z} - \frac{\partial E_{z,0}}{\partial r} = -\mu\mu_0 \frac{\partial H_{\phi,0}}{\partial t}, \tag{5.60}$$

$$\frac{1}{r}\frac{\partial\left(rE_{\phi,0}\right)}{\partial r} = -\mu\mu_0 \frac{\partial H_{z,0}}{\partial t}. \tag{5.61}$$

At an arbitrary integer $k \neq 0$, the amplitudes of the angular harmonics $E^{(1)}$, $E^{(2)}$, $H^{(1)}$ and $H^{(2)}$ from (5.54) and (5.55) are related by the following 12 equations:

$$-\frac{1}{r}kH_{z,k}^{(1)} - \frac{\partial H_{\phi,k}^{(2)}}{\partial z} = \varepsilon\varepsilon_0 \frac{\partial E_{r,k}^{(2)}}{\partial t} + \sigma E_{r,k}^{(2)}, \tag{5.62}$$

$$\frac{\partial H_{r,k}^{(2)}}{\partial z} - \frac{\partial H_{z,k}^{(2)}}{\partial r} = \varepsilon\varepsilon_0 \frac{\partial E_{\phi,k}^{(2)}}{\partial t} + \sigma E_{\phi,k}^{(2)}, \tag{5.63}$$

$$\frac{1}{r}\frac{\partial\left(rH_{\phi,k}^{(2)}\right)}{\partial r} - \frac{1}{r}kH_{r,k}^{(1)} = \varepsilon\varepsilon_0 \frac{\partial E_{z,k}^{(2)}}{\partial t} + \sigma E_{z,k}^{(2)}, \tag{5.64}$$

$$-\frac{1}{r}kE_{z,k}^{(1)} - \frac{\partial E_{\varphi,k}^{(2)}}{\partial z} = -\mu\mu_0 \frac{\partial H_{r,k}^{(2)}}{\partial t}, \tag{5.65}$$

$$\frac{\partial E_{r,k}^{(2)}}{\partial z} - \frac{\partial E_{z,k}^{(2)}}{\partial r} = -\mu\mu_0 \frac{\partial H_{\phi,k}^{(2)}}{\partial t}, \tag{5.66}$$

$$\frac{1}{r}\frac{\partial\left(rE_{\phi,k}^{(2)}\right)}{\partial r} - \frac{1}{r}kE_{r,k}^{(1)} = -\mu\mu_0 \frac{\partial H_{z,k}^{(2)}}{\partial t}, \tag{5.67}$$

$$-\frac{1}{r}kH_{z,k}^{(2)}-\frac{\partial H_{\phi,k}^{(1)}}{\partial z}=\varepsilon\varepsilon_0\frac{\partial E_{r,k}^{(1)}}{\partial t}+\sigma E_{r,k}^{(1)}, \tag{5.68}$$

$$\frac{\partial H_{r,k}^{(1)}}{\partial z}-\frac{\partial H_{z,k}^{(1)}}{\partial r}=\varepsilon\varepsilon_0\frac{\partial E_{\phi,k}^{(1)}}{\partial t}+\sigma E_{\phi,k}^{(1)}, \tag{5.69}$$

$$\frac{1}{r}\frac{\partial\left(rH_{\phi,k}^{(1)}\right)}{\partial r}-\frac{1}{r}kH_{r,k}^{(2)}=\varepsilon\varepsilon_0\frac{\partial E_{z,k}^{(1)}}{\partial t}+\sigma E_{z,k}^{(1)}, \tag{5.70}$$

$$-\frac{1}{r}kE_{z,k}^{(2)}-\frac{\partial E_{\phi,k}^{(1)}}{\partial z}=-\mu\mu_0\frac{\partial H_{r,k}^{(1)}}{\partial t}, \tag{5.71}$$

$$\frac{\partial E_{r,k}^{(1)}}{\partial z}-\frac{\partial E_{z,k}^{(1)}}{\partial r}=-\mu\mu_0\frac{\partial H_{\phi,k}^{(1)}}{\partial t}, \tag{5.72}$$

$$\frac{1}{r}\frac{\partial\left(rE_{\phi,k}^{(1)}\right)}{\partial r}-\frac{1}{r}kE_{r,k}^{(2)}=-\mu\mu_0\frac{\partial H_{z,k}^{(1)}}{\partial t}. \tag{5.73}$$

Note that in [82] of the 12 equations (5.62)–(5.73) only six equations are considered. The number of equations (5.62)–(5.73) could be reduced if the optical element on which diffraction is considered has a cylindrical symmetry and if the an electromagnetic wave with linear, radial or azimuthal polarization falls in the normal direction on this optical element (Fig. 5.28).

Linear polarization of the incident wave (let $E^{\text{inc}} = E_y$) means that at each point in the cross section the electric vector is directed along the axis y (Fig. 5.28a). In the case of azimuthal polarization the electric vector at any point in the incident wave is directed along the tangent to the circles whose centres lie on the optical axis (Fig. 5.28b). In the case of radial polarization the electric vector at any point of the section of the incident wave is directed along the radii of the circles whose centres lie

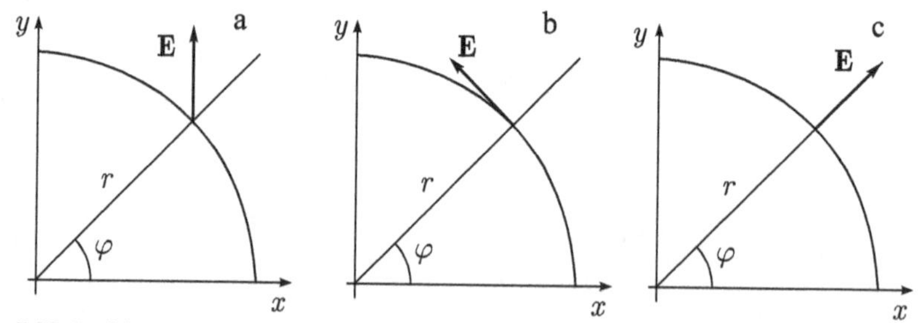

Fig. 5.28. Incident radiation on the optical element has linear (a), azimuthal (b) or radial (c) polarization.

on the optical axis which coincides with the symmetry axis of the optical element (Fig. 5.28c).

5.3.4. Maxwell's equations for the incident wave with linear polarization

Consider the normal incidence of an electromagnetic wave with linear polarization in the initial plane $z = 0$. Then the electric vector of this wave will have only one projection $E^{\text{inc}} = E_y = E_0(r)\cos(\omega t)$, where ω is the angular frequency of a monochromatic wave, $E_0(r)$ is the amplitude of the wave in the plane $z = 0$. We express the incident field E_y through the cylindrical components:

$$E_r = E_y \sin\phi, \quad E_\phi = E_y \cos\phi \tag{5.74}$$

or in the notation (5.54) and (5.55):

$$E_r = E_{r,1}^{(2)} \sin\phi, \quad E_\phi = E_{\phi,1}^{(1)} \cos\phi. \tag{5.75}$$

Thus, from equations (5.62) – (5.73) there are only six equations:

$$-\frac{1}{r}H_{z,1}^{(1)} - \frac{\partial H_{\varphi,1}^{(2)}}{\partial z} = \varepsilon\varepsilon_0 \frac{\partial E_{r,1}^{(2)}}{\partial t} + \sigma E_{r,1}^{(2)}, \tag{5.76}$$

$$\frac{1}{r}\frac{\partial\left(rH_{\varphi,1}^{(2)}\right)}{\partial r} - \frac{1}{r}H_{r,1}^{(1)} = \varepsilon\varepsilon_0 \frac{\partial E_{z,1}^{(2)}}{\partial t} + \sigma E_{z,1}^{(2)}, \tag{5.77}$$

$$\frac{\partial E_{r,1}^{(2)}}{\partial z} - \frac{\partial E_{z,1}^{(2)}}{\partial r} = -\mu\mu_0 \frac{\partial H_{\varphi,1}^{(2)}}{\partial t}, \tag{5.78}$$

$$\frac{\partial H_{r,1}^{(1)}}{\partial z} - \frac{\partial H_{z,1}^{(1)}}{\partial r} = \varepsilon\varepsilon_0 \frac{\partial E_{\phi,1}^{(1)}}{\partial t} + \sigma E_{\phi,1}^{(1)}, \tag{5.79}$$

$$-\frac{1}{r}E_{z,1}^{(2)} - \frac{\partial E_{\varphi,1}^{(1)}}{\partial z} = -\mu\mu_0 \frac{\partial H_{r,1}^{(1)}}{\partial t}, \tag{5.80}$$

$$\frac{1}{r}\frac{\partial\left(rE_{\varphi,1}^{(1)}\right)}{\partial r} - \frac{1}{r}E_{r,1}^{(2)} = -\mu\mu_0 \frac{\partial H_{z,1}^{(1)}}{\partial t}. \tag{5.81}$$

The system of equations (5.76)–(5.81) can be approximately solved by the finite-difference method on the Yee sampling grid [18], but in a cylindrical coordinate system (Fig. 5.29) as in [82].

The finite-difference approximation of equations (5.76)–(5.81) has the form ($\sigma = 0, \mu = 1$):

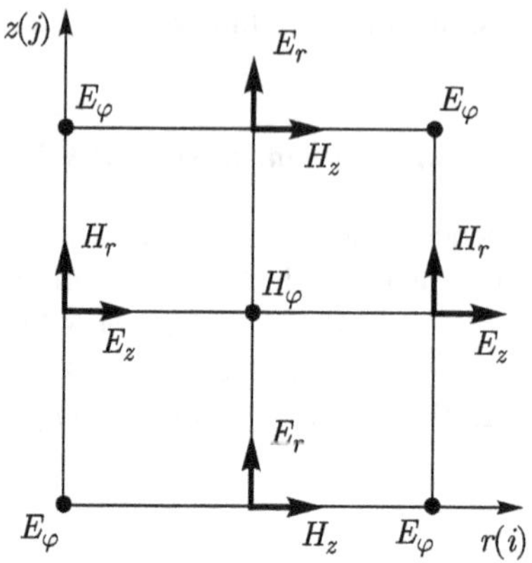

Fig. 5.29. Yee grid with a half-step in the cylindrical coordinates: *h* is the step in the spatial coordinates.

$$\varepsilon(i+\frac{1}{2},j)\varepsilon_0 \frac{E_{r,1}^{(2)n+1}\left(i+\frac{1}{2},j\right)-E_{r,1}^{(2)n}\left(i+\frac{1}{2},j\right)}{\Delta t}=$$

$$=-\frac{1}{r\left(i+\frac{1}{2}\right)}H_{z,1}^{(1)n+\frac{1}{2}}\left(i+\frac{1}{2},j\right)-\frac{H_{\varphi,1}^{(2)n+\frac{1}{2}}\left(i+\frac{1}{2},j+\frac{1}{2}\right)-H_{\varphi,1}^{(2)n+\frac{1}{2}}\left(i+\frac{1}{2},j-\frac{1}{2}\right)}{\Delta z}, \tag{5.82}$$

$$\varepsilon(i,j)\varepsilon_0 \frac{E_{\varphi,1}^{(1)n+1}(i,j)-E_{\varphi,1}^{(1)n}(i,j)}{\Delta t}=$$

$$=\frac{H_{r,1}^{(1)n+\frac{1}{2}}\left(i,j+\frac{1}{2}\right)-H_{r,1}^{(1)n+\frac{1}{2}}\left(i,j-\frac{1}{2}\right)}{\Delta z}-\frac{H_{z,1}^{(1)n+\frac{1}{2}}\left(i+\frac{1}{2},j\right)-H_{z,1}^{(1)n+\frac{1}{2}}\left(i-\frac{1}{2},j\right)}{\Delta r}, \tag{5.83}$$

$$\varepsilon\left(i,j+\frac{1}{2}\right)\varepsilon_0 \frac{E_{z,1}^{(2)n+1}\left(i,j+\frac{1}{2}\right)-E_{z,1}^{(2)n}\left(i,j+\frac{1}{2}\right)}{\Delta t}=$$

$$=\frac{1}{r(i)}\frac{r\left(i+\frac{1}{2}\right)H_{\varphi,1}^{(2)n+\frac{1}{2}}\left(i+\frac{1}{2},j+\frac{1}{2}\right)-r\left(i-\frac{1}{2}\right)H_{\varphi,1}^{(2)n+\frac{1}{2}}\left(i-\frac{1}{2},j+\frac{1}{2}\right)}{\Delta r}$$

$$-\frac{1}{r(i)}H_{r,1}^{(1)n+\frac{1}{2}}\left(i,j+\frac{1}{2}\right), \tag{5.84}$$

$$-\mu_0 \frac{H_{\varphi,1}^{(2)n+\frac{1}{2}}\left(i+\frac{1}{2},j+\frac{1}{2}\right)-H_{\varphi,1}^{(2)n-\frac{1}{2}}\left(i+\frac{1}{2},j+\frac{1}{2}\right)}{\Delta t}=$$

$$=\frac{E_{r,1}^{(2)n}\left(i+\frac{1}{2},j+1\right)-E_{r,1}^{(2)n}\left(i+\frac{1}{2},j\right)}{\Delta z}-\frac{E_{z,1}^{(2)n}\left(i+1,j+\frac{1}{2}\right)-E_{z,1}^{(2)n}\left(i,j+\frac{1}{2}\right)}{\Delta r}, \tag{5.85}$$

$$-\mu_0 \frac{H_{r,1}^{(1)n+\frac{1}{2}}\left(i,j+\frac{1}{2}\right)-H_{r,1}^{(1)n-\frac{1}{2}}\left(i,j+\frac{1}{2}\right)}{\Delta t}=\frac{1}{r(i)}E_{z,k}^{(2)n}\left(i,j+\frac{1}{2}\right)$$
$$-\frac{E_{\varphi,1}^{(1)n}(i,j+1)-E_{\varphi,1}^{(1)n}(i,j)}{\Delta z}, \tag{5.86}$$

$$-\mu_0 \frac{H_{z,1}^{(1)n+\frac{1}{2}}\left(i+\frac{1}{2},j\right)-H_{z,1}^{(1)n-\frac{1}{2}}\left(i+\frac{1}{2},j\right)}{\Delta t}=\frac{1}{r\left(i+\frac{1}{2}\right)}\frac{r(i+1)E_{\varphi,1}^{(1)n}(i+1,j)-r(i)E_{\varphi,1}^{(1)n}(i,j)}{\Delta r}$$
$$-\frac{1}{r\left(i+\frac{1}{2}\right)}E_{r,1}^{(2)n}\left(i+\frac{1}{2},j\right), \tag{5.87}$$

where Δt, Δz, Δr are the discrete steps in the corresponding coordinates: $z = i\Delta z$, $r = j\Delta r$, $t = n\Delta t$. Moreover, the readings of electrical components are calculated in the whole times $t = n\Delta t$, and the readings of the magnetic vectors are computed at half-time points $t = (n + 1/2)\Delta t$. Equations (5.82)–(5.87) are an example of a conditionally stable difference scheme, which is solved by the sweep method with the boundary conditions. For stable convergence of the solutions of the system (5.82)–(5.87) discretization steps should be chosen to satisfy the inequality [82]:

$$c\Delta t \leq \Delta r / k, \quad \Delta r = \Delta z, \tag{5.88}$$

where c is the speed of light in vacuum, k is the number of angular harmonics of the equations (5.54), (5.55).

Note that the system (5.82),(5.87) differs from a similar system in [82] not only by the fact that in the system (5.82)–(5.87) $k = 1$, and in [82] k is arbitrary, but also by the fact that in [82] there are some errors in some signs in the system of equations (5.82)–(5.87).

5.3.5. Maxwell's equations for azimuthal polarization

If the optical element with axial symmetry (optical axis z is the axis of symmetry) received a normally incident electromagnetic monochromatic wave with azimuthal polarization (Fig. 5.28b), then at the electric vector there is only a single projection:

$$E^{\text{inc}} = E_\phi = E_0(r)\cos\omega t. \tag{5.89}$$

In the notations of (5.54) at the electric vector of the incident wave there is only one Fourier component (angular harmonic): $E^{\text{inc}} = E_{\phi,0}$. Therefore, in the system of equations (5.62)–(5.73) for the case of azimuthal polarization there are only three equations:

$$\frac{\partial H_{r,0}}{\partial z} - \frac{\partial H_{z,0}}{\partial r} = \varepsilon\varepsilon_0 \frac{\partial E_{\phi,0}}{\partial t} + \sigma E_{\phi,0}, \tag{5.90}$$

$$-\frac{\partial E_{\phi,0}}{\partial z} = -\mu\mu_0 \frac{\partial H_{r,0}}{\partial t}, \tag{5.91}$$

$$\frac{1}{r}\frac{\partial\left(rE_{\phi,0}\right)}{\partial r} = -\mu\mu_0 \frac{\partial H_{z,0}}{\partial t}. \tag{5.92}$$

The difference approximation of the system (5.90)–(5.92) for the azimuthal polarization takes the form ($\sigma = 0$, $\mu = 1$):

$$\varepsilon(i,j)\varepsilon_0 \frac{E^n_{\varphi,0}(i,j) - E^{n-1}_{\varphi,0}(i,j)}{\Delta t} = \frac{H^{n-\frac{1}{2}}_{r,0}\left(i,j+\frac{1}{2}\right) - H^{n-\frac{1}{2}}_{r,0}\left(i,j-\frac{1}{2}\right)}{\Delta z} - \frac{H^{n-\frac{1}{2}}_{z,0}\left(i+\frac{1}{2},j\right) - H^{n-\frac{1}{2}}_{z,0}\left(i-\frac{1}{2},j\right)}{\Delta r}, \tag{5.93}$$

$$-\mu_0 \frac{H^{n+\frac{1}{2}}_{r,0}\left(i,j+\frac{1}{2}\right) - H^{n-\frac{1}{2}}_{r,0}\left(i,j+\frac{1}{2}\right)}{\Delta t} = -\frac{E^n_{\varphi,0}(i,j+1) - E^n_{\varphi,0}(i,j)}{\Delta z}, \tag{5.94}$$

$$-\mu_0 \frac{H^{n+\frac{1}{2}}_{z,0}\left(i+\frac{1}{2},j\right) - H^{n-\frac{1}{2}}_{z,0}\left(i+\frac{1}{2},j\right)}{\Delta t} = \frac{1}{r\left(i+\frac{1}{2}\right)} \frac{r(i+1)E^n_{\varphi,0}(i+1,j) - r(i)E^n_{\varphi,0}(i,j)}{\Delta r}. \tag{5.95}$$

Note that the equations (5.90)–(5.92) and (5.93)–(5.95) were not considered in [82].

5.3.6. Maxwell's equations for radial polarization

If the optical element the axis of symmetry which coincides with the optical axis, receives a normally incident electromagnetic monochromatic wave with radial polarization (Fig. 5.28c), then the electric vector of the incident wave has only one radial component:

$$E^{\text{inc}} = E_r = E_0(r)\cos\omega t \tag{5.96}$$

or in the notation of equation (5.54): $E^{\text{inc}} = E_{r,0}$. Then only the following three of the six equations (5.76)–(5.81) remain for radial polarization:

$$-\frac{\partial H_{\phi,0}}{\partial z} = \varepsilon\varepsilon_0 \frac{\partial E_{r,0}}{\partial t} + \sigma E_{r,0}, \tag{5.97}$$

$$\frac{1}{r}\frac{\partial\left(rH_{\phi,0}\right)}{\partial r} = \varepsilon\varepsilon_0 \frac{\partial E_{z,0}}{\partial t} + \sigma E_{z,0}, \tag{5.98}$$

$$\frac{\partial E_{r,0}}{\partial z} - \frac{\partial E_{z,0}}{\partial r} = -\mu\mu_0 \frac{\partial H_{\phi,0}}{\partial t}. \tag{5.99}$$

Finite-difference approximation of (5.97)–(5.99) has the form (σ = 0, μ = 1):

$$\varepsilon\left(i+\frac{1}{2},j\right)\varepsilon_0 \frac{E^{n}_{r,0}\left(i+\frac{1}{2},j\right) - E^{n-1}_{r,0}\left(i+\frac{1}{2},j\right)}{\Delta t} = -\frac{H^{n-\frac{1}{2}}_{\varphi,0}(i+\frac{1}{2},j+\frac{1}{2}) - H^{n-\frac{1}{2}}_{\phi,0}\left(i+\frac{1}{2},j-\frac{1}{2}\right)}{\Delta z}, \tag{5.100}$$

$$\varepsilon\left(i,j+\frac{1}{2}\right)\varepsilon_0 \frac{E^{n}_{z,0}\left(i,j+\frac{1}{2}\right) - E^{n-1}_{z,0}\left(i,j+\frac{1}{2}\right)}{\Delta t} =$$
$$= \frac{1}{r(i)} \frac{r(i+\frac{1}{2})H^{n-\frac{1}{2}}_{\varphi,0}\left(i+\frac{1}{2},j+\frac{1}{2}\right) - r\left(i-\frac{1}{2}\right)H^{n-\frac{1}{2}}_{\varphi,0}\left(i-\frac{1}{2},j+\frac{1}{2}\right)}{\Delta r}, \tag{5.101}$$

$$-\mu_0 \frac{H^{n+\frac{1}{2}}_{\varphi,0}\left(i+\frac{1}{2},j+\frac{1}{2}\right) - H^{n-\frac{1}{2}}_{\varphi,0}\left(i+\frac{1}{2},j+\frac{1}{2}\right)}{\Delta t} = \frac{E^{n}_{r,0}\left(i+\frac{1}{2},j+1\right) - E^{n}_{r,0}\left(i+\frac{1}{2},j\right)}{\Delta z}$$
$$-\frac{E^{n}_{z,0}\left(i+1,j+\frac{1}{2}\right) - E^{n}_{z,0}\left(i,j+\frac{1}{2}\right)}{\Delta r}. \tag{5.102}$$

Note that the equations (5.100)–(5.102) are not considered in [82]. Other features of the FDTD method for the case of cylindrical symmetry: calculations of fields on the optical axis at $r = 0$, and compliance with the boundary conditions in the form of perfectly absorbing layers, taken from [82]. The above-described radial FDTD method has been implemented in the programming environment Matlab 7.0.

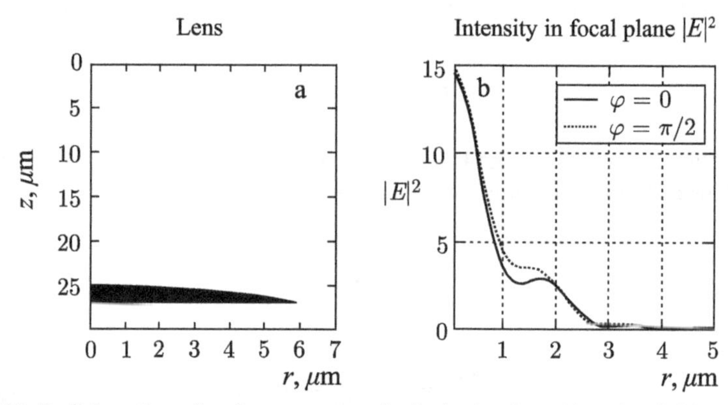

Fig. 5.30. Radial section of a plano-convex spherical microlens (a) and radial intensity distribution in the focal plane (b): curve 1, with $\phi = 0$, curve 2 at $\phi = \pi/2$.

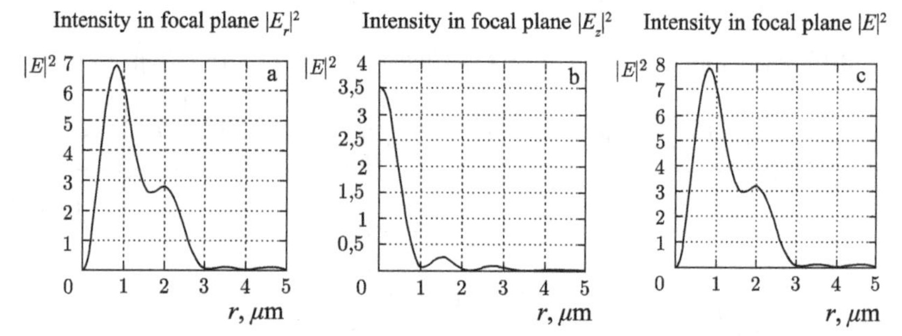

Fig. 5.31. The radial intensity distribution $|E_r|^2$ (a), $|E_z|^2$ (b) and $|E_r|^2 + |E_z|^2$ (c) in the focus of the lens (Fig. 5.30a) upon impact by a plane wave with radial polarization

5.3.7. Modelling the focusing of a plane linearly polarized wave by a spherical microlens

Consider the focus area of a plane axial wave with linear polarization incident on the flat surface of a plano-convex spherical microlens (Fig. 5.30).

Figure 5.30a shows the radial cross section of the microlens: the lens aperture radius of 6 μm, the radius of curvature of the spherical surface 10 μm, the refractive index 1.5, the optical thickness of the lens on the optical axis 2 μm. Wavelength 1 μm. The discreteness of the spatial coordinates of 1 / 20 μm, and the time coordinate 1/40c s. Figure 5.30b shows the radial intensity distribution $I = |E|^2 = |E_r|^2 + |E_\phi|^2$ in the focus on the horizontal x ($\phi = 0$) and the vertical y ($\phi = \pi/2$) axes. The focus is at the distance of 11 μm from the flat surface of the lens. Figure 5.30b shows that the focal spot has a weak ellipticity (eccentricity of the ellipse around 0.97). The long axis of the ellipse is directed along the axis y (polarization axis), and the short axis – along the axis x. The average diameter of the spot intensities at half intensity

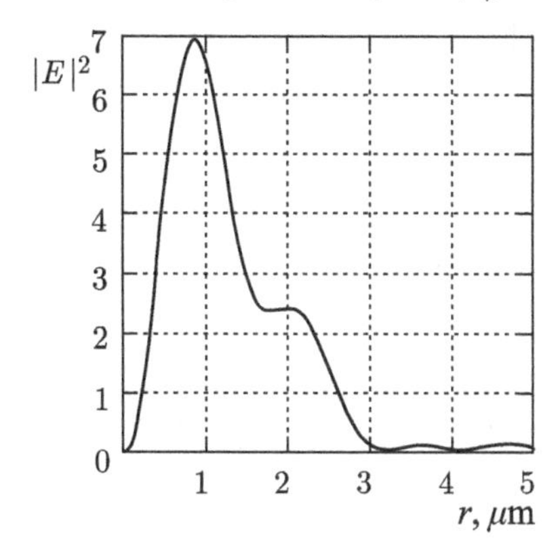

Fig. 5.32. The radial distribution of intensity $|E|^2 = |E_\phi|^2$ in the focal plane of the lens (Fig. 5.30a) upon impact by a plane wave with azimuthal polarization.

is FWHM = $(d_x + d_y)/2 = 1.5\lambda$. The area of the focal spot at half intensity is HMA = $1.77\lambda^2$. The diameter of the spot means the full width of the intensity curve at half intensity.

Figure 5.31 shows the radial intensity distribution at the focus of the same lens (Fig. 5.30a), but upon impact by the radially-polarized plane wave whose electric vector has only one radial component $E^{\text{inc}} = E_r$. Figure 5.31 shows that the numerical aperture of the lens (Fig. 5.30a) is not large enough to ensure that the intensity of the longitudinal component (Fig. 5.31b) exceeds the intensity of the radial component (Fig. 5.31a) to such an extent that the maximum at full intensity distribution (Fig. 5.31c) forms on the optical axis ($r = 0$) . Note that in the case of the radially-polarized light field at focusing through a radially symmetric optical element the amplitude of the radial component of the electric field strength on the optical axis is always zero: E_r $(r = 0) = 0$.

Figure 5.32 shows the radial distribution of the total intensity $|E|^2 = |E_\psi|^2$ of the (there are no other projections at the electric vector in this case) when a plane wave with an azimuthal projection falls on the lens (Fig. 5.30a). It is evident that an annular intensity distribution with a zero on the optical axis forms in the focus.

Plano-convex spherical lenses has a maximum numerical aperture equal to $\text{NA}_0 = (n^2 - 1)^{1/2}/n = 0.745$. This limitation arises due to total internal reflection of light inside the lens. Therefore, half of the maximum angle of convergence of the rays at the focus is 48° (at $n = 1.5$ – refractive index of the lens). To achieve maximum numerical aperture NA_0 it is necessary to ensure that the aperture radius R_0 of the plano-convex spherical lens is equal to $R_0 = R_1/n$, where R_1 is the radius of curvature of the spherical surface. In this case, $R_1 = 10$ μm, so $R_0 = 6.4$ μm. The radius of the lens aperture in Fig. 5.30a is $R = 6$ μm, which is close to the maximum value of R_0. However, when the aperture radius (for a given radius of curvature R_1) approaches the maximum value of R_0 the focal spot is not reduced due to aberrations. In biconvex spherical lenses the numerical aperture can reach unity.

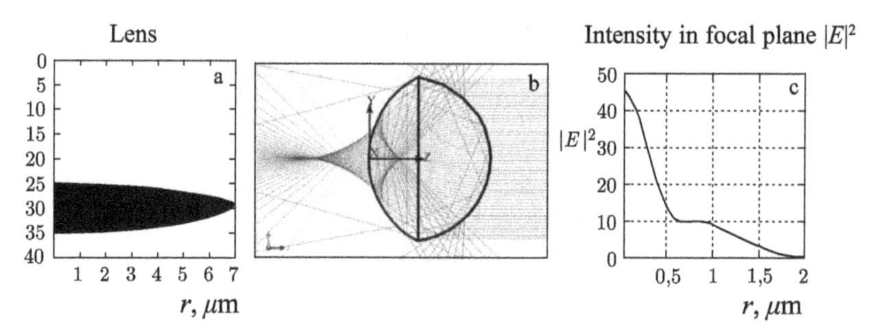

Fig. 5.33. Biconvex spherical microlens (a), the rays in a lens (software TracePro) (b) and radial distribution of intensity $|E|^2 = |E_r|^2 + |E_z|^2$ in the focal plane (c) in incidence of a plane wave with radial polarization,

5.3.8. Focusing the light by biconvex spherical microlenses

Consider focusing of a plane wave with radial polarization by a biconvex spherical microlens. The lens parameters (Fig. 5.33a): the radii of curvature $R_1 = 8.125\ \mu$m, $R_2 = 7.08\ \mu$m, the aperture radius $R = 7\ \mu$m, thickness of the lens at the optical axis $d = 10\ \mu$m, the refractive index $n = 1.5$. The wavelength $\lambda = 1\ \mu$m.

Although the diameter of the focal spot at the chosen radii of curvature R_1 and R_2 of the spherical surfaces (Fig. 5.33c) minimum (while maintaining the aperture radius R), the energy efficiency of such a lens is about 50%. Figure 5.33b shows the rays passing through the lens (Fig. 5.33a), constructed with the help of TracePro commercial software. It is seen that only part the rays that fall inside the lens enter the focal region. The remaining rays, due to total internal reflection, come out of the lens in other directions. The maximum angle at which the optical axis come to the focal point of the rays, is about 60° (half angle). That is the numerical aperture of the lens (Fig. 5.33a) is about NA = sin (60) = 0.86. This is almost two times greater than the NA for the lens shown in Fig. 5.30a. The diameter of the focal spot at half intensity (Fig. 5.33c) is FWHM = 0.78λ. The total diameter (twice the distance from the maximum to first minimum) is equal to 1.4λ, and the area of the focal spot at half intensity is HMA = $0.48\lambda^2$. Recall for comparison that the minimum (with NA = 1) area of the Airy disk in the scalar paraxial case is smaller and equal to HMA = $0.204\lambda^2$. Thus, due to aberrations of the spherical lens it is not possible to achieve the minimum diameter of the focal spot, for example, as in [57, 73].

5.3.9. Focusing of a plane wave with radial polarization by a gradient cylindrical microlens

Consider the focus of a plane wave with radial polarization incident normally on a flat surface of a cylindrical gradient microlens (GL) [14]. The refractive index of the GL depends on the radial variable as follows:

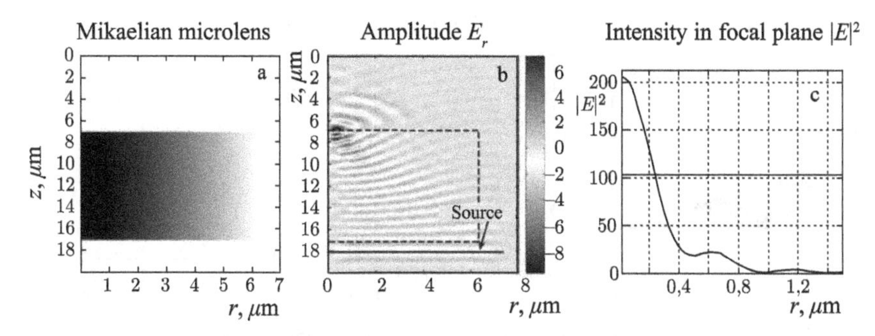

Fig. 5.34. The dependence of the refractive index on the radial coordinate in halftones for the GL (a) and the results of focusing: the instantaneous amplitude distribution E_r on the calculation field 8×20 μm (b), the radial distribution of intensity $|E|^2 = |E_r|^2 + |E_z|^2$ in the focal plane (immediately behind the exit plane of the lens) (c).

$$n(r) = n_0 \operatorname{ch}^{-1}\left[\frac{\pi r}{2L}\right], \tag{5.103}$$

where n_0 is the refraction index on the optical axis, L is thickness of the lens along the optical axis (the lens looks like a cylinder or a piece of the gradient fibre). Equation (5.103) differs from (5.24) only in notation and by the change of variable y to r. All rays parallel to the optical axis and incident on the GL are collected in the focus at the optical axis on the opposite side of the lens. The microlens parameters (Fig. 5.34a): $n_0 = 1.5$, $L = 10$ μm, the aperture radius of the lens $R = 6$ μm. The wavelength of light $\lambda = 1$ μm.

Figure 5.34a shows the radial distribution of the refractive index of the gradient microlens (5.103). The instantaneous field with the amplitude E_r 8×20 μm in size shown in Fig. 5.34b. Figure 5.34c shows the radial distribution of the total intensity of the electric field $|E|^2 = |E_r|^2 + |E_z|^2$ in the focal plane of the GL. The diameter of the focal spot at half intensity is FWHM = 0.44λ, and the radius from the maximum intensity to the first minimum is equal to 0.45λ. The area of the focal spot at half intensity is HMA = 0.152λ_2. The area of this spot (Fig. 5.34c) is slightly lower than that obtained in [57] using a microlens with $N = 0.9$ and a circular aperture (HMA = 0.160λ^2), but slightly larger than the area of the focal spot obtained in [73] using a parabolic mirror (HMA = 0.134λ^2). A gradient microlens (Fig. 5.34a) can be produced in the form of a binary photonic crystal [56].

5.3.10. Focusing of a Gaussian beam with radial polarization using a conical microaxicon

Consider the focusing of a Gaussian beam with radial polarization by a conical microaxicon (Fig. 5.35a). The axicon parameters: the radius of the circular base of the cone (aperture radius) $R = 7$ μm, the thickness of the axicon (cone height) $d = 6$ μm, the refractive index $n = 1.5$. The wavelength $\lambda = 1$ μm, the radius of the waist of the Gaussian beam $w = 7$ μm.

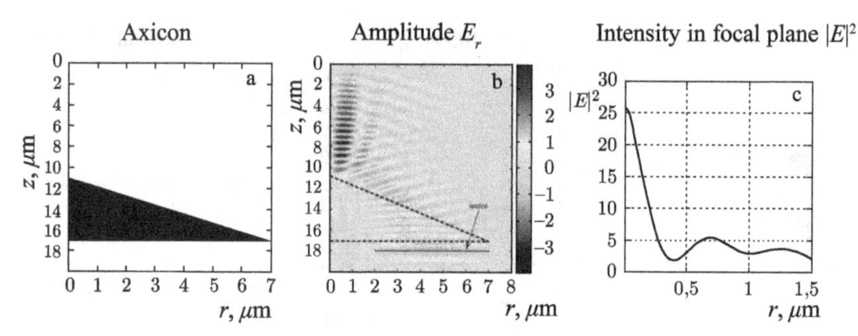

Fig. 5.35. Conical microaxicon (a), the instantaneous amplitude E_r, calculated on the 8×20 μm field (b) and radial distribution of intensity $|E|^2 = |E_r|^2 + |E_z|^2$ in the focal plane of the axicon (next to its apex) (c).

Figure 5.35b shows the calculated instantaneous amplitude of the radial component of the electric field E_r on a platform with the size of 8×20 μm (vertical axis is z, the horizontal axis r). The horizontal segment (source) shows the position of the waist of the Gaussian beam incident on the axicon. Figure 5.35c shows the radial distribution of intensity $|E|^2 = |E_r|^2 + |E_z|^2$ in the focal plane of the axicon (next to its apex). The radius of the focal spot from the maximum intensity to the first minimum is equal to 0.40λ, and the diameter of the focal spot at half intensity is equal to FWHM = 0.36λ. The area of the focal spot in Fig. 5.35c at half intensity is HMA = $0.101\lambda^2$.

This area is less than the record value obtained in [73] (HMA = $0.134\lambda^2$). Note that the numerical aperture of the axicon in this case is not greater than NA = 0.65.

If the axicon (Fig. 5.35a) is illuminated with circular Gaussian beam $\exp(-(r - r_0)^2/w^2)$ with radial polarization, where $r_0 = 4.5$ μm and $w = 2.5$ μm is the radius of the Gaussian beam, we find the record to date parameters of the focal spot: the area at half intensity HMA = $0.096\lambda^2$ and the diameter at half intensity FWHM = 0.35λ. Figure 5.36a shows the calculated radial intensity distribution at the focus of an axicon (Fig. 5.35a), illuminated with a circular Gaussian beam, and Fig. 5.36b shows in halftones the diffraction pattern in the focal plane (immediately after the apex of the cone) in the coordinates (x, y).

One can also note a shortcoming of such sharp focusing of the laser beam using an axicon: a low energy efficiency. Figure 5.36a shows that the maximum intensity at the focus is only 20 relative units, which is 30 times smaller than the intensity at the focus of the GL (Fig. 5.34c) with the same radius of the aperture. This is because the axicon forms a focal region with an extended depth of field. For example, for the axicon in Fig. 5.36a the depth of field at half intensity is 3 μm, and for the GL in Fig. 5.34a it is only 0.5 μm.

5.4. Three-dimensional photonic crystals

The three-dimensional photonic crystal refers to a three-dimensional periodically structured dielectric, which creates a periodically inhomogeneous distribution of dielectric permittivity in the space of the crystal. Such a modulation of dielectric

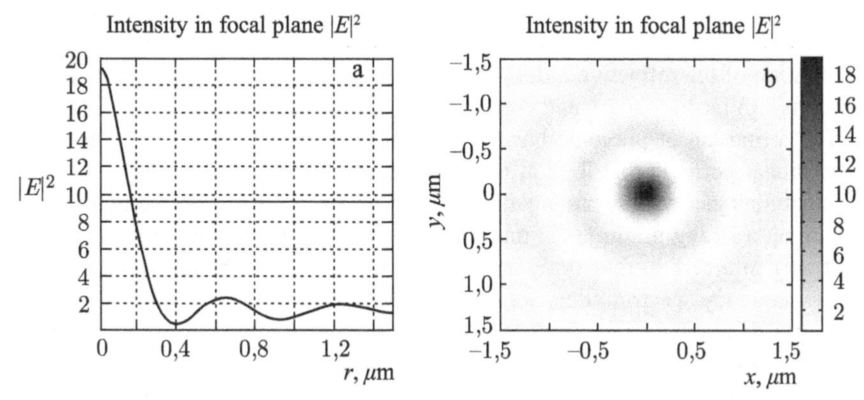

Fig. 5.36. The radial intensity distribution $|E|^2 = |E_r|^2 + |E_z|^2$ in the focus of the microaxicon illuminated with a circular Gaussian beam with radial polarization, and the two-dimensional grayscale diffraction pattern (negative) in the focal plane in the coordinates (x, y) (b).

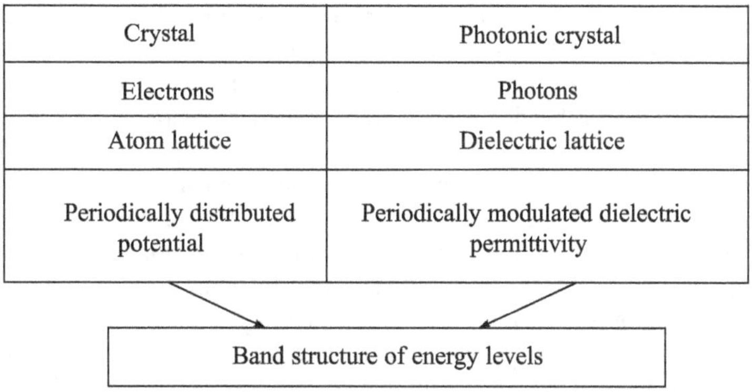

Crystal	Photonic crystal
Electrons	Photons
Atom lattice	Dielectric lattice
Periodically distributed potential	Periodically modulated dielectric permittivity

Band structure of energy levels

Fig. 5.37. Comparison of a conventional (electronic) crystal with a photonic crystal.

permittivity leads to the formation of a band structure of the energy levels of photons. We can trace the analogy between photonic crystals and ordinary crystals. In a typical crystal ions (nuclei) of atoms are arranged in the three-dimensional lattice. This arrangement creates the three-dimensional periodic distribution of the electric potential. Under such a potential the energy levels of electrons are distributed in zones, in other words, the electron energy can only take certain values, corresponding to a certain energy level, just as it does in the potential well field. The potential in the crystal is infinite and periodic in space and in the potential well it is spatially limited, but both leads to the formation of energy levels. The analogue of the three-dimensional periodic potential modulation in a photonic crystal is the modulation of dielectric permittivity or refractive index. This modulation leads to discretization of the energy levels for electromagnetic waves, i.e. the formation of the band structure for photons. For an electron in the potential well the optical analogue is not so close – it is an optical waveguide in which the restriction in the space of wave propagation in two coordinates leads to a discrete spectrum of wave vectors, rather than energy. The discrete spectrum of photon energies requires the three-dimensional localization or three-dimensional periodic modulation of the refractive index. The table in Fig. 5.37 shows some of the basic concepts similar for photon and conventional 'electronic' crystals,

The formation of photonic band gaps can be described as follows. The three-dimensional periodic distribution of the dielectric leads to the situation in which the electromagnetic wave propagating in a certain direction is reflected in the structure as on a Bragg grating or a multilayer dielectric mirror. This 'mirror' reflects light only in a certain range of wavelengths, called the stop-band, whose position in the frequency spectrum depends on the grating period. If for all directions there is a range of overlap of the frequencies of the stop-bands, then this region forms a photonic band gap. Light with a wavelength belonging to this region can not propagate in any direction. Thus, an atom inside a photonic crystal can not emit light at this wavelength. Hence the threshold behaviour of the formation of the band gap on the basis of the magnitude of contrast of the dielectric permittivity of the grating becomes clear. The point is that different directions in the crystal correspond to different periodicity. Thus, the middle parts of the stop bands can be significantly shifted relative to each other in different directions – for example, for a square grating the periodicities along the diagonal of the cube are related as 1 to $\sqrt{2}$. To ensure overlapping of the stop-bands they should be made large enough. This is achieved by creating air–dielectric gratings from materials with a high refractive index. On the other hand, the overlap of stop-bands is best achieved at a more isotropic periodicity, i.e. in the form of the Brillouin zone, which is close to spherical.

The idea of controlling the spontaneous emission of atoms (the suppression of spontaneous emission of atoms) situated in a medium with three-dimensional periodic modulation of the refractive index was expressed in the papers by V.P. Bykov [83] in 1972. This possibility was then realized in 1987 by E. Yablonovich and S. John [9, 84] and the term 'photonic crystal' was suggested. Unlike an ordinary crystal, where the electron probability density wave is scalar, the electromagnetic wave field is of the vector nature. This required the development of new mathematical tools to calculate the band structure and led to a number of distinctive properties, in particular, the threshold nature of the band gap in the depth of modulation of the refractive index. In 1990 the band structure of photonic crystals was calculated for the first time and the photonic band gap was theoretically discovered [85]. Significant technological challenges in making photonic crystal gratings for the optical wavelength range have led to the fact that photonic crystals were first synthesized only in 2000, with these crystals having supposedly the band gap in the near infrared spectrum [86].

The first three-dimensional photonic crystal, in which the band gap was theoretically discovered, was a crystal with the symmetry of the diamond lattice, in which dielectric spheres were placed in the nodes [85]. The band gap was situated between the second and third zones. The band gap was then found in the face-centred cubic (FCC) lattice structure formed by spherical cavities in a dielectric and located between the eighth and ninth bands [87]. The threshold of the existence of

Material	Transmission spectrum boundary, nm	Refractive index
Silicon	1100	3.45
Germanium	1780	4.0
GaAs	870	3.6
AlAs	580	3.1
InP	920	3.55
ZnSe	460	2.5
ZnS	350	2.3
ZnSe	550	2.72

Fig. 5.38. The refractive indices of some semiconductors and dielectrics.

the band gap according to the refractive index of the first diamond lattice was 2.1 for the FCC lattice it was 2.8.

The requirement of a high refractive index of the structure limits the choice of material of the grating and significantly complicates the task of producing a photonic crystal. The table in Fig. 5.38 shows the refractive index for various materials.

Synthesis of three-dimensional photonic crystals is a difficult technological task, due to submicron dimensions of the crystal period and the three-dimensional structure, and also the limited choice of material. To date, there are plenty of ways to solve this problem, each of which, at the same time, has significant drawbacks to the successful implementation of the concept of photonic crystals in practice.

One of the ways is the multiple repetition of well-established methods of traditional (two-dimensional) lithography [88]. Using electron-beam lithography, for example, two-dimensional periodic structures were first formed by selective etching using a template, then new layers were sequentially applied over existing ones. Structures consisting of several periods using compounds A_3B_5 and Si were produced [88–90]. The disadvantage of such structures is the high complexity and difficulty of making a large number of layers (periods). The resulting structure of the 'woodpile' have the symmetry of the diamond and have a complete band gap.

Another method is to use two-photon stereolithography. Using this method, arbitrary three-dimensional structures with a resolution of about 100 nm can be formed in the volume of the photoresist by successive 'pointwise' writing by focused radiation of a femtosecond laser [91]. Absorption of light by the photoresist at a specific wavelength leads to the polymerization reaction. The two-photon absorption mechanism allows to ensure that this process is three-dimensionally localized only by the beam waist area and is not distributed along the beam axis, which ensures the formation of three-dimensional structures. The disadvantages of this approach include the slow recording process and limited resolution, making it difficult to use this method to obtain photonic crystals for near-infrared and visible spectra.

One of the earliest methods for the formation of three-dimensional photonic crystals was the method of synthesizing artificial opals. Artificial colloidal crystals are produced using the monodisperse sol of dielectric particles of latex or silicon oxide. In the first case a suspension of particles is deposited in the solution, placed in an optical cell, forming a close-packed face-centred cubic structure and is studied in this form. In the second case the silica sol is also deposited in the solution, but then heat treatment is carried out, resulting in sintering of particles of silicon oxide to form a solid structure, suitable for further physico-chemical treatments and mechanical polishing.

Such structures are called artificial opals. Several groups of investigators independently proposed to consider artificial opals as prototypes of three-dimensional photonic crystals for the optical region of the electromagnetic spectrum [92, 93]. Sedimentation and agglomeration of the globules of silica lead to the formation of polycrystals with a period ranging over a wide range (100–1500 nm) with single-crystal domains whose size in the best specimens usually does not exceed a few hundred microns. Packing of the globules corresponds to the FCC lattice. However, to obtain the band gap, it is necessary to invert the opal lattice, i.e. the opal pores should be filled with a material with a sufficiently high refractive index ($n > 2.8$) and the initial matrix should be removed.

The three-dimensional periodic structure can be obtained by drilling the material in three directions. In particular, this produced the structure known as yablonovite, proposed by E. Yablonovich, having the symmetry similar to the symmetry of the diamond lattice and the 'stack of wood' structure [94]. Also, similar to these structures, the 'yablonovite' structure has a large fundamental band gap and the lowest threshold for its appearance in contrast to the refractive index (2.0–2.1). For the first time this structure has been implemented for the centimetre wavelength range when drilling ebonite. A row of holes was made in the place and three holes were drilled through each hole symmetrically at an angle of 35° to the normal, and all the resulting holes formed three groups, within each group the axis of the holes were parallel to each other.

For the optical wavelength range the yablonovite structure was produced by drilling of the material with a focused ion beam (FIB) [95]. In this paper, the ions of gallium and the accelerating voltage of 25 keV were used. The diameter of the holes was 350 nm, the diameter of the ion beam was equal to 100 nm. Five periods were obtained over the sample thickness, 25×25 structure periods in the sample plane. The stop band was located near the wavelength 3 μm. Drilling was carried out by an ion beam in two directions and in the third pores formed by electrochemical etching.

The most promising new method is interference lithography, which consists in exposing the photoresist by a three-dimensional interference pattern [96]. As a result, such writing can provide the ideal lattice periodicity of the structure. This method is distinguished by the high rate of production – the entire volume of the lattice is exposed at the same time, low cost – implementation does not require precise positioning systems and the possibility of obtaining large samples. The disadvantages of the method include moderate resolution, which is limited by the

Material	Advantages	Shortcomings
Electron lithography	Freedom in selecting the 'atom' of the lattice type, material, high resolution	Cost, speed. Disruption of long-range order, small lattice size
Two-photon stereolithography	Freedom in selecting the type of lattice	Low speed, limited by large periods
Deposition of opals	Freedom in selecting period, low cost	Restrictions of the type of lattice node ('atom'), structural defects
Ion beam drilling	High resolution, large band gap	High cost, material restrictions
Interference lithography	High degree of periodicity, large lattice size, low cost	Average resolution, restrictions of the type pf lattice

Fig. 5.39. Comparison of methods for the synthesis of three-dimensional photonic crystals.

wavelength of radiation used for exposing the photographic material and the lack of flexibility in choosing the form of the lattice site. More details on this method will be discussed in the next section.

The summary table in Fig. 5.39 compares several methods of synthesis of the photonic crystal structures.

5.5. Interference-lithographic synthesis of photonic crystals

Many methods of synthesis of photonic crystals have been proposed to date. However, the production of macroscopically homogeneous, defect-free crystals is still a difficult technical challenge. One of the most promising synthesis methods for now appears to be the interference lithography technique [96]. This method consists in obtaining a three-dimensional structure of the photopolymer by illuminating the three-dimensional interference pattern by four (or more) coherent beams of light. The advantages of this method are the ideal lattice periodicity, the absence of structural defects, the possibility of obtaining samples of large area and low cost. The interest in this method has increased particularly after a number of studies [97, 98] have shown that it can be used to produce three-dimensional photonic crystals with a band gap at a relatively low refractive index material ($n > 2.0$).

However, in the manufacture of three-dimensional samples the illumination of the film of the photoresist is heterogeneous due to the absorption of light in the photographic material. At the same time, this absorption is necessary to maintain the photosensitivity of the material, since the absorption of photons due to the initiation of polymerization reactions and depolymerization. In order to minimize the impact of absorption, the optimal choice are photoresists with the chemical amplification mechanism of the reaction, when one absorbed photon leads to the generation of a single molecule or atom of the catalyst of the polymerization reaction, each of which, in turn, trigger the emergence of a set of polymer bonds. But when using such photoresist, the thickness of the samples obtained by this method was limited to 10–30 μm [96, 99]. As shown in [100], by choosing the angle between the interfering waves and the normal to the sample surface, this method can be used to synthesize structures corresponding to simple cubic, face- and body-centred lattices.

For the synthesis of three-dimensional lattices of photonic crystals the authors of [101, 102] used the continuous radiation of a helium–cadmium laser with a wavelength of 442 nm, corresponding to low absorption capacity of the material. The photographic material was a photoresist with a cationic polymerization mechanism SU-8. This photoresist has a sharp decline in both absorption and sensitivity in the transition to wavelengths greater than 400 nm. So far this photoresist has been used for recording at the wavelengths of the ultraviolet range [96, 99]. In the transition from 355 nm to 442 nm the required radiation dose per unit area increased by about four orders of magnitude.

5.5.1. The scheme of recording the lattice

In order to form the three-dimensional lattice from the photoresist by interference lithography, the photoresist film was exposed three times by the interference pattern of two waves, similar to how it was done in [99]. The experimental setup is shown in Fig. 5.40.

After each exposure the sample is rotated 120° around the vertical axis. The exposure time should always be the same.

As a result of exposure the following distribution of the absorbed energy is obtained in the volume of the photoresist:

$$I(\vec{r}) = \sum_{i=1}^{N} I_i \cos^2(\vec{b}_i \vec{r} + \phi_i) \qquad (5.104)$$

where $\vec{b}_i = \vec{k}_{1i} - \vec{k}_{2i}, \vec{k}_{1i}, \vec{k}_{2i}$ are the wave vectors of interfering waves at the i-th

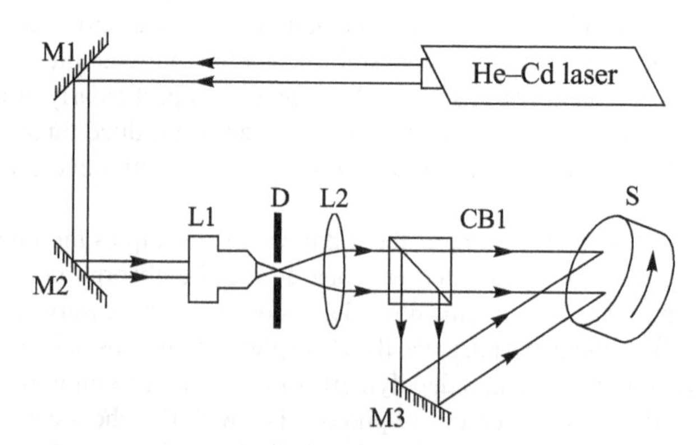

Fig. 5.40. Letters denote: M1, M2, M3 – mirrors, L1, L2 – lenses, D – diaphragm, CB1 – dividing cube, S – substrate of glass coated with a photoresist.

exposure. From (5.104) it is easy to see that the vectors b_i are basis vectors of the reciprocal lattice of the synthesized crystals. At this stage, we implemented only the case when the angle between the interfering beams in each of the three exposures was the same similar, i.e. basic vectors of the reciprocal lattice have the same length. The angle between the basis vectors was also the same. Thus, we have implemented the case of the orthorhombic lattice.

The minimum threshold value of the refractive index for the formation of the band gap occurs at the face-centred lattice and in this case is about $n = 2.5$.

Unlike the case of multiple-beam interference, the triple exposure method can be used to obtain the lattices of the same symmetry but different periods. We have synthesized samples with different lattice constants by changing the angle between the interfering waves. To ensure that the lattice symmetry is not changed, the angle formed by the bisectrix between the vectors $k1_i$, $k2_i$ of the interfering waves normal to the surface of the specimen should remain constant.

5.5.2. Description of experiments and the resulting structure

The synthesis of polymeric matrices of photonic crystals by this method was performed in an experimental setup according to the optical circuit in Fig. 5.40. A helium–cadmium laser GKL-60 (II) with a power of 80 mW was used. The beam was rotated and separated by the mirros M1, M2, M3 and the beam splitter CB1 with bleaching of faces for this wavelength. The sample was rotated using a special holder, which provides both adjustable tilt of the sample in accordance with Fig. 5.40 and its rotation in the perpendicular plane with the desired pitch. Alignment of the sample in the plane of rotation can be achieved by aligning the rotation axis with the optical axis.

The photoresist was deposited on the substrate by centrifugation according to manufacturer‘s recommendations, that is in two stages. Preliminary distribution of the photoresist on the sample surface at 500 rpm for 10 s with formation of the desired thickness at a speed of 3000 rpm for 30 s was achieved. The above procedure produced a layer thickness of 40–50 μm using SU-8-50-40 in accordance with the specifications of the photoresist. The sample was dried in two stages – first 5 min at 60°C, followed by 20 min. at a temperature of 95°C to remove the solvent. The exposed photoresist was baked at a temperature of 95°C for 6 min. After baking, the photoresist was placed in the developer PGMEA (2-(1-methoxy) propylacetate) for 5–7 min, and then washed with isopropyl alcohol. The exposure time was similar and ranged from 10 to 20 min. Experiments showed in some cases insufficient adhesion of the photoresist to the glass at a small area of the irradiated surface and the radiation dose. Therefore, in all experiments an adhesive layer (primer) was initially formed on the glass surface. This layer was a photoresist of the same brand and was deposited and treated by the same technology as the core layer, but before baking the layer was bleached over the entire area until cured. Synthesized photopolymer gratings were investigated using a Quanta 200 scanning electron microscope and a white light interferometer Zygo

Fig. 5.41. Electronic phototograph of the one-dimensional grating.

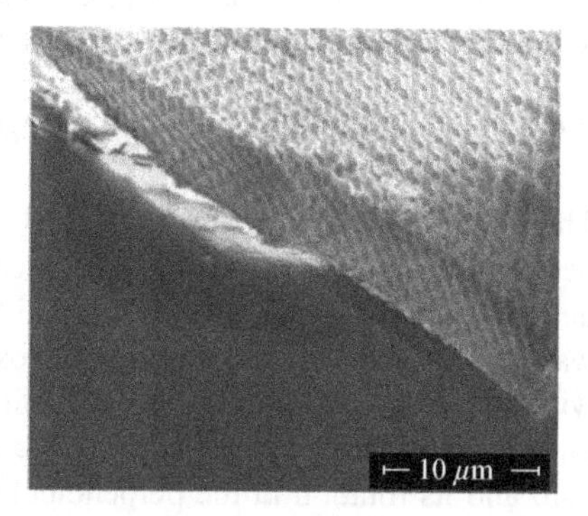

Fig. 5.42. Electronic picture of a three-dimensional grating in normal development.

NewView 5000. Al aluminium layer 10 nm thick was deposited on the surface of the samples prior to investigations..

Trial samples of one-dimensional gratings were obtained by a single exposure and zero angle of inclination of the sample for testing the installation vibrostability and the technological chain of processing the photoresist. The results are shown Fig. 5.41. The grating period is equal to 2.1 μm. The height of the profile, measured by an optical microscope, was 1.2 μm. Here one can see that the profile of the grating is almost rectangular, which indicates the high contrast of the photoresist. This result may also be of independent value for producing the binary-phase diffractive gratings quickly and with minimum expenditure.

In addition, studies were carried out of the recording mode of three-dimensional periodic gratings with different periods and the orthorhombic symmetry type. In addition, exposure, baking and development times were chosen for different periods. The angles of inclination of the sample were also changed to preserve the symmetry of the lattices. These parameters have a complex impact on the quality of the gratings and should also be chosen in the complex. The development time of 8 min produces a grating shown in Fig. 5.42 [101, 102].

5.6. Three-dimensional photonic approximants of quasicrystals and related structures

Quasicrystalline structures, found in metal alloys in the early eighties, have point-symmetry groups, incompatible with the periodicity [103]. In comparison with the crystals, they have a higher rotational symmetry, such as icosahedral, decagonal, etc. This discovery has changed the views on the role of the aperiodic order in condensed matter physics [104] and stimulated the search for physical properties that are typical for aperiodic structures. The largest changes are found in the electronic and phonon properties, as Bloch's theorem does not apply. As a result, the structure of the electronic bands and the quasicrystal lattice vibrations can be very exotic, and they remained a subject of debate for many years, until recently [105–107].

Similar problems arise in the interaction of photons with aperiodic dielectric structures. Photonic quasicrystals are called optical structures which have a quasi-crystalline lattice symmetry. In them, as in the photonic crystals, Bragg diffraction of photons takes place or, in other words, the occurrence of photonic band gaps (PBG). In 1998, it was shown that two-dimensional photonic quasicrystals may possess PBG [108]. The emergence of a full PBG is based on overlapping Bragg stop bands in all directions. At the three-dimensional periodic distribution of the dielectric various directions correspond to different frequency and, consequently, different frequency of the middle of the stop zones. Overlapping of stop-bands can be ensured, on the one hand, by the large size of these zones, i.e. larger amplitude of the appropriate spatial harmonics of the distribution of dielectric permittivity. This is achieved by creating dielectric–air gratings of materials with high refractive index. On the other hand, the overlap of the stop-bands is best achieved at a more isotropic periodicity, i.e. in the form of the Brillouin zone close to spherical [109].

Quasicrystals have a higher rotational symmetry, hence, their band structure can be nearly isotropic, and it can be assumed that such structures are preferred for the emergence of full PBG. In [108] it is shown that two-dimensional photonic quasic-rystals of the 8th order have large PBG for TM (magnetic field in the plane of the structure) and TE (electric field in the plane of the structure) polarizations. In addition, it was noted that the defect states in photonic quasicrystals are more complex and interesting from the standpoint of flexible settings for this condition. The same study indicated the need for investigating three-dimensional photonic quasicrystals.

The first two-dimensional photonic quasicrystal, which has a two-dimensional complete PBG, has been proposed in [110]. Experimentally and theoretically in [110] it is shown that this structure has a low threshold for PBG (the minimum value of the dielectric constant of the material lattice at which there is PBG) is equal to 2.1. In [111] the stereolithography method was used to create a three-dimensional icosahedral quasicrystal, which has stop bands in some directions for microwave range. The transmittance coefficient of microwave radiation, was measured for this crystal but theoretical analysis was not performed. In recent papers the authors reported on the production of three-dimensional photonic quasicrystals for infrared

[112] and visible [113] radiation bands, so that the topic of photonic quasicrystals is of increasing importance [114].

Photonic quasicrystals have no translational symmetry, so reliable methods for calculating their optical properties do not exist, and in any case, they require significant computational resources. In particular, in this case it is difficult to apply the method of expansion in plane waves. The solution to this problem is to study photonic quasicrystals approximants. The approximants of quasicrystals are periodic structures whose geometry is close to quasicrystals with increasing size of the primitive cell. In [115] it is shown that nearly isotropic PBGs can be found in two-dimensional approximants of even the lower orders, and the position and size of the PBGs are essentially independent of the order of the approximant. The authors of [116] have shown that high-order approximant of quasicrystals have a band gap threshold equal to the threshold of the band gap in quasicrystals.

This chapter looks at a three-dimensional approximant 1/0 and two approximants 1/1, with a full PBG [117]. The 'atoms' means the coordinates of lattice sites, which is based photonic quasicrystal. These primitive cells contain a small number of atoms, which greatly reduces the computational volume of the problem, as it allows to keep a relatively small number of plane waves in the solution. The following review does not attempt to find the optimal distribution of the dielectric in the approximant in terms of a complete PBG; this would require an unrealistic computation time. The main thing was to show that the full PBG can exist in technologically achievable structures, to encourage their development and further study, both experimentally and theoretically.

5.6.1. The geometrical structure of the quasicrystal approximants

In [118, 119] it is shown that the phase VS8 and VS32 hypotetical structure, actually existing in silicon and germanium, are respectively 1/0 and 1/1 approximants of the icosahedral quasicrystal with a six-dimensional body-centred cubic (BCC) lattice, so that the coordinates of all atoms can be obtained by projecting a six-dimensional lattice. The approximant 1/0 contains 8 atoms in the primitive rhombohedral unit cell of BCC lattice, and the 1/1 approximant 32 atoms, hence the names of these structures. Consequently, the cubic unit cells 1/0 aand 1/1 contain 64 and 16 atoms, respectively. In the structure of the approximant 1/0 all the atoms are in equivalent crystallographic positions 16 (c) with coordinates xxx, with $x = x_{ic} = \tau^{-2}/4$, where τ is the golden mean. Each atom of the 1/0 approximant has a fourfold coordination (number of nearest 'neighbors'), which makes this structure locally similar to the structure of diamond, and just for the latter the lowest threshold for a full PBG was obtained.

Based on the 1/0 structure, we can build a 1/1 approximant whose unit cell is τ times larger. Its 64 atoms occupy 16 (c) positions with $x = x_{ic}$ and 48 (c) the position with $x = (1-2x_{ic})/2$, $y = (2\tau-1)\, x_{ic}$ $z = x_{ic}$. Atoms in position 16 (c) have a threefold coordination. As shown in [118, 119], we can construct the approximant 1/1 in which all atoms have coordination number 4. In future, such an approximant will be denoted 1/1F.

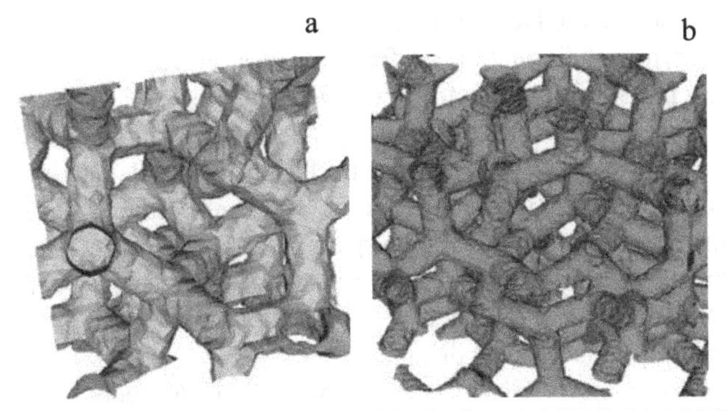

Fig. 5.43. a) graphic representation of the lattice: a) 1/ODR; b) 1/1/FDR.

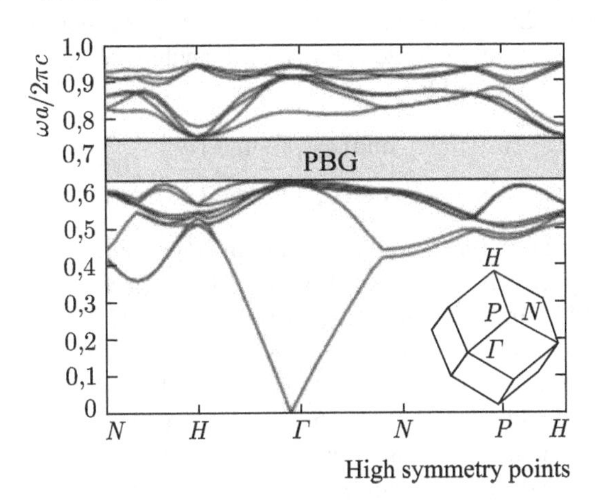

Fig. 5.44. Band structure of approximant 1/0DR (the first 15 zones). Dielectric filling factor $f = 22.7\%$, and its dielectric constant $\varepsilon = 12$. The size of the total band gap $\Delta\omega/\omega_m = 17.6\%$.

5.6.2. Numerical analysis of quasicrystal approximants

For each approximant we studied two types of lattice sites, DR (dielectric rods) with dielectric cylinders, and AS (air spheres) with air spheres. In the first type of structure the dielectric cylinders of a specific radius connect the positions of the neighboring atoms in the lattice. These structures will be referred to hereinafter 1/0DR (see Fig. 5.43.a), 1/1FDR (see Fig. 5.43.b) and 1/1DR. The second type is the air spheres located in the dielectric, and the corresponding approximants are denoted as 1/0AS, 1/1FAS and 1/1AS. For simplicity, we consider the lattice material as non-absorbing, non-magnetic and isotropic. All the considered lattice approximants of quasicrystals are endless in the calculations. To find the eigenmodes of Maxwell's equations with periodic boundary conditions, we used a method of expanding the field with respect to plane waves [120].

Figure 5.44 shows the band structure of approximant 1/0DR at a dielectric constant $\varepsilon = 12$ (corresponding to silicon for near-infrared wavelengths) and the found optimal filling factor of the dielectric $f = 22.7\%$ (the ratio of the dielectric in the unit cell volume to the volume the cell, expressed in percent). The optimal filling factor is called the point at which the size of the band gap reaches its maximum value. The frequency is specified in dimensionless units, where ω is the angular frequency, a is the size of the cubic cell 1/0DR, c is the speed of light in vacuum. On the horizontal there are points of high symmetry of the Brillouin zone of the BCC lattice (inset in Fig. 5.44). Additionally, in the interval between the points of high symmetry there are additional 15 points at cqual distances.

There is a complete PBG with the size $\Delta\omega/\omega_m = 17.6\%$ between 8 and 9 zones (between the frequencies of 0.6293 ($\omega_a/2\pi c$) and 0.7509 ($\omega_a/2\pi c$), at a filling factor of the dielectric $f = 22.7\%$. For a given filling factor by the dielectric the ratio of the radius of the dielectric cylinder to the size of the cubic cell $r/a = 0.09$. The size of PBG $(i) - (i+1)$ is expressed by

$$\frac{\Delta\omega_{i,i+1}}{\omega_m} = 2\frac{\min(\omega_{i+1}) - \max(\omega_i)}{\min(\omega_{i\_1}) + \max(\omega_i)} \cdot 100\%, \tag{5.105}$$

where $\max(\omega_i)$ and $\min(\omega_{i+1})$ are the maximum and minimum frequency for bands (i) and $(i+1)$, respectively. The size of PBG for 1/0AS was found to be $\Delta\omega/\omega_m = 11\%$ between 12 and 13 bands (frequencies between 0.8014 and 0.8948) at $\varepsilon = 12$ and the optimal filling factor by the dielectric $f = 18.9\%$. This filling factor of the dielectric is obtained at the radius of the spheres to the size of the unit cell $r/a = 0.25$. The band structure 1/0AS is shown in Fig. 5.45. It should be noted that the PBG 1/0AS lies at higher frequencies. It is also interesting that the number of bands between which the PBG lies does not coincide for 1/0AS and 1/0DR.

We now consider the photonic quasicrystals approximants of higher order.

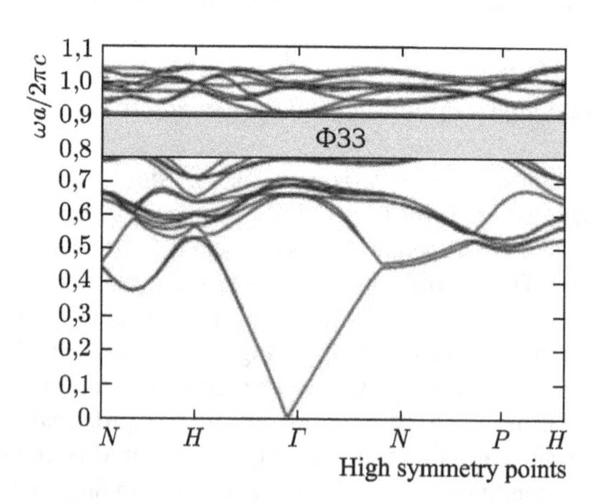

Fig. 5.45. The band structure of approximant 1/0AS (the first 20 zones). Dielectric filling factor $f = 18.9\%$, and its dielectric constant $\varepsilon = 12$. The size of the total gap $\Delta\omega/\omega_m = 11\%$.

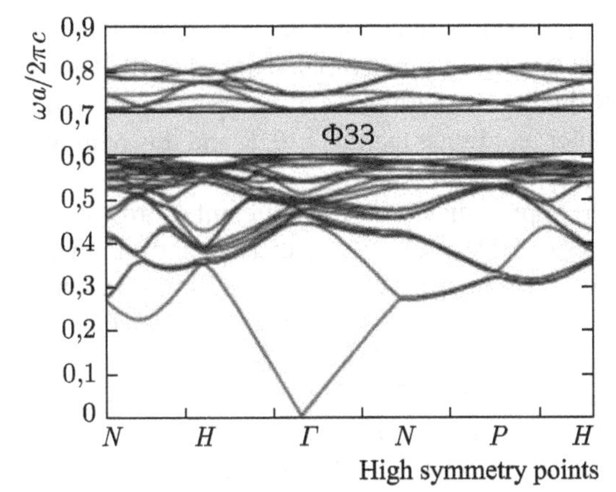

Fig. 5.46. ~Band structure of approximant 1/1FDR (the first 40 bands). Dielectric filling factor $f = 23.8\%$, and its dielectric constant $\varepsilon = 12$. The size of the total gap $\Delta\omega/\omega_m = 10.3\%$.

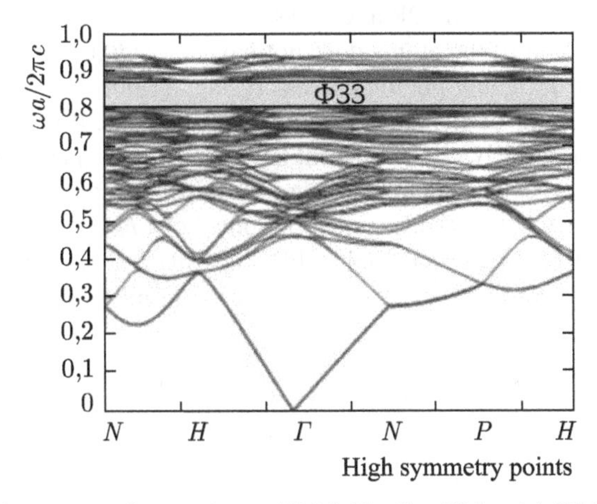

Fig. 5.47. Band structure of approximant 1/1AS (the first 70 bands). Dielectric filling fac tor $f = 20.5\%$, and its dielectric constant $\varepsilon = 12$. The size of the full band gap $\Delta\omega/\omega_m = 7.76\%$.

The structure 1/1FAS at a filling factor by the dielectric $f \sim 22\%$ has an unbound lattice of the dielectric so it cannot be obtained experimentally. In addition, 1/1FAS has no complete PBGs. Therefore, we confined ourselves to the consideration of the structure 1/1FDR. Such gratings are easier to obtain experimentally, as was demonstrated in [109, 110].

The band structure 1/1FDR at $\varepsilon = 12$ and the optimal filling factor by the dielectric $f = 23.8\%$ is shown in Fig. 5.46. Just as in the case of 1/0DR in Fig. 5.46 on the frequency recorded in dimensionless units, and in both cases a is the size of the cubic cell 1/0DR.

The dielectric filling factor $f = 23.8\%$ is obtained for $r/a = 0.095$. The size of the full PBG is equal to $\Delta\omega/\omega_m = 10.3\%$ and it is between 32 and 33 bands (frequencies between 0.6342 ($\omega_a/2\pi c$) and 0.7029 ($\omega_a/2\pi$)).Interestingly, the optimal filling factor is almost identical in both approximants, whereas the size of a full PBG in 1/1FDR was almost two times smaller than that of 1/0DR. This is explained by the fact that in 1/1DR the Fourier harmonics of the spatial distribution of the dielectric constant have smaller amplitude than in 1/0DR and this requires further optimization of structures.

The atoms of the approximant 1/1 have different coordination numbers, so comparing the properties of 1/1DR 1/1FDR we can reveal the influence of the coordination of the atoms on the size and threshold of the band gap. Unlike 1/1FAS, 1/1AS has a bounded lattice of the dielectric. Figure 5.47 shows the band structure 1/1AS at = 12 and f = 20.5%. Full PBG is between 56 and 57 bands (frequencies between 0.8019 and 0.8666) with size = 7.76%. The ratio of the radius of the balloons to the size of the unit cell is obtained 1/0DR r / a = 0.252.

The band structure is shown in 1/1DR Fig. 5.48 at ε = 12 and f = 26%. PBG is between 28 and 29 bands (frequencies between 0.5783 and 0.6379) and size $\Delta\omega/\omega_m$ = 9.8%. For a given filling factor by the dielectric the ratio of the radius of the dielectric cylinder to the size of the cubic cell is r/a = 0.1.

The size of full PBGs in 1/1DR and 1/1FDR is almost the same, therefore, in this case, the coordination of the atoms does not have ant significant impact. This fact may be useful for designing photonic quasicrystals from a six-dimensional simple cubic lattice, where the atoms have different coordinations. It is also interesting to note that the PBG in 1/1AS lies in the higher zones and frequencies, as compared with 1/1DR. This is favourable for lowering the threshold for PBG.

Let us now analyze what indexes of the reflections *hkl* (the coordinates of the reciprocal lattice vector) give a complete PBG in 1/0DR. From the Bragg condition for backward diffraction with the average dielectric permittivity taken into account, we have

$$K^2 = 4\left[1 + f(\varepsilon - 1)\right]\left(\frac{a\omega}{2\pi c}\right)^2,$$

where $K^2 = h^2 + k^2 + l^2$.

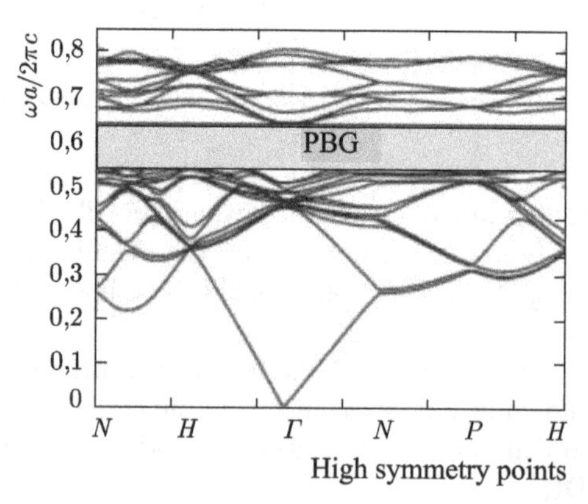

Fig. 5.48. The band structure of the approximant 1/1DR (the first 40 bands). Dielectric filling factor f = 26% and its dielectric constant ε = 12. The size of the full band gap $\Delta\omega/\omega_m$ = 9.8%.

In Fig. 5.44 for the centre of the band $a\omega/2\pi c = 0.7$, which yields $K^2 = 6.85$. It follows from this that the complete PBG is formed mainly by reflections of type 211 ($K^2 = 6$) and 220 ($K^2 = 8$). However, at point P (threefold axis) the band gap is very wide so that this probably contributes to the reflections of the type 222 ($K^2 = 12$). Slightly above the threshold for opening of the full PBG ($\varepsilon = 6, f = 0.296$) calculations give $a\omega/2\pi c = 0.78$, hence $K^2 = 6.04$, i.e. the PBG is due to the reflections 211 and 200, which in a quasicrystal correspond to reflections directed along the twofold axes of the icosahedron (there are thirty such reflections on the sphere). The following results were obtained for 1/0AS. In Fig. 5.47 for the centre of the band $a\omega/2\pi c = 0.85$, which yields $K^2 = 8.9$. So complete PBG is formed mainly by the reflections of type 221 ($K^2 = 9$). If the value of the dielectric constant is close to the threshold $\varepsilon = 6$, for 1/0AS we obtain $a\omega/2\pi c = 1$ at $f = 0.238$, so $K^2 = 8.76$. Thus, the full PBG is formed by the reflections 220 ($K^2 = 8$) and 221 ($K^2 = 9$). As we can see Fig. 5.44 and Fig. 5.45, the PBG in 1/0AS is more isotropic than in 1/0DR, due to the fact that the length of reflections creating the PBGs reflexes is more similar to the length of the reciprocal lattice vectors. This is explained by the fact that in 1/0AS the PBG lies at higher frequencies, compared with 1/0DR.

Let us see what happens to 1/1FDR, in which the Bragg condition looks like $K^2 = 4[1+f(\varepsilon - 1)](\tau a\omega/2\pi c)^2$, where $\tau = \left(1+\sqrt{5}\right)/2$ is the golden mean. If $\varepsilon = 12$, $f = 0.238$ and in the middle of the band $a\omega/2\pi c = 0.67$, then $K^2 = 17.05$. Theoretically, is this approximant there should be strong contributions from the reflections 400 ($K^2 = 16$) and 321 ($K^2 = 14$), but it seems a contribution is also provided by 411 ($K^2 = 18$) and 420 ($K^2 = 20$), which do not correspond to any strong reflections in quasicrystals. At $\varepsilon = 6, f = 0.3123$ and the middle of the band of 0.793, so $K^2 = 16.87$. In percentage terms, the characteristic value K^2 changes not so greatly as in 1/0DR, in transition from $\varepsilon = 6$ to $\varepsilon = 12$. It should also be noted that the PBG in 1/1DR is more isotropic than in 1/0DR, due to the fact that the length of the reflections creating the PBG is more similar to the length of reciprocal lattice vectors. This fact stems from the result that with increasing order of the approximant the structure close to the geometry of the quasicrystal. As can be seen on Fig. 5.44 and Fig. 5.48, the frequency ranges in which the PBGs form overlap. As a result, it turns out that with increasing approximant order the complete PBG is preserved, therefore, this PBG will also be found in the corresponding quasicrystal [116].

The Bragg condition for 1/1DR 1/1AS is similar to the case of 1/1FDR. At $f = 0.26$ and $\varepsilon = 12$ in 1/1DR the middle of the band $a\omega/2\pi c$, so $K^2 = 14.94$. Thus, the main contributions come from the reflections of type 321 ($K^2 = 14$) and 400 ($K^2 = 16$). At $\varepsilon = 6$ 1/1DR does not have complete PBG. This can be explained by the fact that the PBG in 1/1DR is less isotropic than 1/1FDR, due to the smaller absolute value of the reciprocal lattice vector. For 1/1AS at $\varepsilon = 12$ and $f = 0.205$ the middle of the band $a\omega/2\pi c = 0.8342$ so that $K^2 = 23.72$. So the PBG is formed mainly by reflections of type 422 ($K^2 = 24$). For values of $\varepsilon = 6$ and $f = 0.297$ the middle of the band $a\omega/2\pi c = 0.9538$, hence $K^2 = 23.67$. The percentage value of K^2 does not change so much as in 1/0AS, the transition from $\varepsilon = 6$ to $\varepsilon = 12$. Figures 5.45 and 5.47 also show that the frequency bands of the PBGs overlap for 1/0AS 1/1AS.

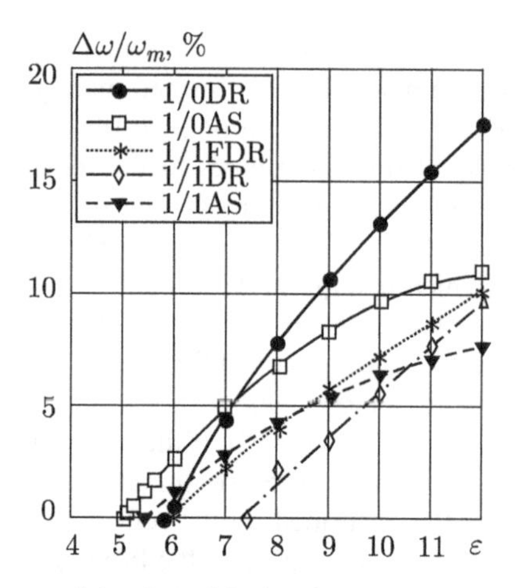

Fig. 5.49. The dependence of the size of the band gap $\Delta\omega/\omega_m$ on the dielectric constant ε.

The optimal filling factor of the dielectric at which the complete PBG has the maximum size was determined for each value of the dielectric constant. These results are shown in Fig. 5.49. The threshold of formation of the PBG as regards the dielectric constant for the approximants 1/0AS, 1/0DR was found to be $\varepsilon_{th} = 5$, $\varepsilon_{th} = 5.8$, respectively. For higher-order approximants 1/1AS, 1/1DR, 1/1FDR, the thresholds were found to be $\varepsilon_{th} = 5.3$, $\varepsilon_{th} = 7.4$, $\varepsilon_{th} = 5.8$, respectively. The threshold of PBG in 1/0DR 1/1FDR turned out to be practically the same, therefore, a further increase in the order of approximants we obtain the same threshold value, or slightly less than [35]. Similar conclusions can be drawn for the approximants 1/0AS and 1/1AS. As can be seen, the threshold of PBG in 1/1DR was significantly greater than that of 1/1FDR. This is explained by the influence of the coordination number of atoms. It turns out that for quasicrystals constructed from dielectric cylinders, the preferred arrangement of the atoms is the one in which they have the same coordination number [121, 122]. In addition, it can be concluded that preferred configuration for the construction of photonic quasicrystals with a low threshold for PBG is the air spheres in the dielectric. In principle, approximants with other sets of strong reflections, which will have a lower threshold for the formation of the complete PBGs, can also be formed.

5.6.3 Photonic crystal with the lattice symmetry of clathrate Si34

Consider a photonic crystal with a large number of nodes in the unit cell which possesses the properties of isotropy, the crystal with the lattice symmetry of clathrate Si34 [123]. This lattice belongs to the class of face-centred cubic (FCC) and contains 34 nodes in the unit cell (see Fig. 5.50). In Si34 each 'atom' is bonded with the neighbouring four 'atoms' distorted by tetrahedral bonds. It is known that the diamond lattice, which has tetrahedral bonds of 'atoms', produces the largest PBG of all photonic crystals. However, the icosahedral structure does not have any tetrahedral bonds.

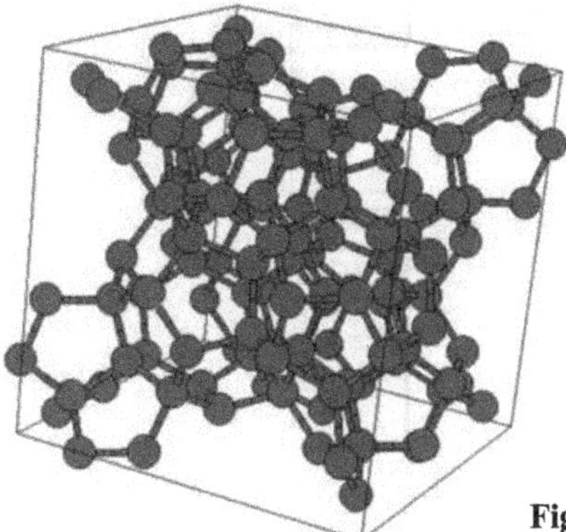

Fig. 5.50. Four primitive lattice cells Si34.

The atomic arrangement in the lattice of the clathrate Si34 is most isotropic, i.e. the shape of the Brillouin zone is closest to the sphere, which is favourable for the opening of a full PBG. The Si34 clathrate lattice is obtained by packing the pentagonal dodecahedron and the tetrakis decahedron. The pentagonal dodecahedron is the right Platonov polyhedron, i.,e. a body in which all vertices are equivalent and all facets are of the same type. It has several axes of symmetry of the fifth order, therefore, under the laws of crystallography, the space cannot be filled only by polyhedra and their combination with other types of polyhedra, which also contain hexagonal faces, is also essential. The tetrakis decahedron has two hexagonal faces. However, as noted above, each atom is bonded to four neighbouring atoms by distorted tetrahedral bonds.

The eigenmodes of Maxwell's equations with periodic boundary conditions were determined by the method of expansion of the field in plane waves [120]. Three cases were considered: 1) lattice sites are dielectric spheres, surrounded by air, 2) lattice sites are spherical cavities in a dielectric, and 3) the neighbouring lattice sites are connected by dielectric rods [124].

As a result of these calculations it was found that the photonic crystal lattice on the basis of Si34, consisting of dielectric spheres in a vacuum, does not have any large band gaps (around 5% for the dielectric constant $\varepsilon = 12$), and the photonic crystal consisting of balloons in the dielectric has no band gap. Therefore, a more detailed study of the photonic crystal consisting of dielectric rods in a vacuum, which connect the neighbouring lattice sites, forming thereby tetrahedral bonds, was carried out.

Figure 5.51 shows the band structure of a photonic crystal at $\varepsilon = 12$ and at the optimum filling factor $f = 22\%$ (the ratio of the volume of the dielectric to the total volume of the cell). In the graphs the frequency is specified in dimensionless units $a\omega/2\pi c$, where ω is the angular frequency, a is the size of the unit cell, and c is the speed of light in vacuum. The x-axis gives high-symmetry points of the Brillouin zone of the FCC lattice. The full PBG is located between 34th and 35th bands and has a size $\Delta\omega_m/\omega = 15.6\%$ (at a filling factor $f = 22\%$). Thus, the photonic crystal with the lattice symmetry of the clathrate contains more than four 'atoms' in the unit

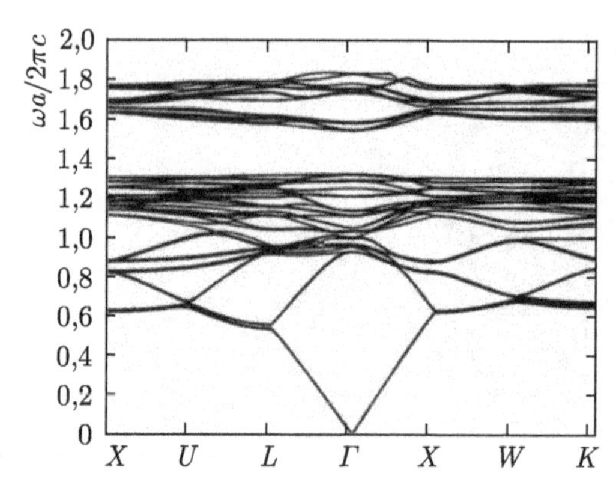

Fig. 5.51. The band structure of Si34. The dielectric constant $\varepsilon = 12$. The filling factor $f = 22\%$. The size of the band gap $\Delta\omega_m/\omega = 15.6\%$.

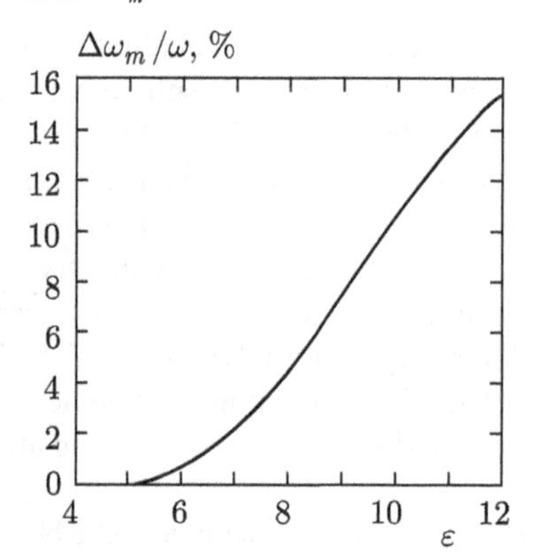

Fig. 5.52. The dependence of the size of the band gap $\Delta\omega_m/\omega$ on the dielectric constant ε.

cell and has a complete PBG. This crystal contains in its lattice a polyhedron, which has the symmetry axes of the fifth order.

To determine the minimum value of the dielectric constant ε_{th} at which there is a complete PBG, calculations were carried out of the dependence of the band gap on the dielectric constant. To calculate the dependence for each value of the dielectric constant ε we determined the filling factor by the dielectric for which the complete PBG has a minimum size. The results are shown in Fig. 5.52. As can be seen from Fig. 5.52, the threshold of a full PBG with respect to the dielectric constant is equal to $\varepsilon_{th} = 5$. The threshold was higher than that of a photonic crystal with the symmetry of the diamond lattice for which the threshold $\varepsilon_{th} = 4.0$, but less than that of a photonic crystal with the symmetry of the inverted opal [125]. This is due to the fact that the PBG of the structure with the Si34 symmetry is more isotropic, that is the frequency of boundary zones (zones restricting PBG) is weakly dependent on the direction of propagation of electromagnetic waves.

The isotropy of the band gap was studied. The parameter of isotropy F of the photonic band is $F = 2(\max(\omega_i) - \min(\omega_i))/(\max(\omega_i) + \min(\omega_i))$ [126]. The resulting dependence of the parameter of isotropy F for bands for 34 ('low' band) and

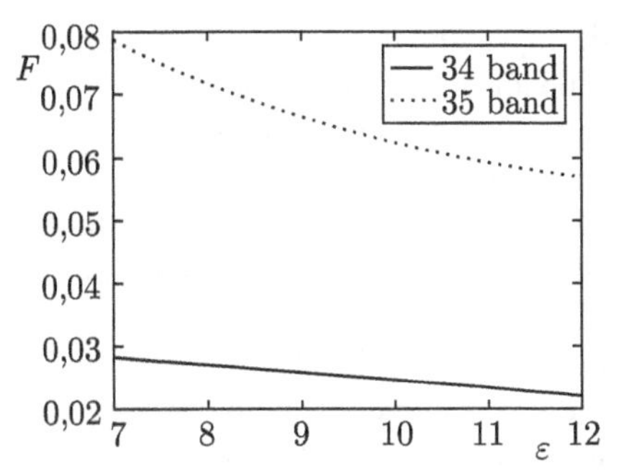

Fig. 5.53. The dependence of the isotropy parameter F for bands 34 (solid line) and 35 (dashed line) on dielectric constant ε, the filling factor $f = 27\%$.

35 ('upper' band) on the dielectric constant ε at the filling factor $f = 27\%$ is shown in Fig. 5.53. Figure 5.53 shows that the parameter F decreases monotonically with increasing dielectric constant. This is due to the fact that the localization of the electromagnetic field increases with increasing dielectric constant.

It is known that the group velocity of propagation of electromagnetic waves can vanish at the boundary of the PBG. The physical significance of this phenomenon lies in the fact that at the boundary of the PBG the scattered waves form a standing wave and therefore generation of coherent radiation is possible [127]. In conventional photonic crystals this is only implemented in some directions, because the position of the boundary of the PBG depends on the direction of propagation of electromagnetic radiation. To produce coherent radiation, no matter the direction of propagation of electromagnetic radiation, it is necessary to use photonic crystals with an isotropic PBG. The isotropy of the photonic bands can also be used for omni-directional negative refraction. This phenomenon was studied for two-dimensional photonic crystals [128] and quasicrystals [129].

5.7. One-dimensional photonic crystal based on a nanocomposite: metal nanoparticles – a dielectric

The first theoretical papers dealt with photonic crystals, obtained from non-absorbing and dispersionless materials [94, 97, 98]. In these studies periodic structures with complete PBG at a high refractive index were investigated. At this moment, the lowest refractive index, allowing a complete PBG to exist, is $n = 2$ for the diamond lattice [130]. This imposes severe restrictions on the choice of material, because in the visible spectrum there is no glass and polymers with the refractive index of this value. In later studies attention was given to materials with significant absorption and dispersion. In particular, photonic crystals composed of ionic material or with metallic inclusions were studied. Using the finite difference method, calculations were carried out to determine the transmission coefficients of a face-centred cubic (FCC) and diamond lattices consisting of perfectly conducting metal spheres [131]. As a result, it was shown in [131] that such a diamond lattice has a complete PBG larger than 45%, which is higher than that of any known dielectric photonic crystal. Using the Drude approximation for dielectric permittivity, which takes into account the dispersion and absorption, it was found that the FCC lattice, consisting of particles of noble metals (e.g. silver) with the size of about 160 nm, has a large full PBG [132]. The authors explain this by saying that in the frequency range in which PBG is formed, the dielectric constant of metals takes large negative values, thereby forming a significant contrast of the dielectric properties, which is favourable for the detection of complete PBGs. In [133] attention was given to a dielectric waveguide coated on one side with a layer of metallic nanoparticles arranged in a square lattice. As a result, it was found that both the waveguide and plasmon modes can propagate in such a structure in the optical range.

Ionic materials in the infrared region have the range of frequencies $\omega_T < \omega < \omega_L$ for which the material is optically like a metal, i.e. the permittivity is negative and has a strong dispersion [134]. The photonic crystals, obtained from the ionic material (polariton photonic crystals), are studied in [135, 136]. It is shown that the photon–phonon interaction can result in the formed of polariton PBG, which should be distinguished from structural PBG, formed as a result of Bragg diffraction of intrinsic electromagnetic states at the edge of the Brillouin zone. The PBG resulting from interaction with the collective optical excitations, such as optical phonons, plasmons, excitons, is called polariton PBG. In [137] the authors observed the effect of the merger of the polariton and structural PBGs, which provides a new tool to obtain structures with PBG.

Composite materials with nanoparticles of noble metals are of great practical interest in the development of various optical devices. The linear and non-linear optical properties of such media are determined by plasmon resonance of metallic nanoparticles and the properties of a transparent matrix. The authors of [138, 139] predicted the appearance of the resonance of the dielectric constant in the nanocomposite consisting of metal nanoparticles suspended in a transparent matrix, and the position of the resonance depends on the dielectric constant of the starting materials and the concentration of nanoparticles. The shape of the resonances of the dielectric constant of such a nanocomposite is identical with the resonances of the ionic material, but the resonance lies in the visible light range. Of great interest, both from applied and the fundamental point of view, the use of such nanocomposite materials as photonic crystals. In this paper, we calculate the transmittance, reflectance, and absorption coefficients for a one-dimensional photonic crystal consisting of a nanocomposite: metallic nanoparticles distributed randomly in a transparent matrix.

To find the dielectric constant of the nanocomposite $\varepsilon_{\text{mix}}(\omega)$, we use the Maxwell–Garnett equation:

$$\frac{\varepsilon_{\text{mix}}(\omega)-\varepsilon_d}{\varepsilon_{\text{mix}}(\omega)+2\varepsilon_d}=f\frac{\varepsilon_m(\omega)-\varepsilon_d}{\varepsilon_m(\omega)+2\varepsilon_d}, \tag{5.106}$$

where f is the relative volume occupied by nanoparticles, $\varepsilon_m(\omega)$ is the dielectric constant of the metal from which the nanoparticles are made, ε_m is the dielectric constant of the matrix in which nanoparticles are immersed, ω is the radiation frequency.

The nanoparticles are distributed in a matrix in a random but uniform manner. We assume that the nanoparticles have spherical shape with a radius within a few nanometers, which is much smaller than the wavelength and the penetration depth into the material. The dielectric constant of the metal of the nanoparticles is determined using the Drude approximation, as follows:

$$\varepsilon_m(\omega) = \varepsilon_0 - \frac{\omega_p^2}{\omega(\omega + i\gamma)} \tag{5.107}$$

where ε_0 is a constant ($\varepsilon_0 = 5$ for silver), ω_p is plasma frequency ($\omega_p = 9$ eV for silver), γ is the relaxation constant ($\gamma = 0.02$eV for silver [140]). To be specific, in all further calculations the metal will be represented by silver. Substituting (5.107) into (5.106) we find as follows:

$$\varepsilon_{\text{mix}}(\omega) = \varepsilon'_{\text{mix}}(\omega) + i\varepsilon''_{\text{mix}}(\omega). \tag{5.108}$$

Figure 5.54 presents the dependences $\varepsilon'_{\text{min}}(\omega)$ and $\varepsilon''_{\text{min}}(\omega)$ at $f = 0.2$, $\varepsilon_d = 2.56$. The curves in Fig. 5.54 have a resonance character (the form of the curves is similar to the case of ionic materials [134]). Neglecting the small factor γ^2 we find that the function vanishes at the points:

$$\omega_{10} = \omega_p \sqrt{\frac{1-f}{\varepsilon_0 + 2\varepsilon_d - f(\varepsilon_0 - \varepsilon_d)}} \tag{5.109}$$

$$\omega_{20} = \omega_p \sqrt{1 + \frac{\varepsilon_0(\varepsilon_0 + 2\varepsilon_d - f(\varepsilon_0 - \varepsilon_d))}{\varepsilon_d(\varepsilon_0 + 2\varepsilon_d + 2f(\varepsilon_0 - \varepsilon_d))}} \tag{5.110}$$

($\omega_{10}/\omega_p = 0.288$ and $\omega_{20}/\omega_p = 0.355$, at $f = 0.2$, $\varepsilon_d = 2.56$). In the gap $[\omega_{10}, \omega_{20}]$ the function $\varepsilon'_{\text{min}}(\omega)$ takes negative values, so in this frequency region the nanocomposite

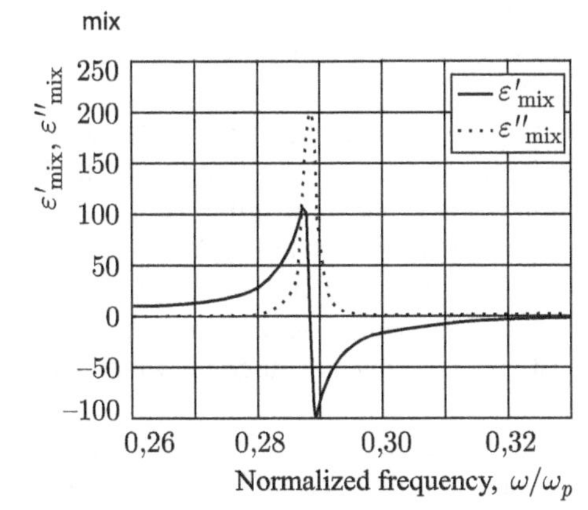

Fig. 5.54. Functions $\varepsilon'_{\text{min}}(\omega)$ and $\varepsilon''_{min}(\omega)$ for silver nanoparticles suspended in a transparent matrix at $f = 0.2$, $\varepsilon_d = 2.56$.

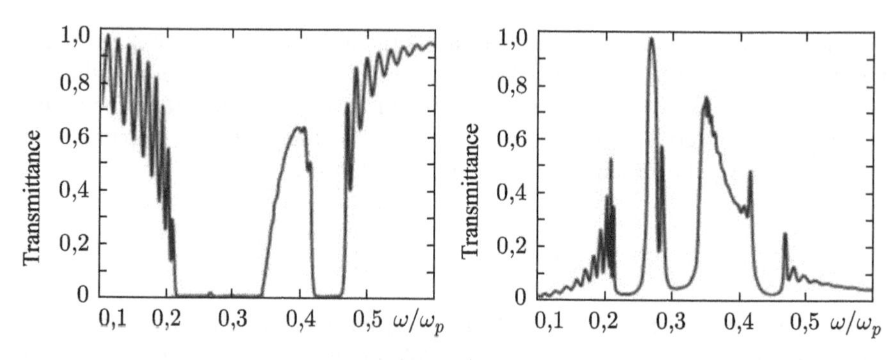

Fig. 5.55. Transmittance and absorbance coefficients at $N = 16, f = 0.2, a/\lambda_p = 1$.

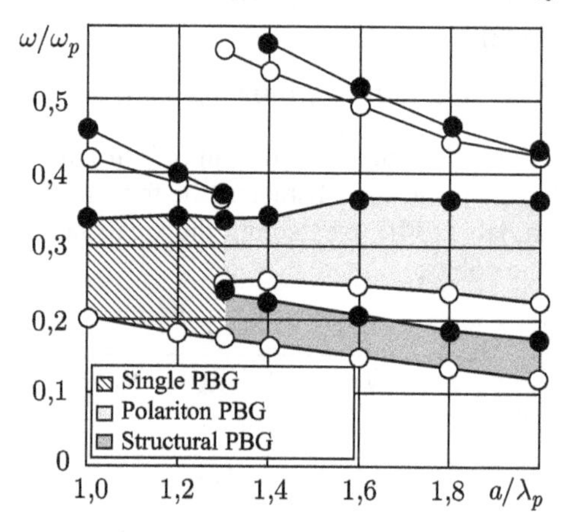

Fig. 5.56. PBG behavior depending on the size of the unit cell a at $N = 16, f = 0.2, d_1/a = 0.5$.

is optically similar to the metal. Consequently, we can expect polariton PBGs in this range of frequencies.

Consider a one-dimensional photonic crystal composed of N unit cells with a period a. Each cell consists of a layer with thickness d_1 and the dielectric constant $\varepsilon'_{\text{mix}}(\omega)$ and a layer with thickness d_2 with a dielectric constant equal to 1. The period of the cell a is a $= d_1 + d_2$. In all cases we consider the normal incidence of electromagnetic waves on a photonic crystal. To find the coefficients of transmission, reflection and absorption, we used the finite difference method [6]. The coefficients of transmission and absorption are shown in Fig. 5.55 at $N = 16, f = 0.2, d_1/a = 0.5$, $a/\lambda_p = 1$. The graph shown the frequency in terms of plasma frequency. Figure 5.56 shows the variation of PBG, depending on the size of the unit cell a at $N = 16, f = 0.2, d_1/a = 0.5$. The PBG is the frequency range for which the transmittance is less than 0.1.

In the graph the size of the unit cell a is marked in units of the plasma wavelength λ_p equal to $\lambda_p = 2\pi c/\omega_p$, where c is the speed of light in vacuum, and the frequency in units of the plasma frequency ω_p. Empty circles marked lower edge of the PBG, and the black circle – the top. Figure 5.56 can be seen that $a/\lambda_p = 1.3$ the single PBG is split into a single polariton PBG and a structural PBG. To prove this, it is necessary to investigate the behaviour of the PBG when the filling factor of the nanocomposite changes, i.e. the ratio d_1/a changes. As shown in [137], the polariton PBG increases in size with increasing filling factor of the material, whereas the structural PBG reaches its maximum and begins to diminish in size. Change of the PBG in the variation of the filling factor of nanocomposites is shown in Fig. 5.57 at $a/\lambda_p = 2$. One can observe that the polariton PBG with the centre between ω_{10}–ω_{20} increases in size, with increasing ratio d_1/a, whereas PBG structure, which lies below, reaches a peak in size and begins to decrease. The splitting of PBG in polariton and structural PBGs gives us a new tool in the design of photonic crystals [142]. This effect gives us the opportunity, without changing the filling factor of nanocomposites, to change the characteristic dimensions of the cell of the photonic crystal dramatically altering the optical properties. In conventional dielectric photonic crystals, changing the cell dimensions (at a constant filling factor), we only shift the PBG, but do not change its structure.

It should also be noted that by reducing the concentration of metallic nanoparticles we reduce the absorption of the nanocomposite. In metallic photonic crystals absorption is very important, due to high metal concentrations. In addition, metal photonic crystals are quite difficult to produce experimentally, because of the need to periodically build submicron metal objects. In the investigated nanocomposite the metal particles are distributed randomly and the uniformity of distribution in space can be implemented using the Coulomb interaction of the charged metal spheres.

Figure 5.58 shows the change in PBG, depending on the concentration of metal nanoparticles f, for $N = 8$, $d_1/a = 0.5$, $a/\lambda_p = 1$. In consequence of the fact that with increasing f the width of the interval $[\omega_{10}, \omega_{20}]$ increases, we observe the growth of the size of the polariton PBG. Structural PBG behaves in a more complicated manner, with $f = 0.15$ it reaches its maximum size, then decreases to zero. It should also be noted that the centre of structural PBG is moved to higher frequencies with increasing f.

The opposite behaviour is observed in the structural PBG at $a/\lambda_p = 2$ as shown in Fig. 5.59. In this case, with increasing f the size of structural PBG does not change and its centre shifts to lower frequencies [137]. This effect, depending on the behaviour of structural PBG on its position relative to the polariton PBG, may be useful in the design of photonic crystals with the desired properties. By varying the concentration of metal nanoparticles, we can achieve that PBG is at the right frequencies. The ability to change the parameter f distinguishes the nanocomposite from the ionic material. In ionic materials the peak position of the resonance can not be changed, which may have a negative impact on the capacity of production of photonic crystals made from these materials. Also important is the fact that the photonic crystals based on the nanocomposite allow to work in the visible frequency range.

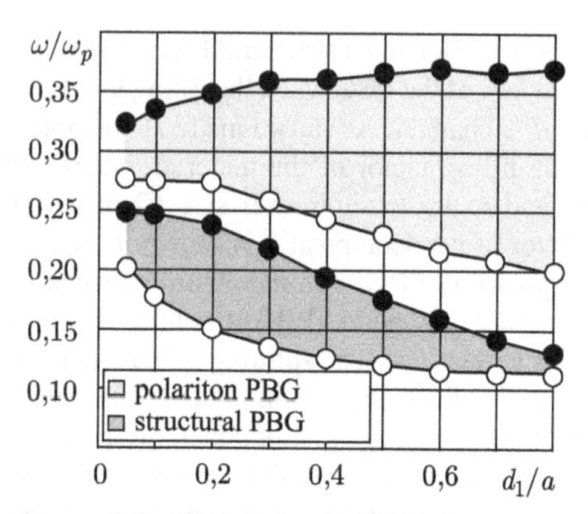

Fig. 5.57. The dependence of the PBG on the the filling factor by the nanocomposite d_1/a at $a/\lambda_p = 2$, $N = 16$, $f = 0.2$.

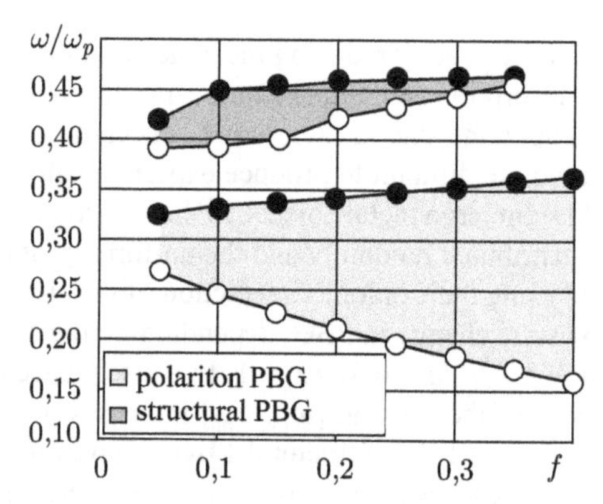

Fig. 5.58. Dependence of PBG on the concentration of metallic nanoparticles f at $a/\lambda_p = 1$, $N = 8$, $d_1/a = 0.5$.

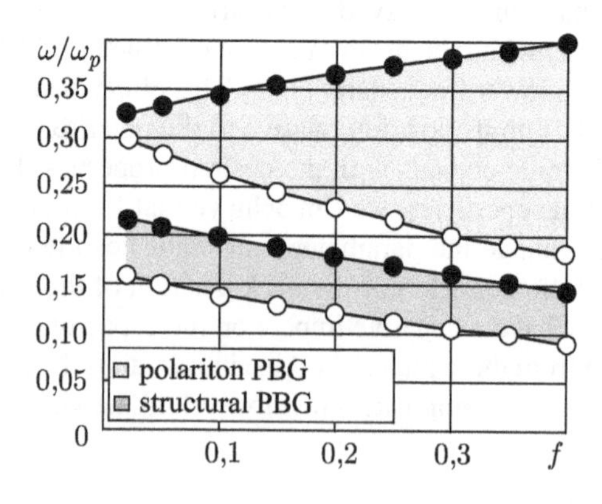

Fig. 5.59. Dependence of PBG on the concentration of metallic nanoparticles f at $a/\lambda_p = 2$, $N = 8$, $d_1/a = 0.5$.

References

1. Joannopoulos J.D., Johonson S.G., Winn J.N. Photonic Crystals: Molding the Flow of Light. Sec. ed. Princeton Univ. Press, 2008. 304 p.
2. Prasad P. N. Nanophotonics. Wiley, 2004. 432 p.
3. Yariv A., Yeh P. Optical wave in crystals: propagation and control of laser radiation. Wiley-Interscience, 2002. 589 p.
4. Fukaya Y., Ohsaki D., Baba T. Two-dimensional photonic cryctal waveguide with 60° bends in a thin slab structure, J. Jap. Soc. Appl. Phys. 2000. V. 39, No. 5A. P. 2619–2623.
5. Frei W.R., Tortorelli D. A., Johnson H. T. Geometry projection method for photonic nanostructures, Opt. Lett. 2007. V. 32, No. 1. P. 77–79.
6. Taflove A., Computational Electrodynamics: the finite-difference time-domain method. MA: Artech House, Inc., 1995. 597 p.
7. Pernice W.H., Payne F. P., Gallagher D. F. Numerical investigation of field enhancement by metal nano-particles using a hybrid FDTD-PSTD algorithm, Optics Express. 2007. V. 15. P. 11433–11443.
8. Kim J.H., Chrostowski L., Bisaillon E., Plant DV DBR, sub-wavelength grating, and photonic crystal slab Fabry-Perot cavity design using phase analysis by FDTD, Optics Express. 2007. V. 15. P. 10330–10339.
9. Yablonovitch E. Inhibited spontaneous emission in solid-state physics and electronic Phys. Rev. 1987. V. 58. P. 2059–2062.
10. Hugonin J. P., Lalanne P., White T. P., Krauss T. F. Coupling into slow-mode photonic crystal waveguides, Opt. Lett. 2007. V. 32. P. 2639–2640.
11. Kwan K.C., Tao X.M., Peng G.D., Transition of lasing modes in disordered active photonic crystals, Opt. Lett. 2007. V. 32. P. 2720–2722.
12. Zabelin V., Dunbar L.A., Thomas N.L., Houndre R., Kotlyar v.V., O'Faolain L., Krauss T.F., Self-collimating photonic crystal polarization beam splitter, Opt. Lett. 2007. V. 32. P. 530–532.
13. Li Y., Jin J., Fast full-wave analysis of large-scale three-dimensional photonic crystal device, J. Opt. Soc. Am. B. 2007. V. 24. P. 2406–2415.
14. Mikaelian A.L. Application of the properties of the medium to focus the waves, Dokl. AN SSSR. 1951. N. 81. P. 569–571.
15. Snyder A.W., Mitchell D. J. Spatial solitons of the power-law nonlinearity, Opt. Lett. 1993. V. 18. P. 101–103.
16. Alimenkov I.V., Exactly solvable mathematical models in nonlinear optics, Komp. Optika. 2005. N. 28. P. 45–54.
17. Alimenkov IV Nonlinear Schrödinger equation in three-dimensional variables, Komp. Optika, 2005. N. 28. P. 55–59.
18. Yee K. S. Numerical solution of initial boundary value problems involving Maxwell's equations in isotropic media, IEEE Trans. Antennas and Propagation. 1966. AP14. P. 302–307.
19. Moore G. Absorbing boundary conditions for the finite-difference approximation of the time-domain electromagnetic field equations, IEEE Trans. Electromagnetic Compatibility. 1981. V. 23. P. 377–382.
20. Berenger J.P., A perfectly matched layer for the absorption of electromagnetic waves, Comput. Physics. 1994. V. 114. P. 185–200.
21. Xu Y., Lee R.K., Yariv A.. Adiabatic coupling between conventional dielectric waveguides with discrete translational symmetry, Opt. Lett. 2000. V. 25 , No. 10. P. 755–757.
22. Mekis A., Joannopoulos J.D., Tapered couplers for efficient interfacing between dielectric and photonic crystal waveguides, J. Light Techn. 2001. V. 19, No. 6. P. 861–865.
23. Happ T.D., Kamp M., Forchel A., Photonic crystal tapers for ultracompact mode conversion, Opt. Lett. 2001. V. 26. P. 14; P. 1102–1104.

24. Talneau A., Lalanne P., Agio M., Soukoulis C.M., Low-reflection photonic crystal taper for efficient coupling between guide sections of arbitrary widths, Opt. Lett. 2002. V. 27, No. 17. P. 1522–1524.
25. Almeida V. R., Panepucci R. R., Lipson M. Nanotaper for compact mode conversion, Opt. Lett. 2003. V. 28, No. 15. P. 1302–1304.
26. Bienstman P., Assefa S., Johson S.G., Joannopoulos J.D., Petrich G.S., Koloziejski L.A., Taper structures for coupling into photonic crystal slab waveguides, J. Opt. Soc. Am. B. 2003. V. 20 , No. 9. P. 1817–1821.
27. MacNab S.J., Moll N., Vlasov Y.A., Ultra-low loss photonic integrated circuit with membrane-type photonic crystal waveguide, Opt. Express. 2003. V. 11, No. 22. P. 2927–2939.
28. Barclay P.E., Srinivasan K., Painter O., Design of photonic crystal waveguide for evanescent coupling to optical fiber tapers and integration with high-Q cavities, J. Opt. Soc. Am. B. 2003. V. 20, No. 11. P. 2274–2284.
29. Orobtchouk R., Layadi A., Gualous H., Pascal D., Koster A., Laval S., High-efficiency light coupling in a submicrometric silicon-on-insulator waveguide, Appl. Opt. 2000. V. 39, No. 31. P. 5773–5777.
30. Lardenois S., Pascal D., Vivien L., Cassan E., Laval S., Orobtchouk R., Heitzmann M., Bonzaida N., Mollard L. Low-loss submicrometer silicon-on-insulator rib waveguides and corner mirrors, Opt. Lett. 2003. V. 28 , No. 13. P. 1150–1153.
31. Taillaert D., Vanlaere F., Ayre M., Bogaerts W., Van Thourhout D., Bienstman P., Baets R., Grating couplers for couping between optical fiber and nanophotonic waveguides, Jap. J. Appl. Phys. 2006. V. 45, No. 8 , P. 6071-6077.
32. Van Laere F., Roelkens G., Ayre M., Taillaert D., Van Thourhout D., Krauss T.F., Baets R., Compact and high efficient grating couplers between optical fiber and nanophotonic waveguides, J. Light. Techn. 2007. V. 25, No. 1. P. 151–156.
33. Bachim B.L., Ogunsola O.O., Gaylord T.K. Optical fiber-to-waveguide coupling using carbon-dioxide-laser-induced long-period fiber gratings, Opt. Lett. 2005. V. 30, No. 16. P. 2080–2082.
34. Prather D.W., Murakowski J., Shi S., Venkataraman S., Sharkawy A., Chen C., Pustai D., High-efficiency coupling structure for a single-line-defect photonic crystal waveguide, Opt. Lett. 2002. V. 27, No. 18. P. 1601–1603.
35. Kim H., Lee S., O B., Park S., Lee E., High efficiency coupling technique for photonic crystal waveguides using a waveguide lens. OSA Techn. Digest: Frontiers in optics, 2003. MT68.
36. Corbett J.C.W., Allington-Smith J.R., Coupling starlight into single-mode photonic crystal fiber using a field lens, Opt. Express. 2005. V. 13, No. 17. P. 6527–6540.
37. Michaelis D., Wachter C., Burger S., Zschiedrich L., Brauer A., Micro-optical assisted high-index waveguide coupling, Appl. Opt. 2006. V. 45, No. 8. P. 1831–1838.
38. Kong G., Kim J., Choi H., Im J.E., Park B., Paek V., Lee B.H., Lensed photonic crystal fiber obtained by use of an arc discharge, Opt. Lett. 2006. V. 31, No. 7. P. 894–896.
39. Pokrovsky A.L., Efros A.L., Lens based on the use of left-handed materials, Appl. Opt. 2003. V. 42, No. 28. P. 5701–5705.
40. Fabre N., Fasquel S., Legrand C., Melique X., Muller M., Francois M., Vanbesien O., Lippens D., Toward focusing using photonic crystal flat lens, Opto-Electronics Rev. 2006. V. 14, No. 3, P. 225–232.
41. Li C., Holt M., Efros A.L., Far-field imagimg by the Veselago lens made of a photonic crystal, J. Opt. Soc. Am. B. 2006. V. 23, No. 3. P. 490–497.
42. Matsumoto T., Eom K., Baba T., Focusing of light by negative refraction in a photonic crystal slab superlens on silicon-on-insulator substrate, Opt. Lett. 2006. V. 31, No. 18. P. 2786–2788.
43. Li C. Y., Holt J.M., Efros A. L. Imaging by the Veselago lens based upon a two-dimensional photonic crystal with a triangular lattice, J. Opt. Soc. Am. B. 2006. V. 23, No. 5. P. 963–968.

44. Geng T., Lin T., Zhuang S., All angle negative refraction with the effective phase index of –1, Chinese Opt. Lett. 2007. V. 5, No. 6. P. 361–363.
45. Asatsume T., Baba T., Abberation reduction and unique light focusing in a photonic crystal negative refractive lens, Opt. Express. 2008. V. 16, No. 12. P. 8711–8718.
46. Fabre N., Lalonat L., Cluzel B., Melique X., Lippens D., deFornel F., Vanbesien O., Measurement of a flat lens focusing in a 2D photonic crystal at optical wavelength. OSA Digest, CLEO / QELS, 2008, CTuDD6, CA.
47. Yang S., Hong C., Yang H. Focusing concave lens photonic crystals with magnetic materials, J. Opt. Soc. Am. A. 2006. V. 23, No. 4. P. 956–959.
48. Luan P., Chang K., Photonic crystal lens coupler using negative refraction, Prog. in Electr. Res. 2007. V. 3, No. 1. P. 91–95.
49. Haxha S., AbdelMalek F., A novel design of photonic crystal lens based on negative refractive index, Prog. in Electr. Res. 2008. V. 4, No. 2. P. 296–300.
50. Lu Z., Shi S., Schuetz C.A., Murakowski J.A., Prather D., Three-dimensional photonic crystal flat lens by full 3D negative refraction, Opt. Express. 2005. V. 13, No. 15. P. 5592–5599.
51. Lu Z., Shi S., Schuetz C.A., Prather D.W., Experimental demonstration of negative refraction imaging in both amplitude and phase, Opt. Express. 2005. V. 13, No. 6. P. 2007–2012.
52. Minin I.V., Minin O.V., Triandafilov Y.R., Kotlyar V.V., Subwavelength diffractive photonic crystal lens, Prog. in Electr. Res. B. 2008. V. 7. P. 257–264.
53. Pshenay-Severin E., Chen C. C., Pertsch T., Augustin M., Chipoline A., Tunnermann A., Photonic crystal lens for photonic crystal waveguide coupling. OSA Techn. Digest: CLEO, 2006, CThK3.
54. Hugonin J.P., Lalanne P., White T.P., Krauss T.F., Coupling into clow-mode photonic crystal waveguide, Opt. Lett. 2007. V. 32, No. 18. P. 2638–2640.
55. Triandafilov Ya.R., Kotlyar V.V., Komp. Optika, 2007. V. 31,N. 3. P. 27–31.
56. Triandafilov Ya.R., Kotlyar V.V., Photonic crystal Mikaelian lens, Opt. Mem. Neur. Net. 2008. V. 17, No. 1. P. 1–7.
57. Dorn R., Quabis S., Leuchs G., Sharper focus for a radially polarized light beam, Phys. Rev. Lett. 2003. V. 91, P. 233901.
58. Davidson N., Bokor N., High-numerical-aperture focusing of radially polarized doughnut beams with a parabolic mirror and a flat diffractive lens, Opt. Lett. 2004. V. 29, No. 12. P. 1318–1320.
59. Borghi R., Santarsiero M., Nonparaxial propagation of spirally polarized optical beams, J. Opt. Soc. Am. A. 2004. V. 21, No. 10. P. 2029–2037.
60. Passilly N., Denis R.S., Ait-Ameur K., Simple interferometric technique for generation of a radially polarized light beam, J. Opt. Soc. Am. A. 2005. V. 22, No. 5. P. 984–991.
61. Jabbour T.G., Kuebler S.M., Vector diffraction analysis of high numerical aperture focused beams modified by two-and three-zone annular multi-phase plates, Opt. Express. 2006. V. 14, No. 3. P. 1033–1043.
62. Kozawa Y, Sato S., Focusing property of a double-ring-shaped radially polarized beam, Opt. Lett. 2006. V. 31, No. 7. P. 867–869.
63. Zhan Q., Properties of circularly polarized vortex beams, Opt. Lett. 2006. V. 31 No. 7. P. 867–869.
64. Deng D., Nonparaxial propagation of radially polarized light beams, J. Opt. Soc. Am. B. 2006. V. 23, No. 6. P. 1228–1234.
65. Salamin Y.I., Fields of a radially polarized Gaussian laser beam beyond the paraxial approximation, Opt. Lett. 2006. V. 31, No. 17. P. 2619–2621.
66. Deng D., Guo Q., Wu L., Yang X., Propagation of radially polarized elegant light beams, J. Opt. Soc. Am. B. 2007. V. 24, No. 3. P. 636–643.
67. Grosjean T., Courjon D., Banier C., Smallest lithographic marks generated by optical focusing systems, Opt. Lett. 2007. V. 32, No. 8. P. 976–978.

68. Kozawa Y, Sato S. Sharper focal spot formed by higher-order radially polarized laser beams, J. Opt. Soc. Am. A. 2007. V. 24, No. 6. P. 1793–1798.
69. Lerman G. M., Levy U., Tight focusing of spatial variant vector optical fields with elliptical symmetry of linear polarization, Opt. Lett. 2007. V. 32, No. 15. P. 2194–2196.
70. Yan S., Yao B. Description of a radially polarized Laguerre–Gauss beam beyondthe paraxial approximation, Opt. Lett. 2007. V. 32, No. 22. P. 3367–3369.
71. Yew E.Y.S., Sheppard C.J.R., Tight focusing radially polarized Gaussian and Bessel-Gauss beams, Opt. Lett. 2007. V. 32, No. 23. P. 3417–3419.
72. Kalosha V.P., Golub I., Toward the subdiffraction focusing limit of optical superresolution, Opt. Lett. 2007. V. 32, No. 24. P. 3540–3542.
73. Stadler J., Stanciu C., Stupperich C., Meixner A.J., Tighter focusing with a parabolic mirror, Opt. Lett. 2008. V. 33, No. 7. P. 681–683.
74. Witkowska A., Leon-Saval S.G., Pham A., Birks T.A., All-fiber LP_{11} mode convertors. Opt. Lett. 2008. V. 33, No. 4. P. 306–308.
75. Nieminen T.A., Heckenberg N.R., Rubinsztein-Dunlop H., Forces in optical tweezers with radially and azimuthally polarized trapping beams, Opt. Lett. 2008. V. 33, No. 2. P. 122–124.
76. Yonezawa K., Kozawa Y., Sato S., Focusing of radially and azimuthally polarized beams through a uniaxial crystal, J. Opt. Soc. Am. A. 2008. V. 25, No. 2. P. 468–472.
77. Ohtaka Y., Ando T., Inone T., Matsumoto N., Toyoda H., Sidelobe reduction of tightly focused radially higher-order Laguerre–Gaussian beams using annular masks, Opt. Lett. 2008. V. 33, No. 6. P. 617–619.
78. Sheppard C.J.R., Alonso M.A., Moore N.J., Localization measures for high-aperture wave fields based on pupil moments, J. Opt. A: Pure Appl. Opt. 2008. V. 10. P. 033001.
79. Minin I.V., Minin O.V., Investigation of the resolution of phase correcting Fresnel lenses with small values of F/D and subwavelength focus, Komp. Optika, 2006. v. 30b. P. 65–68.
80. Richards B., Wolf E., Electromagnetic diffraction in optical systems. II. Structure of the image field in an aplanatic systems, Proc. Roy. Soc. London. Ser. A. 1959. V. 253. P. 358–379.
81. Prudnikov A.P., Brychkov Yu.A., Marichev O.I., Integrals and Series: Special valued functions. Moscow, Nauka, 1983.
82. Prather D.W., Shi S. Formulation and application of the finite-difference time-domain method for the analysis of axially symmetric diffractive optical elements, J. Opt. Soc. Am. A. 1999. V. 16, No. 5. P. 1131–1141.
83. Bykov V.P. Zh. Eksp. Teor. Fiz., 1972. V. 35. P. 269.
84. John S. Strong localization of photons in certain disordered dielectric superlattices, Phys. Rev. Lett. 1987. V. 58. P. 2486–2489.
85. Ho K.M., et al., Existence of a photonic gap in periodic dielectric structures, Phys. Rev. Lett. 1990. V. 65. P. 3152.
86. Blanco A., Chomski E., Grabtchak S. et al., Large-scale synthesis of a silicon photonic crystal with a complete three-dimensional bandgap near 1.5 micrometres, Nature. 2000. V. 405. P. 437–440.
87. Sözüer H.S., Haus J.W., Inguva R., Photonic bands: Convergence problems with the plane-wave method, Phys. Rev. B 45, 13962 (1992).
88. Fleming J.G., Lin Shawn Yu, Three-dimensional photonic crystal with a stop band from 1.35 to 1.95 μm, Optics Lett. 1998. V. 24. P. 1.
89. Noda S., Yamamoto N., Imada M., Kobayashi H., Okano M., Alignment and Stacking of Semiconductor Photonic Bandgaps by Wafer-Fusion, J. Lightwave Technol. 1999. V. 17. P. 1948.
90. Ho K.M., Chan C.T., Soukoulis C.M., et al., Photonic band gaps in three dimensions: New layer-by-layer periodic structures, Solid State Commun. 1994. V. 89. P. 413–416.
91. Maruo S., Nakamura O., Kawata S., Three-dimensional microfabrication with two-photon-absorbed photopolymerization, Optics Lett. 1997. V. 22, No. 2. 132–134.

92. Wijnhoven J.E.G.J., Vos W.L., Preparation of photonic crystals made of air spheres in titania, Science. 1998. V. 281. P. 802–804.
93. Vlasov Yu.A., Xiang-Zheng Bo, Sturm J.C., Norris D.J., On-chip natural assembly of silicon photonic bandgap crystals, Nature. 2001. V. 414. P. 289.
94. Yablonovich E., Gmitter T.J., Leung K.M., Photonic band structure: The face-centered-cubic case employing non-spherical atoms, Phys. Rev. Lett. 1991. V. 67. P. 2295.
95. Cuisin C., Chelnokov A., Lourtioz J.-M., Decanini D., Chen Y., Appl. Phys. Lett. 2000. V. 77. P. 770.
96. Campbell M., Sharp D.N., Harrison M.T., Fabrication of photonic crystals for the visible spectrum by holographic lithography, Nature. 2000. V. 404. P. 53–56.
97. Ullal C.K. et al., Photonic crystals through holographic lithography: Simple cubic, diamond-like, and gyroid-like structures, Appl. Phys. Lett. 2004. V. 84. P. 5434–5436.
98. Toader O., Chan T.Y.M., John S., Photonic Band Gap Architectures for Holographic Lithography, Phys. Rev. Lett. 2004. V. 92. P. 439051–439054.
99. Miklyaev Yu.V., Meisel D.C., Blanco A., et al., Three dimensional face-centered-cubic photonic crystal templates by laser holography: fabrication, optical characterization, and band structure calculations, Appl. Phys. Lett. 2003. V. 82. P. 1284–1286.
100. Pihulya DG, Miklyayev V. Band structures of three-dimensional photonic crystals produced by interference lithography, Izv. RAN. Ser. Fiz., 2006. V. 70. P. 1972–1974.
101. Miklyaev Yu. V., Karpeev S.V., Dyachenko P.N., Pavelyev V.S., Fabrication of three-dimensional photonics crystals by interference lithography with low light absorption, J. of Mod. Opt. 2009. V. 56, Issue 9. P. 1133–1136.
102. Miklyayev Yu.V., et al., Interference lithographic synthesis of three-dimensional photonic crystals using radiation weakly absorbed by the photoresist, Komp. Optika. 2008. V. 32, N. 4. P. 357–360.
103. Shechtman D., Blech I., Gratias D. et al. Metallic phase with long-range orientational order and no translational symmetry, Phys. Rev. Lett. 1984. V. 53. P. 1951–1953.
104. Macia E. The role of aperiodic order in science and technology, Rep. Prog. Phys. 2006. V. 69. P. 397–441.
105. Quilichini M., Janssen T., Phonon excitations in quasicrystals, Rev. Mod. Phys. 1997. V. 69. P. 277–314.
106. Vekilov Yu. Kh., Isaev E. I., Arslanov S. F. Influence of phason flips, magnetic field, and chemical disorder on the localization of electronic states in an icosahedral quasicrystal, Phys. Rev. B. 2000. V. 62. P. 14040–14048.
107. Krajci M., Hafner J. Topologically induced semiconductivity in icosahedral Al-Pd-Re and its approximants, Phys. Rev. B. 2007. V. 75. P. 024116.
108. Chan Y. S., Chan C. T., Liu Z. Y. Photonic Band Gaps in Two Dimensional Photonic Quasicrystals, Phys. Rev. Lett. 1998. V. 80. P. 956–959.
109. Johnson, S. J., Joannopoulos J.D., Photonic Crystals: The Road from Theory to Practice. London: Kluwer Academic Publishers, 2003.
110. Zoorob M.E., Charlton M.D.B., Parker G.J., et al., Complete photonic bandgaps in 12-fold symmetric quasicrystals, Nature. 2000. V. 404. P. 740–743.
111. Man W., Megens M., Steinhardt P. J., et al., Experimental measurement of the photonic properties of icosahedral quasicrystals, Nature. 2005. V. 436. P. 993–996.
112. Lidermann A., Cademartiri L., Hermatschweiler M. et al., Three-dimensional silicon inverse photonic quasicrystals for infrared wavelengths, Nature Mater. 2006. V. 5. P. 942–945.
113. Xu J., Ma R., Wang X., et al., Icosahedral quasicrystals for visible wavelengths by optical interference holography, Opt. Express. 2007. V. 15. P. 4287–4295.
114. Peach M., Quasicrystals step out of the shadows, Materials Today. 2006. V. 9. P. 44–47.
115. Wang K., David S., Chelnokov A., et al., Photonic band gaps in quasicrystal-related approximant structures, J. Mod. Opt. 2003. V. 50. P. 2095–2105.

116. Dyachenko P.N., Miklyaev Yu.V., Band structure calculations of 2D photonic pseudo-quasicrystals obtainable by holographic lithography, Proc. of SPIE. 2006. V. 6182. P. 61822I.
117. Dyachenko P.N., et al., Pis'ma Zh. Eksp. Teor. Fiz., 2007. V. 86, N. 4. P. 270–273.
118. Dmitrienko V.E., Kleman M., Icosahedral order and disorder in semiconductors, Philos. Mag. Lett. 1999. V. 79. P. 359–367.
119. Dmitrienko V.E., Kleman M., Mauri F., Quasicrystal-related phases in tetrahedral semiconductors: Structure, disorder, and ab initio calculations, Phys. Rev. B. 1999. V. 60. P. 9383–9389.
120. Johnson S.G., Joannopoulos J.D., Block-iterative frequency-domain methods for Maxwell's equations in a plane wave basis, Opt. Express. 2001. V. 8. P. 173–190.
121. Dyachenko P.N., Miklyaev Yu. V., Dmitrienko V. E., Pavelyev V. S. Complete photonic band gap in icosahedral quasicrystals with a body-centered six-dimensional lattice, Proc. of SPIE V. 2008. V. 6989. P. 69891T.
122. Dyachenko P.N., et al., Komp. Optika, 2008. V. 32,N 3. P. 216–221.
123. Adams G.B., O'Keeffe M., Demkov A.A., Sankey O.F., Huang Y.M., Wide-bandgap Si in open fourfold-coordinated clathrate structures, Phys. Rev. B. 1994. V. 49. P. 8048.
124. Dyachenko P.N., Kundikova N.D., Miklyaev Yu.V., Band structure of a photonic crystal with the clathrate Si34 lattice, Phys. Rev. B. 2009. V. 79. P. 233102.
125. Bush K., John S. Photonic band gap formation in certain self-organizing systems. Phys. Rev. E. 1998. V. 58. P. 3896.
126. Takeda H., Takashima T., Yoshino K. Flat photonic bands in two-dimensional photonic crystals with kagome lattices, J. Phys.: Condens. Matter. 2004. V. 16. P. 6317.
127. Meier M., Mekis A., Dodabalapur A., Timko A., Slusher R. E., Joannopoulos J.D., Laser action from two-dimensional distributed feedback in photonic crystals. Appl. Phys. Lett. 1999. V. 74. P. 7.
128. Gajic R., Meisels R., Kuchar F., Hingerl K. All-angle left-handed negative refraction in Kagomé and honeycomb lattice photonic crystals, Phys. Rev. B. 2006. V. 73. P. 165310.
129. Feng Z., Zhang X., Wang Y., Li Z.Y., Cheng B., Zhang D.Z., Negative Refraction and Imaging Using 12 -fold-Symmetry Quasicrystals, Phys. Rev. Lett. 2005. V. 94. P. 247402.
130. Sharp D.N., Turberfield A.J., Denning R.G., Holographic photonic crystals with dimond symmetry, Phys. Rev. B. 2003. V. 68, P. 205102–205108.
131. Fan S., Villeneuve P.R., Joannopoulos J.D., Large omnidirectional band gaps in metallo-dielectric photonic crystals, Phys. Rev. B. 1996. V. 54. P. 11245–11251.
132. Wang Z., Chan C.T., Zhang W. et al., Three-dimensional self-assembly of metal nanoparticles: possible photonic crystal with a complete gap below the plasma frequency, Phys. Rev. B. 2001. V. 64. P. 113108–113113.
133. Gantzounis G., Stefanou N., Yannopapas Y. Optical properties of a periodic monolayer of metallic nanospheres on a dielectric waveguide, J. Phys.: Condens. Matter. 2005. V. 17. P. 1791–1802.
134. Kittel C., Introduction to Solid State Physics. 7th ed. N.Y., Wiley, 1966.
135. Siglas M. M., Soukoulis C.M., Chan C.T., et al., Electromagnetic-wave propagation through dispersive and absorptive photonic-band-gap materials, Phys. Rev. B. 1994. V. 49. P. 11080–11087.
136. Huang K.C., Bienstman P., Joannopoulos J.D., et al., Field expulsion and reconfiguration in polaritonic photonic crystals, Phys. Rev. Lett. 2003. V. 90 P. 196402–196406.
137. Runs A., Ribbing C.G., Polaritonic and photonic gap interactions in two-dimensional photonic crystals, Phys. Rev. Lett. 2004. V. 92. P. 123901–123905.
138. Oraevskiy A.N., Protsenko I.E. High refractive index and other optical properties of heterogeneous media, Pis'ma Zh. Eksp. Teor. Fiz., Lett. 2000 T. 72. P. 641–646.
139. Oraevskiy A.N. Protsenko I.E., Optical properties of heterogeneous media, Kvant. Elektronika, 2001. T. 31. P. 252–256.

140. Johnson P.B., Christy R.W. Optical constant of the noble metals, Phys. Rev. B. 1972. V. 6. P. 4370–4379.
141. Dyachenko P.N., Miklyaev Yu.V., One-dimensional photonic crystal based on nanocomposite of metal nanoparticles and dielectric, Optical Memory & Neural Networks (Information Optics). 2007. V. 16, No. 4. P. 198–203.

Chapter 6

Photonic crystal fibres

Photonic crystal fibres (PCF) is a relatively new class of optical fibres, using the properties of photonic crystals [1].

In the cross section the PCF has a quartz or glass microstructure with periodic or aperiodic system of microinclusions, usually cylindrical micro-holes oriented along the fibre axis. The defect of the microstructure, corresponding to the absence of one or more elements in its centre, is the core of the fibre, providing a waveguide mode of propagation of electromagnetic radiation.

The PCFs are divided into two main types according to the mechanism of confinement and the direction of light. There are fibres with a solid core or a core with a higher average refractive index with respect to a microstructural (perforated) cladding that uses the effect of total internal reflection, as well as conventional optical fibres, but unlike the latter, may have a better ability to confine the light in the core as it propagates due to greater local differences in refractive indices of the core and the cladding. Another type of the PCF for making a light effect uses a photonic band gap (Bragg reflection), created by the cladding microstructure; due to this the light can be confined and propagate even in the core with a lower refractive index relative to the cladding, including a hollow core . The photonic band gap that occurs in the transmission spectrum (the dependence of the transmittance on the wavelength) of the two-dimensional periodic shell of this type of fibre provides a high reflection coefficient for the radiation propagating along the hollow core, allowing to significantly reduce the losses inherent in conventional modes of hollow fibres with a solid cladding and quickly increasing [2] with decreasing diameter of the hollow core. Several groups are allocated among the fibres of the second type.

Fibres with a hollow core – the central hole, usually with a larger radius, is surrounded by 'rings' of micro-holes.

The Bragg fibre [3–7] has a hollow core, surrounded by a cladding formed by alternating rings of high and low refractive indices. Study [8] deals with the Bragg fibre with a filled (solid) core in which the working area can be shifted into the visible range by varying the thickness of the cladding formed from the Bragg reflecting rings.

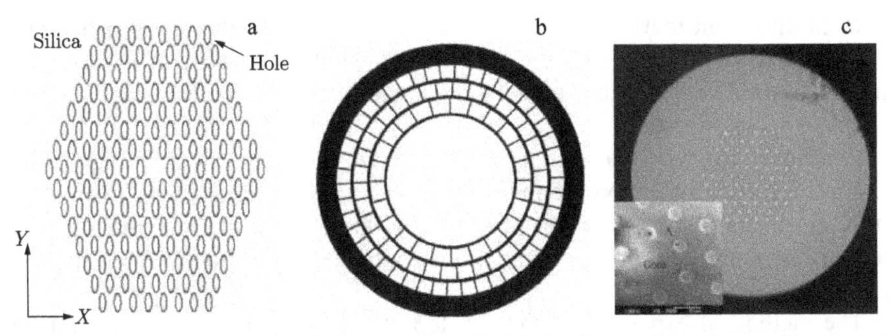

Fig. 6.1. The cross sections for different types of PCF: a) scheme of the flattened hexagonal lattice with elliptical holes [12], b) schematic cross-sections of the Bragg fibre with three concentric cylindrical layers of quartz [6], c) optical micrograph of the cross section solid fibre, light and dark areas in the image correspond to regions high and low refractive indices [10].

Solid-state optical fibres [9,10] whose cladding contains the rods with a high refractive index, are placed in the optical fibre base material with low refractive index.

In addition, in both types of fibres we may vary a number of parameters (Fig. 6.1), such as the form of micro-holes (round, square, elliptical) [11], the scheme of localization of the micro-holes (right hexagonal and square lattices, irregular lattices) [12, 13], the number of cores [14, 15], the number of defects that form the core [16, 17], the material filling the micro-holes [18, 19]. For example, in [20] the cavities in the PCF cladding were filled with liquid crystals, which has led to a hybrid mechanism of propagation of light: the effect of total internal reflection and the effect of photonic band gaps act respectively on orthogonally polarized modes. The application of the PCF, filled with liquid crystals, in optical switches is described in [21–23].

The creation of the PCF resulted new degrees of freedom to control the characteristics of the fibre, which allowed to manage the dispersion, to make fibres with very high or very low non-linearity, with a large or small effective mode area, etc.

Theoretical interest in the PCF is caused to a large extent by the presence in their structure of much larger refractive index contrasts than in conventional weakly guiding fibres, and, consequently, the need for special efficient methods of study. It should be noted that the ability of the PCF to confine light depends on many factors, such as the number of micro-inclusions, their location, the ratio of the diameter and the distance between the centres. This gives scope for designing fibres with predetermined properties.

The PCF modes are electromagnetic waves that can be excited and propagate in the fibre. Any beam of radiation directed into the fibre will bring to it a set of spatial modes.

There are several methods to calculate the modes of the PCF, which can be divided into three groups: the approximately analytical techniques or methods of decomposition, integral methods and finite difference methods.

Decomposition methods. The basic idea exploited in this group of methods is the possibility of representing the mode field of the fibre in the form of an expansion in some basis. As a result, the search mode is reduced to the eigenvalue problem and eigenvectors of a matrix.

Expansion in plane waves [24] with periodic boundary conditions gives the solution for an infinite fibre periodically repeated in the transverse plane, which makes it impossible to obtain in principle by this method the imaginary part of propagation constants corresponding to losses in the propagation of the leaking or non-eigen mode.

The method of expansion in Gauss–Hermite modes [25, 26] is more suitable for describing the complex structure of the cross section of the PCF, rather than the method of expansion in plane waves, but this method is limited to using only the PCF with the holes in the cladding, located at the sites of a regular hexagonal lattice, as the distance between the centres of any two adjacent holes should be fixed, and this value appears in the expression for the basis functions.

Generalization of the multipole method, usually used for optical fibres with multiple cores, to the case of the PCF may be considered [27]. A key aspect of the proposed multipole method is that it uses to advantage the fact of the roundness of inclusions and the symmetry properties inherent in many PCFs in connection with the regular arrangement of micro-holes [28]. The strength of the method also consists in the fact that it allows us to calculate both real and imaginary parts of propagation constants. In contrast to the multipole approach for optical fibres with multiple cores, as demonstrated in [29], using the technique of point-wise coupling of the field at the boundaries of the inclusions, the method [27] handles the boundary conditions by expanding the field components in an orthonormal basis. Application of the method can also be extended to the optical fibres with an arbitrary form of inclusions, in accordance with the strategy proposed in [30].

A characteristic feature of the method of matched sinusoidal modes [31] is the technique of dividing the inhomogeneous cross-section of waveguide structures into rectangular regions with a constant refractive index. In each of these regions the mode field is approximated by superposition of the factorized harmonic functions. The mode propagation constants are determined from the condition of minimization of discrepancy of the representations of the field at the boundaries of neighbouring regions using an integrated approach. The method of matched sinusoidal modes uses the procedure for finding the roots of equations and thus is inferior in speed to the methods based solely on finding the eigenvalues of the matrices.

Integral methods. Integral methods are grid methods, that is, in contrast to the previous group, in this case the solution to the problem of finding the mode field is a grid function and not an analytically defined function.

Among this group we can select the finite element method [32, 33]. It is a powerful tool of vector analysis that can take into account all the features of the geometry of micro-holes and their location in the structure of the section. Sufficiently fast and flexible, it is often used to model the properties of the PCF. Among the shortcomings of the finite element method are the demands on memory resources, so as to describe the structure of the cross section the PCF requires detailed sampling and a large number of variables, as well as the need for human intervention in the algorithm to better define the boundary conditions (Perfectly matched layer) and the sampling grid.

The boundary element method [34], where the cross section is divided into homogeneous areas, and the eigenvalue problem are obtained by applying Green's theorem, are less demanding of memory. However, a significant drawback is the possibility of false solutions.

In the Green function method [35], the problem of finding the propagation constants of modes can also be reduced to the problem of eigenvalues which can be solved by a special fast algorithm developed for this purpose. This method works in the case of complex geometrical forms of the micro-holes, albeit at a slower rate of convergence than in the case of circular holes.

Finite-difference methods. These methods, like the integral methods, give a grid solution.

The finite difference method is widely used for solving various kinds of equations. Due to the simplicity of implementation, this method has become a convenient tool for calculating the modes of optical fibres, particularly those for which no analytical solutions are available, such as the PCF.

The presence of large refractive index contrasts in the structure of the section of the PCF requires the use of the completely vector approach to calculate the modes, instead of the scalar approach commonly used for weakly guiding fibres. However, as was demonstrated in [36], the scalar finite-difference method can be used to get at least a qualitative assessment of the distribution of the modes of the PCF, including those in which the light mechanism is based on the effect of photonic band gaps.

Vector finite-difference schemes have been proposed for a more accurate analysis [37]. Discretization is carried out on differential operators and functions included in the Helmholtz equation or the wave equation. In [38] the authors presented finite-difference time-domain approach (FDTD-method) of calculation of the modes of the PCF using the shifted grids (Yee cells). An improved finite-difference space-domain method that takes into account the dispersion of the material is discussed in [39].

The result of the use of special finite-difference schemes for non-stationary wave equations or Maxwell's equations is a family of methods of propagation of the beam [40–42]. The method consists of simulation of the propagation of a coherent beam of light along the fibre, resulting in modes of the given structure, as it were a posteriori. The method is convenient to investigate the energy loss during the passage of radiation through the optical fibre, although this may be difficult in connection with the convergence of the method.

This chapter details a pair of methods for calculating modes from two fundamentally different groups: the approximately analytical method of matched sinusoidal modes and the grid method based on application of the finite-difference approximations to the stationary wave equations.

The basic idea of the method of matched sinusoidal modes (MSM-method), also known as the transverse resonance technique [43,44], was first formulated in

[45]. Subsequent development of the method was in [46], where it was used to calculate the radiation loss due to outgoing modes in stepped fibres. Then, in [41, 47] the authors introduced the descriptive term 'matched sinusoidal modes', and gave a precise mathematical formulation. The MSM method was modified in [48] with an iterative Krylov method [49] for the most computationally complex stage of solving a non-linear eigenvalue problem of large dimensions of the matrix to which the problem of finding propagation constants of modes is reduced. The MSM method can be used to calculate both scalar and vector modes [50], conventional round fibres [51] and PCF [52].

The basis of the finite-difference method (FD method) considered here was taken the approach proposed in [11], where the modes were calculated using the technique of finite-difference approximations to the stationary vector wave equations for monochromatic light. The FD technique wins in speed of the algorithm in comparison with the MSM method, since the task of finding the propagation constants and sampling grid solutions for the transverse components of electric or magnetic components directly reduces to a linear matrix problem for the eigenvalues and eigenvectors. The FD method also allows full vector analysis modes of the PCF.

6.1. Calculation of modes of photonic crystal fibres by the method of matched sinusoidal modes

The method of matched sinusoidal modes [31, 47] differs advantageously from a variety of other approaches of studying fibres homogeneous in the longitudinal direction by the property of the analytic representation of the field obtained as a result.

The MSM-method is based on the representation of solutions for the spatial mode as a superposition of local sinusoidal modes, which are eigenmodes of homogeneous, with a constant refractive index, rectangular pieces of the fibre with a non-uniform cross section.

6.1.1. Method of matched sinusoidal modes in the scalar case

We formulate the problem of finding the eigenmodes of a dielectric waveguide homogeneous in the longitudinal direction and inhomogeneous in the cross section, surrounded by a perfect conductor, the so-called 'electric walls', or an ideal magnetic material – the 'magnetic walls'.

The fibres homogeneous in the longitudinal direction have constant distribution along the length of the refractive index of the material in the section. The cylindrical symmetry of the fibre allows the separation of variables and the mode field can be presented in the form of

$$\begin{aligned} \overline{E}_j(x,y,z,t) &= \overline{E}_j(x,y)\exp(-ik_{zj}z)\exp(i\omega t), \\ \overline{H}_j(x,y,z,t) &= \overline{H}_j(x,y)\exp(-ik_{zj}z)\exp(i\omega t), \end{aligned} \tag{6.1}$$

where $\overline{E}_j\,(x, y, z, t)$ is the strength of the electric field; $\overline{H}_j\,(x, y, z, t)$ is the magnetic field strength; k_{zj} is the propagation constant, the eigenvalue of j-th mode, or the projection on the longitudinal axis of the wave vector $\overline{k}_j$; ω is the radiation frequency.

Propagation constants of different modes are different.

The spatial component of the field j-th mode can be decomposed into transverse and longitudinal components, which denote the indices t and z, respectively. As a result, we have:

$$\overline{E}_j(x,y,z) = \overline{E}_j(x,y)\exp(ik_{zj}z) = [\overline{e}_{tj}(x,y) + e_{zj}(x,y)\overline{z}]\exp(ik_{zj}z), \quad (6.2a)$$

$$\overline{H}_j(x,y,z) = \overline{H}_j(x,y)\exp(ik_{zj}z) = [\overline{h}_{tj}(x,y) + h_{zj}(x,y)\overline{z}]\exp(ik_{zj}z), \quad (6.2b)$$

where $\overline{z}$ is the unit vector parallel to the fibre axis.

The modes whose longitudinal component of the magnetic vector is zero are called transverse-magnetic modes (TM), and those modes for which the longitudinal component of the electric vector is zero are transverse-electric modes (TE). In general, the TE- and TM-modes are not modes of the optical fibre. The modes of the optical fibres, in general, are hybrid and contain longitudinal components of the electric and magnetic vectors. They are called HE- or EH-modes.

Let us start with the scalar approximation, according to which light propagation is described by the scalar Helmholtz equation:

$$\nabla^2 E(x,y,z) + k_0^2 \varepsilon E(x,y,z) = 0, \quad (6.3)$$

here $E(x, y, z)$ is the complex amplitude of light that can be associated with any component of the electric and magnetic field vectors; $k_0 = 2\pi/\lambda_0$, where λ_0 is the wavelength in vacuum; ε is the dielectric permittivity.

A solution of (6.3) for a rectangular region with a constant value of the refractive index is $n = \sqrt{\varepsilon}$ is a function of the form

$$E = u(x)\varphi(y)e^{-ik_z z}. \quad (6.4)$$

Substituting (6.4) in equation (6.3)

$$\nabla^2 (u(x)\varphi(y)e^{-ik_z z}) + k_0^2 \varepsilon u(x)\varphi(y)e^{-ik_z z} = 0,$$

$$\nabla^2 (u(x)\varphi(y))e^{-ik_z z} - k_z^2 u(x)\varphi(y)e^{-ik_z z} + k_0^2 \varepsilon u(x)\varphi(y)e^{-ik_z z} = 0,$$

$$\nabla^2 (u(x)\varphi(y)) - k_z^2 u(x)\varphi(y) + k_0^2 \varepsilon u(x)\varphi(y) = 0,$$

$$\ddot{u}(x)\varphi(y) + \ddot{\varphi}(y)u(x) + (k_0^2 \varepsilon - k_z^2)u(x)\varphi(y) = 0.$$

We omit the arguments x and y at the functions u and φ, so that records are shorter. Thus, we have

$$\ddot{u}\varphi + \ddot{\varphi}u + (k_0^2 \varepsilon - k_z^2)u\varphi = 0, \quad (6.5)$$

hereinafter referred by the dots above the function denoted derivatives.

After separation of the variables, we divide (6.5) by $u\varphi$

$$\frac{\ddot{\varphi}}{\varphi}+\frac{\ddot{u}}{u}+(k_0^2\varepsilon-k_z^2)=0 \tag{6.6}$$

or

$$\frac{\ddot{\varphi}}{\varphi}+(k_0^2\varepsilon-k_z^2)=-\frac{\ddot{u}}{u}. \tag{6.7}$$

We set

$$\frac{\ddot{u}}{u}=-k_x^2. \tag{6.8}$$

Equation (6.7) can be divided into two equations:

$$\begin{cases}\ddot{\varphi}+(k_0^2\varepsilon-k_z^2-k_x^2)\varphi=0,\\ \ddot{u}+k_x^2u=0.\end{cases} \tag{6.9}$$

We denote

$$k_z^2+k_x^2=k_k^2, \tag{6.10}$$

then

$$\begin{cases}\ddot{\varphi}+(k_0^2\varepsilon-k_k^2)\varphi=0,\\ \ddot{u}+(k_k^2-k_z^2)u=0.\end{cases} \tag{6.11}$$

We also introduce the notation

$$k_0^2\varepsilon-k_k^2=k_y^2, \tag{6.12}$$

that is now

$$k_0^2\varepsilon=k_x^2+k_y^2+k_z^2. \tag{6.13}$$

That is k_x, k_y, k_z are the projections on the respective axes of the wave vector $\overline{k}$, whose modulus is equal $|\overline{k}|=\sqrt{k_0^2\varepsilon}$.

If a section of the fibre can be divided into N rows and M columns so that none of the rectangular cells of this division contains irregularities (Fig. 6.2), then in each of these cells the solution can be represented in a rather simple form.

Each cell located on the intersection of the n-th row and m-th column can be associated with a value of the dielectric permittivity $\varepsilon^{(m,n)}$, constant for a given cell. Let the coordinate axes are arranged as shown in Fig. 6.2. Then the thickness of the i-th row

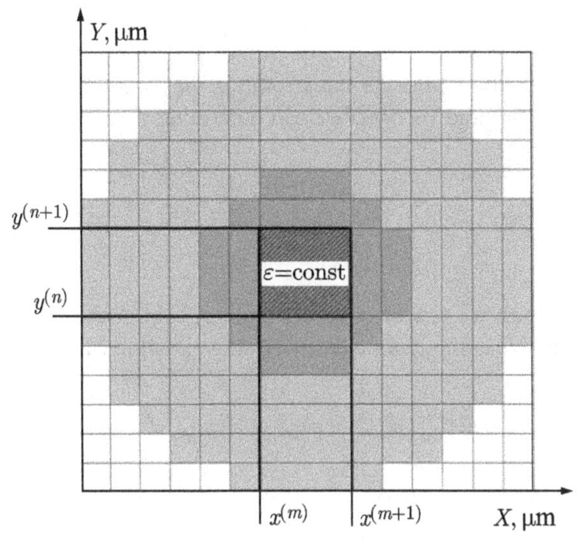

Fig. 6.2. Scheme of the cross section of the fibre, dark gray area shows the value of the refractive index $n_{co} = 1.47$, light gray with $n_{cl} = 1.463$, white – with $n_v = 1$.

$$d_y^{(n)} = y^{(n+1)} - y^{(n)}, \tag{6.14}$$

where $y^{(n)}$ is the coordinate plane, dividing n–1 and n lines. Similarly, the thickness of the column

$$d_x^{(m)} = x^{(m+1)} - x^{(m)}, \tag{6.15}$$

where $x^{(m)}$ is the coordinate plane, dividing m–1 and m columns.

For a homogeneous rectangular portion (cells) of the section (Fig. 6.2), where the value of the dielectric permittivity $\varepsilon(x,y) = \varepsilon^{(m,n)} = \text{const}$, the first equation of the system (6.11) takes the form

$$\ddot{\varphi}(y) + (k_0^2 \varepsilon^{(m,n)} - k_k^2)\varphi(y) = 0, \tag{6.16}$$

and its solution can be written as

$$\varphi(y) = \varphi_s^{(n,l)} \cos[k_y^{(n)}(y - y^{(n)})] + \frac{\varphi_a^{(n,l)}}{k_y^{(n)}} \sin[k_y^{(n)}(y - y^{(n)})], \tag{6.17}$$

where

$$k_y^{(n)} = \sqrt{\varepsilon^{(m,n)} k_0^2 - k_k^2}; \tag{6.18}$$

$$\varphi_s^{(n,l)} = \varphi(y^{(n)} + 0) \tag{6.19a}$$

– the bottom or left value of function $\varphi(y)$ in the given rectangular fragment;

$$\varphi_a^{(n,l)} = \dot{\varphi}(y^{(n)} + 0) \tag{6.19b}$$

– the bottom or left value of the derivative $\dot{\varphi}(y)$ in the same cell section.

Similarly to (6.19a) and (6.19b), we can determine the right value $\varphi(y)$ and its derivative $\dot{\varphi}(y)$:

$$\varphi_s^{(r)} = \varphi(y^{(n+1)} - 0), \tag{6.19c}$$

$$\varphi_a^{(r)} = \dot{\varphi}(y^{(n+1)} - 0). \tag{6.19d}$$

Using these values, we can offer an alternative form of the solution (6.17) and its derivative, which is computationally more convenient in case when $k_y^{(n)}$ is purely imaginary:

$$\varphi(y) = (\varphi_s^{(n,l)} \sin[k_y^{(n)}(y^{(n+1)} - y)] + \varphi_s^{(n,r)} \sin[k_y^{(n)}(y - y^{(n)})]) / \sin(k_y^{(n)} d_y^{(n)}), \tag{6.20a}$$

$$\dot{\varphi}(y) = (\varphi_a^{(n,l)} \sin[k_y^{(n)}(y^{(n+1)} - y)] + \varphi_a^{(n,r)} \sin[k_y^{(n)}(y - y^{(n)})]) / \sin(k_y^{(n)} d_y^{(n)}). \tag{6.20b}$$

Illustration of location $\varphi_s^{(n,l)}$, $\varphi_a^{(n,l)}$, $\varphi_s^{(n,r)}$, $\varphi_a^{(n,r)}$ is given in Fig. 6.3.

The second equation of the system (6.11) in the homogeneous region becomes

$$\ddot{u}_k^{(m)}(x) + (k_k^{(m)2} - k_z^2) u_k^{(m)}(x) = 0, \tag{6.21}$$

and its solution by analogy with (6.17):

$$u_k^{(m)}(x) = u_{sk}^{(m,l)} \cos[k_{xk}^{(m)}(x - x^{(m)})] + \frac{u_{ak}^{(m,l)}}{k_{xk}^{(m)}} \sin[k_{xk}^{(m)}(x - x^{(m)})], \tag{6.22}$$

where

$$k_{xk}^{(m)} = \sqrt{k_k^{(m)2} - k_z^2}, \tag{6.23}$$

$$u_{sk}^{(m,l)} = u_k^{(m)}(x^{(m)} + 0), \tag{6.24a}$$

$$u_{ak}^{(m,l)} = \dot{u}_k^{(m)}(x^{(m)} + 0). \tag{6.24b}$$

Similarly, we introduce the following notations:

$$u_{sk}^{(m,r)} = u_k^{(m)}(x^{(m+1)} - 0), \tag{6.24c}$$

$$u_{ak}^{(m,r)} = \dot{u}_k^{(m)}(x^{(m+1)} - 0). \tag{6.24d}$$

The presence of an index k at the function $u(x)$ in (6.21)–(6.24) is explained below.

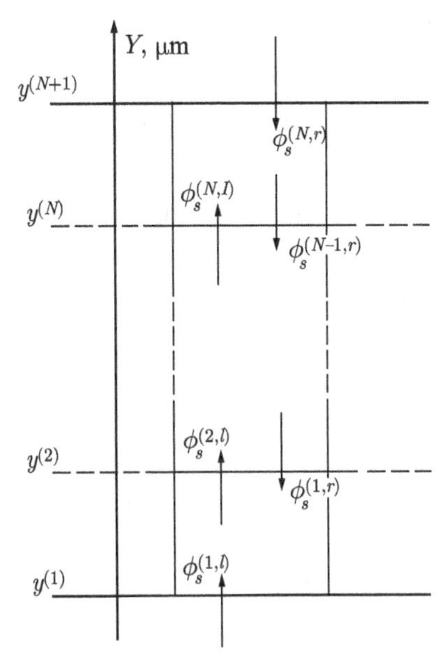

Fig. 6.3. Diagram of notations of fields at the borders of the partition of the section of the fibre.

Thus, in the uniform rectangular cell the Helmholtz equation is satisfied by the product of two harmonic functions.

In accordance with the scalar approximation the modes of the given light guide structure are the solutions of the Helmholtz equation (6.3) in its cross section. Let us assumed that the previously made assumptions about the possibility of dividing the fibre cross section into a finite number of rectangular regions with a constant refractive index are justified, i.e., we consider a waveguide with a piecewise-constant filling of the cross-section. At the edges of the cross section there are assumed to be electric or magnetic walls, ensuring the vanishing of the function of the field or its derivative at $x = 0$, $y = 0$, $x = x^{(M+1)}$, $y = y^{(N+1)}$.

In the scalar case, the field in the m-th column can be written as

$$\xi^{(m)}(x,y) = \sum_{k=1}^{\infty} u_k^{(m)}(x)\varphi_k^{(m)}(y), \tag{6.25}$$

here omitted factor $\exp(-ik_z z)$. Each set of functions $u_k^{(m)}(x)$ satisfies the second, and each of $\varphi_k^{(m)}(y)$ - the first equation of (6.11) in the column, that is for $x^{(m)} \le x < x^{(m+1)}$ and $y^{(1)} \le y < y^{(N+1)}$, and in this area all of these functions are continuous with first derivatives. We call $u_k^{(m)}(x)$ and $\varphi_k^{(m)}(y)$ local and sinusoidal modes, the index k determines the number of the local mode and is directly related to value k_k introduced by (6.10); their relationship will be shown below.

We take into account the proposed form of the solution of equation (6.16) in the form of a sinusoidal mode (6.17). Thus, to solve (6.16) we must determine the value k_k and $2N–2$ and constants $\varphi_s^{(n,l)}$ and $\varphi_a^{(n,l)}$, so that the function $\varphi(y)$ satisfies (6.16) and the boundary conditions $\begin{cases}\varphi(y^{(1)})=0,\\ \varphi(y^{(N+1)})=0,\end{cases}$ or $\begin{cases}\dot{\varphi}(y^{(1)})=0,\\ \dot{\varphi}(y^{(N+1)})=0,\end{cases}$, respectively for electric or magnetic walls.

The requirement of continuity of the solution (6.17) and its derivative leads to the following relations

$$\varphi_s^{(n,r)}=\varphi_s^{(n+1,l)}, \tag{6.26a}$$

$$\varphi_a^{(n,r)}=\varphi_a^{(n+1,l)}. \tag{6.26b}$$

In addition, for $\varphi_s^{(n,l)}$, $\varphi_a^{(n,l)}$, $\varphi_s^{(n,r)}$, $\varphi_a^{(n,r)}$ the following equalities:

$$\varphi_s^{(n,r)}=\varphi_s^{(n,l)}\cos[k_y^{(n)}d_y^{(n)}]+\frac{\varphi_a^{(n,l)}}{k_y^{(n)}}\sin[k_y^{(n)}d_y^{(n)}], \tag{6.27a}$$

$$\varphi_a^{(n,r)}=\varphi_a^{(n,l)}\cos[k_y^{(n)}d_y^{(n)}]-\varphi_s^{(n,l)}k_y^{(n)}\sin[k_y^{(n)}d_y^{(n)}], \tag{6.27b}$$

$$\varphi_s^{(n,l)}=\varphi_s^{(n,r)}\cos[k_y^{(n)}d^{(n)}]-\frac{\varphi_a^{(n,r)}}{k_y^{(n)}}\sin[k_y^{(n)}d^{(n)}], \tag{6.28a}$$

$$\varphi_a^{(n,l)}=\varphi_a^{(n,r)}\cos[k_y^{(n)}d^{(n)}]+\varphi_s^{(n,l)}k_y^{(n)}\sin[k_y^{(n)}d^{(n)}]. \tag{6.28b}$$

We introduce the matrix determinant that is equal to unity

$$P^{(i)}=\begin{bmatrix}\cos[k_y^{(i)}d^{(i)}] & \sin[k_y^{(i)}d^{(i)}]/k_y^{(i)}\\ -\sin[k_y^{(i)}d^{(i)}]k_y^{(i)} & \cos[k_y^{(i)}d^{(i)}]\end{bmatrix}, \tag{6.29}$$

$$Q^{(i)}=\begin{bmatrix}\cos[k_y^{(i)}d^{(i)}] & -\sin[k_y^{(i)}d^{(i)}]/k_y^{(i)}\\ \sin[k_y^{(i)}d^{(i)}]k_y^{(i)} & \cos[k_y^{(i)}d^{(i)}]\end{bmatrix}. \tag{6.30}$$

The equalities $\begin{bmatrix}\varphi_s^{(n,r)}\\ \varphi_a^{(n,r)}\end{bmatrix}=P^{(n)}\begin{bmatrix}\varphi_s^{(n,l)}\\ \varphi_a^{(n,l)}\end{bmatrix}$ and $\begin{bmatrix}\varphi_s^{(n,l)}\\ \varphi_a^{(n,l)}\end{bmatrix}=Q^{(n)}\begin{bmatrix}\varphi_s^{(n,r)}\\ \varphi_a^{(n,r)}\end{bmatrix}$ which are equivalent to the equations (6.27) and (6.28) are obviously justified. Using the matrices $P^{(i)}$ and $Q^{(i)}$ can be easily expressed through the values $\begin{bmatrix}\varphi_s^{(j,r)}\\ \varphi_a^{(j,r)}\end{bmatrix}$ and $\begin{bmatrix}\varphi_s^{(i,l)}\\ \varphi_a^{(i,l)}\end{bmatrix}$, $i\le j$, and the value of $\begin{bmatrix}\varphi_s^{(j,l)}\\ \varphi_a^{(j,l)}\end{bmatrix}$ through $\begin{bmatrix}\varphi_s^{(i,r)}\\ \varphi_a^{(i,r)}\end{bmatrix}$, $i\ge j$. Thus, for example

$$\begin{bmatrix} \varphi_s^{(n',r)} \\ \varphi_a^{(n',r)} \end{bmatrix} = P^{(n')} P^{(n'-1)} \dots P^1 \begin{bmatrix} \varphi_s^{(1,l)} \\ \varphi_a^{(1,l)} \end{bmatrix}, \tag{6.31}$$

$$\begin{bmatrix} \varphi_s^{(n'+1,l)} \\ \varphi_a^{(n'+1,l)} \end{bmatrix} = Q^{(n'+1)} Q^{(n'+2)} \dots Q^{(N)} \begin{bmatrix} \varphi_s^{(N,r)} \\ \varphi_a^{(N,r)} \end{bmatrix}. \tag{6.32}$$

For numerical calculation of k_k, or rather the square of this value since only the square appears in the formulas above, we use the following method. We define a function on the basis of the conditions (6.26) on the boundary:

$$\Delta^{(n')}(k_k^2) = \varphi_s^{(n',r)} \varphi_a^{(n'+1,l)} - \varphi_a^{(n',r)} \varphi_s^{(n'+1,l)}, \tag{6.33}$$

where $n' \in (1, N)$, the values $\varphi_s^{(n',r)}$, $\varphi_a^{(n',r)}$ and $\varphi_s^{(n'+1,l)}$, $\varphi_a^{(n'+1,l)}$ are computed using (6.31) and (6.32) and therefore depend on k_k^2. Zeros of the function $\Delta^{(n')}(k_k^2)$ corresponding to the equality of the left and right values of the local mode at the border of $n' + 1$ n partition rows, are the desired values. Find the zeros (6.33) is equivalent to solving the characteristic equation

$$\varphi_s^{(n',r)} \varphi_a^{(n'+1,l)} - \varphi_a^{(n',r)} \varphi_s^{(n'+1,l)} = 0, \tag{6.34}$$

and n' may be an arbitrary number from the interval $(1,N)$.

It is not difficult to see that the dependence of the solution (6.34) on non-zero values $\begin{bmatrix} \varphi_s^{(1,l)} \\ \varphi_a^{(1,l)} \end{bmatrix}$ and $\begin{bmatrix} \varphi_s^{(N,r)} \\ \varphi_a^{(N,r)} \end{bmatrix}$ does not exist, therefore, solving (6.34), in the case of electric walls it can be assumed that $\begin{bmatrix} \varphi_s^{(1,l)} \\ \varphi_a^{(1,l)} \end{bmatrix} = \begin{bmatrix} 0 \\ 1 \end{bmatrix}, \begin{bmatrix} \varphi_s^{(N,r)} \\ \varphi_a^{(N,r)} \end{bmatrix} = \begin{bmatrix} 0 \\ 1 \end{bmatrix}$, and in the case of magnetic walls $\begin{bmatrix} \varphi_s^{(1,l)} \\ \varphi_a^{(1,l)} \end{bmatrix} = \begin{bmatrix} 1 \\ 0 \end{bmatrix}, \begin{bmatrix} \varphi_s^{(N,r)} \\ \varphi_a^{(N,r)} \end{bmatrix} = \begin{bmatrix} 1 \\ 0 \end{bmatrix}$.

Different values of k_k^2 satisfying (6.34) are infinitely many, but they are all in the interval $(-\infty, k_{\max}^2]$, where they can be sorted in descending order. Each of the values k_k^2 defines a local mode $\varphi_k^{(m)}(y)$ in the m-th column of the cross section of the fibre. The accuracy of construction the field by the formula (6.25) depends on the number of the local modes used for this purpose. But no matter how many modes we would have (much like the roots of (6.34) we found), it will always be some finite number K.

Finding the roots of (6.34) is complicated by the fact that their distribution within the interval $(-\infty, k_{\max}^2]$ is irregular. For some structures of columns (especially with multiple alternation of fragments with different refractive indices), two close roots of the characteristic equation (6.34) can differ by the order 10^{-11}. Skipping the roots and, hence, the corresponding local modes in the expansion (6.25) can lead to significant distortions in the result.

To avoid missing roots and reduce the amount of computation allows the following simple algorithm for adaptive choice of the step in the localization of the roots of the characteristic equation. The initial (arbitrary) value of the step L is defined. Then on the real axis, starting from $k^2_{\max}$, in the direction of decreasing we examine gradually segments length L. On each of these segments at P_1 we calculated values in the left-hand side of equation (6.34). From set the obtained values we determined the statistical characteristics, such as variance and expectation. Also, for the same interval we calculate the values of the left side of the equation still at P_2 points and for P_1+P_2 the values determined by the same statistical characteristics. If the relative change in the variance and expectation does not exceed some predetermined threshold, we carry out the localization of the roots of the P_1+P_2 values found, otherwise the initial segment is divided into parts and all operations are repeated for each part. This recursive procedure can be used to improve accuracy where needed.

When all K values of k^2_k are found, we are interested in the constants $\varphi_s^{(n,l)}$, $\varphi_a^{(n,l)}$, $\varphi_s^{(n,r)}$, $\varphi_a^{(n,r)}$, which can be found from the normalization condition of the function $\varphi(y)$ on unity:

$$\frac{1}{y^{(N+1)}-y^{(1)}}\int\limits_{y^{(1)}}^{y^{(N+1)}}\varphi^2(y)dy=1. \tag{6.35}$$

Here

$$I=\int\limits_{y^{(1)}}^{y^{(N+1)}}\varphi^2(y)dy=\sum_{n=1}^{N}\int\limits_{y^{(n)}}^{y^{(n+1)}}\varphi^2(y)dy=$$

$$=\sum_{n=1}^{N}[\frac{1}{2(k_y^{(n)})^2}(\varphi_s^{(n,l)}\varphi_a^{(n,l)}-\varphi_s^{(n,r)}\varphi_a^{(n,r)})+\frac{d^{(n)}}{2}((\varphi_s^{(n,l)})^2+\left(\frac{\varphi_a^{(n,l)}}{k_y^{(n)}})^2\right)].$$

When k^2_k is known, the value $k_y^{(n)}$ for each row of the partition can be calculated by the formula (6.18). Taking into account the relations

$$\begin{bmatrix}\varphi_s^{(n',r)}\\ \varphi_a^{(n',r)}\end{bmatrix}=P^{(n')}P^{(n'-1)}\dots P^1\begin{bmatrix}\varphi_s^{(1,l)}\\ \varphi_a^{(1,l)}\end{bmatrix}, n'=\overline{1,N}, \tag{6.36}$$

$$\begin{bmatrix}\varphi_s^{(n,l)}\\ \varphi_a^{(n,l)}\end{bmatrix}=\begin{bmatrix}\varphi_s^{(n-1,r)}\\ \varphi_a^{(n-1,r)}\end{bmatrix}, n=\overline{2,N}, \tag{6.37}$$

the value of the integral I can be considered as a function of $\varphi_s^{(1,l)}$ or $\varphi_a^{(1,l)}$ in dependence on the type of boundary conditions. Then (6.35) can be solved by standard numerical methods, such as Newton's method, and has exactly two real roots, differing only by the sign, and any of them can be chosen as the value $\varphi_s^{(1,l)}$ (or $\varphi_a^{(1,l)}$), the other constants are calculated using (6.36), (6.37) and the resultant value.

Acting on the above-described algorithm, for each column of the section of the fibre we can find K values of k_k^2 and for each of them to construct a function that will satisfy (6.16) with this.

When the local y-modes are found, there is a lack of x-modes of the form (6.22) for the construction of the field (6.25). The function $u_k^{(m)}(x)$ is the k-th mode in the column m, it is also consistent with $k_k^{(m)}$ and $\varphi_k^{(m)}(y)$.

The difference between equations (6.16) and (6.21) is that the solution (6.21) is not a continuous function of the form (6.22) but a set of M functions defined and continuous on the corresponding intervals. Equation (6.22) combines them into one discontinuous function, so for $u_{sk}^{(m,l)}, u_{sk}^{(m,r)}$ and $u_{ak}^{(m,l)}, u_{ak}^{(m,r)}$ the relations similar to (6.26) are not satisfied, however, due to the requirement of continuity of the field (6.25) on the boundary of the partition columns, the equalities

$$u_{ak}^{(m,l)} = -\frac{k_{xk}^{(m)} u_{sk}^{(m,l)}}{\operatorname{tg}(k_{kx}^{(m)} d_x^{(m)})} + \frac{k_{xk}^{(m)} u_{sk}^{(m,r)}}{\sin(k_{kx}^{(m)} d_x^{(m)})}, \tag{6.38a}$$

$$u_{ak}^{(m,r)} = -\frac{k_{xk}^{(m)} u_{sk}^{(m,l)}}{\sin(k_{kx}^{(m)} d_x^{(m)})} + \frac{k_{xk}^{(m)} u_{sk}^{(m,r)}}{tg(k_{kx}^{(m)} d_x^{(m)})}. \tag{6.38b}$$

Since the relations (6.38) hold for all $k = \overline{1, K}$, it is convenient to use a matrix notation. We introduce the diagonal matrices $T^{(m)}$ and $S^{(m)}$ with the dimension $K \times K$ with the diagonal elements

$$T_{kk}^{(m)} = k_{xk}^{(m)} / \operatorname{tg}(k_{kx}^{(m)} d_x^{(m)}), \tag{6.39a}$$

$$S_{kk}^{(m)} = k_{xk}^{(m)} / \sin(k_{kx}^{(m)} d_x^{(m)}). \tag{6.39b}$$

Now (6.38) takes the form

$$U_a^{(m,l)} = -T^{(m)} U_s^{(m,l)} + S^{(m)} U_s^{(m,r)}, \tag{6.40a}$$

$$U_a^{(m,r)} = -S^{(m)} U_s^{(m,l)} + T^{(m)} U_s^{(m,r)}. \tag{6.40b}$$

Here

$$U_a^{(m,l)} = [u_{a1}^{(m,l)} \quad u_{a2}^{(m,l)} \quad \cdots \quad u_{aK}^{(m,l)}]^T,$$

$$U_a^{(m,r)} = [u_{a1}^{(m,r)} \quad u_{a2}^{(m,r)} \quad \cdots \quad u_{aK}^{(m,r)}]^T,$$

$$U_s^{(m,l)} = [u_{s1}^{(m,l)} \quad u_{s2}^{(m,l)} \quad \cdots \quad u_{sK}^{(m,l)}]^T,$$

$$U_s^{(m,r)} = [u_{s1}^{(m,r)} \quad u_{s2}^{(m,r)} \quad \cdots \quad u_{sK}^{(m,r)}]^T.$$

To construct the correlations between the left and right values of the x-modes from adjacent columns section, we consider the overlap integral

$$\langle f_1 | f_2 \rangle = \frac{1}{y^{(N+1)} - y^{(1)}} \int_{y^{(1)}}^{y^{(N+1)}} f_1(y) f_2(y) dy \tag{6.41}$$

and form the matrix elements of the form (6.41), where f_1 and f_2 will be represented by the previously found modes $\varphi_k^{(m)}(y)$:

$$O_{pq}^{(m,m')} = <\varphi_q^m | \varphi_p^{m'}> =$$
$$\sum_{n=1}^{N} (\varphi_{sq}^{(n,r)(m)} \varphi_{ap}^{(n,r)(m')} - \varphi_{aq}^{(n,r)(m)} \varphi_{sp}^{(n,r)(m')} - \varphi_{sq}^{(n,l)(m)} \varphi_{ap}^{(n,l)(m')} + \varphi_{aq}^{(n,l)(m)} \varphi_{sp}^{(n,l)(m')}) / (k_{kq}^{(m)2} - k_{kp}^{(m')2}) =$$
$$\frac{\sum_{n=1}^{N} (\varphi_{sq}^{(n,r)(m)} \varphi_{ap}^{(n,r)(m')} - \varphi_{aq}^{(n,r)(m)} \varphi_{sp}^{(n,r)(m')} - \varphi_{sq}^{(n,l)(m)} \varphi_{ap}^{(n,l)(m')} + \varphi_{aq}^{(n,l)(m)} \varphi_{sp}^{(n,l)(m')})}{(k_{kq}^{(m)} + k_{kp}^{(m')})(k_{kq}^{(m)} - k_{kp}^{(m')})} \tag{6.42}$$

We obtain a square matrix $O^{(m,m')}$ with the size $K \times K$ and the property

$$(O^{(m,m')})^T = O^{(m',m)}, \tag{6.43}$$

where the symbol T means the transposition.

In addition, since the orthogonal functions $\varphi_k^{(m)}(y)$ are normalized, then each matrix $O^{(m,m')}$ is unitary:

$$O^{(m,m')} O^{(m',m)} = \begin{bmatrix} 1 & & & & \\ & 1 & & 0 & \\ & & \ddots & & \\ & 0 & & 1 & \\ & & & & 1 \end{bmatrix}. \tag{6.44}$$

In the case of finite values of the dimension $K \times K$ the ratio (6.44) is satisfied with sufficient accuracy only for adjacent columns, that is, if $m' = m + 1$ or $m' = m - 1$. Orthonormality of the functions $\varphi_k^{(m)}(y)$ and requirements of the continuity of the field $\xi(x, y)$ (6.25) together with the derivatives normal to the interface of the columns indicates the fulfilment of the conditions:

$$u_{sk}^{(m,r)} = \sum_{p=1}^{K} O_{kp}^{(m,m+1)} u_{sp}^{(m+1,l)}, \tag{6.45a}$$

$$u_{ak}^{(m,r)} = \sum_{p=1}^{K} O_{kp}^{(m,m+1)} u_{ap}^{(m+1,l)}. \tag{6.45b}$$

Or in the matrix notation

$$U_s^{(m,r)} = O^{(m,m+1)} U_s^{(m+1,l)}, \tag{6.46a}$$

$$U_a^{(m,r)} = O^{(m,m+1)} U_a^{(m+1,l)}. \tag{6.46b}$$

Combining (6.40) and (6.46) and eliminating $U_a^{(m,l)}$, $U_a^{(m,r)}$, $U_s^{(m,r)}$, we obtain

$$(O^{(m,m-1)} T^{(m-1)} O^{(m-1,m)} + T^{(m)}) U_s^{(m,l)} = O^{(m,m-1)} S^{(m-1)} U_s^{(m-1,l)} + S^{(m)} O^{(m,m+1)} U_s^{(m+1,l)} \tag{6.47a}$$

for $2 < m < M$;

$$(O^{(2,1)} T^{(1)} O^{(1,2)} + T^{(2)}) U_s^{(2,l)} = S^{(2)} O^{(2,3)} U_s^{(3,l)} \tag{6.47b}$$

for $m = 2$, when $M > 2$;

$$(O^{(M,M-1)} T^{(M-1)} O^{(M-1,M)} + T^{(M)}) U_s^{(M,l)} = O^{(M,M-1)} S^{(M-1)} U_s^{(M-1,l)} \tag{6.47c}$$

for $m = M$.

In the simplest case, we obtain

$$(O^{(1,2)} T^{(2)} + T^{(1)} O^{(1,2)}) U_s^{(2,l)} = 0. \tag{6.47d}$$

The relations (6.47a)–(6.47d) are satisfied in the case of electric walls where the boundary conditions are as follows:

$$\begin{cases} U_s^{(1,l)} = [0 \quad 0 \quad \cdots \quad 0]^T, \\ U_s^{(M,r)} = [0 \quad 0 \quad \cdots \quad 0]^T. \end{cases}$$

If the walls are magnetic, i.e., the boundary conditions:

$$\begin{cases} U_a^{(1,l)} = [0 \quad 0 \quad \cdots \quad 0]^T, \\ U_a^{(M,r)} = [0 \quad 0 \quad \cdots \quad 0]^T, \end{cases}$$

then for $m = 2$, when $M > 2$:

$$(O^{(2,1)} (T^{(1)} - S^{(1)} (T^{(1)})^{-1} S^{(1)}) O^{(1,2)} + T^{(2)}) U_s^{(2,l)} = S^{(2)} O^{(2,3)} U_s^{(3,l)}, \tag{6.47e}$$

and for $m = M$:

$$(O^{(M,M-1)} T^{(M-1)} O^{(M-1,M)} + T^{(M)} - S^{(M)} (T^{(M)})^{-1} S^{(M)}) \\ U_s^{(M,l)} = O^{(M,M-1)} S^{(M-1)} U_s^{(M-1,l)}. \tag{6.47f}$$

In the simplest case, for $M = 2$ at the magnetic boundary conditions we have:

$$O^{(2,1)} [T^{(1)} - S^{(1)} (T^{(1)})^{-1} S^{(1)}] O^{(1,2)} - S^{(2)} (T^{(2)})^{-1} S^{(2)}) U_s^{(2,l)} = 0. \tag{6.47g}$$

The problem described by (6.47), in a general form can be written as

$$\Lambda(k_z) U = 0, \tag{6.48}$$

where the matrix $\Lambda(k_z)$ consists of blocks $(M-1) \times (M-1)$, each with the dimension $K \times K$. The structure of this matrix has the form:

$$\Lambda(k_z)=\begin{bmatrix} A^{(2)} & C^{(2)} & \mathrm{O} & \mathrm{O} & \mathrm{O} & \mathrm{O} & \mathrm{O} & \mathrm{O} \\ B^{(3)} & A^{(3)} & C^{(3)} & \mathrm{O} & \mathrm{O} & \mathrm{O} & \mathrm{O} & \mathrm{O} \\ \mathrm{O} & B^{(4)} & A^{(4)} & C^{(4)} & \mathrm{O} & \mathrm{O} & \mathrm{O} & \mathrm{O} \\ & & \ddots & & & & & \\ \mathrm{O} & \mathrm{O} & \mathrm{O} & \mathrm{O} & \mathrm{O} & B^{(M-1)} & A^{(M-1)} & C^{(M-1)} \\ \mathrm{O} & \mathrm{O} & \mathrm{O} & \mathrm{O} & \mathrm{O} & \mathrm{O} & B^{(M)} & A^{(M)} \end{bmatrix}. \tag{6.49}$$

Here

$$A^{(m)} = O^{(m,m-1)}T^{(m-1)}O^{(m-1,m)} + T^{(m)}, \tag{6.50a}$$

in the case of magnetic walls the expressions for $A^{(2)}$ and $A^{(M)}$ do not fit into the overall scheme, so they must be brought separately:

$$A^{(2)} = O^{(2,1)}(T^{(1)} - S^{(1)}(T^{(1)})^{-1}S^{(1)})O^{(1,2)} + T^{(2)}, \tag{6.50b}$$

$$A^{(M)} = O^{(M,M-1)}T^{(M-1)}O^{(M-1,M)} + T^{(M)} - S^{(M)}(T^{(M)})^{-1}S^{(M)}. \tag{6.50c}$$

$$B^{(m)} = -O^{(m,m-1)}S^{(m-1)}, \tag{6.51}$$

$$C^{(m)} = -S^{(m)}O^{(m,m-1)}, \tag{6.52}$$

O is the zero matrix of dimension $K \times K$.

In (6.48) the vector

$$U = \begin{bmatrix} U_s^{(2,l)} \\ U_s^{(3,l)} \\ \vdots \\ U_s^{(M,l)} \end{bmatrix}. \tag{6.53}$$

We have a problem (6.48), which has only the trivial solution $U = 0$ if $\det(\Lambda(k_z)) \neq 0$. The value of the parameter k_z for which there is no trivial solution is called the eigenvalue of the matrix $\Lambda(k_z)$. To find these values, we can use the Krylov iterative method discussed in detail in section 6.1.3.

For each of the obtained eigenvalues k_z, the matrix $\Lambda(k_z)$ becomes numerical, defining the eigenvector U, and we get some of the values of the constants required for the construction of K functions of the form (6.22). Others can be found from (6.40) and (6.46) in the following order.

1. For $m = M$

$U_s^{(M,l)}$ known as a part of U.

1.1. If the walls are electrical, then

$U_s^{(M,r)} = [0 \;\; 0 \;\; \cdots \;\; 0]^T$, then

$U_a^{(M,l)} = -T^{(M)} U_s^{(M,l)}$ and $U_a^{(M,r)} = -S^{(M)} U_s^{(M,l)}$.

1.2. If the walls are magnetic, then

$U_a^{(M,r)} = [0 \ \ 0 \ \ \cdots \ \ 0]^T$

$U_a^{(M,l)} = (-T^{(M)} + S^{(M)} S^{(M)} (T^{(M)})^{-1} U_s^{(M,l)}$ and $U_s^{(M,r)} = (T^{(M)})^{-1} S^{(M)} U_s^{(M,l)}$.

2. For $1 \le m < M$.

2.1. $U_s^{(m,r)} = O^{(m,m+1)} U_s^{(m+1,l)}$.

2.2. $U_a^{(m,r)} = O^{(m,m+1)} U_a^{(m+1,l)}$.

2.3. $U_s^{(m,l)} = (S^{(m)})^{-1} (-U_a^{(m,r)} + T^{(m)} U_s^{(m,r)})$ – this holds only for $m =$ 1, as for other values of m the vector $U_s^{(m,l)}$ is known as part of the eigenvector U.

2.4. $U_a^{(m,l)} = -T^{(m)} U_s^{(m,l)} + S^{(m)} U_s^{(m,r)}$.

Thus, for each value k_z we can construct K modes of the form (6.22), combining them with the K modes of the form (6.17) (they are the same for different k_z) by (6.25), we obtain the field propagating in the direction of the z axis with the projection k_z of the wave vector on this axis.

6.1.2. Method of matched sinusoidal modes in the vector case

The principal difference between the vector case and the scalar case is that it is necessary to consider the local modes of two different polarizations – TE and TM, as they both contribute to the formation of a hybrid mode of the fibre. Accordingly, the expression (6.25) is transformed as follows:

$$F^{(m)}(x,y) = \sum_{p=e,h} \sum_{k=1}^{\infty} [u_{pk}^{(m)}(x) F_{spk}^{(m)}(y) + \dot{u}_{pk}^{(m)}(x) F_{apk}^{(m)}(y)], \tag{6.54}$$

where F refers to any of the electric or magnetic field components of the mode field, and the outer sum corresponds to the summation over the polarizations: TE – p = h, and TM – p = e. Let us consider the local modes in the expression (6.54). The local x-mode now has the form:

$$u_{pk}^{(m)}(x) = \left(\frac{k_0}{k_{pk}^{(m)}}\right)^2 u_{spk}^{(m,l)} \cos[k_{xpk}^{(m)}(x - x^{(m)})] + \frac{u_{apk}^{(m,l)}}{k_{xpk}^{(m)}} \sin[k_{xpk}^{(m)}(x - x^{(m)})], \tag{6.55}$$

here we take into account the polarization factor and introduce the multiplier $\left(\frac{k_0}{k_{pk}^{(m)}}\right)^2$, the need for which will be explained below. Expressions for $F_{sp}^{(m)}$ and $F_{ap}^{(m)}$ are defined in Table 6.1, through local y-modes.

In Table 6.1 and later we use the following notation: $B_x = \mu_0 H_x$, $B_y = \mu_0 H_y$, $B_z = \mu_0 H_z$, where μ_0 is the magnetic permeability of free space.

Thus, the expression (6.54) takes a definite form for each of the components of vectors $\bar{E}(x, y)$ and $c\,\bar{B}(x, y) = c\mu_0\,\bar{H}(x, y)$:

$$E_x^{(m)}(x,y) = \sum_{k=1}^{\infty} u_{hk}^{(m)}(x) \frac{k_z}{k_0} \phi_k^{(m)}(y) - \sum_{k=1}^{\infty} \dot{u}_{ek}^{(m)}(x) \frac{\dot{\psi}_k^{(m)}(y)}{k_o^2 \varepsilon^{(m)}(y)}, \quad (6.56a)$$

$$E_y^{(m)}(x,y) = -\sum_{k=1}^{\infty} u_{ek}^{(m)} \left(\frac{k_{ek}^{(m)}}{k_o} \right)^2 \frac{\psi_k^{(m)}(y)}{\varepsilon^{(m)}(y)}, \quad (6.56b)$$

$$E_z^{(m)}(x,y) = -\sum_{k=1}^{\infty} \dot{u}_{hk}^{(m)}(x) \frac{i\varphi_k^{(m)}(y)}{k_0} + \sum_{k=1}^{\infty} u_{ek}^{(m)}(x) \frac{ik_z}{k_0^2} \frac{\dot{\psi}_k^{(m)}}{\varepsilon^{(m)}(y)}, \quad (6.56c)$$

$$cB_x^{(m)}(x,y) = \sum_{k=1}^{\infty} \dot{u}_{hk}^{(m)}(x) \frac{\dot{\varphi}_k^{(m)}(y)}{k_0^2} + \sum_{k=1}^{\infty} u_{ek}^{(m)}(x) \frac{k_z}{k_o} \psi_k^{(m)}(y), \quad (6.56d)$$

$$cB_y^{(m)}(x,y) = \sum_{k=1}^{\infty} u_{hk}^{(m)}(x) \left(\frac{k_{hk}^{(m)}}{k_0} \right)^2 \varphi_k^{(m)}(y), \quad (6.56e)$$

$$cB_z^{(m)}(x,y) = -\sum_{k=1}^{\infty} u_{hk}^{(m)}(x) \frac{ik_z}{k_0^2} \dot{\varphi}_k^{(m)}(y) - \sum_{k=1}^{\infty} \dot{u}_{ek}^{(m)}(x) \frac{i\psi_k^{(m)}(y)}{k_0}. \quad (6.56f)$$

These are the components of the hybrid mode.

In Table 6.1 and the expressions (6.56) $\varphi(y)$ are the local y-modes, in this case corresponding to TE-polarization and reviewed in detail in paragraph 6.1.1.

Modes $\psi(y)$, representing TM-polarization, have the form in the column m similar to (6.17)

$$\psi(y) = \psi_s^{(n,l)} \cos[k_y^{(n)}(y - y^{(n)})] + \frac{\psi_a^{(n,l)}}{k_y^{(n)}} \sin[k_y^{(n)}(y - y^{(n)})], \quad (6.57)$$

but the conditions (6.26) at the interface between the homogeneous regions for them take a different form:

$$\psi_s^{(n,r)} = \psi_s^{(n+1,l)}, \quad (6.58a)$$

$$\frac{\psi_a^{(n,r)}}{\varepsilon^{(n)}} = \frac{\psi_a^{(n+1,l)}}{\varepsilon^{(n+1)}}. \quad (6.58b)$$

This implies changes in the matrix relations (6.31) and (6.32), and for the modes $\psi(y)$, they look like this:

$$\begin{bmatrix} \psi_s^{(n',r)} \\ \psi_a^{(n',r)} \end{bmatrix} = P^{(n')} W^{(n')} P^{(n'-1)} W^{(n'-1)} \ldots W^{(2)} P^1 \begin{bmatrix} \psi_s^{(1,l)} \\ \psi_a^{(1,l)} \end{bmatrix}, \quad (6.59)$$

Table 6.1. Symmetric and antisymmetric components of the field, expressed in terms of local modes

	TE		TM	
F	$F_{sh}^{(m)}$	$F_{ah}^{(m)}$	$F_{se}^{(m)}$	$F_{ae}^{(m)}$
E_x	$\left(\frac{k_z}{k_0}\right)\varphi^{(m)}(y)$	0	0	$-\frac{\dot{\psi}^{(m)}(y)}{k_0^2\varepsilon^{(m)}(y)}$
E_y	0	0	$-\left(\frac{k_{ek}^{(m)}}{k_0}\right)^2\frac{\psi^{(m)}(y)}{\varepsilon^{(m)}(y)}$	0
E_z	0	$-\frac{i\varphi^{(m)}(y)}{k_0}$	$\left(\frac{ik_z}{k_0^2}\right)\frac{\dot{\psi}^{(m)}(y)}{\varepsilon^{(m)}(y)}$	0
cB_x	0	$\frac{\dot{\varphi}^{(m)}(y)}{k_0^2}$	$\left(\frac{k_z}{k_0}\right)\psi^{(m)}(y)$	0
cB_y	$\left(\frac{k_{hk}^{(m)}}{k_0}\right)^2\varphi^{(m)}(y)$	0	0	0
cB_z	$\left(\frac{-ik_z}{k_0^2}\right)\dot{\varphi}^{(m)}(y)$	0	0	$-\frac{i\psi^{(m)}(y)}{k_0}$

$$\begin{bmatrix}\psi_s^{(n'+1,l)}\\ \psi_a^{(n'+1,l)}\end{bmatrix}=Q^{(n'+1)}V^{(n'+1)}Q^{(n'+2)}V^{(n'+2)}\ldots V^{(N-1)}Q^{(N)}\begin{bmatrix}\psi_s^{(N,r)}\\ \psi_a^{(N,r)}\end{bmatrix}, \tag{6.60}$$

Here $P^{(i)}$ and $Q^{(i)}$ are as previously the matrices of the form (6.29) and (6.30), respectively;

$$W^{(i+1)}=\begin{bmatrix}1 & 0\\ 0 & \frac{\varepsilon^{(i+1)}}{\varepsilon^{(i)}}\end{bmatrix}, i=\overline{1,N-1}; \tag{6.61}$$

$$V^{(i)}=\begin{bmatrix}1 & 0\\ 0 & \frac{\varepsilon^{(i)}}{\varepsilon^{(i+1)}}\end{bmatrix}, i=\overline{1,N-1}. \tag{6.62}$$

It is obvious that the matrices $W^{(i+1)}$ and $V^{(i)}$ are inverse and depend solely on the dielectric structure of the fibre cross section. The relations (6.59) and (6.60), as well as their analogues for TE-polarization are used to calculate the quantities involved in the characteristic equation with respect to k_k^2:

$$\varepsilon^{(n')}\psi_s^{(n',r)}\psi_a^{(n'+1,l)} - \varepsilon^{(n'+1)}\psi_a^{(n',r)}\psi_s^{(n'+1,l)} = 0. \quad (6.63)$$

There is no doubt the fact that in general the roots of equations (6.63) and (6.34) are different, and since they characterize the local y-modes of different polarizations, it is necessary to introduce appropriate notations to avoid confusion. Let k_{hk}^2 be the set of solutions of (6.34), corresponding to the TE-case, and k_{ek}^2 the solution of (6.63) describing the local TM-mode.

Also exposed to changes is the formula for calculating the normalization integral

$$\frac{1}{y^{(N+1)} - y^{(1)}} \int_{y^{(1)}}^{y^{(N+1)}} \frac{\psi^2(y)}{\varepsilon(y)} dy = 1. \quad (6.64)$$

Thus, the calculation of vector fields is complicated by the problem exactly by half. Now the algorithm for finding local y-modes is as follows:

– For each column of the section of the fibre it is necessary to find K roots of (6.34), determining the modes $\psi(y)$ with the given boundary conditions (electric or magnetic wall);

– Then calculate the constants $\varphi_s^{(n,l)}$, $\varphi_a^{(n,l)}$, $\varphi_s^{(n,r)}$, $\varphi_a^{(n,r)}$, on the basis of the normalization condition (6.35) to finally form K y-modes of the form (6.17), corresponding to the case of TE-polarization;

– For each column of the section of the fibre we must find exactly the same number of K roots k_{ek}^2 of (6.63) defining the modes $\psi(y)$ with opposite boundary conditions in comparison with the TE-case, that is, if the modes $\varphi(y)$ we use the vanishing function on the boundary, then for $\psi(y)$ we need to use the vanishing of the derivative at the boundary and vice versa;

– Then calculate the constants $\psi_s^{(n,l)}$, $\psi_a^{(n,l)}$, $\psi_s^{(n,r)}$, $\psi_a^{(n,r)}$, on the basis of the normalization condition (6.64) to finally form K y-modes of the form (6.57), corresponding to the case of TM-polarization.

Significant modification is also required in the search algorithm for propagation constants and actually 'cross linking' of local modes into locally continuous functions, which describe the components of the vector field. The following expressions hold for local x-modes of both polarizations

$$U_{ap}^{(m,l)} = -T_p^{(m)} U_{sp}^{(m,l)} + S_p^{(m)} U_{sp}^{(m,r)}, \quad (6.65a)$$

$$U_{ap}^{(m,r)} = -S_p^{(m)} U_{sp}^{(m,l)} + T_p^{(m)} U_{sp}^{(m,r)}. \quad (6.65b)$$

Here

$$T_{pkk}^{(m)} = (k_0 / k_{pk}^{(m)})^2 k_{xpk}^{(m)} / \mathrm{tg}(k_{xpk}^{(m)} d_x^{(m)}), \quad (6.65a)$$

$$S_{pkk}^{(m)} = (k_0 / k_{pk}^{(m)})^2 k_{xpk}^{(m)} / \sin(k_{xpk}^{(m)} d_x^{(m)}). \quad (6.65b)$$

Equalities similar to (6.46) connecting the constants of local x-modes of both polarizations of the adjacent columns are listed below:

$$U_{sh}^{(m,r)} = O_{hh}{}^{(m,m+1)} U_{sh}^{(m+1,l)}, \tag{6.66a}$$

$$U_{ah}^{(m,r)} = O_{hh}^{(m,m+1)} U_{ah}^{(m+1,l)} - k_z O_{he}^{(m,m+1)} U_{se}^{(m+1,l)}, \tag{6.66b}$$

$$U_{se}^{(m,r)} = O_{ee}{}^{(m,m+1)} U_{se}^{(m+1,l)}, \tag{6.66c}$$

$$U_{ae}^{(m,r)} = O_{ee}^{(m+1,m)} U_{ae}^{(m+1,l)} + k_z O_{he}^{(m+1,m)} U_{sh}^{(m+1,l)}, \tag{6.66d}$$

where the matrices $O_{hh}^{(m,m')}, O_{ee}^{(m,m')}$ and $O_{he}^{(m,m')}$ have the following elements:

$$O_{hhkp}^{(m,m')} =< \varphi_k^{(m)} \mid \varphi_p^{(m')} >, \tag{6.67a}$$

$$O_{eekp}^{(m,m')} =< \psi_k^{(m)} \mid \psi_p^{(m')} / \varepsilon^{(m')}(y) >, \tag{6.67b}$$

$$O_{hekp}^{(m,m')} =< \phi_k^{(m)} \mid \psi_p^{(m')} / \varepsilon^{(m')}(y) > / k_{hk}^{(m)2} + < \phi_k^{(m)} \mid \dot{\psi}_p^{(m')} / \varepsilon^{(m')}(y) > / k_{ep}^{(m')2}. \tag{6.67c}$$

The expression for $\left\langle \varphi_k \mid \varphi_p \right\rangle$ through the terms of the values of functions and derivatives at the boundaries of lines (constants $\varphi_s^{(n,l)}$, $\varphi_a^{(n,l)}$, $\varphi_s^{(n,r)}$, $\varphi_a^{(n,r)}$) has been determined previously by (6.42). Here are formulas suitable for calculating the elements of other matrices of overlap integrals:

$$\left\langle \psi_k \mid \psi_p / \varepsilon(y) \right\rangle = \sum_{n=1}^{N} \frac{(\psi_{sk}^{(n,r)} \psi_{ap}^{(n,r)} - \psi_{ak}^{(n,r)} \psi_{sp}^{(n,r)} - \psi_{sk}^{(n,l)} \psi_{ap}^{(n,l)} + \psi_{ak}^{(n,l)} \psi_{sp}^{(n,l)})}{\varepsilon^{(n)} (k_{kk}^2 - k_{kp}^2)}, \tag{6.68}$$

$$\left\langle \dot{\phi}_k \mid \psi_p / \varepsilon(y) \right\rangle = \sum_{n=1}^{N} \frac{(-\phi_{ak}^{(n,l)} \psi_{ap}^{(n,l)} - k_{kk}^2 \phi_{sk}^{(n,l)} \psi_{sp}^{(n,r)} + \phi_{ak}^{(n,r)} \psi_{ap}^{(n,r)} + k_{kk}^2 \phi_{sk}^{(n,r)} \psi_{sp}^{(n,r)})}{\varepsilon^{(n)} (k_{kk}^2 - k_{kp}^2)}, \tag{6.69}$$

$$\left\langle \varphi_k \mid \dot{\psi}_p / \varepsilon(y) \right\rangle = \sum_{n=1}^{N} \frac{(-\varphi_{ak}^{(n,l)} \psi_{ap}^{(n,l)} - k_{kp}^2 \varphi_{sk}^{(n,l)} \psi_{sp}^{(n,l)} + \varphi_{ak}^{(n,r)} \psi_{ap}^{(n,r)} + k_{kp}^2 \varphi_{sk}^{(n,r)} \psi_{sp}^{(n,r)})}{\varepsilon^{(n)} (k_{kk}^2 - k_{kp}^2)}. \tag{6.70}$$

It should be noted that the introduction of the factor $\left(\dfrac{k_0}{k_{pk}^{(m)}} \right)^2$ that distinguishes the expression for the x-modes (6.55) from the scalar analogue of (6.22) allows us, keeping simple expressions for the normalization (6.35) and (6.64) of the local y-modes of the TE-and TM-polarizations, respectively, to obtain an expression for the matrix elements $O_{hhkp}^{(m,m')}$ and $O_{eekp}^{(m,m')}$ which do not depend on the roots of the characteristic equations (6.34) and (6.63).

Similarly to (6.48) for finding propagation constants for the scalar case, we can formulate the problem of solving a homogeneous algebraic system of equations

$$\Xi(k_z) U = 0, \tag{6.71}$$

where matrix $\Xi(k_z)$, as $\Lambda(k_z)$, has the band structure and depends on k_z:

$$\Xi(k_z)=\begin{bmatrix} A_h^{(2)} & C_h^{(2)} & B_h^2 & 0 & 0 & 0 & \dots & \dots & 0 & 0 & 0 & 0 \\ C_e^{(2)} & A_e^{(2)} & 0 & B_e^{(2)} & 0 & 0 & \dots & \dots & 0 & 0 & 0 & 0 \\ D_h^{(3)} & 0 & A_h^{(3)} & C_h^{(3)} & B_h^{(3)} & 0 & \dots & \dots & 0 & 0 & 0 & 0 \\ 0 & D_e^{(3)} & C_e^{(3)} & A_e^{(3)} & 0 & B_e^{(3)} & \dots & \dots & 0 & 0 & 0 & 0 \\ & & & \ddots & & & & & 0 & 0 & 0 & 0 \\ 0 & 0 & 0 & 0 & 0 & 0 & \dots & \dots & D_h^{(M)} & 0 & A_h^{(M)} & C_h^{(M)} \\ 0 & 0 & 0 & 0 & 0 & 0 & \dots & \dots & 0 & D_e^{(M)} & C_e^{(M)} & A_e^{(M)} \end{bmatrix}. \tag{6.72}$$

The components of the block matrix $\Xi(k_z)$, expressed in terms of already defined the matrix of overlap integrals and the diagonal matrices $T_p^{(m)}$ and $S_p^{(m)}$.

$$A_h^{(m)} = O_{hh}^{(m,m-1)} T_h^{(m-1)} O_{hh}^{(m-1,m)} + T_h^{(m)}, \tag{6.73a}$$

$$A_e^{(m)} = O_{ee}^{(m-1,m)T} T_e^{(m-1)} O_{ee}^{(m-1,m)} + T_e^{(m)}. \tag{6.73b}$$

Exactly as stated in paragraph 6.1.1, in the case of magnetic walls the expressions for $A_p^{(2)}$ and $A_p^{(M)}$ differ from the general scheme:

$$A_h^{(2)} = O_{hh}^{(2,1)} (T_h^{(1)} - S_h^{(1)} (T_h^{(1)})^{-1} S_h^{(1)}) O_{hh}^{(1,2)} + T_h^{(2)}, \tag{6.73c}$$

$$A_e^{(2)} = O_{ee}^{(1,2)T} (T_e^{(1)} - S_e^{(1)} (T_e^{(1)})^{-1} S_e^{(1)}) O_{ee}^{(1,2)} + T_e^{(2)}, \tag{6.73d}$$

$$A_h^{(M)} = O_{hh}^{(M,M-1)} T_h^{(M-1)} O_{hh}^{(M-1,M)} + T_h^{(M)} - S_h^{(M)} (T_h^{(M)})^{-1} S_h^{(M)}, \tag{6.73e}$$

$$A_e^{(M)} = O_{ee}^{(M-1,M)T} T_e^{(M-1)} O_{ee}^{(M-1,M)} + T_e^{(M)} - S_e^{(M)} (T_e^{(M)})^{-1} S_e^{(M)}, \tag{6.73f}$$

$$B_h^{(m)} = -S_h^{(m)} O_{hh}^{(m,m+1)}, \tag{6.73g}$$

$$B_e^{(m)} = -S_e^{(m)} O_{ee}^{(m,m+1)}, \tag{6.73h}$$

$$C_h^{(m)} = k_z O_{hh}^{(m,m-1)} O_{he}^{(m-1,m)}, \tag{6.73i}$$

$$C_e^{(m)} = -k_z O_{ee}^{(m-1,m)T} O_{he}^{(m,m-1)T}, \tag{6.73j}$$

$$D_h^{(m)} = -O_{hh}^{(m,m-1)} S_h^{(m-1)}, \tag{6.73k}$$

$$D_e^{(m)} = -O_{ee}^{(m-1,m)T} S_e^{(m-1)}. \tag{6.73l}$$

In constructing the matrix $\Xi(k_z)$ of the problem (6.71), we must also take into account the rule of defining the boundary conditions stated in Table 6.2.

In turn, the vector U in (6.71) contains the constants of local modes with both TE- and TM-polarization.

Table 6.2. Two possible options for setting the boundary conditions

	$\phi(y)$	$\psi(y)$	$u_h(x)$	$u_e(x)$
'Electric walls'	$\begin{cases}\phi_s^{(1,l)}=0,\\ \phi_s^{(N,r)}=0.\end{cases}$	$\begin{cases}\psi_a^{(1,l)}=0,\\ \psi_a^{(N,r)}=0.\end{cases}$	$\begin{cases}u_{ha}^{(1,l)}=0,\\ u_{ha}^{(N,r)}=0.\end{cases}$	$\begin{cases}u_{es}^{(1,l)}=0,\\ u_{es}^{(N,r)}=0.\end{cases}$
'Magnetic walls'	$\begin{cases}\phi_a^{(1,l)}=0,\\ \phi_a^{(N,r)}=0.\end{cases}$	$\begin{cases}\psi_s^{(1,l)}=0,\\ \psi_s^{(N,r)}=0.\end{cases}$	$\begin{cases}u_{hs}^{(1,l)}=0,\\ u_{hs}^{(N,r)}=0.\end{cases}$	$\begin{cases}u_{ea}^{(1,l)}=0,\\ u_{ea}^{(N,r)}=0.\end{cases}$

$$U=\begin{bmatrix} U_{sh}^{(2,l)} \\ U_{se}^{(2,l)} \\ U_{sh}^{(3,l)} \\ U_{se}^{(3,l)} \\ \vdots \\ U_{sh}^{(M,l)} \\ U_{se}^{(M,l)} \end{bmatrix}. \tag{6.74}$$

The problem (6.71) is solved in the same manner as in the scalar case (6.48). This method is discussed in section 6.1.3. As a result, after determining the propagation constants and completion of local x-modes, from the resulting set of functions using the formulas (6.56) we calculated vector field E_x, E_y, E_z, cB_x, cB_y and cB_z.

6.1.3. The Krylov method for solving non-linear eigenvalue problems

In the stage of solving a non-linear eigenvalue problem (6.48), (6.71) it is convenient to use the iterative Krylov method [49], allowing accurately, avoiding gaps, calculate the eigenvalues of matrices of large dimensions.

Interpolating the non-linear matrix operator $\Lambda(k_z)$ between two arbitrary values σ and μ in the following way:

$$\Lambda(k_z)\approx\tilde{\Lambda}(k_z)=\frac{k_z-\sigma}{\mu-\sigma}\Lambda(\mu)+\frac{\mu-k_z}{\mu-\sigma}\Lambda(\sigma), \tag{6.75}$$

obtain a linear eigenvalue problem:

$$\tilde{\Lambda}(k_z)U=0, \tag{6.76}$$

which we solve iteratively as follows:

$$\left[\frac{\mu_{k+1}-\sigma}{\mu_k-\sigma}\Lambda(\mu_k)+\frac{\mu_k-\mu_{k+1}}{\mu_k-\sigma}\Lambda(\sigma)\right]U_k=0, \tag{6.77}$$

where μ_{k+1} is an approximate value of the unknown k_z, obtained in $(k+1)$-th iteration step. We introduce the notation:

$$\theta = \frac{\mu_{k+1} - \mu_k}{\mu_{k+1} - \sigma}. \tag{6.78}$$

Then (6.77) can be represented as:

$$[\Lambda(\sigma)^{-1}\Lambda(\mu_k) - \theta I]U_k = 0. \tag{6.79}$$

Let μ_1 be an initial approximation for the eigenvalue k_z of the matrix $\Lambda(k_z)$, let σ be a fixed value close to μ_1, then the iterative procedure for increasing the accuracy of the eigenvalue k_z consists of the following. At the k-th step we solve with respect to θ:

$$\Lambda(\sigma)^{-1}\Lambda(\mu_k)U = \theta U \tag{6.80}$$

calculate a new estimate for the eigenvalue k_z from the problem (6.48) or (6.71):

$$\mu_{k+1} = \mu_k + \frac{\theta}{1-\theta}(\mu_k - \sigma). \tag{6.81}$$

The iterations are repeated until the sequence of estimates $\{\mu_k\}$ converges. In [53] it is shown that the iterative procedure in the Krylov method converges to the desired eigenvalue.

In [31] the authors considered another method of solving a non-linear eigenvalue problem. It is called the zero function method. The method consists of the following. Choose an arbitrary vector V with non-zero components, such as unit. We solve the inhomogeneous equation

$$\Lambda(k_z)U' = V \tag{6.82}$$

for U'. For different values of the parameter k_z we get different solutions of (6.82). We define the function

$$f(k_z) = 1/U'_p, \tag{6.83}$$

where U_p' is the p-th component of the vector U'. In the vicinity of the desired values of k_z the function $f(k_z)$ is a continuous function of the scalar argument, and its zeros are required values of k_z.

The zeros $f(k_z)$ are found by standard methods.

The discontinuity of the function $f(k_z)$ makes it difficult to find propagation constants in this way. Therefore, even for a sufficiently small sampling step, much smaller than that required for the separation of adjacent zeros of the function $f(k_z)$, there is a chance of missing the roots because of discontinuities close to zero. This problematic situation is shown in Fig. 6.4, where in the root region $k_z = 6.9951$ μm^{-1} the function has a discontinuity.

Numerous discontinuties also prevent us from using statistical estimates of the behaviour of the function in the interval, and the decrease of the discretization step increases the time spent on calculations.

The joint use of the method (6.82)–(6.83) with the Krylov method (6.75)–(6.81) presumably would outperform both of these methods.

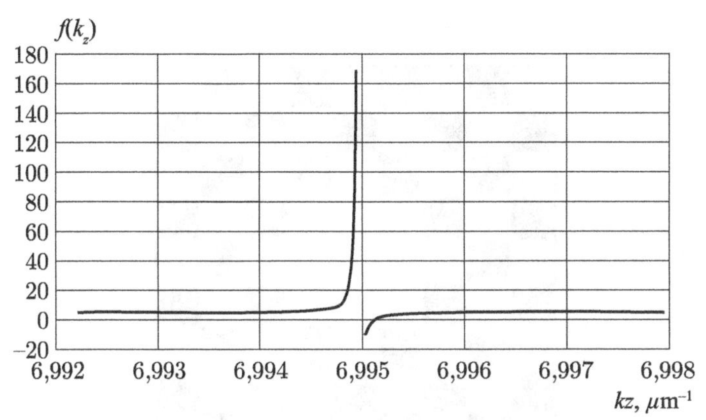

Fig. 6.4. An example of the discontinuity of function $f(k_z)$ near zero.

Let the Krylov method be used to indicate the presence of roots in an interval, where further clarification of values of the roots is carried out using the function (6.83). For a given accuracy of the separation of roots, we choose the sampling step in the Krylov method so as to minimize the computation time

$$t(h_k) = Mt_{ki} + Int_{si} \rightarrow \min. \tag{6.84}$$

In (6.84) $M = L/h_k$ is the number of intervals analyzed by the Krylov method, where L is the length of the interval in which we search for the eigenvalues k_z, t_{ki} is time to execute one iteration of the Krylov method, I is the estimated number of roots in the interval with length L, $n = h_k/h$ is the number of intervals considered in the interval with length h_k when refining roots with a step h, t_{si} is the average time to complete one iteration of zero detection of function $f(k_z)$ in the interval with length h.

The optimal value of the discretization step h_k is the value:

$$h_k^{\text{opt}} = \sqrt{\frac{Lht_{ki}}{It_{si}}}. \tag{6.85}$$

A practical example of calculating several eigenmodes of a photonic waveguide (Fig. 6.5) using these three methods for solving the non-linear eigenvalue problems in the method of matched sinusoidal modes is described below.

Approximation using formula (6.25) was carried out with thirty local modes that, according to a study conducted in [52], suffices to obtain estimates for the propagation constants of modes with an error of not more than 10^{-4}, i.e., up to three decimal places.

In accordance with the MSM-method we considered the problem (6.48), where the matrix elements $\Lambda(k_z)$ depend non-linearly on the parameter k_z, the desired values of which $\Lambda(k_z)$ become a degenerate numerical matrix whose determinant is zero.

For a given structure of the cross section the dimension of the matrix $\Lambda(k_z)$ reduced by the number of local modes was 480×480.

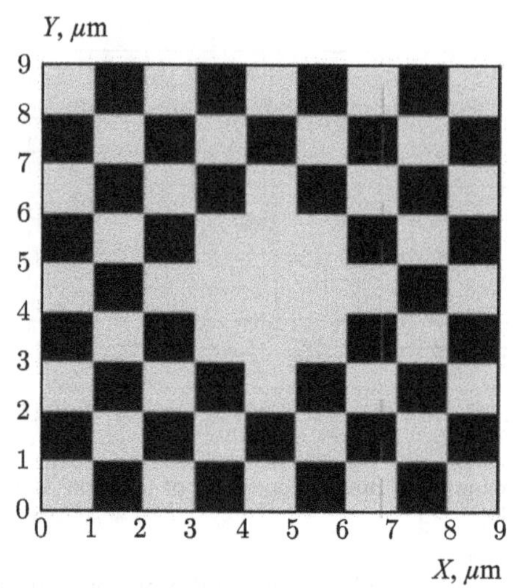

Fig. 6.5. The model of the section of the PCF, light regions show regions with the refractive index $n_1 = 1.47$, dark – with $n_2 = 1$.

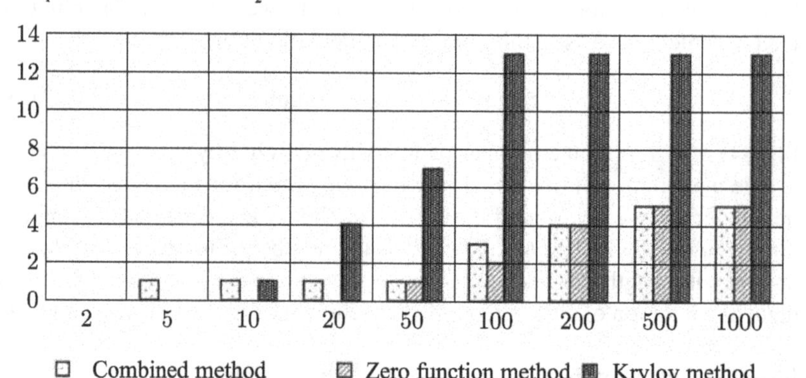

Fig. 6.6. Diagram of dependence of propagation constants found by the first few modes of the PCF (Fig. 6.5) on 1/h, inversely proportional to the accuracy of the separation of roots

In the unit length interval, search for propagation constants was carried out in three ways: by determining the zeros of function (6.83), the Krylov method (6.75) – (6.81) and the combined method with optimal discretization rate (6.85).

The diagram showing the number of roots detected with an accuracy of 10^{-4} for each method with different numbers of points of subdivision – the value inversely proportional to the accuracy of the separation of roots h, is given in Fig. 6.6.

As can be seen from the diagram, the Krylov method is much better than the other two in revealing the unknown parameter k_z values, but, as seen from the graph in Fig. 6.7, is the most time-consuming for a given value of h. For example, the calculation of the thirteen roots by the Krylov method with $h = 10^{-3}$ μm^{-1} on a personal computer takes about three hours. Note (Fig. 6.6) that the Krylov method found the same thirteen roots, but the time was eight times smaller. At the same

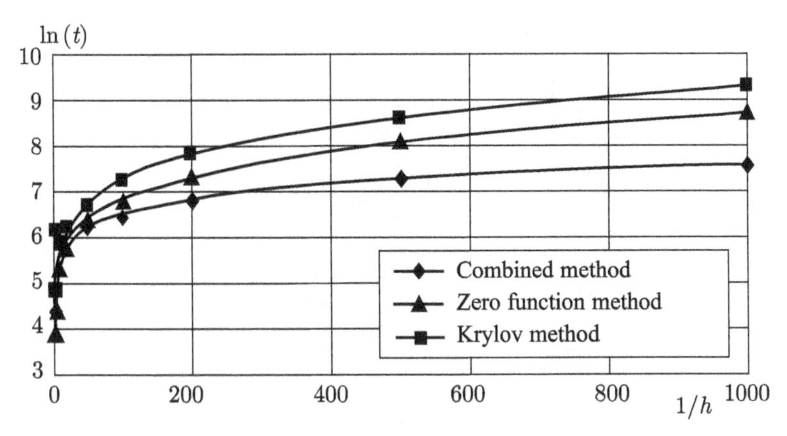

Fig. 6.7. A plot of the dependence of the natural logarithm of time in seconds spent on the calculation of zeros, on the value $1/h$, inversely proportional to the accuracy of the separation of roots.

time, the zero function method and the combined method are approximately the same as regards the frequency of detection of the roots, and the slight advantage of the combined method in the low accuracy range of the separation of roots is due to randomly better positioning of the interval of length h, in which secondary iterations of the zero function method are produced, relative to the root. The combined method is more effective with respect to time than the zero function method, as follows from Fig. 6.7.

Thus, the use in the MSM method of the iterative procedure for finding propagation constants by the Krylov method solves the problem of missing similar values of roots or roots close to the discontinuity of the function (6.83) arising when using the zero function method at the cost of increasing computing time.

6.1.4. Calculation of the modes of the stepped fibre

Consider the use of the MSM method for calculating the modes of stepped optical fibres produced from fused silica SiO_2. The initial radiation is represented by infrared light with a wavelength in vacuum $\lambda_0 = 1.3$ μm, corresponding to the minimum dispersion of the material.

The fibres whose cross sections are shown in Fig. 6.8 (model 1) and Fig. 6.9 (model 2) represent the approximate model of a stepped waveguide with a circular cross section, with a refractive index of the material in the core and the cladding of $n_{co} = 1.47$ and $n_{cl} = 1.463$, respectively. Since the refractive indices of the core and the cladding differ slightly

$$n_{co} \cong n_{cl}, \tag{6.86}$$

this fibre is weakly guiding.

In addition, the fibre is not a multi-mode fibre [54], since the condition

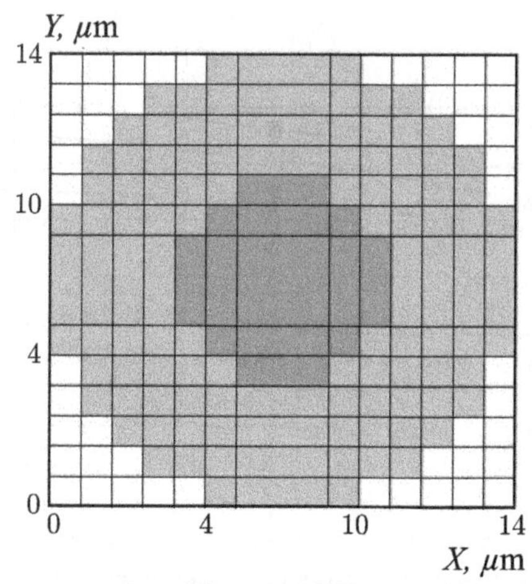

Fig. 6.8. Scheme of the cross section of the optical fibre (model 1), dark gray area shows the value of the refractive index $n_{co} = 1.47$, light gray – with $n_{cl} = 1.463$, white – with $n_v = 1$.

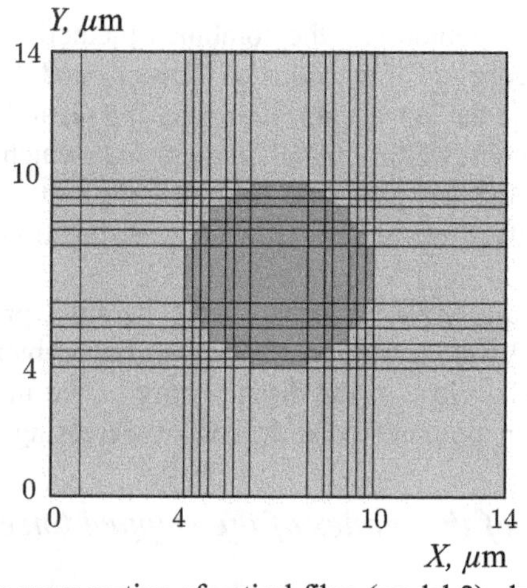

Fig. 6.9. Scheme of the cross section of optical fibre (model 2), dark gray area shows the value of the refractive index $n_{co} = 1.47$, light gray – with $n_{cl} = 1.463$.

$$V = \frac{2\pi\rho\sqrt{n_{\text{co}}^2 + n_{\text{cl}}^2}}{\lambda_0} >> 1 \tag{6.87}$$

is not satisfied. In this case, the characteristic core size (radius) is equal to 3 μm, and fibre optic parameter [54]:

$$V = \frac{2\pi\sqrt{1.47^2 + 1.463^2} \cdot 3\mu\text{m}}{1.3\,\mu\text{m}} \approx 2.078. \tag{6.88}$$

Therefore, in the calculations attention will be paid to only a few lower-order modes. Figure 6.10 shows the intensity distribution of the fundamental modes

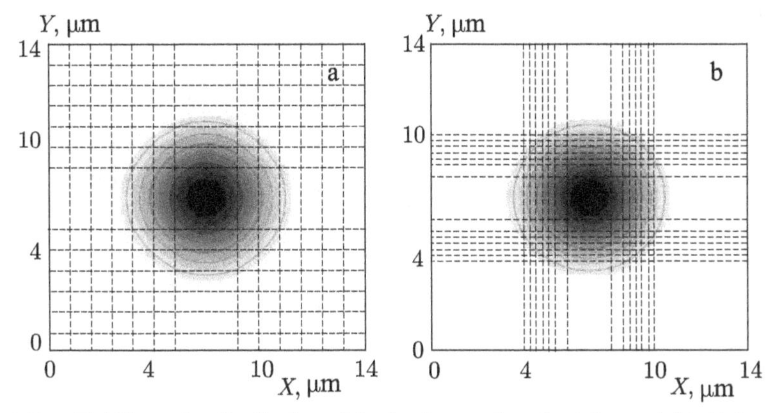

Fig. 6.10. Field intensity distribution of the fundamental mode a) for model 1, b) for model 2.

Table 6.3. The values of the propagation constants of the first ten modes of model 2

Number of mode	k_z, mm^{-1}	$n_{\text{eff}} = k_z/k_0$
1	7.0879	1.4665
2	7.0617	1.4611
3	7.0447	1.4576
4	7.0389	1.4564
5	7.0301	1.4545
6	7.0102	1.4504
7	6.9890	1.4460
8	6.9846	1.4451
9	6.9551	1.4390
10	6.9512	1.4382

for both models derived from the scalar version of the MSM method, function $|\xi(x, y)^2|$ from (6.25). Fig. 6.10 shows the modes in haltones: black corresponds to the maximum intensity of the mode, white to the minimum (zero).

Since there are no fundamental qualitative differences between these two results, for further calculations we use a more convenient model 2 with the cross section shown in Fig. 6.9. At the same number of rows and columns in the models 1 and 2, the accuracy of the approximation of the form of the core for model 2 is higher, at the same time, the radius of the cladding can be considered much greater than the radius of the core, because this usually happens in practice. The intensity distributions of the first ten modes of model 2 are shown in Fig. 6.11 of which only the fundamental mode is guided, as only its propagation constant satisfies the cutoff condition $n_{\text{cl}} k_0 < k_z$ [54] (Table 6.3).

The found spatial modes are a system of mutually orthogonal functions in the section. The matrix of the values of the integrals

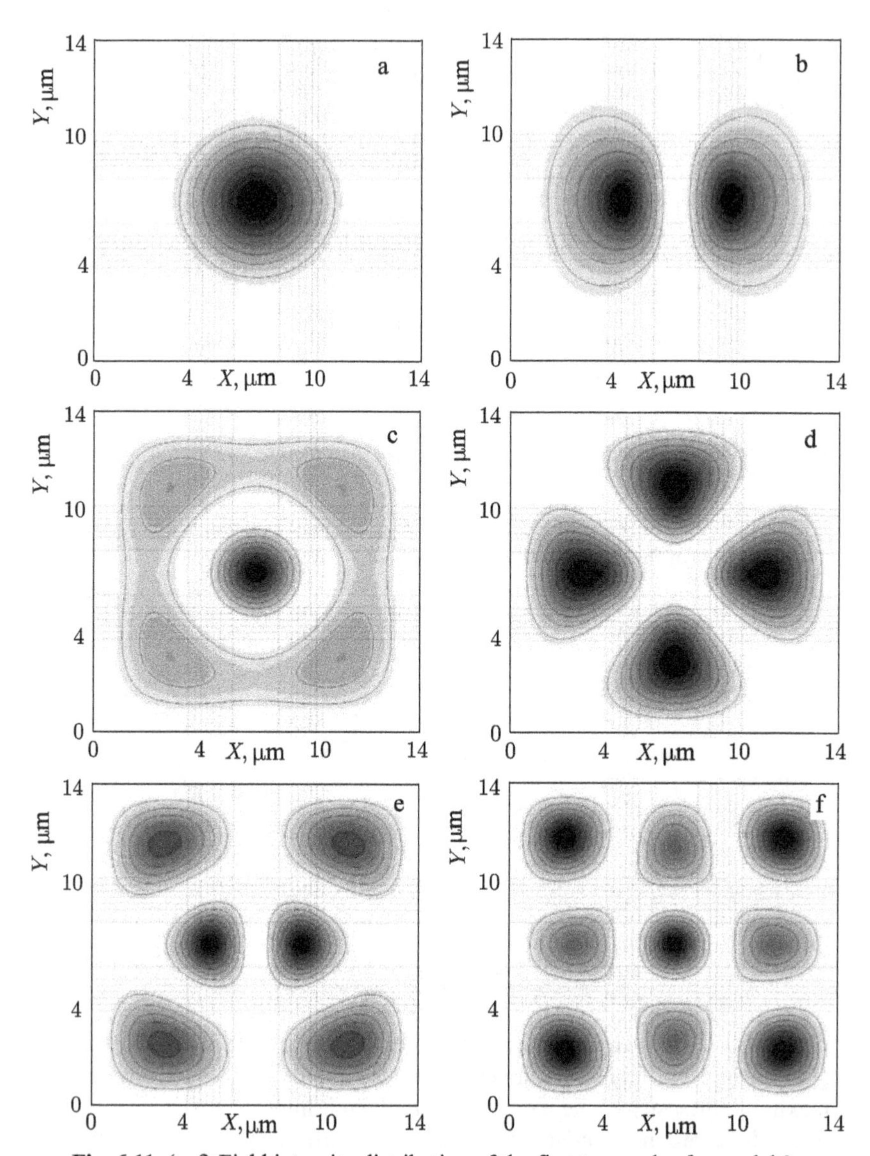

Fig. 6.11. (a–f) Field intensity distribution of the first ten modes for model 2.

$$\int_{x^{(1)}}^{x^{(M+1)}} \int_{y^{(1)}}^{y^{(N+1)}} \xi_{k1}(x,y)\xi_{k2}(x,y)dxdy, \text{ where } k1,k2=\overline{1,10} \tag{6.89}$$

is shown in Table 6.4. which shows that the calculated modes are orthogonal to within four decimal places.

After normalization of each mode, we obtain a system of orthonormal functions.

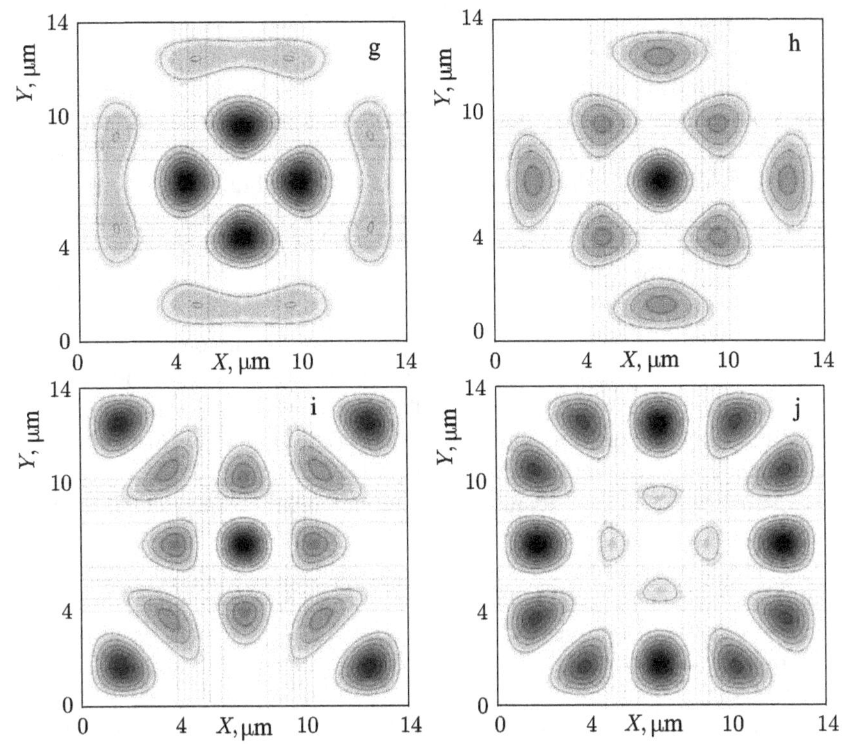

Fig. 6.11. (g–j) Field intensity distribution of the first ten modes for model 2.

Table 6.4. Matrix of values of the overlap integrals for the amplitudes of the first ten modes of model 2

Mode No	1	2	3	4	5	6	7	8	9	10
1	8.4803	0	0	0	0	0	0	0	0	0
2	0	7.7895	0	0	0	0	0	0	0	0
3	0	0	14.7646	0	0	0	0	0	0	0
4	0	0	0	13.3078	0	0	0	0	0	0
5	0	0	0	0	10.5809	0	0	0	0	0
6	0	0	0	0	0	40.0445	0	0	0	0
7	0	0	0	0	0	0	9.9666	0	0	0
8	0	0	0	0	0	0	0	10.4592	0	0
9	0	0	0	0	0	0	0	0	19.4402	0
10	0	0	0	0	0	0	0	0	0	21.7969

6.1.5. Calculation of modes of the photonic-crystal fibre

The use of vector methods for calculating the eigenmodes is especially important for the study of light guides sections of which contain strong (20%) variations in the refractive index, such as the PCF. Further, the vector and scalar modes of the PCF, calculated by the SCM, whose cross section is shown in Fig. 6.5, are compared.

The intensity distribution of the component cB_y of the fundamental mode with effective refractive index $n_{\text{eff}} = 1.4473$ and the intensity of the corresponding scalar field with $n_{\text{ef}} = 1.4491$, calculated for the radiation with a wavelength of $\lambda_0 = 1.3$ μm are shown in Fig. 6.12.

There is quite a similarity between the configuration of the scalar field and vector component cB_y. The standard deviation between the two solutions, normalized to unity, in the field $W_z \times W_y = 9 \times 9$ μm is 0.0000014 or 0.00014%. For the other three modes, also notes the distribution of the square of the amplitude of one component of the vector and scalar field intensity (Figs. 6.13–6.15).

The similarity of the configurations of the individual components of the vector mode and scalar mode is not coincidental. The corresponding vector component is the most powerful of the six vector field components, i.e., has the greatest value of the integral in the cross section (Table 6.5).

Characteristically, the integrals of the longitudinal components are insignificant in relation to the dominant component of the number of transverse magnetic components.

Thus, the product calculation shows that the scalar MSM-method provides a fairly good approximation of the most intense component of the vector modes, even in the case of fibres with a strong variation of the refractive index. Nevertheless, the presence of several non-zero vector components, making use of the vector approach appropriate in the case of the PCF.

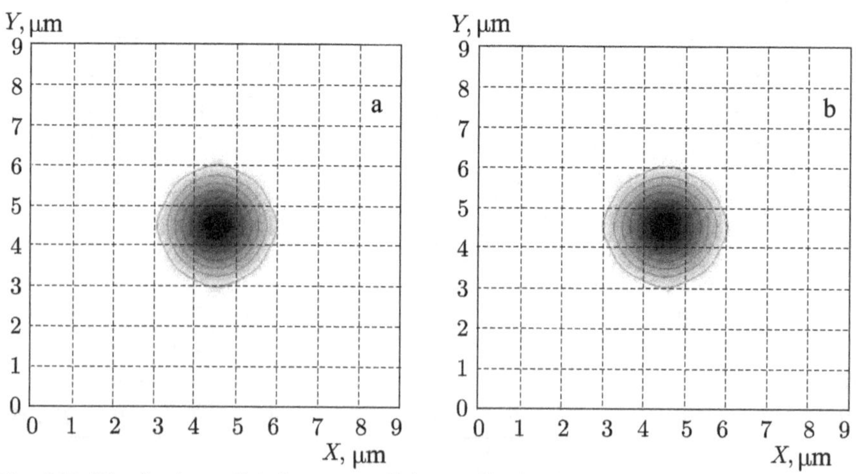

Fig. 6.12. Distributions of a) the square of the amplitude of the fundamental component of the vector mode cB_y, and b) the intensity of the corresponding scalar mode.

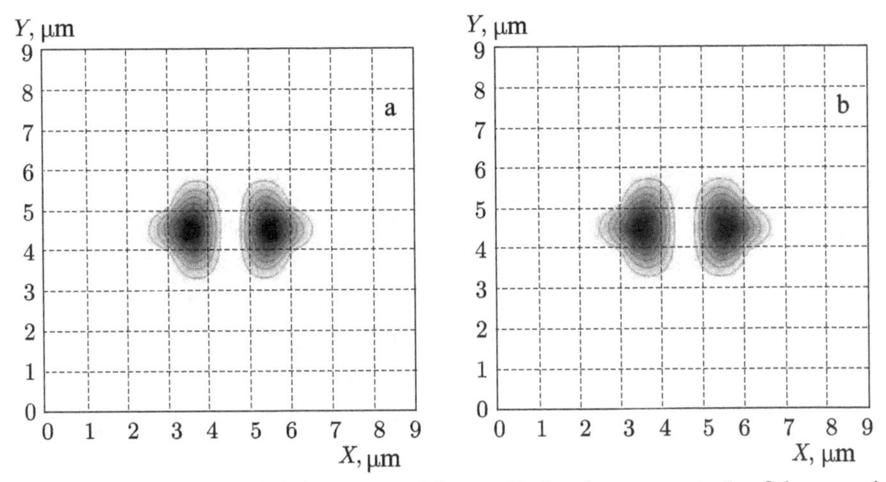

Fig. 6.13. Distributions of a) the square of the amplitude of component cB_y of the second vector mode, and b) the intensity of the corresponding scalar mode.

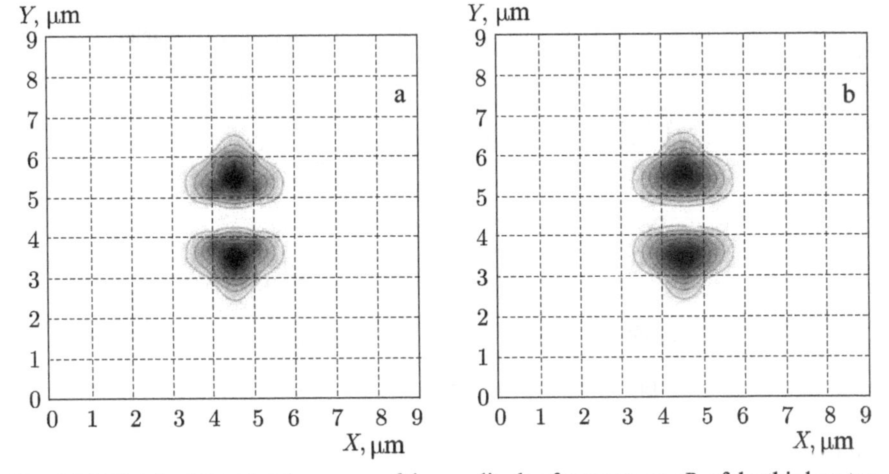

Fig. 6.14. Distributions of a) the square of the amplitude of component cB_y of the third vector mode, b) the intensity of the corresponding scalar modes

6.1.6. Calculation of modes using Fimmwave software

Commercial software simulation of light propagation FIMMWAVE v.4.6 http://www.photond.com/products/fimmwave.htm allows the calculation of dielectric waveguides with modes of an arbitrary cross section uniform in the longitudinal direction, including the PCF, through the implementation of the vector method of matched sinusoidal modes (FMM Solver (real)) and the effective index method (Eff. Idx. Solver (real)).

For the model of the PCF, whose cross section is shown in Fig. 6.5, and the wavelength $\lambda = 1.3$ μm in FIMMWAVE with FMM Solver (real), with a number of local y-modes is equal to sixty, was obtained from the fundamental mode effective index of 1.4477.

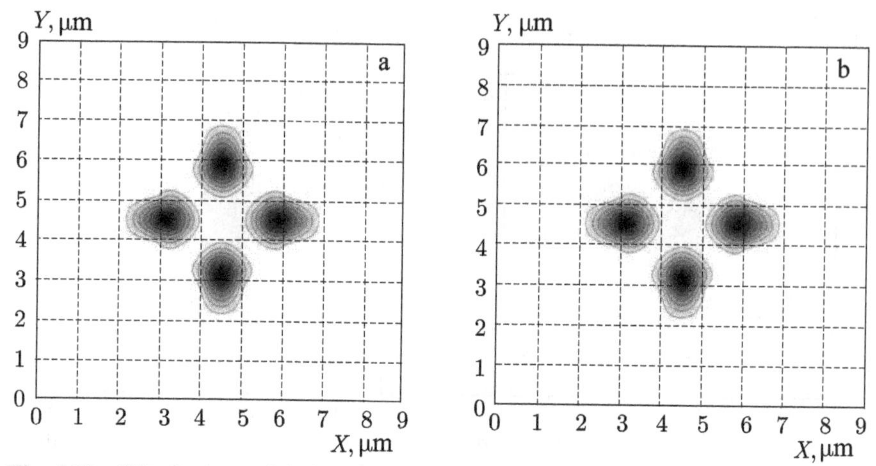

Fig. 6.15. Distributions of a) the square of the amplitude of component cB_y of the fourth vector mode, b) the intensity of the corresponding scalar modes.

Table 6.5. The values of the integrals of the square of the amplitude component of the first four modes

F	$I(\lvert F_1\rvert^2)$	$I(\lvert F_2\rvert^2)$	$I(\lvert F_3\rvert^2)$	$I(\lvert F_4\rvert^2)$
E_x	4.7671	4.7263	0.2488	5.9050
E_y	0.0662	0.0421	4.8538	0.2406
E_z	0.1364	0.2805	0.2899	0.9118
cB_x	0.2799	0.1442	10.6741	1.7101
cB_y	10.1416	10.3738	0.4344	11.1618
cB_z	0.2481	0.2208	0.3118	1.8900

The results obtained in [51] using MSM and FD (finite difference) techniques in the Matlab medium for this model are compared with the results of FIMMWAVE in Table 6.6.

As shown in Table 6.6, the calculated values differ from the results FMM Solver (real) not more than one-tenth of one percent. The intensity distribution of modes (Fig. 6.16), resulting in FIMMWAVE also agrees well with the intensity distribution of the scalar modes and the main components of the vector shown in Fig. 6.12.

In the accompanying documentation FIMMWAVE, recommended minimum number of local y-modes in the calculation using the FMM Solver (real) equal to thirty, which is consistent with studies of convergence of the method carried out in [52].

Table 6.6. Absolute and relative deviation of the effective index of fundamental mode of the PCF (Fig. 6.5), calculated by different methods, from the result obtained by FIMMWAVE

Method	The effective index n_{eff}	The absolute deviation Δ	The relative deviation δ
FD method[1], $n_x \times n_y$ =5252	1.4480	0.0003	0.02%
The MSM-method (vector)	1.4473	0.0004	0.03%
The MSM-method (scalar)	1.4491	0.0014	0.10%
FMM Solver (real)	1.4477	0	0

[1]FD-method will be discussed in section 6.2

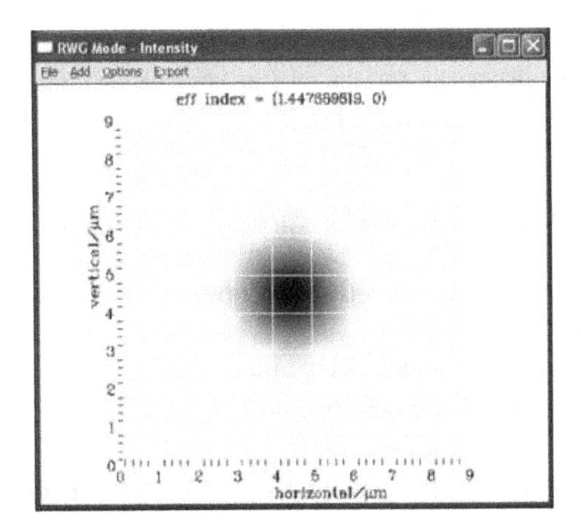

Fig. 6.16. The intensity distribution of the fundamental mode of the PCF model (Fig. 6.5), resulting in FIMMWAVE software.

We compare two implementations of MSM methods: proposed in [50] and the commercial software FIMMWAVE – the convergence of the relative number of local modes. Changing the relative errors in the calculation of one of the output parameters – constant propagation, the number of local modes in the range of 10 to 60 for the PCF (Fig. 6.5) is shown in Fig. 6.17.

From a comparison of plots of the relative errors in the calculation of the propagation constants of the local modes for the two implementations of the SMS-method that was proposed in [50] implementation in Matlab provides a more stable and monotonic convergence, as well as a significantly lower value of the error with a small number of local modes than a commercial program FIMMWAVE.

6.2. Calculation of modes of photonic-crystal light guides by the finite difference method

The basis of the method considered in this chapter is the approach proposed in [11], where sufficiently detailed calculations of solutions of the stationary wave equation

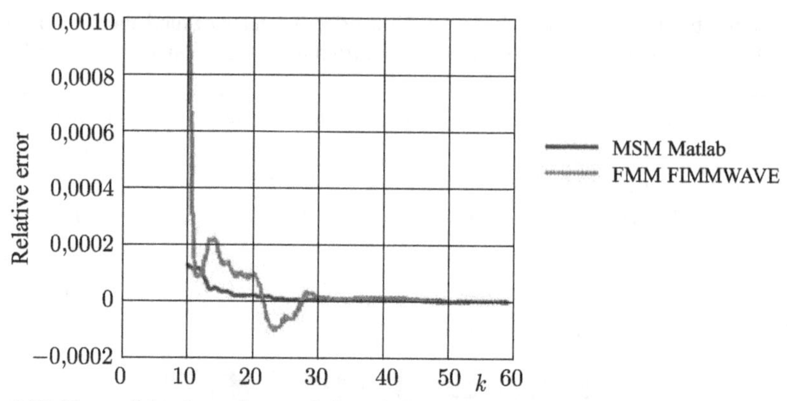

Fig. 6.17. Plots of the dependence of the relative error of the propagation constants of the fundamental mode of the PCF (Fig. 6.5) on the number of local modes for the two variants of the MSM method.

for the electric field component were carried out by the finite-difference approximations. Appropriate formulation for the calculation of the magnetic component can be found in [55, 56]. This section provides formulas for the two components of the electromagnetic field. Particular attention is paid to a full description of the matrix of the difference problem for the eigenvalues and eigenvectors for the intensity of the electric component of the electromagnetic field [57].

6.2.1. A difference method for calculating the modes for electric fields

Consider the homogeneous wave equations for monochromatic radiation in a dielectric medium without sources [11, 58]:

$$\nabla^2 \bar{E} + \nabla\left(\nabla \ln n^2 \cdot \bar{E}\right) + n^2 k_0^2 \bar{E} = 0, \tag{6.90}$$

$$\nabla^2 \bar{H} - (\nabla \times \bar{H}) \times \nabla \ln n^2 + n^2 k_0^2 \bar{H} = 0, \tag{6.91}$$

where n is the refractive index, which depends on the transverse coordinates (x,y); $k_0 = 2\pi / \lambda_0$ is the wave number in vacuum; λ_0 is the wavelength of radiation.

Next, we apply equation (6.90) for optical fibres, which are homogeneous along the longitudinal axis z. In this case the electric field component can be represented as $\bar{E}(x, y, z) = \bar{E}(x, y) \exp(-ik_z z)$, where $\bar{E}(x, y)$ is the vector of the strength of the electric component of electromagnetic field in the transverse plane, k_z is the propagation constant. Taking into account the invariance of the optical fibre along the longitudinal axis z and the equalities $\dfrac{\partial \ln n^2}{\partial z} = 0$, $\dfrac{\partial \bar{E}}{\partial z} = -ik_z \bar{E}$, the vector equation (6.90) can be represented in the matrix form:

$$\begin{bmatrix} P_{xx} & P_{xy} \\ P_{yx} & P_{yy} \end{bmatrix} \begin{bmatrix} E_x \\ E_y \end{bmatrix} = k_z^2 \begin{bmatrix} E_x \\ E_y \end{bmatrix}, \tag{6.92}$$

where the continuous differential operators are defined as follows:

$$P_{xx}E_x = \frac{\partial}{\partial x}\left[\frac{\partial(\ln n^2 \cdot E_x)}{\partial x}\right] + \frac{\partial^2 E_x}{\partial y^2} + n^2 k_0^2 E_x, \tag{6.93}$$

$$P_{yy}E_y = \frac{\partial^2 E_y}{\partial x^2} + \frac{\partial}{\partial y}\left[\frac{\partial(\ln n^2 \cdot E_y)}{\partial y}\right] + n^2 k_0^2 E_y, \tag{6.94}$$

$$P_{xy}E_y = \frac{\partial}{\partial x}\left[\frac{\partial(\ln n^2 \cdot E_y)}{\partial y}\right] - \frac{\partial^2 E_y}{\partial x \partial y}, \tag{6.95}$$

$$P_{yx}E_x = \frac{\partial}{\partial y}\left[\frac{\partial(\ln n^2 \cdot E_x)}{\partial x}\right] - \frac{\partial^2 E_x}{\partial y \partial x}. \tag{6.96}$$

Replacing the continuous differential operators by finite-difference ones and using the following approximation:

$$\left.\frac{\partial f(x,y)}{\partial x}\right|_{x_i} = \frac{f(x_{i+1},y) - f(x_{i-1},y)}{2h_x}, \tag{6.97}$$

$$\left.\frac{\partial f(x,y)}{\partial y}\right|_{y_j} = \frac{f(x,y_{j+1}) - f(x,y_{j-1})}{2h_y}, \tag{6.98}$$

$$\left.\frac{\partial^2 f(x,y)}{\partial x^2}\right|_{x_i} = \frac{f(x_{i-1},y) - 2f(x_i,y) + f(x_{i+1},y)}{h_x^2}, \tag{6.99}$$

$$\left.\frac{\partial^2 f(x,y)}{\partial y^2}\right|_{y_j} = \frac{f(x,y_{j-1}) - 2f(x,y_j) + f(x,y_{j+1})}{h_y^2}, \tag{6.100}$$

$$\left.\frac{\partial f(x,y)}{\partial x \partial y}\right|_{x_i y_j} = \frac{[f(x_{i+1},y_{j+1}) - f(x_{i+1},y_{j-1})] - [f(x_{i-1},y_{j+1}) - f(x_{i-1},y_{j-1})]}{4h_x h_y}, \tag{6.101}$$

we obtain a linear algebraic eigenvalue problem with respect to the square of the propagation constant:

$$RE_t^h = k_z^2 E_t^h, \tag{6.102}$$

The dimensions of the problem (6.102) are $2M \times 2\,M$, where $M = n_x n_y$, here n_x, n_y is the number of nodes in a uniform rectangular grid on the relevant Cartesian axes.

The matrix of (6.102) has the following structure:

$$R = \begin{bmatrix} \tilde{A}_1^1 & B_1^2 & O & & O & O & \tilde{A}_2^{n_x+1} & B_2^{n_x+2} & O & & O & O \\ C_1^1 & A_1^2 & B_1^3 & & \vdots & \vdots & C_2^{n_x+1} & A_2^{n_x+2} & B_2^{n_x+3} & & \vdots & \vdots \\ O & C_1^2 & A_1^3 & & O & O & O & C_2^2 & A_2^{n_x+3} & & O & O \\ O & O & C_1^3 & \ddots & B_1^{n_x-1} & O & O & O & C_2^{n_x+3} & \ddots & B_2^{2n_x-1} & O \\ \vdots & \vdots & \vdots & & A_1^{n_x-1} & B_1^{n_x} & \vdots & \vdots & \vdots & & A_2^{2n_x-1} & B_2^{2n_x} \\ O & O & O & & C_1^{n_x-1} & \tilde{A}_1^{n_x} & O & O & O & & C_2^{2n_x-1} & \tilde{A}_2^{2n_x} \\ \tilde{A}_3^1 & B_3^2 & O & & O & O & \tilde{A}_4^{n_x+1} & B_4^{n_x+2} & O & & O & O \\ C_3^1 & O & B_3^3 & & \vdots & \vdots & C_4^{n_x+1} & A_4^{n_x+2} & B_4^{n_x+3} & & \vdots & \vdots \\ O & C_3^2 & O & & O & O & O & C_4^{n_x+2} & A_4^{n_x+3} & & O & O \\ O & O & C_3^3 & \ddots & B_3^{n_x-1} & O & O & O & C_4^{n_x+3} & \ddots & B_4^{2n_x-1} & O \\ \vdots & \vdots & \vdots & & O & B_3^{n_x} & \vdots & \vdots & \vdots & & A_4^{2n_x-1} & B_4^{2n_x} \\ O & O & O & & C_3^{n_x-1} & \tilde{A}_3^{n_x} & O & O & O & & C_4^{2n_x-1} & \tilde{A}_4^{2n_x} \end{bmatrix}, \qquad (6.103)$$

where O is the zero matrix with the dimension $n_y n_y$. The exact formulas for calculating the non-zero elements of the matrix (6.103) are presented below.

Everywhere below in the calculation of discrete values of permittivity and its derivative we used the approximation

$$\varepsilon_{\text{discr}}^{(i,j)} = \varepsilon(x_i, y_j) + \frac{\varepsilon(x_{i+1}, y_j) + \varepsilon(x_{i-1}, y_j) + \varepsilon(x_i, y_{j+1}) + \varepsilon(x_i, y_{j-1}) - 4\varepsilon(x_i, y_j)}{24}. \qquad (6.104)$$

Elements of the upper-left quadrant of the matrix (6.103), marked by unity in the lower index, are defined as follows:

$$\tilde{A}_1^1 = \begin{bmatrix} T_2^1 + T_y + T_4^1 & T_y & 0 & & 0 & 0 \\ T_y & T_2^1 + T_4^1 & T_y & & \vdots & \vdots \\ 0 & T_y & T_2^1 + T_4^1 & & 0 & 0 \\ 0 & 0 & T_y & \ddots & T_y & 0 \\ \vdots & \vdots & \vdots & & T_2^1 + T_4^1 & T_y \\ 0 & 0 & 0 & & T_y & T_2^1 + T_4^1 + T_y \end{bmatrix}_{n_y n_y}, \qquad (6.105)$$

$$A_1^i = \begin{bmatrix} T_y + T_4^i & T_y & 0 & & 0 & 0 \\ T_y & T_4^i & T_y & & \vdots & \vdots \\ 0 & T_y & T_4^i & & 0 & 0 \\ 0 & 0 & T_y & \ddots & T_y & 0 \\ \vdots & \vdots & \vdots & & T_4^i & T_y \\ 0 & 0 & 0 & & T_y & T_4^i + T_y \end{bmatrix}_{n_y n_y}, \quad i = 2,3,\ldots,n_x - 1, \tag{6.106}$$

$$\tilde{A}_1^{n_x} = \begin{bmatrix} T_y + T_4^{n_x} + T_5^{n_x} & T_y & 0 & & 0 & 0 \\ T_y & T_2^{n_x} + T_4^{n_x} + T_5^{n_x} & T_y & & \vdots & \vdots \\ 0 & T_y & T_2^{n_x} + T_4^{n_x} + T_5^{n_x} & & 0 & 0 \\ 0 & 0 & T_y & \ddots & T_y & 0 \\ \vdots & \vdots & \vdots & & T_2^{n_x} + T_4^{n_x} + T_5^{n_x} & T_y \\ 0 & 0 & 0 & & T_y & T_4^{n_x} + T_5^{n_x} + T_y \end{bmatrix}_{n_y n_y}, \tag{6.107}$$

$$B_1^i = \mathrm{diag}\left\{T_5^i\right\}_{n_y}, \quad i = 1,2,\ldots,n_x, \tag{6.108}$$

$$C_1^i = \mathrm{diag}\left\{T_2^i\right\}_{n_y}, \quad i = 1,2,\ldots,n_x, \tag{6.109}$$

Let (l_1, l_2) be the position of T_j^i in the matrix X_j^i, where $X = A, B, C$, $i = 1, 2,\ldots, n_x$, and let its position in the resulting matrix (6.103) is (g_1, g_2) then the relationship between them is expressed as

$$\begin{cases} g_1 = (i-1)n_y + l_1, \\ g_2 = (i-1)n_y + l_2. \end{cases} \tag{6.110}$$

In (6.105)–(6.109):

$$T_2^i = -a_x(i_x,i_y)e(i_x-1,i_y)/2/h_x + e(i_x-1,i_y)/e(i_x,i_y)/h_x^2, \tag{6.111}$$

where $\begin{cases} g_1 = (i_x-1)n_y + i_y, \\ g_2 = (i_x-2)n_y + i_y. \end{cases}$

$$T_4^i = e(i_x,i_y)k_0^2 - 2/h_x^2 - 2/h_y^2, \tag{6.112}$$

where $\begin{cases} g_1 = (i_x-1)n_y + i_y, \\ g_2 = (i_x-1)n_y + i_y. \end{cases}$

$$T_5^i = a_x(i_x, i_y)e(i_x+1, i_y)/2/h_x + e(i_x+1, i_y)/e(i_x, i_y)/h_x^2, \quad (6.113)$$

where $\begin{cases} g_1 = (i_x - 1)n_y + i_y, \\ g_2 = i_x n_y + i_y. \end{cases}$

Elements of the right lower quadrant of the matrix (6.103), designated by the number 4 in the subscript, are given by:

$$\tilde{A}_4^{n_x+1} = \begin{bmatrix} T_x + V_3^{n_x+1} + T_4^{n_x+1} & V_4^{n_x+1} & 0 & & 0 & 0 \\ V_3^{n_x+1} & T_x + T_4^{n_x+1} & V_4^{n_x+1} & & \vdots & \vdots \\ 0 & V_3^{n_x+1} & T_x + T_4^{n_x+1} & & 0 & 0 \\ 0 & 0 & V_3^{n_x+1} & \ddots & V_4^{n_x+1} & 0 \\ \vdots & \vdots & \vdots & & T_x + T_4^{n_x+1} & V_4^{n_x+1} \\ 0 & 0 & 0 & & V_3^{n_x+1} & T_x + T_4^{n_x+1} + V_6^{n_x+1} \end{bmatrix}_{n_y n_y}, \quad (6.114)$$

$$A_4^{n_x+i} = \begin{bmatrix} V_3^{n_x+i} + T_4^{n_x+i} & V_4^{n_x+i} & 0 & & 0 & 0 \\ V_3^{n_x+i} & T_4^{n_x+i} & V_4^{n_x+i} & & \vdots & \vdots \\ 0 & V_3^{n_x+i} & T_4^{n_x+i} & & 0 & 0 \\ 0 & 0 & V_3^{n_x+i} & \ddots & V_4^{n_x+i} & 0 \\ \vdots & \vdots & \vdots & & T_4^{n_x+i} & V_4^{n_x+i} \\ 0 & 0 & 0 & & V_3^{n_x+i} & T_4^{n_x+i} + V_4^{n_x+i} \end{bmatrix}_{n_y n_y}, \quad i = 2,3,\ldots,n_x - 1, \quad (6.115)$$

$$\tilde{A}_4^{2n_x} = \begin{bmatrix} V_3^{2n_x} + T_4^{2n_x} + T_x & V_4^{2n_x} & 0 & & 0 & 0 \\ V_3^{2n_x} & T_4^{2n_x} + T_x & V_4^{2n_x} & & \vdots & \vdots \\ 0 & V_3^{2n_x} & T_4^{2n_x} + T_x & & 0 & 0 \\ 0 & 0 & V_3^{2n_x} & \ddots & V_4^{2n_x} & 0 \\ \vdots & \vdots & \vdots & & T_4^{2n_x} + T_x & V_4^{2n_x} \\ 0 & 0 & 0 & & V_3^{2n_x} & T_4^{2n_x} + T_x + V_6^{2n_x} \end{bmatrix}_{n_y n_y}, \quad (6.116)$$

$$B_4^{n_x+i} = \operatorname{diag}\{T_x\}_{n_y}, \quad i = 1,2,\ldots,n_x, \quad (6.117)$$

$$C_4^{n_x+i} = \operatorname{diag}\{T_x\}_{n_y}, \quad i = 1,2,\ldots,n_x, \quad (6.118)$$

Let (l_1, l_2) be the position of Y_j^i, $Y = T, V$ in the matrix $X_4^{n_x+i}$, where $X = \mathrm{A}$, $i = 1,2,\ldots, n_x$, and let its position in the resulting matrix (6.103) be (g_1, g_2), then the relationship between them is expressed as

$$\begin{cases} g_1 = n_x n_y + (i-1)n_y + l_1, \\ g_2 = n_x n_{y+}(i-1)n_y + l_2. \end{cases} \tag{6.119}$$

In (6.114)–(6.116):

$$V_3^{n_x+i} = -a_y(i_x,i_y)e(i_x,i_y-1)/2/h_y + e(i_x,i_y-1)/e(i_x,i_y)/h_y^2, \tag{6.120}$$

where $\begin{cases} g_1 = n_x n_y + (i_x-1)n_y + i_y, \\ g_2 = n_x n_y + (i_x-1)n_y + i_y - 1. \end{cases}$

$$T_4^{n_x+i} = e(i_x,i_y)k_0^2 - 2/h_x^2 - 2/h_y^2, \tag{6.121}$$

where $\begin{cases} g_1 = n_x n_y + (i_x-1)n_y + i_y, \\ g_2 = n_x n_y + (i_x-1)n_y + i_y. \end{cases}$

$$V_6^{n_x+i} = a_y(i_x,i_y)e(i_x,i_y+1)/2/h_y + e(i_x,i_y+1)/e(i_x,e_y)/h_y^2, \tag{6.122}$$

where $\begin{cases} g_1 = n_x n_y + (i_x-1)n_y + i_y, \\ g_2 = n_x n_y + (i_x-1)n_y + i_y + 1. \end{cases}$

Elements of the right upper quadrant of the matrix (6.103), designated by number 2 in the subscript, are given by:

$$\tilde{A}_2^{n_x+1} = \begin{bmatrix} T_1^{n_x+1}+U_3^{n_x+1} & U_6^{n_x+1}+T_9^{n_x+1} & 0 & & 0 & 0 \\ T_1^{n_x+1}+U_3^{n_x+1} & 0 & U_6^{n_x+1}+T_9^{n_x+1} & & \vdots & \vdots \\ 0 & T_1^{n_x+1}+U_3^{n_x+1} & 0 & & 0 & 0 \\ 0 & 0 & T_1^{n_x+1}+U_3^{n_x+1} & \ddots & U_6^{n_x+1}+T_9^{n_x+1} & 0 \\ \vdots & \vdots & \vdots & & 0 & U_6^{n_x+1}+T_9^{n_x+1} \\ 0 & 0 & 0 & & T_1^{n_x+1}+U_3^{n_x+1} & U_6^{n_x+1}+T_9^{n_x+1} \end{bmatrix}_{n_y n_y}, \tag{6.123}$$

$$A_2^{n_x+i} = \begin{bmatrix} U_3^{n_x+i} & U_6^{n_x+i} & 0 & & 0 & 0 \\ U_3^{n_x+i} & 0 & U_6^{n_x+i} & & \vdots & \vdots \\ 0 & U_3^{n_x+i} & 0 & & 0 & 0 \\ 0 & 0 & U_3^{n_x+i} & \ddots & U_6^{n_x+i} & 0 \\ \vdots & \vdots & \vdots & & 0 & U_6^{n_x+i} \\ 0 & 0 & 0 & & U_3^{n_x+i} & U_6^{n_x+i} \end{bmatrix}_{n_y n_y}, \quad i = 2,3,\ldots,n_x-1, \tag{6.124}$$

$$\tilde{A}_2^{2n_x} = \begin{bmatrix} U_3^{2n_x}+T_8^{2n_x} & U_6^{2n_x}+T_7^{2n_x} & 0 & & 0 & 0 \\ U_3^{2n_x}+T_8^{2n_x} & 0 & U_6^{2n_x}+T_7^{2n_x} & & \vdots & \vdots \\ 0 & U_3^{2n_x}+T_8^{2n_x} & 0 & & 0 & 0 \\ 0 & 0 & U_3^{2n_x}+T_8^{2n_x} & \ddots & U_6^{2n_x}+T_7^{2n_x} & 0 \\ \vdots & \vdots & 0 & & 0 & U_6^{2n_x}+T_7^{2n_x} \\ 0 & 0 & 0 & & U_3^{2n_x}+T_8^{2n_x} & U_3^{2n_x}+T_8^{2n_x} \end{bmatrix}_{n_y n_y}, \tag{6.125}$$

$$B_2^{n_x+i} = \begin{bmatrix} T_8^{n_x+i} & T_7^{n_x+i} & 0 & & 0 & 0 \\ T_8^{n_x+i} & 0 & T_7^{n_x+i} & & \vdots & \vdots \\ 0 & T_8^{n_x+i} & 0 & & 0 & 0 \\ 0 & 0 & T_8^{n_x+i} & \ddots & T_7^{n_x+i} & 0 \\ \vdots & \vdots & \vdots & & 0 & T_7^{n_x+i} \\ 0 & 0 & 0 & & T_8^{n_x+i} & T_8^{n_x+i} \end{bmatrix}_{n_y n_y}, \quad i = 1, 2, \ldots, n_x, \tag{6.126}$$

$$C_2^{n_x+i} = \begin{bmatrix} T_1^{n_x+i} & T_9^{n_x+i} & 0 & & 0 & 0 \\ T_1^{n_x+i} & 0 & T_9^{n_x+i} & & \vdots & \vdots \\ 0 & T_1^{n_x+i} & 0 & & 0 & 0 \\ 0 & 0 & T_1^{n_x+i} & \ddots & T_9^{n_x+i} & 0 \\ \vdots & \vdots & \vdots & & 0 & T_9^{n_x+i} \\ 0 & 0 & 0 & & T_1^{n_x+i} & T_1^{n_x+i} \end{bmatrix}_{n_y n_y}, \quad i = 1, 2, \ldots, n_x, \tag{6.127}$$

Let (l_1, l_2) be the position Y_j^i, $Y = T, U$ in the matrix $X_2^{n_x+i}$, where $X = A, B, C$, $i = 1, 2, \ldots, n_x$, and let its position in the resulting matrix be (g_1, g_2) then the relationship between them is expressed as

$$\begin{cases} g_1 = (i-1)n_y + l_1, \\ g_2 = n_x n_y + (i-1)n_y + l_2. \end{cases} \tag{6.128}$$

In (6.123)–(6.127):

$$T_1^{n_x+i} = \left[e(i_x-1, i_y-1)/(e(i_x, i_y)-1)\right]/4/h_x/h_y, \tag{6.129}$$

where $\begin{cases} g_1 = (i_x-1)n_y + i_y, \\ g_2 = n_x n_y + (i_x-2)n_y + i_y - 1. \end{cases}$

$$U_3^{n_x+i} = -a_x(i_x,i_y)e(i_x,i_y-1)/2/h_y, \tag{6.130}$$

where $\begin{cases} g_1 = (i_x-1)n_y + i_y, \\ g_2 = n_x n_y + (i_x-1)n_y + i_y - 1. \end{cases}$

$$U_6^{n_x+i} = a_x(i_x,j_y)e(i_x,i_y+1)/2/h_y, \tag{6.131}$$

where $\begin{cases} g_1 = (i_x-1)n_y + i_y, \\ g_2 = n_x n_y + (i_x-1)n_y + i_y + 1. \end{cases}$

$$T_7^{n_x+i} = \left[e(i_x+1,i_y+1)/(e(i_x,i_y)-1\right]/4/h_x/h_y, \tag{6.132}$$

where $\begin{cases} g_1 = (i_x-1)n_y + i_y, \\ g_2 = n_x n_y + i_x n_y + i_y + 1. \end{cases}$

$$T_8^{n_x+i} = \left[(1-e(i_x+1,i_y-1))/e(i_x,i_y)\right]/4/h_x/h_y, \tag{6.133}$$

where $\begin{cases} g_1 = (i_x-1)n_y + i_y, \\ g_2 = n_x n_y + i_x n_y + i_y - 1. \end{cases}$

$$T_9^{n_x+i} = \left[(1-e(i_x-1,i_y+1))/e(i_x,i_y)\right]/4/h_x/h_y, \tag{6.134}$$

where $\begin{cases} g_1 = (i_x-1)n_y + i_y, \\ g_2 = n_x n_y + (i_x-2)n_y + i_y + 1. \end{cases}$

Elements of the left lower quadrant of the matrix (6.103), designated by number 3 in the subscript, are given by:

$$\tilde{A}_3^1 = \begin{bmatrix} T_1^1 + U_2^1 & T_9^1 & 0 & & 0 & 0 \\ T_1^1 & U_2^1 & T_9^1 & & \vdots & \vdots \\ 0 & T_1^1 & U_2^1 & & 0 & 0 \\ 0 & 0 & T_1^1 & \ddots & T_9^1 & 0 \\ \vdots & \vdots & \vdots & & U_2^1 & T_9^1 \\ 0 & 0 & 0 & & T_1^1 & U_2^1 + T_9^1 \end{bmatrix}_{n_y n_y}, \tag{6.135}$$

$$\tilde{A}_3^{n_x} = \begin{bmatrix} U_5^{n_x}+T_8^{n_x} & T_7^{n_x} & 0 & & 0 & 0 \\ T_8^{n_x} & U_5^{n_x} & T_7^{n_x} & & \vdots & \vdots \\ 0 & T_8^{n_x} & U_5^{n_x} & & 0 & 0 \\ 0 & 0 & T_8^{n_x} & \ddots & T_7^{n_x} & 0 \\ \vdots & \vdots & \vdots & & U_5^{n_x} & T_7^{n_x} \\ 0 & 0 & 0 & & T_8^{n_x} & U_5^{n_x}+T_7^{n_x} \end{bmatrix}_{n_y n_y}, \tag{6.137}$$

$$B_3^i = \begin{bmatrix} U_5^i+T_8^i & T_7^i & 0 & & 0 & 0 \\ T_8^i & U_5^i & T_7^i & & \vdots & \vdots \\ 0 & T_8^i & U_5^i & & 0 & 0 \\ 0 & 0 & T_8^i & \ddots & T_7^i & 0 \\ \vdots & \vdots & \vdots & & U_5^i & T_7^i \\ 0 & 0 & 0 & & T_8^i & U_5^i+T_7^i \end{bmatrix}_{n_y n_y}, \quad i=1,2,\ldots,n_x, \tag{6.138}$$

$$C_3^i = \begin{bmatrix} T_1^i+U_2^i & T_9^i & 0 & & 0 & 0 \\ T_1^i & U_2^i & T_9^i & & \vdots & \vdots \\ 0 & T_1^i & U_2^i & & 0 & 0 \\ 0 & 0 & T_1^i & \ddots & T_9^i & 0 \\ \vdots & \vdots & \vdots & & U_2^i & T_9^i \\ 0 & 0 & 0 & & T_1^i & U_2^i+T_9^i \end{bmatrix}_{n_y n_y}, \quad i=1,2,\ldots,n_x, \tag{6.139}$$

Let (l_1, l_2) be the position Y_j^i, $Y = T, U$ in the matrix X_3^i, where $X = A, B, C$, $i = 1,2,\ldots, n_x$, then its position in the resulting matris is (g_1, g_2) and the relationship between them is expressed as

$$\begin{cases} g_1 = n_x n_y + (i-1)n_y + l_1, \\ g_2 = (i-1)n_y + l_2. \end{cases} \tag{6.140}$$

In (6.135)–(6.139):

$$T_1^i = \left[e(i_x-1, i_y-1)/(e(i_x, i_y)-1)\right]/4/h_x/h_y, \tag{6.141}$$

where $\begin{cases} g_1 = n_x n_y + (i_x-1)n_y + i_y, \\ g_2 = (i_x-2)n_y + i_y - 1. \end{cases}$

$$U_2^i = -a_y(i_x,i_y)e(i_x-1,i_y)/2/h_x, \tag{6.142}$$

where $\begin{cases} g_1 = n_x n_y + (i_x-1)n_y + i_y, \\ g_2 = (i_x-2)n_y + i_y. \end{cases}$

$$U_5^i = a_y(i_x,i_y)e(i_x+1,i_y)/2/h_x, \tag{6.142}$$

where $\begin{cases} g_1 = n_x n_y + (i_x-1)n_y + i_y, \\ g_2 = i_x n_y + i_y. \end{cases}$

$$T_7^i = \left[e(i_x+1,i_y+1)/e(i_x,i_y)-1\right]/4/h_x/h_y, \tag{6.143}$$

where $\begin{cases} g_1 = n_x n_y + (i_x-1)n_y + i_y, \\ g_2 = i_x n_y + i_y + 1. \end{cases}$

$$T_8^i = \left[1-e(i_x+1,i_y-1)/e(i_x,i_y)\right]/4/h_x/h_y, \tag{6.144}$$

where $\begin{cases} g_1 = n_x n_y + (i_x-1)n_y + i_y, \\ g_2 = i_x n_y + i_y - 1. \end{cases}$

$$T_9^i = \left[1-e(i_x-1,i_y+1)/e(i_x,i_y)\right]/4/h_x/h_y, \tag{6.145}$$

where $\begin{cases} g_1 = n_x n_y + (i_x-1)n_y + i_y, \\ g_2 = (i_x-2)n_y + i_y + 1. \end{cases}$

In the above formulas h_x is the step of the sampling grid on axis x, and h_y is the step of the sampling grid along the axis y. Below is the definition of other parameters used in the description of the matrix (6.103).

$$T_x = 1/h_x^2, \tag{6.146}$$

$$T_y = 1/h_y^2, \tag{6.147}$$

$$a_x(i_x,i_y) = \frac{\partial}{\partial x}\left(\frac{1}{n^2(x,y)}\right)\Bigg|_{x=(i_x-1)h_x,\,y=(i_y-1)h_y}, \tag{6.148}$$

$$a_y(i_x, i_y) = \frac{\partial}{\partial y}\left(\frac{1}{n^2(x,y)}\right)\Bigg|_{x=(i_x-1)h_x, y=(i_y-1)h_y}, \tag{6.149}$$

$$e(i_x, i_y) = n^2(x,y)\Big|_{x=(i_x-1)h_x, y=(i_y-1)h_y}. \tag{6.150}$$

A column vector E_t^h is an eigenvector of the problem (6.102) and contains samples of the two transverse components of the electric component of the mode.

Let us consider the structure of the vector

$$E_t^h = \left(E_x^h \quad E_y^h\right)^T, \tag{6.151}$$

Here T denotes transposition, and

$$\begin{cases} E_x^h = (E_1^h \quad E_2^h \quad E_3^h \quad \ldots \quad E_M^h), \\ E_y^h = (E_{M+1}^h \quad E_{M+2}^h \quad E_{M+3}^h \quad \ldots \quad E_{2M}^h). \end{cases} \tag{6.152}$$

Thus, this vector contains samples of both of the resulting grid functions E_x^h and E_y^h.

To find the longitudinal component in the known cross-correlation exists [54]:

$$E_z = -\frac{i}{k_z}\{\nabla_t \cdot \bar{E}_t + (\bar{E}_t \cdot \nabla_t)\ln n^2\}, \tag{6.153}$$

where $\nabla_t = \left(\frac{\partial}{\partial x}, \frac{\partial}{\partial y}\right)^T$, and $\bar{E}_t = \left(E_x, \quad E_y\right)^T$. Thus, using the grid solutions found for the transverse components E_x^h and E_y^h we can determine E_z^h.

Having a matrix of (6.102) we must find its eigenvalues. Each of the eigenvalues is a square of the propagation constant and defines a single mode. The corresponding eigenvector stores in the expanded form two-dimensional sets of values of the two transverse electric components.

6.2.2. The difference method for calculating the modes for magnetic fields

Similar to the procedure used in section 6.2.1, we apply equation (6.91) to an optical fibre invariant with respect to the longitudinal axis, where $\bar{H}(x, y, z) = \bar{H}(x, y)\exp(-ik_z z)$, $\bar{H}(x, y)$ is the vector of the strength of the magnetic component of the electromagnetic field in the transverse plane, k_z is the propagation constant. Given the equality $\frac{\partial \ln n^2}{\partial z} = 0$ and $\frac{\partial \bar{H}}{\partial z} = -ik_z\bar{H}$, we write the matrix form of equation (6.91):

$$\begin{bmatrix} Q_{xx} & Q_{xy} \\ Q_{yx} & Q_{yy} \end{bmatrix} \begin{bmatrix} H_x \\ H_y \end{bmatrix} = k_z^2 \begin{bmatrix} H_x \\ H_y \end{bmatrix}, \tag{6.154}$$

where the corresponding continuous differential operators Q_{ij} have the form:

$$Q_{xx}H_x = \frac{\partial^2 H_x}{\partial x^2} + (1+\ln n^2)\frac{\partial H_x}{\partial y^2} - \frac{\partial}{\partial y}(\ln n^2 \frac{\partial H_x}{\partial y}) + n^2 k_0^2 H_x, \tag{6.155}$$

$$Q_{yy}H_y = (1+\ln n^2)\frac{\partial^2 H_y}{\partial x^2} + \frac{\partial^2 H_y}{\partial y^2} - \frac{\partial}{\partial y}(\ln n^2 \frac{\partial H_y}{\partial y}) + n^2 k_0^2 H_y, \tag{6.156}$$

$$Q_{xy}H_y = \frac{\partial}{\partial y}(\ln n^2 \frac{\partial H_y}{\partial x}) - \ln n^2 \frac{\partial^2 H_y}{\partial y \partial x}, \tag{6.157}$$

$$Q_{yx}H_x = \frac{\partial}{\partial x}(\ln n^2 \frac{\partial H_x}{\partial y}) - \ln n^2 \frac{\partial^2 H_x}{\partial x \partial y}. \tag{6.158}$$

Using finite-difference approximations (6.97)–(6.101), we obtain another linear eigenvalue problem with respect to the square of the propagation constants:

$$SH_t^h = k_z^2 H_t^h. \tag{6.159}$$

The matrix of (6.159) is similar to the matrix (6.103) and the structure of the same size. In view of its cumbersome nature it is not given in explicit form.

The expression for the calculation of continuous longitudinal magnetic component in terms of known cross has the following form [54]:

$$H_z = -\frac{i}{k_z}\nabla_t \cdot \bar{H}_t, \tag{6.160}$$

where $\bar{H}_t = \left(H_x, \ H_y\right)^T$.

As in the MSM-method, in the problems (6.102) and (6.159) we may use two types of boundary conditions, the so-called 'electric' and 'magnetic walls' on the boundaries of the calculation domain. At the boundary of a perfect conductor the the component of the electric vector tangent to the boundary and the component of the magntic vector normal to the boundary should convert to zero. Therefore, in the first case, components E_x and H_y convert on the horizontal boundaries parallel to the x axis, and on the vertical boundaries parallel to the y axis, – E_y and H_x. In the case of the 'magnetic walls' on the contrary, at the horizontal boundaries the component E_y and H_x turn zero and on the vertical boundaries – E_x and H_y.

The problems (6.102) and (6.159) can be solved independently of each other, and it is clear that the values of propagation constants of modes resulting from solving these problems, should coincide with some accuracy.

6.2.3. Calculation of modes of photonic-crystal fibres with a filled core

The object of the study was selected the PCF model, whose cross section is shown in Fig. 6.18. The original scheme of the PCF has been taken from [59] and modified.

This fibre has a solid core with a refractive index $n = 1.46$. The distance between the centres of holes Λ and the wavelength λ was selected to ensure that their ratio λ/Λ is equal to 0.6. The filling coefficient, equal to the ratio of the hole diameter d to the distance between centres of adjacent holes Λ was 0.85. The calculations were performed throughout the entire cross section, where a uniform grid with the size of 200×220 was determined, which led to the construction of the resulting components of the electromagnetic field of the mode with a sampling step of the order of 0.05 Λ. The transverse as well as longitudinal electric and magnetic components of the fundamental mode, obtained by solving (6.102) and (6.159), are shown in Figs. 6.19a and 6.19b, respectively. The solutions of both problems yielded two values of the effective mode index $n_{\text{eff}} = k_z/k_0$. The deviation between the values of the effective index of the fundamental mode in these two cases was 0.008%. Thus, we can accept as the true value of this average $n^{\text{m}}_{\text{eff}} = (n^{\text{el}}_{\text{eff}} + n^{\text{mag}}_{\text{eff}})/2 = 1.4509$.

The view of the mode in Fig. 6.19a agrees qualitatively with the kind of mode, calculated in [59] by another method. In [59], the problem was solved by expanding the transverse components of the strength of the magnetic field into a series of orthogonal Hermite–Gaussian functions.

The graph of the dispersion parameter $D = -(\lambda / c) d^2 n_{\text{eff}} / d\lambda^2$ [11] is shown in Fig. 6.20.

This parameter characterizes the group velocity dispersion of the mode pulse during its propagation in the fibre. Figure 6.20 shows that the optical fibre, shown in Fig. 6.18, has a normal dispersion.

The dispersion curve in Fig. 6.20 agrees qualitatively with the analogous curve obtained in [59], and with the curves obtained in [60]. In [60] dispersion curves were calculated using the MIT Photonic-Bands package. The package, developed by the University of Massachusetts, is an OpenSource application with source code available for UNIX-like operating systems. The method that is embedded in the software is based on an approach similar to the MSM method. In a unit cell, the

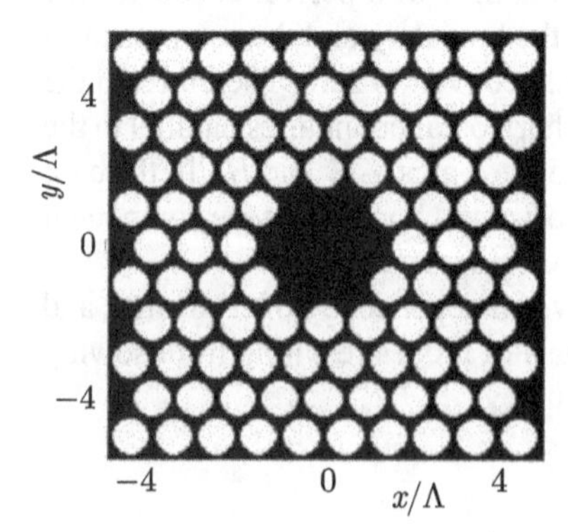

Fig. 6.18. The cross section for the PCF, dark areas correspond to material with a refractive index $n_1 = 1.46$, light – the holes filled with air $n_2 = 1$.

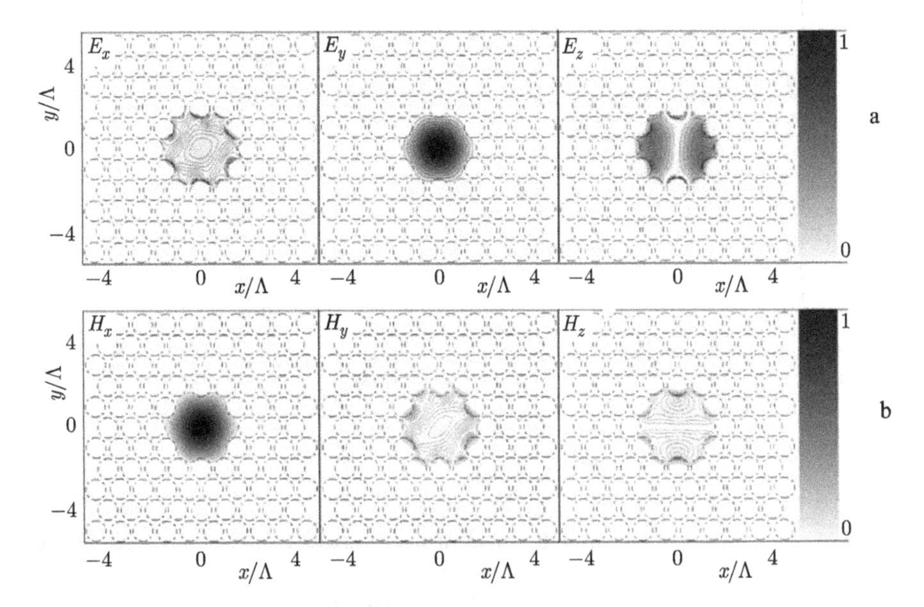

Fig. 6.19. Distribution of the absolute values of the amplitudes: a) the electrical component of the fundamental mode, from left to right: E_x, E_y, E_z, b) the magnetic components of the fundamental mode, from left to right: H_x, H_y, H_z.

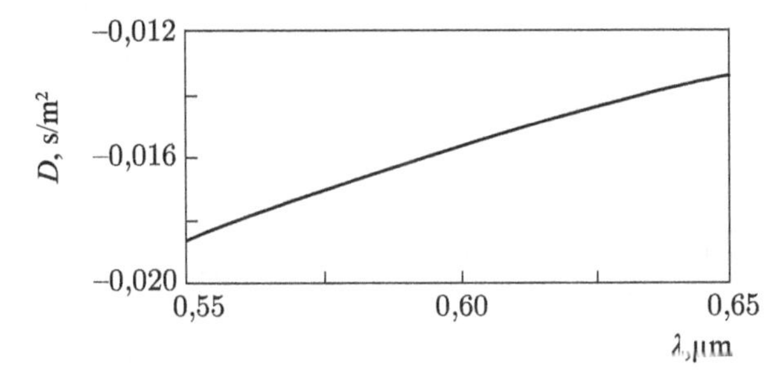

Fig. 6.20. Graph of the dispersion parameter for the PCF model (Fig. 6.18).

solution of Maxwell's equations is an expansion of plane waves in the basis. The dimensions of the problem are determined by the accuracy of an expansion in plane waves and by the number of eigenvalues (eigenfrequencies) or, equivalently, by the number of modes to be found.

For the PCF considered (Fig. 6.18) the distribution of the components of higher-order modes was obtained in [61].

Figures 6.21 and 6.22 show that the modes of higher orders of the PCF do not extend beyond the core, despite the fact that the cladding consists of only 4 rows of round holes located in the 'chess-like' or triangular manner.

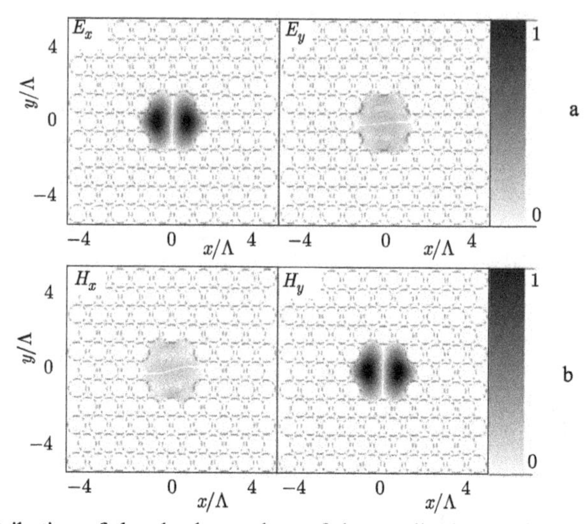

Fig. 6.21. Distribution of the absolute values of the amplitudes: a) the transverse electric components, from left to right: E_x, E_y; b) transverse magnetic components, from left to right: H_x, H_y – higher mode of PCF (Fig. 6.18) with the effective index n^{m}_{eff} = 1.4385.

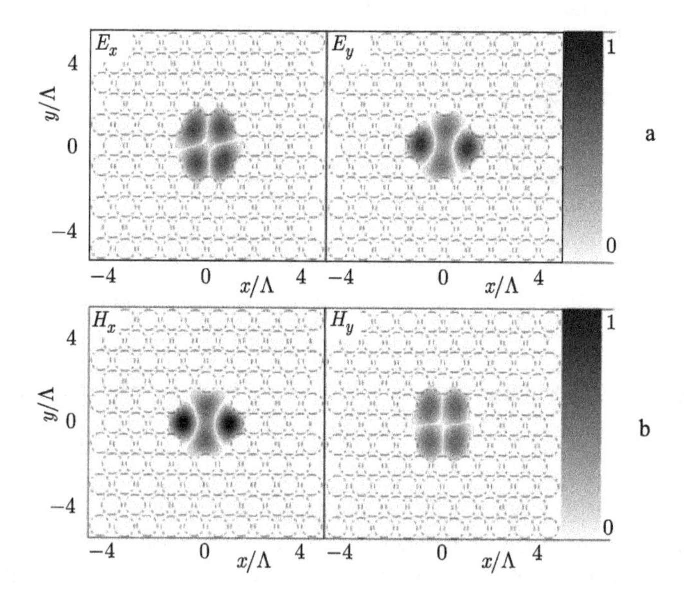

Fig. 6.22. Distribution of the absolute values of the amplitudes: a) the transverse electric components, from left to right: E_x, E_y; b) transverse magnetic components from left to right: H_x, H_y – higher mode of the PCF (Fig. 6.18) with an effective index n^{m}_{eff} = 1.4205.

6.2.4. Calculation of modes of photonic-crystal fibre with hollow core

Using the FD method described above, we calculate the distribution of the transverse components of the electric vector (vector of the strength of the electric field) for the fundamental mode of the hollow PCF, shown in Fig. 6.23a. The distance between the centres of the holes Λ and wavelength λ_0 are selected so that their ratio λ_0/Λ is equal to 0.6. The fill factor, equal to the ratio of diameter of the hole d to the distance between centres of adjacent holes Λ, is 0.85. Let the area of section $10\Lambda \times 11\Lambda$ be defined as the uniform rectangular grid of size 200×220 nodes with a sampling step of 0.05Λ.

The main feature of the PCF with a hollow core consists in the fact that the propagation of light is realized in them by improper modes, i.e. modes with effective indices satisfying the condition $n_{\text{eff}} < 1$. In general, the propagation constant k_z is a complex quantity whose real part is responsible for the dispersion characteristics of the fibre, and the imaginary – for its modal loss. Therefore, the search for the eigenvalues of the matrix in the case of the PCF with a hollow core must be done among the set of complex numbers, and not among the set of real numbers, as was previously discussed in the examples.

The modal loss is measured in dB per unit length (meter), and calculated using the formula [62]:

$$L = 20\log_{10}(\exp(\text{Im}[k_{zj}])) \sim 8.69\,\text{Im}[k_{zj}], \tag{6.161}$$

where k_{zj} is the propagation constant of j-th mode in m^{-1}.

Figure 6.23b shows the transverse electric components of the fundamental mode, calculated for the given model. The share of the intensity of each component, concentrated in the core, is 40% and 60% for E_x and E_y respectively.

A plot of the dependence of the real part of the effective index n_{eff} of the fundamental mode of the ratio λ_0/Λ is shown in Fig. 6.24.

In this case, according to calculations [63], in the range of λ_0/Λ from 0.55 to 0.65 the imaginary part of the complex propagation constants is of the order of $10^{-10} \div 10^{-9}$ m^{-1}, which corresponds to a fairly low for hollow-core fibres [64] energy

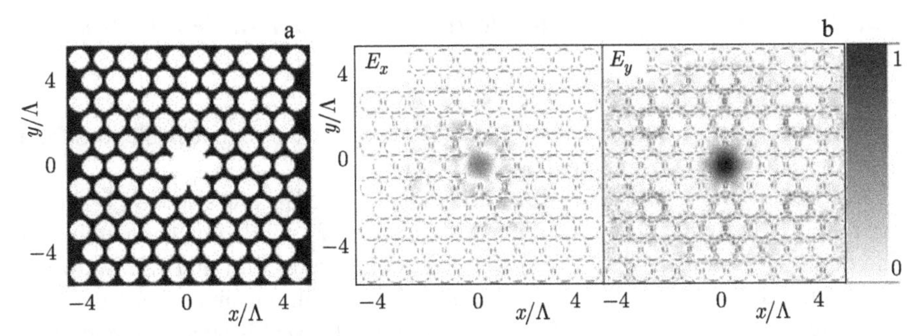

Fig. 6.23. a) Structure of the PCF section of a hollow core, the light areas correspond to the refractive index $n_1 = 1$, dark – $n_2 = 1.46$, b) distribution of the absolute values of the normalized electrical components of the fundamental mode.

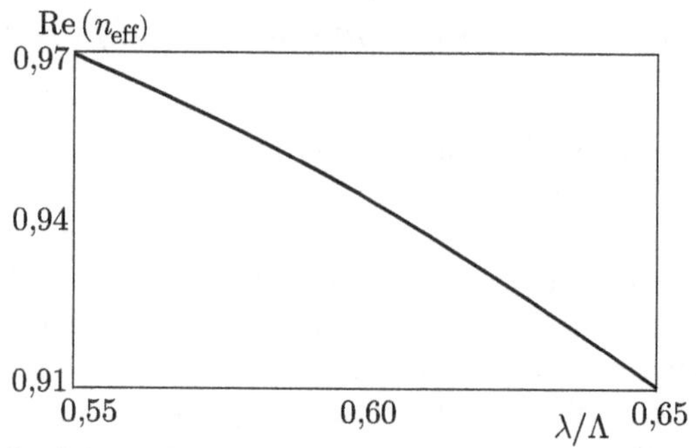

Fig. 6.24. The graph of the real part of the effective index of the fundamental mode for a model of a hollow PCF (Fig. 6.23).

losses of the order of 0.1÷1 dB/km. In particular, for $\Lambda = 1$ μm, and $\lambda_0 = 0.6$ μm the energy loss is 0.1 dB/km, which is less than the fundamental limit for conventional silica fibres of 0.2 dB/km.

6.2.5. Calculation of modes of Bragg fibres

Consider the example of a Bragg fibre with a hollow core (Fig. 6.25a), the mechanism of light direction in which is based on the effect of the phonon band gaps created by the cladding in the form of a one-dimensional photonic crystal with quartz ($n = 1.46$) and air layers ($n = 1$). The geometrical parameters of the fibre were chosen so as to be consistent with values in [17], namely, the radius of the core, the thickness of silica layer and the thickness of the layer of air equalled to 10 μm, 0.28 μm and 2.27 μm respectively. A rectangular grid with a sampling step h = 0.28 μm was set in the 60×60 μm cross section.

The FD method was employed to obtain the intensity distribution of the transverse electric and magnetic field components of the fundamental mode (Fig. 6.25 b, c) for the wavelength $\lambda_0 = 1.55$ μm, and the effective index of the mode was $n_{eff} = 0{,}99 + i7 \cdot 10^{-14}$.

6.2.6. Comparison of the calculation of the waveguide modes by different methods

In order to compare the two methods described above, we present calculations for a pair of models of microstructured optical fibres.

As a first example, consider the PCF, the structure of its cross section is shown in Fig. 6.5. The radiation wavelength was chosen at $\lambda_0 = 1.3$ μm. The number of grid points (FD method) is given equal to 52×52. Figures 6.21 and 6.22 show the distribution of the square of the the modulus of the amplitude and cross section on the Cartesian axes for the scalar field, calculated by the MSM-method (Fig. 6.26a), and the main vector component H_y of the fundamental mode, calculated by the MSM-method (Fig. 6.26b) and FD method (Fig. 6.26c).

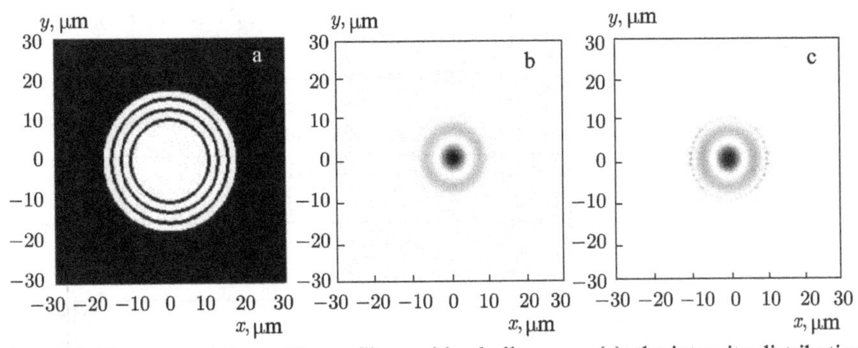

Fig. 6.25. The cross section of Bragg fibres with a hollow core (a); the intensity distribution of the transverse field components of the fundamental mode $\bar{E}_t$ (b) and $\bar{H}_t$ (c).

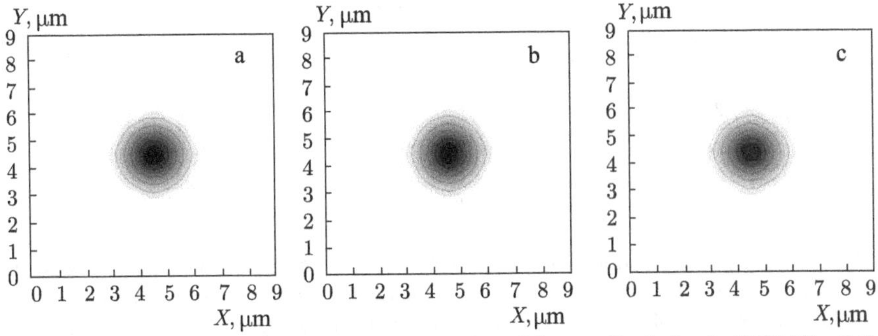

Fig. 6.26. Distribution of the square of the modulus of the amplitude for the PCF (Fig. 6.5): a) the scalar field of the fundamental mode, calculated by MSM method; b) major vector component H_y of the fundamental mode, calculated by MSM method; c) the main vector component H_y of the fundamental mode, calculated by the FD method.

Both methods give good concordant results. The standard deviation between the two solutions, normalized by the square of the modulus of the amplitude per unit, in the region $W_x \times W_y$ is 0.86%. The field obtained by using the MSM method in the scalar approximation describes qualitatively quite accurately the distribution of the main vector components of the fundamental mode. The values of the effective mode index for the considered three methods of calculating are shown in Table 6.7.

We see that the difference in the values obtained by vector methods forms only the third decimal place.

The second example – a Bragg fibre filled with a quartz (n = 1.5) core radius of 1 μm, surrounded by alternating quartz and air nanolayers of the cladding (Fig. 6.28a). Suppose that in the MSM-method the cross section of the fibre is described by a system of 37×37 homogeneous rectangular cells in the area 6.8 μm×6.8 μm. To calculate the fundamental mode of the fibre by the FD method at a wavelength λ_0 = 1.55 μm using two uniform grids with sampling steps on the axes of h = $0.1\lambda_0$ and $h/2 = 0.05\lambda_0$. The distributions of the dominant electric component E_y of the fundamental mode with effective index n_{eff} = 1.4166 obtained by different

Table 6.7. Parameter values for the fundamental mode of the PCF (see Fig. 6.5), calculated by different methods

Parameter	FD method, $n_x \times n_y = 52 \times 52$	MSM-method (vector)	SMS-method (scalar)
k_z, μm^{-1}	6.9985	6.9951	7.0038
n_{eff}	1.4480	1.4473	1.4491

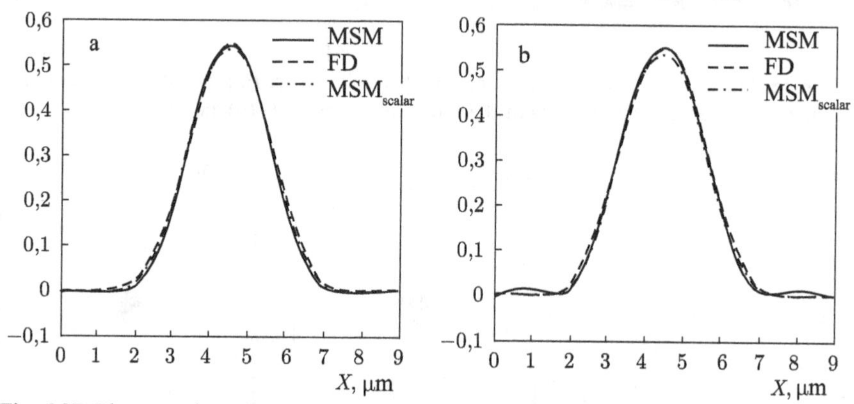

Fig. 6.27. The central section along the axes a) x and b) y of the distributions of the square of the amplitude of the scalar field and the main vector components H_y of the fundamental mode, calculated MSM and FD techniques for the PCF (see Fig. 6.5).

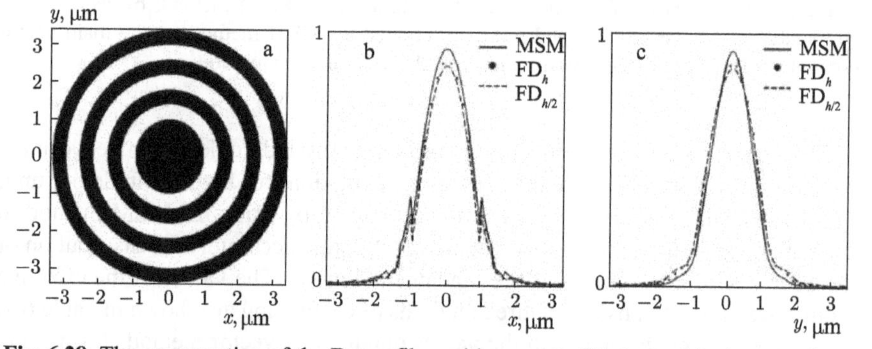

Fig. 6.28. The cross section of the Bragg fibre with a quartz-filled core (a); the central section of the distributions of the component E_y of the fundamental mode along axes x (b) and y (c).

methods agree well (Fig. 6.28b, c). The standard deviation between the MSM and FD solutions does not exceed 2%.

Thus, Figs. 6.27 and 6.28 show that both discussed methods of calculating the modes of optical fibres inhomogeneous in the section (MSM method and FD method) give almost identical results that differ by an average of 1%–2%.

Literature

1. Knight J.C., Birks T.A., Russel P.S.J., Atkin D.M., All-silica single mode optical fibre-with photonic crystal cladding, Opt. Lett. 1996. V. 21 (19). P. 1547–1549.
2. Adams M.J., An Introduction to Optical Waveguides,New York: Wiley, 1981.
3. Yeh P., Yariv A., Marom E. Theory of Bragg fiber, J. Opt. Soc. Am. 1978. V. 68. P. 1196–1201.
4. Ibanescu M., Fink Y., Fan S., Thomas E. L., Joannopoulos J.D. All-dielectric coaxial waveguide, Science. 2000. V. 289. P. 415–419.
5. Cojocaru E. Dispersion analysis of hollow-core modes in ultralarge-band with all-silica Bragg fibers, with nanosupports, Appl. Opt. 2006. V. 45 (9). P. 2039–2045.
6. Foroni M. et al. Confinement loss spectral behavior in hollow-core Bragg fiber, Opt. Lett. 2007. V. 32 (21). P. 3164–3166.
7. Zhelticov A.M. Ray-optic analysis of the (bio) sensing ability of ring-cladding hollow waveguides, Appl. Opt. 2008. V. 47 (3). P. 474-479.
8. Dupuis A. et al. Guiding in the visible with «colorful» solid-core Bragg fiber, Opt. Lett. 2007. V. 32 (19). P. 2882–2884.
9. Fang Q., Despersion design of all-solid photonic bandgap fiber, J. Opt. Soc. Am. A. 2007. V. 24 (11). P. 2899–2905.
10. Ren G., et al., Low-loss all-solid photonic bandgap fiber, Opt. Lett. 2007. V. 32 (9). P. 1023–1025.
11. Yang R., et al., Research of the effects of air hole shape on the properties of microstructured optical fibers, Opt. Eng. 2004. V. 43 (11). P. 2701–2706.
12. Yue Y. et al., Highly birefringent elliptical-hole photonic crystal fiber with squeezed hexagonal lattice, Opt. Lett. 2007. V. 32 (5). P. 469–471.
13. Choi H.-G., et al., Dispersion and birefringence of irregularly microstructured fiber with elliptical core, Appl. Opt. 2007. V. 46 (35). P. 8493-8498.
14. Mafi A., Moloney J.V., Shaping Modes in Multicore Photonic Crystal Fiber. IEEE Photonics Tech. Lett. 2005. V. 17. P. 348-350.
15. Michaille L. et al., Characteristics of a Q-switched multicore photonic crystal fiber laser with a very large mode field area, Opt. Lett. 2008. V. 33 (1). P. 71–73.
16. Szpulak M., et al., Experimental and theoretical investigations of birefringent holey fibers with a triple defect, Appl. Opt. 2005. V. 44. P. 2652–2658.
17. Eguchi M., Tsuji Y., Geometrical birefringence in square-lattice holey fibers having a core consisting of multiple defect, J. Opt. Soc Am. A. 2007. V. 24 (4).
18. Zhang Ch., et al., Design of tunable bandgap guidance in high-index filled microstructure fibers, J. Opt. Soc. Am. A. 2006. V. 23 (4). P. 782-786.
19. Sun J., et al., Refractive index measurement using photonic crystal fiber, Opt. Eng. 2007. V. 46 (1).
20. Sun J., Chan Ch.Ch., Hybrid guiding in liquid-crystal photonic crystal fibers, J. Opt. Soc. Am. A. 2007. V. 24 (10). P. 2640-2646.
21. Larsen T., et al., Optical devices based on liquid crystal photonic bandgap fibres, Opt. Express. 2003. V. 11 (20). P. 2589–2596.
22. Haakestad M.W., et al., Electrically tunable photonic bandgap guidance in a liquid crystal– filled photonic crystal fiber, IEEE Photon. Technol. Lett. 2005. V. 17 (4). P. 819–821.
23. Domachuk P., Nguyen H.C., Eggleton B.J., Transverse probed microfluidic switchable photonic crystal fiber devices, Photon. Technol. Lett. 2004. V. 16 (8). P. 1900–1902.
24. Ferrando A., et al., Nearly zero ultra-flattened dispersion in photonic crystal fibers, Opt. Lett. 2000. V. 25. P. 790–792.
25. Broderick AN.G.R., et al., Modeling Large Air Fraction Holey Optical Fiber, J. Opt. Tech. 2000. V. 18. P. 50–56.
26. Broderick AN.G.R., et al., Nonlinearity in holey optical fibers: measurement and future opportunities, Opt. Lett. 1999. V. 24. P. 1395.

27. White T.P. et al., Multipole method for microstructured optical fibers, J. Opt. Soc. Am. A. 2002. V. 19 (10). P. 2322–2330.
28. Steel M.J., et al., Symmetry and degeneracy in microstructured optical fibers, Opt. Lett. 2001. V. 26. P. 488–490.
29. Yamashita E., Ozeki S., Atsuki K., Modal analysis method for optical fibers with symmetrically distributed multiple cores, J. Lighhtwave Techn. 1985. V. 3. P. 341–346.
30. Tayed G., et al., Scattering by a random set of parallel cylinders, J. Opt. Soc. Am. A. 1994. V. 11. P. 2526–2538.
31. Sudbo A.S., Film mode matching: A versatile method for mode field calculations in dielectric waveguides, Pure Appl. Opt. (J. Europ. Opt. Soc. A). 1993. V. 2. P. 211–233.
32. Cucinotta A., et al., Holey fiber analysis through the finite element method, IEEE Photon. Technol. Lett. 2002. V. 14. P. 1530–1532.
33. Brechet F., et al., Complete analysis of characteristics of propagation into photonic crystal fibers by the finite element method, Opt. Fiber Technol. 2000. V. 6 (2). P. 181–191.
34. Guan N., et al., Boundary Element Method for Analysis of Holey Optical Fibers, J. Lightwave Technol. 2003. V. 21 (8).
35. Cheng H., et al., Fast, accurate integral equation methods for the analysis of photonic crystal fibers, Opt. Express. 2004. V. 12 (16). P. 3791–3805.
36. Riishede J., Mortensen N.S., Legsgaard J., A Poor Man's Approach to Modeling Micro–Structured Optical Fibers, J. Opt. A: Pure Appl. Opt. 2003. V. 5. P. 534–538.
37. Hardley G.R., Smith R.E., Full-vector waveguide modeling using an iterative finite-difference method with transparent boundary conditions, J. Lightwave Technol. 1994. V. 13. P. 465–469.
38. Zhu Z., Brown T.G.. Full-vectorial finite-difference analysis of microstructured optical fibers, Opt. Express. 2002. V. 10 (17). P. 853–864.
39. Jiang W., et al., An Extended FDTD Method With Inclusion of Material Dispersion for the Full-Vectorial Analysis of Photonic Crystal Fibers, J. Lightwave Technol. 2006. V. 24 (11). P. 4417–4423.
40. Xu C.L., et al., Full-vectorial mode calculations by finite difference method, Inst. Electr. Eng. Proc. J. 1994. V. 141. P. 281–286.
41. Huang W.P., et al., The finite–difference vector beam propagation method. Analysis and Assessment, J. Lightwave Technol. 1992. V. 10. P. 295–305.
42. Xu C.L., Efficient and accurate vector mode calculations by beam propagation method, J. Lightwave Technol. 1993. V. 11 (9). P. 1209–1215.
43. Itoh T., Numerical techniques for microwave and millimeter-wave passive structures. New York: Wiley, 1988.
44. Sorrentino R., Transverse resonance technique, Ch. 11 in Itoh's book [43].
45. Schlosser W., Unger H.G., Partially filled waveguides and surface waveguides of rectangular cross section, New York: Advances in Microwaves. Academic Press, 1966.
46. Peng S.T., Oliner A.A., Guidance and leakage properties of a class of open dielectric waveguides. Part I: Mathematical formulations, IEEE Trans. Microwave Theory Techn. 1981. V.MTT–29. P. 843–855.
47. Sudbo A.S., Improved formulation of the film mode matching method for mode field calculations in dielectric waveguides, Pure Appl. Opt. (J. Europ. Opt. Soc. A). 1994. V. 3. P. 381–388.
48. Kotlyar V.V. and Shuyupova Ya.O., Komp. Optika, ISOI RAN, 2007. V. 31. P. 27–30.
49. Ruhe A. A Rational Krylov Algorithm For Nonlinear Matrix Eigenvalue Problems, Zapiski Nauchnih Seminarov, Steklov Mathem. Inst., 2000. V. 268. P. 176–180.
50. Kotlyar V.V. and Shuyupova Ya.O., Vector mode calculation of the optical waveguide modes, Komp. Optika, ISOI RAN, 2005. V. 27. C. 89–94.
51. Kotlyar V.V. and Shuyupova Ya.O., Komp. Optika, ISOI RAN, 2005. V. 27. P. 84–88.
52. Kotlyar V.V. and Shuyupova Ya.O., Komp. Optika, ISOI RAN, 2003. V. 25. C. 41–48.
53. Jarlebring E., Voss H., Rational Krylov for nonlinear eigenproblems, an iterative projection method, Applications of Mathematics. 2005. V. 50. P. 543–554.

54. Snyder A., Love G.,. Optical Waveguide Theory, Moscow, Radio i Svyaz', 1987, 655.
55. Kotlyar V.V., Shuyupova Ya.O., Calculating spatial modes in photonic crystal fibers based on applying finite–difference method to wave equations, Proc. of ICO Topical Meeting on Optoinformatics, Information Photonics 2006, St. Petersburg, Russia, Sept. 4–7, 2006. P. 483–485.
56. Kotlyar V.V., Shuyupova Ya.O., Komp. Optika, ISOI RAN, 2005. V. 28. P. 41–44.
57. Kotlyar V.V., Shuyupova Ya.O., Opt. Zhurnal, 2007. V. 74, N. 9. P. 600–608.
58. Dangui V., Digonnet M.J.F., Kino G.S., A fast and accurate numerical tool to model the modal properties of photonic bandgap fibers, Opt. Express. 2006. V. 14 No. 7. P. 2979–2993.
59. Mogilevtsev D., Birks T.A., Russell P.St.J., Group-velocity dispersion in photonic crystal fibers, Opt. Lett. 1998. V. 23. P. 1662–1664.
60. Sokolov V.O., et al., Kvant. Elektronika, 2006. V. 36, N. 1. P. 67–72.
61. Kotlyar V.V., Shuyupova Ya.O., Opt. Zhurnal, 2007. V. 74, N. 9. P. 600–608.
62. Chen M., Yu R., Design of defect-core in highly birefringent photonic crystal fibers with anisotropic claddings, Opt. Commun. 2006. V. 258, No. 2. P. 164–169.
63. Kotlyar V.V., Shuyupova Ya.O., Izv. SNTs RAN, 2007. V. 9, N. 3. P. 592–597.
64. Sakai J. Optical loss estimation in Bragg fiber, J. Opt. Soc. Am. A. 2007. V. 24 (4).

Chapter 7

Singular optics and superresolution

Objects with a vortex structure exist in the various spheres of the material world, in the macrocosm (the spiral shape of galaxies and nebulae), in the microcosm (elementary particles, optical fields) and in our daily lives (cyclones and anticyclones, tornadoes and typhoons). Their structure and behaviour have still not been exhaustively studied and represent a vast field for research. Thus, in recent years a separate section ('singular optics') was formed by the branch of optics dealing with the study of light beams with screw phase singularities (i.e. the vortex laser beams).

At the point of singularity the intensity of the light field is zero, and the phase is undefined. There are abrupt phase changes in the vicinity of this point.

Singular features in light fields may appear as they pass through randomly inhomogeneous and nonlinear media. It is also possible to excite the vortex fields in laser resonators and multimode optical fibres. The simplest and most controllable method of forming the vortex fields is to use spiral diffractive optical elements (DOE), and dynamic liquid-crystal transparants (energy efficiency of the latter is still quite low). The simplest of such DOEs are spiral phase plates (SPP) and helical axicon (HA).

Vortex laser beams are the subject of numerous studies and publications by Russian scientists and their foreign counterparts. The properties of such beams on the basis of the Bessel, Laguerre–Gaussian (LG), Hermite–Gaussian (HG) modes, etc. are being actively studied [1–3].

The scope of optical vortices is constantly expanding. In particular, in nanophotonics problems it is proposed for use them to manipulate dielectric micro- and nano-objects. For example, a recent paper [4] studied the motion of gold nanoparticles (100 nm to 250 nm), captured in the central part of the optical vortex. Also, more attention is paid to exploring the possibilities of using plasmon effects in nano-tweezers [5, 6].

The use of optical vortices in photolithography can achieve a resolution of $\lambda/10$ (λ is the wavelength of light). It is possible to use efficiently optical spiral structures, even with a small number of quantization levels [7].

Other applications include optical vortices, such as interferometry: using the SPP, placed in the plane of the spatial spectrum of a 4*f*-optical system (*f* is the focal length of the spherical lens) a method was proposed for producing spiral interferograms which can be used to easily distinguish convex and concave sections of the wavefront [8].

Spiral filters are used for contrasting and relief imaging of nanosized phase objects [9].

SPPs are also used in stellar coronagraphs, in which the light from a bright star is converted into the ring and stopped down, and the faint light from the planets of the star passes through the aperture and registers. It is known that vortex waves in a coherent system have a well-defined phase, which, however, are poorly defined in the partially-coherent system. In the limit, for a fully incoherent case neither the helical phase nor zero intensity is observed. This allows the optical vortices to be excluded from the observation of coherent radiation in order to enhance the incoherent signal; this effect is used in coronagraphs [10, 11].

SPPs are also used for the optical realisation of the radial Hilbert transform [12]. Hilbert-optics, as well as shadows optics with transformations taking place in a similar frequency domain, has been used successfully for pre-processing of images and phase analysis. Hilbert spectroscopy allows to achieve nanoresolution for spectral analysis. Using the radial Hilbert transform, including fractional-based transform, SPP opens up new possibilities in solving the problems mentioned above.

Phase dislocations that define zero intensity, represent a promising tool in metrology. Since the accuracy of determining the position of the dislocation is not limited to the classical diffraction limit (the gradient of the phase change in this case tends to infinity), and is limited only by the signal / noise ratio, the geometry of an object subject to the availability of a priori information about the object can be determined with very high accuracy [13]. This approach is based on the method of optics-vortex metrology [14], successfully applied in the optical vortex interferometer, which allows to track objects with nanometer displacement accuracy [15].

The sensitivity of singular beams to wavefront changes and all kinds of defects can be used to test surfaces [16] and for the analysis of optical systems [17].

With the help of optical vortex interferometers, which are based on the generation of light fields, representing regular gratings or grids of optical vortices [18, 19] (i.e. measurements are carried out of the status of nodes with not the maximum but minimum light intensity) we can determine the angles rotation with an accuracy of 0.03 arc seconds [20] and measure the angles of inclination of the wavefront with an accuracy of 0.2 arc seconds [21].

In non-linear optical media, optical vortices can be used to form waveguide structures [22] and 'labyrinths' [23], and also to study various physical phenomena [24, 25].

This chapter describes the main types of paraxial optical vortices and their formation by diffractive optical elements in the scalar theory of diffraction. Vector diffraction is studied for the SPPs.

7.1. Optical elements that form wavefronts with helical phase singularities

Consider the light fields having a wavefront with a helical phase singularity. The complex amplitude of such fields is as follows:

$$E(r,\varphi,z) = A(r,z)\exp(in\varphi), \tag{7.1}$$

where (r, φ, z) are the cylindrical coordinates, n is the order of the phase singularity or topological charge.

These fields are called optical vortices, they have an orbital angular momentum, and the Umov–Poynting vector is directed along the helix (Fig. 7.1).

Optical vortices are formed by the spiral optical elements, which include a spiral phase plate (SPP) (Fig. 7.2a). Figure 7.2b shows a conical axicon. When combining the axicon and the SPP we obtain a helical or spiral axicon.

7.1.1. The spiral phase plate (SPP)

The spiral phase plate (SPP) as an optical element, whose transmittance function is proportional to exp $(in\varphi)$, φ is the polar angle, n is an integer (the order of the SPP),

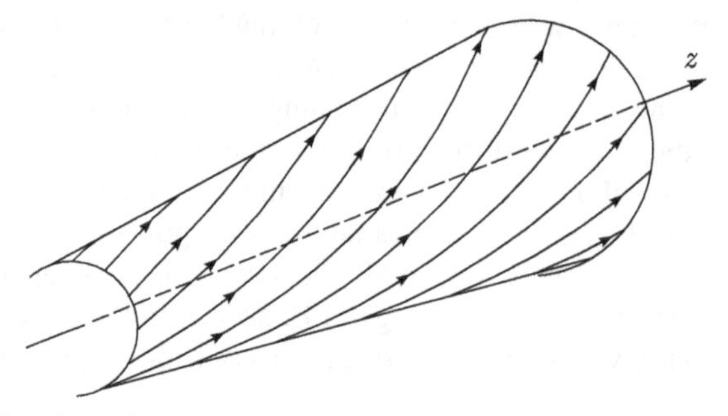

Fig. 7.1. The direction of energy flux in optical vortices (z is the axis along which light propagates).

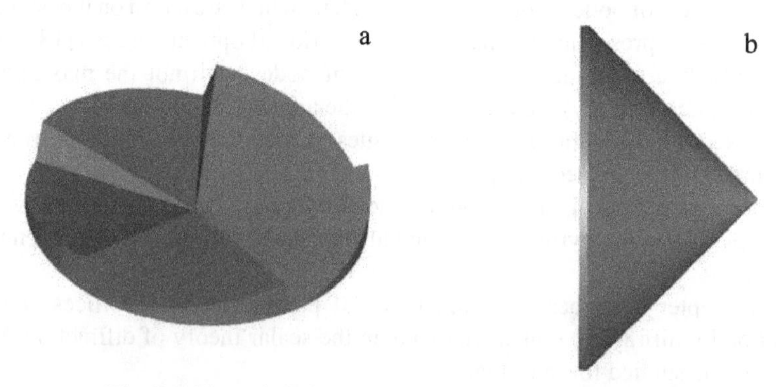

Fig. 7.2. The spiral phase plate (a) and conical axicon (b).

was first produced and analyzed in [26]. In recent years, particularly with respect to the optical manipulation of microparticles, the interest in the SPP increased [27-32]. In [30] SPP few millimetres in diameter with $n = 3$ were produced and characterized at a wavelength of 831 nm using the moulding technology with a maximum height of the microrelief of 5 μm. The accuracy of manufacturing the surface relief of the SPP on a polymer was very high (~ 3% error). In [27] using a conventional scanning electron microscope, converted into an electron-beam lithograph, writing directly on the negative photoresist SU-8 was carried out to record an SPP with a diameter 500 μm with $n = 1$ and with a continuous profile of relief with a maximum height of the step of 1.4 μm for a helium–neon laser. The formed diffraction pattern of a Gaussian beam on the SPP differed from the ideal 'doughnut' form by only 10%. The SPP made in this way was used for simultaneous optical trapping of latex beads of diameter 3 μm each, with the refractive index $n' = 1.59$. In [29] the same authors showed that the displacement of the centre of the SPP from the axis of the Gaussian beam results in the formation of an off-axis vortex and its transverse intensity distribution is rotated around the optical axis during beam propagation.

In [31] using standard photolithographic techniques with four binary amplitude masks the authors constructed and studied a 16-level and 16-sector SPP for a pulsed solid-state laser with a wavelength of 789 nm. SPP was made on SiO_2 100 mm in diameter with a maximum height of the step of the relief of 928 nm. In [28] using direct writing by the electron beam on a negative photoresist at a wavelength of 514 nm the authors produced an SPP with a diameter of 2.5 mm and a relied depth of 1082 nm. In addition, in [28] a theoretical analysis of Fresnel diffraction for a plane wave and Gaussian beam on the SPP was carried out. In [32], using a liquid-crystal spatial light modulator and a neodymium laser with doubled frequency and a wavelength of 532 nm an SPP with a high order singularity $n = 80$ was produced. In addition, in [32] analytical expressions were derived for Fraunhofer diffraction of the Gaussian beam on the SPP with large orders of singularity $n \gg 1$.

7.1.2. Spiral zone plates

The wavefronts with a helical phase singularity can also be produced by methods of digital holography [33]. When encoding a spiral phase plate with a circular carrier spatial frequency we obtain a function of the form sgn $[\cos(n\varphi + kr^2)]$. The transmission function of such a hologram is shown in Fig. 7.3.

7.1.3. Gratings with a fork

When using the carrier frequency the hologram will have the form shown in Fig. 7.4.

7.1.4. Screw conical axicon

The optical element called the axicon has been known for a long time [34]. It is a glass cone, which is illuminated from the base, and its optical axis passes along the height of the cone. It is usually used in optics to create a narrow 'diffraction-free'

Fig. 7.3. The formation of optical vortices by methods of digital holography.

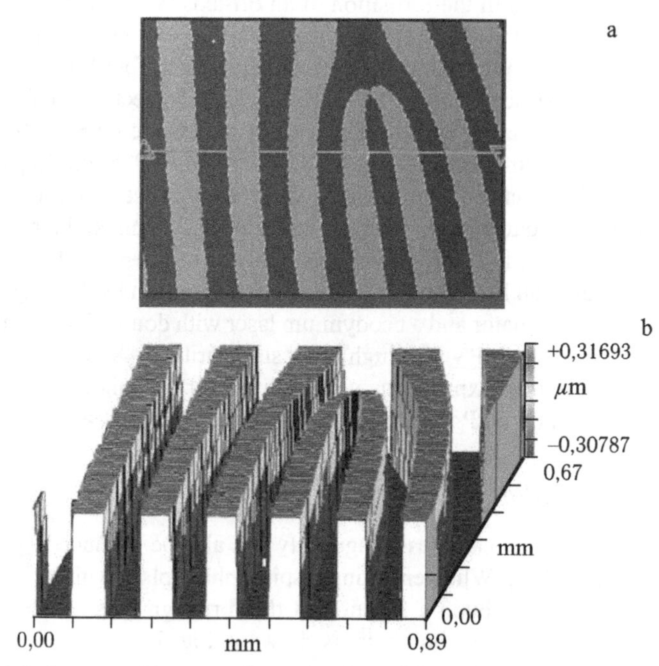

Fig. 7.4. A hologram with a one-dimensional spatial carrier frequency for generation of optical vortices.

laser beam [35, 36] or in conjunction with a lens to form a narrow annular light intensity distribution [37–39]. The diffractive axicon is shown in Fig. 7.5.

Axicons are also used in imaging systems to increase the depth of field, which can be 10–100 times greater than in a traditional lens.

In [40] grayscale photolithography was used to made on a low-contrast photoresist an optical element whose transmittance is proportional to the product of the transmission function of the axicon and SPP. In [40] this element is named

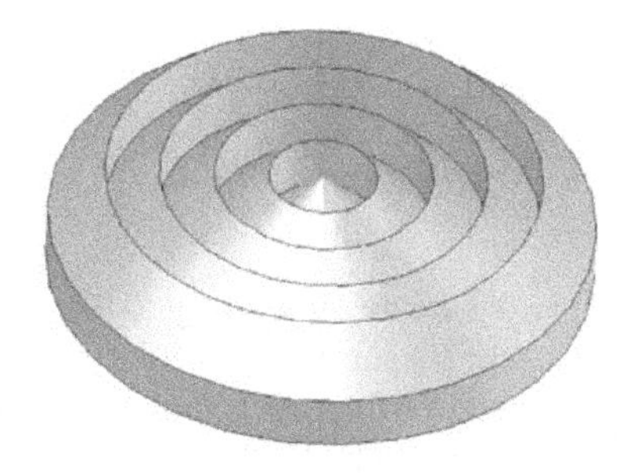

Fig. 7.5. Diffractive axicon.

the trochoson, i.e. forming a light pipe. Such an optical element is sometimes called the helical axicon [41]. Diffraction of a plane wave on such an diffraction element is identical to the diffraction of a conical wave on the SPP. In [42] using a 16-level helical axicon of the 5-th order and 6 mm diameter, made by direct writing with the electron beam and a He–Ne laser experiments were carried out with optical trapping and rotation with a period of 2 s of yeast particles and polystyrene beads with a diameter of 5 μm.

7.1.5. Helical logarithmic axicon

In the geometric approximation, a special feature of the classical (conical) axicon is a linear increase of intensity on the optical axis [34, 43, 44].

In [45], attention was paid to a generalized axicon that generates a given intensity distribution along the optical axis. In particular, a logarithmic axicon is suitable for producing uniform intensity along the optical axis (Fig. 7.6).

7.2. The spiral phase plate

In this section, we discuss some of the early uses of spiral phase plates for optical information processing, namely the optical performance of the Hankel transform and radial Hilbert transform. The remainder of the section is devoted to the theory of diffraction of light by the SPP, it gives expressions for the diffraction of a

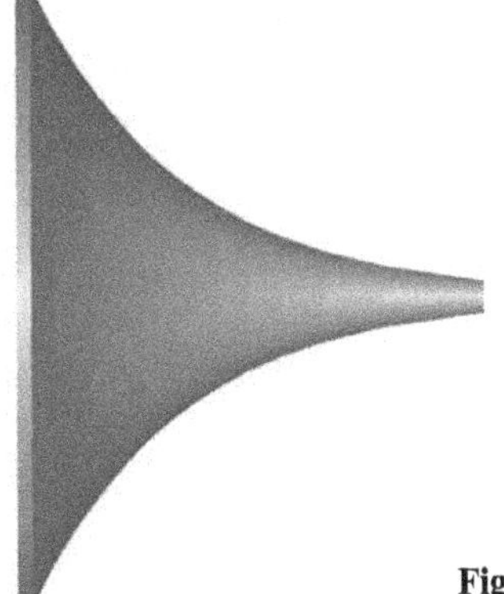

Fig. 7.6. Logarithmic axicon.

Gaussian beam and plane wave on SPP in both the scalar approximation and taking into account the vector nature of the electromagnetic field.

7.2.1. Hankel transform

In problems with cylindrical symmetry, for example, to create such images, like a circle, ring, or a set of rings [46] or by focusing on the 3D surface of revolution [47], to generate the laser modes [48], and in the formation of vortex beams [42], the calculations can be greatly accelerated by reducing the integral expressions to the Hankel transform.

For example, the Fourier transform in polar coordinates:

$$F(\rho,\theta) = -\frac{ik}{2\pi}\int_0^{\infty}\int_0^{2\pi} f(r,\varphi)\exp\left[-i\frac{k}{f_L}r\rho\cos(\theta-\varphi)\right]rdrd\varphi, \tag{7.2}$$

where (r, φ) and (ρ, θ) are the polar coordinates in the front and back focal plane of the lens, f_L is the focal length of the spherical lens used for the Fourier transform for the input function that is presented in the form $f(r, \varphi) = t(r) \exp(im\varphi)$, reduces to the Hankel transform of the m-th order:

$$H_m(\rho,\theta) = \exp(im\theta)\int_0^{\infty} t(r)J_m\left(\frac{kr\rho}{f}\right)rdr, \tag{7.3}$$

where $J_m(x)$ is the Bessel function of the first kind and m-th order:

$$J_m(z) = \frac{(-i)^m}{2\pi}\int_0^{2\pi}\exp(iz\cos\varphi + im\varphi)d\varphi. \tag{7.4}$$

A similar expression is obtained for the transformation of the Fresnel function describing the propagation of function $f(r, \varphi)$ in free space at distance z:

$$F_m(\rho,\theta,z) = \frac{ik}{z}\exp(ikz)\exp(im\theta)\exp\left(\frac{ik\rho^2}{2z}\right)\int_0^{\infty} t(r)\exp\left(\frac{ikr^2}{2z}\right)J_m\left(\frac{kr\rho}{z}\right)rdr. \tag{7.5}$$

For a quick calculation of the Hankel transform (7.3) we can consider using an exponential change of variables [49]. This method assumes that with the exponential change of variables the Hankel transforms reduces to convolution, which can be calculated using the fast Fourier transform algorithm.

Indeed, after the change of variables

$$r = r_0 e^x, \qquad \rho = \rho_0 e^y, \tag{7.6}$$

where r_0 and ρ_0 are constant, instead of (7.3) we obtain

$$\bar{H}(y) = r_0^2\int_{-\infty}^{\infty}\bar{t}(x)S(x+y)e^{2x}dx, \tag{7.7}$$

where

$$\overline{t}(x) = t(r_0 e^x), S(x+y) = J_m\left(\frac{k}{f} r_0 \rho_0 e^{x+y}\right),$$
$$\overline{H}(y) = H(\rho_0 e^y) = H_m(\rho, \theta)\exp(-im\theta). \quad (7.8)$$

For the function $S(x)$ to tend to 0 at $x \to \pm\infty$, it can be multiplied by $\exp(x/4)$, and to ensure that the integrand in (7.7) is not changed, function $\overline{t}(x)$ should also be multiplied by $\exp(-x/4)$.

Since the transmission function $t(r)$ is limited by the aperture $r \in [0, R]$, where R is the radius of the aperture, no problems arise for function $\overline{t}(x)$ at $x \to \infty$.
We introduce the following notation:

$$t_c(x) = \overline{t}(x) r_0^2 e^{7x/4}, \; S_c(x+y) = S(x+y) e^{x+y/4}. \; H_c(y) = \overline{H}(y) e^{y/4}, \quad (7.9)$$

and instead of (7.7) we obtain the convolution

$$H_c(y) = \int_{-\infty}^{a} h_c(x) S_c(x+y)\, dx, \; a = \ln\frac{R}{r_0}. \quad (7.10)$$

The integral (7.10) can be expressed in terms of 1D Fourier transform:

$$H(\rho) = \left[\frac{\rho}{\rho_0}\right]^{-\frac{1}{4}} \int_{-\infty}^{\infty} T(-u) U(u) e^{iu\ln\frac{\rho}{\rho_0}} du, \quad (7.11)$$

where $T(u)$ is the Fourier transform of the functions $h_c(x)$, $U(u)$ is the Fourier transform of $S_c(y)$.

There are other methods for rapid calculation of the Hankel transform [50].

When using the SPP as a filter in the spatial frequency plane of the Fourier correlator, we obtain the following chain of expressions. Before the filter frequency distribution of the radially symmetric function is a Hankel transform of zero order:

$$F(\rho) = 2\pi \int_0^{\infty} f(r) J_0(2\pi r\rho) r\, dr. \quad (7.12)$$

At the output of the correlator taking into account (7.4) and $J_m(-x) = (-1)^m J_m(x)$ we obtain:

$$E_c(r', \varphi') = 2\pi \int_0^{\infty}\int_0^{2\pi} F(\rho) \exp(im\theta) \exp\left[-i2\pi r' \rho \cos(\theta - \varphi')\right] \rho\, d\rho\, d\theta =$$
$$= 4\pi^2 i^m e^{im\varphi} \int_0^{\infty} F(\rho) J_m(2\pi r' \rho) \rho\, d\rho = E_c(r'). \quad (7.13)$$

This is a radially symmetric function, equal to the Hankel transform of m-th order of the Fourier spectrum $F(\rho)$ of the original function $f(r)$.

7.2.2. Radial Hilbert transform

The Hilbert transform is used in image processing since it emphasizes the contour of objects. The disadvantage of this transformation is the one-dimensionality, and the detection of contours takes place along a single direction. In [51], attention was paid to the radially symmetric version of the Hilbert transform, allowing a two-dimensional detection of the contours of objects of arbitrary shape.

Let $g(x, y)$ be a function describing the light field in the input plane of the Fourier correlator. The convolution operation is performed in this correlator, and the light field with the complex amplitude of the form in the output plane is obtained

$$\tilde{g}(x,y) = g(x,y) * h(x,y), \tag{7.14}$$

where $h\ (x, y)$ is the Fourier transform of the function of the masks in the frequency domain of the correlator $H(u, v)$.

The mask for a one-dimensional Hilbert transform of the P-th order is as follows:

$$H_P(u) = \exp\left(\frac{iP\pi}{2}\right)S(u) + \exp\left(-\frac{iP\pi}{2}\right)S(-u), \tag{7.15}$$

where $S(u)$ is the Heaviside function (step function).

The function (7.15) can be rewritten as follows:

$$H_P(u) = \cos\left(\frac{P\pi}{2}\right) + i\sin\left(\frac{P\pi}{2}\right)\operatorname{sgn}(u), \tag{7.16}$$

where sgn (u) is the function of the sign.

Since the Fourier transform of the sign function sgn (u) has the form $1/(i\pi x)$, then by (7.14) in the output plane of the correlator is a field with amplitude

$$\tilde{g}(x,y) = g(x,y)\cos\left(\frac{P\pi}{2}\right) + i\sin\left(\frac{P\pi}{2}\right)\left[g(x,y) * \frac{1}{i\pi x}\right], \tag{7.17}$$

Such a transformation is still one-dimensional. To generalize to the two-dimensional case we can use a mask type $H_P(u)\ H_Q(v)$. But in this case we will emphasize the contour along the axes x and y. To avoid this, we can make a mask and transmission at each point of this mask is equal to the transmission in the opposite point, but with a phase difference πP. This mask has a transmittance of exp $(iP\varphi)$ and hence is a spiral phase plate.

In [51] the authors used a liquid-crystal spatial light modulator (SLM) in place of the SPP. The light from an argon laser passed through a lens with a focal length of 36.8 cm and illuminated either a slit with a width of 200 mm or a circular aperture with a diameter of 300 mm. The SLM was placed in the Fourier plane of the lens, where the Fraunhofer diffraction pattern from the input field formed.

Figure 7.7 shows profiles of the diffraction pattern in the output plane of the Fourier correlator, when $P = 0$ (a), $P = 1/2$ (b), $P = 1$ (c), $P = 3/2$ (d).

Figure 7.8 shows the diffraction pattern in the output plane of the Fourier correlator when the depicted object is a circular aperture ($P = 0$ (a), $P = 1/2$ (b), $P = 1$ (c), $P = 3/2$ (d)) .

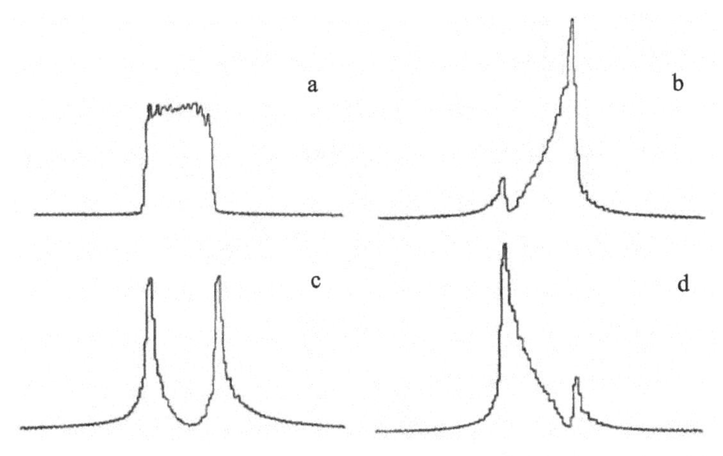

Fig. 7.7. Slit image (the mask was not used in the correlator) (a), the results of the Hilbert transform of the order of $P = 1/2$ (b), $P = 1$ (c), $P = 3/2$ (d).

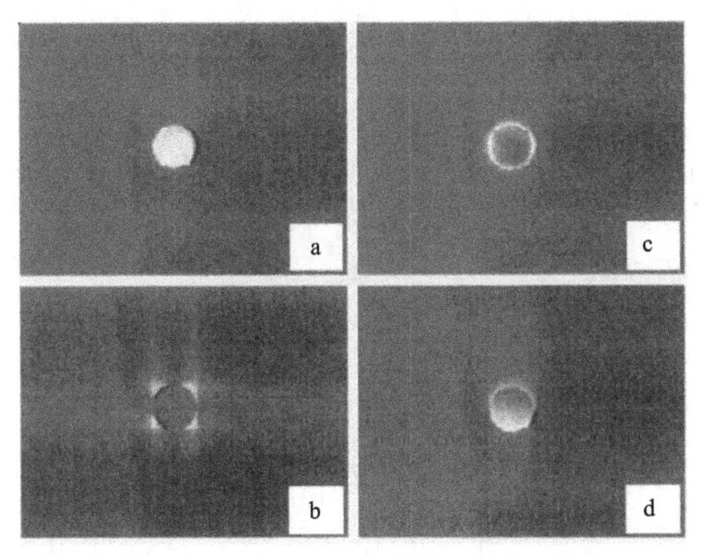

Fig. 7.8. Diffraction pattern in the output plane of the Fourier correlator in the absence of a mask (a), with a mask $H_1(u)\,H_1(v)$ (b), with a mask exp $(i\varphi)$ (c) with a mask exp $(i\varphi/2)$ (d).

7.2.3. Diffraction of a Gaussian beam on SPP: scalar theory Fresnel diffraction of Gaussian beam on SPP

In [28, 52] explicit analytical expressions describing the Fresnel diffraction of Gaussian beam on the SPP were derived. At a distance z, the complex amplitude of the light field in the paraxial approximation has the form:

$$E_n(\rho,\theta,z)=\frac{(-i)k}{2\pi z}\int_0^{\infty}\int_0^{2\pi}E_n^0(r,\varphi)\exp\left\{\frac{ik}{2z}\left[r^2+\rho^2-2r\rho\cos(\varphi-\theta)\right]\right\}r\,\mathrm{d}r\,\mathrm{d}\varphi=$$

$$=\frac{(-i)^{n+1}\sqrt{\pi}}{2}\left(\frac{z_0}{z}\right)^2\left(\frac{\rho}{w}\right)\left[1+\left(\frac{z_0}{z}\right)^2\right]^{-\frac{3}{4}}\exp\left[i\frac{3}{2}\mathrm{arctg}\left(\frac{z_0}{z}\right)+i\frac{k\rho^2}{2R(z)}-\frac{\rho^2}{w^2(z)}+in\theta\right]\times \quad (7.18)$$

$$\times\left\{I_{\frac{n-1}{2}}\left[\rho^2\left(\frac{1}{w^2(z)}+\frac{ik}{2R_0(z)}\right)\right]-I_{\frac{n+1}{2}}\left[\rho^2\left(\frac{1}{w^2(z)}+\frac{ik}{2R_0(z)}\right)\right]\right\},$$

where

$$w^2(z)=2w^2\left[1+\left(\frac{z}{z_0}\right)^2\right],\quad R(z)=2z\left[1+\left(\frac{z_0}{z}\right)^2\right]\cdot\left[2+\left(\frac{z_0}{z}\right)^2\right]^{-1},$$

$$R_0(z)=2z\left[1+\left(\frac{z}{z_0}\right)^2\right],\quad z_0=\frac{kw^2}{2},\quad E_n^0(r,\varphi)=\exp\left[-\left(\frac{r}{w}\right)^2+in\varphi\right],$$

$I_\nu(x)$ is the Bessel function of second kind and the ν-th order.

In [52] by a limiting transition from diffraction in the Fresnel zone to the far field an expression was derived for the Fraunhofer diffraction of the Gaussian beam on the SPP.

When $z \gg z_0$

$$w^2(z)\approx 2w^2\frac{z^2}{z_0^2},\quad R(z)\approx z,\quad R_0(z)\approx\frac{2z^3}{z_0^2}.$$

$$E_n(\rho,\theta,z\to\infty)=\frac{(-i)^{n+1}\sqrt{\pi}}{2}\left(\frac{z_0}{z}\right)^2\left(\frac{\rho}{w}\right)\times$$

$$\times\exp(in\theta)\exp\left[-\frac{\rho^2}{w^2(z)}\right]\left\{I_{\frac{n-1}{2}}\left(\frac{\rho^2}{w^2(z)}\right)-I_{\frac{n+1}{2}}\left(\frac{\rho^2}{w^2(z)}\right)\right\}. \quad (7.19)$$

(7.19) yields an expression for the intensity of the Gaussian beam with a phase singularity in the far diffraction zone

$$\hat{I}_n(\rho, z \to \infty) \approx \frac{\pi}{4}\left(\frac{z_0}{z}\right)^4 \left(\frac{\rho}{w}\right)^2 \exp\left[-\frac{2\rho^2}{w^2(z)}\right] \left\{ I_{\frac{n-1}{2}}\left(\frac{\rho^2}{w^2(z)}\right) - I_{\frac{n+1}{2}}\left(\frac{\rho^2}{w^2(z)}\right) \right\}^2 =$$

$$= \frac{\pi}{2}\left(\frac{z_0}{z}\right)^2 x \exp(-2x) \left\{ I_{\frac{n-1}{2}}(x) - I_{\frac{n+1}{2}}(x) \right\}^2, \quad (7.20)$$

where $x = \left(\frac{\rho z_0}{\sqrt{2}\, z w}\right)^2$

Fraunhofer diffraction of a Gaussian beam on the SPP

Above we obtained explicit analytical expressions describing the Fresnel diffraction of the Gaussian beam on the SPP. By limiting transition from diffraction in the Fresnel zone to the far zone, an expression was also obtained for the Fraunhofer diffraction of a Gaussian beam on the SPP (the expression (7.20)). In this section, we derive analytical formulas to describe the Fraunhofer diffraction of the Gaussian beam on the SPP, located in its waist. The Fraunhofer diffraction pattern is formed in the focal plane of a spherical lens.

Consider the initial function in the form of:

$$f_n'(r, \theta) = \exp\left(-\frac{r^2}{w^2} + in\theta\right), \quad (7.21)$$

where w is the waist radius of the Gaussian beam. Then the complex amplitude of Fraunhofer diffraction of a Gaussian beam at the waist on the SPP will be described by the expression:

$$F_n'(\rho, \varphi) = \frac{(-i)^{n+1} k}{f} \exp(in\varphi) \int_0^\infty \exp\left(-\frac{r^2}{w^2}\right) J_n\left(\frac{k}{f} r\rho\right) r\mathrm{d}r. \quad (7.22)$$

The known reference integral [53]:

$$\int_0^\infty \exp(-px^2) J_n(cx) x dx = \frac{c\sqrt{\pi}}{8p^{3/2}} \exp\left(-\frac{c^2}{8p}\right) \left[I_{(n-1)/2}\left(\frac{c^2}{8p}\right) - I_{(n+1)/2}\left(\frac{c^2}{8p}\right) \right], \quad (7.23)$$

where $I_\nu(x)$ is the modified Bessel function or the Bessel function of second kind. In view of (7.23) the expression (7.22) can be rewritten as:

$$F_n'(\rho, \varphi) = (-i)^{n+1} \exp(in\varphi) \left(\frac{kw^2}{4f}\right) \sqrt{2\pi x} \exp(-x) \left[I_{(n-1)/2}(x) - I_{(n+1)/2}(x) \right], \quad (7.24)$$

where

$$x = \frac{1}{2}\left(\frac{kw\rho}{2f}\right)^2.$$

The function of the intensity of the Fraunhofer diffraction pattern of the Gaussian beam on the SPP has the form:

$$\overline{I}'_n(\rho) = \left|F'_n(\rho,\varphi)\right|^2 = 2\pi\left(\frac{kw^2}{4f}\right)^2 x\exp(-2x)\left[I_{(n-1)/2}(x) - I_{(n+1)/2}(x)\right]^2. \quad (7.25)$$

From equation (7.25) we can see that for $x = 0$ at the centre of the Fourier plane intensity will be zero ($n \neq 0$): $\overline{I}'_n(0) = 0$. The factors $x \exp(-2x)$ in equation (7.25) show that an annular intensity distribution forms in the far zone. The radius of the ring can be found from the equation [28]:

$$(n-4x)I_{(n-1)/2}(x) + (n+4x)I_{(n+1)/2}(x) = 0. \quad (7.26)$$

We find the form of the function of the intensity on the outer side of the ring at $\rho \to \infty$ (or $x \to \infty$). For this we use the asymptotics of the Bessel function:

$$I_\nu(x) \approx \frac{\exp(x)}{\sqrt{2\pi x}}\left(1 - \frac{4\nu^2 - 1}{8x}\right), \quad x >> 1. \quad (7.27)$$

Then instead of (7.25) at $x \to \infty$ we obtain:

$$\overline{I}'_n(\rho) \approx \left(\frac{nf}{k\rho^2}\right)^2. \quad (7.28)$$

It is interesting that equation (7.28) does not depend on the radius of the Gaussian beam waist. From this match we can conclude that the asymptotic behaviour of the intensity at $\rho \to \infty$ is determined only by the number of the SPP, the size of the focus of a spherical lens and the wavelength of the radiation and does not depend on the amplitude and phase parameters of the beam illuminating the SPP.

Note that the expression (7.28) can be obtained from equation (7.25), letting go to infinity the radius of the Gaussian beam $w \to \infty$ at a fixed ρ.

We find the form of the function of intensity inside the ring. When ρ tends to zero (for fixed w) the argument of the Bessel function x also tends to zero, and we can use the first terms of expansion of the cylindrical function into a series:

$$I_\nu(x) \approx \left(\frac{x}{2}\right)^\nu \Gamma^{-1}(\nu+1), \quad x << 1, \quad (7.29)$$

where $\Gamma(x)$ is the gamma function. Then instead of (7.25) when we obtain:

$$\overline{I}'_n(\rho) \approx \pi\,\Gamma^{-2}\left(\tfrac{n+1}{2}\right)\left(\frac{kw^2}{f}\right)\left(\frac{kw\rho}{4f}\right)^{2n}. \quad (7.30)$$

From equation (7.30) we see that the intensity near the centre of the Fourier plane increases as the degree $2n$ of the radial coordinate:

$$\overline{I}'_n(\rho) \approx (w\rho)^{2n}, \quad \rho \ll 1. \tag{7.31}$$

If in addition to ρ tending to zero the Gaussian beam radius w should tend to infinity so that their product $w\rho$ remained constant, from equation (7.30) it follows that the intensity near the centre of the Fourier plane will tend to infinity as the square of the radius of the waist:

$$\overline{I}'_n(\rho \to 0, w \to \infty) \approx w^2, \quad \rho w = \text{const}, \tag{7.32}$$

but in the most central point at $\rho = 0$, the intensity will be zero $\overline{I}'_n(\rho = 0) = 0$, for any w.

For the experiments made with a 32-level SPP we generated a light field with the singularity of the second order. The size of the element is equal to 2.5 × 2.5 mm^2, and the size of the frame 5 × 5 μm^2. These SPPs were designed for wavelength $\lambda = 633$ nm. The depth of the microrelief, measured using a contact profilometer was 1320 nm. The optimum depth of the 32-level microrelief was 1341 nm on the assumption that the refractive index of the resist is $n_r = 1.457$ (exact value unknown). Thus, the deviation from the optimum depth is only about 1%. In [28] SPP was manufactured by the same technology, but for a wavelength of 514 nm, and the experiments were carried out at a wavelength of 543 nm. This led to the formation of a low quality tubular beam. The design of the SPP and the experiments were conducted using the same wavelength of a helium–neon laser at 633 nm. Therefore, the intensity distribution of the generated beam actually has a circular symmetry.

Figure 7.9a shows the estimated distribution of the phase (white colour indicates zero phase, black 2π $(1-1/N)$, where N is the number of quantization levels). Figure 7.9b shows the microrelief of the SPP obtained using the interferom eter NEWVIEW 5000 Zygo (200-fold magnification).

The annular intensity distribution in Fig. 7.10a was obtained as a result of diffraction of a Gaussian beam with a waist radius of $\sigma = 0.8$ mm on the SPP of the second order ($n = 2$). As a result of inexact matching the centre of the Gaussian beam and the centre of the SPP the circular symmetry on the diffraction pattern is violated.

Figure 7.10b shows a comparison of theoretical and experimental profiles of annular intensity distributions shown in Fig. 7.10a. The graph of the intensity in

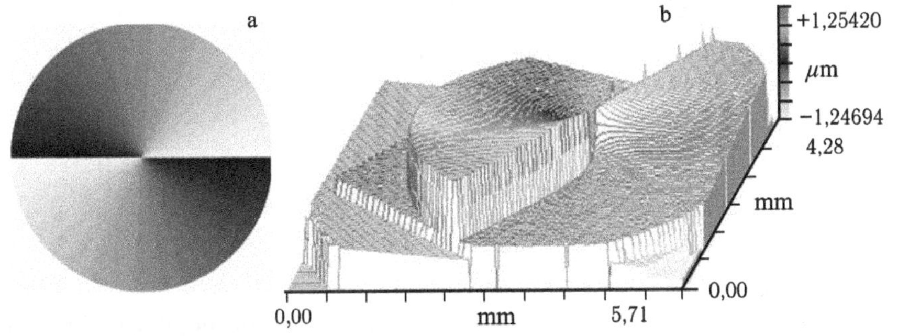

Fig. 7.9. The generation of the laser field with a phase singularity of the second order: (a) theoretical phase distribution, (b) the central part of the microrelief of the SPP.

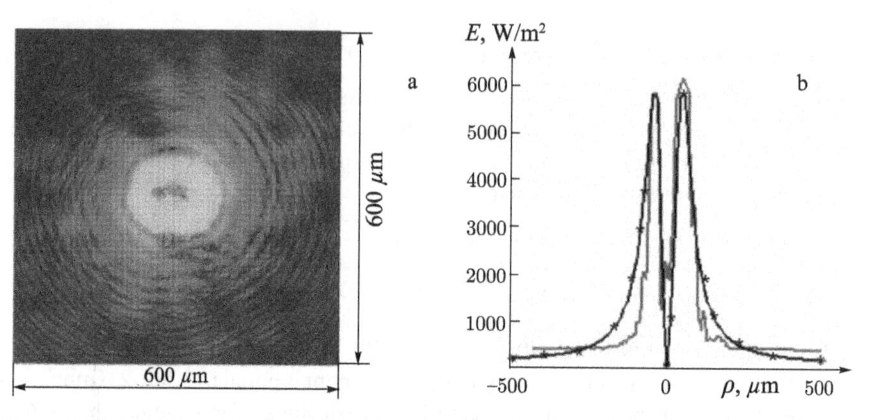

Fig. 7.10. Fraunhofer diffraction of the Gaussian beam (waist radius $\sigma = 0.8$ mm) on the SPP of the order $n = 2$; a) diffraction pattern; b) solid curve – experimental, and the curve with asterisks – theoretical intensity distribution.

Fig. 7.10b is calibrated taking into account the power of the illuminating laser beam, measured by a wattmeter with an accuracy of 15%.

The radius of the ring in Fig. 7.10b can be obtained by using the ratio from [28]: $\rho_2 = 0.46\,\lambda f/\sigma = 45$ μm. Figure 7.10b shows that the experimental and theoretical curves agree quite well.

7.2.4. Diffraction of a Gaussian beam on SPP: vector theory

This section analyzes the diffraction of Gaussian beam on SPP in the vector theory, and analytical expressions are derived for the longitudinal field component, which, as shown numerically, in some cases makes a significant contribution.

It is known that the propagation of light in free space is described by the Rayleigh–Sommerfeld diffraction integrals [54, 55]:

$$\begin{cases} E_x(u,v,z) = -\dfrac{1}{2\pi}\displaystyle\iint_{R^2} E_x(x,y,0)\dfrac{\partial}{\partial z}\left[\dfrac{\exp(ikR)}{R}\right]\mathrm{d}x\mathrm{d}y, \\ E_y(u,v,z) = -\dfrac{1}{2\pi}\displaystyle\iint_{R^2} E_y(x,y,0)\dfrac{\partial}{\partial z}\left[\dfrac{\exp(ikR)}{R}\right]\mathrm{d}x\mathrm{d}y, \\ E_z(u,v,z) = \dfrac{1}{2\pi}\displaystyle\iint_{R^2} \left\{E_x(x,y,0)\dfrac{\partial}{\partial x}\left[\dfrac{\exp(ikR)}{R}\right] + E_y(x,y,0)\dfrac{\partial}{\partial y}\left[\dfrac{\exp(ikR)}{R}\right]\right\}\mathrm{d}x\mathrm{d}y, \end{cases} \tag{7.33}$$

where $R = [(u-x)^2 + (v-y)^2 + z^2]^{1/2}$, (x, y) are Cartesian coordinates in the SPP plane $z = 0$, (u, v) are the Cartesian coordinates in the plane, at a distance z from the plane of the CPP, $k = 2\pi/\lambda$ is the wave number.

In the calculation of these integrals the factors containing derivatives of the function R^{-1} exp (ikR) are normally replaced by approximate expressions. In the paraxial approximation this is done as follows: the following change is made in the exponent of rapidly oscillating functions

$$R \approx z + \frac{(u-x)^2 + (v-y)^2}{2z}, \tag{7.34}$$

but in other cases it is assumed that $R \approx z$. After these transformations, instead of (7.33) we obtain the following expressions:

$$\begin{cases} E_{x,y}(u,v,z) \approx -\dfrac{ik}{2\pi z}\exp(ikz)\displaystyle\iint_{R^2} E_{x,y}(x,y,0)\exp\left\{\dfrac{ik}{2z}\left[(u-x)^2 + (v-y)^2\right]\right\}\mathrm{d}x\mathrm{d}y, \\ E_z(u,v,z) \approx \dfrac{ik}{2\pi z^2}\exp(ikz)\displaystyle\iint_{R^2}\left[(x-u)E_x(x,y,0) + (y-v)E_y(x,y,0)\right]\times \\ \qquad \times\exp\left\{\dfrac{ik}{2z}\left[(u-x)^2 + (v-y)^2\right]\right\}\mathrm{d}x\mathrm{d}y, \end{cases} \tag{7.35}$$

where $E_{x,y}$ is either E_x, or E_y. From (7.35) it can be seen that for the transverse components we obtain the well-known Fresnel transformation. In [56] the authors used a less rough approximation: the following change is made in the exponent of rapidly oscillating functions

$$R \approx \sqrt{u^2 + v^2 + z^2} + \frac{(u-x)^2 + (v-y)^2}{2\sqrt{u^2 + v^2 + z^2}}, \tag{7.36}$$

but in other cases it is considered that $R \approx (u^2 + v^2 + z^2)^{1/2}$. After these transformations, instead of (7.33) we can write approximately:

$$\begin{cases} E_{x,y}(u,v,z) \approx -\dfrac{ikz\exp\left(ik\sqrt{u^2+v^2+z^2}\right)}{2\pi\left(u^2+v^2+z^2\right)}\times \\ \qquad \times\displaystyle\iint_{R^2} E_{x,y}(x,y,0)\exp\left[\dfrac{ik}{2\sqrt{u^2+v^2+z^2}}\left(x^2+y^2-2ux-2vy\right)\right]\mathrm{d}x\mathrm{d}y, \\ E_z(u,v,z) \approx \dfrac{ik}{2\pi\left(u^2+v^2+z^2\right)}\exp\left(ik\sqrt{u^2+v^2+z^2}\right)\times \\ \qquad \times\displaystyle\iint_{R^2}\left[(x-u)E_x(x,y,0) + (y-v)E_y(x,y,0)\right] \\ \qquad \times\exp\left[\dfrac{ik}{2\sqrt{u^2+v^2+z^2}}\left(x^2+y^2\right)\right]\exp\left[-\dfrac{ik}{\sqrt{u^2+v^2+z^2}}(ux+vy)\right]\mathrm{d}x\mathrm{d}y. \end{cases} \tag{7.37}$$

We can see that in formula $(u^2 + v^2)^{1/2} \ll z$ (7.37) becomes (7.35).

In the case when the beam in the input plane has a vortical component, i.e.

$$\begin{cases} E_x(r,\varphi,0) \equiv A_x(r)\exp(in\varphi), \\ E_y(r,\varphi,0) \equiv A_y(r)\exp(in\varphi), \end{cases} \tag{7.38}$$

where (r, φ) are the polar coordinates in the plane $z = 0$, the double integrals in (7.35) and (7.37) after transition to the polar coordinates can be reduced to single.

In the case of the paraxial approximation (7.35) we obtain the following expression:

$$\begin{cases} E_{x,y}(\rho,\theta,z) = (-i)^{n+1}\dfrac{k}{z}\exp\left(\dfrac{ik\rho^2}{2z} + in\theta + ikz\right)\displaystyle\int_0^\infty A_{x,y}(r)\exp\left(\dfrac{ikr^2}{2z}\right)J_n\left(\dfrac{k\rho r}{z}\right)r\mathrm{d}r, \\ E_z(\rho,\theta,z) = (-i)^n\dfrac{k}{z^2}\exp\left(\dfrac{ik\rho^2}{2z} + in\theta + ikz\right)\times \\ \times\left[\exp(i\theta)\displaystyle\int_0^\infty \dfrac{A_x(r) - iA_y(r)}{2}\exp\left(\dfrac{ikr^2}{2z}\right)J_{n+1}\left(\dfrac{k\rho r}{z}\right)r^2\mathrm{d}r - \right. \\ -\exp(-i\theta)\displaystyle\int_0^\infty \dfrac{A_x(r) + iA_y(r)}{2}\exp\left(\dfrac{ikr^2}{2z}\right)J_{n-1}\left(\dfrac{k\rho r}{z}\right)r^2\mathrm{d}r - \\ \left. -i\rho\displaystyle\int_0^\infty \left[A_x(r)\cos\theta + A_y(r)\sin\theta\right]\exp\left(\dfrac{ikr^2}{2z}\right)J_n\left(\dfrac{k\rho r}{z}\right)r\mathrm{d}r\right], \end{cases} \tag{7.39}$$

where (ρ, θ) are the polar coordinates in the plane at distance z from the plane of the SPP; $J_n(x)$ is the Bessel function of n-th order.

In the case of non-paraxial approximation (7.37) we obtain expressions similar in form but more accurate:

$$\begin{cases} E_{x,y}(\rho,\theta,z) = (-i)^{n+1} \dfrac{kz\exp\left(in\theta + ik\sqrt{\rho^2+z^2}\right)}{\rho^2+z^2} \times \\ \times \displaystyle\int_0^\infty A_{x,y}(r)\exp\left(\frac{ikr^2}{2\sqrt{\rho^2+z^2}}\right) J_n\left(\frac{k\rho r}{\sqrt{\rho^2+z^2}}\right) r\mathrm{d}r, \\ E_z(\rho,\theta,z) = (-i)^n \dfrac{k}{\rho^2+z^2}\exp\left(ik\sqrt{\rho^2+z^2}+in\theta\right)\times \\ \times\left[\exp(i\theta)\displaystyle\int_0^\infty \frac{A_x(r)-iA_y(r)}{2}\exp\left(\frac{ikr^2}{2\sqrt{\rho^2+z^2}}\right)J_{n+1}\left(\frac{k\rho r}{\sqrt{\rho^2+z^2}}\right)r^2\mathrm{d}r - \right. \\ -\exp(-i\theta)\displaystyle\int_0^\infty \frac{A_x(r)+iA_y(r)}{2}\exp\left(\frac{ikr^2}{2\sqrt{\rho^2+z^2}}\right)J_{n-1}\left(\frac{k\rho r}{\sqrt{\rho^2+z^2}}\right)r^2\mathrm{d}r - \\ \left. -i\rho\displaystyle\int_0^\infty\left[A_x(r)\cos\theta + A_y(r)\sin\theta\right]\exp\left(\frac{ikr^2}{2\sqrt{\rho^2+z^2}}\right)J_n\left(\frac{k\rho r}{\sqrt{\rho^2+z^2}}\right)r\mathrm{d}r\right]. \end{cases} \tag{7.40}$$

If a Gaussian beam falls on the SPP, i.e. $A_{x,y}(r) = B_{x,y}\exp(-r^2/w^2)$, the integrals in (7.39) and (7.40) can be calculated using the following reference integral [53]:

$$\int_0^\infty \exp\left(-px^2\right)J_\nu(cx)x\mathrm{d}x = \frac{c\sqrt{\pi}}{8p^{3/2}}\exp(-y)\left[I_{\frac{\nu-1}{2}}(y) - I_{\frac{\nu+1}{2}}(y)\right], \mathrm{Re}\,\nu > -2, \tag{7.41}$$

where $y = c^2/(8p)$, $I_n(x)$ is the Bessel function of second kind, and with the help of another integral, which can be obtained from (7.41):

$$\int_0^\infty \exp\left(-px^2\right)J_\nu(cx)x^2\mathrm{d}x = \frac{\sqrt{\pi}}{8p^{3/2}}\exp(-y)\times \\ \times\left\{(\nu+2-3y)\left[I_{\frac{\nu}{2}}(y) - I_{\frac{\nu+2}{2}}(y)\right] + y\left[I_{\frac{\nu-2}{2}}(y) - I_{\frac{\nu+4}{2}}(y)\right]\right\}. \tag{7.42}$$

After application of the integrals (7.41) and (7.42) to the expressions (7.40), we get:

$$\begin{cases} E_{x,y}(\rho,\theta,z)=(-i)^{n+1}\dfrac{B_{x,y}kz\exp\left(in\theta+ik\sqrt{\rho^2+z^2}\right)}{\rho^2+z^2}\dfrac{c\sqrt{\pi}}{8p^{3/2}}\exp(-y)\left[I_{\frac{n-1}{2}}(y)-I_{\frac{n+1}{2}}(y)\right], \\ E_z(\rho,\theta,z)=(-i)^n\dfrac{k}{\rho^2+z^2}\exp\left(ik\sqrt{\rho^2+z^2}+in\theta\right)\dfrac{\sqrt{\pi}}{8p^{3/2}}\exp(-y)\times \\ \times\left(\dfrac{B_x-iB_y}{2}\exp(i\theta)\left\{(n+3-3y)\left[I_{\frac{n+1}{2}}(y)-I_{\frac{n+3}{2}}(y)\right]+y\left[I_{\frac{n-1}{2}}(y)-I_{\frac{n+5}{2}}(y)\right]\right\}-\right. \\ -\dfrac{B_x+iB_y}{2}\exp(-i\theta)\left\{(n+1-3y)\left[I_{\frac{n-1}{2}}(y)-I_{\frac{n+1}{2}}(y)\right]+y\left[I_{\frac{n-3}{2}}(y)-I_{\frac{n+3}{2}}(y)\right]\right\}- \\ \left.-i\left(B_x\cos\theta+B_y\sin\theta\right)c\rho\left[I_{\frac{n-1}{2}}(y)-I_{\frac{n+1}{2}}(y)\right]\right). \end{cases} \quad (7.43)$$

In equation (7.43) the notation:

$$p=\frac{1}{w^2}-\frac{ik}{2\sqrt{\rho^2+z^2}},\qquad c=\frac{k\rho}{\sqrt{\rho^2+z^2}},\qquad y=\frac{c^2}{8p}. \quad (7.44)$$

The Cartesian components of the vector of the strength of the electric field (7.43) in cylindrical coordinates describe the non-paraxial Gaussian beam diffraction on the SPP with a topological charge n. Note that for $B_z = \pm iB_y$ a Gaussian beam has a circular polarization, and with $B_x \neq 0$, $B_y = 0$ – linear polarization.

In the numerical simulation the integrals (7.39) and (7.40) are calculated by the method of rectangles and compared with the values obtained with the formulas (7.43) and the formulas obtained from (7.43) for the paraxial approximation. Thus, the resulting expression (7.42) was verified.

This was followed by numerical comparison of the paraxial approximation (7.39) and the more accurate non-paraxial approximation (7.40). The simulation results are shown in Fig. 7.11. We used the following parameters: wavelength λ = 633 nm, the radius of the Gaussian beam waist w = 1 μm, the order of SPP n = 3, and the distance along the optical axis z = 10 mm, the amplitudes of the Gaussian beam B_x = 1 and B_y = 0.2i (elliptical polarization).

Figure 7.11 shows that the transverse components of the vector of the strength of the electric field obtained with the paraxial and non-paraxial approximations differ from each other (the maximum error was 14%). The longitudinal component in this case is small.

Figure 7.12 shows the diffraction of the same Gaussian beam, but at a distance z = 10 μm.

Figure 7.12 shows that under these conditions it is already important to consider the effect on the intensity of the longitudinal projection of the vector of the electric field, as it is about 3% of the transverse projection.

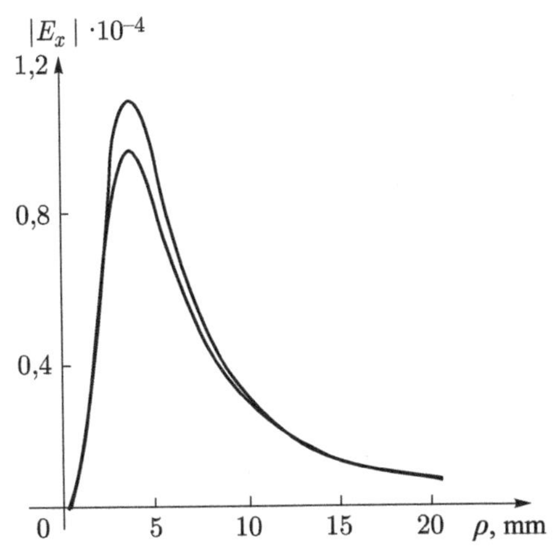

Fig. 7.11. Diffraction of Gaussian beam on SPP: the transverse component $|E_x|$ (top line – in the paraxial approximation, bottom line – in non-paraxial approximation).

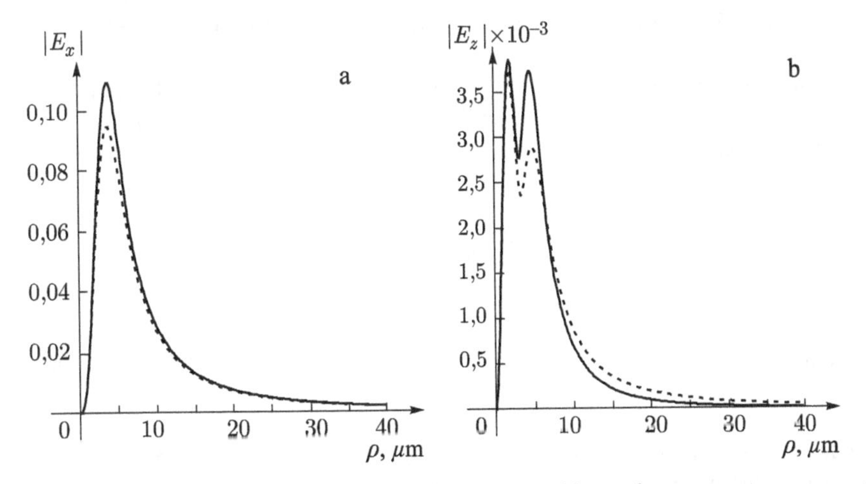

Fig. 7.12. Diffraction of a Gaussian beam on the SPP at $z = 10$ μm: the transverse component $|E_x|$ (a) and the longitudinal component $|E_z|$ (b) (solid line – in the paraxial approximation, dashed – in non-paraxial approximation).

Figure 7.13 shows the radial distribution of the modulus of the electric vector obtained by the formula (7.43) (solid line) and using the Rayleigh–Sommerfeld diffraction integral (7.33) (mean computational complexity of the individual values are shown by dots). Calculation parameters are the same as that for Fig. 7.11 and 7.12. The distance was taken equal to $z = 10$ mm.

Figure 7.13 shows that the formula (7.43) yields results virtually identical with the exact formula (7.33).

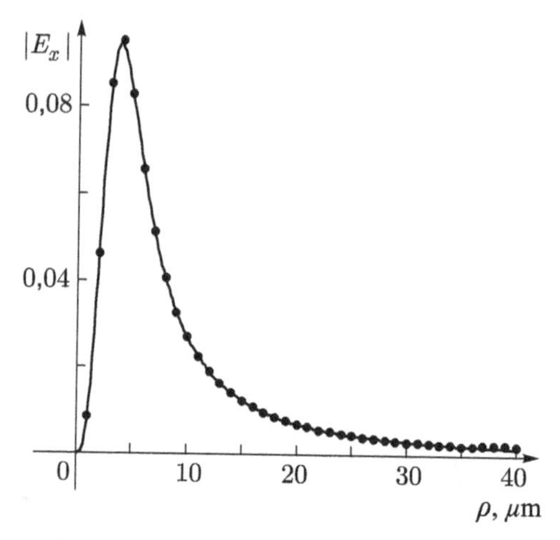

Fig. 7.13. Comparison of the calculation results for the diffraction pattern of the Gaussian beam, using the approximate formula (7.43) (solid line) and the exact formula (7.33) (graph is shown in dots).

7.2.5. Fresnel diffraction of a restricted plane wave on SPP

Consider the Fresnel diffraction of a restricted plane wave on SPP. Paraxial wave diffraction on the SPP will be described by the following transformation (derived from the Fresnel transform):

$$E_n(\rho,\theta,z)=\frac{(-i)^{n+1}k}{z}\exp\left(\frac{ik\rho^2}{2z}+in\theta\right)\int_0^R\exp\left(\frac{ikr^2}{2z}\right)J_n\left(\frac{k}{z}r\rho\right)r\,\mathrm{d}r=$$

$$=\exp\left(\frac{iz_0\bar{\rho}^2}{z}+in\theta\right)\frac{2\left(\frac{-iz_0}{z}\right)^{n+1}\bar{\rho}^n}{n!}\sum_{m=0}^{\infty}\frac{\left(\frac{iz_0}{z}\right)^m}{(2m+n+2)m!}\,{}_1F_2\left[\frac{2m+n+2}{2},\frac{2m+n+4}{2},n+1;-\left(\frac{z_0\bar{\rho}}{z}\right)^2\right],\tag{7.45}$$

where $z_0 = kR^2/2$ is the Rayleigh length, $\bar{\rho} = \rho/R$. Equation (7.45) shows that at $n \neq 0$ in the centre of the beam at $\rho = 0$ the amplitude is zero $E_n(\rho = 0, \theta, z)$ for all z, except $z = 0$. Equation (7.45) also shows that with increasing z in the series of hypergeometric functions contributions are provided only the first few terms in the series, and if $z \to \infty$ ($z \gg z_0$, far field) the contribution to the amplitude will come only from the first term at $m = 0$. Note that in (7.45) the integer part of ratio z_0 / z is equal to the Fresnel number. Note also that the expression (7.45) at $n = 0$ (no SPP) describes the Fresnel diffraction of a plane wave on a circular aperture of radius R:

$$E_0(\rho,z)=(-1)\exp\left(\frac{iz_0\bar{\rho}^2}{z}\right)\sum_{m=0}^{\infty}\frac{\left(\frac{iz_0}{z}\right)^{m+1}}{(m+1)!}\,{}_1F_2\left[m+1,m+2,1;-\left(\frac{z_0\bar{\rho}}{z}\right)^2\right].\tag{7.46}$$

From (7.46) can be a simple dependence of the complex amplitude of the light field on the optical axis ($\rho = 0$) of the distance z to the diaphragm:

$$E_0(\rho = 0, z) = 1 - \exp\left(\frac{iz_0}{z}\right). \tag{7.47}$$

The expression (7.47) coincides with that obtained previously [57].

Figure 7.14 shows the results of the comparison of experiment and calculation. In Fig. 7.14 shows a surface profile of the SPP with the number $n = 3$ and a diameter of 2.5 mm, visualized with an interferometer Newview 5000 Zygo (increase by 200 times). The SPT profile differs from the ideal of 4.3%, while the SPP itself has 32 gradations of relief and was produced by a low-contrast negative resist XAR-N7220 by direct write electron beam with the lithographer Leica LION LV1 with a resolution of 5 microns.

Figure 7.14 b, c shows the experimental and calculated diffraction pattern of a plane wave at the SPP with a radius $R = 1.25$ mm and a wavelength $\lambda = 0.633$ mm at a distance z = 80 mm. Both diffraction patterns have the same number of rings (8 rings).

Figure 7.15 shows the result of registering with the CCD-camera pictures of the Fraunhofer diffraction at the lens focus (f = 150 mm) obtained for a plane wave with a radius of 1.25 mm, 0.633 mm wavelength and SPP $n = 3$.

The relative standard deviation of the theoretical and experimental curves in Fig. 7.15 b was 14.3%.

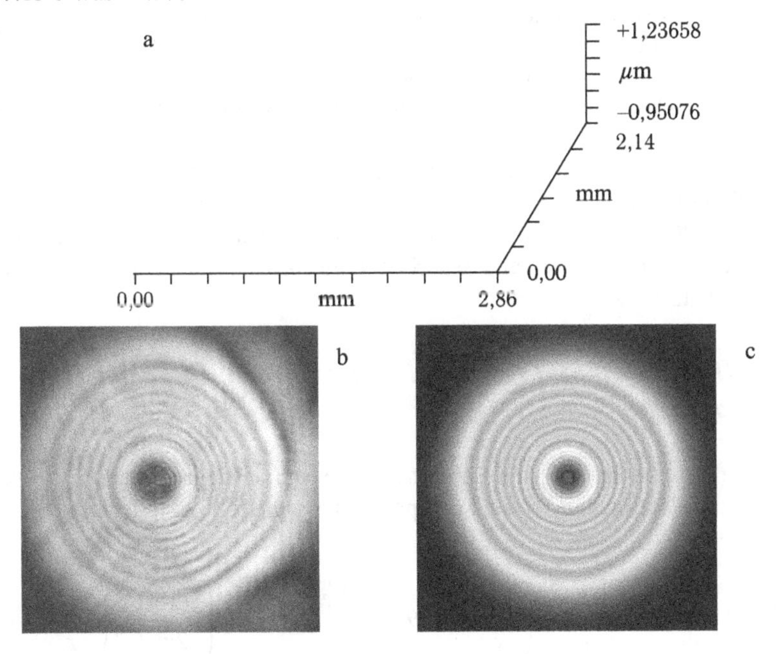

Fig. 7.14. Profile of the surface of SPP ($n = 3$) (a), Fresnel diffraction pattern of a plane wave with radius $R = 1.25$ mm and wavelength $\lambda = 0.633$ μm at distance $z = 80$ mm from the SPP: experiment (b) and theory (c).

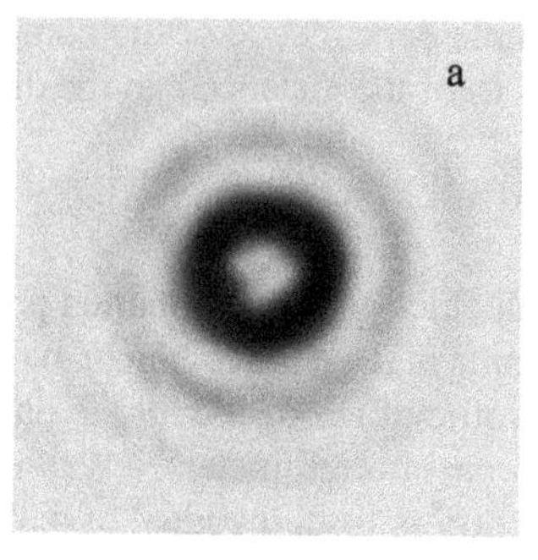

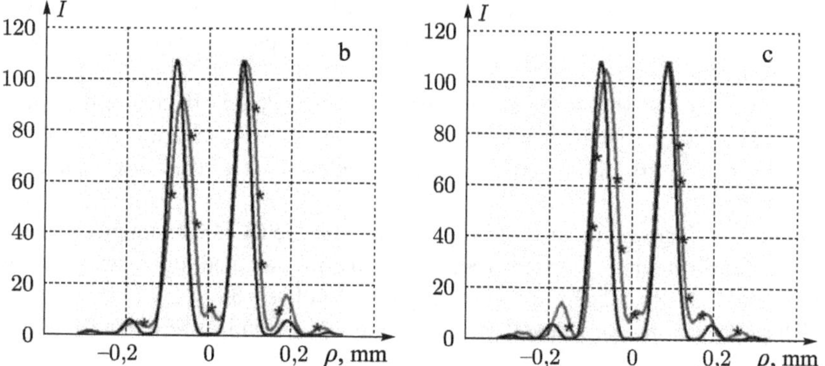

Figure 7.15. Fraunhofer diffraction pattern (negative) on SPP with the number $n = 3$, a plane wave with a radius of 1.25 mm and wavelength 0.633 μm, formed in the focal plane of Fourier lens with a focal length of 150 mm: the intensity distribution of (negative) (a), vertical (b) and horizontal (c) in the intensity section (solid curves – theory, * – * – experiment).

In this section we derived analytical expressions that describe the paraxial diffraction of a restricted plane wave on the SPP. Using a SPP produced with high accuracy with the number $n = 3$ we obtained Fresnel and Fraunhofer experimental diffraction patterns. Theory and experiment are consistent with an average error of not more than 15%.

7.2.6. Diffraction of a restricted plane wave on SPP: paraxial vectorial theory

From the expression (7.39) it follows that in the paraxial approximation, the expressions of the electromagnetic field component have the following form:

$$E_x(\rho,\theta,z) = (-i)^{n+1} \frac{k}{z} \exp\left(\frac{ik\rho^2}{2z} + in\theta + ikz\right) \int_0^\infty A_x(r) \exp\left(\frac{ikr^2}{2z}\right) J_n\left(\frac{k\rho r}{z}\right) r dr, \tag{7.48}$$

$$E_y(\rho,\theta,z)=(-i)^{n+1}\frac{k}{z}\exp\left(\frac{ik\rho^2}{2z}+in\theta+ikz\right)\int_0^\infty A_y(r)\exp\left(\frac{ikr^2}{2z}\right)J_n\left(\frac{k\rho r}{z}\right)r\mathrm{d}r, \tag{7.49}$$

$$\begin{gathered}E_z(\rho,\theta,z)=\frac{(-i)^n k}{2z^2}\exp\left(\frac{ik\rho^2}{2z}+in\theta+ikz\right)\times\\ \times\left\{\exp(i\theta)\int_0^\infty\left[A_x(r)-iA_y(r)\right]\exp\left(\frac{ikr^2}{2z}\right)J_{n+1}\left(\frac{k\rho r}{z}\right)r^2\mathrm{d}r-\right.\\ -\exp(-i\theta)\int_0^\infty\left[A_x(r)+iA_y(r)\right]\exp\left(\frac{ikr^2}{2z}\right)J_{n-1}\left(\frac{k\rho r}{z}\right)r^2\mathrm{d}r-\\ \left.2i\rho\int_0^\infty\left[A_x(r)\cos\theta+A_y(r)\sin\theta\right]\exp\left(\frac{ikr^2}{2z}\right)J_n\left(\frac{k\rho r}{z}\right)r\mathrm{d}r\right\},\end{gathered} \tag{7.50}$$

where $J_n(x)$ is the Bessel function of n-th order.

In the case where in the plane $z = 0$ there is a spiral phase plate (SPP) of radius R and n-th order, and a lens with a focal length f, we obtain the expression:

$$A_x(r)\equiv A_x\,\mathrm{circ}\left(\frac{r}{R}\right)\exp\left(-\frac{ikr^2}{2f}\right), \tag{7.51}$$

$$A_y(r)\equiv A_y\,\mathrm{circ}\left(\frac{r}{R}\right)\exp\left(-\frac{ikr^2}{2f}\right), \tag{7.52}$$

where Ax and A_y are the complex amplitudes of a plane wave incident on the SPP with a lens. Then, at a distance z an electromagnetic field with the following components ($E_{x,y}$ – it is either E_x, or E_y) will form:

$$E_{x,y}(\rho,\theta,z)=(-i)^{n+1}\frac{kA_{x,y}}{z}\exp\left(\frac{ik\rho^2}{2z}+in\theta+ikz\right)\int_0^R\exp\left[\frac{ikr^2}{2}\left(\frac{1}{z}-\frac{1}{f}\right)\right]J_n\left(\frac{k\rho r}{z}\right)r\mathrm{d}r, \tag{7.53}$$

$$\begin{gathered}E_z(\rho,\theta,z)=(-i)^n\frac{k}{2z^2}\exp\left(\frac{ik\rho^2}{2z}+in\theta+ikz\right)\times\\ \times\left\{(A_x-iA_y)\exp(i\theta)\int_0^R\exp\left[\frac{ikr^2}{2}\left(\frac{1}{z}-\frac{1}{f}\right)\right]J_{n+1}\left(\frac{k\rho r}{z}\right)r^2\mathrm{d}r-\right.\\ -(A_x+iA_y)\exp(-i\theta)\int_0^R\exp\left[\frac{ikr^2}{2}\left(\frac{1}{z}-\frac{1}{f}\right)\right]J_{n-1}\left(\frac{k\rho r}{z}\right)r^2\mathrm{d}r-\end{gathered} \tag{7.54}$$

$$-2i\rho\left(A_x\cos\theta+A_y\sin\theta\right)\int\limits_0^R\exp\left[\frac{ikr^2}{2}\left(\frac{1}{z}-\frac{1}{f}\right)\right]J_n\left(\frac{k\rho r}{z}\right)r\mathrm{d}r\Bigg\}.$$

In the geometric focus of the lens, i.e. at $z=f$, the expressions can be simplified [58]:

$$\begin{gathered}E_{x,y}\left(\rho,\theta,z=f\right)=\left(-i\right)^{n+1}\frac{kA_{x,y}}{f}\exp\left(\frac{ik\rho^2}{2f}+in\theta+ikf\right)\int\limits_0^R J_n\left(\frac{k\rho r}{f}\right)r\mathrm{d}r=\\=\left(-i\right)^{n+1}\frac{kA_{x,y}}{f}\exp\left(\frac{ik\rho^2}{2f}+in\theta+ikf\right)\times\\\times\begin{cases}n\left[1-J_0\left(y\right)-2\sum\limits_{m=1}^{n/2-1}J_{2m}\left(y\right)\right]-yJ_{n-1}\left(y\right),n=2p,\\n\left[\int\limits_0^y J_0\left(t\right)\mathrm{d}t-2\sum\limits_{m=1}^{(n-1)/2}J_{2m-1}\left(y\right)\right]-yJ_{n-1}\left(y\right),n=2p+1,\end{cases}\end{gathered}\tag{7.55}$$

where $y=kR\rho/f$.

$$\begin{gathered}E_z\left(\rho,\theta,z=f\right)=\frac{(-i)^n k}{2f^2}\exp\left(\frac{ik\rho^2}{2f}+in\theta+ikf\right)\times\\\Bigg\{\left(A_x-iA_y\right)\exp\left(i\theta\right)\int\limits_0^R J_{n+1}\left(\frac{k\rho r}{f}\right)r^2\mathrm{d}r-\\-\left(A_x+iA_y\right)\exp\left(-i\theta\right)\int\limits_0^R J_{n-1}\left(\frac{k\rho r}{f}\right)r^2\mathrm{d}r-\\-2i\rho\left(A_x\cos\theta+A_y\sin\theta\right)\int\limits_0^R J_n\left(\frac{k\rho r}{f}\right)r\mathrm{d}r\Bigg\}.\end{gathered}\tag{7.56}$$

The last integral in (7.56) is calculated as in (7.55). For the first two integrals we can also obtain analytical expressions for even values of the order of SPP n, so as for $p=N\pm1$

$$\int x^2J_p\left(cx\right)\mathrm{d}x=\frac{1-p^2}{c^3}\left[J_0\left(cx\right)+2\sum_{q=1}^{(p-3)/2}J_{2q}\left(cx\right)\right]-\frac{x^2}{c}J_{p-1}\left(x\right)-\frac{p+1}{c^2}xJ_{p-2}\left(x\right).\tag{7.57}$$

For small orders of SPP we obtain simple formulas. In particular, for $n=2$:

$$E_{x,y}\left(\rho,\theta,z=f\right)=-\frac{2ifA_{x,y}}{k\rho^2}\exp\left(\frac{ik\rho^2}{2f}+i2\theta+ikf\right)\left[J_0\left(\frac{kR\rho}{f}\right)+\frac{kR\rho}{2f}J_1\left(\frac{kR\rho}{f}\right)-1\right],$$

(7.58)

$$E_z\left(\rho,\theta,z=f\right)=-\frac{1}{2k\rho}\exp\left(\frac{ik\rho^2}{2f}+i2\theta+ikf\right)\times$$
$$\left(-\frac{2kR^2}{f}\left(A_x\cos\theta+A_y\sin\theta\right)J_2\left(\frac{kR\rho}{f}\right)+\right.$$
$$+\left[\frac{4}{\rho}\left(A_x-iA_y\right)\exp\left(i\theta\right)-\frac{2ik\rho}{f}\left(A_x\cos\theta+A_y\sin\theta\right)\right]\times$$
$$\left.\times\left\{\frac{f}{k\rho}\left[2-2J_0\left(\frac{kR\rho}{f}\right)\right]-RJ_1\left(\frac{kR\rho}{f}\right)\right\}\right). \tag{7.59}$$

Let us consider two special cases of the circular polarization of the field in the initial plane.

At $A_y=-iA_x$:

$$E_z\left(\rho,\theta,z\right)=\frac{iA_xR}{f\rho}\exp\left(\frac{ik\rho^2}{2f}+i\theta+ikf\right)\times$$
$$\left\{\frac{f}{kR}\left[2-2J_0\left(\frac{kR\rho}{f}\right)\right]-\rho J_1\left(\frac{kR\rho}{f}\right)-iRJ_2\left(\frac{kR\rho}{f}\right)\right\}. \tag{7.60}$$

At $A_y=iA_x$:

$$E_z\left(\rho,\theta,z\right)=\frac{iA_xR}{f\rho}\exp\left(\frac{ik\rho^2}{2f}+i3\theta+ikf\right)\times$$
$$\left(-iRJ_2\left(\frac{kR\rho}{f}\right)+\left(1-\frac{4f}{ik\rho^2}\right)\left\{\frac{f}{kR}\left[2-2J_0\left(\frac{kR\rho}{f}\right)\right]-\rho J_1\left(\frac{kR\rho}{f}\right)\right\}\right). \tag{7.61}$$

Figure 7.16 shows the distribution of the amplitude of the z-component of the electromagnetic field along the optical axis. Calculation parameters: wavelength $\lambda=514.5$ nm, the aperture radius: $R=2$ mm, the order of the SPP: $n=1$.

Figures 7.17 and 7.18 show the amplitude distribution of the x- and z-components of the electromagnetic field along the radial coordinate. Calculation parameters: wavelength: $\lambda=514.5$ nm, focal length of the lens: $f=500$ mm, the order of the SPP: $n=1$.

Figure 7.18 shows that the z-component of the amplitude can be several percent, so in some cases it makes sense to consider its existence, even in the paraxial case.

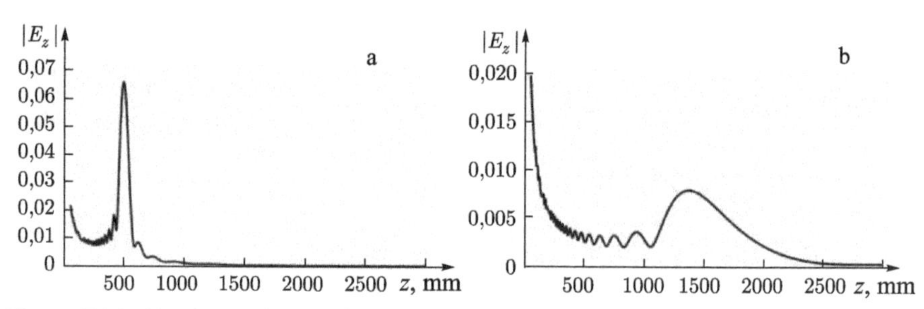

Figure 7.16. Absolute value of the z-component on the optical axis, calculated for f = 500 mm (a) and f = 1500 mm (b).

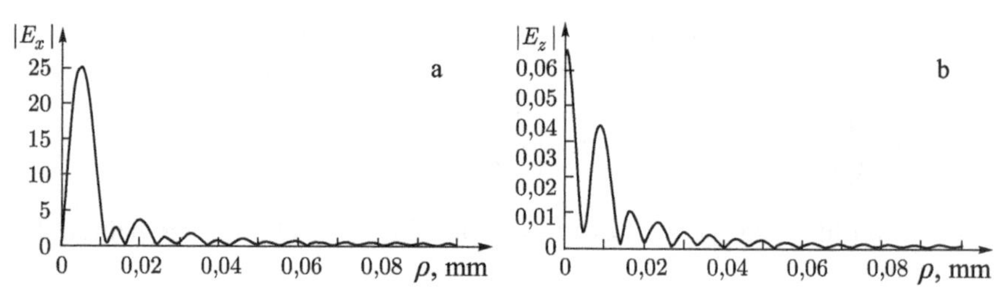

Fig. 7.17. The value of the amplitude in the plane z = 500 mm at an aperture radius of 2 mm. x-component (a) and z-component (b).

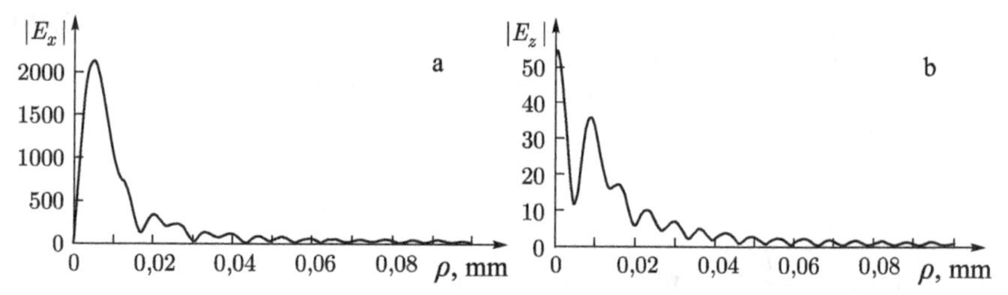

Fig. 7.18. The value of the amplitude in the plane z = 500 mm at an aperture radius of 20 mm. x-component (a) and z-component (b).

7.3. Quantized SPP with a restricted aperture, illuminated by a plane wave

There are many ways of making the SPP, such as multistage etching of silicon [59] or by ablation of polyamide substrates using an excimer laser [60]. The microrelief of the resultant SPP is stepped or quantized.

Multilevel SPPs were studied in [31, 61]. In [31] the efficiency of conversion of a Gaussian beam to the Laguerre–Gaussian mode (0,1) was theoretically calculated, and experiments were also carried out with a 16-level SPP produced by photolithography.

In [61] the authors found theoretically the minimum number of levels of the SPP phase (for the numbers $n < 8$), in which the finite-level SPPs slightly differ from the continuous SPPs. With the help of the finite-level SPP, produced on the basis of a liquid crystal cell, vortex laser beams with indices of singularity to 6 were formed in [61].

In [62, 63] attention was paid to the achromatic SPP, which forms almost the same vortex fields, if the wavelength of the illuminating radiation varies in a relatively wide range (140 nm). In these studies [31, 61–63] the SPP analyzed using a series expansion of angular harmonics:

$$\exp\left[i \operatorname{mod}\left(\frac{P\varphi}{2\pi}\right)\frac{2\pi n}{P}\right] = \sum_{m=-\infty}^{\infty} C_m \exp(im\varphi), \tag{7.62}$$

where mod (...) is an integer, P is the total number of phase levels of SPP, φ is the azimuthal angle of the polar coordinate system, n is the number of SPP, C_m are complex coefficients, exp $(in\varphi)$ are the angular harmonics describing the transmission of the continuous SPP with the number m.

In this section, the finite-level SPP, bounded by a polygonal aperture (i.e., having the shape of the polygon) is considered. Moreover, the number of quantization levels of the SPP phase equals the number of sides of a regular polygon, bounding the aperture of the SPP. In this case it was possible to obtain analytical expressions as a finite sum of plane waves for the complex amplitude, which describes the Fraunhofer diffraction of a plane wave on a finite-level SPP, bounded by a regular polygon.

Note that the possibility of the formation of vortex fields using non-spiral phase plates was already considered [64]. In our case, unlike in [64], with an increase in the number of phase quantization levels (or the number of sides), the diffraction pattern in the far field tends to the diffraction pattern formed by a continuous SPP with a circular aperture.

The equation of the polygonal aperture

Let Ω be the polygon defined by the coordinates of its vertices $A_p(x_p, y_p)$, $p = \overline{0, P-1}$ where P is the number of vertices (see Fig. 7.19).

Let the equation of the polygon connecting the p-th and (p +1)-th vertex is given by:

$$y = a_p x + b_p. \tag{7.63}$$

Let $f(x, y)$ be a function of two variables defined in R^2 as follows:

$$f(x,y) = \begin{cases} 1, (x,y) \in \Omega, \\ 0, (x,y) \notin \Omega. \end{cases} \tag{7.64}$$

It is known that the Fourier transform of such function $f(x, y)$ is calculated using the equation of the polygonal aperture [65]:

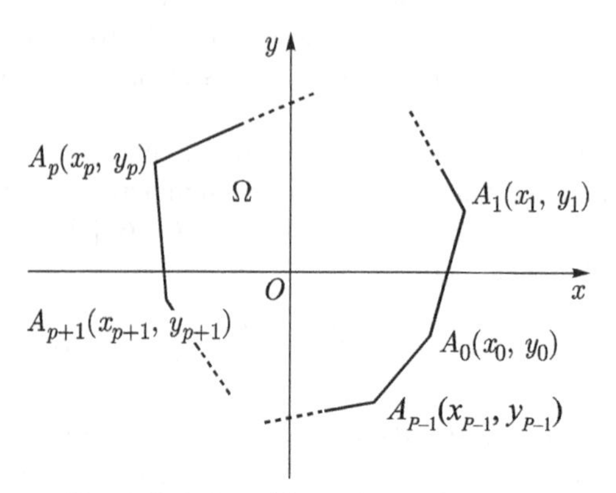

Fig. 7.19. DOE with a polygonal aperture.

$$\iint_{\Omega} \exp\left[\pm i\left(x\xi + y\eta\right)\right]dxdy = \sum_{p=1}^{P} \frac{a_p - a_{p-1}}{\left(\xi + \eta a_{p-1}\right)\left(\xi + \eta a_p\right)} \exp\left[\pm i\left(\xi x_p + \eta y_p\right)\right] =$$

$$= \sum_{p=1}^{P} \frac{\left(y_{p+1} - y_p\right)\left(x_p - x_{p-1}\right) - \left(y_p - y_{p-1}\right)\left(x_{p+1} - x_p\right)}{\left[\xi\left(x_{p+1} - x_p\right) + \eta\left(y_{p+1} - y_p\right)\right]\left[\xi\left(x_p - x_{p-1}\right) + \eta\left(y_p - y_{p-1}\right)\right]} \exp\left[\pm i\left(\xi x_p + \eta y_p\right)\right], \tag{7.65}$$

where p refers to value of the modes (p, P), i.e. $(x_P, y_P) = (x_0, y_0)$, $(x_{-1}, y_{-1}) = (x_{P-1}, y_{P-1})$, etc.

Then the complex amplitude describing Fraunhofer diffraction at polygonal apertures (Fig.7.19) of a plane wavelength λ at a focal length spherical lens is equal to f, is given by:

$$E(\xi,\eta) = -\frac{if}{2\pi k} \sum_{p=1}^{P} \frac{\left(y_{p+1} - y_p\right)\left(x_p - x_{p-1}\right) - \left(y_p - y_{p-1}\right)\left(x_{p+1} - x_p\right)}{\left[\xi\left(x_{p+1} - x_p\right) + \eta\left(y_{p+1} - y_p\right)\right]\left[\xi\left(x_p - x_{p-1}\right) + \eta\left(y_p - y_{p-1}\right)\right]} \exp\left[\pm i\frac{k}{f}\left(\xi x_p + \eta y_p\right)\right], \tag{7.66}$$

where $k = 2\pi/\lambda$ is the wave number.

Fraunhofer diffraction of a plane wave on the DOE with the form of a regular polygon and a piecewise constant microrelief

Consider the diffractive optical element having the shape of a regular polygon $\Omega = A_0 A_1 ... A_{P-1}$, inscribed in a circle of radius R and containing the origin O. Then,

$\Omega = \bigcup_{p=0}^{P-1} \Omega_p$, where Ω_p are the triangles OA_pA_{p+1}, and each vertex has the coordinates of A_p (Fig. 7.20)

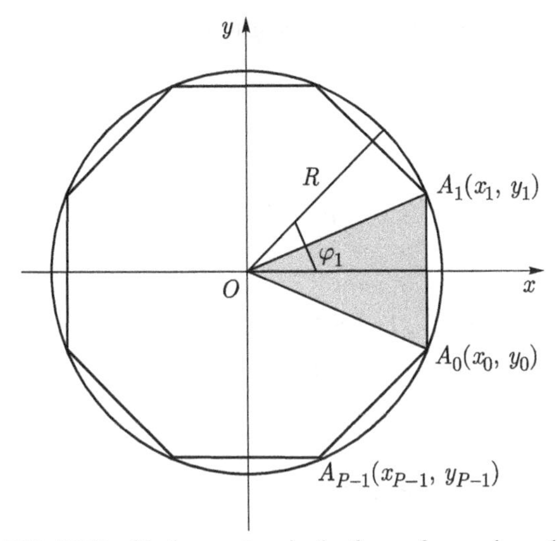

Fig. 7.20. DOE with the aperture in the form of a regular polygon.

$$\begin{cases} x_p = R\cos\left(\varphi_p - \dfrac{\pi}{P}\right), \\ y_p = R\sin\left(\varphi_p - \dfrac{\pi}{P}\right), \end{cases} \qquad \varphi_p = \frac{2\pi}{P}p. \tag{7.67}$$

Let the depth of the microrelief inside of each triangle Ω_p be constant, then inside Ω_p and the complex transmission function of the DOE is constant:

$$\tau(x,y) = \exp\left(i\Psi_p\right). \tag{7.68}$$

Then, using the equation for the polygonal aperture, we can obtain an expression for the complex amplitude, which describes the Fraunhofer diffraction of a plane wave length λ at a DOE (Figure 7.20):

$$\begin{aligned} E(\xi,\eta) &= \frac{if}{2\pi k}R^2\sin\left(\frac{2\pi}{P}\right)\sum_{p=0}^{P-1}\frac{\exp\left(i\Psi_p\right)}{\left(\xi x_{p+1}+\eta y_{p+1}\right)\left(\xi x_p+\eta y_p\right)} - \\ &-\frac{if}{2\pi k}R^2\sin\left(\frac{2\pi}{P}\right)\sum_{p=0}^{P-1}\frac{\exp\left(i\Psi_p\right)\exp\left[-i\frac{k}{f}\left(\xi x_p+\eta y_p\right)\right]}{\left[\xi\left(x_{p+1}-x_p\right)+\eta\left(y_{p+1}-y_p\right)\right]\left(\xi x_p+\eta y_p\right)} + \\ &+\frac{if}{2\pi k}R^2\sin\left(\frac{2\pi}{P}\right)\sum_{p=0}^{P-1}\frac{\exp\left(i\Psi_p\right)\exp\left[-i\frac{k}{f}\left(\xi x_{p+1}+\eta y_{p+1}\right)\right]}{\left(\xi x_{p+1}+\eta y_{p+1}\right)\left[\xi\left(x_{p+1}-x_p\right)+\eta\left(y_{p+1}-y_p\right)\right]}. \end{aligned} \tag{7.69}$$

In the transition to the polar coordinates instead of (7.68) we obtain the following expression:

$$E(\rho,\theta)=\frac{if}{2\pi k\rho^2}\sin\left(\frac{2\pi}{P}\right)\sum_{p=0}^{P-1}\frac{\exp\left(i\Psi_p\right)}{\cos\left(\varphi_p+\frac{\pi}{P}-\theta\right)\cos\left(\varphi_p-\frac{\pi}{P}-\theta\right)}+$$

$$+\frac{if}{2\pi k\rho^2}\cos\frac{\pi}{P}\sum_{p=0}^{P-1}\left[\frac{\exp\left(i\Psi_p\right)}{\sin\left(\varphi_p-\theta\right)}-\frac{\exp\left(i\Psi_{p-1}\right)}{\sin\left(\varphi_{p-1}-\theta\right)}\right]\frac{\exp\left[-i\frac{kR\rho}{f}\cos\left(\varphi_p-\frac{\pi}{P}-\theta\right)\right]}{\cos\left(\varphi_p-\frac{\pi}{P}-\theta\right)}. \tag{7.70}$$

In the case of quantized SPP, i.e. $\Psi_p=n\varphi_p$, from (7.69) we get:

$$E_n^P(\rho,\theta)=\frac{if}{2\pi k\rho^2}\sin\left(\frac{2\pi}{P}\right)\sum_{p=0}^{P-1}\frac{\exp\left(in\varphi_p\right)}{\cos\left(\varphi_p+\frac{\pi}{P}-\theta\right)\cos\left(\varphi_p-\frac{\pi}{P}-\theta\right)}+$$

$$+\frac{if}{2\pi k\rho^2}\cos\frac{\pi}{P}\sum_{p=0}^{P-1}\left[\frac{\exp\left(in\varphi_p\right)}{\sin\left(\varphi_p-\theta\right)}-\frac{\exp\left(in\varphi_{p-1}\right)}{\sin\left(\varphi_{p-1}-\theta\right)}\right]\frac{\exp\left[-i\frac{kR\rho}{f}\cos\left(\varphi_p-\frac{\pi}{P}-\theta\right)\right]}{\cos\left(\varphi_p-\frac{\pi}{P}-\theta\right)}. \tag{7.71}$$

Figure 7.21 shows a picture of the Fraunhofer diffraction of a plane wave on a continuous SPP limited by a circular aperture, obtained by the mean sum of Bessel functions [66]:

$$E_n(\rho,\theta)=\frac{(-i)^{n+1}k\exp(in\theta)}{f\bar{\rho}^2}\begin{cases}n\left[1-J_0(y)-2\sum\limits_{m=1}^{(n-2)/2}J_{2m}(y)\right]-yJ_{n-1}(y), n=2m,\\ n\left[\int\limits_0^y J_0(t)dt-2\sum\limits_{m=1}^{(n-1)/2}J_{2m-1}(y)\right]-yJ_{n-1}(y), n=2m+1,\end{cases} \tag{7.72}$$

where $y=R\bar{\rho}=kR\rho/f$, $J_n(x)$ is the Bessel function of the n-th order

$$\int\limits_0^y J_0(t)dt=\frac{y}{2}\left\{\pi J_1(y)H_0(y)+J_0(y)\left[2-\pi H_1(y)\right]\right\}, \tag{7.73}$$

$H_{0,1}(y)$ is the Struve function of zero and first orders.

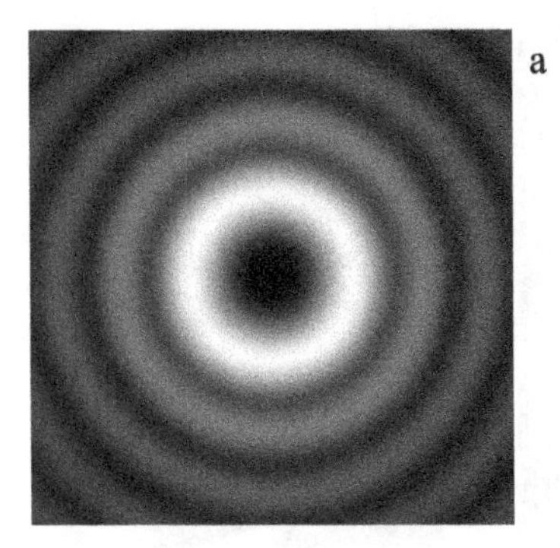

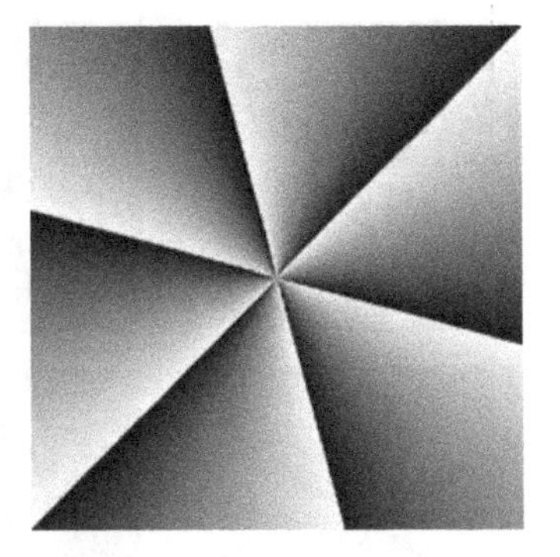

Fig. 7.21. Fraunhofer diffraction pattern of a plane wave on a continuous restricted SPP: amplitude (a) and phase (b).

Table 7.1

The number of sectors	SD
18	19.1411
30	1.9003
42	0.1320
54	0.0479

Table 7.2

Number of SPP	Minimum number of sectors
2	19
4	25
6	29
8	35
10	39

In the calculation we used the following parameters: wavelength 633 nm, the focal length of the spherical lens 150 mm, the radius of the aperture 2 mm, the order of the SPP 6.

Figure 7.22 shows the Fraunhofer diffraction pattern of a plane wave on a quantized limited SPP, obtained by the formula (7.71).

Table 7.1 shows the dependence of the standard deviation of the Fraunhofer diffraction pattern of a plane wave on a limited quantized spiral phase plate on the diffraction patterns from a limited continuous SPP for different numbers of sectors.

Table 7.2 shows, for several numbers of SPP, the minimum number of sectors of the multilevel SPP, in which the standard deviation of the Fraunhofer diffraction pattern from the diffraction pattern for continuous SPP does not exceed 2%.

7.4. Helical conical axicon

The spiral phase plate is the simplest optical element intended to generate wave fronts with a helical phase singularity. The transmission function of the SPP has only one parameter – topological charge n. By varying it we can change the radius

Fig. 7.22. Fraunhofer diffraction patterns of a plane wave on a quantized limited SPP: DOE phase (a, d, g, j), amplitude (b, d, h, k) and phase (c, f, i, l) in the zone of Fraunhofer diffraction. The number of sectors: 18 (a, b, c), 30 (c, d, e), 42 (g, h, i), 54 (j, k, l).

of the main ring of the diffraction pattern, however, to control other properties of the beam we do not have enough degrees of freedom. This leads to the idea of using combined optical elements which also include SPP. The simplest such element is a helical axicon whose phase depends linearly on both the angular and radial polar coordinates.

7.4.1. Diffraction of Gaussian beam in an aperture-limited helical axicon

Consider the scalar paraxial diffraction of a collimated Gaussian beam with a complex amplitude

$$E_0\left(r\right)=\exp\left(-\frac{r^2}{w^2}\right). \tag{7.74}$$

The helical axicon (HA), which in the approximation of a thin transparant is described by the transmission function of the form

$$\tau_n(r,\varphi)=\exp(i\alpha r+in\varphi), \tag{7.75}$$

where w is the Gaussian beam waist radius, (r, φ) are the polar coordinates in the plane of the HA at $z = 0$, z is the optical axis, α is the axicon parameter; $n = 0, \pm1, \pm2,...$, is the number of SPP.

Then paraxial diffraction of the wave (7.74) on HA (7.75) is described by the Fresnel transformation:

$$F_n(\rho,\theta,z)=-\frac{ik}{2\pi z}\exp\left(ikz+\frac{ik\rho^2}{2z}\right)\times$$
$$\times\int_0^R\int_0^{2\pi}\exp\left[-\frac{r^2}{w^2}+i\alpha r+in\varphi+\frac{ikr^2}{2z}-\frac{ik}{z}\rho r\cos(\varphi-\theta)\right]rdrd\varphi, \tag{7.76}$$

where (ρ, θ) are the polar coordinates in the plane z (z is the optical axis), $k = 2\pi/\lambda$ is the wave number. Using the background integral [53]

$$\int_0^\infty x^{\lambda+1}\exp(-px^2)J_\nu(cx)dx=\frac{c^\nu p^{-(\nu+\lambda+2)/2}}{2^{\nu+1}\nu!}\Gamma\left(\frac{\nu+\lambda+2}{2}\right){}_1F_1\left[\frac{\nu+\lambda+2}{2},\nu+1,-\left(\frac{c}{2\sqrt{p}}\right)^2\right], \tag{7.77}$$

instead of (7.76) we get:

$$F_n(\rho,\theta,z)=\frac{(-i)^{n+1}k}{z}\exp\left[in\theta+ikz+\frac{ik\rho^2}{2z}\right]\left(\frac{k\rho}{2z}\right)^n\frac{\gamma^{-(n+2)/2}}{2^{n+1}n!}\times$$
$$\times\sum_{m=0}^{\infty}\frac{(i\alpha)^m\gamma^{-m/2}}{m!}\Gamma\left(\frac{m+n+2}{2}\right){}_1F_1\left[\frac{m+n+2}{2},n+1,-\left(\frac{k\rho}{2z\sqrt{\gamma}}\right)^2\right], \tag{7.78}$$

where $\gamma=1/w^2-ik/(2z)$, ${}_1F_1(a,b,x)$ is the degenerate or confluent hyper-geometric function:

$${}_1F_1(a,b,x)=\sum_{m=0}^{\infty}\frac{(a)_m x^m}{(b)_m m!}, \tag{7.79}$$

$(a)_m = \Gamma(a + m)/\Gamma(a)$, $(a)_0 = 1$, and $\Gamma(x)$ is the gamma function.

From the expression (7.78) it follows that the diffraction pattern is a set of concentric rings. When $\rho = 0$ the intensity in the centre of the diffraction pattern at any $n \neq 0$ zero. Since the complex amplitude (7.78) depends on the combination of variables $k\rho/(2z\sqrt{\gamma})$ then the radii ρ_l of the local maxima and minima of the diffraction pattern must satisfy the following expression:

$$\rho_l = \frac{wza_l}{z_0}\left(1+\frac{z_0^2}{z^2}\right)^{1/4}, \tag{7.80}$$

where a_l is a constant depending only on the number of the rings $l = 1,2,...$ of diffraction patterns and the parameter α, $z_0 = kw^2/2$ is the Rayleigh length.

At $\alpha = 0$ (i.e. no axicon), from (7.78) we obtain the relationshp for the complex amplitude of Fresnel diffraction of the Gaussian beam on the SPP:

$$F_n(\rho,\theta,z,\alpha=0) = \frac{(-i)^{n+1}k}{z}\exp\left[i(n\theta+kz)+\frac{ik\rho^2}{2z}\right]\left(\frac{k\rho}{2z}\right)^n \times$$
$$\times\frac{\gamma^{-(n+2)/2}}{2^{n+1}n!}\Gamma\left(\frac{n+2}{2}\right){}_1F_1\left[\frac{n+2}{2},n+1,-\left(\frac{k\rho}{2z\sqrt{\gamma}}\right)^2\right]. \tag{7.81}$$

Given the connection between the hypergeometric and Bessel functions

$$J_{(n-1)/2}(x) = \frac{\left(\frac{x}{2}\right)^{(n-1)/2}\exp(-ix)}{\Gamma\left(\frac{n-1}{2}\right)}{}_1F_1\left(\frac{n}{2},n;2ix\right) \tag{7.82}$$

and the recurrence relation for the hypergeometric functions

$${}_1F_1\left(\frac{n}{2},n+1;2ix\right) = \left(i\frac{d}{dx}+2\right){}_1F_1\left(\frac{n}{2},n;2ix\right), \tag{7.83}$$

we can replace (7.81) to obtain a well-known relation for the Fresnel diffraction of Gaussian beam on the SPP [28, 52]:

$$E_n(\rho,\theta,z,\alpha=0) = \frac{(-i)^{n+1}\sqrt{\pi}}{2}\left(\frac{z_0}{z}\right)^2\left(\frac{\rho}{w}\right)\left[1+\left(\frac{z_0}{z}\right)^2\right]^{-3/4} \times$$
$$\times\exp\left[i\frac{3}{2}\tan^{-1}\left(\frac{z_0}{z}\right)-i\frac{k\rho^2}{2R_0(z)}+i\frac{k\rho^2}{2z}-\frac{\rho^2}{w^2(z)}+in\theta+ikz\right]\times$$
$$\times\left\{I_{\frac{n-1}{2}}\left[\rho^2\left(\frac{1}{w^2(z)}+\frac{ik}{2R_0(z)}\right)\right]-I_{\frac{n+1}{2}}\left[\rho^2\left(\frac{1}{w^2(z)}+\frac{ik}{2R_0(z)}\right)\right]\right\}, \tag{7.84}$$

where $w^2(z) = 2w^2[1+(z/z_0)^2]$, $R_0(z) = 2z[1+(z/z_0)^2]$, $I_\nu(x)$ is the Bessel function of second kind and ν-th order.

When $z\to\infty$ ($z \gg z_0$) the expression (7.78) yields the following formula for the complex amplitude of Fraunhofer diffraction of the Gaussian beam on the HA ($\gamma = 1/w^2$)

$$F_n\left(\rho,\theta,z\to\infty\right)=\frac{\left(-i\right)^{n+1}z_0}{2^n n!z}\exp\left(in\theta+ikz+\frac{ik\rho^2}{2z}\right)\left(\frac{z_0\rho}{zw}\right)^n\times$$
$$\times\sum_{m=0}^{\infty}\frac{\left(i\alpha w\right)^m}{m!}\Gamma\left(\frac{m+n+2}{2}\right){}_1F_1\left[\frac{m+n+2}{2},n+1,-\left(\frac{z_0\rho}{zw}\right)^2\right]. \tag{7.85}$$

At $\alpha=0$ (i.e. no axicon) and $z\to\infty$ ($z\gg z_0$) from (7.78) follows the expression for the complex amplitude of Fraunhofer diffraction of the Gaussian beam on the SPP:

$$F_n\left(\rho,\theta,z\to\infty,\alpha=0\right)=$$
$$=\frac{\left(-i\right)^{n+1}z_0}{2^n n!z}\exp\left(in\theta+ikz+\frac{ik\rho^2}{2z}\right)\left(\frac{z_0\rho}{zw}\right)^n\Gamma\left(\frac{n+2}{2}\right){}_1F_1\left[\frac{n+2}{2},n+1,-\left(\frac{z_0\rho}{zw}\right)^2\right]. \tag{7.86}$$

It is interesting to compare the expression (7.86) with the complex amplitude of Fraunhofer diffraction of a restricted plane wave of radius R on the SPP, when the focal length of the spherical lens is equal to f [67]:

$$E_n\left(\rho,\theta\right)=\frac{\left(-i\right)^{n+1}\exp\left(in\theta+ikz\right)}{\left(n+2\right)n!}\left(\frac{kR^2}{f}\right)\left(\frac{kR\rho}{2f}\right)^n{}_1F_2\left[\frac{n+2}{2},\frac{n+4}{2},n+1,-\left(\frac{kR\rho}{2f}\right)^2\right], \tag{7.87}$$

where ${}_1F_2\,(a, b, c, x)$ is the hypergeometric function:

$${}_1F_2\left(a,b,c,x\right)=\sum_{m=0}^{\infty}\frac{\left(a\right)_m x^m}{\left(b\right)_m\left(c\right)_m m!}. \tag{7.88}$$

Figure 7.23 shows the calculated distribution of the amplitude $|F_n(\rho,\theta)|$ in relative units as a function of the radial variable. These curves represent the radial profile of the Fresnel diffraction pattern ($z = 200$ mm) of the Gaussian beam with the waist radius $w = 1$ mm and a wavelength $\lambda = 633$ nm on the HA ($n = 8$) with parameter $\alpha = 0$ mm^{-1} (a), $\alpha = 20$ mm^{-1} (b), $\alpha = 50$ mm^{-1} (c).

Figure 7.23 shows that the radius of the main peak of the amplitude increases with increasing values of α.

Figure 7.24 shows two calculated radial Fresnel diffraction patterns (amplitude $|F_n(\rho,\theta)|$) for a Gaussian beam ($w = 1$ mm, $\lambda = 633$ nm) for HA ($n = 8$) with parameter $\alpha = 20$ mm^{-1} at a distance $z = 400$ mm (a) and $z = 500$ mm (b). From Fig. 7.24 it can be seen that with increasing distance z the radius of the first bright ring in the diffraction pattern, characterized by the maximum amplitude, also increases. Comparing Figures 7.23 and 7.24 gives reason to conclude that the radius of the first ring can be changed either by changing the parameter α of the axicon at

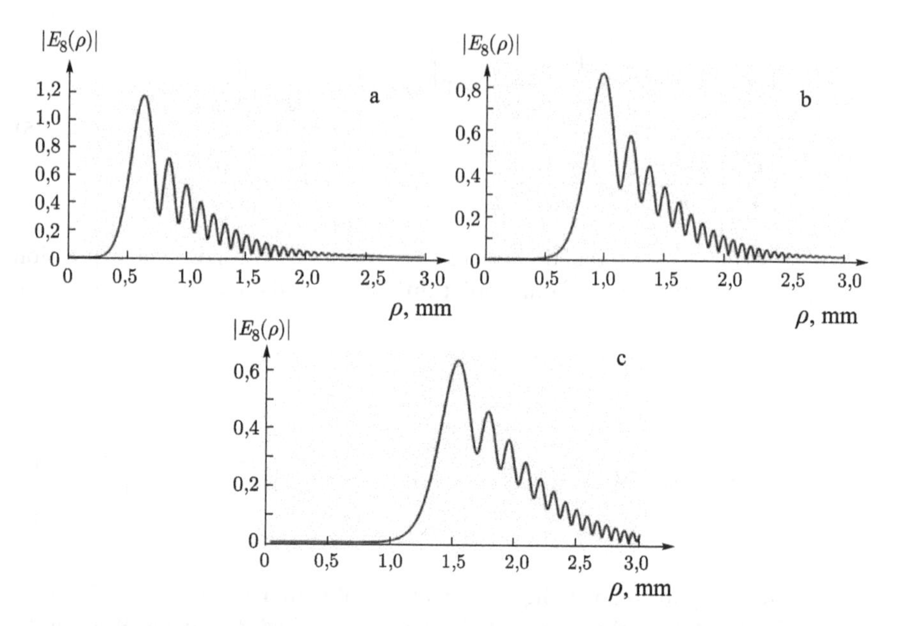

Fig. 7.23. The radial profile of the Fresnel diffraction pattern (the amplitude at a distance $z = 200$ mm) for a Gaussian beam ($\lambda = 633$ nm, $w = 1$ mm) on the HA ($n = 8$): $\alpha = 0$ mm^{-1} (a), $\alpha = 20$ mm^{-1} (b), $\alpha = 50$ mm^{-1} (c).

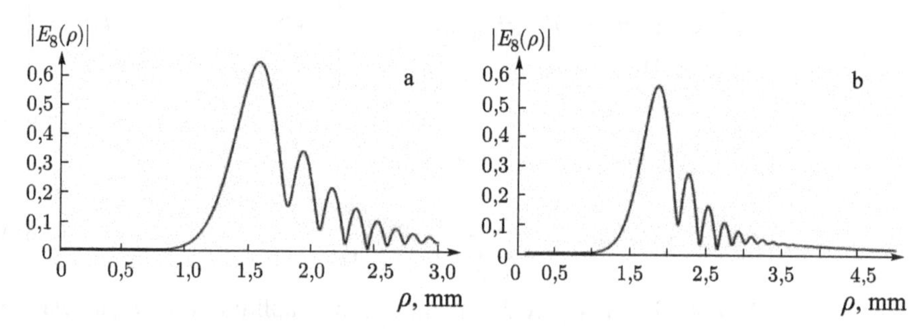

Fig. 7.24. The radial profile of the Fresnel diffraction pattern (amplitude $|F_n(\rho, \theta)|$) of a Gaussian beam ($\lambda = 633$ nm, $w = 1$ mm) on the HA ($n = 8$, $\alpha = 20$ mm^{-1}): $z = 400$ mm (a), $z = 500$ mm (b).

a constant distance z, or by changing the distance z from the axicon to the plane of observation. The difference will be in the amount of the peripheral rings (sidelobes) in the diffraction pattern. From Fig. 7.23 it can be seen that 13 peripheral diffraction rings are stacked in a radial range from 1.5 mm to 3 mm. At the same time in Fig. 7.24a in the same radial range from 1.5 mm to 3 mm there are only seven lateral lobes, despite the fact that the radius of the first ring is the same in both patterns.

Figure 7.25 shows two calculated radial Fresnel diffraction pattern (the amplitude $|F_n(\rho, \theta)|$ at a distance $z = 200$ mm) Gaussian beam ($w = 1$ mm, $\lambda = 633$ nm) at HA ($\alpha = 20$ mm^{-1}) of different orders n: 40 (a) and 50 (b).

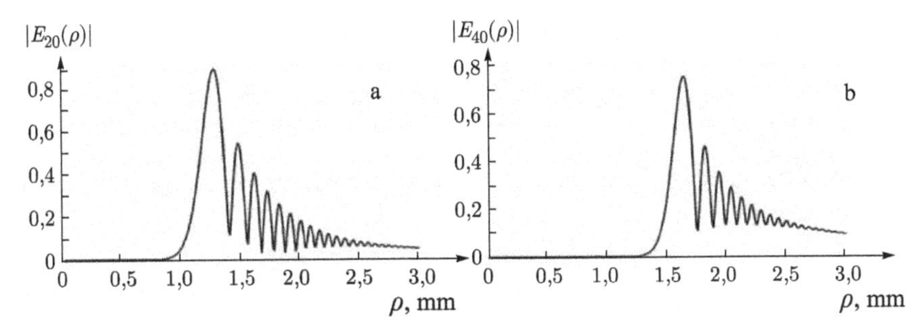

Fig. 7.25. The radial profile of the Fresnel diffraction pattern (amplitude $|F_n(\rho, \theta)|$ at distance $z =$ 200 mm) of a Gaussian beam (λ = 633 nm, w = 1 mm) on the HA (α = 20 mm^{-1}): a) n = 20 mm, b) n = 40.

Figure 7.25 shows that the radius of the first ring in the diffraction pattern can be changed by varying both the order of HA n, and the parameter α. Note, however, that in the case of Fig. 7.25, in addition to an increase in the radius of the first ring, an increase of order n leads to thinning of the first ring, to a larger number of peripheral rings and an increased contrast of the rings.

7.4.2. Diffraction of a restricted plane wave on a helical axicon

Diffraction of an unbounded plane wave is considered in [68]. Fraunhofer diffraction of a plane wave by a helical axicon of a finite radius with the transmission function circl(r/R)exp(iar) is described by the following expression:

$$F(\bar{\rho}) = \int_0^R \exp(i\alpha r) J_n\left(\frac{k}{f}\bar{\rho}r\right) r dr. \tag{7.89}$$

We consider the integral:

$$I = \int_0^R \exp(i\alpha r) J_n(\rho r) r dr, \quad \rho = \frac{k}{f}\bar{\rho}. \tag{7.90}$$

Using the integral representation of Bessel functions

$$J_n(x) = \frac{(-i)^n}{2\pi}\int_0^{2\pi} \exp(in\varphi)\exp(ix\cos\varphi)d\varphi, \tag{7.91}$$

we obtain:

$$I = \frac{(-i)^{n+2}}{2\pi}\frac{\partial}{\partial\alpha}\left[\exp(i\alpha R)\int_0^{2\pi}\exp(in\varphi)\frac{\exp(iR\rho\cos\varphi)}{\alpha+\rho\cos\varphi}d\varphi - \int_0^{2\pi}\frac{\exp(in\varphi)d\varphi}{\alpha+\rho\cos\varphi}\right]. \tag{7.92}$$

Using the known relation for the Bessel functions

$$\exp(ix\cos\varphi) = \sum_{m=-\infty}^{+\infty} i^m \exp(-im\varphi) J_m(x), \tag{7.93}$$

instead of the integral (7.92) we can obtain an expression for the diffraction in the form of a series:

$$I = \frac{(-i)^n}{2\pi}\frac{\partial I_1^n}{\partial\alpha} - \frac{\exp(i\alpha R)}{2\pi}\sum_{m=-\infty}^{+\infty} i^m \left(iRI_1^m + \frac{\partial I_1^m}{\partial\alpha}\right) J_{m+n}(R\rho), \tag{7.94}$$

where

$$I_1^n = \int_0^{2\pi} \frac{\exp(in\varphi)\,d\varphi}{\alpha+\rho\cos\varphi}. \tag{7.95}$$

The integrals (7.95) and their derivatives are computed by applying the theory of residues. The expressions for the integrals I_1^n and $\partial I_1^n/\partial\alpha$ are given below.

Case 1. $0<\rho<|\alpha|$.

$$I_1^n = \frac{2\pi\,\mathrm{sgn}\,\alpha}{\sqrt{\alpha^2-\rho^2}}\left(\frac{-\alpha+\mathrm{sgn}\,\alpha\sqrt{\alpha^2-\rho^2}}{\rho}\right)^{|n|}, \tag{7.96}$$

$$\frac{\partial I_1^n}{\partial\alpha} = -2\pi\,\mathrm{sgn}\,\alpha\left(\frac{-\alpha+\mathrm{sgn}\,\alpha\sqrt{\alpha^2-\rho^2}}{\rho}\right)^{|n|}\frac{\alpha+\mathrm{sgn}\,\alpha\,|n|\sqrt{\alpha^2-\rho^2}}{\left(\alpha^2-\rho^2\right)^{\frac{3}{2}}}. \tag{7.97}$$

Case 2. $\rho>|\alpha|$:

$$I_1^n = \frac{\pi i}{\sqrt{\rho^2-\alpha^2}}\left[\left(\frac{-\alpha-i\sqrt{\rho^2-\alpha^2}}{\rho}\right)^{|n|} - \left(\frac{-\alpha+i\sqrt{\rho^2-\alpha^2}}{\rho}\right)^{|n|}\right] = \pi i\frac{\chi^{*|n|}-\chi^{|n|}}{\sqrt{\rho^2-\alpha^2}}, \tag{7.98}$$

where $\chi = \left[-\alpha+i\left(\rho^2-\alpha^2\right)^{1/2}\right]/\rho$.

$$\frac{\partial I_1^n}{\partial\alpha} = \pi i\left[\alpha\left(\rho^2-\alpha^2\right)^{-\frac{3}{2}}\left(\chi^{*|n|}-\chi^{|n|}\right) - i\left(\rho^2-\alpha^2\right)^{-1}|n|\left(\chi^{*|n|}+\chi^{|n|}\right)\right]. \tag{7.99}$$

Diffraction of a restricted plane wave on a spiral phase plate

We obtain a formula for the Fraunhofer diffraction of a plane wave on a limited spiral phase plate (i.e. $\alpha = 0$). We shall also assume that $n \geq 0$, as for $n < 0$ it is sufficient to multiply the complex amplitude in the output plane by $(-1)^n$ (it can be seen from (7.89)).

At $\alpha = 0$ the expressions for the integrals I_1^n and $\partial I_1^n / \partial \alpha$ can be significantly simplified:

$$I_1^n = \begin{cases} 0, n = 2m, \\ -2\pi i^{|2m+1|+1} \rho^{-1}, n = 2m+1; \end{cases} \tag{7.100}$$

$$\frac{\partial I_1^n}{\partial \alpha} = \begin{cases} 2\pi \, i^{|n|} \, |n| \, \rho^{-2}, n = 2m, \\ 0, n = 2m+1. \end{cases} \tag{7.101}$$

Substituting these expressions in (7.94), dividing the sum $m \in (-\infty, +\infty)$ by the sums $m \in [0, +\infty)$ and $m \in (-\infty, -1]$, to get rid of the modules, using the recurrence relation for Bessel functions $2\nu J_\nu(z) = z[J_{\nu-1}(z) + J_{\nu+1}(z)]$ and, given that $\lim\limits_{\nu \to \infty} J_\nu(z) = 0$ we can reduce (7.94) to the following form:

$$I = \frac{(-i)^n}{2\pi} \frac{\partial I_1^n}{\partial \alpha} + n\rho^{-2} \left[\sum_{m=0}^{+\infty} J_{2m+n}(R\rho) - \sum_{m=1}^{+\infty} J_{n-2m}(R\rho) \right] - R\rho^{-1} J_{n-1}(R\rho). \tag{7.102}$$

For even $n \geq 0$:

$$I = n\rho^{-2} - n\rho^{-2} \left[J_0(R\rho) + 2\sum_{m=1}^{n/2-1} J_{2m}(R\rho) \right] - R\rho^{-1} J_{n-1}(R\rho). \tag{7.103}$$

Multiplying by k/f and substituting ρ for $k\rho/f$, we obtain a formula for the Fraunhofer diffraction of a plane wave on a restricted SPP of the even non-negative integer order n:

$$F(\rho) = \frac{f}{k\rho^2} \left\{ n \left[1 - J_0\left(\frac{k}{f} R\rho \right) \right] - \frac{k}{f} R\rho J_{n-1}\left(\frac{k}{f} R\rho \right) - 2n \sum_{m=1}^{n/2-1} J_{2m}\left(\frac{k}{f} R\rho \right) \right\}. \tag{7.104}$$

For odd n, $n > 0$:

$$I = n\rho^{-2} \left[\int_0^{R\rho} J_0(x)\,dx - 2\sum_{m=1}^{(n-1)/2} J_{2m-1}(R\rho) \right] - R\rho^{-1} J_{n-1}(R\rho). \tag{7.105}$$

Multiplying by k/f and substituting ρ for $k\rho/f$, we obtain a formula for the Fraunhofer diffraction of a plane wave on a restricted SPP of the positive odd integer order n:

$$F(\rho)=\frac{f}{k\rho^2}\left[n\int\limits_0^{\frac{k}{f}R\rho}J_0(x)dx-2n\sum_{m=1}^{(n-1)/2}J_{2m-1}\left(\frac{k}{f}R\rho\right)-\frac{k}{f}R\rho J_{n-1}\left(\frac{k}{f}R\rho\right)\right].$$

(7.106)

The use of a conical axicon provides an additional degree of freedom (the parameter α) as compared to a plane wave. For example, we can achieve a smooth radial distribution of the amplitude.

Figures 7.26 and 7.27 show the results of numerical simulation of Fraunhofer diffraction of a plane wave by a helical axicon with a finite radius. We used the following settings:

- Wavelength: $\lambda = 633$ nm.
- Focal length of spherical lens: $f = 140$ mm.
- The order of the SPP: $n = 4$.
- Parameter of the axicon: $\alpha = 0$ mm^{-1} (i.e. no axicon) (a) and $\alpha = 1$ mm^{-1} (b).
- Aperture radius: $R = 2$ mm.

It is seen that the graph in Fig. 7.26b, obtained using an axicon, is 'smoother'.

If we increase the value of the axicon parameter α, then increase of the value of the radial coordinate ρ is accompanied by an increase in the number of 'lobes' (Fig. 7.28).

Experiments on the formation of a ring of light with the help of the HA are given in [58].

Thus, the use of the helical axicon raises the possibility of formation of optical vortices with desired characteristics. This is of practical importance for the problems of nanophotonics, in particular the optical manipulation of micro- and nano-objects. Due to the pressure of light these objects tend to be drawn into the area with the greatest intensity, but the presence of side lobes in the diffraction pattern shown in Figs. 7.26a and 7.27a means that an object can be drawn into the side instead of the main ring. The values of the radius and the speed of rotation of

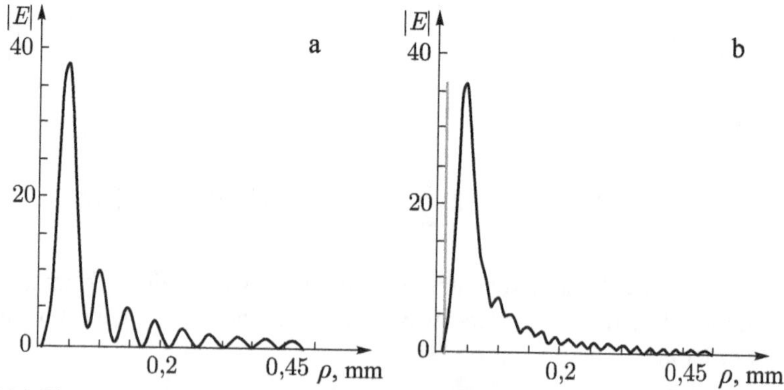

Fig. 7.26. The result of numerical simulation of Fraunhofer diffraction of a plane wave on a helical axicon with a finite radius (dependence of the modulus of the amplitude on the radial coordinate) without the axicon $\alpha = 0$ (a) and with an axicon $\alpha = 1$ mm^{-1} (b).

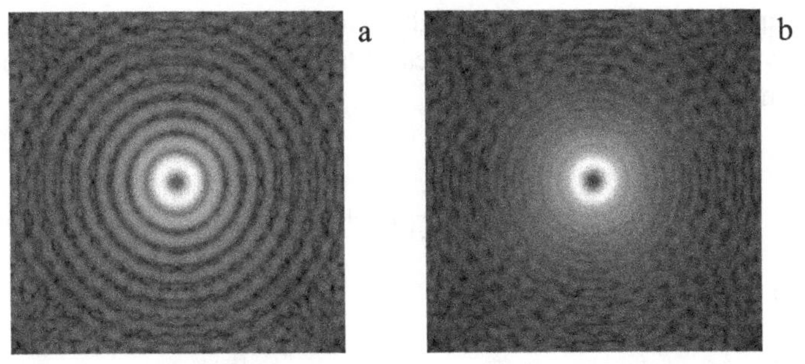

Fig. 7.27. The result of numerical simulation of Fraunhofer diffraction of a plane wave on a helical axicon finite radius (two-dimensional diffraction pattern) with no axicon $\alpha = 0$ (a) and with an axicon $\alpha = 1$ mm^{-1} (b).

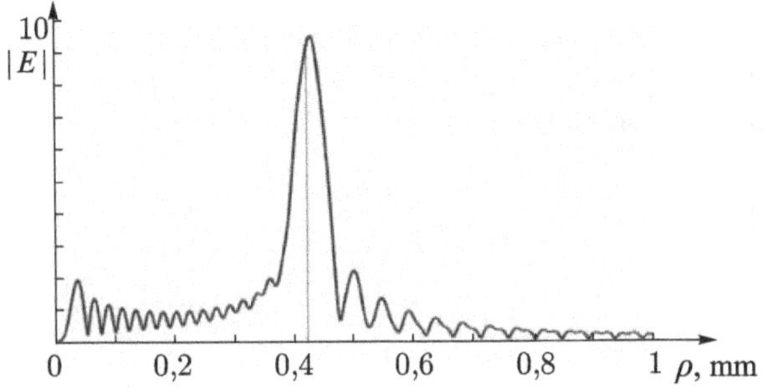

Fig. 7.28. The result of numerical simulation of Fraunhofer diffraction of a plane wave on a helical axicon with a finite radius at the axicon parameter $\alpha = 30$ mm^{-1}.

the object will differ from the target. The use of an axicon for the formation of the diffraction patterns, shown in Figs. 7.26b and 7.27b, is designed to eliminate this problem.

7.5. Helical logarithmic axicon

7.5.1. General theory of hypergeometric laser beams

Let us consider a light field with the initial function of the complex transmittance of the form:

$$E_{\gamma nm}(r,\varphi) = \frac{1}{2\pi}\left(\frac{r}{w}\right)^m \exp\left(-\frac{r^2}{2\sigma^2} + i\gamma \ln\frac{r}{w} + in\varphi\right), \tag{7.107}$$

where (r, φ) are the polar coordinates in the initial plane ($z = 0$), w and γ are the actual parameters of the logarithmic axicon, σ is the Gaussian beam waist radius, n is the integer order of a spiral phase plate, m is a parameter. The complex amplitude (7.107) describes a light field with infinite energy and a singularity at $r = 0$ and $m < 0$. Despite this, in any transverse plane at a distance z from the initial plane the complex amplitude of the light field, generated by the function (7.107), will not have any singularities and will be finite.

In the paraxial propagation of the light field (7.107), its complex amplitude at a distance z will be determined by the Fresnel transform, which in polar coordinates has the form:

$$E(\rho,\theta,z) = -\frac{ik}{2\pi z}\iint_{R^2} E(r,\varphi,0)\exp\left\{\frac{ik}{2z}\left[\rho^2 + r^2 - 2\rho r\cos(\varphi-\theta)\right]\right\} r\mathrm{d}r\mathrm{d}\varphi. \tag{7.108}$$

We have the reference integral:

$$\int_0^\infty x^{\alpha-1}\exp(-px^2)J_\nu(cx)\mathrm{d}x = c^\nu p^{-\frac{\nu+\alpha}{2}} 2^{-\nu-1}\Gamma\left(\frac{\nu+\alpha}{2}\right)\Gamma^{-1}(\nu+1)\,{}_1F_1\left(\frac{\nu+\alpha}{2},\nu+1,-\frac{c^2}{4p}\right), \tag{7.109}$$

where ${}_1F_1(a, b, x)$ is the confluent hypergeometric function or Kummer's function $\Gamma(x)$ is the gamma function.

Then the transformation from the Fresnel (7.108) has the form:

$$E_{\gamma nm}(\rho,\theta,z) = \frac{(-i)^{n+1}}{2\pi n!}\left(\frac{z_0}{zq^2}\right)\left(\frac{\sqrt{2}\sigma}{wq}\right)^{m+i\gamma}\left(\frac{k\sigma\rho}{\sqrt{2}qz}\right)^n \exp\left(\frac{ik\rho^2}{2z}+in\theta\right)\times$$

$$\times\Gamma\left(\frac{n+m+2+i\gamma}{2}\right){}_1F_1\left(\frac{n+m+2+i\gamma}{2},n+1,-\left(\frac{k\sigma\rho}{\sqrt{2}qz}\right)^2\right), \tag{7.110}$$

where $z_0 = k\sigma^2$, $q = (1-iz_0/z)^{1/2}$. Laser beams with a complex amplitude (7.110) are termed hypergeometric beams (HyG-beams).

The modulus of the complex amplitude (7.110) is proportional to Kummer's function:

$$\left|E_{\gamma nm}(\rho,\theta,z)\right| \sim x^{\frac{n}{2}}\,{}_1F_1(a,b,-x), \tag{7.111}$$

where x is a complex argument:

$$x = \left(\frac{k\sigma\rho}{\sqrt{2}qz}\right)^2. \tag{7.112}$$

Since the Kummer function is represented in the form of a series:

$${}_1F_1(a,b,-x) = \sum_{l=0}^\infty C_l(-1)^l x^l, \tag{7.113}$$

where

$$C_l = \frac{\Gamma(a+l)}{\Gamma(a)}\frac{\Gamma(b)}{\Gamma(b+l)l!}, \tag{7.114}$$

then

$$x^l = \left(\frac{k\sigma\rho}{\sqrt{2}z}\right)^l \frac{1}{q^{2l}} = \left\{\frac{1}{q^{2l}} = \frac{1}{\left(1 - iz_0/z\right)^l} = \left[\frac{\exp\left(i\text{arctg}\frac{z_0}{z}\right)}{\sqrt{1+z_0^2/z^2}}\right]^l\right\} =$$

$$= \left[\frac{k\sigma\rho}{\sqrt{2}z\left(1+z_0^2/z^2\right)^{1/4}}\right]^{2l} \exp\left(il\text{arctg}\frac{z_0}{z}\right). \tag{7.115}$$

Then

$${}_1F_1\left(a,b,-x\right) = \sum_{l=0}^{\infty}\left(-i\right)^l C_l \left[\frac{k\sigma\rho}{\sqrt{2}z\left(1+z_0^2/z^2\right)^{1/4}}\right]^{2l} \exp\left(-il\text{arctg}\frac{z}{z_0}\right), \tag{7.116}$$

From (7.116) it follows that the function $|{}_1F_1|$ and the amplitude and phase of each term of the series varies with changes in z. This means that each 'partial' light field at l = const in (7.116) will be propagating in space with its phase velocity determined by the factor $\exp[-il\,\text{tg}^{-1}(z/z_0)]$. As a result of the longitudinal interference of all terms in (7.116), the modulus of function (7.116), and hence the modulus of the complex amplitude of the light field (7.110), will change its appearance during propagation.

Hypergeometric beams in the near zone

At $z \ll z_0 = k\sigma^2$ the dependence on σ is lost, as

$$\frac{1}{2\sigma^2} - \frac{ik}{2z} \approx -\frac{ik}{2z}. \tag{7.117}$$

Then $q \approx \sqrt{-iz_0/z}$, $q^2 \approx iz_0/z$, and

$$E_{\gamma nm}\left(\rho,\theta,z << z_0\right) = \frac{\left(-i\right)^{\frac{n-m-i\gamma}{2}}}{2\pi n!}\left(\frac{kw^2}{2z}\right)^{-\frac{m+i\gamma}{2}}\left(\frac{k\rho^2}{2z}\right)^{\frac{n}{2}} \exp\left(\frac{ik\rho^2}{2z} + in\theta\right) \times$$

$$\times\Gamma\left(\frac{n+m+2+i\gamma}{2}\right){}_1F_1\left(\frac{n+m+2+i\gamma}{2}, n+1, -\frac{ik\rho^2}{2z}\right). \tag{7.118}$$

From (7.118) it follows that at $z \ll z_0$ the modulus of the complex amplitude $|E_{\gamma nm}(\rho, \theta, z \ll z_0)|$ will maintain its form and vary only on a large scale. Note that at $\sigma \to\infty$ (Gaussian beam is replaced by a plane wave) and instead of (7.110) we obtain the equation (7.118) describing the paraxial mode beams, the generalized hypergeometric modes [66].

Hypergeometric beams in the far zone
At $z \gg z_0 = k\sigma^2$ $q = (1-iz_0/z)^{1/2} \approx 1$. Then

$$E_{\gamma nm}\left(\rho,\theta,z >> z_0\right) = \frac{(-i)^{n+1}}{2\pi n!}\left(\frac{z_0}{z}\right)\left(\frac{\sqrt{2}\sigma}{w}\right)^{m+i\gamma}\left(\frac{k\sigma\rho}{\sqrt{2}z}\right)^n \exp\left(\frac{ik\rho^2}{2z}+in\theta\right)\times$$

$$\times\Gamma\left(\frac{n+m+2+i\gamma}{2}\right){}_1F_1\left(\frac{n+m+2+i\gamma}{2},n+1,-\left(\frac{k\sigma\rho}{\sqrt{2}z}\right)^2\right). \quad (7.119)$$

The dependence of the diffraction pattern on z changes qualitatively. And in the near- and far-field zones the diffraction pattern has a set of concentric rings of light with increasing spatial frequency, since the distribution of the amplitude is proportional to ρ^2. But in the near-field the diffraction pattern does not change (up to a factor) at a constant ratio ρ^2/z, while in the far field – at a constant ratio ρ/z. That is, in propagation of near-field the light ring radii grow more slowly than in the far-field: in the near-field the radii of the rings grow in proportion to $\sqrt{z}$, and in the far-field in proportion to z.

7.5.2. Hypergeometric modes

The Helmholtz equation, which describes the propagation of a non-paraxial monochromatic light wave in a homogeneous space permits eleven solutions with separable variables in different coordinate systems [69]. This means that there are light fields, which propagate without changing their structure. Examples are the well-known Bessel modes [70]. The paraxial analogue of the Helmholtz equation is the parabolic equation of Schrödinger type, which describes the propagation of paraxial optical fields. This equation permits seventeen solutions with separable variables in the coordinate systems [71]. Light fields, which are described by such solutions, retain their structure during propagation up to scale. Example include the well-known Hermite–Gaussian and Laguerre–Gaussian modes [71].

In recent years there has been a dramatic increase in the number of papers in which solutions with separable variables for the Helmholtz equation and Schrödinger were used in optics [72–80]. New non-paraxial light beams that retain their structure during the propagation were considered in [72–74]. These are parabolic bundles [72], Helmholtz–Gauss waves [73] and Laplace–Gauss waves [74]. New paraxial light beams that retain their structure up to scale were considered in [75–80]. These are Ince–Gaussian modes [75], elegant Ince–Gaussian beams [76], Hermite-Laguerre–Gaussian modes [77] and the pure optical vortices [28]. Some of these beams have been realized with laser resonators [77, 78], diffractive optical elements [28] and liquid crystal displays [79].

This section deals with another family of laser modes, which are an orthonormal basis and are solutions with separated variables of the paraxial parabolic equation in a cylindrical coordinate system. In this coordinate system, the Schrödinger equation except for solutions in the form of Bessel and Laguerre–Gaussian modes, also has a solution in the form of confluent hypergeometric functions. These solutions are special cases of the considered hypergeometric beams of general form. Intensity distribution in the cross section of such beams is close to the intensity distribution for the Bessel modes. It is also a set of concentric light rings, but their intensity decreases with increasing radial variable as r^{-2}, i.e. faster than that for the Bessel modes. Like the Bessel modes, the hypergeometric modes have infinite energy. In contrast to the Bessel modes, the light ring radii of the hypergeometric modes increases with increasing longitudinal coordinate z as $\sqrt{z}$. Experiments with the generation of such laser modes using liquid crystal microdisplays are also described.

The complex amplitude of the paraxial optical field $E(r, \varphi, z)$ in a cylindrical coordinate system (r, φ, z) satisfies the equation of Schrödinger type:

$$\left(2ik\frac{\partial}{\partial z}+\frac{\partial^2}{\partial r^2}+\frac{1}{r}\frac{\partial}{\partial r}+\frac{1}{r^2}\frac{\partial^2}{\partial \varphi^2}\right)E(r,\varphi,z)=0, \tag{7.120}$$

where $k = 2\pi/\lambda$ is the wave number of light with the wavelength λ. Equation (7.120) is satisfied by the functions that form an orthonormal basis:

$$E_{\gamma,n}(r,\varphi,z)=\frac{1}{2\pi n!}\left(\frac{z_0}{z}\right)^{\frac{1}{2}}\Gamma\left(\frac{n+1+i\gamma}{2}\right)\times$$

$$\times\exp\left[\frac{i\pi}{4}(3n+i\gamma-1)+\frac{i\gamma}{2}\ln\frac{z_0}{z}+in\varphi\right]x^{\frac{n}{2}}\,{}_1F_1\left(\frac{n+1-i\gamma}{2},n+1,ix\right), \tag{7.121}$$

where $-\infty < \gamma < \infty$, $n = 0, \pm 1, \pm 2,...$ are continuous and discrete parameters that affect the functions (7.121) and which will be called the mode numbers; $z_0 = kw^2/2$ is an analog of Rayleigh length, w is the mode parameter, similar to the radius of the Gaussian beam, although it has a different meaning here; $x = kr_2/(2z)$; $\Gamma(x)$ is the gamma function; ${}_1F_1(a, b, y)$ is the degenerate or confluent hypergeometric function [81]:

$${}_1F_1(a,b,y)=\frac{\Gamma(b)}{\Gamma(a)\Gamma(b-a)}\int\limits_0^1 t^{a-1}(1-t)^{b-a-1}\exp(yt)\,\mathrm{d}t, \tag{7.122}$$

where $\mathrm{Re}(b) > \mathrm{Re}\,(a) > 0$. From (7.122) we see that ${}_1F_1(a, b, y)$ it is an entire analytic function. In the case of (7.121) $\mathrm{Re}\,(y) = 0$ and then (7.122) is a one-dimensional Fourier transform of a bounded function on the interval [0, 1]. According to Shannon's theorem asymptotically at $r \to \infty$ the modulation period of function (7.121) (i.e. the distance between adjacent maxima or minima) is 2π. For large values of the argument $x >> 1$ we have the asymptotic behaviour $x^{n/2}\left|{}_1F_1\left[(n+1-i\gamma)/2,\, n+1, ix\right]\right| \approx 1/x$. This behaviour of the modulus of the function (7.121) leads to a more rapid decline than that of the Bessel function. In addition, the zeros of the confluent hypergeometric functions are ${}_1F_1\ (a, b, y_{0m})$ are close to the zeros of Bessel functions $J_{b-1}\ (y_{b-1,\,m})$ [81]:

$$y_{0,m} \approx \left| \frac{y_{b-1,m}^2}{2b-4a} \right|. \tag{7.123}$$

The light beams (7.121), which will be called hypergeometric (HyG) modes, can be generated using an optical element having a transmission function:

$$E_{\gamma,n}(\rho,\theta,z) = \frac{1}{2\pi}\left(\frac{w}{\rho}\right)\exp\left[i\gamma\ln\left(\frac{\rho}{w}\right)+in\theta\right]. \tag{7.124}$$

In illuminating the optical element (7.124), located in the plane $z = 0$, by an unbounded plane wave a light field with the complex amplitude (7.121) forms at distance z. The energy of the light fields (7.121) and (7.124) is unbounded, as in the the Bessel mode

$$E_{\beta,n}(r,\varphi,z) = J_n(\beta r)\exp\left[i\frac{\beta^2 z}{2k}+in\varphi\right], \tag{7.125}$$

which also satisfies (7.120). Therefore, to produce the mode (7.121) in practice, the optical element (7.124) should be limited by a circular aperture. At the same time, the mode (7.121) will form effectively at a finite distance $z_0 < R$ tg (γ/R), where R is the large radius of the circular aperture.

In propagation, the light field (7.121) retains its structure and only its scale changes. The transverse intensity distribution of the HyG mode (7.121) is a set of concentric light rings, whose radii satisfy the condition:

$$r_m = \left(\alpha_m z\lambda/\pi\right)^{1/2}, \tag{7.126}$$

where α_m is a constant depending on the number of rings m and the number of modes (γ, n). Therefore, the ring radii increase with increasing z as $\sqrt{z}$. From the relation [81]:

$${}_1F_1\left(\frac{n+1+i\gamma}{2}, n+1, -ix\right) = \exp(-ix)\,{}_1F_1\left(\frac{n+1-i\gamma}{2}, n+1, ix\right) \tag{7.127}$$

it follows that the phase of the hypergeometric function is equal to $x/2$ (up to π):

$$\arg\left\{{}_1F_1\left(\frac{n+1+i\gamma}{2}, n+1, -ix\right)\right\} = -\frac{x}{2}. \tag{7.128}$$

Interestingly, this phase does not depend on the number of the mode (γ, n). Then we can write the expression for the phase of the HG mode:

$$\arg\left\{E_{\gamma,n}\left(r,\varphi,z\right)\right\} = \frac{\gamma}{2}\ln\frac{z}{z_0} + n\varphi + \frac{kr^2}{4z} + \frac{\pi}{4}\left(3n-1\right) + \arg\Gamma\left(\frac{n+1+i\gamma}{2}\right), \tag{7.129}$$

where the first term has the meaning of the Gouy phase.

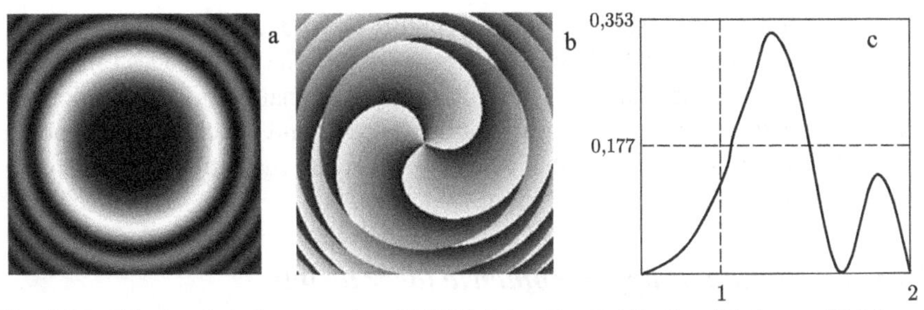

Fig. 7.29. Calculated (using equation (7.121)) intensity distribution $|E_{\gamma,n}(r, \varphi, z)|^2$ (a) and phase arg $\{E_{\gamma,n}(r, \varphi, z)\}$(b) and the intensity radial section (c) of the HyG-mode $(\gamma, n) = (2, 3)$ at a distance $z = 1000$ mm. The size of frames (a) and (b) is equal to 4 mm × 4 mm.

Figure 7.29 shows the intensity (a), phase (b) and the radial section of intensity (c) for the HyG-mode with the number (γ, n) calculated by the formula (7.121) at a distance $z = 1000$ mm. The wavelength $\lambda = 633$ nm, $w = 1$ mm.

Experimentally, the HyG-modes were produced with the help of liquid crystal microdisplays CRL OPTO with a resolution of 1316×1024 elements. The microdisplay formed a binary diffractive optical element (DOE) with a diameter of 6.5 mm, which was illuminated by a linearly polarized plane wave from a solid-state laser with a wavelength of 532 nm and 500 mW. Figure 7.30a shows the binary phase of the DOE $S(\rho, \theta)$ which satisfies the equation:

$$S(\rho,\theta,z) = \operatorname{sgn}\left\{\cos\left[\gamma \ln\frac{\rho}{w} + n\theta + \frac{k\rho^2}{2f}\right]\right\}, \tag{7.130}$$

where sign (ξ) is the sign function, f is the focal length of the spherical lens. The amplitude of the function ρ^{-1} (7.124) was replaced by a constant value. Figure 7.30b shows the intensity distribution formed with the help of a liquid crystal microdisplay with the phase in Fig. 7.30a and registered at a distance of 700 mm from the display.

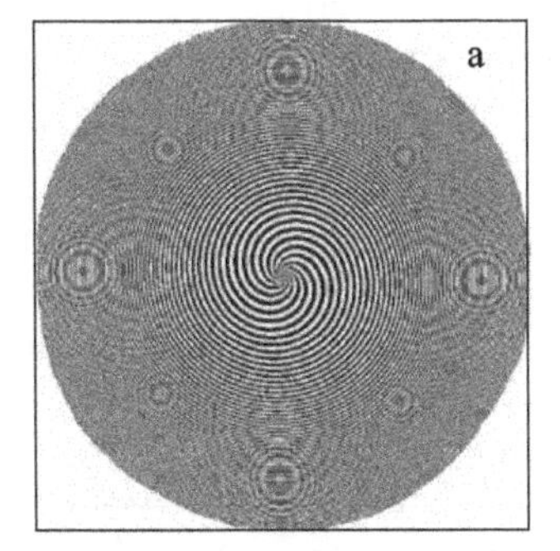

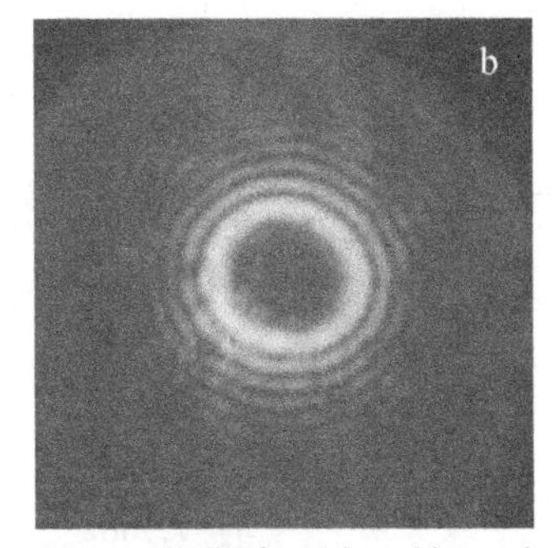

Fig. 7.30. The binary phase formed on the liquid crystal display (a) and intensity distribution of the HyG-mode $(\gamma, n) = (5, 10)$, registered with the CCD-camera at a distance $z = 700$ mm from the display (b).

In this section, we examine, both theoretically and experimentally, a new family of optical modes, called hypergeometric (HyG). The HyG modes satisfy the paraxial equation of Schrödinger type, form an orthogonal basis of the functions and in propagation they retain the structure up to scale and are close to the known modes of Bessel functions. Note that when $\gamma = -i$ the HG-mode transfer to the one-parameter family of pure optical vortices [28].

7.5.3. Formation of hypergeometric laser beams

The formation of HyG-modes [82] using the diffractive optical elements is not an easy task. First, similar to the Bessel modes [70], HyG-modes have infinite energy, and secondly, HyG modes are generated by the initial light field (7.124), which has a singularity at the origin ($\sigma \to \infty$, $m = -1$):

$$E_{\gamma,n,-1}\left(r,\varphi,z=0\right)=\frac{1}{2\pi}\left(\frac{\omega}{r}\right)\exp\left[i\gamma\ln\left(\frac{r}{\omega}\right)+in\varphi\right]. \tag{7.131}$$

In practice, therefore, to form the HyG-mode, the light field (7.131) should be restricted by a circular aperture with radii R_1 and R_2 ($R_1 < R_2$). However, such a restriction of the aperture of the initial field at some of the parameters does not lead to significant distortions of the HyG-mode. Figure 7.31 shows a view of the radial intensity distribution of the field (7.131), bounded by a circular aperture and the intensity at a distance $z = 100$ mm.

At the following parameters: λ = 532 nm, R_1 = 0.05 mm, R_2 = 1 mm, w = 1 mm, number of pixles $N = 512$; HyG-mode parameters: $n = 4$, $\gamma = -10$, the standard deviation of the exact intensity obtained on the basis of equation (7.121) from that calculated taking into account the limited aperture (Fig. 7.31b) is 5.5%.

Implementation of the amplitude distribution, shown in Fig. 7.31a, for the formation of HyG-mode is not an effective way. More energy-efficient and technologically advanced is the formation of HyG-mode by using a binary phase DOE [83].

Transmission function of the DOE might look like this:

$$\tau_{\gamma,n}(r,\varphi)=\operatorname{sgn}\left\{\cos\left[\gamma\ln\left(\frac{r}{w}\right)+n\varphi+cr\cos\varphi\right]\right\}, \tag{7.132}$$

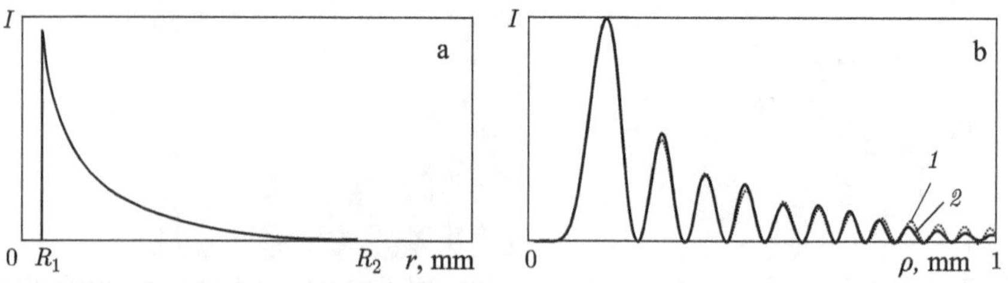

Fig. 7.31. The radial intensity distribution of the HyG-mode (n = 4, γ = −10, m = −1) at z = 0 (a) and z = 100 mm (b): exact HyG-mode (7.121) and calculated after limiting by the aperture (7.124).

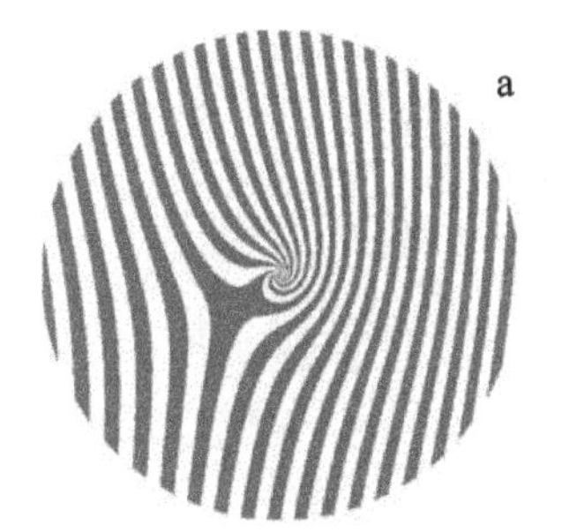

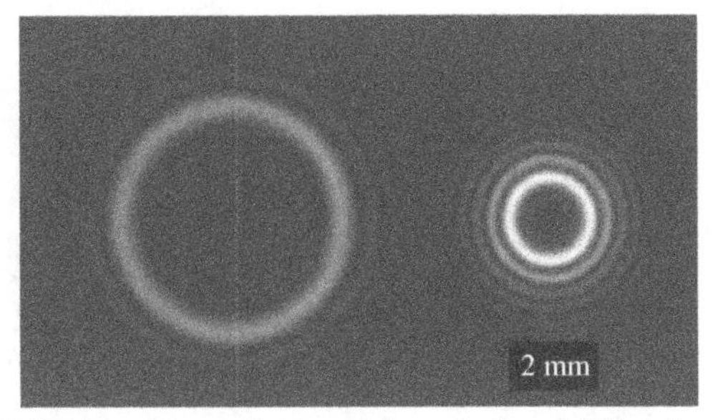

Fig. 7.32. The binary phase of the DOE ($c = 10$ mm^{-1}) (a) and the calculated diffraction pattern at a distance $z = 700$ mm (b).

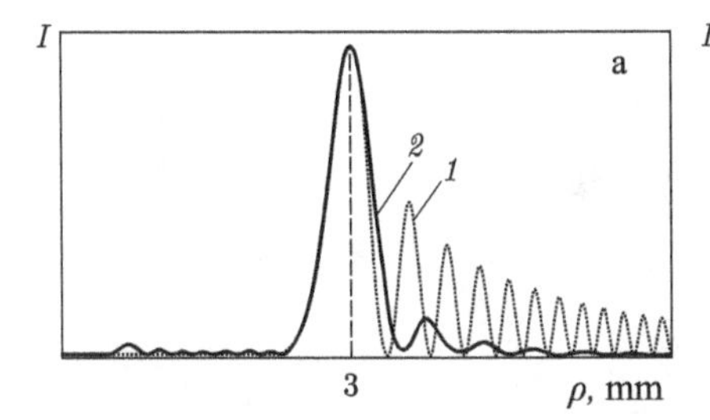

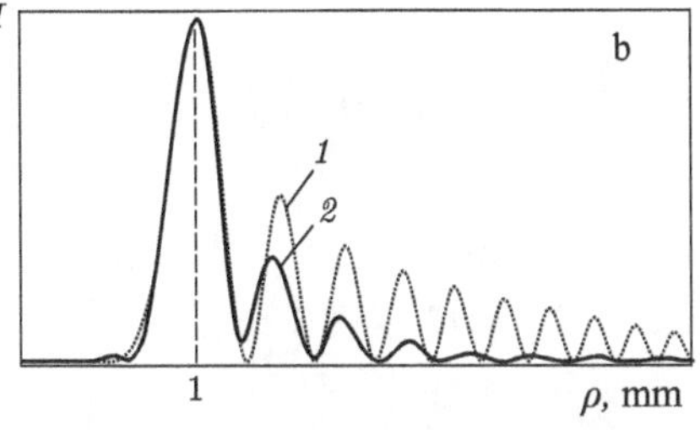

Fig. 7.33. The radial intensity distribution of exact HyG modes (curves 1) and calculated behind the binary DOE (Fig. 7.32a) (curves 2) at a distance $z = 2000$ mm: $n = 7$, $\gamma = 10$ (a) and $n = -7$, $\gamma = -10$ (b).

where c is the carrier spatial frequency.

Figure 7.32 shows (a) the binary phase of the DOEs (7.132) (diameter 5 mm, $n = 7$, $\gamma = 10$, $c = 10$ mm^{-1}, $w = 1$ mm) and (b) the calculated diffraction pattern at a distance $z = 700$ mm from the DOE.

Figure 7.32b shows that in illumination of the DOE (Fig. 7.32a) by a plane wave two circular diffraction patterns form mainly at some distance and are similar to the HyG-modes with numbers $n = 7$, $\gamma = 10$ (large ring) and $n = -7$, $\gamma = -10$ (small ring). In each of the two beams there is about 40% of light energy.

Figure 7.33 shows the radial intensity distribution calculated for the ideal HyG-modes (curves 1) and formed with a binary DOE (Fig. 7.32a) (curves 2) at a distance $z = 2000$ mm: $n = 7$, $\gamma = 10$ (a) and $n = -7$, $\gamma = -10$ (b).

The standard deviation of the exact HyG-modes from those calculated at a distance $z = 2000$ mm from the binary DOE (Fig. 7.32a) was 43% (a) and 35% (b). Thus, the replacement of the function of the amplitude decreasing from r (7.131) by a constant one in (7.132) leads to a noticeable error in the formation of the HyG-beam. However, the differences concern only the side lobes of the diffraction patterns and almost do not affect the main ring.

Electron-beam lithography was used to fabricate the binary phase DOEs, 5×5 mm in size with a resolution of 10 μm for a wavelength of 532 nm. Figure 7.34 shows the image of the central part of the DOE microrelief, obtained with an interferometer Zygo NewView 5000, ×100 magnification. The required height of the microrelief of a substrate of fused silica (SiO_2) is equal to 578.3 nm and the height of the produced

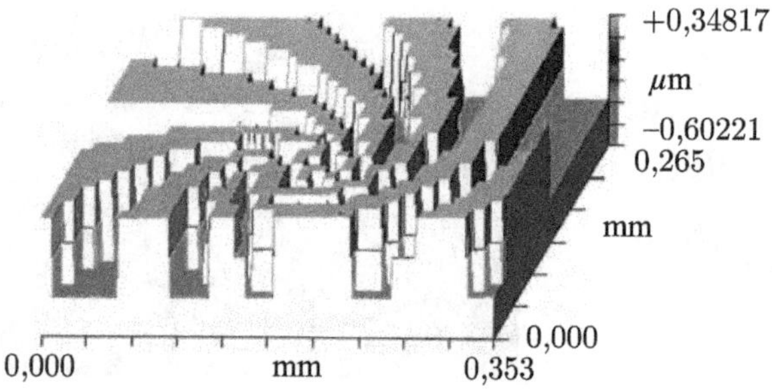

Fig. 7.34. Microrelief of the central part of the binary DOE (Fig. 7.32a), size 353 × 265 μm in a fused silica substrate.

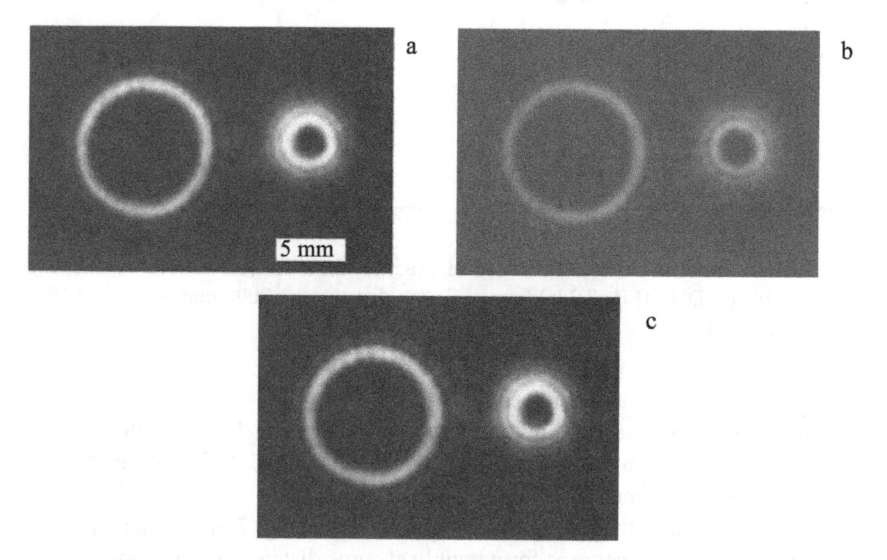

Fig. 7.35. Diffraction pattern formed by the DOE (Figure 7.34) in illumination with a plane wave with a diameter of 4 mm (λ = 532 nm) and recorded with a CCD-camera at distances of 2000 (a), 2300 (b) and 3000 mm (c).

relief ranged from 572 to 583 nm. That is, the binary DOE was made with high accuracy, about 1%.

Figure 7.35 shows the diffraction patterns formed after illumination of the DOE with a plane wave with a diameter of 4 mm from a solid-state laser with a wavelength of 532 nm and 500 mW and measured at different distances from the DOE: 2000 mm (a), 2300 mm (b) and 3000 mm (c). Figure 7.35 shows that the diffraction patterns coincide in form with the calculated pattern (Fig. 7.32c), and that at a sufficient distance from the DOE both light beams $n = 7,\ \gamma = 10$ and $n = -7,\ \gamma = -10$ change little in propagation.

Figure 7.36 shows the experimental (a) and calculated (c) diffraction patterns of a plane wave on the DOE for the HyG-mode ($n = 7,\ \gamma = 10$) and the intensity of the radial cross-section (b and d, respectively).

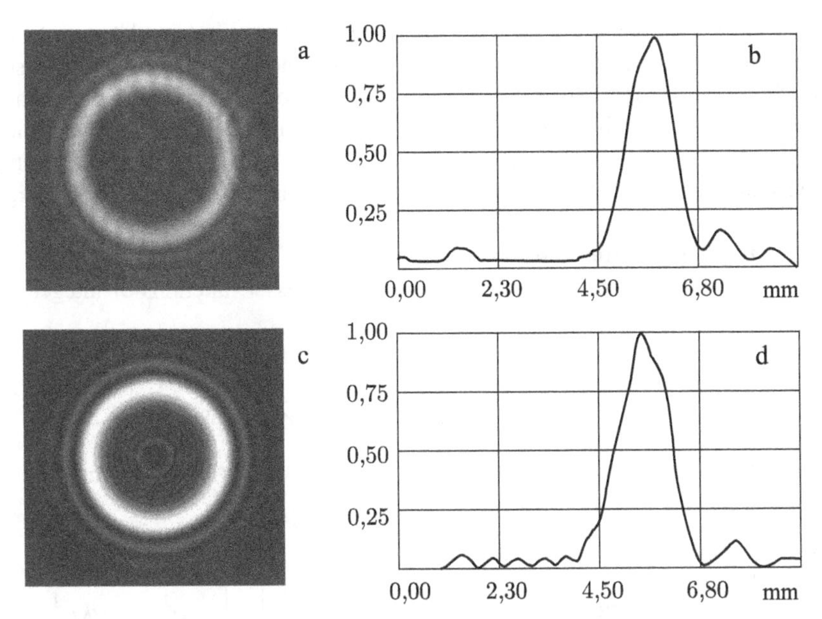

Fig. 7.36. Experimental (a) and calculated (c) diffraction pattern of a plane wave on the DOE for the HyG-mode ($n = 7, \lambda = 10$) and the radial cross-section of intensity: experimental (b) and calculated (d).

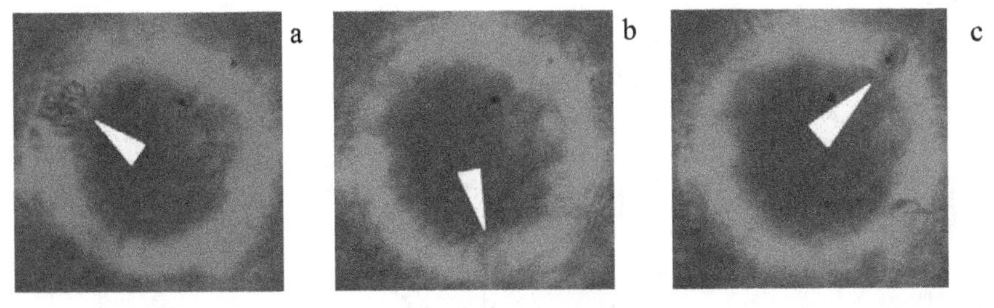

Fig. 7.37. Rotation of the polystyrene particle with a diameter of 5 μm (the location of the particle is shown with white triangle) in the main ring of the diffraction pattern for HyG-beam ($n = 7, \gamma = 10$), formed by the diffraction of a plane filled by the DOE (Figure 7.34).

Figure 7.36 shows that the two diffraction patterns and their radial cross-sections agree satisfactorily with each other, and the rms error is 27%.

Figure 7.37 shows three fragments (separated by a time interval of 15 seconds) of rotation of a polystyrene bead with a diameter 5 μm, rotating on the main ring of the HG beam ($n = 7,\ \gamma = 10$) formed by a binary DOE (Fig. 7.34). The light ring in Fig. 7.37 after focusing with a ×40 microobjective had a diameter of 39 μm.

7.5.4. Special cases of hypergeometric beams

As noted above, the most important special case of hypergeometric beams are hypergeometric modes. This section discusses some more special cases for which analytical expressions were derived. All these beams are not modes and do not retain their structure during the propagation. The diffraction pattern in the cross section of these beams is a collection of bright and dark concentric rings, among which the brightest is the main ring, while the rest of the ring are side lobes. From these beams only the modified Bessel–Gaussian quadratic beams have infinite energy, such as ordinary Bessel modes [70], while the remaining beams have finite energy.

The modified quadratic Bessel–Gaussian beams

The connection between Kummer's function and Bessel functions of integer and half-integer orders is well known [81]:

$$ {}_1F_1\left(\frac{n+1}{2}, n+1, x\right) = \Gamma\left(\frac{n}{2}+1\right)\exp\left(\frac{x}{2}\right)\left(-\frac{ix}{4}\right)^{-\frac{n}{2}} J_{\frac{n}{2}}\left(-\frac{ix}{2}\right), \tag{7.133} $$

where $J_\nu(x)$ is the Bessel function. Then, from (7.121) with (7.133) and on the condition that $\gamma = i(m+1)$, we obtain:

$$ E_{i(m+1),n,m}(\rho,\theta,z) = E_{0,n,-1}(\rho,\theta,z) = $$
$$ = \frac{(-i)^{n+1}}{\sqrt{2\pi}}\left(\frac{k\sigma\omega}{2zq}\right)\exp\left[in\theta + \frac{ik\rho^2}{2R_1(z)} - \frac{\rho^2}{2\sigma^2(z)}\right] I_{\frac{n}{2}}\left[\frac{\rho^2}{2\sigma^2(z)} + \frac{ik\rho^2}{2R(z)}\right], \tag{7.134} $$

where

$$ \begin{cases} \sigma^2(z) = 2\sigma^2\left(1+\dfrac{z^2}{z_0^2}\right), \\ R(z) = 2z\left(1+\dfrac{z^2}{z_0^2}\right), \\ R_1(z) = R(z)\left(1+\dfrac{2z^2}{z_0^2}\right)^{-1}, \end{cases} \tag{7.135} $$

$I_\nu(x)$ is the modified Bessel function.

The light beams, described by the complex amplitude (7.134), have as the cofactors the Gaussian exponent and the Bessel function. Therefore, they are similar to the well-known Bessel–Gaussian beams (BG) [84]. The dependence of the argument of the Bessel function in (7.134) on the radial coordinate is quadratic, and therefore the beams (7.134) are similar to the quadratic Bessel–Gaussian beams [85] (QBG-beams). However, QBG-beams are generated by the initial light field at $z = 0$, described by the function $J_{n/2}(ar^2)\exp(-br^2 + in\varphi)$, a and b are constants, and in any other plane $z > 0$ the expression for the amplitude of the QBG-beams can be obtained by the reference integral [53]:

$$\int_0^\infty J_{\frac{n}{2}}\left(ar^2\right)\exp\left(-br^2\right)J_n\left(cr\right)r\mathrm{d}r=\frac{1}{2\left(a^2+b^2\right)^{\frac{1}{2}}}J_{\frac{n}{2}}\left[\frac{c^2a}{4\left(a^2+b^2\right)}\right]\exp\left[-\frac{c^2b}{4\left(a^2+b^2\right)}\right].$$

(7.136)

The beams (7.134) are generated by using another light field in the initial plane:

$$E_{0,n,-1}\left(r,\varphi,z=0\right)=\frac{1}{2\pi}\left(\frac{\omega}{r}\right)\exp\left(-\frac{r^2}{2\sigma^2}+in\varphi\right) \quad (7.137)$$

and therefore differ from the QBG-beams. The light fields (7.134) can be called modified quadratic Bessel–Gauss beams (mQBG-beams).

When $\sigma\to\infty$ (Gaussian beam is replaced by a plane wave) from (7.134) we obtain diverging Bessel beams described in [86]:

$$\tilde{E}_{0,n,-1}\left(\rho,\theta,z\right)=\frac{\omega}{2}\sqrt{\frac{ik}{2\pi z}}\left(-i\right)^{\frac{n}{2}+1}\exp\left(in\theta+\frac{ik\rho^2}{4z}\right)J_{\frac{n}{2}}\left(\frac{k\rho^2}{4z}\right). \quad (7.138)$$

The light field (7.137) has a singularity at the centre of the initial plane $z = 0$ and $r = 0$ and has infinite energy. We next consider the case without such features.

Gaussian optical vortices

This section provides an explicit form of the complex amplitude for another special case of HG-beams – Gaussian optical vortices (GOV) [28, 87]. The recurrence relation between Kummer functions is known [81]:

$${}_1F_1\left(a,b,-x\right)=\exp\left(-x\right){}_1F_1\left(b-a,b,x\right). \quad (7.139)$$

In view of (7.139) a general view of the HyG-beams (7.121) is convenient for further consideration written as:

$$F_{\gamma,n,m}\left(\rho,\theta,z\right)=\frac{\left(-i\right)^{n+1}}{2\pi n!}\left(\frac{z_0}{zq^2}\right)\left(\frac{\sqrt{2}\sigma}{\omega q}\right)^{m+i\gamma}\times\left(\frac{k\sigma\rho}{\sqrt{2}qz}\right)^n\exp\left[in\theta+\frac{ik\rho^2}{2z}-\left(\frac{k\sigma\rho}{\sqrt{2}qz}\right)^2\right]\times$$
$$\Gamma\left(\frac{n+m+2+i\gamma}{2}\right){}_1F_1\left[\frac{n-m-i\gamma}{2},n+1,\left(\frac{k\sigma\rho}{\sqrt{2}qz}\right)^2\right].$$

(7.140)

In [86] a connection was found between the Kummer function and the modified Bessel functions:

$${}_1F_1\left(\frac{n}{2},n+1,x\right)=\Gamma\left(\frac{n+1}{2}\right)2^{\frac{n-1}{2}}\left(\frac{x}{2}\right)^{-\left(\frac{n-1}{2}\right)}\exp\left(\frac{x}{2}\right)\left[I_{\frac{n-1}{2}}\left(x\right)-I_{\frac{n+1}{2}}\left(x\right)\right]. \quad (7.141)$$

Using (7.141) and setting $\gamma = im$, from the general form of (7.140) for HyG-beams we obtain a special case explicitly:

$$E_{im,n,m}(\rho,\theta,z) = E_{0,n,0}(\rho,\theta,z) =$$
$$\times \frac{(-i)^{n+1}}{4\sqrt{\pi}}\left(\frac{z_0}{zq^2}\right)\left(\frac{k\sigma\rho}{\sqrt{2}qz}\right)\exp\left[in\theta + \frac{ik\rho^2}{2R_1(z)} - \frac{\rho^2}{2\sigma^2(z)}\right]\left[I_{\frac{n-1}{2}}(y) - I_{\frac{n+1}{2}}(y)\right], \tag{7.142}$$

where

$$y = \frac{1}{2}\left(\frac{k\sigma\rho}{\sqrt{2}qz}\right)^2 = \frac{\rho^2}{2\sigma^2(z)} + \frac{ik\rho^2}{2R(z)}.$$

The light field (7.142) is generated by the initial field of the form:

$$E_{0,n,0}(r,\varphi,z=0) = \frac{1}{2\pi}\exp\left(-\frac{r^2}{2\sigma^2} + in\varphi\right), \tag{7.143}$$

which can be implemented by diffraction of a Gaussian beam by the SPP. Therefore, light beams, described by the complex amplitude (7.142) can be called GOV [28, 87].

Hollow Gaussian optical vortices

We can obtain an explicit analytic form in terms of modified Bessel functions for the complex amplitude of the light beams that are similar in shape to the beams (7.142). Such beams can be formed by diffraction of a hollow Gaussian beam on the SPP:

$$E_{0,n,1}(r,\varphi,z=0) = \frac{1}{2\pi}\left(\frac{r}{\omega}\right)\exp\left(-\frac{r^2}{2\sigma^2} + in\varphi\right). \tag{7.144}$$

To obtain the complex amplitude at $z > 0$, we first obtain an intermediate ratio. To do this, we compare the two reference integrals, one of which is proportional to the Kummer's function [53]:

$$\int_0^\infty r^2 \exp(-pr^2) J_n(cr)\,\mathrm{d}r = c^n p^{-(n+3)/2} 2^{-(n+1)} \Gamma\left(\frac{n+3}{2}\right)\Gamma^{-1}(n+1)\,{}_1F_1\left(\frac{n+3}{2}, n+1, -\frac{c^2}{4p}\right), \tag{7.145}$$

and the second integral is obtained by differentiating both sides [53]

$$\int_0^\infty \exp(-pr^2) J_n(cr)\,\mathrm{d}r = \frac{1}{2}\sqrt{\frac{\pi}{p}}\exp\left(-\frac{c^2}{8p}\right) I_{\frac{n}{2}}\left(\frac{c^2}{8p}\right), \tag{7.146}$$

with respect to the parameter p:

$$\int_0^\infty r^2 \exp(-pr^2) J_n(cr)\,\mathrm{d}r = \frac{\sqrt{\pi}}{2} p^{-\frac{3}{2}} \exp\left(-\frac{c^2}{8p}\right)\left[\left(\frac{1-n}{2} - \frac{c^2}{8p}\right) I_{\frac{n}{2}}\left(\frac{c^2}{8p}\right) + \left(\frac{c^2}{8p}\right) I_{\frac{n-2}{2}}\left(\frac{c^2}{8p}\right)\right]. \tag{7.147}$$

Comparing the right-hand sides of (7.145) and (7.147), we obtain the relation between Kummer's function and modified Bessel functions:

$$ {}_1F_1\left(\frac{n+3}{2}, n+1, -x\right) = x^{-\frac{n}{2}} \exp\left(-\frac{x}{2}\right)\sqrt{\pi}\, n!\,\Gamma^{-1}\left(\frac{n+3}{2}\right)\left[\left(\frac{1-n}{2}-\frac{x}{2}\right)I_{\frac{n}{2}}\left(\frac{x}{2}\right)+\frac{x}{2}I_{\frac{n-2}{2}}\left(\frac{x}{2}\right)\right]. \tag{7.148}$$

Using (7.148), from the general equation (7.121), provided $\gamma = i(m-1)$, we obtain an explicit expression for the new light beams:

$$ E_{i(m-1),n,m}(\rho,\theta,z) = E_{0,n,1}(\rho,\theta,z) = $$
$$ = \frac{(-i)^{n+1}}{\sqrt{2\pi}}\left(\frac{k\sigma^3}{z\omega q^3}\right)\exp\left[in\theta+\frac{ik\rho^2}{2R_1(z)}-\frac{\rho^2}{2\sigma^2(z)}\right]\left[\left(\frac{1-n}{2}-y\right)I_{\frac{n}{2}}(y)+yI_{\frac{n-2}{2}}(y)\right], \tag{7.149}$$

where y is the same as in (7.142). The light fields (7.149) for the type of field generating them (7.144) can be called hollow Gaussian optical vortices (HGOV).

The modified elegant Laguerre–Gaussian beams

These beams are generated by the initial field of the form:

$$ E_{\gamma,n,2p+n}(r,\varphi,z=0) = \frac{1}{2\pi}\left(\frac{r}{\omega}\right)^{2p+n}\exp\left(-\frac{r^2}{2\sigma^2}+in\varphi\right). \tag{7.150}$$

We obtain a new kind of light beams, as a special case of HyG-beams (7.121) under the condition $\gamma = -i(2p - m + n)$, p is an integer ($p \geq -n/2$). For this we use the known relationship between the Kummer function and the associated Laguerre polynomials [81]:

$$ {}_1F_1(-p, n+1, x) = \frac{p!\,n!}{(n+p)!}L_p^n(x), \tag{7.151}$$

where $L_p^n(x)$ is the associated Laguerre polynomial.

In view of (7.151) from (7.150) we obtain

$$ E_{0,n,2p+n}(\rho,\theta,z) = \frac{(-i)^{n+1}p!}{2\pi}\left(\frac{z_0}{zq^2}\right)\left(\frac{\sqrt{2}\sigma}{\omega q}\right)^{n+2p} t^{\frac{n}{2}}\exp\left(in\theta+\frac{ik\rho^2}{2z}-t\right)L_p^n(t), \tag{7.152}$$

where $t = 2y = \left[k\sigma\rho/\left(\sqrt{2}qz\right)\right]^2$.

The light beams, described by the complex amplitude (7.152), can be called modified elegant Laguerre–Gaussian beams (meLG-beams). These beams are referred to in [88], but their explicit form is not given. We call this new light beams elegant, as the argument of the Laguerre polynomial is complex, as in the usual elegant Laguerre–Gaussian (eLG-beams) [89]. But the dependence of the argument of the Laguerre polynomial in (7.152) on variable z is different from the analogous dependence in the normal eLG-beams [89]. For comparison, we give an explicit form of the eLG-beams in the notation adopted here:

$$E_{\mathrm{eLG}}(\rho,\theta,z)=(-i)^{p+1}\left(\frac{z_0}{zq^2}\right)^{p+1}\left(\frac{-2i\sigma^2 z_0}{\omega^2 q^2 z}\right)^{\frac{n}{2}} s^{\frac{n}{2}}\exp(in\theta-s)L_p^n(s), \tag{7.153}$$

where $s=-ik\rho^2/(2q^2z)$. Comparison of (7.152) and (7.153) shows that the arguments s and t are characterized by their dependence on the coordinate z. This difference arises from the fact that the meLG-beams (7.152) are generated by the initial field (7.150), and eLG-beams (7.153) are generated by the initial field of the form:

$$E_{\mathrm{eLG}}(r,\varphi,z=0)=\left(\frac{r}{\omega}\right)^n \exp\left(-\frac{r^2}{2\sigma^2}+in\varphi\right)L_p^n\left(\frac{r^2}{2\sigma^2}\right). \tag{7.154}$$

Hypergeometric gamma beams

Here we give another explicit form of the complex amplitude, which describes a special case of HyG-beams (7.121). For this we use the relation between the Kummer function and the incomplete gamma function [81]:

$${}_1F_1(n,n+1,-x)=nx^{-n}\gamma(n,x), \tag{7.155}$$

where $\gamma(\nu, x)$ is the incomplete gamma function,

$$\gamma(\nu,x)=\int_0^x \xi^{\nu-1}\exp(-\xi)\,\mathrm{d}\xi. \tag{7.156}$$

In view of (7.155) from (7.121) we obtain a particular form of HyG-beams at $\gamma = i(m+2)$:

$$\begin{gathered}E_{i(m+2-n),n,m}(\rho,\theta,z)=E_{0,n,n-2}(\rho,\theta,z)=\\ =\frac{(-i)^{n+1}}{2\pi}\left(\frac{k\omega^2}{2z}\right)\left(\frac{k\rho\omega}{2z}\right)^{-n}\exp\left(\frac{ik\rho^2}{2z}+in\theta\right)\gamma\left[n,\left(\frac{k\sigma\rho}{\sqrt{2}qz}\right)^2\right].\end{gathered} \tag{7.157}$$

The light beams (7.157) are described by the complex amplitude proportional to the incomplete gamma function, and so we called them γHG-beams. The light beams (7.157) are generated by the initial light field of the form:

$$E_{0,n,n-2}(r,\varphi,z=0)=\frac{1}{2\pi}\left(\frac{r}{\omega}\right)^{n-2}\exp\left(-\frac{r^2}{2\sigma^2}+in\varphi\right) \tag{7.158}$$

Note that the meLG-beams (7.152) become γHG-beams (7.157) for $p = -1$.

7.5.5. Non-paraxial hypergeometric beams

The above hypergeometric beams and modes are considered in the paraxial approximation which in some cases cannot be used, such as in tasks that require sharp focusing of laser radiation (sharp focusing can be used, for example, to seal the information in laser writing, in lithography, in surgery, laser deposition of silicon vapours, for welding in confined spaces.)

This section discusses the HyG-modes in a non-paraxial case. An analytical expression, which is an exact solution of the Helmholtz equation in cylindrical coordinates, is obtained. This solution is proportional to the product of two Kummer functions. Further, this solution is represented as a sum of two terms describing direct non-paraxial hypergeometric (nHyG$^+$) mode and inverse non-paraxial hypergeometric (nHyG$^-$) modes. These light beams propagate along the optical axis in the forward and reverse directions. It is shown that at large distances from the initial plane (much larger than the wavelength) the nGG$^+$-mode coincides up to a constant with the paraxial HG-mode [82, 86].

The angular spectrum of plane waves for non-paraxial hypergeometric modes

It is known that any solution of the Helmholtz equation

$$(\Delta+k^2)E(x,y,z)=0, \tag{7.159}$$

where k is the wave number, can be represented in the form of the angular spectrum of plane waves

$$E(x,y,z)=\int_{-\pi}^{\pi}\int_{0}^{\pi} f(\theta,\varphi)\exp\left[-ik\left(x\sin\theta\cos\varphi+y\sin\theta\sin\varphi+z\cos\theta\right)\right]\sin\theta\,\mathrm{d}\theta\,\mathrm{d}\varphi, \tag{7.160}$$

where (θ, φ) are the Euler angles that define a point on the sphere which defines the direction of propagation of a plane wave. Consider the specific form of the angular spectrum

$$f(\theta,\varphi)=\frac{1}{2\pi}\left(\tan\frac{\theta}{2}\right)^{\beta}\sin^{-1}(\theta)\exp(2in\varphi), \tag{7.161}$$

where n is an integer and β is the real number. Substituting (7.161) into (7.160), we obtain:

$$E(r,\phi,z)=(-1)^n\exp(i2n\phi)\int_0^{\pi}\exp(-ikz\cos\theta)\left(\tan\frac{\theta}{2}\right)^{\beta}J_{2n}(kr\sin\theta)\,\mathrm{d}\theta, \tag{7.162}$$

where $J_v(x)$ is the Bessel function. With the reference integral [53] (the integral 2.12.27.3)

$$\int_0^{\pi}\exp(p\cos x)\left(\tan\frac{x}{2}\right)^{2m}J_{2n}(c\sin x)\,\mathrm{d}x=\frac{1}{c}\Gamma\left[\begin{matrix}m+n+1/2,n-m+1/2\\2n+1,2n+1\end{matrix}\right]M_{m,n}(z_+)M_{m,n}(z_-), \tag{7.163}$$

where $z_{\pm} = p \pm (p^2 - c^2)^{1/2}$, $M_{\chi,\mu}(z)$ is the Whittaker's confluent hypergeometric function:

$$M_{\chi,\mu}(z) = z^{\mu+1/2} \exp\left(-\frac{z}{2}\right) {}_1F_1\left(\mu - \chi + \frac{1}{2}, 2\mu + 1, z\right), \tag{7.164}$$

and ${}_1F_1(a, b, x)$ is a confluent hypergeometric function (Kummer function) [53], instead of (7.162) we can obtain an explicit analytical expression:

$$E(r,\phi,z) = \frac{(-1)^n}{[(2n)!]^2} \exp(i2n\phi + ikz) \Gamma\left(\frac{2n+\beta+1}{2}\right) \Gamma\left(\frac{2n-\beta+1}{2}\right) (kr)^{2n} \times$$
$$\times {}_1F_1\left(\frac{2n-\beta+1}{2}, 2n+1, x_+\right) {}_1F_1\left(\frac{2n-\beta+1}{2}, 2n+1, x_-\right), \tag{7.165}$$

where $x_{\pm} = -ikz\left\{1 \pm \left[1 + (r/z)^2\right]^{1/2}\right\}$, (r, φ, z) are cylindrical coordinates, $\Gamma(x)$ is the gamma function. Equation (7.165) is an exact solution of equation (7.159), and describes the sum of two non-paraxial hypergeometric beams:

$$E(r,\varphi,z) = E^+(r,\varphi,z;\beta) + E^-(r,\varphi,z;\beta), \tag{7.166}$$

where E^+ is direct nHyG$^+$ mode, which is described by (7.162), in which the integral over θ is calculated from 0 to $\pi/2$, and E^- – inverse nHyG$^-$-mode, which is described by (7.162), in which the integral over θ is calculated from the $\pi/2$ to π. One can show that $E^-(r, \varphi, z; \beta) = E^+(r, \varphi, -z; -\beta)$. Hence, in particular, it follows that at $z = \beta = 0$ the direct and inverse nHyG-modes coincide and are equal to the expression:

$$E^-(r,\varphi,0;0) = E^+(r,\varphi,0;0) = 0.5E(r,\varphi,z) = (-1)^n (\pi/2) \exp(i2n\varphi) J_n^2(kr/2). \tag{7.167}$$

From this expression it follows that the main nHyG-mode at $z = \beta = n = 0$ is generated by the square of the Bessel function of zero order and has a diameter of the central light spot 1.53λ, where λ is the wavelength (the diameter is twice the distance from the maximum to the first root of the Bessel function).

Direct and inverse non-paraxial hypergeometric modes

In the general case, when $z \neq 0$ or $\beta \neq 0$, in (7.165) we must identify explicitly the components describing direct and inverse modes. For the confluent functions we know the asymptotic expansion (at $z \to \infty$) [81]:

$$\frac{{}_1F_1(a,b,z)}{\Gamma(b)} = \frac{\exp(\pm i\pi a) z^{-a}}{\Gamma(b-a)} \left[\sum_{n=0}^{R-1} \frac{(a)_n (1+a-b)_n}{n!} (-z)^{-n} + O\left(|z|^{-R}\right)\right] +$$
$$+ \frac{\exp(z) z^{a-b}}{\Gamma(a)} \left[\sum_{n=0}^{S-1} \frac{(b-a)_n (1-a)_n}{n!} z^{-n} + O\left(|z|^{-S}\right)\right], \tag{7.168}$$

where the upper sign is taken for the case $-\pi/2 < \arg z < 3\pi/2$ and the bottom sign for $-3\pi/2 \le \arg z \le \pi/2$. Tending R and S to infinity, we find that

$$\frac{{}_1F_1(a,b,z)}{\Gamma(b)} = \frac{\exp(\pm i\pi a) z^{-a}}{\Gamma(b-a)}\, {}_2F_0\left(a, 1+a-b, -\frac{1}{z}\right) + \frac{\exp(z) z^{a-b}}{\Gamma(a)}\, {}_2F_0\left(b-a, 1-a, \frac{1}{z}\right). \tag{7.169}$$

Substituting (7.169) into (7.165) instead of the confluent function with an argument x_+:

$$\begin{aligned} E(r,\phi,z) &= (-1)^n \exp(i2n\phi)\Gamma\left(\frac{2n+\beta+1}{2}\right)\Gamma\left(\frac{2n-\beta+1}{2}\right)[(2n)!]^{-1}(kr)^{2n}\, {}_1F_1\left(\frac{2n-\beta+1}{2}, 2n+1, x_-\right) \times \\ &\times \left\{ \frac{1}{\Gamma\left(\frac{2n+\beta+1}{2}\right)} \exp(ikz)\left[+ikz + ik\left(z^2+r^2\right)^{1/2}\right]^{\frac{2n-\beta+1}{2}} {}_2F_0\left(\frac{2n-\beta+1}{2}, \frac{-2n-\beta+1}{2}, -\frac{1}{x_+}\right) + \right. \\ &\left. + \frac{1}{\Gamma\left(\frac{2n-\beta+1}{2}\right)} \exp\left[-ik\left(z^2+r^2\right)^{1/2}\right]\left[-ikz - ik\left(z^2+r^2\right)^{1/2}\right]^{\frac{-2n-\beta-1}{2}} {}_2F_0\left(\frac{2n+\beta+1}{2}, \frac{-2n+\beta+1}{2}, \frac{1}{x_+}\right) \right\}. \end{aligned} \tag{7.170}$$

where ${}_2F_0(a, b, x)$ is the hypergeometric function [53]. Applying Kummer's transformation

$${}_1F_1(a,b,z) = \exp(z)\, {}_1F_1(b-a, b, -z), \tag{7.171}$$

we obtain:

$$\begin{aligned} E(r,\phi,z) &= (-1)^n \exp(i2n\phi)\Gamma\left(\frac{2n+\beta+1}{2}\right)\Gamma\left(\frac{2n-\beta+1}{2}\right)[(2n)!]^{-1}(kr)^{2n} \times \\ &\times \left\{ \frac{\exp(+ikz)}{\Gamma\left(\frac{2n+\beta+1}{2}\right)}(-x_+)^{\frac{2n-\beta+1}{2}} {}_2F_0\left(\frac{2n-\beta+1}{2}, \frac{-2n-\beta+1}{2}, -\frac{1}{x_+}\right) {}_1F_1\left(\frac{2n-\beta+1}{2}, 2n+1, x_-\right) + \right. \\ &\left. + \frac{\exp(-ikz)}{\Gamma\left(\frac{2n-\beta+1}{2}\right)}(+x_+)^{\frac{2n+\beta+1}{2}} {}_2F_0\left(\frac{2n+\beta+1}{2}, \frac{-2n+\beta+1}{2}, +\frac{1}{x_+}\right) {}_1F_1\left(\frac{2n+\beta+1}{2}, 2n+1, -x_-\right) \right\}. \end{aligned} \tag{7.172}$$

Taking into account that $z^{-a}{}_2F_0(a, 1 + a-b, -1/z) = U(a, b, z)$, where $U(a, b, z)$ is another solution of the Kummer equation [81, expression 13.1.3], the expression (7.172) becomes:

$$E(r,\phi,z)=\exp(i2n\phi)\Gamma\left(\frac{2n+\beta+1}{2}\right)\Gamma\left(\frac{2n-\beta+1}{2}\right)[(2n)!]^{-1}(kr)^{2n}\times$$
$$\times\left\{\frac{\exp(+ikz)}{\Gamma\left(\frac{2n+\beta+1}{2}\right)}(-1)^{-\frac{-\beta-1}{2}}U\left(\frac{2n-\beta+1}{2},2n+1,x_+\right){}_1F_1\left(\frac{2n-\beta+1}{2},2n+1,x_-\right)+\right.$$
$$\left.+\frac{\exp(-ikz)}{\Gamma\left(\frac{2n-\beta+1}{2}\right)}(-1)^{-\frac{\beta+1}{2}}U\left(\frac{2n+\beta+1}{2},2n+1,-x_+\right){}_1F_1\left(\frac{2n+\beta+1}{2},2n+1,-x_-\right)\right\}. \quad (7.173)$$

We write the expression for the complex amplitudes of nHyG modes:

$$E(r,\phi,z)=\frac{(-1)^{-\frac{\mp\beta-1}{2}}}{(2n)!}\Gamma\left(\frac{2n\mp\beta+1}{2}\right)\exp(i2n\phi\pm ikz)(kr)^{2n}\times$$
$$\times U\left(\frac{2n\mp\beta+1}{2},2n+1,\pm x_+\right){}_1F_1\left(\frac{2n\mp\beta+1}{2},2n+1,\pm x_-\right). \quad (7.174)$$

The expression for $E(r, \phi, z)$ from (7.173) is the sum $E^+(r, \phi, z)$ and $E^-(r, \phi, z)$ from (7.174).

Next, we consider only the field $E^+(r, \phi, z)$, propagating from the $z = 0$ plane to the $z\to\infty$ plane.

If in the expression (7.174) for the direct wave $E^+(r, \phi, z)$ the distance z tends to infinity, we obtain an asymptotic expression:

$$E^+(r,\phi,z\gg\lambda)\simeq\left[(2n)!\right]^{-1}\exp(i2n\phi+ikz+it)\Gamma\left(\frac{2n-\beta+1}{2}\right)\times$$
$$\times(2ikz)^{(\beta-1)/2}t^n{}_1F_1\left(\frac{2n+\beta+1}{2},2n+1,-t\right), \quad (7.175)$$

where $t = ikr^2/(2z)$. Equation (7.175) coincides (up to a constant factor) with the expression for the complex amplitude of the HyG-paraxial modes [82, 86], provided that $m = -1$, $i\gamma = \beta$, $w = k^{-1}$ and n is replaced by $2n$.

Modelling

For the distribution near the initial plane $z = 0$ the intensity distribution for the nHyG$^+$-beam varies mainly in the side lobes (peripheral light rings of the diffraction pattern) (Fig. 7.38a, b). When $z \gg \lambda$, where nHyG$^+$-beam coincides with the HyG-mode, changes in intensity occur only on a large scale, and the view of the diffraction pattern of the beam is preserved (Fig. 7.38c). Figure 7.38 shows the intensity $I = |E_x|^2$ in relative units at $\lambda = 633$ nm, $\beta = 0$, $n = 1$ at different distances from z = 0,

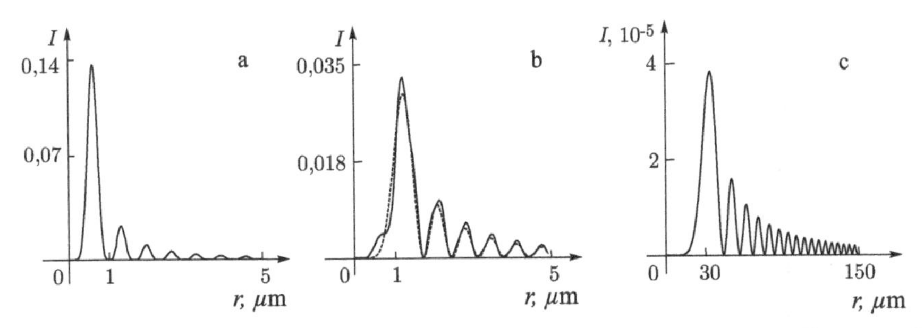

Fig. 7.38. Intensity distribution of nHG$^+$-mode at $\lambda = 633$ nm, $\beta = 0$, $n = 1$ at a distance z: 0 (a), 1 μm (b), 1 mm (c),

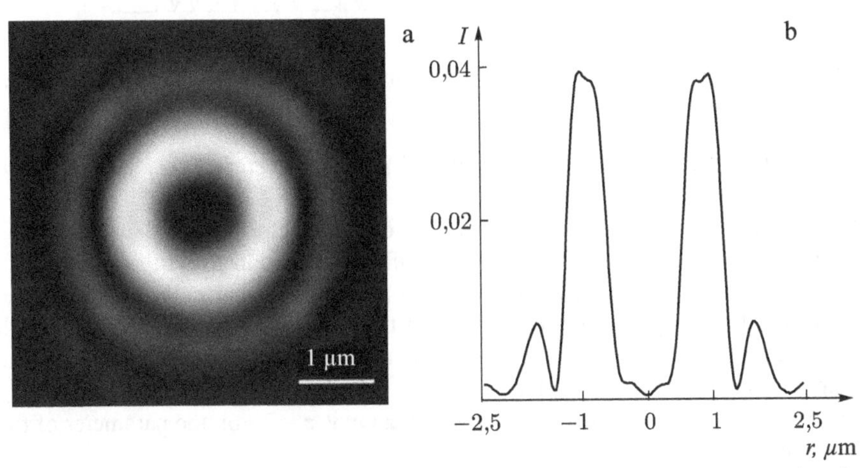

Fig. 7.39. Diffraction pattern $I = |E_x|^2 + |E_y|^2 + |E_z|^2$ (a) and its horizontal cross section (by the plane $y = 0$) (b) for a non-paraxial hypergeometric beam with an initial ($z = 0$) complex amplitude (7.167) at $z = 1$ μm.

calculated using equation (7.162) with integration from 0 to $\pi/2$. In Fig. 7.38b the dotted line shows the result of calculation by formula (7.174).

To verify the calculations (Fig. 7.38b) numerical simulation was carried out using the program FullWave 6.0 (RSoft Design, USA, http://www.rsoftdesign.com), designed to solve the Maxwell equations by the FDTD method (finite-difference time-domain). In the plane $z = 0$ we specified the electromagnetic linearly polarized (along the x-axis) field (7.167) for $n = 1$, $\lambda = 633$ nm, with sampling $\lambda/20$. Figure 7.39a shows the diffraction pattern of such a field in the plane $z = 1$ μm. The size of the pattern is 5×5 μm. Figure 7.39b shows a section of the diffraction pattern. A comparison of Figs. 7.38b and 7.39b show that they are in good agreement with each other, although in Fig. 7.38 shows the value of $|E_x|^2$, and Fig. 7.39b $I = |E_x|^2 + |E_y|^2 + |E_z|^2$.

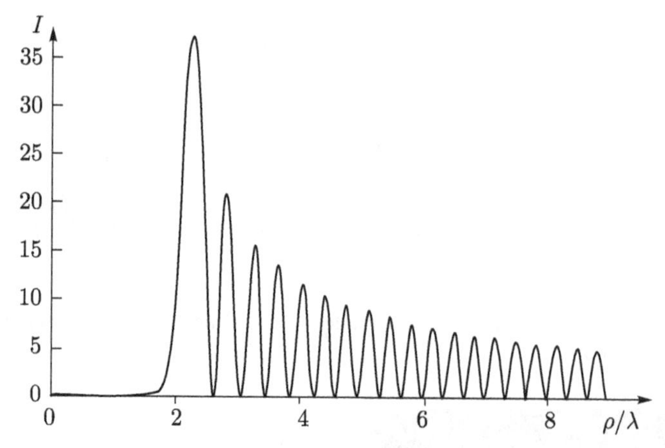

Fig. 7.40. Radial section of the intensity of the hypergeometric beam (7.118) at $z = 7$ μm.

7.5.6 Superresolution by means of hypergeometric laser beams

Formula (7.118) shows that the amplitude of the electric vector depends on the value ρ^2/z. This means that the frequency of the diffraction pattern increases with distance from the optical axis. Figure 7.40 shows the radial cross section of the intensity of the hypergeometric beam with the following parameters: wavelength $\lambda = 532$ nm, a Gaussian beam waist radius of $\sigma = 5$ mm, the power index of the amplitude component $m = 0$, the order of a spiral phase plate $n = 20$, the scaling factor $w = 5$ μm, the distance along the optical axis $z = 7$ μm, the parameter of the logarithmic axicon $\gamma = -200$.

Figure 7.40 shows that at one wavelength we can fit about two maxima. First, the frequency really increases with distance from the optical axis (the distance between the first two peaks is greater than the distance between the last two), but the formula (7.118) was obtained in the paraxial approximation. In the non-paraxial approximation, similar to that used in section 7.2.4 to analyze the diffraction vector of a Gaussian beam, using the formulas (7.40) the argument of the hypergeometric function $-ik\rho^2/(2z)$ is replaced by $-ik\rho^2\big/\sqrt{z^2+\rho^2}$, tending for large z to $-ik\rho^2/(2z)$, but for large values of ρ depends on the radial coordinate is not square and almost linear. Therefore, increase the spatial frequency really takes place, but not indefinitely, as it may seem from (7.118).

7.6. Elliptic vortex beams

7.6.1. Astigmatic Bessel beams

In recent papers [91, 92], attention was given to the oblique incidence of a plane wave on a conical axicon [91] and on a binary diffractive axicon [92].

In [91] the authors studied theoretically and experimentally the diffraction pattern in the case of oblique incidence (angle of 8–16°) of a beam of a helium-neon laser on a conical axicon with an angle of 0.01 radians at the base and the base diameter of 40 mm. The method of the stationary phase and using a Taylor expansion of functions in the exponent of a spherical wave included in the Kirchhoff–Fresnel integral [91] was used to obtain an approximate formula describing the diffraction pattern on the axicon at oblique incidence (equation (30) in [91]). The results of numerical simulation are in good agreement with experiment.

In [92] an equation was derived describing the caustic surface (shape and size of the focal spot) in the light beam formed by a diffraction axicon at oblique incidence of the light beam. It is also shown that at oblique incidence (the angle of incidence 0–10°) of the beam of a helium–neon laser with a wavelength of 0.6328 μm on a binary diffractive axicon with a diameter of 16.4 mm and the axicon parameter $\alpha = 0.036$ (axicon transmittance is described by the function $\exp(ik\alpha r)$, $k = 2\pi/\lambda$ is the wave number, α is the axicon parameter, r is the transverse radial coordinate), a characteristic astigmatic diffraction pattern was produced at a distance of 100 mm behind the axicon. The numerical results are in good agreement with experiment. Numerical simulation was carried using the two-dimensional integral transformation (equation (17) in [92]), obtained using a Taylor expansion of the exponent in the Kirchhoff–Fresnel integral.

In this section, we examine both theoretically and experimentally the oblique incidence of a beam of a helium-neon laser on a phase diffractive optical element (DOE) that generates simultaneously multiple single-mode Bessel beams of different orders that propagate at different angles to the optical axis. The complex amplitude of a single-mode Bessel beam is described by the function $J_n(\alpha r) \exp(in\varphi)$, (r, φ) are the polar coordinates, α is the beam parameter, n is the order of the Bessel modes, $J_n(x)$ is the Bessel function. Methods of calculation of phase formers of laser modes are described in [83]. The same paper describes a method of calculation of binary multi-order DOEs forming multiple modes of laser radiation.

If in [91] modelling of the diffraction of the oblique beam on the axicon was performed using the approximate equation obtained by both the stationary phase method and by the expansion of the phase of the impulse response function of free space in a Taylor series, in [92] a similar simulation of the diffraction of an inclined beam on the axicon was performed using a two-dimensional integral transform obtained by using only the Taylor expansion of the phase of the impulse response function of free space.

In this section, the numerical simulation of diffraction of the inclined beam on the phase DOE is carried out using only the stationary phase method, which reduces the two-dimensional Kirchhoff–Fresnel integral transformation to the one-dimensional Taylor series expansion. Therefore, the results of numerical modelling more accurately correspond to the experimental diffraction pattern.

In addition, this section shows that the oblique incidence on the DOE allows the visualization of the orbital angular momentum which the non-zero-order Bessel modes have [93]. The cross-sections of the beam of the Bessel mode with a right-handed rotation $J_n(ar)$ exp $(+in\varphi)$ is rotated by 45° to the right and for the modes with left rotation $J_n(ar)$ exp $(-in\varphi)$ is rotated by 45° to the left compared with the diffraction pattern of the zero Bessel modes $J_0(ar)$. Moreover, the higher the order of the mode n, the greater the number of local maxima in the cross section of an astigmatic Bessel beam, if we consider a fixed angle of incident radiation and a fixed distance from the DOE.

Analytical calculation of the cross section of single-mode astigmatic Bessel beam

The complex amplitude of the light behind the DOE, which is illuminated by an inclinded plane wave with amplitude exp(ikr cos φ × sin η), where η is the angle of the beam relative to the axis x, and generates a single-mode Bessel beam with an amplitude $J_n(ar)$ exp $(in\varphi)$, is described in the scalar diffraction theory by the Kirchhoff integral:

$$E(\rho,\psi,z)=\frac{z}{i\lambda}\int_0^{2\pi}\int_0^{\infty}J_n(\alpha r)\exp(in\varphi)\exp(ikr\cos\varphi\sin\eta)\frac{\exp(ikR)}{R^2}r\,\mathrm{d}r\,\mathrm{d}\varphi, \quad (7.176)$$

where

$$R=\sqrt{z^2+r^2+\rho^2-2r\rho\cos(\varphi-\psi)} \quad (7.177)$$

is the distance from the current point in the plane of the DOE to the point in the cross-section of the beam at a distance z from the DOE, (ρ, ψ) are the polar coordinates in the beam.

We assume that $z \gg r$, $z \gg \rho$ and, therefore, in equation (7.176) we can assume in the denominator $R^2 \approx z^2$. Next, following [91], we expand the square root of (7.177) in a Taylor series:

$$\sqrt{z^2+r^2+\rho^2-2r\rho\cos(\varphi-\psi)}\approx z+\frac{r^2+\rho^2-2r\rho\cos(\varphi-\psi)}{2z}-$$
$$-\frac{r^4+\rho^4+2r^2\rho^2-(4r^3\rho+4r\rho^3)\cos(\varphi-\psi)+4r^2\rho^2\cos^2(\varphi-\psi)}{8z^3}+\ldots \quad (7.178)$$

We neglect in (7.178) the terms:

$$\frac{r^4}{8z^3}<<z,\quad \frac{4r^3\rho}{8z^3}<<z,\quad \frac{r^2+\rho^2}{2z}<<z, \quad (7.179)$$

Then instead of (7.178), we obtain:

$$\sqrt{z^2+r^2+\rho^2-2r\rho\cos(\varphi-\psi)}\approx\left(z+\frac{\rho^2}{2z}-\frac{\rho^4}{8z^3}\right)+\left(\frac{r^2}{2z}-\frac{r^2\rho^2}{2z^3}\right)-$$
$$-\frac{r\rho}{z}\cos(\varphi-\psi)-\frac{r^2\rho^2}{4z^3}\cos 2(\varphi-\psi). \quad (7.180)$$

In this approximation, the integral (7.176) can be written as:

$$E(\rho,\psi,z)=\frac{\exp\left[ik\left(z+\frac{\rho^2}{2z}-\frac{\rho^4}{8z^3}\right)\right]}{i\lambda z}\int_0^{\infty}J_n(\alpha r)\exp\left[ik\frac{r^2}{2z}\left(1-\frac{\rho^2}{z^2}\right)\right]\times$$
$$\times\left\{\int_0^{2\pi}\exp(in\phi)\exp(ikr\cos\phi\sin\eta)\exp\left[-ik\frac{r\rho}{z}\cos(\phi-\psi)\right]\exp\left[-ik\frac{r^2\rho^2}{4z^3}\cos2(\phi-\psi)\right]\mathrm{d}\phi\right\}r\,\mathrm{d}r. \tag{7.181}$$

We consider separately the inner integral in φ (7.181), and make the change of variables, $\varphi-\psi=\phi$:

$$I_n=\exp(in\psi)\int_0^{2\pi}\exp(in\phi)\exp\left[-ik\frac{r\rho_0}{z}\cos(\phi-\nu)\right]\exp\left[-ik\frac{r^2\rho^2}{4z^3}\cos2\phi\right]\mathrm{d}\phi, \tag{7.182}$$

where

$$\rho_0=\sqrt{(\rho-z\sin\eta\cos\psi)^2+(z\sin\eta\sin\psi)^2}, \tag{7.183}$$

$$\nu=\mathrm{arctg}\left(\frac{z\sin\eta\sin\psi}{\rho-z\sin\eta\cos\psi}\right). \tag{7.184}$$

Introducing the notation:

$$N=-k\frac{r^2\rho^2}{4z^3},\quad M=-k\frac{r\rho_0}{z}. \tag{7.185}$$

and using the known formula of the series expansion of Bessel functions:

$$\exp(ix\sin t)=\sum_{m=-\infty}^{\infty}J_m(x)\exp(imt), \tag{7.186}$$

instead of (7.182) can be obtained from a series of Bessel functions:

$$I_n=2\pi\exp(in\psi)\sum_{m=-\infty}^{\infty}i^{n-m}J_m(N)J_{n+2m}(M)\exp(i2m\nu)=$$
$$=2\pi i^n\exp(in\psi)\left\{J_0(N)J_n(M)-iJ_1(N)\left[J_{n+2}(M)\exp(i2\nu)+J_{n-2}(M)\exp(-i2\nu)\right]+\ldots\right\}. \tag{7.187}$$

It is known that the Bessel beams are formed using conical waves. Therefore, the intensity distribution in the cross section of a Bessel beam at a fixed distance z from the DOE is determined near the beam axis by the contribution of the secondary light waves propagating from the DOE and located at a radius

$$r_0=\frac{z\alpha}{k}. \tag{7.188}$$

Equation (7.188) is the equation of stationary points, if the integral (7.181) is calculated over r by the stationary phase method. Then the integral over r in (7.181) will be equal to the integrand, in which the variable r should be replaced by the constant r_0 from (7.188). Therefore, the intensity distribution in the cross section of a Bessel beam at oblique incidence of a plane wave on the DOE is described approximately by the expression (7.187).

Let us analyze the equation (7.187). For sufficiently large z we have $|N| \ll |M|$, so the series (7.187) can be approximated by the first three terms. If there is no inclination ($\sin \eta = 0$), then $\nu = 0$ and function $I(\rho, \psi, z)$ is radially symmetric with respect to the modulus:

$$I_n(\rho,\psi,z) \approx 2\pi(-i)^n \exp(in\psi) J_n\left(k\frac{r_0\rho}{z}\right),$$
$$|I_n(\rho,\psi,z)| \approx 2\pi |J_n(\alpha\rho)|. \tag{7.189}$$

To analyze the structure of the intensity pattern in the cross-section of the astigmatic Bessel beam, we find the square of the modulus of the function I from (7.187), selecting only the first three terms in the sum. We obtain:

$$|I_0(\rho,\psi,z)|^2 \approx J_0^2(N)J_0^2(M) + J_1^2(N)J_2^2(M)\cos^2 2\nu, \tag{7.190}$$

$$|I_n(\rho,\psi,z)|^2 \approx J_0^2(N)J_n^2(M) + 2J_0(N)J_1(N)J_n(M)\left[J_{n+2}(M) - J_{n-2}(M)\right]\sin 2\nu. \tag{7.191}$$

Equation (7.190) was obtained for $n = 0$, and equation (7.191) for $n \neq 0$. From (7.191) it follows that for different signs of the number n we obtained different forms of the function I_n. For example, when $n = \pm 1$, from (7.191) we get:

$$|I_{\pm 1}(\rho,\psi,z)|^2 \approx J_0^2(N)J_1^2(M) \pm 2J_0(N)J_1(N)J_1(M)\left[J_1(M) + J_3(M)\right]\sin 2\nu. \tag{7.192}$$

Since $z \sin \eta \gg \rho$, then $|M| \approx \alpha z \sin \eta$ and $\nu \approx \psi$, then the Bessel function in the equation (7.192) does not depend on the azimuthal angle. The astigmatic pattern in the beam cross section is formed as a result of the addition of two terms in the equations (7.190)–(7.192). For a Bessel beam with $n = 0$ equation (7.190) shows that the maximum contribution of the second term to the overall diffraction pattern will occur at the points with $\cos 2\psi = \pm 1 \Rightarrow \psi = \pi m/2$, i.e., the diffraction pattern will be the axis of symmetry, coinciding with the axes x and y. For beams with $n \neq 0$ the angles at which both terms in (7.191) and (7.192) will give the maximum contribution will be different. From (7.192) it follows that for $n = 1$, the maximum contribution of the second term to the overall pattern will be provided if $\sin 2\psi = 1$, that is, if $\psi = \pi/4 + \pi m$, and for the Bessel beam with $n = -1$ at $\psi = -\pi/4 + \pi m$. That is, the axis of symmetry of the patterns of intensity distribution in the cross section of astigmatic Bessel beams with the right rotating phase ($n = 1$) and left-rotating phase ($n = -1$) are perpendicular to each other and lie on the axes at 45° to the axes x and y.

Thus, it appears that at oblique incidence of the light field on the DOE (or with the DOE inclined in relation to the optical axis), instead of radially symmetric intensity patterns in the cross section of Bessel beams we obtain not radially symmetrical astigmatic pattern, the axes of symmetry which are directed in different ways for the Bessel beams with the left and right rotation of the phase and without rotation. That is, by simple tilting of the DOE we can determine the sign of the orbital angular momentum of the Bessel beam.

Note also that, since the argument of the Bessel functions in (7.189)–(7.192) includes a combination of variables $\alpha\, z\sin\eta$, the degree of astigmatism of the Bessel beam can be varied not only by changing the angle η but also by changing the distance z at which is the diffraction pattern is examined.

Numerical simulation of the formation of astigmatic Bessel beams

Numerical simulation of the formation of the single-mode astigmatic Bessel beam at oblique incidence of a plane wave on the DOE was carried out using the equation (7.176), in which the integral over r was calculated in advance by the stationary phase method. Then the complex amplitude in the cross secion of the beam can be approximately calculated using the one-dimensional integral transform:

$$E_n(\rho,\psi,z)=\frac{zJ_n(\alpha r_0)r_0}{i\lambda}\int_0^{2\pi}\exp(in\varphi)\exp(ikr_0\cos\varphi\sin\eta)\frac{\exp(ikR_1)}{R_1^2}\mathrm{d}\varphi, \tag{7.193}$$

where

$$R_1=\sqrt{z^2+r_0^2+\rho^2-2r_0\rho\cos(\varphi-\psi)}, \tag{7.194}$$

r_0 is determined from the equation of the stationary point (7.188).

Calculation by formula (7.193) was performed with the following parameters: λ = 0.63 μm, η = 0.115π, D = 10 mm – diameter of the DOE, α = 36 mm^{-1}, 550 mm < z < 1210 mm. The number of samples in the angular variable φ – 100, the angular variable ψ – 250, the radial variable ρ – 250.

Figures 7.41a, 7.42a and 7.43a show the results of numerical simulation. The figures show the calculated intensity in the cross section of the Bessel beams with indices n = 0 (Fig. 7.41, a1–a3), n = +1 (Fig. 7.42, a1–a3), n = –2 (Fig. 7.43, a1–a3) at different distances z = 550 mm (a1), z = 785 mm (a2), z = 1210 mm (a3).

It is seen that the zero Bessel beam (n = 0) has an astigmatic diffraction pattern, the axes of symmetry lying along the axes x and y, as follows from (7.190). Increasing z increases the degree of astigmatism and the number of local maxima (Figs. 7.41–7.43 show the negatives).

For a Bessel beam with n = +1 the axis of symmetry of the diffraction pattern is rotated clockwise by an angle of 45°, as predicted by the formula (7.192). If we compare the diffraction patterns in Fig. 7.41a and 7.42a, we can see that the number of local maxima in the extreme row of the appropriate patterns differs for the two beams: at z = 550 mm – 2 peaks (Fig. 7.41, a1) and 3 peaks (Fig. 7.42, a1), at z = 785 mm – 3 peaks (Fig. 7.41, a2) and 4 peaks (Fig. 7.42, a2), at z = 1210 mm – 4

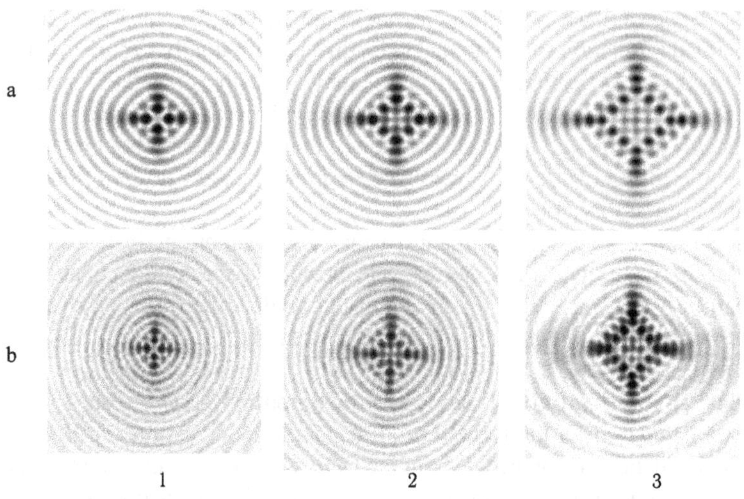

Fig. 7.41. Numerical (a) and experimental (b) intensity distribution (negative) at distances of (a) 550 mm, (2) 785 mm, (3) 1210 mm from the DOE, which forms a zero-order Bessel beam J_0 (αr) at oblique incidence (at an angle of 21°) of the illuminating beam ($\alpha = 36$ mm^{-1}).

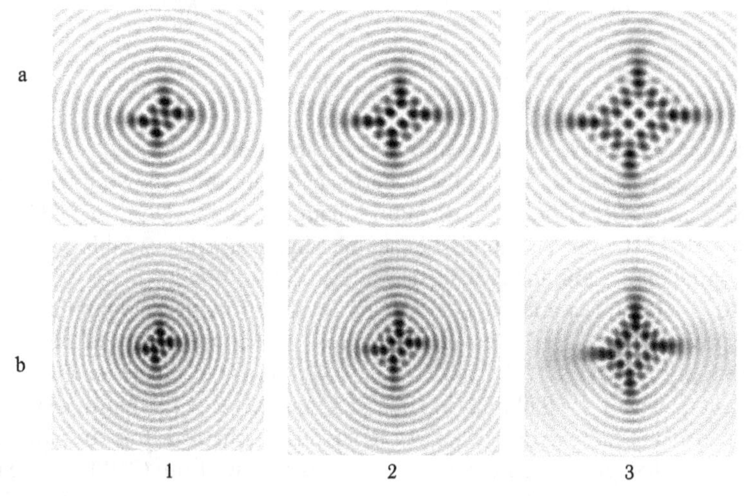

Fig. 7.42. Numerical (a) and experimental (b) intensity distribution (negative) at distances of (1) 550 mm, (2) 785 mm, (3) 1210 mm from the DOE, which forms a first-order Bessel beam $J_1(\alpha r)\exp(i\varphi)$at oblique incidence (at an angle of 21°) of the illuminating beam ($\alpha = 36$ mm^{-1}).

peaks (Fig. 7.41, a3) and 5 peaks (Fig. 7.42, a3). Thus, with the order of the Bessel beam increasing by 1 the number of local maxima in the far field increases by one.

Figure 7.43a shows that the symmetry axes of the diffraction pattern for a Bessel beam with index $n = -2$ are rotated counterclockwise by 45° to the axes x and y, as predicted by the formulas (7.191) and (7.192).

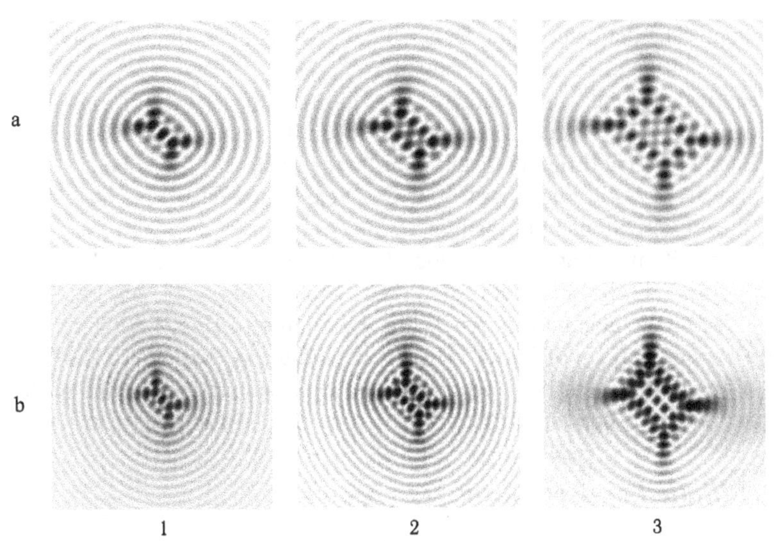

Fig. 7.43. Numerical (a) and experimental (b) intensity distribution (negative) at distances of (1) 550 mm, (2) 785 mm, (3) 1210 mm from the DOE, which forms a second-order Bessel beam $J_2(\alpha r)$ exp $(i\varphi)$ at oblique incidence (at an angle of 21°) of the illuminating beam ($\alpha = 36$ mm^{-1}).

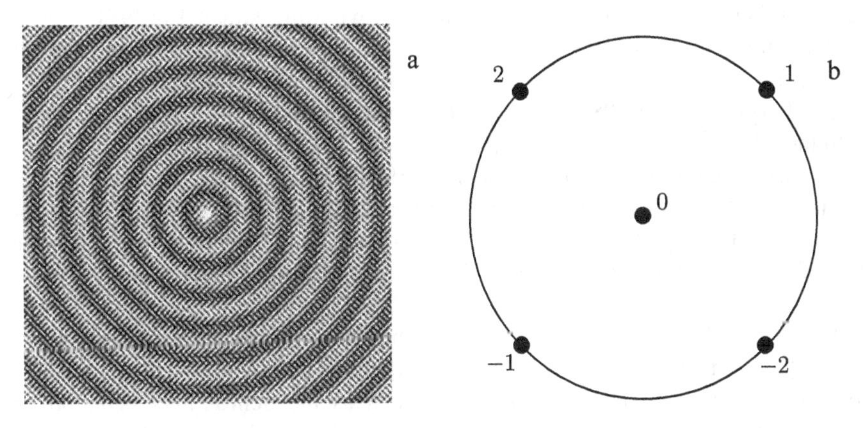

Fig. 7.44. (a) The central part of the binary phase of the complex transmission function of a five-order DOE that forms Bessel beams $J_m(\alpha r)$ exp $(im\varphi)$, $m = 0, \pm 1, \pm 2$, (b) the layout of the diffraction orders in the Fourier plane ($\alpha = 36$ mm^{-1}).

Experimental formation of astigmatic Bessel beams

To investigate the astigmatic Bessel beams, the University of Joensuu (Finland) produced by the electron-beam lithography technology a binary phase five-order DOE that forms simultaneously five Bessel beams, the complex amplitude of which is proportional to the functions $J_n(\alpha r)$ exp$(in\varphi)$, $\alpha = 36$ mm^{-1}, $n = 0, \pm1, \pm2$. Figure 7.44 shows a fragment of the central binary mask for producing this DOE, and Fig. 7.44b shows the arrangement of the five diffraction orders, which are formed in the far zone. The diameter of the DOE 10 mm, the number of pixels 2000 × 2000 with the resolution of 5 × 5 μm^2 per pixel.

The optical element was illuminated by a collimated beam of a He–Ne laser with a wavelength of 0.633 μm at different angles and the intensity distribution of each of the Bessel beam at different distances from the DOE was recorded.

Figures 7.41c–7.43c (negative) show CCD-camera-recorded intensity distributions in the cross sections of the astigmatic Bessel beams at distances from the DOE equal to $z = 550$ mm (Fig. 7.41, b1, Fig. 7.42, b1, Fig. 7.43, b1), $z = 785$ mm (Fig. 7.41, b2, Fig. 7.42, b2, Fig. 7.43, b2), $z = 1210$ mm (Fig. 7.41, b3, Fig. 7.42, b3, Fig. 7.43, b3) with the numbers: $n = 0$ (Fig. 7.41b), $n = -1$ (Fig. 7.42b), $n = +2$ (Fig. 7.43b). The comparison of Figs. 7.41a, 7.42a, 7.43a and Figs. 7.41b, 7.42b, 7.43b shows a good agreement between theory and experiment, at least, the number of local maxima in the corresponding diffraction pattern is exactly the same.

7.6.2. Elliptic Laguerre–Gaussian beams

Recently, interest has grown to elliptic Gaussian laser beams. For example, in [80] the decentred elliptical Gaussian beam, propagating in an axially-asymmetric optical system, was studied. Such a beam is described using the tensor technique. In [94] the authors investigated elliptical decentred Hermite–Gaussian beams. Elliptical Gaussian beams of partially coherent light were investigated in [95]. Fresnel diffraction of elliptical (astigmatic) Gaussian beams by a diffraction grating is discussed in [96]. In [97] attention is given to the propagation of such a beam in a uniaxial crystal, and in [98] the second-harmonic generation in a non-linear crystal with the elliptical beam was considered.

Gaussian beams with different degrees of ellipticity can be used to align the beam shape [99], that is, for elliptical laser beams with uniform intensity across the beam. With the help of a linear combination of the elliptical Hermite–Gaussian it is possible to form 'tubular' (hollow) elliptical laser beams with zero intensity [100] on their axis.

In [75, 78, 82, 101, 102] Ince–Gaussian beams were investigated theoretically and in [78] both theoretically and experimentally. Such light fields are particular solutions of the paraxial wave equation (such as the Schrödinger equation) in elliptic coordinates. In these coordinates, the equation can be solved by separation of variables, and the solution is obtained as the product of a Gaussian function by Ince polynomials. The Ince polynomials are solutions of the differential Whittaker-Hill equation. The Ince–Gaussian beams are an orthogonal basis which generalizes the well-known Hermite–Gaussian and LG modal bases. When the ellipse becomes a circle (eccentricity $\varepsilon = 1$), the Ince–Gaussian modes become the LG modes and when ε tends to infinity (the ellipse becomes a straight line) Ince–Gaussian modes transform into Hermite–Gaussian modes.

Elliptical beams may also be formed by oblique incidence of the axisymmetric beam on the optical element.

In [91, 92], oblique incidence of a laser beam with a plane wavefront on a conical axicon and a binary diffractive axicon was studied. In [91] experimentally and numerically studies were made of the diffraction pattern which is formed at oblique incidence (angle of 8–16°) of a collimated beam from a helium–neon laser on a conical axicon with an angle of 0.01 radians at the base and the base diameter of 40 mm. It was shown that if the axial illumination of the base of the axicon results in the formation by a light diffraction-free zero-order Bessel beam, then in oblique incidence the diffraction pattern loses its radial symmetry, the beam begins to diverge and change the structure of the transverse distribution of intensity.

Similar studies were carried out in [92], but a binary diffractive axicon was used instead of a conventional axicon. The angle of incident light with a wavelength λ = 632.8 nm was up to 10°, the diameter of the diffractive optical element (DOE) was 16.4 mm, and the axicon parameter $\alpha = 0.036$. The transmission function of the axicon is exp $(-ik\alpha r)$, where k is the wave number, r the radial coordinate.

In [103], both theoretically and experimental studies were carried out of the diffraction pattern, which can be obtained at oblique incidence of a plane wave on a multichannel binary DOE, whose transmission is proportional to the transmission function which describes five Bessel beams propagating under different angles to the optical axis with amplitudes $J_m(\alpha r)$ exp $(im\varphi)$, $m = 0, +1, +2$.. It has been shown that this is accompanied by the formation of divergent astigmatic Bessel beams whose diffraction patterns are rotated by 45° to the right when $m > 0$, and by 45° to the left at $m = 0$. Moreover, the greater the number of the Bessel mode $|m|$, the larger is the number of local minima and maxima observed in the diffraction pattern at a fixed distance from the DOE.

This section describes the distribution of an elliptical LG beam, which is no longer a mode. It is shown that at oblique incidence of a plane wave on the DOE, whose transmission function is proportional to a function that describes the LG mode of the order (m, n), there is a diffraction pattern which is identical to the diffraction pattern formed by an elliptical LG beam. In the propagation in a homogeneous space the elliptical LG beam forms a diffraction pattern which is rotated by +45° at $m > 0$ and –45° for $m < 0$. Moreover, the number of local maxima initially increases with increasing distance z and then decreases, and at $z \to \infty$ (in the far field) a diffraction pattern appears, consisting of a set of concentric ellipses rotated by 90° with respect to the original diffraction pattern with $z = 0$.

Fresnel transformation of the LG mode

With the reference integral [53]:

$$\int_0^\infty x^{\frac{m}{2}} \exp(-px) J_m\left(b\sqrt{x}\right) L_n^m(cx)\,\mathrm{d}x = \left(\frac{b}{2}\right)^m \frac{(p-c)^n}{p^{m+n+1}} \exp\left(-\frac{b^2}{4p}\right) L_n^m\left(\frac{b^2 c}{4pc - p^2}\right), \tag{7.195}$$

where $J_m(x)$ is the Bessel function of the m-th order and the first kind; $L_n^m(x)$ is the associated Laguerre polynomial [81, Table 22.3], (n, m) are integers, we can find an expression for the Fresnel transformation of the LG mode:

$$\Psi_{mn}(r,\varphi)=\left(\frac{r\sqrt{2}}{w_0}\right)^m \exp\left(-\frac{r^2}{w_0^2}\right)L_n^m\left(\frac{2r^2}{w_0^2}\right)\exp(im\varphi), \tag{7.196}$$

where (r, φ) are the polar coordinates, w_0 is the Gaussian beam waist radius.

Using (7.195) and (7.196), we obtain an expression for the amplitude of the LG mode at a distance z from the waist:

$$\Psi_{mn}(\rho,\theta,z)=\frac{-ik}{2\pi z}\exp\left(\frac{ik\rho^2}{2z}+im\theta\right)\int_0^\infty\left(\frac{r\sqrt{2}}{w}\right)^m \exp\left(-\frac{r^2}{w^2}+\frac{ikr^2}{2z}\right)L_n^m\left(\frac{2r^2}{w^2}\right)r\,\mathrm{d}r=$$

$$=\exp\left[\frac{-\rho^2}{w^2(z)}+\frac{ik\rho^2}{2R(z)}+im\theta-i(2n+m+1)\operatorname{arctg}\left(\frac{z}{z_0}\right)\right]\frac{i^m}{2\pi}\frac{w_0}{w(z)}\left(\frac{\rho\sqrt{2}}{w(z)}\right)^m L_n^m\left(\frac{2\rho^2}{w^2(z)}\right), \tag{7.197}$$

where $w(z)=w_0\sqrt{1+z^2/z_0^2}$, $R(z) = z\,(1 + z_0^2/z^2)$, $z_0 = kw_0^2/2$, k is the wave number of light.

Equation (7.197) shows that the structure of the LG mode is retained and the intensity is radially symmetric function: $I_{mn}(\rho, z) = |\Psi_{mn}(\rho, \theta, z)|^2$.

With the help of the light field of the form (at $z = 0$):

$$\Psi_{mn}^{(1)}(r,\varphi)=r^{2n+m}\exp\left(-\frac{r^2}{w_0^2}+im\varphi\right) \tag{7.198}$$

we can create generic LG modes which will also be radially symmetric and will retain their structure during the propagation up to scale. We use the reference integral [53]:

$$\int_0^\infty r^{2n+m}\exp\left(-pr^2\right)J_m(cr)r\,\mathrm{d}r=\frac{n!c^m}{2^{m+1}p^{m+n+1}}\exp\left(-\frac{c^2}{4p}\right)L_n^m\left(\frac{c^2}{4p}\right). \tag{7.199}$$

Then the Fresnel transformation from the initial light field (7.198) with (7.199) has the form:

$$\Psi_{mn}^{(1)}(\rho,\theta,z)=\frac{-ik}{2\pi z}\exp\left(\frac{ik\rho^2}{2z}+im\theta\right)\int_0^\infty r^{2n+m}\exp\left(-\frac{r^2}{w^2}+\frac{ikr^2}{2z}\right)J_m\left(\frac{kr\rho}{z}\right)r\,\mathrm{d}r=$$

$$=\frac{kn!2^{m-1}}{2\pi z}\left(\frac{1}{w_0^2}-\frac{ik}{2z}\right)^{-n-1}\exp\left[\frac{ik\rho^2}{2z}+im\theta\right]x^m\exp(-x)L_n^m(x), \tag{7.200}$$

where $x=\left[\frac{1}{w^2(z)}+\frac{ik}{2\hat{R}(z)}\right]\rho^2$, $\hat{R}(z)=z\left(1+\frac{z^2}{z_0^2}\right)$, $w(z)=w_0\sqrt{1+\frac{z^2}{z_0^2}}$.

These light fields are called elegant LG beams.

Oblique paraxial beams

Let us illuminate with an inclined plane wave exp ($ikr\cos\varphi\ \sin\gamma$), γ is the angle to the axis $x = r\cos\varphi$, a plane optical element whose transmission is proportional to the function, which describes the LG mode $\Psi_{mn}(r,\varphi)$ whose radial part is denoted as follows:

$$\hat{\Psi}_{mn}(r) = \left(\frac{r\sqrt{2}}{w_0}\right)^m \exp\left(-\frac{r^2}{w_0^2}\right) L_n^m\left(\frac{2r^2}{w_0^2}\right). \tag{7.201}$$

Then the Fresnel transformation from the LG mode (7.196) for oblique illumination will look like:

$$F_\gamma(\rho,\theta,z) = \frac{2\pi(-i)^m k}{z}\exp\left[\frac{ik\rho^2}{2z} + im\,\mathrm{arctg}\left(\frac{\rho\sin\theta}{z\sin\gamma - \rho\cos\theta}\right)\right]\times$$
$$\times\int_0^\infty \hat{\Psi}_{mn}(r)\exp\left(\frac{ikr^2}{2z}\right)J_m\left(\frac{kr}{z}\sqrt{z^2\sin^2\gamma + \rho^2 - 2\rho z\cos\theta\sin\gamma}\right)r\,\mathrm{d}r. \tag{7.202}$$

In the polar coordinates, with the slope:

$$\begin{cases}\xi = \rho\cos\theta - z\sin\gamma,\\ \eta = \rho\sin\theta,\end{cases} \tag{7.203}$$

the argument of the Bessel functions in the integral of equation (7.202) will depend only on the radial component:

$$\xi^2 + \eta^2 = z^2\sin^2\gamma + \rho^2 - 2\rho z\cos\theta\sin\gamma. \tag{7.204}$$

That is, from equation (7.202) with (7.203) and (7.204) it follows that the intensity of the oblique paraxial LG modes is radially symmetric:

$$I_\gamma(\xi^2 + \eta^2, z) = \left|F_\gamma(\rho,\theta,z)\right|^2. \tag{7.205}$$

For oblique incidence of the illuminating beam to lead to a distortion of the LG mode it is required to consider the non-paraxial propagation of light, i.e., to transfer from the Fresnel to Kirchhoff transformation.

Kirchhoff's transformation from an inclined LG beam is:

$$F(\xi,\eta,z) = \frac{-ik}{2\pi}\int_{-\infty}^{\infty}\int_{-\infty}^{\infty}\Psi_{mn}(x,y)\frac{\exp(ikR)}{R}\,\mathrm{d}x\,\mathrm{d}y, \tag{7.206}$$

where $R^2 = (\xi - x)^2 + (\eta - y)^2 + z^2$.

In the polar coordinates (7.206) becomes:

$$F(\rho,\theta,z) = \frac{-ik}{2\pi z}\int_0^\infty\int_0^{2\pi}\Psi_{mn}(r,\varphi)\exp\left[ik\sqrt{r^2 + \rho^2 + z^2 - 2r\rho\cos(\theta - \varphi)}\right]r\,\mathrm{d}r\,\mathrm{d}\varphi, \tag{7.207}$$

where $R \approx z$ in the denominator of (7.206).

Then the non-paraxial propagation of the inclined LG beam will be described by the expression:

$$F_\gamma(\rho,\theta,z)=\frac{-ik}{2\pi z}\int_0^\infty\int_0^{2\pi}\Psi_{mn}(r,\varphi)\exp\times \\ \times\left[ikr\cos\varphi\sin\gamma+ik\sqrt{r^2+\rho^2+z^2-2r\rho\cos(\theta-\varphi)}\right]r\,\mathrm{d}r\,\mathrm{d}\varphi. \tag{7.208}$$

Suppose that $z \gg r$ and $z \gg \rho$ and expand the square root in the exponent in (7.208) into a Taylor series:

$$\left[z^2+r^2+\rho^2-2r\rho\cos(\theta-\varphi)\right]^{\frac{1}{2}}\approx \\ \approx z+\frac{r^2+\rho^2-2r\rho\cos(\theta-\varphi)}{2z}-\frac{1}{8z^3}\left[r^4+\rho^4+2r^2\rho^2-\right. \\ \left.-\left(4r^3\rho+4r\rho^3\right)\cos(\theta-\varphi)+4r^2\rho^2\cos^2(\theta-\varphi)\right]+... \tag{7.209}$$

Suppose that in (7.209), $\frac{r^4}{8z^3}<<z$, $\frac{4r^3\rho}{8z^3}<<z$ and $\frac{4r\rho^3}{8z^3}<<z$. Then instead of (7.209) we get:

$$\left[z^2+r^2+\rho^2-2r\rho\cos(\theta-\varphi)\right]^{\frac{1}{2}}\approx\left(z+\frac{\rho^2}{2z}-\frac{\rho^4}{8z^3}\right)+\left(\frac{r^2}{2z}-\frac{r^2\rho^2}{8z^3}\right)- \\ -\frac{r\rho}{z}\cos(\theta-\varphi)-\frac{r^2\rho^2}{4z^3}\cos 2(\theta-\varphi). \tag{7.210}$$

In view of (7.208) instead of (7.210) we get:

$$F_\gamma(\rho,\theta,z)=\frac{-ik}{2\pi z}\exp\left[ik\left(z+\frac{\rho^2}{2z}-\frac{\rho^4}{8z^3}\right)\right]\times \\ \times\int_0^\infty\hat{\Psi}_{mn}(r)\exp\left[\frac{ikr^2}{2z}\left(1-\frac{\rho^2}{z^2}\right)\right]\times\left\{\int_0^{2\pi}\exp\left[im\varphi+ikr\cos\varphi\sin\gamma-\right.\right. \\ \left.\left.-\frac{ikr\rho}{z}\cos(\theta-\varphi)-\frac{ikr^2\rho^2}{4z^3}\cos 2(\theta-\varphi)\right]\mathrm{d}\varphi\right\}r\,\mathrm{d}r. \tag{7.211}$$

We rewrite the integral over φ in the curly brackets in (7.211) separately as follows:

$$I_0=\exp(im\theta)\int_0^{2\pi}\exp\left[im\psi-\frac{ikr\rho_0}{z}\cos(\psi-\nu)-\frac{ikr^2\rho^2}{4z^3}\cos 2\psi\right]\mathrm{d}\psi, \tag{7.212}$$

where

$$\begin{cases} \psi = \theta - \varphi, \\ \rho_0^2 = \left(\rho - z\sin\gamma\cos\theta\right)^2 + \left(z\sin\gamma\sin\theta\right)^2, \\ \nu = \operatorname{arctg}\left(\dfrac{z\sin\gamma\sin\theta}{\rho - z\sin\gamma\cos\theta}\right). \end{cases} \tag{7.213}$$

In (7.212) we denote:

$$P = \frac{kr\rho_0}{z},\quad Q = \frac{kr^2\rho^2}{4z^3}. \tag{7.214}$$

Then the integral (7.212) becomes:

$$\begin{aligned} I_0 &= \exp(im\theta)\int_0^{2\pi} \exp\left[im\psi - iP\cos(\psi - \nu) - iQ\cos 2\psi\right]\mathrm{d}\psi = \\ &= \exp(im\theta)\sum_{p=-\infty}^{\infty}(-i)^p J_p(Q)\int_0^{2\pi}\exp\left[i2p\psi + im\psi - iP\cos(\psi - \nu)\right]\mathrm{d}\psi = \\ &= 2\pi(-i)^m \exp[im(\theta+\nu)]\sum_{p=-\infty}^{\infty}(-i)^p J_p(Q) J_{m+2p}(P)\exp(i2p\nu). \end{aligned} \tag{7.215}$$

Note that the same expression as (7.215) was obtained in [95] to describe the astigmatic Bessel beam.

In view of (7.215), the amplitude of the light field (7.211), which describes the astigmatic LG beam (similar to the astigmatic Bessel beam [95]) takes the form:

$$\begin{aligned} F_\gamma(\rho,\theta,z) &= \frac{(-i)^{m+1}k}{z}\exp\left[im(\theta+\nu) + ik\left(z + \frac{\rho^2}{2z} - \frac{\rho^4}{8z^3}\right)\right]\times \\ &\times\sum_{p=-\infty}^{\infty}(-i)^p\exp(i2p\nu)\int_0^{\infty}\hat{\Psi}_{mn}(r)\exp\left[\frac{ikr^2}{2z}\left(1 - \frac{\rho^2}{z^2}\right)\right]J_p\left(\frac{kr^2\rho^2}{4z^3}\right)J_{m+2p}\left(\frac{kr\rho_0}{z}\right)r\,\mathrm{d}r. \end{aligned} \tag{7.216}$$

Equation (7.216) shows that the astigmatic LG beam is not radially symmetric and does not retain its structure during propagation, as azimuthal angle θ is included in the integrand in equation (7.216) in ρ_0 and ν (see (7.213)).

Elliptical paraxial LG beams

We show that if we replace the inclined LG beam by an elliptical one, then the expression similar to equation (7.216) can be obtained using the Fresnel transform, rather than the Kirchhoff transform. That is, the paraxial elliptical LG beam also will not retain its structure and lose ellipticity in the Fresnel diffraction zone.

The elliptical LG beam at $z = 0$ is described by:

$$\Psi_{mn}(x,y;\alpha) = \left(\frac{2x^2 + 2\alpha^2 y^2}{w_0^2}\right)^{\frac{m}{2}} \exp\left(-\frac{x^2 + \alpha^2 y^2}{w_0^2}\right) L_n^m\left(\frac{2x^2 + 2\alpha^2 y^2}{w_0^2}\right) \exp\left[im \operatorname{arctg}\left(\frac{\alpha y}{x}\right)\right]. \tag{7.217}$$

In the elliptical coordinates:

$$\begin{cases} x = \alpha r \cos\varphi, \\ y = r\sin\varphi, \quad 0 \le \alpha \le 1, \end{cases} \tag{7.218}$$

instead of (7.217) we get:

$$\Psi_{mn}(r,\varphi;\alpha) = \left(\frac{\alpha r\sqrt{2}}{w_0}\right)^m \exp\left(-\frac{\alpha^2 r^2}{w_0^2}\right) L_n^m\left(\frac{2\alpha^2 r^2}{w_0^2}\right) \exp(im\varphi). \tag{7.219}$$

At $\alpha = 1$, expression (7.219) coincides with (7.196). The Fresnel transformation in the elliptical coordinates (7.218) for the beam (7.219) becomes:

$$F_\alpha(\rho,\theta,z) = \frac{-ik}{2\pi z} \exp\left[\frac{ik\rho^2}{2z}\left(\cos^2\theta + \alpha^2 \sin^2\theta\right)\right] \times$$
$$\times \int_0^\infty \int_0^{2\pi} \hat{\Psi}_{mn}(\alpha r) \exp\left[im\varphi + \frac{ikr^2}{2z}\left(\alpha^2 \cos^2\varphi + \sin^2\varphi\right) - \frac{ik\alpha r\rho}{z}\cos(\theta - \varphi)\right] r\,\mathrm{d}r\,\mathrm{d}\varphi, \tag{7.220}$$

where $\Psi_{mn}(r,\varphi;\alpha) = \hat{\Psi}_{mn}(\alpha r)\exp(im\varphi)$.

In equation (7.220) we used the elliptical coordinates in a plane rotated by 90 ° with respect to the coordinates in the plane z = 0:

$$\begin{cases} \xi = \rho\cos\theta, \\ \eta = \alpha\rho\sin\theta. \end{cases} \tag{7.221}$$

We distinguish in equation (7.220) the integral over the angle φ:

$$F_\alpha(\rho,\theta,z) = \frac{-ik}{2\pi z} \exp\left[\frac{ik\rho^2}{2z}\left(\cos^2\theta + \alpha^2 \sin^2\theta\right)\right] \int_0^\infty \hat{\Psi}_{mn}(\alpha r) \exp\left[\frac{ikr^2}{4z}\left(1 + \alpha^2\right)\right] \times$$
$$\times \left\{ \int_0^{2\pi} \exp\left[im\varphi - \frac{ikr^2}{4z}\left(1 - \alpha^2\right)\cos^2\varphi - \frac{ik\alpha r\rho}{z}\cos(\theta - \varphi)\right] \mathrm{d}\varphi \right\} r\,\mathrm{d}r. \tag{7.222}$$

We write separately to the integral over φ in the curly brackets in equation (7.222):

$$\hat{I}_0 = \int_0^{2\pi} \exp\left[im\varphi - \frac{ikr^2}{4z}\left(1-\alpha^2\right)\cos^2\varphi - \frac{ik\alpha r\rho}{z}\cos(\theta-\varphi) \right] d\varphi. \quad (7.223)$$

The integral (7.223) up to notation and the factor before the integral coincides with the expression (7.212). We introduce the following notations:

$$A = \frac{k\alpha r\rho}{z}, \quad B = \frac{kr^2(1-\alpha^2)}{4z}. \quad (7.224)$$

Then instead of (7.223) we get:

$$\hat{I}_0 = \sum_{p=-\infty}^{\infty} (-i)^p J_p(B) \int_0^{2\pi} \exp\left[i2p\varphi + im\varphi - A\cos(\varphi-\theta)\right] d\varphi = \\ = 2\pi \sum_{p=-\infty}^{\infty} (-i)^{p+m} J_p(B) J_{m+2p}(A) \exp[i(m+2p)\theta]. \quad (7.225)$$

It is evident that the series in (7.215) and (7.225) coincide up to notation. Finally, from (7.221) and (7.224) we obtain an expression for Fresnel diffraction of the elliptical LG beam:

$$F_\alpha(\rho,\theta,z) = \frac{(-i)^{m+1}k}{z} \exp\left[\frac{ik\rho^2}{2z}\left(\cos^2\theta + \alpha^2\sin^2\theta\right)\right] \sum_{p=-\infty}^{\infty} (-i)^p \exp[i(2p+m)\theta] \times \\ \times \int_0^{\infty} \hat{\Psi}_{mn}(\alpha r) \exp\left[\frac{ikr^2}{4z}\left(1+\alpha^2\right)\right] J_p\left[\frac{kr^2\left(1-\alpha^2\right)}{4z}\right] J_{m+2p}\left(\frac{k\alpha r\rho}{z}\right) r\,dr. \quad (7.226)$$

Note that when $\alpha = 1$, equation (7.226) coincides with equation (7.197) for the Fresnel transformation of the LG mode. Indeed, when $\alpha = 1$, all terms of a series in p, except $p = 0$, will be zero, since $J_p(0) = 0$ at $p \neq 0$, and $J_0(0) = 1$.

Note also that at $z \to \infty$ in such a way that $r^2/z \ll r\rho/z$, instead of equation (7.226) we obtain the Fourier transform of the elliptical LG beam:

$$F_\alpha(\rho,\theta,z\to\infty) \approx \frac{(-i)^{m+1}k}{z} \exp(im\theta) \int_0^{\infty} \hat{\Psi}_{mn}(\alpha r) J_m\left(\frac{kr\alpha\rho}{z}\right) r\,dr = \\ = (-i)^m (-1)^n \left(\frac{w_0}{\alpha^2\sigma}\right)\left(\frac{\rho\sqrt{2}}{\sigma}\right)^m \exp\left(-\frac{\rho^2}{\sigma^2}\right) L_n^m\left(\frac{2\rho^2}{\sigma^2}\right), \quad (7.227)$$

where $\sigma = \dfrac{2z}{kw_0}$, $\rho^2 = \xi^2 + \left(\dfrac{\eta}{\alpha}\right)^2$.

Equation (7.227) shows that the elliptical LG beam at $z = 0$ in the far-field diffraction zone at $z \to \infty$ once again takes the elliptical symmetry, but is rotated by 90° with respect to the beam at $z = 0$.

At final z the elliptical symmetry of the LG beam is lost. At finite z at the centre of the diffraction pattern ($\rho = 0$) for an elliptical LG beam with even $m = 2l \neq 0$ the intensity is different from zero, while at $z = 0$ and $z \to \infty$ the intensity at the centre of the beam at $\rho = 0$ is equal to zero. Indeed, when $\rho = 0$ all the terms of the series in (7.226) are zero except the term with index $p = -m/2$, since $J_{m+2p}(0) = J_0(0) = 1$. Then from equation (7.226) for even $m = 2l$ we obtain:

$$F_\alpha(\rho = 0, \theta, z) \sim \int_0^\infty \hat{\Psi}_{mn}(\alpha r) \exp\left[\frac{ikr^2}{4z}\left(1+\alpha^2\right)\right] J_{\frac{m}{2}}\left[\frac{kr^2}{4z}\left(1-\alpha^2\right)\right] r\,\mathrm{d}r, \tag{7.228}$$

where the symbol ~ means proportionality.

At low beam ellipticity $\alpha \to 1$ the Bessel function in equation (7.228) can be replaced by the approximate expression:

$$J_p(x) \cong \frac{\left(\frac{x}{2}\right)^p}{\Gamma(p+1)}, \quad x \to 0, \tag{7.229}$$

where $\Gamma(p + 1)$ is the gamma function, $p > 0$. If $p < 0$, we have to use the equality $J_{-p}(x) = (-1)^p J_p(x)$ if p is an integer.

Then the integral in (7.228) can be written as:

$$\begin{aligned} F_\alpha(\rho = 0, z) &\sim \Gamma^{-1}\left(\frac{m}{2}+1\right)\left[\frac{k\left(\alpha^2-1\right)}{8z}\right]^{\frac{m}{2}}\left(\frac{\alpha\sqrt{2}}{w_0}\right)^m \times \\ &\times \int_0^\infty r^{2m} \exp\left[-\frac{\alpha^2 r^2}{w_0^2} + \frac{ikr^2\left(1+\alpha^2\right)}{4z}\right] L_n^m\left(\frac{2\alpha^2 r^2}{w_0^2}\right) r\,\mathrm{d}r. \end{aligned} \tag{7.230}$$

Using the background integral [53]:

$$\int_0^\infty x^m \exp(-px) L_n^m(cx)\,\mathrm{d}x = \frac{\Gamma(m+n+1)(p-c)^n}{n!\,p^{m+n+1}} \tag{7.231}$$

for the integral in (7.230) we get:

$$\begin{aligned} &\int_0^\infty r^{2m} \exp\left[-\frac{\alpha^2 r^2}{w_0^2} + \frac{ikr^2\left(1+\alpha^2\right)}{4z}\right] L_n^m\left(\frac{2\alpha^2 r^2}{w_0^2}\right) r\,\mathrm{d}r = \\ &= \frac{(-1)^n}{n!}\Gamma(m+n+1)\exp\left[i(2n+m+1)\eta\right]\left[\frac{\alpha^4}{w_0^4} + \frac{k^2\left(1+\alpha^2\right)^2}{8z^2}\right]^{-\frac{m+1}{2}}, \end{aligned} \tag{7.232}$$

where $\eta = \text{arctg}\left[\frac{k\left(1+\alpha^2\right)w_0^2}{4z^2\alpha^2}\right]$.

Equation (7.232) shows that the modulus of the integral (7.232) is always different from zero. Only when $\alpha = 1$ it follows from (7.230) that the field $F_\alpha(\rho = 0, z) = 0$. Thus, we have shown that even a small ellipticity of the LG mode violates the conditions under which in the centre of the beam with $\rho = 0$ for any z the intensity is zero. But the vortex character of the light field in the presence of ellipticity does not change, and the phase singularity at the centre of the order of $m = 2l$ 'breaks up' into p singularities with m/p numbers (orders), and p points of zero intensity occur near the centre of the beam. The number p depends on the degree of ellipticity. As shown below, with weak ellipticity $p = 2$ and at 45° there are two zero-order intensities of the $m/2$-th order. For odd $m = 2l+1$ in the centre of the diffraction pattern ($\rho = 0$) the intensity is always zero.

Consider the propagation characteristics of an elliptical LG beam with weak ellipticity ($\alpha^2 \approx 1$).Under the integral in equation (7.226) we replace the Bessel function of order p by the approximate expression (7.229). This is possible, first, because the argument of the Bessel function tends to zero for small ellipticity, and secondly, because even though the integral in (7.226) has an infinite upper integration limit, but the Gaussian exponent, included in the LG mode, limits the integration scope by the final effective radius of the mode. Then instead of (7.226) we write:

$$F_{\alpha\to 1}(\rho,\theta,z) \approx S(\rho,\theta)\sum_{p=-\infty}^{\infty}\left(\frac{-i\varepsilon}{2}\right)^p \Gamma^{-1}(p+1)\exp(i2p\theta)\,\delta(p)\times$$

$$\times\int_0^\infty\left(\frac{\alpha r\sqrt{2}}{w_0}\right)^m r^{2p}\exp\left[-\frac{\alpha^2 r^2}{w_0^2}-\frac{ikr^2\left(1+\alpha^2\right)}{4z}\right]L_n^m\left(\frac{2\alpha^2 r^2}{w_0^2}\right)J_{m+2p}\left(\frac{k\alpha r\rho}{z}\right)r\,\mathrm{d}r, \tag{7.233}$$

where

$$\delta(p) = \begin{cases} 1, \ p \ge 0, \\ (-1)^{|p|}, \ p < 0, \end{cases} \tag{7.234}$$

$$S(\rho,\theta) = \frac{(-i)^{m+1}k}{z}\exp\left[im\theta + \frac{ik\rho^2}{2z}\left(\cos^2\theta + \alpha^2\sin^2\theta\right)\right]. \tag{7.235}$$

Note that in (7.233) the terms with positive and negative numbers p give a different contribution to the total amount. If $m > 0$, in the formula (7.233) the term with $p > 0$ will have a factor $\exp[i\theta(m + 2p)]r^{|m|+2|p|}J_{m+2p}(x)$, and the term with $p < 0$ – another factor $\exp[i\theta(m - 2p)]r^{|m|+2|p|}J_{m-2p}(x)$. It is seen that in the first case

($p > 0$) the exponent of the radial variable in the factor coincides with the order of the Bessel function. Conversely, at $m < 0$ the contribution to the sum in (7.233) will be provided by the terms with $p < 0$.

The reference integral from [53]:

$$\int_0^\infty x^{\frac{m+p}{2}} \exp(-cx) L_n^m(cx) J_{m+p}\left(b\sqrt{x}\right) \mathrm{d}x = \frac{(p-c)^n}{p^{m+n+1}} \left(\frac{b}{2}\right)^m \exp\left(-\frac{b^2}{4p}\right) L_n^m\left(\frac{b^2 c}{4pc-p^2}\right), \tag{7.236}$$

indirectly confirms the ellipticity: only if the radial variable $x = r^2$ and the order of the Bessel function are equal to each other can self-reproduction of the modified LG beam take place. Although the direct use of the integral (7.236) for calculations using (7.233) is not possible, as the exponent in the integral (7.233) differs from the argument of the associated Laguerre polynomial. Thus, leaving only the terms with $p > 0$ in the expression (7.233) (assuming that $m > 0$), and taking into account the weak ellipticity of the LG beam, we save only the first two terms:

$$F_{\alpha\to1}(\rho,\theta,z) \approx S(\rho,\theta) \int_0^\infty \left(\frac{\sqrt{2}\alpha r}{w}\right)^m L_n^m\left(\frac{2r^2\alpha^2}{w^2}\right) \times$$
$$\times \exp\left[-\frac{r^2\alpha^2}{w^2} + \frac{ikr^2\left(1+\alpha^2\right)}{4z}\right] \left\{J_m\left(\frac{k\alpha r\rho}{z}\right) - \frac{i\varepsilon r^2}{2} e^{i2\theta} J_{m+2}\left(\frac{k\alpha r\rho}{z}\right) + O(\varepsilon^2)\right\} r\,\mathrm{d}r, \tag{7.237}$$

where $\varepsilon = k(1-\alpha^2)/(4z) << 1$ is a small parameter.

According to the series in [91]:

$$\sum_{p=0}^\infty \frac{t^{2p} J_{m+2p}(x)}{(2p)!} = \frac{1}{2} x^{-\frac{m}{2}} J_m(\sqrt{x^2-2tx}) \times \left\{(x-2t)^{\frac{m}{2}} + (x+2t)^{-\frac{m}{2}}\right\}. \tag{7.238}$$

Suppose that $t \ll 1$, then instead of (7.238) we can write approximately:

$$\sum_{p=0}^\infty \frac{t^{2p} J_{m+2p}(x)}{(2p)!} \approx J_m(x) + \frac{t^2}{2} J_{m+2}(x) + O(t^4). \tag{7.239}$$

Comparing the expression in curly brackets in (7.237) on the right side of (7.239) and with (7.238), we obtain instead of (7.237):

$$F_{\alpha\to1}(\rho,\theta,z) \approx S'(\rho,\theta) \int_0^\infty \left(\frac{\sqrt{2}\alpha r}{w}\right)^m L_n^m\left(\frac{2r^2\alpha^2}{w^2}\right) \times$$
$$\times \exp\left[-\frac{r^2\alpha^2}{w^2} + \frac{ikr^2\left(1+\alpha^2\right)}{4z}\right] J_m\left(r\sqrt{\left(\frac{k\alpha\rho}{z}\right)^2 - \sqrt{-i\varepsilon} e^{i\theta}\left(\frac{k\alpha\rho}{z}\right)}\right) r\,\mathrm{d}r, \tag{7.240}$$

where

$$S'(\rho,\theta)=S(\rho,\theta)\frac{1}{2}\left(\frac{k\alpha\rho}{z}\right)^{\frac{m}{2}}\left\{\left(\frac{k\alpha\rho}{z}-2\sqrt{-i\varepsilon}e^{i\theta}\right)^{\frac{-m}{2}}+\left(\frac{k\alpha\rho}{z}+2\sqrt{-i\varepsilon}e^{i\theta}\right)^{\frac{-m}{2}}\right\}. \tag{7.241}$$

Equation (7.240) shows that at $\rho\to\infty$ the expression (7.240) changes to equation (7.197) for Fresnel transformation from the LG mode:

$$F_{\alpha\to1}(\rho\to\infty,\theta,z)\approx S(\rho,\theta)\int_0^\infty\left(\frac{\sqrt{2}\alpha r}{w}\right)^m L_n^m\left(\frac{2r^2\alpha^2}{w^2}\right)\times$$
$$\times\exp\left[-\frac{r^2\alpha^2}{w^2}+\frac{ikr^2\left(1+\alpha^2\right)}{4z}\right]J_m\left(\frac{k\alpha\rho r}{z}\right)r\,\mathrm{d}r, \tag{7.242}$$

which is calculated using the reference integral (7.195).

Equation (7.242) shows that for the weak ellipticity the LG beam at the periphery behaves like a normal LG mode, but with elliptical symmetry, i.e. the diffraction pattern is not a set of concentric rings but a set of ellipses. In the central part of the diffraction pattern of the elliptical LG beam at small ρ, as follows from equation (7.240), isolated zeros of the intensity will be located at the points where the argument of the Bessel function in equation (7.240) turns to zero:

$$\rho=\frac{1}{2\alpha}\sqrt{\frac{z}{k}(1-\alpha^2)e^{i\left(\theta-\frac{\pi}{4}\right)}}. \tag{7.243}$$

It follows from (7.243) that the two real zeros of the $m/2$-th order of intensity lie on a straight line $\theta=\pi/4$ at a distance from the centre $\rho=0$ equal to:

$$\rho_0=\frac{1}{2\alpha}\sqrt{\frac{z(1-\alpha^2)}{k}}. \tag{7.244}$$

Equation (7.244) shows that the greater the ellipticity of the beam $(1-\alpha^2)$ and the distance from the waist z, the higher ρ_0 and the greater the distance of the zeros of intensity from the centre $\rho_0=0$.

Note that from (7.240) it follows that at $\rho=0$ the intensity should also be zero but it is not. The point is that equation (7.240) does not take into account the terms in equation (7.233) with negative $\rho<0$, which at $\alpha\approx1$ are small but not zero. Previously it was shown that the main contribution to the intensity value of the elliptic LG beam is determined by the term of the form (7.228), which is always different from zero at finite z.

To understand the physical meaning of why the original zero intensity of the m-th order in the centre of the diffraction pattern at $z=0$ 'disappears' at $z>0$, we represent the intensity of the elliptical LG beam in the form of interference between two fields.

From equation (7.237) it follows that

$$\left|F_{\alpha\to 1}(\rho,\theta,z)\right|^2 \approx \left|F_0(\rho)-\frac{ik(1-\alpha^2)}{8z}e^{2i\theta}F_1(\rho)\right|^2, \tag{7.245}$$

where

$$F_0(\rho)=\frac{k}{z}\int_0^\infty\left(\frac{\sqrt{2}\alpha r}{w}\right)^m L_n^m\left(\frac{2r^2\alpha^2}{w^2}\right)\exp\left[-\frac{r^2\alpha^2}{w^2}+\frac{ikr^2\left(1+\alpha^2\right)}{4z}\right]J_m\left(\frac{k\alpha\rho r}{z}\right)r\,\mathrm{d}r, \tag{7.246}$$

$$F_1(\rho)=\frac{k}{z}\int_0^\infty\left(\frac{\sqrt{2}\alpha r}{w}\right)^m L_n^m\left(\frac{2r^2\alpha^2}{w^2}\right)\exp\left[-\frac{r^2\alpha^2}{w^2}+\frac{ikr^2\left(1+\alpha^2\right)}{4z}\right]J_{m+2}\left(\frac{k\alpha\rho r}{z}\right)r\,\mathrm{d}r. \tag{7.247}$$

From (7.245)–(7.247) it follows that the structure of the diffraction pattern of the LG beam with a low degree of ellipticity has neither radial nor elliptical symmetry. The integral (7.246) can be calculated using equation (7.195). From (7.245) it can also be seen that increasing z decreases the contribution of the second term and at $z \to \infty$ instead of (7.245) we obtain (7.227), i.e. the elliptical LG beam rotated by 90° with respect to the input LG beam at $z = 0$. From (7.245) it follows that the two light fields $F_0(\rho)$ and $F_1(\rho)$ would be added up in 'phase' on line $\theta = -\pi/4$. If we consider the expansion (7.233) for $m < 0$, we would obtain an equation similar to (7.245), but the two light fields $F_0(\rho)$ and $F_1(\rho)$ would add up in 'phase' when $\theta = \pi/4$. That is, the elliptical LG beam during propagation has the diffraction pattern in its cross section which allows one to define the left ($m < 0$) or right ($m > 0$) 'twisting of the phase', has the initial LG beam at $z = 0$. Moreover, the rotation of the diffraction pattern through the angle $\pm\pi/4$ in the Fresnel zone is independent of the number $|m|$. In the far diffraction field the elliptical LG beam is rotated by ±90° in relation to the original, and we can not distinguish the direction of rotation and the sign of the number $\pm|m|$.

From the intensity patterns in Fig. 7.45 we can trace the validity of theoretical calculations: with formation of the light beam on the DOE, the latter acquires a slope of 45°, and in the far zone the whole pattern rotates by 90°. With increasing distance z the number of local maxima first increases and then begins to decrease, as predicted. In addition, it is clear that for finite z the intensity in the centre of the diffraction pattern is not zero and two intensity zeros occur along a line at an angle of 45°.

The propagation pattern of the LG mode (2, 2) also confirms the theoretical conclusions. We have two predicted local minima near the centre of the image as in the case of LG modes (5, 2) in Fig. 7.45. Also, in the centre of the pattern when $z \neq 0$ there is non-zero intensity.

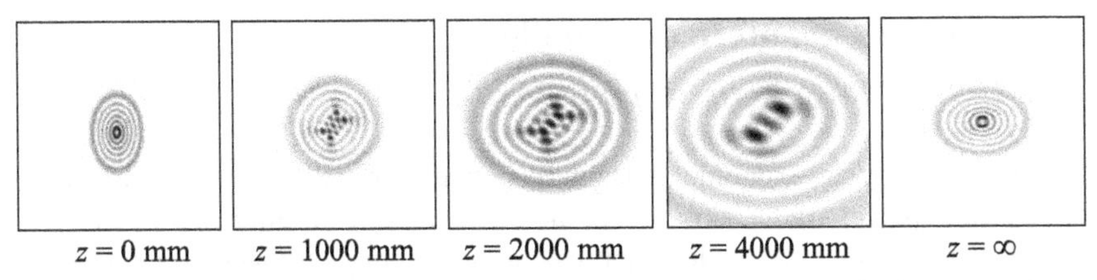

$z = 0$ mm $z = 1000$ mm $z = 2000$ mm $z = 4000$ mm $z = \infty$

Fig. 7.45. Propagation of LG mode (5,2) with an elliptical distortion in free space (negative). Image size 5×5 mm, 256×256 pixles. Typical Gaussian beam radius $\sigma = 0.391$ mm, the wavelength $\lambda = 0.63$ μm, ellipticity $\alpha = 0.66$.

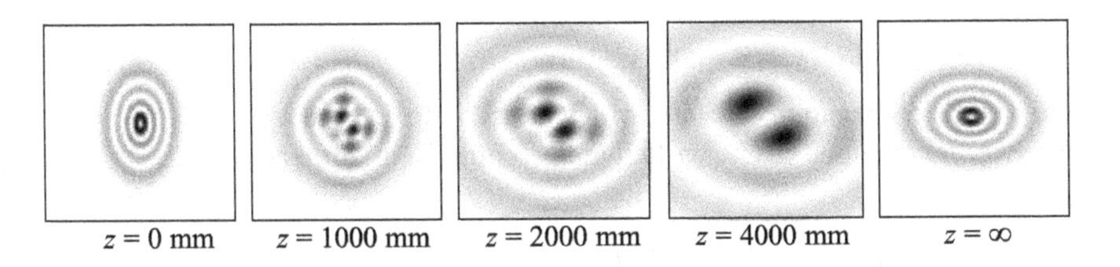

$z = 0$ mm $z = 1000$ mm $z = 2000$ mm $z = 4000$ mm $z = \infty$

Fig. 7.46. Propagation of LG mode (3, –1) with an elliptical distortion in free space (negative). Ellipticity $\alpha = 0.66$.

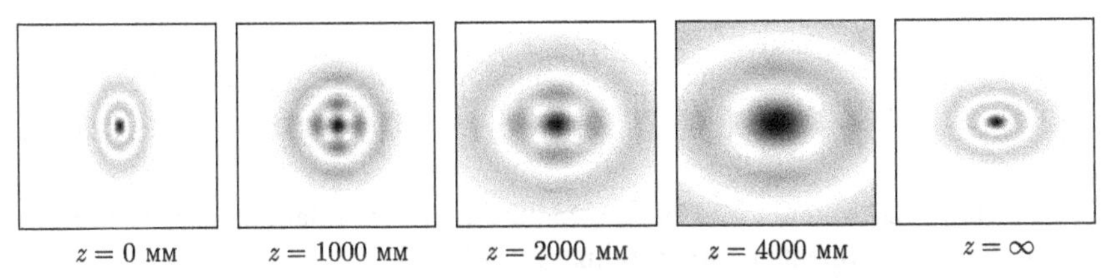

$z = 0$ мм $z = 1000$ мм $z = 2000$ мм $z = 4000$ мм $z = \infty$

Fig. 7.47. Propagation of LG modes (2,0) with an elliptical distortion in free space (negative). Ellipticity $\alpha = 0.66$.

The behaviour of the beam is shown in Fig. 7.46 (LG mode (3, –1)), in general, similar to Fig. 7.45, with the only difference being that we have only one central local minimum, because the mode index $m = -1$. In addition, in this case there is a rotation through –45 ° in the middle area of propagation.

An interesting example is the propagation of the LG mode (2,0) (Fig. 7.47). In this case there are no local minima in the central part, however, there are some interesting effects in the first ring of the mode. For $m = 0$ the central part of the pattern is subjected to the greatest transformation (takes the form of a square) at a distance close to $z_0 \approx 1000$ mm. At the same, the outer rings are circles instead of ellipses. A similar effect is observed with decreasing size of the elliptical distortion. As seen from the patterns of modelling of LG modes, the number of near-centre local maxima-minima depends on the order (n, m) of LG modes and on the coefficient of ellipticity. Thus, their number varies with changes in the degree of elliptical distortion, i.e. the angle of the DOE for the same LG modes.

7.7. The vortex beams in optical fibres

7.7.1. Optical vortices in a step-index fibre

Consider a dielectric waveguide in which the energy is distributed in a certain direction, for example, along the axis z, with the propagation constant β. We assume that the dielectric constant $\varepsilon(x, y)$ does not depend on the longitudinal coordinate z, and changes only in the transverse direction. This model of an inhomogeneous medium gives a good description of the optical fibre.

Then the electromagnetic field, propagating in the waveguide, can be written as:

$$\mathbf{E}(x,y,z)=\mathbf{E}_0(x,y)\exp(-i\beta z), \tag{7.248}$$

$$\mathbf{H}(x,y,z)=\mathbf{H}_0(x,y)\exp(-i\beta z), \tag{7.249}$$

where β is the propagation constant, which needs to be determined.

Maxwell's equations can be written in the 'component' form as:

$$\left(\frac{\partial E_z}{\partial y}-\frac{\partial E_y}{\partial z}\right)\mathbf{e}_x-\left(\frac{\partial E_z}{\partial x}-\frac{\partial E_x}{\partial z}\right)\mathbf{e}_y+\left(\frac{\partial E_y}{\partial x}-\frac{\partial E_x}{\partial y}\right)\mathbf{e}_z=-i\omega\mu_0\left(H_x\mathbf{e}_x+H_y\mathbf{e}_y+H_z\mathbf{e}_z\right), \tag{7.250}$$

$$\left(\frac{\partial H_z}{\partial y}-\frac{\partial H_y}{\partial z}\right)\mathbf{e}_x-\left(\frac{\partial H_z}{\partial x}-\frac{\partial H_x}{\partial z}\right)\mathbf{e}_y+\left(\frac{\partial H_y}{\partial x}-\frac{\partial H_x}{\partial y}\right)\mathbf{e}_z=i\omega\varepsilon\left(E_x\mathbf{e}_x+E_y\mathbf{e}_y+E_z\mathbf{e}_z\right). \tag{7.251}$$

Substituting (7.248) and (7.249), we obtain the known relations expressing the transverse field components by longitudinal ones [104]:

$$E_x=\frac{-i}{k^2-\beta^2}\left(\omega\mu_0\frac{\partial H_z}{\partial y}+\beta\frac{\partial E_z}{\partial x}\right), \tag{7.252}$$

where $k^2=\omega^2\mu_0\varepsilon$. Similarly, for other components:

$$E_y=\frac{-i}{k^2-\beta^2}\left(\beta\frac{\partial E_z}{\partial y}-\omega\mu_0\frac{\partial H_z}{\partial x}\right), \tag{7.253}$$

$$H_x=\frac{-i}{k^2-\beta^2}\left(\beta\frac{\partial H_z}{\partial x}-\omega\varepsilon\frac{\partial E_z}{\partial y}\right), \tag{7.254}$$

$$H_y=\frac{-i}{k^2-\beta^2}\left(\beta\frac{\partial H_z}{\partial y}+\omega\varepsilon\frac{\partial E_z}{\partial x}\right). \tag{7.255}$$

Next, we obtain the equations only for the longitudinal component. From (7.254), (7.255) and from Maxwell's equations we obtain independent equations with respect to one component E_z or H_z:

$$\frac{\partial^2 E_z}{\partial x^2}+\frac{\partial^2 E_z}{\partial y^2}+(k^2-\beta^2)E_z=0, \tag{7.256}$$

$$\frac{\partial^2 H_z}{\partial x^2}+\frac{\partial^2 H_z}{\partial y^2}+(k^2-\beta^2)H_z=0. \tag{7.257}$$

These expressions can be obtained from the Helmholtz wave equation by substituting the field in the form (7.248).

Consider a step-index optical fibre of circular cross section, where the core with radius a has a refractive index n_1, and the cladding of radius b has a refractive index n_2 (Fig. 7.48a). We assume that the radius of the fibre cladding is large enough and the field in the cladding, which decreases exponentially, is close to zero at the junction of the fibre cladding with air. This assumption allows, as shown in Fig. 7.48b, to consider a fibre with a single boundary surface.

Given the shape of the fibre, we consider a cylindrical coordinate system for the components of the electromagnetic field. We assume that the field propagates in the direction of axis z. To derive the expression for the modes of the step-index fibre, it is necessary to solve the modified wave equations (7.256) and (7.257) for the z-components of the electric and magnetic vectors in the core and the cladding, and then get from (7.252)–(7.255) other (transverse) components.

Given the circular cross section of the optical fibre, it is better to use the transverse components in the polar coordinates (r, ϕ). In this case, equations (7.252) and (7.253) will be as follows [105]:

$$E_r=\frac{-i}{\alpha^2}\left(\beta\frac{\partial E_z}{\partial r}+\omega\mu_0\frac{1}{r}\frac{\partial H_z}{\partial\varphi}\right),\qquad H_r=\frac{-i}{\alpha^2}\left(\beta\frac{\partial H_z}{\partial r}-\omega\varepsilon\frac{1}{r}\frac{\partial E_z}{\partial\varphi}\right),$$
$$E_\varphi=\frac{-i}{\alpha^2}\left(\beta\frac{1}{r}\frac{\partial E_z}{\partial\varphi}-\omega\mu_0\frac{\partial H_z}{\partial r}\right),\qquad H_\varphi=\frac{-i}{\alpha^2}\left(\beta\frac{1}{r}\frac{\partial H_z}{\partial\varphi}+\omega\varepsilon\frac{\partial E_z}{\partial r}\right), \tag{7.258}$$

$$\frac{\partial^2 E_z}{\partial r^2}+\frac{1}{r}\frac{\partial E_z}{\partial r}+\frac{1}{r^2}\frac{\partial^2 E_z}{\partial\varphi^2}+\alpha^2E_z=0,\qquad \frac{\partial^2 H_z}{\partial r^2}+\frac{1}{r}\frac{\partial H_z}{\partial r}+\frac{1}{r^2}\frac{\partial^2 H_z}{\partial\varphi^2}+\alpha^2H_z=0, \tag{7.259}$$

where $\alpha^2=k^2-\beta^2$.

To obtain the solution of equation (7.259), we apply the method of separation of variables. We seek a solution in the form:

$$E_z(r,\varphi)=A\cdot R(r)\cdot\Phi(\varphi). \tag{7.260}$$

Given the radial symmetry of the optical fibre, we choose as $\Phi(\phi)$ the angular harmonics $\Phi(\phi)=\exp(im\phi)$, where m is the positive or negative integer. Then the differential equation for the radial part is as follows:

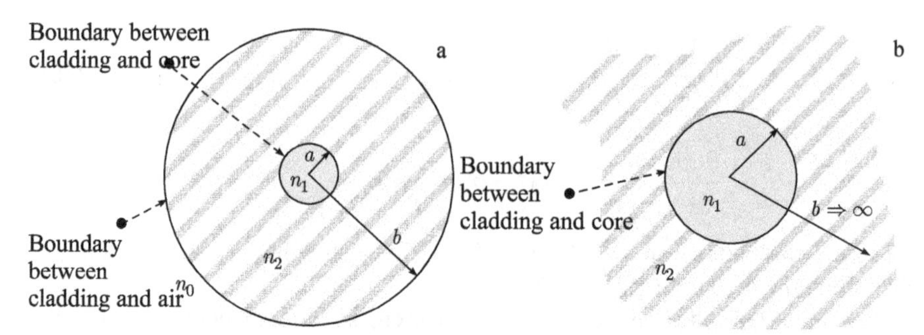

Fig. 7.48. The geometry of the circular step-index fibre.

$$\frac{d^2R(r)}{dr^2}+\frac{1}{r}\frac{dR(r)}{dr}+\left(\alpha^2-\frac{m^2}{r^2}\right)R(r)=0. \tag{7.261}$$

This is the well-known differential equation whose solution are numerous cylindrical functions.The specific form of the functions is chosen from the following considerations:

a) the field in the *fibre core* must be finite, including in the centre at $r = 0$,
b) in the *cladding of the fibre* the field should decay exponentially at large distances from the centre.

Then, in the fibre core we should use the Bessel function of the first kind where $J_m(\alpha r)$, and in the cladding – the modified Bessel function $K_m(\gamma r)$, which decays as $\exp(-\alpha r)$ for $\alpha = i\gamma$.

Given that the core of the fibre, $r \leq a$, has a refractive index n_1, we need to use the parameter $\alpha_1=\sqrt{\omega^2\mu_0\varepsilon_1-\beta^2}=\sqrt{(k_0n_1)^2-\beta^2}$, $k_0=2\pi/\lambda_0$ is the wave number in the air. Similarly, in the cladding, $r \geq a$, having a refractive index n_2.

Since we also have to 'match' the two solutions, obtained in the core and the cladding, on the border between the two media, i.e. at $r = a$, we introduce the following parameters:

$$u=a\sqrt{(k_0n_1)^2-\beta^2}, \tag{7.262}$$

$$w=a\sqrt{\beta^2-(k_0n_2)^2}, \tag{7.263}$$

$$u^2+w^2=V^2, \tag{7.264}$$

$$V=k_0a\sqrt{n_1^2-n_2^2}. \tag{7.265}$$

The parameter V is the the cut-off number.

Then the components of the electromagnetic field will be as follows:

– in the *fibre core*, i.e. for $0 \leq r \leq$ a:

$$E_z = A_1 J_m\left(u\frac{r}{a}\right)\exp(im\varphi), \qquad H_z = B_1 J_m\left(u\frac{r}{a}\right)\exp(im\varphi), \tag{7.266}$$

– in the *fibre cladding*, i.e. with $a \le r \le b$:

$$E_z = A_2 K_m\left(w\frac{r}{a}\right)\exp(im\varphi), \qquad H_z = B_2 K_m\left(w\frac{r}{a}\right)\exp(im\varphi), \tag{7.267}$$

The transverse components are obtained from (7.258). For example, the

$$E_r = \frac{-ia^2}{u^2}\left[A_1\beta\frac{u}{a}J_m^{'}\left(u\frac{r}{a}\right) + A_2\frac{im\omega\mu_0}{r}J_m\left(u\frac{r}{a}\right)\right]\exp(im\varphi). \tag{7.268}$$

Satisfying the boundary conditions at the interface between two dielectric media Γ in general terms means the continuity of the tangential and normal components of electric and magnetic fields at this boundary. From this, in particular, it follows that

$$\left(E_{z1} - E_{z2}\right)\Big|_{r=a} = 0, \tag{7.269}$$

$$\left(E_{\phi1} - E_{\phi2}\right)\Big|_{r=a} = 0. \tag{7.270}$$

$$\left(H_{z1} - H_{z2}\right)\Big|_{r=a} = 0, \tag{7.271}$$

$$\left(H_{\phi1} - H_{\phi2}\right)\Big|_{r=a} = 0. \tag{7.272}$$

The four equations (7.269)–(7.272) allow us to uniquely determine the coefficients A_1, B_1, A_2, B_2:

$$A_1 J_m(u) - A_2 K_m(w) = 0, \tag{7.273}$$

$$A_1\frac{im\beta}{u^2}J_m(u) - B_1\frac{\omega\mu_0}{u}J_m^{'}(u) + A_2\frac{im\beta}{w^2}K_m(w) - B_2\frac{\omega\mu_0}{w}K_m^{'}(w) = 0, \tag{7.274}$$

$$B_1 J_m(u) - B_2 K_m(w) = 0, \tag{7.275}$$

$$A_1\frac{\omega\varepsilon_1}{u}J_m^{'}(u) + B_1\frac{im\beta}{u^2}J_m(u) + A_2\frac{\omega\varepsilon_2}{w}K_m^{'}(w) + B_2\frac{im\beta}{w^2}K_m(w) = 0. \tag{7.276}$$

Equations (7.273)–(7.276) have a non-trivial solution if the determinant of these equations is equal to zero:

$$\begin{vmatrix} J_m(u) & 0 & -K_m(w) & 0 \\ \frac{im\beta}{u^2}J_m(u) & -\frac{\omega\mu_0}{u}J_m^{'}(u) & \frac{im\beta}{w^2}K_m(w) & -\frac{\omega\mu_0}{w}K_m^{'}(w) \\ 0 & J_m(u) & 0 & -K_m(w) \\ \frac{\omega\varepsilon_1}{u}J_m^{'}(u) & \frac{im\beta}{u^2}J_m(u) & \frac{\omega\varepsilon_2}{w}K_m^{'}(u) & \frac{im\beta}{w^2}K_m(w) \end{vmatrix} = 0. \tag{7.277}$$

Disclosure of equation (7.277) gives an equation which is called the equation for the eigenvalues or the characteristic equation of the waveguide:

$$\left[\frac{J_m'(u)}{uJ_m(u)}+\frac{K_m'(w)}{wK_m(w)}\right]\left[\frac{n_1^2}{n_2^2}\cdot\frac{J_m'(u)}{uJ_m(u)}+\frac{K_m'(w)}{wK_m(w)}\right]=m^2\left(\frac{1}{u^2}+\frac{1}{w^2}\right)\left(\frac{n_1^2}{n_2^2}\cdot\frac{1}{u^2}+\frac{1}{w^2}\right). \tag{7.278}$$

Equation (7.278) yields a set of valid values u_{mq}, w_{mq}, β_{mq}, determining the number of modes propagating in a step-index fibre. Expressing the coefficients B_1, A_2, B_2 through A_1 with the help of (7.273)–(7.276) and setting $A_1 = 1$, we can finally write [106]:

$$E_{mq}^z=\exp(im\phi)\begin{cases}J_m\left(\dfrac{u_{mq}r}{a}\right), & 0\le r\le a,\\[2ex] \dfrac{J_m\left(u_{mq}\right)}{K_m\left(w_{mq}\right)}K_m\left(\dfrac{w_{mq}r}{a}\right), & a\le r\le b,\end{cases} \tag{7.279}$$

$$E_{mq}^r=\exp(im\phi)\begin{cases}-i\dfrac{a\beta_{mq}}{u_{mq}}\left\{\dfrac{1-P_{mq}}{2}J_{m-1}\left(\dfrac{u_{mq}r}{a}\right)-\dfrac{1+P_{mq}}{2}J_{m+1}\left(\dfrac{u_{mq}r}{a}\right)\right\}, & 0\le r\le a,\\[2ex] -i\dfrac{a\beta_{mq}}{w_{mq}}\dfrac{J_m(u_{mq})}{K_m(w_{mq})}\left\{\dfrac{1-P_{mq}}{2}K_{m-1}\left(\dfrac{w_{mq}r}{a}\right)+\dfrac{1+P_{mq}}{2}K_{m+1}\left(\dfrac{w_{mq}r}{a}\right)\right\}, & a\le r\le b,\end{cases} \tag{7.280}$$

$$E_{mq}^\varphi=\exp(im\phi)\begin{cases}-i\dfrac{a\beta_{mq}}{u_{mq}}\left\{\dfrac{1-P_{mq}}{2}J_{m-1}\left(\dfrac{u_{mq}r}{a}\right)+\dfrac{1+P_{mq}}{2}J_{m+1}\left(\dfrac{u_{mq}r}{a}\right)\right\}, & 0\le r\le a,\\[2ex] -i\dfrac{a\beta_{mq}}{w_{mq}}\dfrac{J_m(u_{mq})}{K_m(w_{mq})}\left\{\dfrac{1-P_{mq}}{2}K_{m-1}\left(\dfrac{w_{mq}r}{a}\right)-\dfrac{1+P_{mq}}{2}K_{m+1}\left(\dfrac{w_{mq}r}{a}\right)\right\}, & a\le r\le b,\end{cases} \tag{7.281}$$

$$H_{mq}^z=\exp(im\phi)\begin{cases}\dfrac{\beta_{mq}P_{mq}}{\omega\mu_0}J_m\left(\dfrac{u_{mq}r}{a}\right), & 0\le r\le a,\\[2ex] \dfrac{\beta_{mq}P_{mq}}{\omega\mu_0}\dfrac{J_m\left(u_{mq}\right)}{K_m\left(w_{mq}\right)}K_m\left(\dfrac{w_{mq}r}{a}\right), & a\le r\le b,\end{cases} \tag{7.282}$$

$$H_{mq}^{r} = \exp(im\phi)\begin{cases} i\dfrac{a\beta_{mq}^2}{\omega\mu_0 u_{mq}}\left\{\dfrac{\gamma_{mq}^1 - P_{mq}}{2}J_{m-1}\left(\dfrac{u_{mq}r}{a}\right) + \dfrac{\gamma_{mq}^1 + P_{mq}}{2}J_{m+1}\left(\dfrac{u_{mq}r}{a}\right)\right\}, 0 \le r \le a, \\ i\dfrac{a\beta_{mq}^2}{\omega\mu_0 w_{mq}}\dfrac{J_m(u_{mq})}{K_m(w_{mq})}\left\{\dfrac{\gamma_{mq}^2 - P_{mq}}{2}K_{m-1}\left(\dfrac{w_{mq}r}{a}\right) - \dfrac{\gamma_{mq}^2 + P_{mq}}{2}K_{m+1}\left(\dfrac{w_{mq}r}{a}\right)\right\}, a \le r \le b, \end{cases} \tag{7.283}$$

$$H_{mq}^{\phi} = \exp(im\phi)\begin{cases} -i\dfrac{a\beta_{mq}^2}{\omega\mu_0 u_{mq}}\left\{\dfrac{\gamma_{mq}^1 - P_{mq}}{2}J_{m-1}\left(\dfrac{u_{mq}r}{a}\right) - \dfrac{\gamma_{mq}^1 + P_{mq}}{2}J_{m+1}\left(\dfrac{u_{mq}r}{a}\right)\right\}, 0 \le r \le a, \\ -i\dfrac{a\beta_{mq}^2}{\omega\mu_0 w_{mq}}\dfrac{J_m(u_{mq})}{K_m(w_{mq})}\left\{\dfrac{\gamma_{mq}^2 - P_{mq}}{2}K_{m-1}\left(\dfrac{w_{mq}r}{a}\right) + \dfrac{\gamma_{mq}^2 + P_{mq}}{2}K_{m+1}\left(\dfrac{w_{mq}r}{a}\right)\right\}, a \le r \le b, \end{cases} \tag{7.284}$$

where

$$P_{mq} = \frac{m\left(\dfrac{1}{u_{mq}^2} + \dfrac{1}{w_{mq}^2}\right)}{\dfrac{J_m'(u_{mq})}{u_{mq}J_m(u_{mq})} + \dfrac{K_m'(w_{mq})}{w_{mq}K_m(w_{mq})}}, \gamma_{mq}^1 = \left(\frac{k_0 n_1}{\beta_{mq}}\right)^2, \gamma_{mq}^2 = \left(\frac{k_0 n_2}{\beta_{mq}}\right)^2, \tag{7.285}$$

and we also used the well-known recurrence relations and the relations for the derivatives of Bessel functions.

In general, the modes propagating in a step-index fibre are described by the 6-component electromagnetic field (7.279)–(7.285). The mode, in which there are longitudinal components $E_z \neq 0$ and $H_z \neq 0$ are called hybrids and designated by HE, if $E_z > H_z$ and EH, if $E_z < H_z$.

In the case where $m = 0$ the right-hand side of equation (7.278) is reset to zero, and there are equations for the eigenvalues for the transverse TE- and TM-modes:

$$\frac{J_m'(u)}{uJ_m(u)} + \frac{K_m'(w)}{wK_m(w)} = 0, \tag{7.286}$$

$$\frac{n_1^2}{n_2^2} \cdot \frac{J_m'(u)}{uJ_m(u)} + \frac{K_m'(w)}{wK_m(w)} = 0. \tag{7.287}$$

Equation (7.286) corresponds to the TE-mode ($E_z = 0$), while equation (7.287) to the TM-mode ($H_z = 0$).

In terms of the beam model of light propagation in a step-index optical fibre the hybrid modes correspond to oblique rays and the transverse modes to meridian rays.

An important parameter for each propagating mode is the cut-off frequency. The mode will be 'leaking away', i.e. not propagating, if its field in the cladding is not decaying. For the modified Bessel function we know [81] the following approximation for large values of the argument:

$$K_m\left(\frac{wr}{a}\right)\sim\sqrt{\frac{\pi a}{2wr}}\exp\left(-\frac{wr}{a}\right). \tag{7.288}$$

For large values of w the field is concentrated in the core of an optical fibre. With decreasing w the field is beginning to emerge more and more into the cladding, and when $w = 0$, it 'leaves' the waveguide. The frequency at which this occurs is the cut-off frequency. At cut-off, $w_c = a\sqrt{\beta_c^2-(k_0n_2)^2}=0$ or $\beta_c = k_0n_2 = \omega_c\sqrt{\mu_0\varepsilon_2}$. On the other hand, at a 'cut-off' the following equality is satified in the fibre core,

$$u_c = a\sqrt{(k_0n_1)^2-\beta_c^2} = a\omega_c\sqrt{\mu_0(\varepsilon_1-\varepsilon_2)} \text{ or } \omega_c = \frac{u_c}{a\sqrt{\mu_0(\varepsilon_1-\varepsilon_2)}}.$$

The parameter u_c is the fibre cut-off number:

$$u_c \equiv V = k_0a\sqrt{n_1^2-n_2^2}. \tag{7.289}$$

In the optical fibre with $\omega_c = 0$ ($u_c = 0$) only one mode extends – hybrid mode HE_{11}, which exists for all frequencies. Such a single-mode fibre has a very small core diameter a and a small difference in refractive indices of the fibre core and the cladding.

Linearly-polarized modes of a weakly guiding step-index fibre

For the most common commercial fibres the difference between the refractive indices of the core and the cladding $\Delta n = n_1-n_2$ is less than 1%. Such fibres are called weakly guiding and for them, assuming $n_1 \cong n_2$, we can greatly simplify expressions for the propagating electromagnetic field.

Assuming a weakly guiding approximation, the eigenvalue equation takes the form:

$$\frac{J_m'(u)}{uJ_m(u)}+\frac{K_m'(w)}{wK_m(w)}=\pm m\left(\frac{1}{u^2}+\frac{1}{w^2}\right). \tag{7.290}$$

Using the known relations for Bessel functions, equation (7.290) can be simplified. If you use the '+', we obtain the equation:

$$\frac{uJ_m(u)}{J_{m+1}(u)}=-\frac{wK_m(w)}{K_{m+1}(w)}, \tag{7.291}$$

corresponding to the hybrid modes EH, and for the sign '–':

$$\frac{uJ_{m-2}(u)}{J_{m-1}(u)}=-\frac{wK_{m-2}(w)}{K_{m-1}(w)}, \tag{7.292}$$

corresponding to the hybrid modes HE.

Comparing (7.291) and (7.292) we can see that the HE modes of the order $m = m_0 + 1$ degenerate into the EH mode of the order $m = m_0-1$. Then more than one mode will have the same eigenvalues (and the velocity of propagation). We can produce a linear combination of hybrid modes which is linearly polarized (LP mode) and is predominantly transverse [107, 108].

If we enter for the LP mode indes p instead of m:

$$p = \begin{cases} 1, \ \text{TE, TM}, \\ m+1, \ \text{EH}, \\ m-1, \ \text{HE}, \end{cases} \tag{7.293}$$

then the equation for the eigenvalues will be:

$$\frac{uJ_{p-1}(u)}{J_p(u)} = -\frac{wK_{p-1}(w)}{K_p(w)}. \tag{7.294}$$

In the approximation of a weakly guiding optical fibre it is more convenient to switch from the polar to Cartesian coordinates. Given that the transverse field for the LP modes is substantially linearly polarized, a complete set of modes forms when only one electric and one magnetic component are dominant. Then the field can be regarded as a scalar [109, 110]:

$$\Psi_{pq}(r,\varphi,z) = \exp\left(-i\beta_{pq}z\right)\begin{Bmatrix} \cos(p\varphi) \\ \sin(p\varphi) \end{Bmatrix} \begin{cases} \dfrac{J_p(u_{pq}r/a)}{J_p(u_{pq})}, & 0 \le r \le a, \\ \dfrac{K_p(w_{pq}r/a)}{K_p(w_{pq})}, & a \le r \le b. \end{cases} \tag{7.295}$$

We can choose the electric vector, as described by (7.295), lying along any radius, for example, along the axis x, E_x, and the magnetic vector perpendicular to it $H_y = -\frac{k_0 n_{1,2}}{\omega\mu_0} E_x$. E_y and H_x are very small compared with the E_x and H_y. In this case, we can always choose to change the polarization and select a second pair of electric and magnetic vectors perpendicular to the first. Each pair of independent polarizations can be taken either with cos $(p\phi)$ or with sin $(p\phi)$, which arise in linear combinations of hybrid modes. Thus, the four types of LP-modes reflect the four types of hybrid modes.

The intensity of the scalar field (7.295) expresses the distribution of energy propagating in a weakly guiding step-index waveguide [108].

Calculation of the LP-modes

To calculate the LP-modes we need to:

– From the specified characteristics of the optical fibre (a, b, n_1, n_2) and the laser radiation wavelength λ, determine the cut-off number V (7.289);

– For the determined V to determine the set of solutions $\{u_{pq}\}$ of the system of equations (7.264) and (7.294) with respect to the parameter u;

– For each u_{pq} we can determine β_{pq} and w_{pq} and using (7.262) and (7.263);

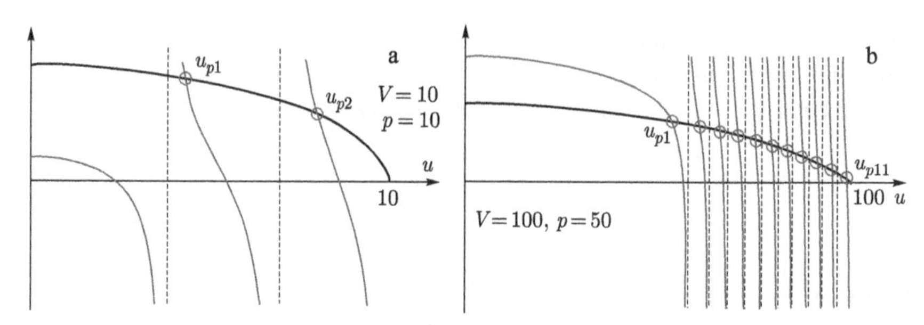

Fig. 7.49. Graphical solution of the eigenvalue equation (7.294) for fixed values of V and p. The thin line shows the function representing the left-hand side of equation (7.294), and the bold line the right side of the equation.

- Using the expression (7.295) for the determined parameters u_{pq}, w_{pq}, β_{pq} we can determine the complex distribution of the mode with indices (p, q) at any point in the step-index optical fibres.

Usually the greatest problem is the definition of the set of solutions $\{upq\}$.

It should be noted that the entire domain of $u \in [0, V]$ is divided into intervals by the roots of the Bessel function $J_p(u)$, located in the denominator of the left side of equation (7.294). In each interval the expressions presented in the left and right side are unimodal functions.Therefore, the interval can not have more than one solution u_{pq}, where the index p corresponds to the order of the Bessel functions in the denominator, and the index q – to the number of the interval (starting from one). For any fixed p the inequality $u_{p,q} < u_{p,q+1}$ is satisfied.

Figure 7.49 shows a graphical solution of the eigenvalue equation (7.294) for (a) $V = 10$ and (b) $V = 100$ for fixed values of p. The number of solutions is determined by the number of intersections of the functions representing the left and right side of equation (7.294).

To determine the $u_{p,q}$ we tabulate the functions in equation (7.294), but knowledge of the behaviour of these functions can reduce the amount of computations, which becomes significant with increasing cut-off number V.

Self-reproduction of multimode laser fields in weakly guiding step-index fibres

We consider the propagation of a linear superposition of LP-modes in an ideal step-index optical fibre:

$$U_0(r,\varphi) = \sum_{p,q \in \Omega} C_{pq} \Psi_{pq}(r,\varphi), \tag{7.296}$$

where C_{pq} are complex coefficients, $\Psi_{pq}(r, \varphi)$ are the modes from (7.295) at $z = 0$, the angular part of which is presented without loss of generality in a somewhat different form:

$$\Psi_{pq}(r,\phi,z)=\exp\left(-i\beta_{pq}z\right)T_p\left(\phi\right)R_{pq}\left(r\right)=$$
$$=\exp\left(-i\beta_{pq}z\right)\exp\left(ip\phi\right)\begin{cases}\dfrac{J_p(u_{pq}r/a)}{J_p(u_{pq})}, 0\le r\le a,\\ \dfrac{K_p(w_{pq}r/a)}{K_p(w_{pq})}, a\le r\le b.\end{cases} \tag{7.297}$$

Although the expressions (7.295) and (7.297) are related to each other by a simple relation, they describe the modes with slightly different properties. For example, the modes (7.295) are valid at $z = 0$, but do not have the orbital angular momentum.

The linear density of projection on the z axis of the orbital angular momentum of a linearly polarized field, given in polar coordinates, can be calculated using the following formula [111]:

$$J_{z0}=-\frac{i}{2\omega}\frac{\int\limits_0^\infty\int\limits_0^{2\pi}\left(U\dfrac{\partial U^*}{\partial\varphi}-U^*\dfrac{\partial U}{\partial\varphi}\right)r\,\mathrm{d}r\,\mathrm{d}\varphi}{\int\limits_0^\infty\int\limits_0^{2\pi}UU^*\,r\,\mathrm{d}r\,\mathrm{d}\varphi}, \tag{7.298}$$

where ω is the angular frequency of light in vacuum.

For the field (7.296) with the modes of the form (7.297), expression (7.298) can be written as follows:

$$\omega J_{z0}=-\frac{\sum\limits_{p,q\in\Omega}\left|C_{pq}\right|^2 p}{\sum\limits_{p,q\in\Omega}\left|C_{pq}\right|^2}. \tag{7.299}$$

Equation (7.299) is valid for modes with the normalized radial part:

$$\int\limits_0^b R_{pq}^2(r)r\,dr=1, \tag{7.300}$$

and this can be satified in the calculations.

Thus, each mode of the form (7.297) has a linear density of the z-projection of the orbital angular momentum proportional to the first index p. At the same time, the expression (7.298) for all modes of the form (7.295) is equal to zero.

Note that the basic property of the modes – the invariance of the operator to propagation in the medium. That is, the propagating mode does not change its structure, acquiring only a phase shift. In particular, the transverse intensity distribution of the field (7.297) at any distance is the same as at $z = 0$:

$$\left|\Psi_{pq}(r,\varphi,z)\right|^2=\left|R_{pq}(r)\exp(ip\varphi)\exp(-i\beta_{pq}z)\right|^2=\left|R_{pq}(r)\right|^2=\left|\Psi_{pq}(r,\varphi)\right|^2. \tag{7.301}$$

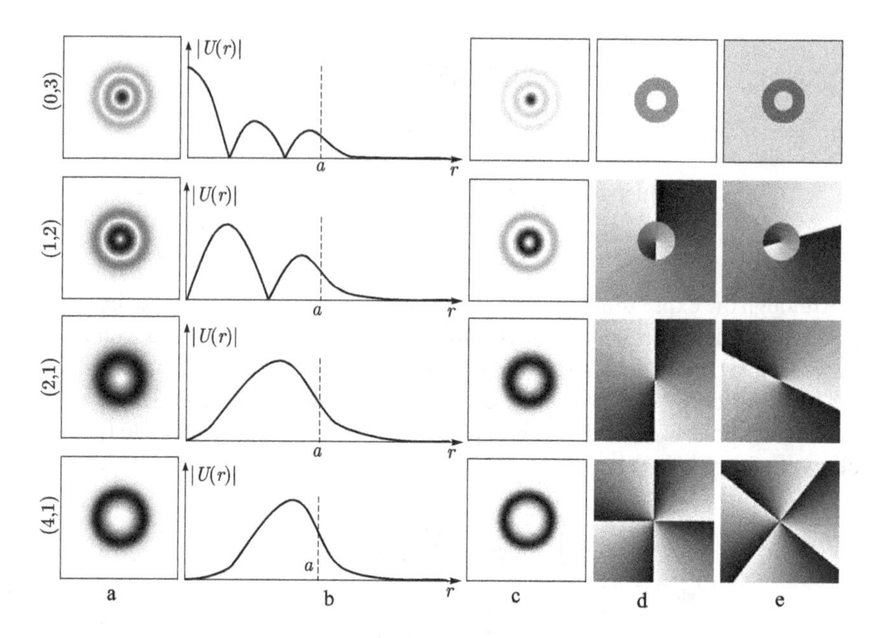

Fig. 7.50. Modes (p, q): (0,3), (1,2), (2,1), (4,1): transverse distribution of the (negative) amplitude (a), the radial cross section of amplitude (b), the transverse distribution (negative) intensity (c) in the planes $z = 0$ and $z = 100$ μm, the transverse distribution of the phase (white colour corresponds to the value 0, and black to 2π) in the planes $z = 0$ (d) and $z = 100$ μm (d).

Figure 7.50 shows the transverse distribution of the (negative) amplitude (a) and intensity (c), as well as the the radial cross section of amplitude (b) for some modes (7.297) of the step-index fibre with the cut-off number $V = 8.4398$. These characteristics of the mode do not change during propagation in an ideal fibre. Changes occur only in the phase. For comparison, Fig. 7.50d shows the phase at $z = 0$, and Fig. 7.50e – after 100 μm.

For numerical simulation we used the following parameters: the core radius $a = 5$ μm, the radius of the cladding $b = 62.5$ µm, the refractive indices of the core and the cladding, respectively $n_1 = 1.45$, $n_2 = 1.44$. Optical waveguides with such parameters are commonly used for the single-mode emission at a wavelength $\lambda = 1.31$ μm, and $\lambda = 1.55$ μm. However, for the emission of a helium-neon laser, $\lambda = 0.633$ μm, the regime becomes low-mode [112, 113], i.e. several propagating modes appear.

If the cut-off number is equal to $V = 8.4398$, there remain 11 LP-modes (the rest will come from the fibre): (0, 1), (0, 2), (0, 3), (±1, 1), (±1, 2), (±2, 1), (±2, 2), (±3, 1), (±3, 2), (±4, 1), (±5, 1).

In general, the field (7.296) will not have the property of invariance of the individual mode (7.301). However, we can choose a mode structure in (7.296) for which the intensity of such superposition will have some special properties of self-reproduction.

The superposition of LP-modes (7.296) in the ideal fibre would have the following distribution of the complex amplitude at a distance of z:

$$U(r,\varphi,z)=\sum_{p,q\in\Omega} C_{pq}\Psi_{pq}(r,\varphi,z), \tag{7.302}$$

where $\Psi_{pq}(r,\varphi,z)=\Psi_{pq}(r,\varphi)\cdot\exp(-i\beta_{pq}z)$, β_{pq} are the propagation constants.

For any pair of modes the intensity at a distance z:

$$\begin{aligned}&\left|C_{p_iq_i}\Psi_{p_iq_i}(r,\varphi,z)+C_{p_jq_j}\Psi_{p_jq_j}(r,\varphi,z)\right|^2=\left|C_{p_iq_i}\right|^2 R^2_{|p_i|q_i}(r)+\left|C_{p_jq_j}\right|^2 R^2_{|p_j|q_j}+\\&+2\left|C_{p_iq_i}\right|\left|C_{p_jq_j}\right|R_{|p_i|q_i}(r)R_{|p_j|q_j}(r)\cos\left[(\arg C_{p_iq_i}-\arg C_{p_jq_j})+(p_i-p_j)\varphi+(\beta_{|p_i|q_i}-\beta_{|p_j|q_j})z\right]\end{aligned} \tag{7.303}$$

different from the intensity at $z = 0$:

$$\begin{aligned}&\left|C_{p_iq_i}\Psi_{p_iq_i}(r,\varphi)+C_{p_jq_j}\Psi_{p_jq_j}(r,\varphi)\right|^2=\left|C_{p_iq_i}\right|^2 R^2_{|p_i|q_i}(r)+\left|C_{p_jq_j}\right|^2 R^2_{|p_j|q_j}+\\&+2\left|C_{p_iq_i}\right|\left|C_{p_jq_j}\right|R_{|p_i|q_i}(r)R_{|p_j|q_j}(r)\cos\left[(\arg C_{p_iq_i}-\arg C_{p_jq_j})+(p_i-p_j)\varphi\right]\end{aligned} \tag{7.304}$$

through the term containing the cosine.

If we impose certain conditions on all pairs of the modes included in the superposition (7.296), it is possible to obtain fields with special properties of reproduction (repetition) of the intensity distribution.

Changes in the complex field distribution during propagation were evaluated using the following criteria:

– normalized standard deviation of the transverse distribution of the field amplitude at a distance z $|U(r, \varphi, z)|$ from the initial distribution $|U_0(r, \varphi)|^2$:

$$\delta_A=\left(\int_0^{2a}\int_0^{2\pi}\left(\left|U(r,\varphi,z)\right|-\left|U_0(r,\varphi)\right|\right)^2 r\,\mathrm{d}r\,\mathrm{d}\varphi\right)^{1/2}\left(\int_0^{2a}\int_0^{2\pi}\left|U_0(r,\varphi)\right|^2 r\,\mathrm{d}r\,\mathrm{d}\varphi\right)^{-1/2}, \tag{7.305}$$

– normalized standard deviation of the transverse intensity distribution at a distance z $|U(r, \varphi, z)|^2$ from the initial distribution $|U_0(r, \varphi)|^2$:

$$\delta_I=\left(\int_0^{2a}\int_0^{2\pi}\left(\left|U(r,\varphi,z)\right|^2-\left|U_0(r,\varphi)\right|^2\right)^2 r\,\mathrm{d}r\,\mathrm{d}\varphi\right)^{1/2}\left(\int_0^{2a}\int_0^{2\pi}\left|U_0(r,\varphi)\right|^4 r\,\mathrm{d}r\,\mathrm{d}\varphi\right)^{-1/2}, \tag{7.306}$$

– the normalized overlap integral of the complex field at a distance z $U(r, \varphi, z)$ with the original field $U_0(r, \varphi)$:

$$\eta=\left|\int_0^{2a}\int_0^{2\pi}U(r,\varphi,z)U_0^*(r,\varphi)r\,\mathrm{d}r\,\mathrm{d}\varphi\right|^2\left(\int_0^{2a}\int_0^{2\pi}\left|U(r,\varphi,z)\right|^2 r\,\mathrm{d}r\,\mathrm{d}\varphi\int_0^{2a}\int_0^{2\pi}\left|U_0(r,\varphi)\right|^2 r\,\mathrm{d}r\,\mathrm{d}\varphi\right)^{-1}. \quad (7.307)$$

The invariance of the whole propagation area

The change of the transverse distribution of intensity of the light field in propagation is due to the intermode dispersion, determined by the difference in propagation constants of modes β_{pq}. For the functions of the form (7.297) the same velocity of propagation is recorded only for the modes with the same index ($|p|$, q). That is, invariant in any interval (in an ideal fibre) is the superposition of pairs of modes of the form:

$$C_{|p|q}\Psi_{|p|q}(r,\varphi)+C_{-|p|q}\Psi_{-|p|q}(r,\varphi). \quad (7.308)$$

In this case, in expression (7.303)

$$\cos\left[(\arg C_{|p|q}-\arg C_{-|p|q})+(|p|+|p|)\varphi+(\beta_{|p|q}-\beta_{-|p|q})z\right]=$$
$$=\cos\left[(\arg C_{|p|q}-\arg C_{-|p|q})+2|p|\varphi\right]$$

and the intensity in the cross section of the field becomes independent of z, i.e. remains without changes. The shape of the intensity distribution is completely determined by the complex coefficients C_{pq} (see Fig. 7.51).

In the particular case at $|C_{|p|q}| = \pm|C_{-|p|q}|$ we obtain the 'classic' LP-modes (the first row of Fig. 7.51). Interestingly, the arguments of the complex coefficients do not affect the value of the orbital angular momentum (7.298) of superposition of (7.296), i.e. if the amplitude of the coefficients remain the same, we will get the rotated 'classic' LP-mode whose orbital angular momentum is also equal to zero (the second row of Fig. 7.51).

Changing the amplitude of the coefficients leads to both a change in the structure of the cross-section and in the projection of the orbital angular momentum (7.298). For the cases shown in Fig. 7.51c in the third and bottom rows, the expressions (7.299) for the orbital angular momentum are different and equal 0.6 and 0.923, respectively.

The invariance of the whole propagation area with the accuracy to rotation

If we assume invariance with the accuracy up to rotation, then the following conditions must be satified for the pairs of modes in superposition:

$$\cos\left[(p_i-p_j)\varphi+(\beta_{p_i|q_i}-\beta_{p_j|q_j})z\right]=\cos\left[(p_i-p_j)(\varphi+\varphi_0)\right], \quad (7.309)$$

where φ_0 is a certain angle.

Equation (7.309) implies the rotation condition:

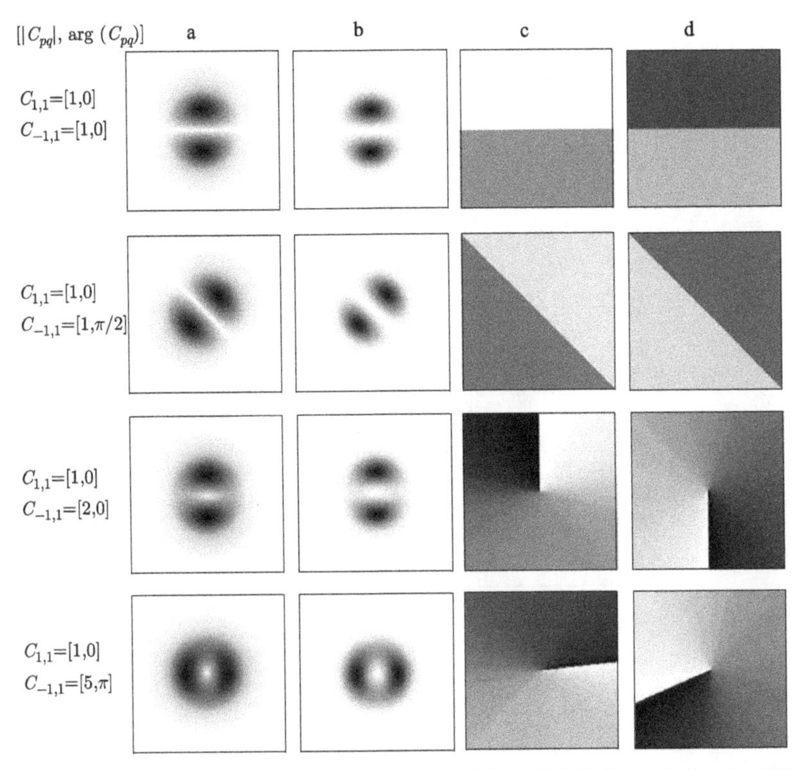

Fig. 7.51. The superposition of the modes (p, q): (1,1) + (–1,1) (negative) with different complex factors: the transverse distribution of the amplitude (a), intensity (b) and phase (c) in the plane $z = 0$, and phase distribution at a distance $z = 200$ m??? (d).

$$\frac{\beta_{|p_i|q_i} - \beta_{|p_j|q_j}}{p_i - p_j} z = \varphi_0, \tag{7.310}$$

for each pair constituting the superposition.

The exact condition (7.310) satisfies any superposition consisting of two modes, if $|p_i| \neq |p_j|$, because with $|p_i| = |p_j|$ rotation will take place through an angle $\varphi_0 = 0$, i.e. complete invariance discussed in the previous section. Thus, exciting different pairs of modes we can produce fields preserving their structure (up to rotation) in the interval of any length. There may be 154 such superpositions, which is many times greater than the number of invariant superpositions equal to 8. For illustration, Fig. 7.52 shows examples of propagation to a distance of 150 m of invariant rotating pairs of modes (p, q): (1,2) + (–2,1) – the top column, (3,2) + (5,1) – the middle column (4.1) + (5.1) – the bottom column.

As seen in Fig. 7.52, the superpositions, consisting of two modes, have the symmetry of the order

$$s = |p_1 - p_2|, \tag{7.311}$$

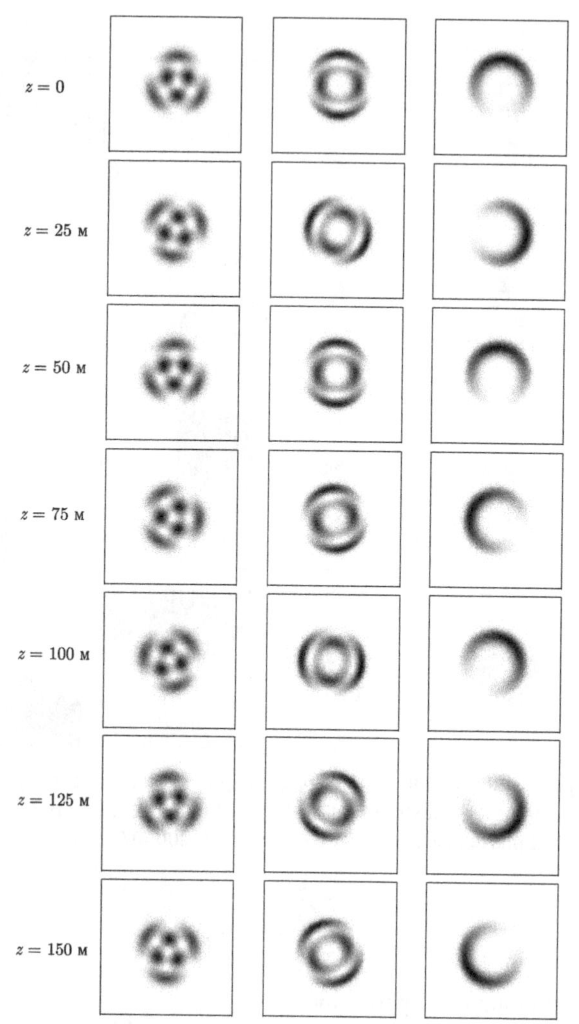

Fig. 7.52. Propagation of rotating mode pairs (p, q): (1,2) + (–2,1) – the left column, (3,2) + (5,1) – the middle column (4,1) + (5, 1) – the right column; the distribution of the intensity (negative) at different distances z is shown.

while the total rotation distance the transverse intensity distribution is reproduced s times due to symmetry.

The speed of rotation of such a pair is given by:

$$\theta = \frac{\beta_{|p_1|q_1} - \beta_{|p_2|q_2}}{p_1 - p_2}, \tag{7.312}$$

and the direction of rotation corresponds to the sign of (7.312).

Interestingly, the rotational speed of the interference pattern of modes, included in the superposition (7.312), is not associated with the expression for the orbital angular momentum (7.299), as it does not depend on the coefficients of the modes but depends on their propagation constants. In particular, for the pairs of modes shown in Fig. 7.52, given the equality of the coefficients, the values (7.299) and (7.312) are as follows: for (1,2) + (-2,1) $\omega J_{z0} = 0.5$, $\theta = 0.54$; for (3, 2) + (5, 1) $\omega J_{z0} = -4$, $\theta = -0.35$; for (4, 1) + (5, 1) $\omega J_{z0} = -4.5$, $\theta = 1.02$.

Note that the transverse distribution of intensity of a beam, consisting of two modes, can be varied by changing the coefficients at the modes. In this case the intensity distribution itself will be preserved during the propagation of the beam in an ideal fibre.

Periodic self-reproduction with the accuracy up to rotation

Similarly to the previous section, given a certain distance z_L (e.g. fibre length), we consider a superposition of modes, replicated at a given distance (in a given period) with the accuracy up to a rotation with some acceptable mismatch (otherwise the set will consist of two-mode superpositions). In this case, the modes, included in the superposition, must pairwise satisfy the condition:

$$\max \Delta_{ij}^{kl} - \min \Delta_{ij}^{kl} \leq \varphi_\varepsilon, \tag{7.313}$$

where $\Delta_{ij}^{kl} = \left| \left[\varphi_{ij} - \varphi_{kl} \right]_\pi \right|$, $\varphi_{ij} = \dfrac{(\beta_{p_i q_i} - \beta_{p_j q_j}) z_L}{p_i - p_j}$, φ_ε is the allowable misalignment angle in the plane z_L.

When choosing z_L = 1 m and defining the permissible error in the angle $\varphi_\varepsilon \leq \pi/9$, we can obtain a set of 173 possible superpositions containing from 2 to 3 modes. Figure 7.53 shows the distribution of one of these superpositions from three modes (p, q): (2,1) + (3,1) + (4,1) in the interval from $z = 0$ to z_L (in this case the angle of misalignment at point $z_L = 1$ m is $\varphi_\varepsilon \leq \pi/30$).

Figure 7.53 shows that the intensity of the superposition is reproduced (with some accuracy) also in other planes. However, this study does not attempt to find all points of reproduction for a particular superposition. Another problem was solved – on the basis of the physical characteristics of a given step-index optical fibre (transverse and longitudinal dimensions, material properties) to determine all propagating modes and the set of superpositions of these modes, with various properties of self-reproduction with a given accuracy.

7.7.2. *Optical vortices in gradient fibres*

The formation and propagation of beams with helical singularity in free space is well studied [22, 26, 28, 33, 67, 68, 70, 114–117]. Excitation of the individual vortex modes or their superpositions in optical fibres is a difficult task [118–125]. But at the same time it is an urgent task as, for example, the use of vortex beams in the near-field probe scanning optical microscope is promising for further improving the resolution (for the moment the resolution is from tens to a few nanometers).

It is especially interesting to study the excitation and propagation of optical vortices which are not modes of step-index or gradient fibres formed by, for example,

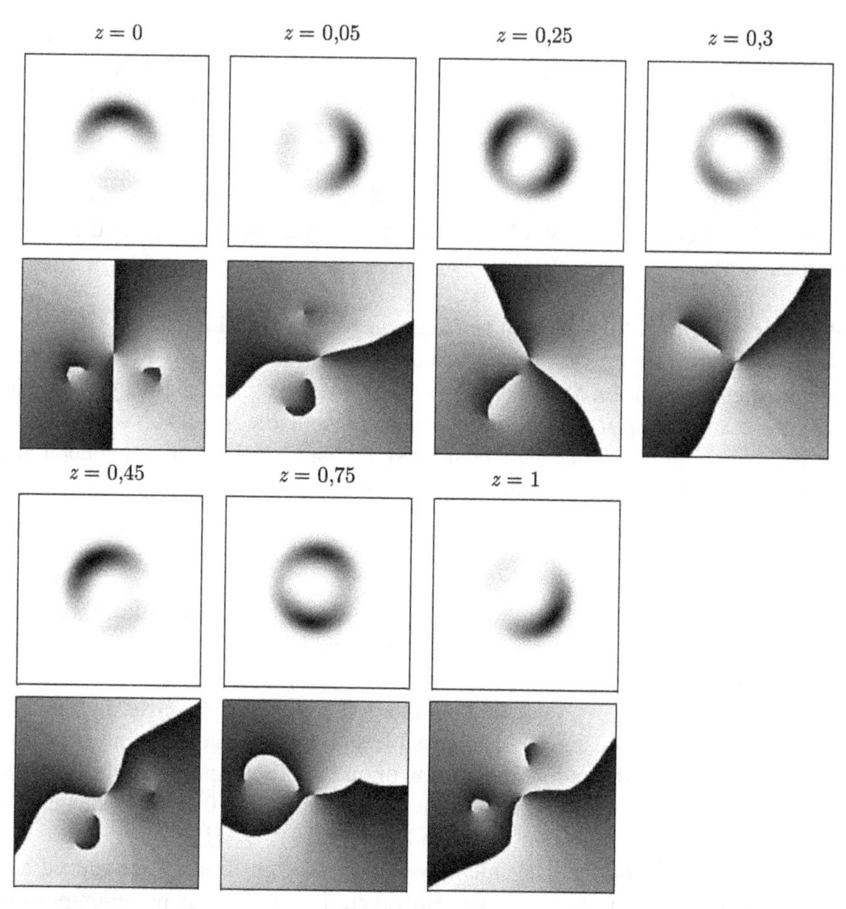

Fig. 7.53. Distribution of the superposition (p, q): (2, 1) + (3, 1) + (4, 1), virtually reproducing at a distance z_L = 1 m; the distributions of the intensity (negative) and the phase at different distances z is given.

the introduction of a helical phase dislocation (with, in particular, the diffractive optical element) in the Gaussian beam – the fundamental mode of laser radiation.

There are several ways of modelling electromagnetic wave propagation in a medium.The most common way to describe this propagation is the system of Maxwell's equations, from which we can obtain the vector wave equations for determining the electric and magnetic field components. If the relative change in the refractive index at a wavelength is much smaller than unity for any of the scalar component of the vector field, we can write the Helmholtz equation [127].

In this section, we study the propagation of monochromatic light beams in a medium with a parabolic refractive index profile. The solutions of the Helmholtz equation for a given medium in cylindrical coordinates are the Gauss–Laguerre modes, which form a complete orthonormal basis which allows decomposition of any field with any desired accuracy [107, 127].

Also, media with small inhomogeneities can be approximated by a system of periodically repeating identical optical elements mounted in a homogeneous medium. In particular, for a parabolic medium these elements are large collecting lens. In the limiting case when the number of the lenses (Fig. 7.54) is infinite, and the distance between them is infinitely small, we obtain an integral operator, which describes the propagation of light in a medium with a parabolic refractive index profile in the paraxial approximation in the scalar theory. This integral operator is analogous to the Fresnel transform describing in the same approximation the propagation of light in a homogeneous medium.

The vortex light fields were simulated by the two methods described above. It can be seen their good agreement in the paraxial region. The properties of an integral operator and its effect on the singular laser beams, including the Gaussian vortex beam, were investigated analytically. The periodic behaviour of the transverse distribution of the light beam in propagation, if it is not an eigenmode of the waveguide, is shown. The above described operator can be used to simulate the propagation of beams with an arbitrary configuration, but at distances that are multiples of a half period.

We also discuss an alternative way of modelling the propagation through the decomposition of the input light beam with respect to the eigenmodes of the medium. Analytical results were obtained for the effect of the integral operator on the non-paraxial Gauss–Laguerre modes with an arbitrary initial effective radius. The expansion coefficients for the Gaussian vortex beam were obtained in the analytical form and can be used for non-paraxial modelling.

In [128] it was shown that the propagation of the light beam in an inhomogeneous medium satisfying the condition $\delta_n/n_0 \ll 1$, up to the error $O(\delta z^3)$, is equivalent to its propagation through a periodic system of identical thin optical element with the transmission function $\tau(x, y) = \exp\{i\delta z\chi(x, y)\}$ in a homogeneous medium with refractive index n_0. The first element is located at a distance $\delta z/2$ from the beginning of penetration, and the two adjacent elements are arranged at a distance δz from each other.

It is also shown that in this case, the propagation of the light field along the optical axis is described by the following integral transformation:

$$E(x,y,z) \approx \left(\frac{k}{2\pi}\right)^2 \int_{-\infty}^{\infty}\int_{-\infty}^{\infty} E_0(\xi,\eta) H(\xi,\eta,x,y,z)\, d\xi d\eta + O\left(z^3\right), \quad (7.314)$$

where $E_0(\xi,\eta)$ is the complex amplitude in the initial plane $z = 0$,

$$\begin{aligned} H(\xi,\eta,x,y,z) = \left(\frac{k}{2\pi}\right)^2 \int_{-\infty}^{\infty}\int_{-\infty}^{\infty} H_0\left(u-\xi, v-\eta, \frac{z}{2}\right) \times \\ \times \exp\{iz\chi(u,v)\} H_0\left(x-u, y-v, \frac{z}{2}\right) du dv. \end{aligned} \quad (7.315)$$

and $H_0(x, y, z)$ is the function of the pulse response of a homogeneous medium with refractive index n_0

$$H_0(x,y,z)=\int_{-\infty}^{\infty}\int_{-\infty}^{\infty}\exp\left\{ikz\sqrt{1-a^2-b^2}\right\}\exp\left\{ik(xa+yb)\right\}dadb. \quad (7.316)$$

Under the condition $k_x^2 + k_y^2 \ll k^2$, or $a^2 + b^2 \ll 1$, the propagation of the light wave can be described by simple paraxial operators:

$$H(\xi,\eta,x,y,z)=-\frac{4}{z^2}\exp\{ikz\}\int_{-\infty}^{\infty}\int_{-\infty}^{\infty}\exp\left\{\frac{ik}{z}\left[(u-\xi)^2+(v-\eta)^2\right]\right\}\times$$
$$\times\exp\left\{iz\chi(u,v)\right\}\exp\left\{\frac{ik}{z}\left[(x-u)^2+(y-v)^2\right]\right\}dudv. \quad (7.317)$$

Consider in particular a parabolic medium with the following dependence of the refractive index:

$$n^2(r)=n_0^2\left(1-2\Delta\frac{r^2}{r_0^2}\right)=n_0^2\left(1-\alpha^2r^2\right). \quad (7.318)$$

For it we can write

$$\chi(x,y)=k\frac{\delta n(x,y)}{n_0}=k\left(\sqrt{1-2\Delta\frac{x^2+y^2}{r_0^2}}-1\right)=-k\Delta\frac{x^2+y^2}{r_0^2}+O\left(\Delta^2\right). \quad (7.319)$$

It is known that the $\tau(x,y)=\exp\left\{-i\frac{k}{2f}\left(x^2+y^2\right)\right\}$ is a transmission function of a thin circular converging lens in the paraxial approximation, where f the focal length.

Thus, in the case of a parabolic medium optical elements are

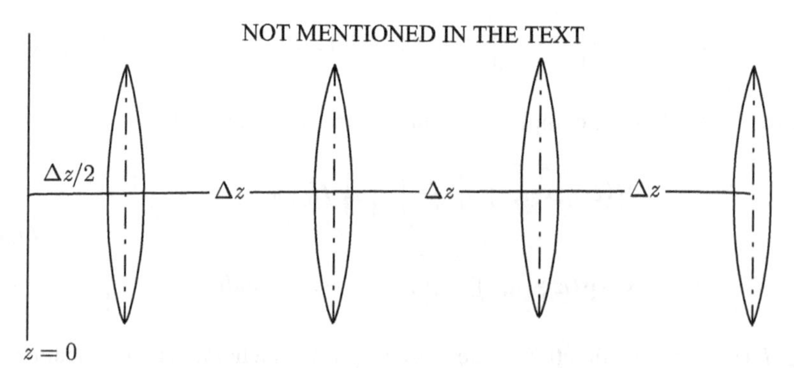

Fig. 7.54. The system of collecting lenses.

converging lens with the focal length $f = \frac{r_0^2}{2\Delta}\frac{1}{\delta z} = \frac{1}{\alpha^2 \delta z}$.

Furthermore, using the integral equations (7.314) and (7.317), we obtain a paraxial integral operator describing the propagation of a light beam in a medium with a parabolic refractive index profile.

Substituting (7.338) for the function $\chi(x, y)$ in (7.317), we obtain

$$H(\xi,\eta,x,y,\delta z) = -\frac{2\pi i}{k\delta z}\frac{1}{\gamma(\delta z)}\exp\{ik\delta z\}\times$$
$$\times\exp\left\{\frac{ik}{2\delta z}\left[2-\frac{1}{\gamma(\delta z)}\right]\left[\xi^2+\eta^2+x^2+y^2\right]\right\}\exp\left\{-\frac{ik}{\delta z}\frac{1}{\gamma(\delta z)}\left[\xi x+\eta y\right]\right\}, \quad (7.320)$$

where $\gamma(z) = 1 - \alpha^2 z^2/4$.

Given this expression, equation (7.314) can be written as follows

$$E(x,y,\delta z) \approx -\frac{ik}{2\pi\delta z}\frac{1}{\gamma(\delta z)}\exp\{ik\delta z\}\exp\left\{\frac{ik}{2\delta z}\left[2-\frac{1}{\gamma(\delta z)}\right]\left[x^2+y^2\right]\right\}\times$$
$$\times\int_{-\infty}^{\infty}\int_{-\infty}^{\infty} E_0(\xi,\eta)\exp\left\{\frac{ik}{2\delta z}\left[2-\frac{1}{\gamma(\delta z)}\right]\left[\xi^2+\eta^2\right]\right\}\times \quad (7.321)$$
$$\exp\left\{-\frac{ik}{\delta z}\frac{1}{\gamma(\delta z)}\left[\xi x+\eta y\right]\right\}d\xi d\eta + O\left(\delta z^3\right).$$

Equation (7.321) describes the complex amplitude distribution obtained after the passage through a thin lens at a distance δz in the paraxial approximation under the condition $\Delta \ll 1$.

One can show that after passage of the light beam through N identical thin lenses its complex amplitude at a distance $N\delta z$ will be described by the following integral relation

$$E(x,y,N\delta z) \approx -\frac{ik}{2\pi\delta z}\frac{1}{\gamma_N^{(2)}(\delta z)}\exp\{ikN\delta z\}\times$$
$$\times\exp\left\{\frac{ik}{2\delta z}\left[2-\frac{1}{\gamma_N^{(1)}(\delta z)}\right]\left[x^2+y^2\right]\right\}\times$$
$$\times\int_{-\infty}^{\infty}\int_{-\infty}^{\infty} E_0(\xi,\eta)\exp\left\{\frac{ik}{2\delta z}\left[2-\frac{1}{\gamma_N^{(1)}(\delta z)}\right]\left[\xi^2+\eta^2\right]\right\}\times \quad (7.322)$$
$$\times\exp\left\{-\frac{ik}{\delta z}\frac{1}{\gamma_N^{(2)}(\delta z)}\left[\xi x+\eta y\right]\right\}d\xi d\eta + O\left(N\delta z^3\right),$$

where $\gamma_N^{(1)}(z)$ and $\gamma_N^{(2)}(z)$ are sequences, which satisfy the following recurrence relations

$$\gamma_{N+1}^{(1)}(z)=\gamma(z)-\cfrac{1}{4-\cfrac{1}{\gamma_N^{(1)}(z)}},$$

$$\gamma_{N+1}^{(2)}(z)=\gamma_N^{(2)}(z)\left[\left(4-\frac{1}{\gamma_N^{(1)}(z)}\right)\gamma(z)-1\right],\gamma_1^{(1)}(z)=\gamma_1^{(2)}(z)=\gamma(z). \quad (7.323)$$

Let us pass to the limit: $\delta z \to 0$, $N \to \infty$, $N\delta z = z$.

One can show that in this transition, the following limit relations hold

$$\frac{1}{\delta z}\left[2-\frac{1}{\gamma_N^{(1)}(\delta z)}\right] \to \frac{\alpha}{\tan(\alpha z)}, \frac{1}{\delta z \gamma_N^{(2)}(\delta z)} \to \frac{\alpha}{\sin(\alpha z)}. \quad (7.324)$$

Then, using these limits and the integral (7.322), we obtain the desired integral

$$E(x,y,z) \approx -\frac{ik\alpha}{2\pi\sin(\alpha z)}\exp\{ikz\}\exp\left\{\frac{ik\alpha}{2\tan(\alpha z)}\left[x^2+y^2\right]\right\}\times$$

$$\times\int_{-\infty}^{\infty}\int_{-\infty}^{\infty} E_0(\xi,\eta)\exp\left\{\frac{ik\alpha}{2\tan(\alpha z)}\left[\xi^2+\eta^2\right]\right\}\exp\left\{-\frac{ik\alpha}{\sin(\alpha z)}\left[\xi x+\eta y\right]\right\}d\xi d\eta. \quad (7.325)$$

The integral (7.325) is the paraxial operator of the propagation of an electromagnetic wave in a fibre with a parabolic refractive index profile under the condition $\Delta \ll 1$. This integral determines the distribution of complex amplitude $E(x, y, z)$ in the plane located at an arbitrary distance z from the origin, knowing the initial distribution $E_0(\xi, \eta)$ on the plane with a value of $z = 0$.

Let us analyze the paraxial operator (7.325) at different distances z from the initial plane.

1. The distance determined by the condition $z \ll \alpha^{-1}$.

Under this condition, it can be assumed that $\sin(\alpha z) \approx \alpha z$, $\cos(\alpha z) \approx 1$ and the integral (7.325) takes the form of Fresnel transformation, which describes the propagation of electromagnetic waves in a homogeneous medium in the paraxial approximation. It should be noted that the paraxial operator also takes the form of the Fresnel integral in the limit $\alpha \to 0$, which corresponds to the transition from a heterogeneous to a homogeneous parabolic medium.

2. The distance defined by the relation $z_s = \frac{\pi}{\alpha}\left[s-\frac{1}{2}\right]$, $s \in N$..

In this case $\sin(\alpha z_s) = (-1)^{s-1}$, $\cos(\alpha z_s) = 0$ and the paraxial integral can be rewritten as

$$E(x,y,z_s)=-\frac{ik\alpha(-1)^{s-1}}{2\pi}\exp\{ikz_s\}\int_{-\infty}^{\infty}\int_{-\infty}^{\infty}E_0(\xi,\eta)\exp\left\{-ik\alpha(-1)^{s-1}[\xi x+\eta y]\right\}d\xi d\eta. \tag{7.326}$$

Consider the following options:

1) for odd values s the integral (7.326) takes the form of the Fourier transform

$$E(x,y,z_s)=-\frac{ik\alpha}{2\pi}\exp\{ikz_s\}\int_{-\infty}^{\infty}\int_{-\infty}^{\infty}E_0(\xi,\eta)\exp\left\{-ik\alpha[\xi x+\eta y]\right\}d\xi d\eta; \tag{7.327}$$

2) for even values of s the integral (7.326) is the inverse Fourier transform

$$E(x,y,z_s)=\frac{ik\alpha}{2\pi}\exp\{ikz_s\}\int_{-\infty}^{\infty}\int_{-\infty}^{\infty}E_0(\xi,\eta)\exp\left\{ik\alpha[\xi x+\eta y]\right\}d\xi d\eta; \tag{7.328}$$

3. The distance defined by the relation $z_s=\frac{\pi}{\alpha}s,\ s\in N.$.

In this situation, the integral (7.325) should be considered in the limit at $z\to z_s$. At the same time $\sin(\alpha z_s)\to 0$, $\cos(\alpha z_s)\to(-1)^s$

$$E(x,y,z_s)=-\frac{ik\alpha}{2\pi}\exp\{ikz_s\}\lim_{t\to 0}\left[\frac{1}{t}\exp\left\{\frac{ik\alpha(-1)^s}{2t}\left[x^2+y^2\right]\right\}\int_{-\infty}^{\infty}\int_{-\infty}^{\infty}E_0(\xi,\eta)\times\right.$$

$$\left.\times\exp\left\{\frac{ik\alpha(-1)^s}{2t}\left[\xi^2+\eta^2\right]\right\}\exp\left\{-\frac{ik\alpha}{t}[\xi x+\eta y]\right\}d\xi d\eta\right]. \tag{7.329}$$

Further, there are two possible situations:

1) if the value of s is odd, then after substitution $u=\frac{k\alpha}{2t}$, we obtain the following

$$E(x,y,z_s)=-\exp\{ikz_s\}\lim_{u\to\infty}\left[\frac{iu}{\pi}\int_{-\infty}^{\infty}\int_{-\infty}^{\infty}E_0(\xi,\eta)\exp\left\{-iu\left[(\xi+x)^2+(\eta+y)^2\right]\right\}d\xi d\eta\right], \tag{7.330}$$

using the limit of $\lim_{u\to\infty}\left\{\sqrt{\frac{iu}{\pi}}\exp\left(-iux^2\right)\right\}=\delta(x)$ [115], we finally have

$$E(x,y,z_s)=-\exp\{ikz_s\}\int_{-\infty}^{\infty}\int_{-\infty}^{\infty}E_0(\xi,\eta)\delta(\xi+x)\delta(\eta+y)d\xi d\eta= \tag{7.331}$$

$$=-\exp\{ikz_s\}E_0(-x,-y),$$

where $\delta(x)$ is the Dirac delta function;

2) if the value of s is even, then after substitution $u=-\frac{k\alpha}{2t}$, we obtain the same

$$E(x,y,z_s)=\exp\{ikz_s\}\lim_{u\to\infty}\left[\frac{iu}{\pi}\int_{-\infty}^{\infty}\int_{-\infty}^{\infty}E_0(\xi,\eta)\times\right.$$
$$\left.\times\exp\left\{-iu\left[(\xi-x)^2+(\eta-y)^2\right]\right\}d\xi d\eta\right]=\exp\{ikz_s\}\times \tag{7.332}$$
$$\times\int_{-\infty}^{\infty}\int_{-\infty}^{\infty}E_0(\xi,\eta)\delta(\xi-x)\delta(\eta-y)d\xi d\eta=\exp\{ikz_s\}E_0(x,y).$$

Thus, according to the steps of calculations, the paraxial integral (7.325) gives the periodically repetitive resulting distribution of the complex amplitude $F(x, y, z) = E(x, y, z)\exp\{-ikz\}$ with a period $z_T = 2\pi/\alpha$. In the interval $z\in[0, z_T]$, this integral has the following special results:

1) when $z=\frac{1}{4}z_T=\frac{\pi}{2\alpha}$ distribution $F(x, y, z)$ is the Fourier transform;

2) with $z=\frac{1}{2}z_T=\frac{\pi}{\alpha}$ result $F\left(x,y,\frac{1}{2}z_T\right)=-E_0(-x,-y)$;

3) with $z=\frac{3}{4}z_T=\frac{3\pi}{2\alpha}$ distribution is the inverse Fourier transform;

4) when the distribution is exactly equal to the initial distribution.

If the function $E_0(x, y, z)$ has the form $A(r)\exp(in\varphi)$ in the polar coordinates, it is obvious that the intensity in the transverse plane will be repeated through the period π/α instead of $2\pi/\alpha$, as shown in Figure 7.55.

7.8. Matrices of optical vortices

The ability to create a large number of optical traps, each of which has its own structure, and move them in the three-dimensional space, provides additional control of micro- and nanosized particles for biomedical research and micro- and nanoengineering.

Spatial light modulators (SLM), characterized by still low diffraction efficiency, are very promising for the development and dynamic reconfiguration of the matrix of optical traps [129]. Modern SLMs have a resolution of 2 megapixels, frame rate 60 Hz, 8-bit addressing, the pixel size 8 μm. The large size of the pixel is one of the main drawbacks of the SLM compared with the DOE.

In this section iterationless coding techniques [130] are used for dynamic calculations of the phase of the SLM generating sets of optical 'tweezers' or 'spanners' of different numbers in any place of the three-dimensional space.

In the methods of digital holography each sample of the amplitude–phase functions to be coded is replaced by a $K\times K$ cell of readings of digital holograms. In the methods oriented to SLMs, each sample of the encoded amplitude–phase

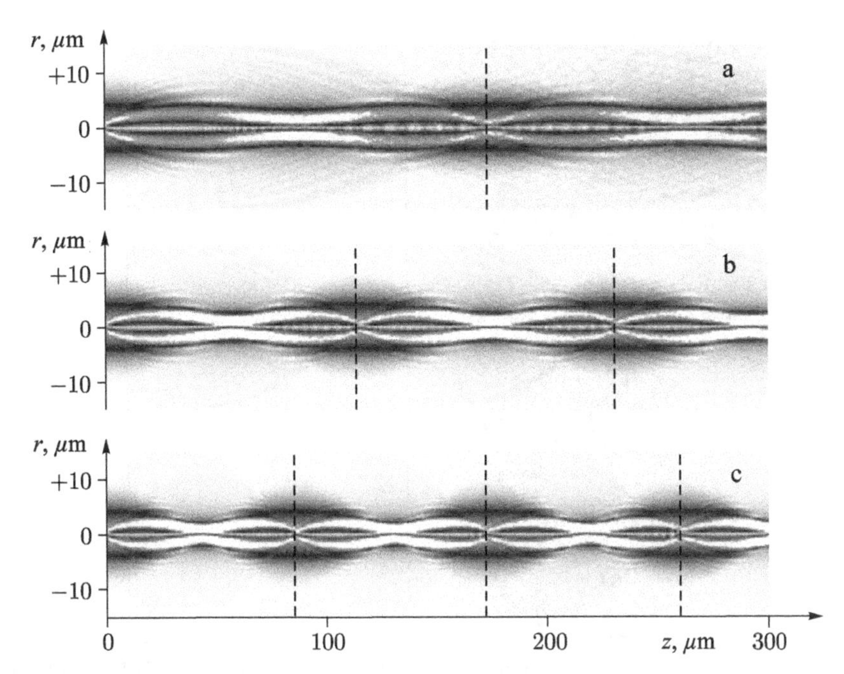

Fig. 7.55. Propagation of a Gaussian optical vortex in a fibre with a parabolic profile for different values of α: $\alpha = 17.88\ \text{mm}^{-1}$ (upper graph), $\alpha = 26.82\ \text{mm}^{-1}$ (middle graph) and $\alpha = 35.76\ \text{mm}^{-1}$ (lower graph). Dotted lines indicate periods, after which the diffraction pattern in the transverse plane begins to repeat.

function is replaced by a sample of the phase function. In addition, for the highest local concentration of energy it is desirable to form each light distribution with a minimum diffraction radius [130]. These limits determine the use in the calculation of the SLM phase of the superposition method [131] as follows:

$$f(x,y)=\sum_{n=1}^{N} C_n \exp\left[im_n \text{arctg}\left(\frac{y}{x}\right)\right]\exp\left[i\left(\alpha_n x+\beta_n y\right)\right], \tag{7.333}$$

where N is the number of optical traps in the matrix, m_n is the order of the vortex singularity of each of the traps (if $m_n = 0$ the vortex component is absent and a focused spot of light energy will form), α_n, β_n are the parameters proportional to the coordinates of each trap in the focal plane.

Equation (7.333) involves the addition of a spherical lens to the SLM.

The intensity of focused radiation in each of the traps is determined by the square of the modulus of the corresponding coefficient $|C_n|^2$, the phase of the coefficients $\arg C_n$ – the free parameters.

To encode the amplitude –phase function (7.333) we used the method of pseudo-random coding [132], which consists in replacing the function $f(x, y)$ by the phase function by the rule:

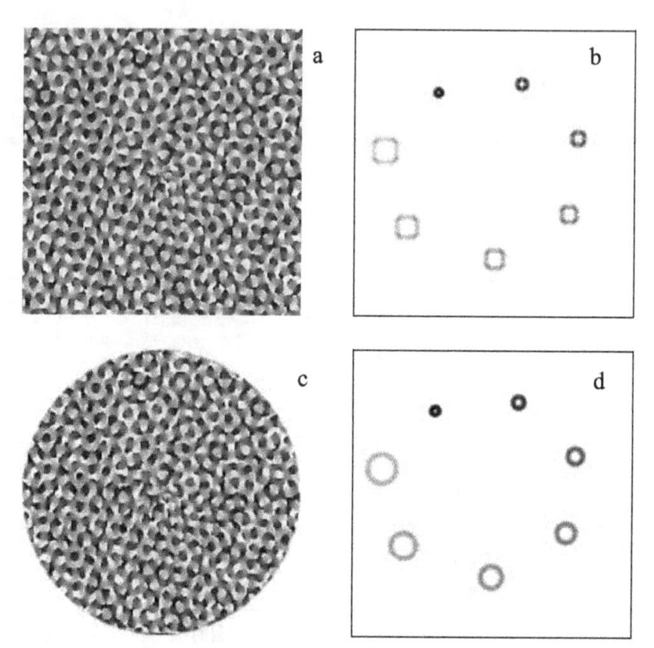

Fig. 7.56. The results of the formation of seven traps: halftone phase of SLM (a, c) and intensity distribution (negative) in the focal plane of a spherical lens in illumination of the SLM with a plane wave, limited by the square (b) or a circular aperture (d).

$$\begin{aligned} \hat{f}(x,y) &= \exp\left[i\psi(x,y)\right], \\ \psi(x,y) &= \arg f(x,y) - \operatorname{sgn} S(x,y) \arccos\left|f(x,y)\right|, \end{aligned} \tag{7.334}$$

where $S(x, y)$ is the pseudorandom value generated by the sensor of pseudorandom numbers in the range [–0.5, 0.5].

Figure 7.56 shows the results of the formation of seven traps (the order of an optical vortex varies from 1 to 7), distributed around the circumference in illumination of the SLM with a halftone phase with a flat beam bounded by a square (a, b) and a circular aperture (c, d).

Binary SLMs are more common and have better resolution than the halftone ones. Figure 7.57 shows the result of a simple binarization of the phase (7.333):

$$\psi_b(x,y) = \frac{\pi}{2}\left\{1 - \operatorname{sgn}\cos\left[\alpha_n x + \beta_n y + m_n \operatorname{arctg}\left(\frac{y}{x}\right)\right]\right\}. \tag{7.335}$$

As can be seen from Fig. 7.57, in the focal plane of the lens we obtained additional associated optical vortices (with opposite directions of rotation), located symmetrically about the centre of origin.

The conversion phase (7.334) or (7.335) for different locations and numbers of optical vortices is produced on a computer very quickly – convert images with the inscription DOE to the image with the words SLM (see Fig. 7.58) took less than 1 sec. on the frame.

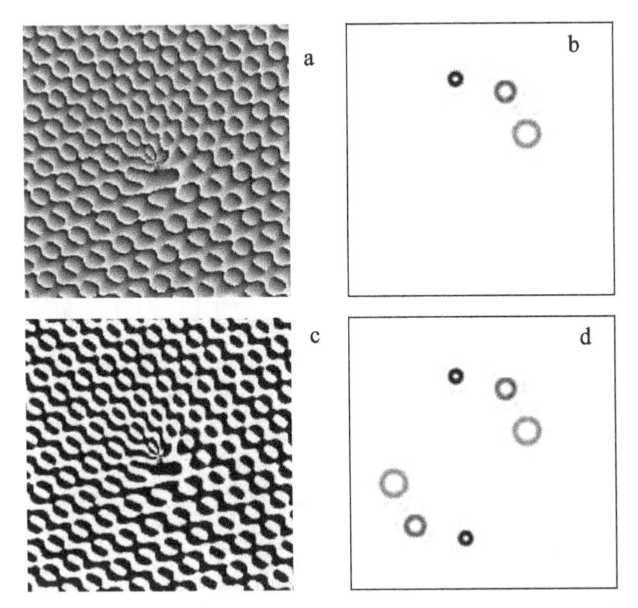

Fig. 7.57. The results of formation of a set of traps with a halftone SLM (a, b) and a binary SLM (c, d).

Fig. 7.58. Dynamic change in the position of traps in the matrix; some frames are transformations from a sequence in which each trap is shifted with a step of the order of the minimum diffraction spot.

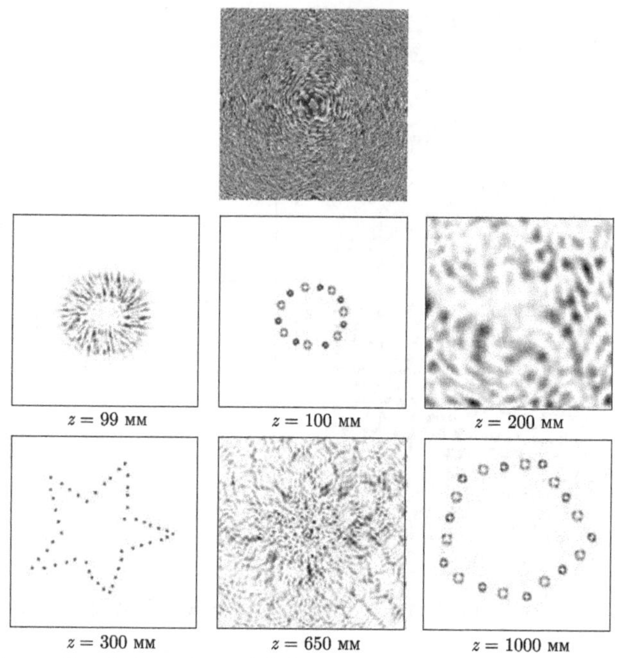

Fig. 7.59. A set of optical vortices of different orders in the field of 3D: halftone phase of the PMS and the intensity distribution at different distances from the plane of the PSPP

It should be noted that when using the SLM the coordinates of the traps should not be too far away from the optical axis, as in this case larger values of α_n and β_n result in a high frequency phase change, which will not be prescribed by a limited number of samples of SLM.

Also, using the superposition method [130], we can form a 3D mesh of optical vortices each having its own spatial location and order of the vortex:

$$f(x,y)=\sum_{p=1}^{P}\left(\sum_{n=1}^{N_p}C_{np}\exp\left[im_{np}\operatorname{arctg}\left(\frac{y}{x}\right)\right]\exp\left[i\left(\alpha_{np}x+\beta_{np}y\right)\right]\right)\exp\left[-i\frac{\pi}{\lambda z_p}\left(x^2+y^2\right)\right], \tag{7.336}$$

where α_{np}, β_{np}, z_p correspond to the three-dimensional coordinates of a single optical vortex in a 3D structure, which has an individual vortex order m_{np}.

No additional optical elements are required when using the formula (7.336) for the formation of a set of optical vortices in the 3D field.

Figure 7.59 shows the formation of different sets of optical vortices at a distance: $z = 100$ mm, $z = 300$ mm, $z = 1000$ mm from the plane of the SLM.

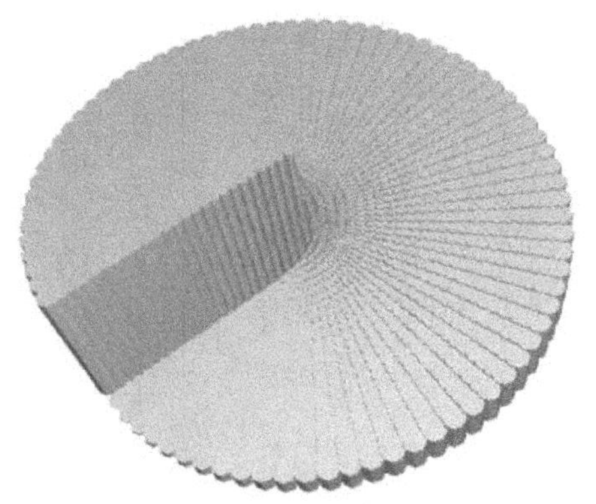

Fig. 7.60. Three-dimensional model of the SLM ($n = 1$).

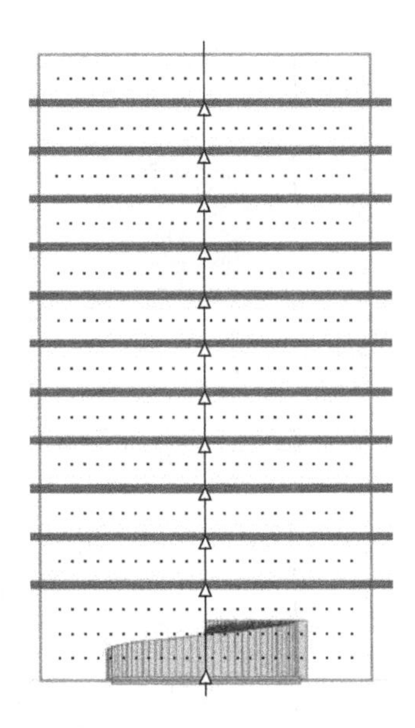

Fig. 7.61. The scheme for numerical simulation: orange colour – a source of green light ($\lambda = 633$ nm), a plane wave with the size of 4 μm passes through the SLM of 1st order, the green coloir – screens for measurement located in planes at $z = (m + 1)\Delta z$, $\Delta z = 1$ μm, $m = 1, ..., 11$.

7.9. Simulation of an optical vortex generated by a plane wave diffracted by a spiral phase plate

When light passes through a real SPP (Fig. 7.60), the diffraction pattern as a set of concentric rings does not appear immediately. Figure 7.61 shows a scheme for modelling the passage of a plane wave ($\lambda = 633$ nm) through a SLM with a radius of 2 μm radius ($n = 1$). Green colour represents the screens to measure the intensity, located in the planes $z = (m + 1)\ \Delta z$, $\Delta z = 1\ \mu$m, $m = 1, ..., 11$.

Figure 7.62 shows the time-averaged intensity values in the planes of several screens.

Figure 7.62 shows that at a distance of approximately 20λ the diffraction pattern is indeed a bright ring (secondary rings are not visible in the picture).

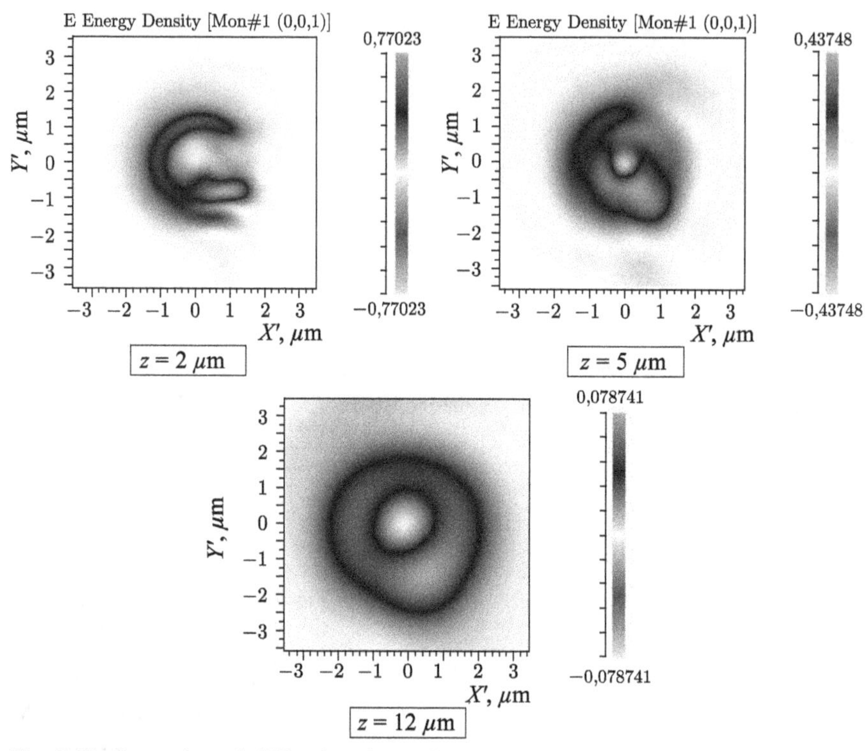

Fig. 7.62. Generation of diffraction rings after passage of a plane wave through a spiral phase plate.

References

1. Litvin I. A., McLaren M.G., Forbes A., A conical wave approach to calculating Bessel–Gauss beam reconstruction after complex obstacles, Opt. Commun. 2009. V. 282. P. 1078–1082.
2. Sato S., Kozawa Y. Hollow Vortex Beams, J. Opt. Soc. Am. A. 2009. V. 26 (1). P.142–146.
3. Anand S. Generation of a hollow Gaussian beam and its anomalous behavior, Opt. Commun. 2009. V. 282. P. 1335–1339.
4. Dienerowitz M., Mazilu M., Reece P. J., Krauss T. F., Dholakia K., Optical vortex trap for resonant confinement of metal nanoparticles, Opt. Express. 2008. V. 16 (7). P. 4991–4999.
5. Zelenina A., Quidant R., Nieto-Vesperinas M. Tunable optical sorting and manipulation of nanoparticles via plasmon excitation, Opt. Lett. 2006. V. 31. P. 2054–2056.
6. Burresi M., Engelen R. JP, Opheij A., van Oosten D., Mori D., Baba T., Kuipers L., Observation of Polarization Singularities at the Nanoscale, Phys. Rev. Lett., 2009. V. 102. P. 033902.
7. Levenson M., Tan S.M., Dai G., Morikawa Y., Hayashi N., Ebihara T., The vortex via process: analysis and mask fabrication for contact CDs < 80 nm, SPIE. 2003. V. 5040. P. 344–370.
8. Jesacher A., Fürhapter S., Bernet S., Ritsch-Marte M., Spiral interferogram analysis, J. Opt. Soc. Am. A. 2006. V. 23 (6). P. 1400–1409.

9. Bernet S., Jesacher A., Fürhapter S., Maurer C., Ritsch-Marte M. Quantitative imaging of complex samples by spiral phase contrast microscopy, Opt. Express. 2006. V. 14 (9). P. 3792–3805.
10. van Dijk T., Visser T.D., Evolution of singularities in a partially coherent vortex beam, J. Opt. Soc. Am. A. 2009. V. 26 (4). P. 741–744.
11. Mawet D., Serabyn E., Liewer K., Hanot Ch., McEldowney S., Shemo D., O'Brien N. Optical Vectorial Vortex Coronagraphs using Liquid Crystal Polymers: Theory, Manufacturing and Laboratory Demonstration, Opt. Express. 2009. V. 17 (3). P. 1902–1918.
12. Davis J.A., Smith D.A., McNamara D.E., Cottrell D.M., Campos J., Fractional derivatives – analysis and experimental implementation, Appl. Opt. 2001. V. 40 (32). P. 5943–5948.
13. Tychinskii V.P. Superresolution and singularities in phase images, Usp. Fiz. Nauk, 2008. V. 178 (11). P. 1205–1214.
14. Wang W., Ishii N., Hanson SG, Miyamoto Y., Takeda M., Phase singularities in analytic signal of white-light speckle pattern with application to micro- displacement measurement, Opt. Commun. 2005. V. 248. P. 59–68.
15. Wang W., Yokozeki T., Ishijima R., Wada A., Miyamoto Y., Takeda M., Optical vortex metrology for nanometric speckle displacement measurement, Opt. Express. 2006. V. 14 (1). P. 120–127.
16. Angelsky O. V., Burkovets D. N., Maksimyak P.P., Hanson S.G., Applicability of the singular optics concept for diagnostics of random and fractal rough surfaces. Appl. Opt. 2003. V. 42 (22). P. 4529–4540.
17. Singh R.K., Senthilkumaran P., Singh K. Structure of a tightly focused vortex beam in the presence of primary coma, Opt. Commun. 2009. 2009. V. 282. P. 1501–1510.
18. Otsuka K., Chu S.-C. Generation of vortex array beams from a thin-slice solid-state laser with shaped wide-aperture laser–diode pumping, Opt. Lett. 2009. V. 34 (1). P. 10–12.
19. Wang L.-G., Wanga L.-Q., Zhu S.-Y. Formation of optical vortices using coherent laser beam arrays, Opt. Commun. 2009. V. 282. P. 1088–1094.
20. Masajada J., Small-angle rotations measurement using optical vortex interferometer, Opt. Commun. 2004. V. 239. P. 373–381.
21. Popiolek-Masajada A., Kurzynowski P., Wozniak W.A., Borwinska M., Measurements of the small wave tilt using the optical vortex interferometer with the Wollaston compensator, Appl. Opt. 2007. V. 46 (33). P. 8039–8044.
22. Desyatnikov A.S., Kivshar Yu.S., Torner L., Optical vortices and vortex solitons, Progress in Optics. V. 47. Ed. by E.Wolf. Amsterdam: North-Holand, 2005. P. 291–391.
23. O'Holleran K., Dennis M.R., Flossmann F., Padgett M.J., Fractality of Light's Darkness, Phys. Rev. Lett. 2008. V. 100. P. 053902.
24. Zagrodzinski J. Vortices in different branches of physics, Physica C. 2002. V. 369. P. 45–54.
25. Bliokh K.Y., Shadrivov I.V., Kivshar Y.S., Goos–Hänchen and Imbert–Fedorov shifts of polarized vortex beams, Opt. Lett. 2009. V. 34 (3). P. 389–391.
26. Khonina S.N., Koltyar V.V., Shinkarev M.V., Soifer V.A., Uspleniev G.V., The rotor phase filter, J. Mod. Opt. 1992. V. 39 (5). P. 1147–1154.
27. Cheong W.C., Lee W.M., Yuan X.-C., Zhang L.-S., Dholakia K., Wang H., Direct electron-beam writing of continuous spiral phase plates in negative resist with high power efficiency for optical manipulation, Appl. Phys. Lett. 2004. V. 85 (23). P. 5784–5786.
28. Koltyar V.V., Almazov A.A., Khonina S.N., Soifer V.A., Elfstrom H., Turunen J., Generation of phase singularity through diffracting a plane or Gaussian beam by a spiral phase plate, J. Opt. Soc. Am. A. 2005. V. 22 (5). P. 849–861.

29. Lee W.M., Ahluwalia B.P.S., Yuan X.-C., Cheong W.C., Dholakia K., Optical steering of high and low index microparticles by manipulating an off-axis optical vortex, J. Opt. A: Pure Appl. Opt. 2005. V. 7 (1). P. 1–6.
30. Oemrawsingh S.S.R., van Houwelinger J.A.W., Eliel E.R., Woerdman J.R., Vestegen E. J.K., Kloosterboer J.G., Hooft G.W., Production and characterization of spiral phase plates for optical wavelengths, Appl. Opt. 2004. V. 43 (3). P. 688–694.
31. Sueda K., Miyaji G., Miyanaga N., Nakatsura M., Laguerre–Gaussian beam generated with a multilevel spiral phase plate for high intensity laser pulses, Opt. Expr. 2004. V. 12 (15). P. 3548–3553.
32. Sundbeck S., Gruzberg I., Grier D.G., Structure and scalling of helical modes of light, Opt. Lett. 2005. V. 30 (5) P. 1–13.
33. Heckenberg N.R., McDuff R., Smith C.P., White A.G., Generation of optical phase singularities by computer-generated holograms, Opt. Lett. 1992. V. 17 (3). P. 221.
34. McLeod J.H., The axicon: a new type optical element, J. Opt. Soc. Am. 1954. V. 44 (8). P. 592–597.
35. Tremblay R., D'Astons Y., Roy G., Blanshard M., Laser plasma-optically pumped by focusing with axicon a CO_2–TEA laser beam in a high-pressure gas, Opt. Commun. 1979. V. 28 (2). P. 193–194.
36. Michaltsova I.A., Nalivaiko V.I., Soldatenkov I.S.. Kinoform axicon, Optik, 1984. V. 67 (3). P. 267.
37. Belanger P., Rioux M., Ring pattern of a lens–axicon doublet illuminated by a Gaussian beam, Appl. Opt. 1978. V. 17 (7). P. 1080–1086.
38. Fedotovsky A., Lehovec H., Optimal filter design for annular imaging, Appl. Opt., 1974. V. 13 (12). P. 2919–2923.
39. Koronkiewicz V.P., Palchikova I.G., Paleshchuk A.G., et al., kinoform optical elements with ring impulse response. Preprint IAiE CO RAN, Number 265. Novosibirsk, 1985.
40. Khonina S.N., Kotlyar V.V., Soifer V.A., Shinkarev M.V., Uspleniev G.V., Trochoson, Opt. Commun. 1992. V. 91 (3–4). P. 158–162.
41. Paterson C., Smith R. Higher–order Bessel waves produced by axicon-type computer–generated holograms, Opt. Commun. 1996. V. 124 (1–2). P. 123–130.
42. Khonina S.N., Kotlyar V.V., Skidanov R.V., Soifer V.A., Jefimos K., Simonen J., Turunen J., Rotation of microparticles with Bessel beams generated by diffractive elements, J. Mod. Opt. 2004. V. 51 (14). P. 2167–2184.
43. McLeod J.H., Axicons and their uses, J. Opt. Soc. Am. 1960. V. 50 (2). P. 166–169.
44. Perez M.V., Gomez-Reino C., Cuadrado J.M., Diffraction pattern and zone plates produced by thin linear axicons, J. Mod. Opt. 1986. V. 33 (9). P. 1161–1176.
45. Sochacki J., Koiodziejczyk A., Jaroszewicz Z., Bar'a S., Nonparaxial design of generalized axicons, Appl. Opt. 1992. V. 31 (25). P. 5326–5330.
46. Kotlyar V.V., Soifer V.A., Khonina S.N., Diffraction calculation of focusers using a fast Hankel transform, Optika Spektroskopiya. 1991. V. 71 (2). P. 372–377.
47. Kotlyar V.V., Khonina S.N., Soifer V.A., Iterative calculation of diffractive optical elements focusing into a three dimensional domain and the surface of the body of rotation, J. Modern Optics. 1996. V. 43 (7). P. 1509–1524.
48. Khonina S.N., Kotlyar V.V., Soifer V.A., Diffraction optical elements matched to the Gauss–Laguerre modes, Optics and Spectroscopy. 1998. V. 85 (4). P. 636–644.
49. Khonina S.N., Kotlyar V.V., Soifer V.A., Fast Hankel transform for focusator synthesis, Optik. 1991. V. 88 (4). P. 182–184.
50. Zhang D.W., Yuan X.–C., Ngo N.Q., Shum P., Fast Hankel transform and its application for studying the propagation of cylindrical electromagnetic fields, Optics Express. 2002. V. 10 (12). P. 521–525.

51. Davis J.A., McNamara D. E., Cottrell D.M., Campos J., Image processing with the radial Hilbert transform: theory and experiments, Opt. Lett. 2000. V. 25 (2). P. 99–101.
52. Sacks Z.S., Rozas D., Swartzlander G.A., Jr., Holographic formation of optical vortex filaments, J. Opt. Soc. Am. B. 1998. V. 15 (8). P. 2226–2234.
53. Prudnikov A.P., Brychkov Yu.A., Marichev O.I. Integrals and series. Special valued functions, Moscow, Nauka, 1983.
54. Lü B., Duan K., Non-paraxial propagation of vectorial Gaussian beams diffracted at a circular aperture, Opt. Lett. 2003. V. 28 (24). P. 2440–2442.
55. Luneburg R.K., Mathematical Theory of Optics. Berkeley, Univ. of California Press, 1996.
56. Mei Z., Zhao D., Non-paraxial analysis of vectorial Laguerre–Bessel–Gaussian beams, Opt. Express. 2007. V. 15 (19). P. 11942–11951.
57. Teng S., Liu L., Liu D., Analytical expression of the diffraction of a circular aperture, Optik. 2005. V. 116 (12). P. 568–572.
58. Kotlyar V.V., et al., Komp. Optika, 2006. V. 30. P. 36–43.
59. Oron R., Davidson N., Friesem A.A., Hasman E., Efficient formation of pure helical laser beams, Opt. Commun. 2000. V. 182 (1–3). P. 205–208.
60. Peele A.G., McMahon P. J., Paterson D., Tran C. Q., Mancuso A. P., Nugent K.A., Hayes J. P., Harvey E., Lai B., McNulty I., Observation of an x–ray vortex, Opt. Lett. 2002. V. 27 (20). P. 1752–1754.
61. Wang Q., Sun X.W., Shum P., Yin X.J., Dynamic switching of optical vortices with dynamic gamma-correction liquid crystal spiral phase plate, Opt. Express. 2005. V. 13 (25). P. 10285–10291.
62. Swartzlander G.A., Jr., Broadband nulling of a vortex phase mask, Opt. Lett. 2005. V. 30 (21). P. 2876–2878.
63. Swartzlander G.A., Jr. Achromatic optical vortex lens, Opt. Lett. 2006. V. 31 (13). P. 2042–2044.
64. Kim G.-H., Jeon J.-H., Ko K.-H., Moon H.-J., Lee J.-H., Chang J.-S. Optical vortices produced with a non-spiral phase plate, Appl. Opt. 1997. V. 36 (33). P. 8614–8621.
65. Saga N. New line integral expressions for Fraunhofer diffraction, Optics Commun. 1987. V. 64 (1). P. 4–8.
66. Kotlyar V.V., Kovalev A.A., Skidanov R.V., Moiseev O.Y., Soifer V.A., Diffraction of a finite-radius plane wave and a Gaussian beam by a helical axicon and a spiral phase plate, J. Opt. Soc. Am. A. 2007. V. 24 (7). P. 1955–1964.
67. Kotlyar V.V., Khonina S.N., Kovalev A.A., Soifer V.A., Elfstrom H., Turunen J., Diffraction of a plane, finite-radius wave by a spiral phase plate, Opt. Lett. 2006. V. 31 (11). P. 1597–1599.
68. Kotlyar V.V., Kovalev A.A., Khonina S.N., Skidanov R.V., Soifer V.A., Elfstrom H., Tossavainen N., Turunen J., Diffraction of conic and Gaussian beams by a spiral phase plate, Appl. Opt. 2006. V. 45 (12). P. 2656–2665.
69. Miller W. Symmetry and Separation of Variables, Addison-Wesley Publ. Comp., MA, 1977.
70. Durnin J., Miceli J.J., Jr., Eberly J.H.. Diffraction–free beams, Phys. Rev. Lett. 1987. V. 58 (15), P. 1499–1501.
71. Sigman A.E., Lasers, Mill Valley, California, Univ. Science, 1986.
72. Bandres M.A., Gutiérrez-Vega J. C., Chávez-Cerda S., Parabolic non-diffracting optical wave fields, Opt. Lett. 2004. V. 29 (1). P. 44–46.
73. Gutiérrez-Vega J.C., Bandres M.A., Helmholtz–Gauss waves, J. Opt. Soc. Am. A. 2005. V. 22 (2). P. 289–298.

74. Bandres M.A., Gutiérrez-Vega J.C., Vector Helmholtz–Gauss and vector Laplace–Gauss beams, Opt. Lett. 2005. V. 30 (16). P. 2155–2157.
75. Bandres M.A., Gutiérrez-Vega J.C., Ince Gaussian beams, Opt. Lett. 2004. V. 29 (2). P. 144–146.
76. Bandres M.A., Elegant Ince–Gaussian beams, Opt. Lett. 2004. V. 29 (15). P. 1724–1726.
77. Abramochkin E.G., Volostnikov V.G., Generalized Gaussian beams, J. Opt. A: Pure Appl. Opt. 2004. V. 6 (5). P. 5157–5161.
78. Schwarz U.T., Bandres M.A., Gutiérrez-Vega J.C., Observation of Ince–Gaussian modes in stable resonators, Opt. Lett. 2004. V. 29 (16). P. 1870–1872.
79. Bentley, J.B., et al., Generation of helical Ince–Gaussian beams with a liquid–crystal display, Opt. Lett. 2006. V. 31 (5). P. 649–651.
80. Cai Y., Lin Q., Decentered Elliptical Gaussian Beam, Appl. Opt. 2002. V. 41 (21). P. 4336–4340.
81. Handbook of mathematical functions, Ed. by M.Abramovitz, I.A. Stegun, NBS, Appl. Math. Ser. 1964. V. 55.
82. Kotlyar V.V., Skidanov R.V., Khonina S.N., Soifer V.A., Hypergeometric modes, Opt. Lett. 2007. V. 32 (7). P. 742–744.
83. Methods for computer design of diffractive optical elements, Ed. by V.A. Soifer. New York: John Wiley & Sons, Inc. 2002.
84. Gori F., Guattari G., Padovani C., Bessel–Gauss beams, Opt. Commun. 1987. V. 64 (6). P. 491–495.
85. Caron C.F.R., Potvliege R.M., Bessel–modulated Gaussian beams with quadratic radial dependence, Opt. Commun. 1999. V. 164 (1–3). P. 83–93.
86. Kotlyar V.V., Kovalev A.A., Family of hypergeometric laser beams, J. Opt. Soc. Am. A. 2008. V. 25 (1). P. 262–270.
87. Rozas D., Law C.T., Swartzlander G.A., Jr., Propagation dynamics of optical vortices, J. Opt. Soc. Am. B. 1997. V. 14 (11). P. 3054–3065.
88. Karimi E., Zito G., Piccirillo B., Marrucci L., Santamato E., Hypergeometric Gaussian modes, Opt. Lett. 2007. V. 32 (21). P. 3053–3055.
89. Takenaka T., Yokota M., Fukumitsu O., Propagation of light beams beyond the paraxial approximation, J. Opt. Soc. Am. A. 1985. V. 2 (6). P. 826–829.
90. Kotlyar V.V., Kovalev A.A., Non-paraxial hypergeometric beams, J. Opt. A Pure Appl. Opt. 2009. V. 11 (4). P. 045,711.
91. Bin Z., Zhu L., Diffraction property of an axicon in oblique illumination, App. Opt. 1998. V. 37 (13). P. 2563–2568.
92. Thaning A., Jaroszewicz Z., Friberg A.T., Diffractive axicons in oblique illumination: Analysis and experiments and comparison with elliptical axicons, App. Opt. 2003. V. 42 (1). P. 9–17.
93. Khonina S.N., Kotlyar V.V., Soifer V.A., Paakkonen P., Turunen J., Measuring the light field orbital angular momentum using DOE, Opt. Mem. and Neur. Networks. 2001. V. 10 (4). P. 241–255.
94. Cai Y., Lin Q., Decentered elliptical Hermite–Gaussian beam, J. Opt. Soc. Am. A. 2003. V. 20 (6). P. 1111–1119.
95. Cai Y., Lin Q., A partially coherent elliptical flattened Gaussian beam and its propagation, J. Opt. A: Pure and Appl. Opt. 2004. V. 6 (12). P. 1061–1066.
96. Mitreska Z., Diffraction of elliptical Gaussian light beams on rectangular profile grating of transmittance, J. Opt. A: Pure and Appl. Opt. 1994. V. 3 (6). P. 995–1004.
97. Seshadri S.R., Basic elliptical Gaussian wave and beam in a uniaxial crystal, J. Opt. Soc. Am. A. 2003. V. 20 (9). P. 1818–1826.

98. Steinbach A., Rauner M., Cruz F.C., Bergquist J.C., CW second harmonic generation with elliptical Gaussian beam, Opt. Commun. 1996. V. 123 (1–3). P. 207–214.
99. Cai Y., Lin Q., Light beams with elliptical flat–topped profiles, J. Opt. A: Pure and Appl. Opt. 2004. V. 6 (4). P. 390–395.
100. Cai Y., Lin Q., Hollow elliptical Gaussian beam and its propagation through aligned and misaligned paraxial optical systems, J. Opt. Soc. Am. A. 2004. V. 21 (6). P. 1058–1065.
101. Bandres M.A., Gutiérrez-Vega J.C., Ince–Gaussian modes of the paraxial wave equation and stable resonators, J. Opt. Soc. Am. A. 2004. V. 21 (5). P. 873–880.
102. Bandres M.A., Gutiérrez-Vega J. C. Higher-order complex source for elegant Laguerre–Gaussian waves, Opt. Lett. 2004. V. 29 (19). P. 2213–2215.
103. Khonina S.N., Kotlyar V.V., Soifer V.A., Jefimovs K., Paakkonen P., Turunen J. Astigmatic Bessel laser beams, J. of Modern Optics. 2004. V. 51 (5). P. 677–686.
104. Khonina S.N., Volotovskiy S.G., Self-reproduction of multimode laser fields in step weakly guiding optical fibers, Izv. Samarsk. Nauchn. Tsentra RAN. 2004. V. 6 (1). P. 53–64.
105. Cherin A.H., An introduction to optical fibers. Singapore: McGraw-Hill book Co., 1987.
106. Koshiba M., Optical waveguide analysis, Tokyo: McGraw-Hill Inc., 1990.
107. Snyder A., Love D., Theory of optical waveguides. Moscow, Radio i Svyaz', 1987.
108. Levi L. Applied Optics, New York: John Wiley & Sons Inc., 1980.
109. Gloge D., Weakly Guiding Fibers, Appl. Opt. 1971. V. 10 (10). C. 2252–2258.
110. Yeh C., Handbook of Fiber Optics. Theory and Applications, New York: Academic Press Inc., 1990.
111. Kotlyar V.V., et al., Avtometriya, 2002. V. 38 (3). P. 33–44.
112. Khonina S.N., Skidanov R.V., Kotlyar V.V., Soifer V.A., Komp. Optika, 2003. V. 25. P. 89–94.
113. Karpeev S.V, Paul V.S., Khonina S.N., Komp. Optika, 2003. V. 25. P. 95–99.
114. Soskin M.S., Vasnetsov M.V., Singular optics, Progress in Optics. V. 42. Ed. by E.Wolf. Amsterdam: North-Holand, 2001. P. 219–276.
115. Bereznyy A.E. et al., Bessel optics, Dokl. AN SSSR. 1984. V. 274 (4). P. 802–805.
116. Abramochkin E., Losersky N., Volostnikov V., Generation of spiral-type laser beams, Opt. Commun. 1997. V. 141 (1–2). P. 59–64.
117. Arlt J., Dholakia K., Generation of high-order Bessel beams by use of an axicon, Opt. Commun. 2000. V. 177 (1–6). P. 297–301.
118. Berdagué S., Facq P., Mode division multiplexing in optical fibers, Appl. Opt. 1982. V. 21 (11). P. 1950–1955.
119. Mikaelian A.L.. Optical Methods in Informatics, Moscow, Nauka, 1990.
120. Soifer V.A., Golub M.A., Laser beam mode selection by computer-generated holograms. Boca Raton, CRC Press, 1994.
121. Thornburg W.Q., Corrado B.J., Zhu X.D., Selective launching of higher-order modes into an optical fiber with an optical phase shifter, Opt. Lett. 1994. V. 19 (7). P. 454–456.
122. Dubois F., Emplit Ph., Hugon O., Selective mode excitation in graded–index multimode fiber by a computer–generated optical mask, Opt. Lett. 1994. V. 19 (7). P. 433–435.
123. Bolshtyansky M.A., Savchenko A.Yu., Zel'dovich B.Ya.m Use of skew rays in multimode fibers to generate speckle field with nonzero vorticity, Opt. Lett. 1999. V. 24 (7). P. 433–435.
124. Volyar A.V., Fadeeva T.A., Dynamics of topological multipoles. II: Creation, annihilation, and evolution of nonparaxial optical vortices, Optics and Spectr., 2002. V. 92 (2). P. 253–262.

125. Karpeev S.V., Khonina S.N., Experimental excitation and detection of angular harmonics in a step–index optical fiber, Optical Memory & Neural Networks (Information Optics). Allerton Press, 2007. V. 16 (4). P. 295–300.
126. Agrawal G.P. Fiber-Optic Communication Systems, 3rd Ed., John Wiley, 2002.
127. Methods of Computer Optics, Ed. by. V.A. Soifer, Moscow, Fizmatlit, 2003.
128. Strilets A.S., Khonina S.N., Komp. Optika, 2008. V. 32 (1). P. 33–38.
129. Curtis J.E., Koss B.A., Grier D.G., Dynamic holographic optical tweezers, Optics Commun. 2002 . V. 207 (1-6). P. 169-175 .
130. Kotlyar V.V., Khonina S.N., Soifer V.A., Coding techniques of compositional DOEs-Komp. Optika, 2001. V. 21. P. 36-39 .
131. Khonina S.N., Kotlyar V.V., Lushpin V.V., Soifer V.A., A method for design of composite DOEs for the generation of letter image, Optical Memory and Neural Networks (Allerton Press). 1997. V. 6 (3). P. 213-220 .
132. Duelli M., Reece M., Cohn R.W., Modified minimum-distance criterion for blanded random and non-random encoding, J. Opt. Soc. Am. A. 1999. V. 16 (10). P. 2425–2438 .

Chapter 8

Optical trapping and manipulation of micro- and nano-objects

The optical trapping and rotation of microobjects are based on the well-known phenomenon of light pressure. After developing lasers it became possible to generate the radiation pressure force sufficient for acceleration, deceleration, deflection, direction and stable trapping of microscopic objects, whose dimensions range from tens of nanometers to tens of micrometers. If the refractive index is greater than the refractive index of the medium, the force resulting from changes in the direction of the light acts on the micro-object so that it moves into the region of the highest light intensity.

The first experiments with the trapping and acceleration of micro-objects, suspended in liquid and gas, are described in [30]. In 1977, changes were detected in the force of radiation pressure on the transparent dielectric spherical objects, depending on the wavelength and size [31].

If in the first studies it was shown that a micro-object can be trapped and moved linearly, then subsequent studies considered the possibility of rotating and orienting micro-objects in space. Optical rotation allows non-contact drives for micromechanical systems [58], also has many applications in biology [85].

There are three main ways of rotation of microscopic objects.

– Due to the spin angular momentum, which exists in the fields with circular polarization. In this case only the birefringent micro-objects, such as micro-objects made of Iceland spar, rotate [35, 36]. The main drawback of this method is the restriction on the material from which the micro-object is made.

– Due to the orbital angular momentum, which arises due to the spiral shape of the wave front, such as Laguerre–Gaussian (LG) and Bessel beams of higher orders. Transfer of the orbital angular momentum is due to partial absorption of light in the micro-object. This method is presented in [37, 50, 58]. In these studies, Bessel and LG beams formed with the use of amplitude holograms, which is unprofitable from the standpoint of energy efficiency. It is much more efficient to use pure phase DOEs, for example to create Bessel beams (BB) [65]. There are works in which the micro-objects move along paths other than the circle, for example, a light triangle, square, spiral [5, 23–26].

– By changing the phase shift in the interference pattern (in trapping of a microscopic object in the interference pattern) between the beam having a helical wavefront (i.e., LG beam) and the Gaussian beam. This pattern is rotated by changing the optical path length of one of the beams. This method is described in [82].The main drawback of this method is the need to use a fairly complex optical circuit. In this case it is also easier to use the DOE that forms a superposition of Bessel of LG modes [11]. Rotating BB or LG which in propagation along the optical axis is accompanied by the rotation of the intensity distribution in the beam cross section can be used to rotate microscopic objects with variable speed by using the linear displacement of the radiation source or a focusing lens. In this case, the optical system is reduced, in fact, to one DOE.

2D-arrays of traps (micro-objects are pressed to the table of the microscope) have potential application for building elements of micro-optomechanical systems [49, 59], the formation of different micro-configurations [60], the sorting of biological cells [54] and other applications that do not require manipulation of longitudinal objects.

A system of two traps was realized with the help of a beam splitter and refractive optics [46, 84]. However, this approach is very complicated if a larger number of traps is required. An alternative and more promising approach is the separation and guiding of the laser beam with the DOE [37, 42, 43, 53, 56].

In [38] the authors proposed to supplement a dynamic diffractive element, which is a matrix of $N \times N$ programmable phase gratings, with a $N \times N$ matrix of microlenses. In [37] the iterative method was used for calculating the phase DOE for creating 2D and 3D arrays of optical traps. In the experiments, the matrix was formed from eight Gaussian beams. The main disadvantage of spatial light modulators based on liquid crystals remain low diffraction efficiency (strong diffraction noise due to the high discreteness of the modulators) and the insufficient resolution of the matrix of pixels to handle the complex phase distributions. Also, the final pixel size limits the maximum variation of the diffraction orders (high carrier spatial frequencies are accompanied by the binarization of the phase profile and diffraction efficiency decreases).

Measurements have shown [79] that 15% of the energy of the incident beam remains after the liquid crystal modulator. Energy losses are due to several reasons:

1) the opaque part of the panel (core loss, up to 65%),
2) the structure of the liquid crystal modulator is similar to the grating, generating high orders (54% loss),
3) the inability to concentrate all the energy in a useful manner, because the modulator has a maximum phase shift of less than 2π (maximum ratio achieved between the first and zero order of 2:1) [79],
4) the discrepancy between the square aperture of the panel and the round profile of the incident beam (8%). Thus, the use of DOE for the formation of multiorder light beams for micro-object rotation problems if no dynamics is required, is preferred to the use of dynamic light modulators.

There are many studies in which solutions with separable variables for the Helmholtz and Schrödinger equations are used in optics. These studies examined multimode Bessel beams [57], multiorder LG beams [21], non-paraxial light beams that retain their structure during propagation [33], parabolic beams, Gauss-Helmholtz waves, paraxial light beams that retain their structure up to scale, Ince-Gaussian modes [33], elegant Ince–Gaussian beams [33], Hermite–Laguerre-Gaussian modes [22], optical vortices [66]. Some of these beams have been realized with laser resonators [66], liquid crystal displays [34, 87], phase DOEs [57, 67, 68]. These beams can provide additional new features in the problem of 'optical tweezers'.

There is a considerable number of works [13, 29, 32, 41, 51, 72, 74, 77, 78, 80, 83] concerned with the calculation of forces acting on the micro-object. In the well-known papers on the calculation of forces acting on the micro-object using the geometric optics approach, restrictions are imposed on the shape of the micro-object and the shape of the beam, and, as a rule, the motion parameters of the micro-object are not considered. For example, in [29] the authors considered only spherical micro-objects in a Gaussian beam. In [51], the force was calculated for the non-spherical micro-objects, but the authors consider the case of a Gaussian beam. In [13] the spherical and elliptical micro-objects in Gaussian and LG beams were considered.

8.1. Calculation of the force acting on the micro-object by a focused laser beam

This section describes the derivation of the expressions for the force acting on a two-dimensional dielectric cylindrical object from a monochromatic electromagnetic wave.

8.1.1. Electromagnetic force for the three-dimensional case

In [1] a formula is derived which expresses the conservation of the total momentum of the system of the electromagnetic field plus the object V, bounded by the surface S:

$$\frac{\partial}{\partial t}\int\limits_{V_1} P_i dV + \frac{\partial}{\partial t} P_{0i} = -\oint\limits_{S_1} \sigma_{ik} n_k dS, \tag{8.1}$$

where P_i are the coordinates of the vector of the momentum of the electromagnetic field (V_1 and S_1 are the volume and the surface restricting it, which include an object $V \in V_1$) that is associated with the Umov–Poynting vector by the relation:

$$\mathbf{P} = \frac{\mathbf{S}}{c} = \frac{1}{4\pi c}\left[\mathbf{E}\times\mathbf{H}\right], \tag{8.2}$$

P_{0i} are the coordinates of the momentum vector of the object, $\delta P_{0i}/\delta t$ are the coordinates of the force vector of the light on the object ($\mu = 1$):

$$\sigma_{ik} = \frac{1}{4\pi}\left(\frac{|\mathbf{E}|^2 + |\mathbf{H}|^2}{2}\delta_{ik} - E_i E_k - H_i H_k\right); \tag{8.3}$$

σ_{ik} is the Maxwell stress tensor of the electromagnetic field ($\sigma_{ik} = \sigma_{ki}$); **E**, **H** are the vectors of the stress of the electric and magnetic fields in a vacuum.

After averaging over the time period of $T = 2\pi/\omega$ of the monochromatic light:

$$\mathbf{E}(\mathbf{x},t) = \mathrm{Re}\left\{\mathbf{E}(\mathbf{x})e^{i\omega t}\right\}, \quad \mathbf{H}(\mathbf{x},t) = \mathrm{Re}\left\{\mathbf{H}(\mathbf{x})e^{i\omega t}\right\} \tag{8.4}$$

instead of equation (8.1) we obtain:

$$F_i = \left\langle \frac{\partial P_{0i}}{\partial t} \right\rangle = -\oint \langle \sigma_{ik} \rangle n_k dS, \tag{8.5}$$

as

$$\left\langle \frac{\partial}{\partial t}\int_V P_i dV \right\rangle = \int_V \left\langle \frac{\partial}{\partial t} P_i \right\rangle dV = 0. \tag{8.6}$$

It can be shown that:

$$\begin{aligned}\left\langle \frac{\partial P_x}{\partial t} \right\rangle = \frac{1}{4\pi c}\Big\{ & \left\langle \mathrm{Re}\left(i\omega E_y(\vec{x})e^{i\omega t}\right)\mathrm{Re}\left(H_z(\vec{x})e^{i\omega t}\right)\right\rangle + \\ & + \left\langle \mathrm{Re}\left(E_y(\vec{x})e^{i\omega t}\right)\mathrm{Re}\left(i\omega H_z(\vec{x})e^{i\omega t}\right)\right\rangle - \\ & - \left\langle \mathrm{Re}\left(i\omega E_z(\vec{x})e^{i\omega t}\right)\mathrm{Re}\left(H_y(\vec{x})e^{i\omega t}\right)\right\rangle - \\ & - \left\langle \mathrm{Re}\left(E_z(\vec{x})e^{i\omega t}\right)\mathrm{Re}\left(i\omega H_y(\vec{x})e^{i\omega t}\right)\right\rangle\Big\} = 0,\end{aligned} \tag{8.7}$$

where Re (...) is the real part of complex number $\langle f(t) \rangle = \frac{1}{T}\int_0^T f(t)dt$. Similarly to (8.7) for the other projections of the momentum vector of the electric field it can be shown that $\left\langle \frac{\partial P_y}{\partial t} \right\rangle = \left\langle \frac{\partial P_z}{\partial t} \right\rangle = 0$.

To obtain expressions for the time-averaged stress tensor (8.3) we take into account that

$$\left\langle \mathrm{Re}\left(E_i(\vec{x})e^{i\omega t}\right)\mathrm{Re}\left(E_j(\vec{x})e^{i\omega t}\right)\right\rangle = \frac{1}{2}\mathrm{Re}\left[E_i(\vec{x})E_j^*(\vec{x})\right]. \tag{8.8}$$

Then instead of (8.5) we obtain (ε_2 is the dielectric constant of the medium):

$$F_x = \frac{1}{8\pi}\oint_S \left\{ \frac{1}{2}\left[\varepsilon_2 |E_x|^2 + |H_x|^2 - \varepsilon_2 |E_y|^2 - \right.\right.$$

$$\left. - |H_y|^2 - \varepsilon_2 |E_z|^2 - |H_z|^2 \right] dS_x +$$

$$\left. + \mathrm{Re}\left(\varepsilon_2 E_x E_y^* + H_x H_y^*\right) dS_y + \mathrm{Re}\left(\varepsilon_2 E_x E_z^* + H_x H_z^*\right) dS_z \right\},$$

$$F_y = \frac{1}{8\pi}\oint_S \left\{ \frac{1}{2}\left[\varepsilon_2 |E_y|^2 + |H_y|^2 - \varepsilon_2 |E_x|^2 - \right.\right.$$

$$\left. - |H_x|^2 - \varepsilon_2 |E_z|^2 - |H_z|^2 \right] dS_y +$$

$$\left. + \mathrm{Re}\left(\varepsilon_2 E_y E_z^* + H_y H_z^*\right) dS_z + \mathrm{Re}\left(\varepsilon_2 E_y E_x^* + H_y H_x^*\right) dS_x \right\},$$

$$F_z = \frac{1}{8\pi}\oint_S \left\{ \frac{1}{2}\left[\varepsilon_2 |E_z|^2 + |H_z|^2 - \varepsilon_2 |E_x|^2 - \right.\right.$$

$$\left. - |H_x|^2 - \varepsilon_2 |E_y|^2 - |H_y|^2 \right] dS_z$$

$$\left. + \mathrm{Re}\left(\varepsilon_2 E_z E_x^* + H_z H_x^*\right) dS_x + \mathrm{Re}\left(\varepsilon_2 E_z E_y^* + H_z H_y^*\right) dS_y \right\}, \qquad (8.9)$$

where $dS_x = -\frac{\partial z}{\partial x} dxdy$, $dS_y = \frac{\partial z}{\partial y} dxdy$, $dS_z = dxdy$, $E_1 = E_x$, $E_2 = E_y$, $E_3 = E_z$ (and similarly for H_i and F_i).

8.1.2. Electromagnetic force for the two-dimensional case

We rewrite the expression (8.9) for the force of the action of light on the micro-object in the 2D case. For the TE-polarization ($H_x = E_y = E_z = 0$) the electric field is directed along the axis X:, $E_x \neq 0$, Z is the optical axis, the 2D-object has the form of a cylinder with the arbitrary cross-sectional shape and has an infinite length along the axis X. The plane YOZ is the plane of incidence. In this case the relation (8.9) takes the form:

$$F_x = 0,$$

$$F_y = \frac{1}{8\pi}\oint_{S_1} \left\{ \frac{1}{2}\left[|H_y|^2 - \varepsilon_2 |E_x|^2 - |H_z|^2\right] dS_y + \mathrm{Re}\left(H_y H_z^*\right) dS_z \right\}, \qquad (8.10)$$

$$F_z = \frac{1}{8\pi}\oint_{S_1} \left\{ \frac{1}{2}\left[|H_z|^2 - \varepsilon_2 |E_x|^2 - |H_y|^2\right] dS_z + \mathrm{Re}\left(H_z H_y^*\right) dS_y \right\},$$

Here S_1 is already a contour enclosing a section of the object in the plane *YOZ*. Force F_z is directed along the optical axis and is analogous to the scattering force for the Rayleigh particles [2], and F_y is directed across the optical axis and is analogous to the gradient force [2].The relationship between the projections H_y, H_z and E_x (TE-polarization) follows from Maxwell's equations:

$$H_y = \frac{i}{k_0\mu}\frac{\partial E_x}{\partial z}, H_z = \frac{1}{ik_0\mu}\frac{\partial E_x}{\partial y}, \tag{8.11}$$

and between the projections E_y, E_z and H_x (TM-polarization):

$$E_y = \frac{1}{ik_0\varepsilon}\frac{\partial H_x}{\partial z}, E_z = \frac{i}{k_0\varepsilon}\frac{\partial H_x}{\partial y}, \tag{8.12}$$

where $k_0 = 2\pi/\lambda$ is the wave number of light with a wavelength λ, ε is the dielectric constant of the medium, μ is the magnetic permeability of the medium. Similar to (8.10), the force of light pressure with TM-polarization for the 2D object will have the following projections ($E_x = H_y = H_z = 0$):

$$F_x = 0,$$

$$F_y = \frac{1}{8\pi}\oint_{S_1}\left\{\frac{1}{2}\left[\varepsilon_2\left|E_y\right|^2 - \varepsilon_2\left|E_z\right|^2 - \left|H_x\right|^2\right]dS_y + \varepsilon_2 \operatorname{Re}\left(E_y E_z^*\right)dS_z\right\},$$

$$F_z = \frac{1}{8\pi}\oint_{S_1}\left\{\frac{1}{2}\left[\varepsilon_2\left|E_z\right|^2 - \varepsilon_2\left|E_y\right|^2 - \left|H_x\right|^2\right]dS_z + \varepsilon_2 \operatorname{Re}\left(E_z E_y^*\right)dS_y\right\}, \tag{8.13}$$

where (as in equation (8.10)) $dS_y = n_y dl = \sin\phi\, dl = dz$ and $dS_z = n_z dl = \cos\phi\, dl = dy$ and dl is the element of the arc.

8.1.3. Calculation of force for a plane wave

To calculate the force exerted by the light field on a cylindrical object, we must calculate the integral over the contour within which the object resides. As follows from the formulas for calculating the force projections (8.10), (8.13), the force should not change when the radius of integration R_i changes, if the object is completely enclosed in the integration contour: $R_i > R$.

We calculate the iterative algorithm of the diffraction field of a plane wave on a cylindrical object, and we also calculate the force acting on it at various radii of integration. Simulation parameters: the incident wave is flat, the entire calculated diffraction field 10×10 μm, the wavelength 1 μm. The object is a cylinder with a circular cross-section, a diameter of 1 μm, or a square with 1 μm side. The refractive index of the cylinder n_1 = 1.4 (ε_1 = 1.96). The density of the light energy flux is 100 mW/m over the entire diffraction field.

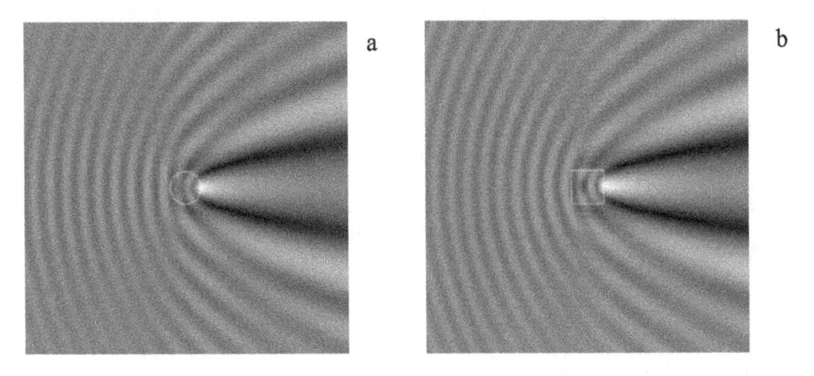

Fig. 8.1. The diffraction field $|E_x|$ of a TE-polarized plane wave on a) a cylinder with a circular cross section and b) a cylinder with a square cross-sectional shape.

Table 8.1. Dependence of the projection of force from the radius of integration

	Cylindrical object with a circular cross-section			
R_i, μm	5	3.75	2.5	1
$F_z \cdot 10^{-10}$ N/m	0.33176	0.32036	0.33213	0.31781
	Cylindrical object with a square cross-section			
R_i, μm	5	3.75	2.5	1
$F_z \cdot 10^{-10}$ N/m	0.32137	0.31588	0.31792	0.31688

Figure 8.1 shows the amplitude of the diffraction field ($|E_x|$, TE-polarization) of a plane wave on a cylindrical object with the above parameters.

Table 8.1 shows the dependence of the projection of the force F_z acting on the cylinder along the axis of light propagation Z, on the radius of integration R_i. Since the incident wave is flat and extends along the axis Z, the projection of force along the Y axis must be zero. Simulation shows that the projection of force on the Y-axis in this case is three orders of magnitude less than the projection of the force along the axis Z. For example, at the radius of integration $R_i = 5$ μm for a cylinder with a circular cross-section the projection of force $F_z = 0.33176 \cdot 10^{-10}$ N/m and projection $F_y = 0.0007617 \cdot 10^{-10}$ N/m.

As shown in Table 8.1, the fluctuations of the results of calculation of force are less than 5%. The number of samples over the entire diffraction field 256×256. This result proves that the force is calculated correctly using formulas (8.10) (to within 5%).

Let us consider the dependence of the force, calculated by formulas (8.10), on the resolution of the diffraction field.

Table 8.2. Dependence of the projection of force F_z on the number of counts in the entire diffraction field

$K \times K$	64×64	128×128	256×256	512×512
	Cylindrical object with a circular cross-section			
$F_z \cdot 10^{-10}$ N/m	0.4058	0.3523	0.3479	0.3454
	Cylindrical object with a square cross-section			
$F_z \cdot 10^{-10}$ N/m	0.3220	0.3259	0.3216	0.3324

Table 8.3. Dependence of the projection of force F_z on the number of samples K taken at the diameter of the circular cylinder

K	12	25	50	100
$F_z \cdot 10^{-10}$ N/m	0.4046	0.3594	0.3497	0.3327

Table 8.2 shows the dependence of the projection of force F_z on the Z axis on the number of counts in the entire diffraction field for the above parameters. The force was calculated for an integration radius of 2.5 μm. All of the diffraction field was 5×5 μm in size.

Table 8.2 shows that when the number of counts in the entire diffraction field is 64×64, calculation of the force in the case of a circular cylinder is less accurate due to an error in the description of the boundary of the circular cylinder by a broken line. This does not apply to the last three values of force for a cylinder with a circular cross-section of 2% and 1.5% for the square.

Table 8.3 shows the projection of force F_z in the Z-axis under the same conditions on the number of samples K, taken for the diameter of a cylinder with a circular cross section, at a fixed resolution of the diffraction field – 256×256 pixels.

Table 8.3 shows that at low resolution of the object (in this case 12 samples per diameter of the circular cylinder), the value of the projection of the force acting on the cylinder is considered to be inaccurate. The difference for the last three values of the forces in Table 8.3 is 8%. When taking less than 12 samples per wavelength the iterative algorithm ceases to converge for the given parameters.

8.1.4. Calculation of force for a non-paraxial Gaussian beam

In this section we calculate the projection of force by the formulas (8.10) acting from the non-paraxial Gaussian beam on a dielectric cylinder with a circular cross section, depending on the displacement L of the centre of the cylinder from the centre of the beam waist.

Projections of the force, calculated by formula (8.10), acting on a cylinder with a circular cross-section in the case of a TE-polarized wave are shown in Fig. 8.2. The parameters of the experiment: $D = \lambda = 2\omega_0 = 1$ μm, $\varepsilon_2 = 1$ (centre), $\varepsilon_1 = 2$, (object), the power of incident radiation per unit length is $P =$ 0.1 W/m. The offset from the centre of the waist L has the dimension of a μm.

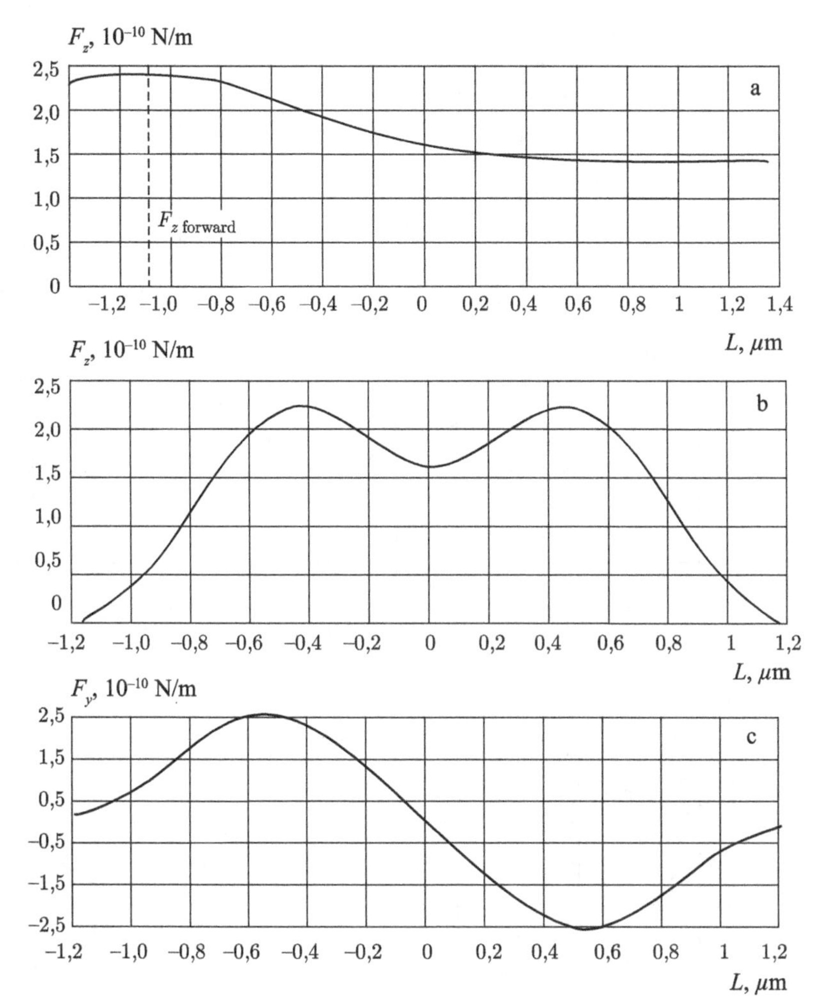

Fig. 8.2. TE-polarization: a) the dependence of the projection of force F_z on the displacement L of the object along the Z axis through the centre of the waist ($Y = 0$), the dependence of the projections of the forces F_z (b) and F_y (c) on displacement L of the object along the Y axis through the centre of the waist ($Z = 0$).

Similar projections of force in the case of TM-polarization, calculated by the formulas (8.13), are shown in Fig. 8.3.

Figures 8.2b and 8.3b shows that at the transverse displacement of the cylinder along the Y axis there is a projection of force F_y tending to return the cylinder to the centre of the waist. Moreover, the maximum projection of the force F_y and F_z *is* obtained for the transverse displacement of the cylinder L approximately equal to the radius of the waist of the Gaussian beam: $L \approx \omega_0$.

In [3] the results are presented of numerical simulation of the force acting on a Kerr microsphere in the 3D case. Simulation parameters: the refractive index of the

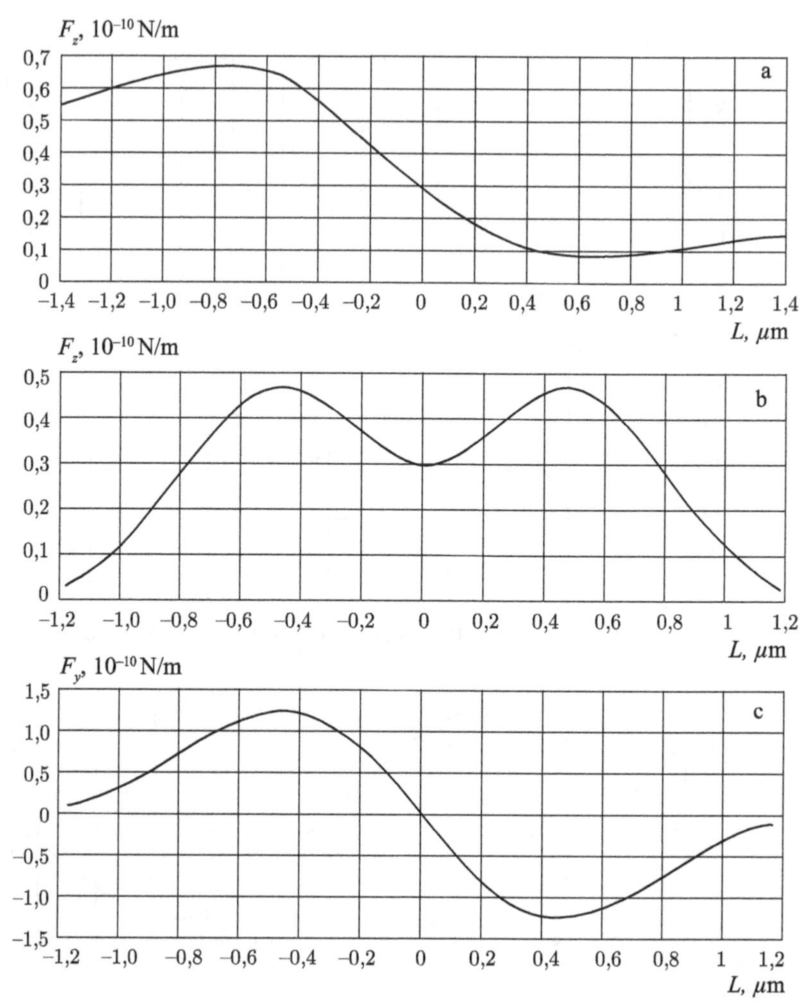

Fig. 8.3. TM-polarization: a) the dependence of the projection of force F_z on the displacement L of the object along the Z axis through the centre of the waist ($Y = 0$), the dependence of the projections of the forces F_z (b) and F_y (c) on displacement L of the object along the Y axis through the centre of the waist ($Z = 0$).

sphere $n_1 = 1.4$, refractive index $n_2 = 1.33$, sphere diameter $D = 2$ μm, wavelength $\lambda = 1.06$ μm, relative aperture (the ratio of the aperture of the lens to the focal length) NA = 1.4, the shift from the focus along the axis Z $L = 1$ μm. The force acting perpendicular to the propagation of light when a subject moves from the centre in a plane perpendicular to the propagation of radiation at the given parameters, $F = 0.3 \cdot 10^{-10}$ N. Figures 8.2b and 8.3b show that the projection of force is of the same order of magnitude per unit length of the cylinder $(0.5–1)\cdot 10^{-10}$ N/m.

Figure 8.4 shows the interference pattern of two Gaussian beams directed against each other with a waist at the origin, creating a standing wave. Figure 8.4a shows

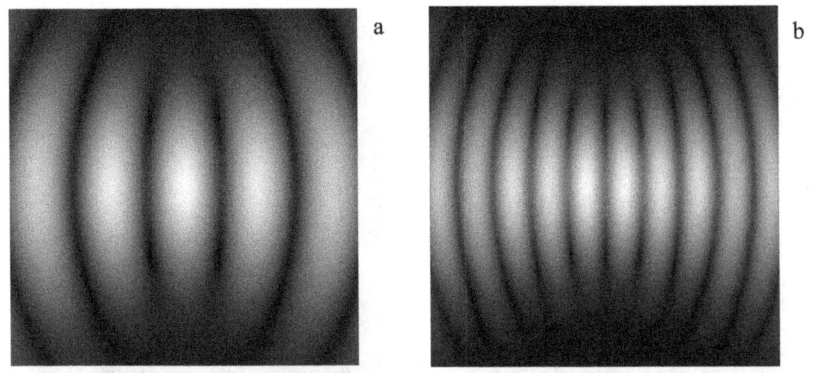

Fig. 8.4. The interference pattern of two non-paraxial Gaussian beams propagating in opposite directions along the axis Z: a) the total amplitude of the electrical field vector $|E_x|$, and b) the projection on the Z axis of the Umov–Poynting vector $|S_z|$.

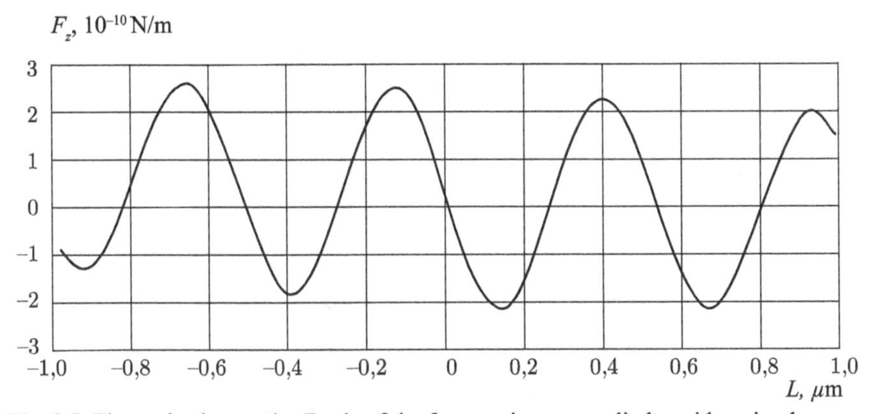

Fig. 8.5. The projection on the Z axis of the force acting on a cylinder with a circular cross section with $\varepsilon_1 = 2$, depending on the displacement of the centre circle of the cylinder along the axis Z.

the amplitude of the total field $|E_x|$ (TE-polarization), Fig. 8.4b is the modulus of the projection of the Umov–Poynting vector on the axis of light propagation Z. The first Gaussian beam is directed along the axis Z, the second beam in the opposite direction of the axis Z. For the first Gaussian beam the wavelength is 1 μm, the radiation power of 50 mW/m, the Gaussian beam waist is at the origin, its diameter is 1 μm. The radiation power of the second beam 50 mW/m, the wavelength is also equal to 1 μm and the diameter of the waist is 1.5 μm. If a dielectric object with the size of the order of the wavelength is placed in such a field, then this field will be a trap for it: the object is drawn into the intensity maxima of the field. Figure 8.5 is a plot of the dependence of the projection of force F_z directed along the Z axis on the displacement L from the axis Z. The object is a cylinder with a circular cross section with a diameter of 1 μm, dielectric constant $\varepsilon_1 = 2$. The diffraction field has a size of 2.5×2.5 μm. Figure 8.5 shows that near the waist along the Z axis almost periodically over a distance of about 0.25 μm there are points at which the force is

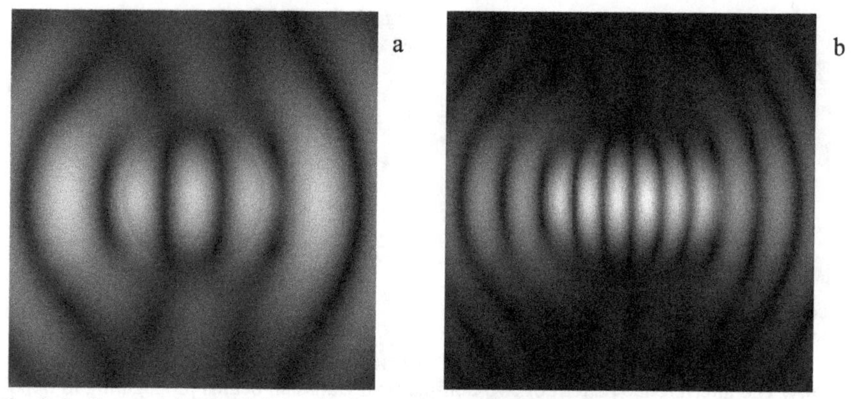

Fig. 8.6 Same as in Fig. 8.4, but in the presence of a cylinder with a circular cross section in the centre of the waist.

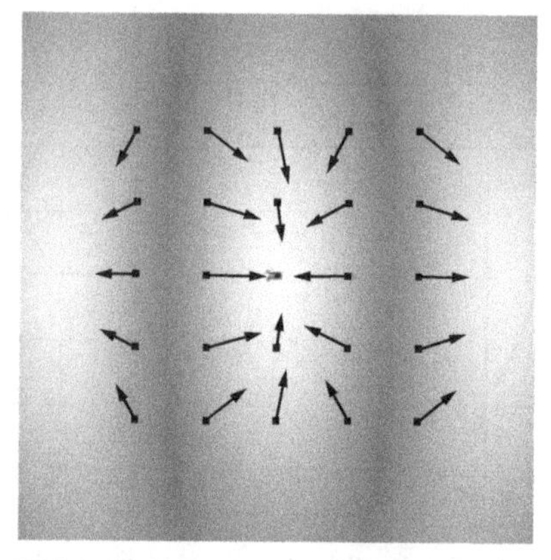

Fig. 8.7. The field vectors of the forces acting on the part of two colliding Gaussian beams on a cylinder with a circular cross section, which centre is located at different points in the interference pattern: the centre circle of the cylinder coincides with the beginning of each hand, and the length of each arrow is proportional to the modulus of strength at this point.

zero. If the centre of the cylinder coincides with these points, then the cylinder will be in a stable or unstable equilibrium. The points of stable and unstable equilibria alternate, that is, approximately every 0.5 μm the cylinder will be in the 'optical trap' (at the point of stable equilibrium).

Figure 8.6 shows the diffraction of Gaussian beams directed against each other, shown in Fig. 8.4, on a cylinder with a circular cross section, as described above. Figure 8.6a represents the strength of the electric field $|E_x|$ (TE-polarization), Fig. 8.6b – a projection of the Umov–Poynting vector on the axis Z. The object is located in the centre of the waist ($z = 0$). For visualization, the object in Fig. 8.6a is slightly obscured itself.

Figure 8.7 shows the central part of the diffraction pattern in Fig. 8.6a with the size of 0.31×0.31 μm. The arrows displayed the direction of the force acting on this cylinder by radiation, with the object placed in each specific point in space. One can see that the object is 'drawn' into the maxima of the interference pattern. The length of the arrows is proportional to the absolute force value.

If the refractive index is less than the refractive index of the particles, under certain conditions one can observe the 'trapping' of the particle along the *Z* axis, not only in the case of two colliding beams, but also in the case of a focused Gaussian beam.

Figure 8.8 shows a graph of the projection of force F_z in the displacement of the cylinder over distance *L* along the axis *Z*. The parameters of the experiment: the wavelength 1 μm, the diameter of the Gaussian beam waist $2\omega_0 = 1$ μm, the dielectric constant of the particles $\varepsilon_1 = 1.2$, the medium $\varepsilon_2 = 1$, the particle diameter $D = 2$ μm. From the graph we can see the trapping mechanism: the projection of force F_z in front of the focus is positive and directed towards the focus, behind the focus it is negative and pushes the particle back into focus. From numerical experiments it was determined that the ability to trapping depends on the dielectric constant of the particle. For the given parameters 'trapping' occurs when $1 < \varepsilon_1 < 1.35$. A plot of the force F_z under these parameters and the dielectric constant of the particle $\varepsilon_1 = 1.35$ is shown in Fig. 8.8.

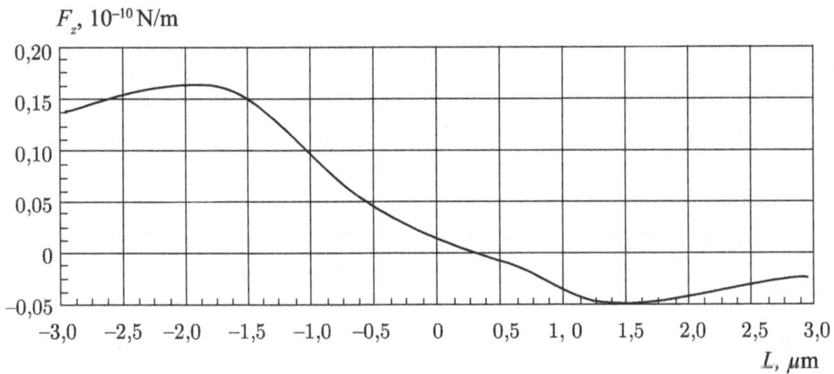

Fig. 8.8. The projection of force in the Z axis for a Gaussian beam acting on a cylinder with a circular cross section with $\varepsilon_1 = 1.2$ (medium $\varepsilon_2 = 1$).

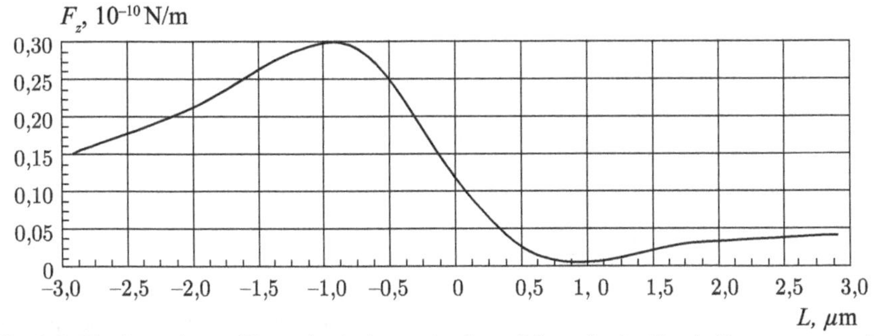

Fig. 8.9. The boundary of 'trapping': the projection of force in the Z axis for a non-paraxial Gaussian beam and a cylinder with a circular cross section with $\varepsilon_1 = 1.35$.

Figure 8.8 shows that in the displacement of the cylinder along the optical axis at a distance $L \approx 0.4$ μm the force exerted on it by the light beam goes to zero: $F_y = F_x = 0$. The existence of such an equilibrium point for the cylinder can be explained in terms of the two forces (scattering and gradient) acting on the cylinder near the waist of the Gaussian beam. Indeed, when the centre of the cylinder is exactly in the centre of the beam waist, then it is subjected only to the scattering force (photons push the cylinder forward), which is proportional to the intensity $|E_x|^2$. At the offset from the centre of the cylinder along the optical axis a gradient force arises due to the presence of the gradient of intensity $\Delta|E_x|^2$, which is aimed at the centre of the beam. At displacement $L \approx 0.4$ μm those forces are equivalent and the cylinder is in equilibrium.

A real cylindrical object has a finite length. But the two-dimensional approximation, which we consider here, can be applied to the description of the real situation if the length of the cylinder will be much larger than the diameter of its cross section. Indeed, consider the case where a three-dimensional dielectric cylinder of finite length is located near the waist of a cylindrical Gaussian beam (see Fig. 8.10).

Let the cylinder axis tilted at an angle θ, in the plane XY. Then the maximum deviation from the stable equilibrium point of the cylinder cross-section in the YZ plane will be equal to $\Delta y = l \cdot \mathrm{tg}(\theta)$, $y/a \ll 1$, where l is the length of the cylinder, a is the radius of its cross section. This results in a projection force F_y directed to the point of maximum intensity on the beam axis. That is, small rotations and displacements of the three-dimensional finite-length cylinder near a stable equilibrium of the cylindrical Gaussian beam waist will give rise to forces seeking to return the cylinder to the 'optical trapping' position.

8.1.5. Calculation of forces for the refractive index of the object smaller less than the refractive index of the medium

It is interesting to calculate the light force and field, acting on a dielectric 2D object whose refractive index is smaller than that of the medium.

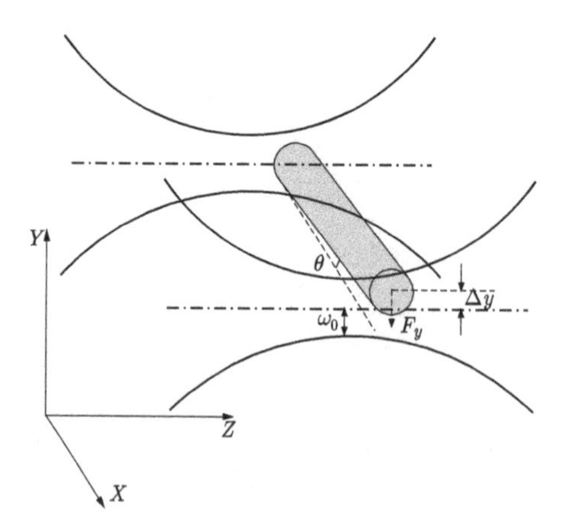

Fig. 8.10. Defining of the slope of a three-dimensional finite cylinder in the two-dimensional model.

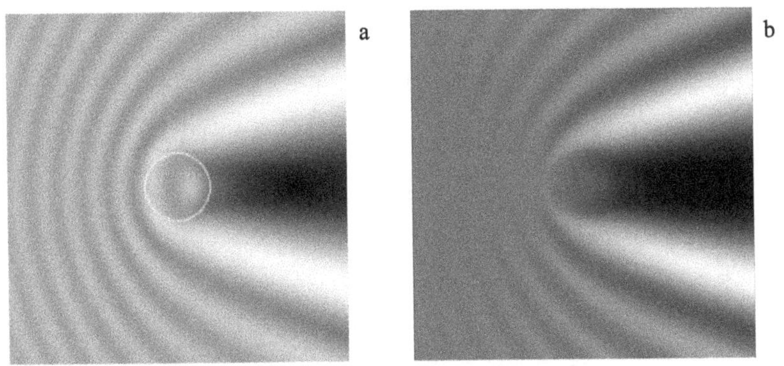

Fig. 8.11. Modulus of the strength of the electric field $|E_x|$ (a) and modulus of the projection of the Umov–Poynting vector on the axis Z $|S_z|$ (b) for diffraction of a plane wave on an air cylinder with a circular cross section in water.

Figure 8.11 shows the diffraction pattern of a plane wave in a medium with a refractive index $n_2 = 1.33$ (water) on a cylindrical object with a circular cross section with a refractive index $n_1 = 1$ (cylindrical air bubble). The diameter of the cylinder is equal to the wavelength and is equal to 1 μm. Figure 8.11a is the modulus of the strength of the electric field $|E_x|$ (TE-polarization), Fig.8.11b is a projection of the Umov–Poynting vector on the axis Z. The energy does not propagate behind the 'air bubble' which can be clearly seen on the sections shown in Fig. 8.12, taken along the Z axis through the point $Y = 0$.

Figure 8.12a shows the value of the amplitude $|E_x|$, Fig. 8.12b – the projection of the Umov–Poynting vector on the axis Z.

If such an object is placed near the focus of a Gaussian beam, it will be pushed out of it, as illustrated in the graphs in Fig. 8.13.

Figure 8.13a is a plot of the dependence of the force F_z along the Z axis on the displacement L along the axis Z, Fig. 8.13b – the dependence of force F_y on the displacement L along the Y axis through the focus.The Gaussian beam has a wavelength of 1 μm, the diameter of the waist is 1 μm, the radiation power 100 mW/m. It is seen that in deviation in either direction from the focus in the transverse direction the force, directed toward the deflection, increases, which leads to the movement of the focus in this direction. In deviation along the axis Z of light the force exerted on an object in front of the focus is less in the absolute value than behind the focus.

8.2. Methods for calculating the torque acting on a micro-object by a focused laser beam

In this section, we calculate the torque acting on a cylindrical micro-object with an elliptical cross-section from the side of the focused non-paraxial Gaussian beam. Calculation of the moment of the force was conducted depending on the size and shape of the integration region encompassing the micro-object under study. We

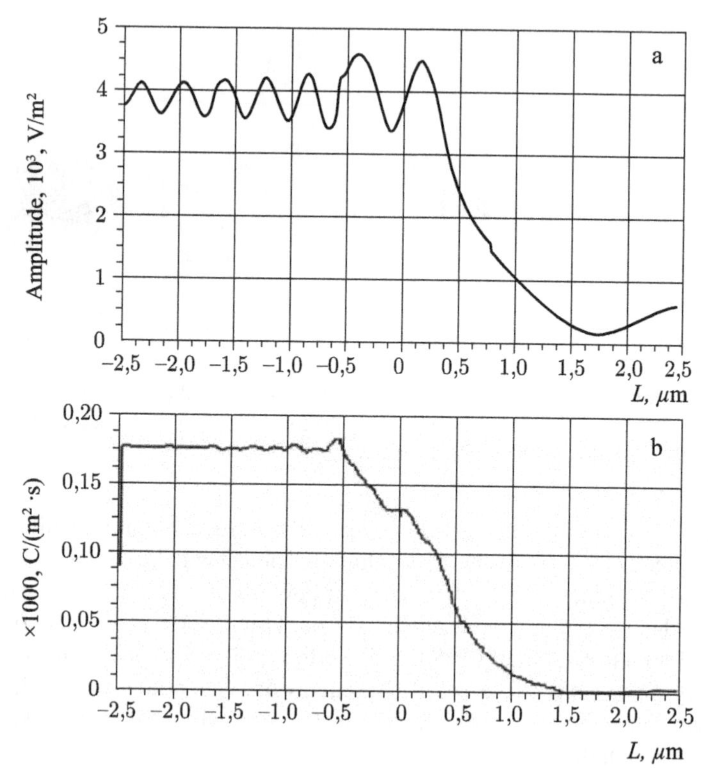

Fig. 8.12. Section of Fig. 8.11a (a) and 8.11b (b) along the Z axis through the point $Y = 0$.

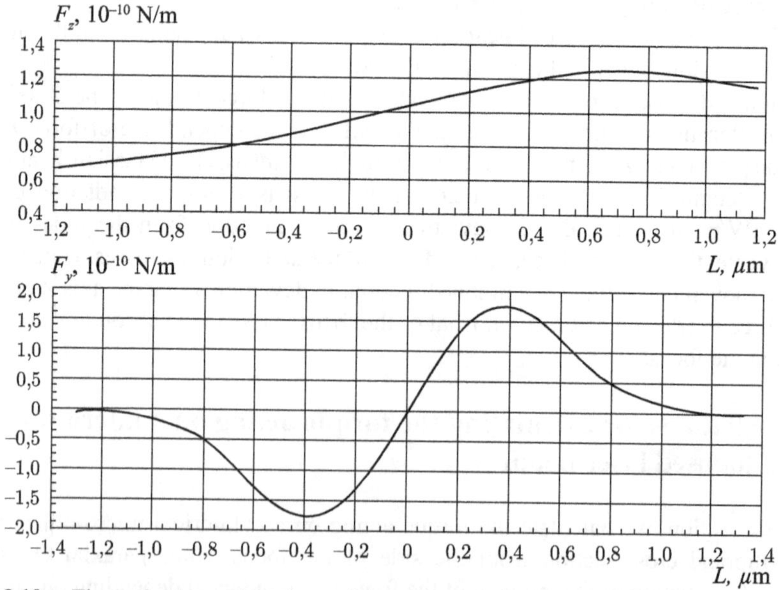

Fig. 8.13. The projections of the force of the non-paraxial Gaussian beam on 'a cylindrical air bubble' in the water: on the longitudinal axis F_z (a) and transverse axis F_y (b).

consider the torque on the micro-object, located at the point where the force exerted by light is zero. We also consider a torque acting on the micro-object with an elliptical section in the standing wave.

8.2.1. The orbital angular momentum in cylindrical microparticles

Figure 8.14 shows the scheme of the problem.

Light propagates along the Z axis in a medium with permittivity ε_1 and falls on an object with a dielectric constant ε_2. Then the torque $\mathbf{M}$ at any point A can be calculated using the formula [4]:

$$\mathbf{M} = \oint_S \left[\mathbf{r} \times \left(\tilde{T} \cdot \mathbf{n}\right)\right] dS, \tag{8.14}$$

where $\mathbf{n}$ is the normal to the surface S, covering the object in question, A is the point at which we calculated torque $\mathbf{M}$, $\mathbf{r}$ is the radius vector from point A to the integration surface S, $\tilde{T}$ is the Maxwell's stress tensor of the electromagnetic field.

The product of the Maxwell stress tensor on the normal can be written as:

$$\begin{gathered}\left(\tilde{T} \cdot \mathbf{n}\right) = \begin{bmatrix} T_{ii} & T_{ij} & T_{ik} \\ T_{ji} & T_{jj} & T_{jk} \\ T_{ki} & T_{kj} & T_{kk} \end{bmatrix} \begin{pmatrix} n_x \\ n_y \\ n_z \end{pmatrix} = \\ = \begin{bmatrix} T_{ii}n_x + T_{ij}n_y + T_{ik}n_z \\ T_{ji}n_x + T_{jj}n_y + T_{jk}n_z \\ T_{ki}n_x + T_{kj}n_y + T_{kk}n_z \end{bmatrix} = \begin{pmatrix} t_x \\ t_y \\ t_z \end{pmatrix}. \end{gathered} \tag{8.15}$$

Then

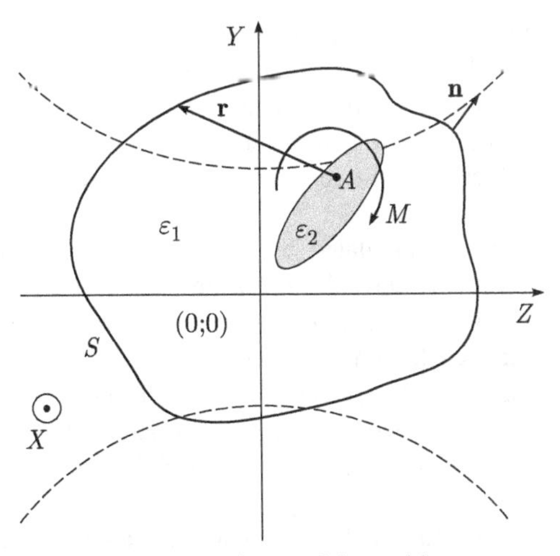

Fig. 8.14. Scheme of the problem.

$$\mathbf{r}\times\left(\tilde{T}\cdot\mathbf{n}\right)=\begin{bmatrix} i & j & k \\ r_x & r_y & r_z \\ t_x & t_y & t_z \end{bmatrix}=\mathbf{i}(r_y t_z - r_z t_y)- \\ -\mathbf{j}\left(r_x t_z - r_z t_y\right)+\mathbf{k}\left(r_x t_y - r_y t_x\right). \tag{8.16}$$

For the two-dimensional case (cylindrical object) this makes sense only in relation to the axis X, i.e.

$$M_x = \oint_S \left(r_y t_z - r_z t_y\right) dS. \tag{8.17}$$

As $E_y = E_z = H_x = 0$, for TE-polarization, we finally obtain:

$$M_x = \frac{1}{4}\oint_S \left[\varepsilon_0\varepsilon_1\left|E_x\right|^2 + \mu\mu_0\left|H_y\right|^2 - \mu\mu_0\left|H_z\right|^2\right]\times \\ \times r_y dy - \frac{1}{2}\oint_S \mu\mu_0 \operatorname{Re}\left(H_z H_y^*\right) r_y dz - \\ -\frac{1}{4}\oint_S \left[\varepsilon_0\varepsilon_1\left|E_x\right|^2 - \mu\mu_0\left|H_y\right|^2 + \mu\mu_0\left|H_z\right|^2\right] r_z dz - \\ -\frac{1}{2}\oint_S \mu\mu_0 \operatorname{Re}\left(H_y H_z^*\right) r_z dy. \tag{8.18}$$

Calculation of the final formula for the moment for TM-polarization differs by the substitution in (8.17) $E_x = H_y = H_z = 0$.

8.2.2. The results of numerical simulation of the torque

To verify the correctness of the formula (8.18), we calculated the torque acting on an elliptical microparticle by a non-paraxial Gaussian beam with different sizes of square integration contour S. Figure 8.15b shows a plot of the changes of torque M_x on the integration contour, whose parameters are shown in Fig. 8.15a. The waist of the Gaussian beam is located in the centre of coordinates, the light propagates in the positive direction along the axis Z.

The diffraction field is calculated by the method described in the previous section. The integration was carried out on the square contour S, the number of counts in Fig. 8.15a 254×254. This is convenient because it is not necessary to calculate the normal $\mathbf{n}$ at each point of the circuit – the contour integral (8.18) splits into a sum of integrals, some of which are taken on the contour sides parallel to the axis Y, and another part on the sides of the contour parallel to the axis Z. The power of the incident radiation was equal to P = 100 mW/m, the wavelength λ = 1 μm, the dielectric constant of the medium ε_1 = 1, the dielectric constant of the particle ε_2= 2. Point A, in relation to which the moment is calculated, coincides with the centre of an elliptical particle.

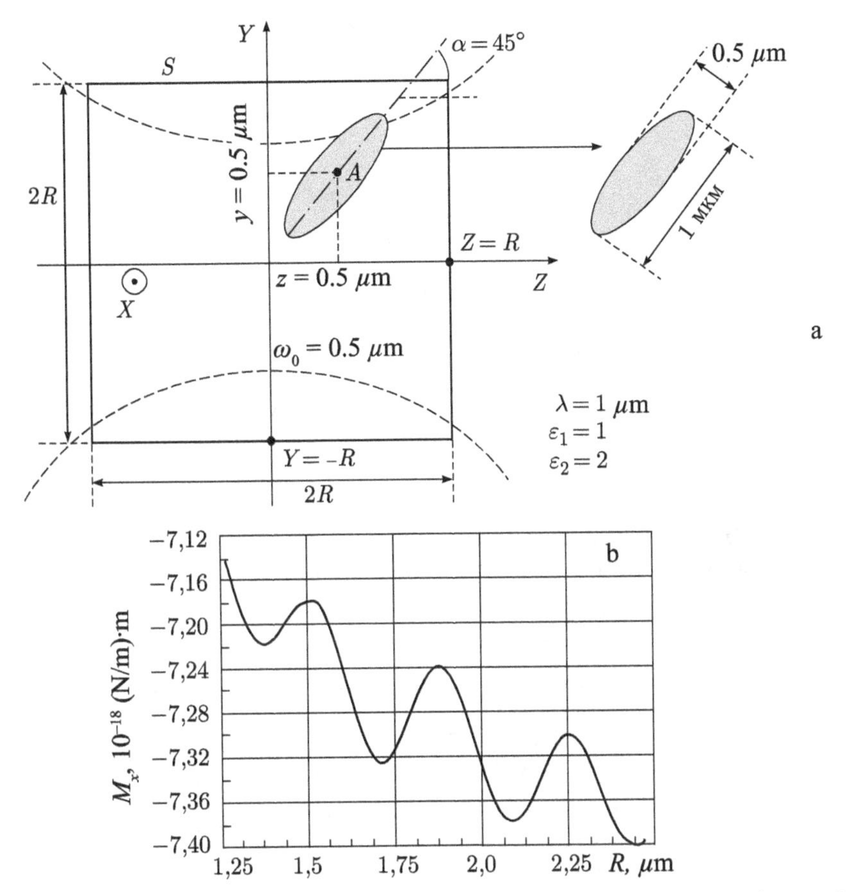

Fig. 8.15. a) The calculation of the torque and b) the results of the calculation of torque M_x for different R.

Note that the torque **M** is measured in units [N ·m]. However, in this case the dimension of M_x is [(N/m)·m] because the force in the two dimensional case is expressed, due to an infinitely long cylindrical object along the X axis, in units N/m and is unit force per unit length of the micro-object.

Figure 8.15b shows that the fluctuations in the value of M_x with a change in R are less than 4%.

Note that the magnitude of the torque **M** depends on the location of point A, relative to which it is calculated, if the force acting on the particle is not zero. If we calculate the torque M_x for the scheme in Fig. 8.15a at the same location of the cylinder, but when the point A has coordinates (0.25 μm, 0.25 μm), the value will be $M_x = 8.8 \cdot 10^{-17}$ (N/m)·m.

Figure 8.16 shows the scheme of calculating the moment of the force M_x for the case when the force of light acting on a particle is zero. An elliptical cylinder with dielectric constant $\varepsilon_2 = 2$ is in the water with $\varepsilon_1 = 1.77$.

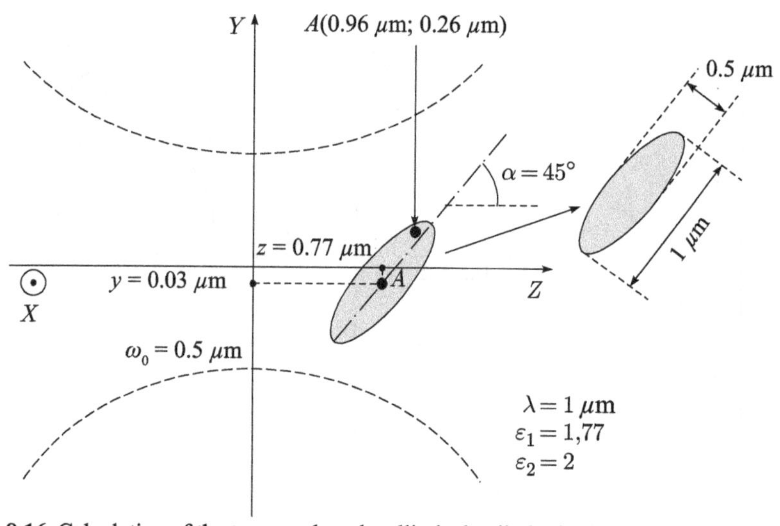

Fig. 8.16. Calculation of the torque when the elliptical cylinder is situated at the equilibrium point (the force of the light to an object is equal to zero).

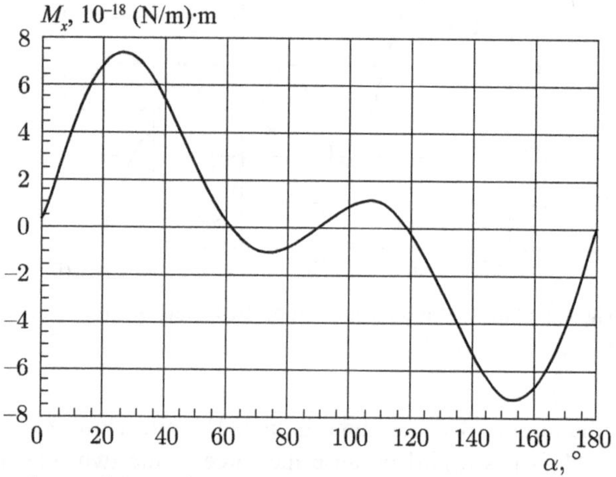

Fig. 8.17. Dependence of the projection on the X axis of the torque acting on the elliptical cylinder in the focus of a Gaussian beam, on angle α.

The torque in finding the elliptical cylinder at the focus with point A, located in the centre of the particle, is $M_x = 7.11 \cdot 10^{-8}$ (N/m)·m. If we move the observation point A to a point with the coordinates (0.96 μm, 0.26 μm), then the moment will be equal to $M_x = 7.11 \cdot 10^{-8}$ (N/m)·m. Thus, we see that the torque acting on the particle is almost independent of the point relative to which it is calculated, if the particle is at the point where the force exerted by the light field is zero.

Figure 8.17 shows a plot of the dependence of the torque on angle α. The simulation parameters are the same as in Fig. 8.15a, but the object is located in the centre of the waist, point A is located in the centre of the object, the major diameter

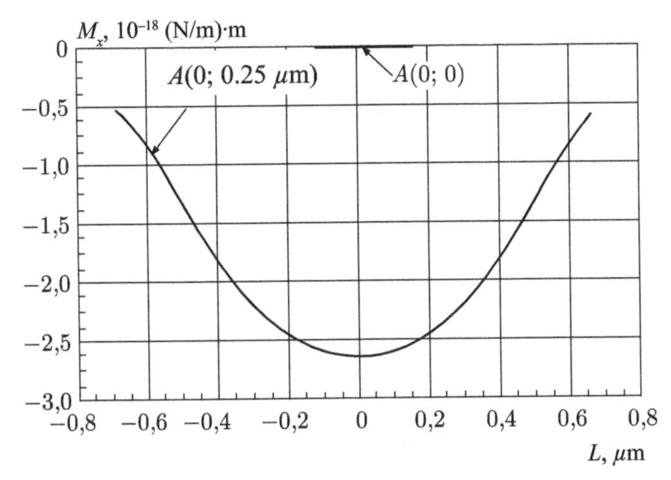

Fig. 8.18. The dependence of the torque M_x on the displacement L of a spherical particle along the Y axis for the two positions of point A relative to the centre of the particle.

of the ellipse is 1.2 μm, minor 0.3 μm, the dielectric constant of the object $\varepsilon_1 =$ 2.25. The angle α was determined as shown in Figure 8.16 (measured from the Z-axis counter-clockwise).

Figure 8.17 shows that the positions of the elliptical particle in the focus when its long axis lies along and perpendicular to the optical axis, are stable equilibria ($M_x > 0$ for a particle rotating in the clockwise direction). When α is approximately equal to 60° and 120°, there are two points of unstable equilibrium, when the moment vanishes. This agrees well with the data in [4], where the same plot of the dependence the torque on the rotation angle of the particle α was obtained.

Figure 8.18 is a plot of the dependence of the torque M_x on the displacement L of a circular particle with a radius of 0.25 μm on the Y axis through the centre of the waist. Point A is located in the centre of particle (0; 0) and 0.25 μm above the centre of the Y axis. The other parameters are the same as in Fig. 8.15.

As can be seen from Fig. 8.18 the torque acting on a circular non absorbing particle in relation to the centre is equal to zero when it is displaced along the Y axis, and varies with the displacement of point A up by 0.25 μm from the centre of the particle along the Y axis. If to the function, describing the dielectric permittivity of the particles, we add an imaginary component (the absorbing particle), the torque on the circular particle in relation to the centre will be different from zero at the displacement of the particle identical to that in Fig. 8.18. Figure 8.19 shows a plot of the dependence of the torque acting on a circular particle with $\varepsilon_2 = 2 + 1i$ with respect to the centre of the particle. The other parameters are the same as in the calculation of the graph in Fig. 8.18.

As can be seen from Fig. 8.19, the torque acting on a circular absorbing particles relative to the centre is not equal to zero when the particle is displaced from the optical axis.

Figure 8.20 is a plot of the dependence of the torque M_x acting on an elliptical dielectric ($\varepsilon_2 = 2 + 0i$) and absorbing microparticles ($\varepsilon_2 = 2 + 0i$), on the angle of

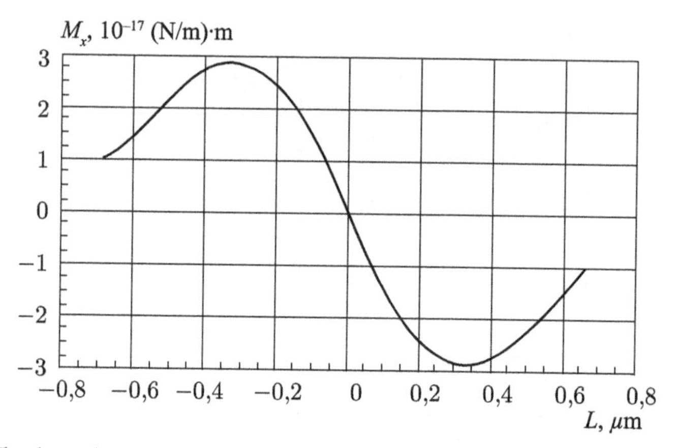

Fig. 8.19. The dependence of the torque on circular displacement L of the absorbing particles along the Y axis with respect to point A (0, 0).

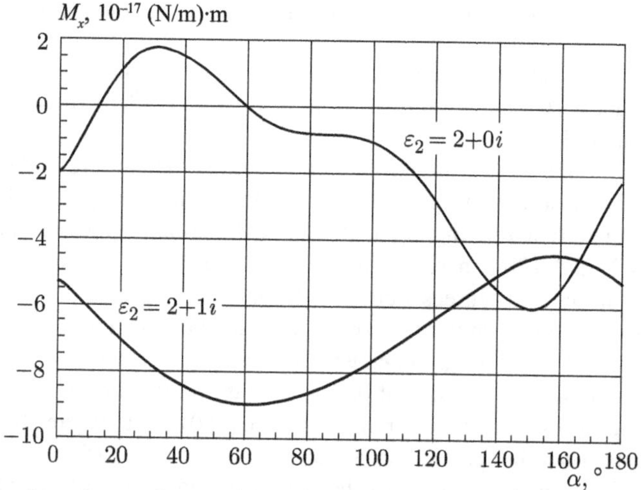

Fig. 8.20. The dependence of torque M_x on the elliptical particle on the rotation angle α of the particle.

rotation of the particle α. The particle is located in the coordinates (0, 0.25) with respect to the beam waist, the moment was calculated relative to the centre of the ellipse. The other simulation parameters are the same as in Fig. 8.15a.

As can be seen from Fig. 8.20, the addition of the imaginary part to the function of dielectric permittivity of the particle increases the scattering strength, due to which at any angle α the torque is non-zero and is directed counterclockwise ($M_x < 0$).

Figure 8.21a shows the results of the calculation of the dependence of the torque M_x acting on the elliptical particle, located at the centre of the Gaussian beam, on the angle of rotation α. All parameters are the same as in Fig. 8.15a, the particle is taken only half the size (the smaller diameter of the ellipse 0.25 μm, larger diameter 0.5

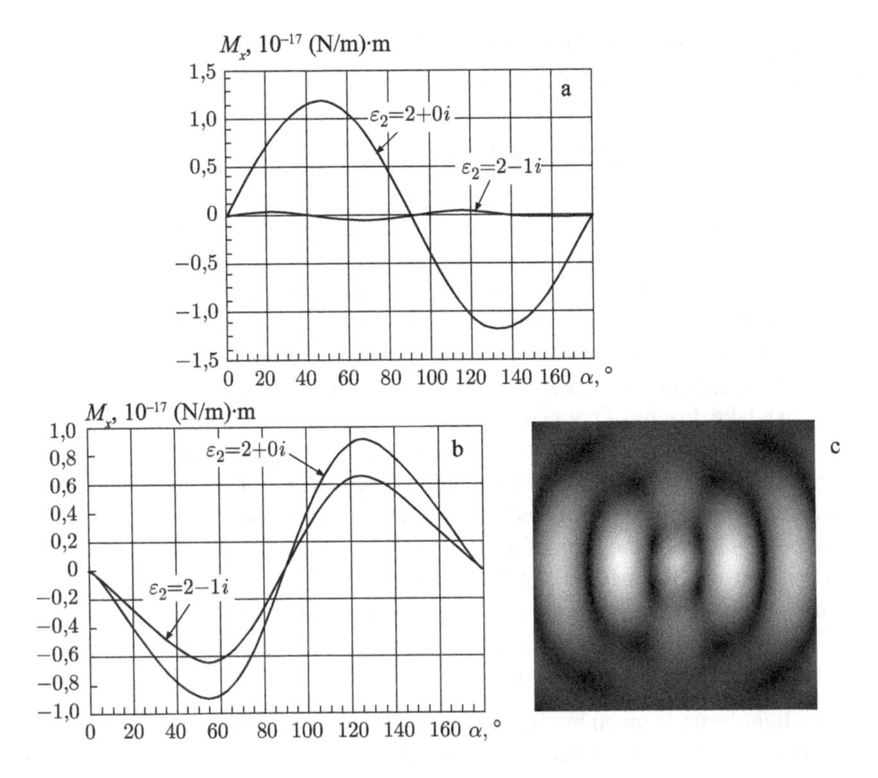

Fig. 8.21. a) The dependence of the torque M_x on the angle α for a single Gaussian beam b) the dependence of torque M_x on angle α for the two oppositely directed Gaussian beams, c) the amplitude of the field $|E_x|$ of two oppositely directed Gaussian beams on an elliptical particle.

μm), the dielectric constant and the particle is taken. Figure 8.21b shows the same chart, but for the two Gaussian beams, the first waist radius $\sigma_1 = 0.5$ μm, second $\sigma_2 = 0.6$ μm, the intensity of the two Gaussian beams was the same and equal to 50 mW/m. Diffraction of the oppositely directed beams on a particle at $\alpha = 45°$ is shown in Fig. 8.21c.

As can be seen from Fig. 8.21, by adding the imaginary part of dielectric permittivity if the cylinder is located in the centre of the waist of a Gaussian beam, the torque is strongly attenuated and has several zero points. It is also clear that the position of stable equilibrium when adding a second oppositely directed Gaussian beam will be observed at the location of the major axis of an elliptical particle along the axis Y, i.e. the elliptical particle tends to settle along the line of maximum intensity.

8.3. A geometrical optics method for calculating the force acting by light on a microscopic object

8.3.1. Description of the method

As can be seen in the previous section in the problem of calculating the electromagnetic forces in the approach is very capacious as regards the volume of calculations, as well as in the problems of calculating the forces acting on the micro-object in the light beam it is generally required to calculate the forces in some areas; consequently, the amount of computation grows as the square of the size of the area. In this case, for example, the calculation of forces to simulate the motion of the micro-object can take dozens of hours on a PC. In this regard, we use a simple geometrical optics approach, which gives approximate results. The geometrical optics approach has been used in the calculation of the simplest optical traps [32]. However, this and other studies generally considered some of the simpler cases in which restrictions are imposed either on the form of the micro-object [32, 86], or the shape of the beam [78]. We consider the method of calculating the forces acting on the micro-object of arbitrary shape in a light beam with a given distribution of intensity and phase [15].

Consider the micro-object of arbitrary shape in the light beam. We assume that the observed number of conditions:

1. The light beam is given by functions of intensity $I(x, y)$ and phase $\varphi(x, y)$.
2. The micro-object is bounded by two surfaces: the top, which is given by the function $f_1(x, y)$ and the bottom, which is given by the function $f_2(x, y)$ (Fig. 8. 22). The functions $f_1(x, y)$ and $f_2(x, y)$ are unique.
3. Micro-object is moving in the plane xy (however, this method is easy to calculate the force extends to three-dimensional motion).
4. The light beam is incident on the micro-object vertically from top to bottom.

Unit vectors $\vec{\mathbf{a}}\left(a_x,a_y,a_z\right)$, $\vec{\mathbf{b}}\left(b_x,b_y,b_z\right)$, $\vec{\mathbf{c}}\left(c_x,c_y,c_z\right)$ define the direction of the incident and refracted rays. The vector $\vec{\mathbf{a}}\left(a_x,a_y,a_z\right)$ is determined by the function $\varphi(x, y)$. This vector must always be perpendicular to the wavefront.

The components of the force **F** of a single beam on the micro-object are determined by the formula

$$F_x = \frac{N}{c}\left(a_x - c_x\right),$$
$$F_y = \frac{N}{c}\left(a_y - c_y\right), \tag{8.19}$$

where N is the power of the beam, c is the velocity of light.

For the whole beam, this formula takes the form

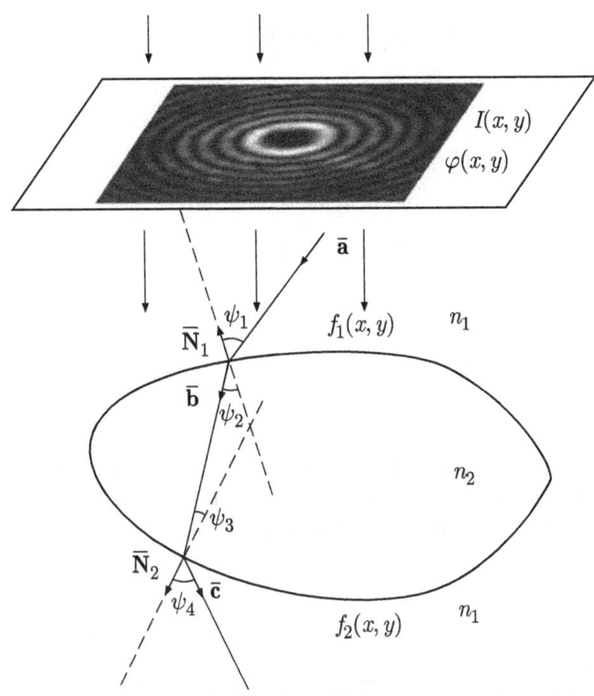

Fig. 8.22. Scheme of refraction of light rays on a micro-object.

$$F_x = \frac{1}{c}\iint_{\Omega} I(x,y)\left(a_x - c_x\right) dx\,dy,$$
$$F_y = \frac{1}{c}\iint_{\Omega} I(x,y)\left(a_y - c_y\right) dx\,dy, \tag{8.20}$$

where Ω is the region of maximum cross-sectional area of a microscopic object in a plane perpendicular to the direction of propagation of the light beam.

It should be borne in mind that the direction vector for the refracted output beam depends on the direction vector of the incident beam

$$F_x = \frac{1}{c}\iint_{\Omega} I(x,y)\left(a_x - c_x(\vec{\mathbf{a}})\right) dx\,dy,$$
$$F_y = \frac{1}{c}\iint_{\Omega} I(x,y)\left(a_y - c_y(\vec{\mathbf{a}})\right) dx\,dy. \tag{8.21}$$

The dependence $\vec{\mathbf{c}}(\vec{\mathbf{a}})$, i.e. direction of the beam after the micro-object, depending on the initial beam direction, can be determined from the following relations

$$
\begin{aligned}
&\left(\vec{\mathbf{N}}_1,-\vec{\mathbf{a}}\right)=\cos\left(\psi_1\right),\\
&\left(\vec{\mathbf{N}}_1,-\vec{\mathbf{b}}\right)=\cos\left(\psi_2\right),\\
&\left(\vec{\mathbf{a}},\vec{\mathbf{b}}\right)=\cos\left(\psi-\psi_{21}\right),\\
&n_1\sin\psi_1=n_2\sin\psi_2,\\
&\left(\vec{\mathbf{N}}_2,\vec{\mathbf{b}}\right)=\cos\left(\psi_3\right),\\
&\left(\vec{\mathbf{N}}_2,\vec{\mathbf{c}}\right)=\cos\left(\psi_4\right),\\
&\left(\vec{\mathbf{b}},\vec{\mathbf{c}}\right)=\cos\left(\psi_3-\psi_4\right),\\
&n_2\sin\psi_3=n_1\sin\psi_4,
\end{aligned}
\tag{8.22}
$$

where n_1, n_2 are the refractive indices of the medium and micro-object, $\vec{\mathbf{N}}_1$ and $\vec{\mathbf{N}}_2$ are the normal vectors to the surfaces $f_1(x, y)$ and $f_2(x, y)$, ψ_1, ψ_2, ψ_3, ψ_4 are the angles of incidence and refraction at the surfaces. By simple transformations we obtain from (8.22)

$$
b_x=\frac{A_1^2}{A_1^2+A_2^2}\left[\left(\frac{A_3A_2}{A_1}+K_1K_2\right)+\sqrt{\left(\frac{A_3A_2}{A_1^2}+K_1K_2\right)^2-\left(1+\frac{A_2^2}{A_1^2}+K_1^2\right)\left(\frac{A_3^2}{A_1^2}+K_1^2-1\right)}\right].
\tag{8.23}
$$

Here we have introduced a number of intermediate symbols which can significantly simplify the writing of the formula:

$$
A_1=a_y-\frac{N_{1y}a_z}{N_{1z}},\qquad A_2=a_x-\frac{N_{1x}a_z}{N_{1z}},
$$

$$
A_3=\frac{a_z\cos\psi_2}{N_{1z}}+\cos(\psi_1+\psi_2),
$$

$$
K_1=\frac{\cos\psi_2}{N_{1z}}-\frac{N_{1y}A_3}{A_1},\qquad K_2=\frac{N_{1x}}{N_{1z}}-\frac{N_{1y}A_2}{A_1}.
$$

The same calculation procedure is used to determine other components of the vectors $\vec{\mathbf{b}}\left(b_x,b_y,b_z\right)$ and $\vec{\mathbf{c}}\left(c_x,c_y,c_z\right)$. Since the cumulative record of these formulas is very large and is similar to the formula (8.23), it will not be presented here.

To determine the components of the direction vector $\vec{\mathbf{a}}(a_x,a_y,a_z)$ we need to use the phase $\varphi(x, y)$ of the light beam. As mentioned above, the direction vector must always be perpendicular to the wavefront. Then the direction vector is described by the following relation

$$\vec{\mathbf{a}} = \frac{\operatorname{grad}\varphi(x,y)}{|\operatorname{grad}\varphi(x,y)|}. \tag{8.24}$$

For normal vectors we can write the relation

$$\vec{\mathbf{N}}_1 = \left(\frac{\frac{\partial f_1}{\partial x}}{\sqrt{\left(\frac{\partial f_1}{\partial x}\right)^2 + \left(\frac{\partial f_1}{\partial y}\right)^2 + 1}}, \frac{\frac{\partial f_1}{\partial y}}{\sqrt{\left(\frac{\partial f_1}{\partial x}\right)^2 + \left(\frac{\partial f_1}{\partial y}\right)^2 + 1}}, -\frac{1}{\sqrt{\left(\frac{\partial f_1}{\partial x}\right)^2 + \left(\frac{\partial f_1}{\partial y}\right)^2 + 1}} \right),$$

$$\vec{\mathbf{N}}_2 = \left(\frac{\frac{\partial f_2}{\partial x}}{\sqrt{\left(\frac{\partial f_2}{\partial x}\right)^2 + \left(\frac{\partial f_2}{\partial y}\right)^2 + 1}}, \frac{\frac{\partial f_2}{\partial y}}{\sqrt{\left(\frac{\partial f_2}{\partial x}\right)^2 + \left(\frac{\partial f_2}{\partial y}\right)^2 + 1}}, -\frac{1}{\sqrt{\left(\frac{\partial f_2}{\partial x}\right)^2 + \left(\frac{\partial f_2}{\partial y}\right)^2 + 1}} \right). \tag{8.25}$$

Similarly, we determine the parameters of the reflected light rays (Fresnel reflection). Using (8.21)–(8.25), we can calculate the force acting on an arbitrarily shaped micro-object from an arbitrary light beam. Knowing the mechanical properties of the medium in which the micro-object is located, we can also simulate the motion of a microscopic object. In modelling the motion of a microscopic object in the light beam was solved by a system of equations of motion.

$$\begin{cases} m\dfrac{d\mathbf{v}}{dt} = \mathbf{F}_l + \mathbf{F}_f, \\ \dfrac{d\mathbf{r}}{dt} = \mathbf{v}, \end{cases}$$

where $\mathbf{F}_l$ is the force acting on the micro-object from the light beam, $\mathbf{F}_f$ is the force of viscous friction. Based on the above proposed method software has been developed that allows one not only to calculate the force of the light on the microscopic object in a given light field, but also to simulate the motion of a microscopic object in a given environment. Figure 8.23 shows the interface of the software.

However, the geometrical optics approach for the micro-objects comparable in size to the wavelength usually gives very inaccurate results where we need to calculate, for example, the intensity behind a microscopic object. It is necessary to verify how we apply this approach to calculate the force of light. To do this, we compare the values of force obtained by two methods: geometrical optics, which has been described in this section, and the electromagnetic method, which was described in Section 8.1.

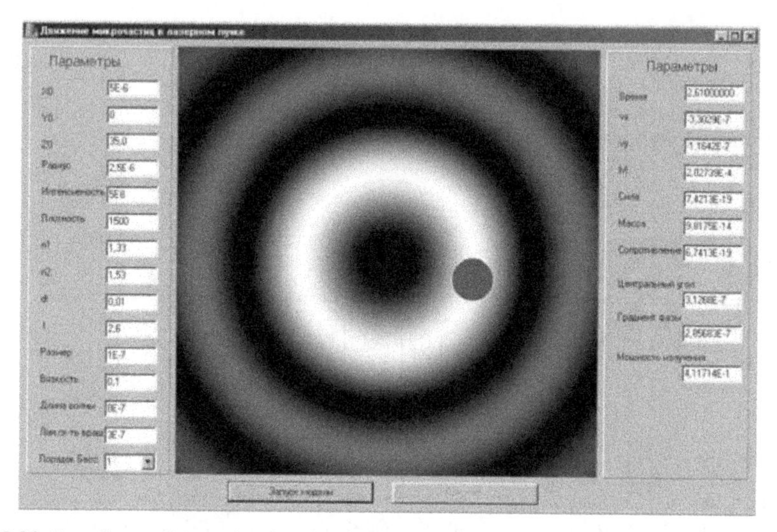

Fig. 8.23. Interface software for the simulation of the motion of microscopic objects in the light field.

8.3.2. Comparison of results of calculations by geometrical optics and electromagnetic methods

The method of calculating the force (8.19)–(8.25) allows us to calculate the effect of light also for the cylindrical micro-object [15]. A non-paraxial Gaussian beam was approximated by a system of rays with the parameters on the same beam shown in Fig. 8.24. A cylindrical object with a diameter of 1 μm and a refractive index of 1.41 was shifted across the axis of the Gaussian beam, with the values of the forces F_x and F_y calculated in each position. Figure 8.24 shows approximation of a Gaussian beam by a system of rays and refraction of the rays by this system at different positions of the cylindrical micro-object with respect to the optical axis.

For a cylindrical micro-object we calculated the force of light in the Gaussian beam. Figure 8.25a is a plot of the dependence of the projections of the force F_z on the displacement with respect to the waist along the propagation axis of the beam, superimposed on the graph presented in Section 8.1. Figure 8.25b is a plot of the dependence of projections of force F_y on the displacement relative to the axis of beam propagation, superimposed on the graph presented in Section 8.1 [7–9].

For the longitudinal force (Fig. 8.25a) there is an area where the standard deviation of force, obtained in the geometrical optics approximation, from the force calculated in the framework of the electromagnetic approach is no more than 0.1. As seen in Fig. 8.25b, for the shear force there is also a region in which the standard deviation is not more than 0.1.

To determine the accuracy, a standard formula was used for calculating the standard deviation (S.K.O.):

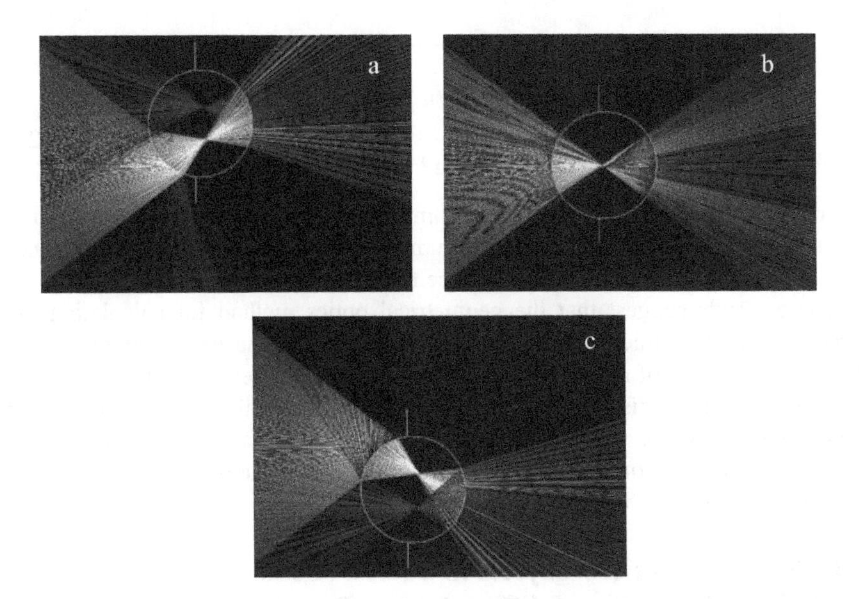

Fig. 8.24. Refraction of beams on a cylindrical micro-object 1 μm in diameter in different positions relative to the beam axis, (a) the displacement by 0.5 μm upwards, (b) exactly on the beam axis, (c) the displacement by 0.5 μm downward.

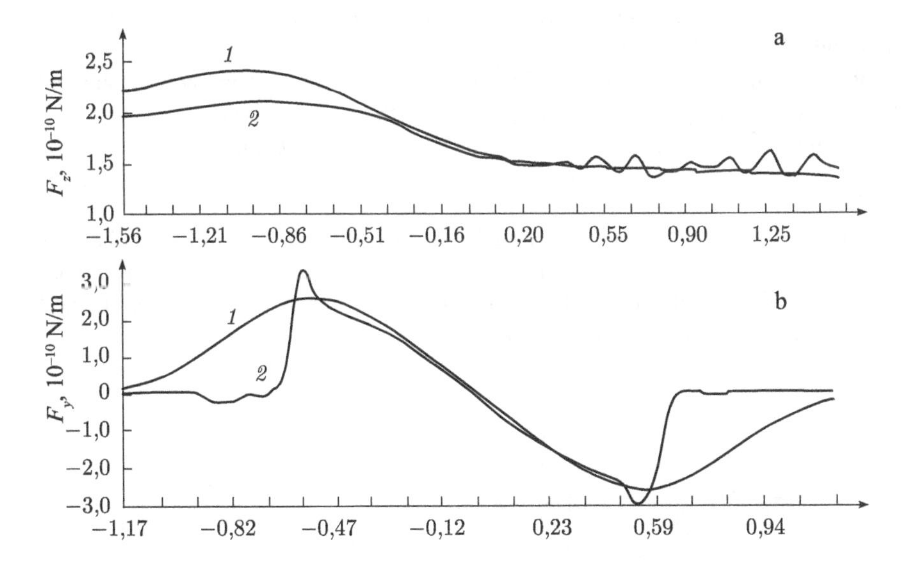

Fig. 8.25. a) The dependence of force F_z on the displacement L of the object along the Z axis through the centre of the waist ($Y = 0$), b) dependence of force F_y on the displacement L of the object along the Y axis through the centre of the waist ($Z = 0$) (1 – exact calculation, 2 – calculation by geometrical optics .)

$$\sigma = \frac{\sqrt{\frac{1}{N}\sum_{i=1}^{N}(f_{1i} - f_{2i})^2}}{\langle f_2 \rangle}, \tag{8.26}$$

where f_{1i} is the sampling point of the compared function, f_{2i} is the discretization point of the reference function, N is the number of discretization points of the function, $\langle f_2 \rangle$ is the mean value of the reference function.

These results suggest that the geometrical optics method for calculating the forces acting on the micro-object from the light beam has the error of no more than 0.1, except in cases of no practical interest, such as displacement of more than 0.5 μm from the beam axis in Fig. 8.25b. In this case, the method has a much lower computational complexity than the exact methods of calculation, which allows not only to calculate the forces acting on the micro-object in the light field, but also to simulate the motion of the micro-object.

8.4. Rotation of micro-objects in a Bessel beam

8.4.1. Transformation of diffractionless Bessel beams

In optical systems for the rotation of micro-objects with the help of a Bessel beam (BB) [28, 52, 73, 92] the light energy is concentrated using a ring BB 'squeezed' through a spherical lens. However, it appears that BB has the property to maintain its diameter near the axicon or DOE, loses this property in imaging using a spherical lens and begins to disperse.

We show that the image of the diffractionless BB produced using with a spherical lens leads to a divergent BB. Figure 8.26 shows the optical scheme.

As an initial function we choose the zero-order Bessel beam:

$$\Psi_0(r) = J_0(\alpha r). \tag{8.27}$$

To find how the function Ψ_0 is transformed by a lens, we need to simulate the propagation of the beam in free space over distance a by Fresnel transform, then multiply by the complex transmission function of the lens with a focal length f, and again apply the Fresnel transform at distance z:

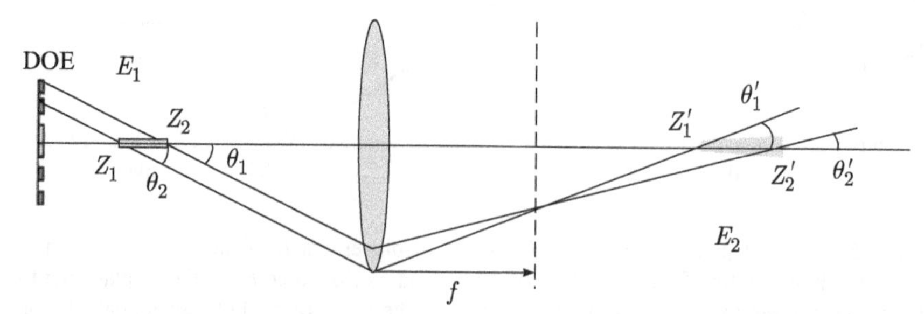

Fig. 8.26. The optical circuit for imaging the Bessel beam, used to manipulate microscopic objects.

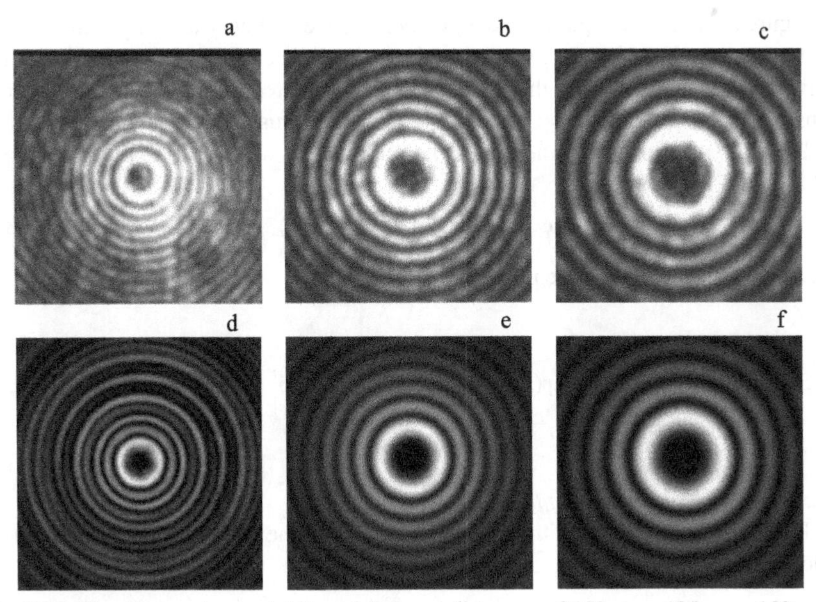

Fig. 8.27. The divergent paraxial Bessel beam at distances of 100 mm, 125 mm, 150 mm, respectively (a), (b), (c) (experiment), (d), (e), (f) (theory).

$$\Psi(u,v,z)=\left(\frac{k}{2\pi i}\right)^2\frac{1}{az}$$
$$\times\int_{-\infty}^{\infty}\int_{-\infty}^{\infty}\int_{-\infty}^{\infty}\int_{-\infty}^{\infty}\Psi_0(r)\exp\left\{\frac{ik}{2}\left[\frac{(x-\xi)^2}{a}+\frac{(y-\eta)^2}{a}-\frac{\xi^2+\eta^2}{f}+\frac{(\xi-y)^2}{z}+\frac{(\eta-v)^2}{z}\right]\right\}\mathrm{d}x\,\mathrm{d}y\,\mathrm{d}\xi\,\mathrm{d}\eta. \tag{8.28}$$

The distance z is related to the distance a by the lens formula. Equation (8.28) uses the transmission function of the paraxial lens in the form of:

$$\tau(\xi,\eta)=\exp\left[-\frac{ik}{2f}\left(\xi^2+\eta^2\right)\right]. \tag{8.29}$$

The evaluation of the integral (8.28) we get:

$$\Psi(\rho,z)=\frac{f}{(z-f)}J_0\left(\frac{\alpha f\rho}{f-z}\right)\exp\left[-i\frac{\alpha^2}{2k}\left(a+\frac{fz}{f-z}\right)+i\frac{k\rho^2}{2(f-z)}\right], \tag{8.30}$$

where $\rho^2 = u^2 + v^2$.

Equation (8.30) shows that the BB (8.27) like the non-paraxial beam diverges for $z > f$. This is due to the fact that the lens produces a divergent parabolic wave front in the BB. On the contrary, the beam diverges before the focal plane.

Figure 8.27 shows the results of the experiment compared with the theoretical results for the Bessel function of the fifth order; the experiments were caried out used a lens with a focal length f = 50 mm, the distance from the DOE to the lens 200 mm, a He–Ne laser with a wavelength of 0.633 μm.

Figure 8.27 shows a qualitative agreement between theory and experiment.

It may seem that when imaging with a lens, the diffractionless BB (8.27) becomes a divergent BB (8.30) due to the fact that it was calculated using the paraxial Fresnel transformation. But it is not so. It can be shown that the use of Fresnel transformation for BB (8.27) retains its diffractionless nature. Indeed, we choose as the initial light field the BB in the form:

$$\Psi_0(r,\varphi,z=0)=J_n(\alpha r)\exp(in\varphi), \tag{8.31}$$

Then at a distance z from the plane $z = 0$, we obtain:

$$\Psi(\xi,\eta,z)=\frac{k}{2\pi iz}\int\limits_0^{\infty}\int\limits_0^{2\pi}J_n(\alpha r)\exp(in\varphi)\exp\left[\frac{ik}{2z}\left(r^2+\rho^2\right)\right]\exp\left[-\frac{ik}{z}r\rho\cos(\theta-\varphi)\right]r\,\mathrm{d}r\,\mathrm{d}\varphi, \tag{8.32}$$

where $\rho^2 = \xi^2 + \eta^2$, $=$ arctg (η/ξ).

Replacing the integral with respect to φ(8.32) by the Bessel function of the n-th order, instead of (8.32) we get:

$$\Psi(\rho,\theta,z)=\frac{(-i)^{n+1}k}{z}\exp\left(\frac{ik}{2z}\rho^2\right)\exp(in\theta)\int\limits_0^{\infty}J_n(\alpha r)J_n\left(\frac{kr\rho}{z}\right)\exp\left(\frac{ik}{2z}r^2\right)r\,\mathrm{d}r. \tag{8.33}$$

The integral in (8.33) can be calculated, then instead of (8.33) we get:

$$\Psi(\rho,\theta,z)=\exp\left(-i\frac{z\alpha^2}{2k}\right)J_n(\alpha\rho)\exp(in\theta). \tag{8.34}$$

From (8.34) shows that the Frenel transforma preserves the original paraxial non-diverging BB (with the accuracy up to a phase factor) [19, 62]:

$$\Psi(\rho,\theta,z)=\exp\left(-i\frac{z\alpha^2}{2k}\right)\Psi_0(\rho,\theta). \tag{8.35}$$

Thus, in this section it is shown that the paraxial BB keeps its diameter in propagation in space, but converts into convergent and divergent light BB with a spherical lens. This property of BB should be considered in optical manipulation of micro-objects.

8.4.2. Umov–Poynting vector for a non-paraxial 2D vector Bessel beam

In the two-dimensional case we obtain a simpler relation between the scalar and vector BBs. In the two-dimensional case, the expansion of a complex function satisfying the Helmholtz equation in plane waves has the form [75]:

$$\Psi(x,z)=\int_{-\infty}^{\infty}\Psi_0(t)\exp\left[ik\left(xt+z\sqrt{1-t^2}\right)\right]\mathrm{d}t, \tag{8.36}$$

where z is optical beam axis (on the y-axis there are no changes $\partial/\partial y=0$).

If we select $\Psi_0(t)$ in the form of:

$$\Psi_0(t)=\frac{(-i)^n\exp\left[in\arccos\left(\frac{t}{a}\right)\right]}{2\pi\sqrt{a^2-t^2}}\operatorname{rect}\left(\frac{t}{a}\right), \tag{8.37}$$

then substituting (8.37) into (8.36) we obtain an expression for the non-paraxial two-dimensional beam which at $z=0$ coincides with the Bessel beams:

$$\Psi(x,z)=\frac{(-i)^n}{2\pi}\int_{-a}^{a}\frac{\exp\left[in\arccos\left(\frac{t}{a}\right)\right]}{\sqrt{a^2-t^2}}\exp\left[ik\left(xt+z\sqrt{1-t^2}\right)\right]\mathrm{d}t. \tag{8.38}$$

From Eq. (8.38) at $z=0$ and after the change $t=a\cos\varphi$, we obtain

$$\Psi(x,z=0)=\frac{(-i)^n}{2\pi}\int_{0}^{\pi}\exp(in\varphi)\exp(ika\cos\varphi)\mathrm{d}\varphi=J_n(kax). \tag{8.39}$$

To get a compact notation for the Bessel beam at any z we can write Eq. (8.36) as

$$\Psi(x,z)=\int_{-\pi}^{\pi}\Psi_0(\theta)\exp\left[ik(x\cos\theta+z\sin\theta)\right]d\theta. \tag{8.40}$$

Then, for

$$\Psi_0(\theta)=\frac{(-i)^n}{2\pi}\exp(in\theta), \tag{8.41}$$

instead of (8.38) we obtain

$$\Psi(x,z)=\frac{(-i)^n}{2\pi}\int_{-\pi}^{\pi}\exp(in\theta)\exp\left[ik(x\cos\theta+z\sin\theta)\right]d\theta=J_n(kr)\exp(in\varphi), \tag{8.42}$$

where $x=r\sin\varphi$, $y=r\cos\varphi$.

At $z=0$ from Eq. (8.42) we get:

$$\Psi(x,z=0)=J_n(kr)(i\operatorname{sgn}x)^n. \tag{8.43}$$

The scalar two-dimensional Bessel beam (8.42) can be regarded as a vector beam, assuming that $\Psi(x, z)$ is the projection on the y-axis of the vector of the electric field $E_y(x, z) = \Psi(x, z)$ for the TE-polarized monochromatic electromagnetic wave. This field is described by three quantities E_y, H_x, H_z, where H_x and H_z are the projections on the x and z axes of the vector of the strength of the magnetic field of the wave. Projections of the magnetic vector can be found through E_y:

$$H_x = \frac{1}{ik}\frac{\partial E_y}{\partial z},$$
$$H_z = \frac{i}{k}\frac{\partial E_y}{\partial x}. \tag{8.44}$$

With the help of equations (8.42) and (8.44) we can find an expression for the Umov–Pointing vector of the two-dimensional Bessel beam. Indeed, the Umov–Poynting vector is defined for complex vector fields in the form:

$$\mathbf{S} = \frac{c}{4\pi}\mathrm{Re}\left[\mathbf{E}\times\mathbf{H}^*\right], \tag{8.45}$$

where c is the speed of light.

In two dimensions, taking into account (8.42), instead of (8.44) we get:

$$S_x = \frac{ic}{4\pi k}\left(E_y\frac{\partial E_y^*}{\partial x} - E_y^*\frac{\partial E_y}{\partial x}\right) = \frac{c}{4\pi k}\mathrm{Im}\left(E_y\frac{\partial E_y^*}{\partial x}\right), \tag{8.46}$$

$$S_z = \frac{ic}{4\pi k}\left(E_y\frac{\partial E_y^*}{\partial z} - E_y^*\frac{\partial E_y}{\partial z}\right) = \frac{c}{4\pi k}\mathrm{Im}\left(E_y\frac{\partial E_y^*}{\partial z}\right). \tag{8.47}$$

Substituting (8.42) into (8.46) and (8.47), we obtain the projection of the Umov–Poynting vector for the two-dimensional Bessel beam with TE-polarization

$$S_x(x,z) = \frac{cnz}{4\pi k r^2}J_n^2(kr), \tag{8.48}$$

$$S_z(x,z) = \frac{-cnx}{4\pi k r^2}J_n^2(kr). \tag{8.49}$$

From equations (8.48) and (8.49) it follows that at $z = 0$ $S_x(x, z = 0) = 0$ and

$$S_z(x, z=0) = \frac{-cn}{4\pi\, kx}J_n^2(kx), \tag{8.50}$$

and at $x = 0$, $S_z(x = 0, z) = 0$ and

$$S_x(x=0, z) = \frac{cn}{4\pi\, kz}J_n^2(kz). \tag{8.51}$$

Arrows in Fig. 8.28 arrows indicate the direction of the Umov–Poynting vector, which follows from equations (8.48)–(8.51).

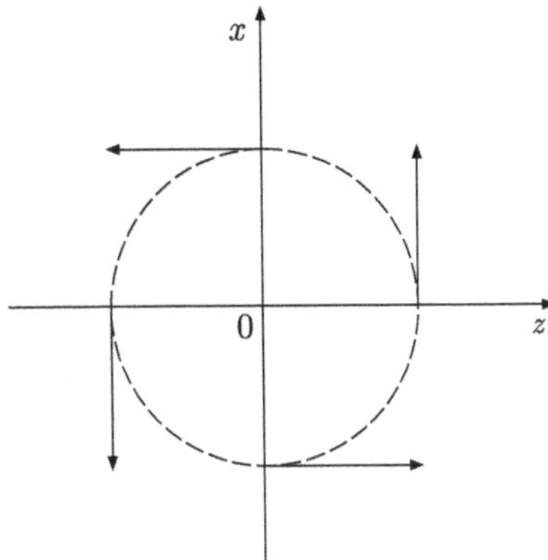

Fig. 8.28. The arrows indicate the direction of the Umov–Poynting vector at r = const for a 2D Bessel beam of the n-th order.

8.4.3. Umov–Poynting vector for the paraxial 3D vector Bessel beam

Let a monochromatic Bessel beam be linearly polarized along axis x:

$$\mathbf{E} = \mathbf{e}_x U(x,y,z) = \mathbf{e}_x J_n(\alpha r)\exp\left[i\left(\beta z + n\varphi\right)\right], \tag{8.52}$$

where $\alpha = k \sin\theta$, $\beta = \mathrm{k}\cos\theta$, θ is the angle of the conical wave to the optical axis z, (r, φ) are the polar coordinates. From Maxwell's equations

$$\begin{cases} \operatorname{rot}\mathbf{E} = ik\mathbf{H}, \\ \operatorname{rot}\mathbf{H} = -ik\varepsilon\mathbf{E}. \end{cases} \tag{8.53}$$

We find the rest of the projection of the electric and magnetic vectors:

$$\begin{cases} E_x = U, \\ E_y = \dfrac{1}{k^2\varepsilon}\dfrac{\partial^2 U}{\partial y \partial x}, \\ E_z = \dfrac{1}{k^2\varepsilon}\dfrac{\partial^2 U}{\partial z \partial x}, \end{cases} \quad \begin{cases} H_x = 0, \\ H_y = \dfrac{1}{ik}\dfrac{\partial U}{\partial z}, \\ H_z = \dfrac{i}{k}\dfrac{\partial U}{\partial y}. \end{cases} \tag{8.54}$$

It is seen that the projection H_y and H_z of the order k^{-1}, and E_y and E_z of the order k^{-2}, i.e. small compared with E_x.

The Umov–Poynting vector is defined by equation (8.45) with (8.54) takes the form:

$$S_z = \frac{c}{4\pi k}\operatorname{Im}\left[U^* \frac{\partial U}{\partial z}\right], \tag{8.55}$$

$$S_y = \frac{c}{4\pi k}\,\mathrm{Im}\left[U^*\frac{\partial U}{\partial y}\right], \tag{8.56}$$

$$S_x = \frac{c}{4\pi\varepsilon k}\,\mathrm{Im}\left[\frac{\partial^2 U}{\partial y \partial x}\frac{\partial U^*}{\partial y} + \frac{\partial^2 U}{\partial z \partial x}\frac{\partial U^*}{\partial y}\right]. \tag{8.57}$$

We substitute the expression for U from (8.52) into (8.55)–(8.57) and obtain the projection of the Umov–Poynting vector of the 3D vector paraxial Bessel beam:

$$S_z = \frac{c\beta}{4\pi k} J_n^2(\alpha r), \tag{8.58}$$

$$S_y = \frac{cnx}{4\pi k r^2} J_n^2(\alpha r), \quad x = r\cos\varphi, \tag{8.59}$$

$$\begin{aligned} S_x = \frac{-cn}{4\pi k\varepsilon}\Bigg\{ & \left(\frac{\beta^2 y}{k^2 r^2} + \frac{n^2 x^2 y}{k^2 r^6}\right) J_n^2(\alpha r) - \\ & -\left(\frac{\alpha y}{k^2 r^3} - \frac{2\alpha x^2 y}{k^2 r^5} + \frac{\alpha x y^2}{k^2 r^5}\right) J_n(\alpha r)\frac{\partial J_n(t)}{\partial t} + \\ & +\left(\frac{\alpha^2 y^3}{k^2 r^4} - \frac{\alpha^2 x^2 y}{k^2 r^4}\right)\left(\frac{\partial J_n(t)}{\partial t}\right)^2 + \frac{\alpha^2 x^2 y}{k^2 r^4} J_n(\alpha r)\frac{\partial^2 J_n(t)}{\partial t^2}\Bigg\}. \end{aligned} \tag{8.60}$$

Note that equation (8.58) and (8.59) are similar and almost identical with the equations (8.49) and (8.48), respectively.

If in (8.60) we leave only the terms proportional to k^{-1}, and the terms with k^{-2} and k^{-3} ignored, instead of (8.60), we obtain a simple expression:

$$S_x = \frac{-cn}{4\pi k\varepsilon}\left\{\frac{\beta^2 y}{k^2 r^2} J_n^2(\alpha r) + \left(\frac{\alpha^2 y^3}{k^2 r^4} - \frac{\alpha^2 x^2 y}{k^2 r^4}\right)\left(\frac{\partial J_n(t)}{\partial t}\right)^2 + \frac{\alpha^2 x^2 y}{k^2 r^4} J_n(\alpha r)\frac{\partial^2 J_n(t)}{\partial t^2}\right\}. \tag{8.61}$$

From Eq. (8.58)–(8.60) we see that at $x = 0$:

$$\begin{cases} S_y = 0, \\ S_x = \dfrac{-cn}{4\pi k\varepsilon}\left[\dfrac{\beta^2 y}{k^2 r^2} J_n^2(\alpha r) + \dfrac{\alpha^2 y^3}{k^2 r^4}\left(\dfrac{\partial J_n(t)}{\partial t}\right)^2\right]. \end{cases} \tag{8.62}$$

The sign S_x is determined by the product ny, with $n > 0$ and at $r = \text{const}$ projection S_x is aimed at in Fig. 8.29 (z-axis is directed toward the observer).

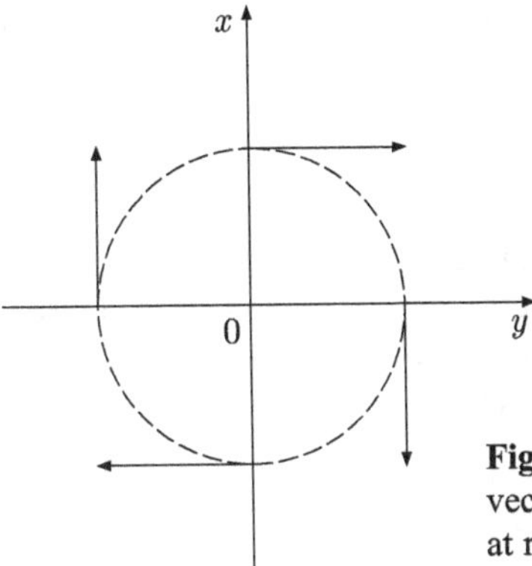

Fig. 8.29 The arrows indicate the direction of the Poynting vector in the cross section (x, y) 3D paraxial Bessel beam at r = const. The light is linearly polarized along the axis x.

At $y = 0$:

$$\begin{cases} S_x = 0, \\ S_y = \dfrac{cnx}{4\pi k r^2} J_n^2(\alpha r). \end{cases} \tag{8.63}$$

The sign S_y is determined by the product nx and for $n > 0$ the projection is shown in Fig. 8.29.

8.4.4. *The orbital angular momentum for a Bessel beam*

The orbital angular momentum of the electromagnetic field is given by [27]:

$$\mathbf{M} = [\mathbf{r} \times \mathbf{S}] = \left[\mathbf{r} \times \left\{\frac{c}{4\pi}\mathrm{Re}[\mathbf{E} \times \mathbf{H}]\right\}\right]. \tag{8.64}$$

The projection on the optical axis of the orbital angular momentum for a linearly polarized electromagnetic field calculated in the paraxial approximation has the form:

$$M_z = \frac{1}{4\pi kc}\left[y\,\mathrm{Im}\left(E\frac{\partial E^*}{\partial x}\right) - x\,\mathrm{Im}\left(E\frac{\partial E^*}{\partial y}\right)\right]. \tag{8.65}$$

For a linearly polarized Bessel beam

$$E_x = J_n(\alpha r)\exp(in\varphi)\exp(i\beta z), \quad \alpha^2 + \beta^2 = k^2 \tag{8.66}$$

projection on the z axis of the orbital angular momentum will be:

$$M_z = \frac{nJ_n^2(\alpha r)}{4\pi kc}. \tag{8.67}$$

The expression (8.67) up to a constant coincides with the first term in the equation obtained in [93].

8.4.5. DOE to form a Bessel beam

Diffractive optical elements can generate Bessel beams which retain the modal nature at a great distance along the propagation axis. Based on geometrical considerations, the distance at which the single-modal nature of Bessel light fields $J_n(\alpha r)$ exp $(in\varphi)$, is estimated by the following formula [45]:

$$z_{\max} = R\left[\left(\frac{2\pi}{\alpha\lambda}\right)^2 - 1\right]^{1/2}, \tag{8.68}$$

where R is the radius of the DOE, α is the parameter of the Bessel function.

In [92] it is shown that the formation of BBs by holographic optical elements, the maximum distance over which they maintain the character of their mode increases by about two times compared with the method of forming a BB with a narrow gap [44]. However, we need some distance from the plane of the holographic optical element so that the beam can form. Thus, the segment of the optical axis in which BB, formed by the final phase DOE, retains its modal nature, begins at some $z_{\min}$, required for a beam to form, and ends at $z_{\max}$, defined by the DOE radius R and the BB parameter α.

In [81] it is proposed to form a Bessel mode beam with a spiral zone plate, the transmission function of which is a function of:

$$\tau(r,\varphi) = \operatorname{sgn}\left(J_n\left(\alpha r\right)\right)\exp\left(in\varphi\right). \tag{8.69}$$

A helical DOE with transmittance (69) effectively forms a light field whose amplitude is proportional to the Bessel functions $J_n(\alpha r)$ exp $(in\varphi)$, near the optical axis in the interval $0 < z < Rk/\alpha$ [81]. At the same time, the DOE with transmittance (69) forms a light ring in the Fourier plane with a maximum intensity [48].

In calculating the phase of the DOE, for the formation of BB of the 5-th order we used the following parameters: $R = 3$ mm, $\lambda = 633$ nm, $\alpha = 44.5$ mm^{-1}. Figure 8.30a shows a template (600 × 600 samples), used at the University of Joensuu (Finland) to made a 16-gradation DOE (discretization step 10 μm). Figure 8.30b shows the central part of the DOE microrelief at a magnification of 50 (top view), and Fig. 8.30c – at a magnification of 200 (oblique vies). Pictures of the microrelief are obtained with an interferometer NEWVIEW 5000 of the firm Zygo.

The results of comparing the experimental formation of a Bessel beam of he 5-th and the numerical simulation based on the Fresnel integral transform are shown in Fig. 8.31. The fabricated phase DOE was illuminated by a collimated beam of an He–Ne laser. The resulting intensity distribution at different distances behind the DOE was recorded by a CCD-camera. Figure 8.31a–d (top row) shows the experimentally recorded intensity distributions at the following distances from the plane of the DOE: 300 mm (a), 400 mm (b), 500 mm (c) 600 mm (d). Figure 8.31e–h (bottom row) shows the corresponding patterns of numerical simulation [19].

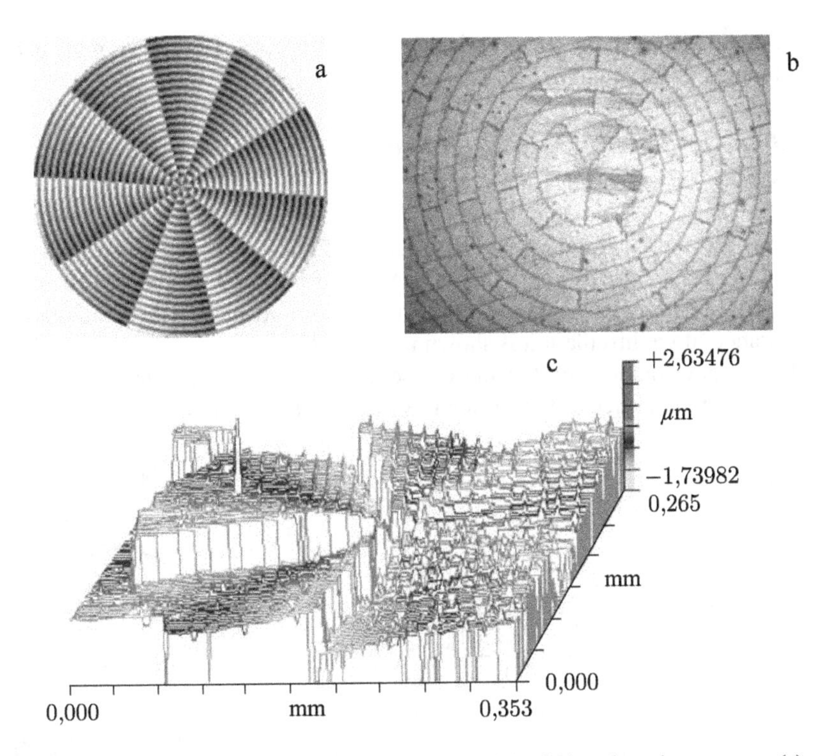

Fig. 8.30. The phase DOE that forms a Bessel beam of the fifth-order: phase pattern (a) and the form of the central part of the microrelief at a magnification of 50 times (b) and 200 times (s).

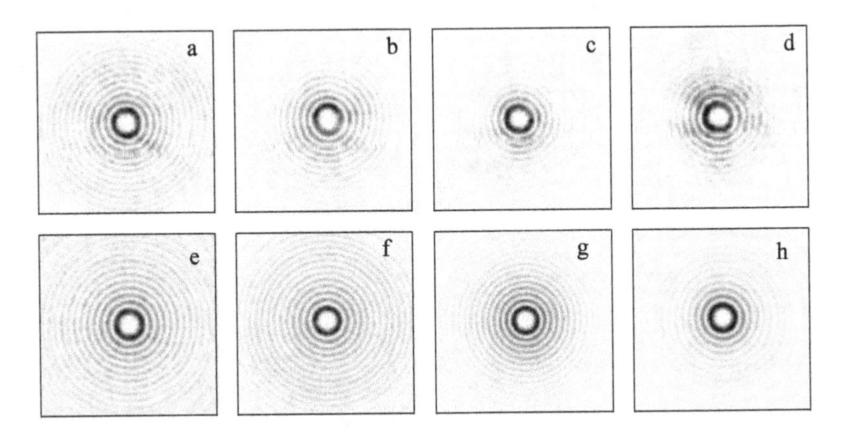

Fig. 8.31. Experimentally recorded intensity distribution (negative) in the cross-section at the following distances from the plane of the DOE: 300 mm (a), 400 mm (b), 500 mm (c) 600 mm (d), and the corresponding results of numerical simulation (e–h).

A comparison of the patterns in Fig. 8.31 shows a good agreement between theory and experiment.

8.4.6. Experimental study of movements of the micro-object in the Bessel beam

Experiments were carried out by the rotation of microscopic objects in the optical setup [14, 19] shown in Fig. 8.32a. The basis of the set is a modified microscope Biolam – M. A standard optical circuit was used for generating laser radiation. The appearance of the installation is shown in Fig. 8.32b.

In the development of the optical setup it was necessary to satisfy several conflicting requirements: firstly, for the most efficient focusing it was necessary

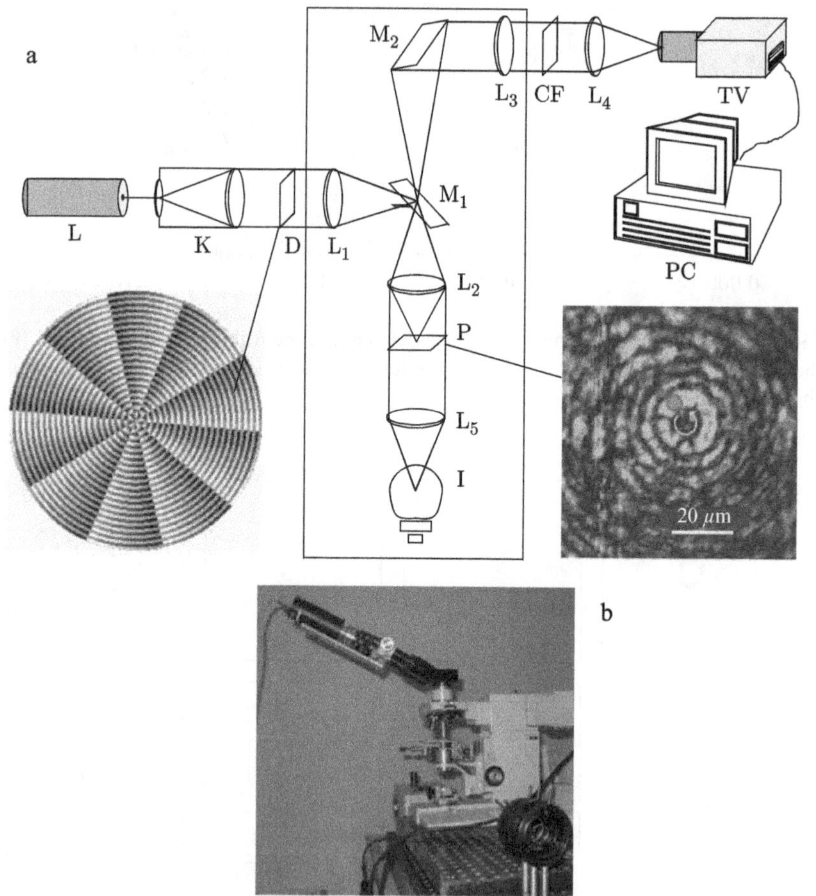

Fig. 8.32. Optical diagram of the experimental setup (a): L – argon laser, K – collimator, D – DOE, L_1 – corrective lens, M_1 – a semitransparent mirror of the microscope, M_2 – rotating mirror, L_2 – microscope objective, P – cell with microobjects, L_3 – the eyepiece of the microscope, CF – a red filter, TV – TV camera, L_4 – camera lens, L_5 – condenser of the illuminator, I – light fixture, picture of the experimental setup (b).

to use a microscope objective with high magnification, and secondly, the size of the DOE determined the size of the beam incident on the microscope objective, and for example, for a 90× microscope the beam size was significantly greater than the entrance aperture, which inevitably leads to a decrease in beam energy. In addition, the use of a microscope for focusing and image formation leads to the need to combine the focal and working planes of the microscope objective. Both of these problems were successfully solved with the help of corrective lenses L_1. To determine the minimum required beam power, the developed method for calculating the forces was used to define the minimum intensity of 2×10^8 W/m^2 at which motion is possible of a micro-object with a diameter of 5 μm, with a refractive index of 1.5 in the Bessel beam of the 5-th order. When using the 16× microscope objective, the power of the beam in the working plane is 90 mW. Given that the losses by reflection from the refractive surfaces of the focusing system are 55–60% (experimentally obtained value), it follows that the power of the beam at the output of the laser should be about 200 mW. At the same time it is taken into account that the central ring of BB receives no more than 30% of the energy (experimentally determined value).

The work of the installation will now be described. The argon laser beam travels from collimator K to the DOE D, which forms the fifth-order Bessel beam. The correction lens L_1 then forms the final beam, which then enters the optical system of the microscope (lenses L_1, L_2). The generated beam is imaged by a decrease in the cell with an aqueous suspension of micro-objects. Background illumination is provided by the lamp I, through the lens L_5. Lens L_2 (microscope objective 16×, 20×, 90×) is used for focusing and at the same time to form an image of the cell. Yeast cells were used as micro-objects. Figure 8.33 shows the different stages of movement of yeast cells trapped by the first light ring of the Bessel beam. Light filter CF in the experiment was chosen so that the micro-object could be seen but the beam not. The micro-object made a total of eight revolutions, and then was stuck to the bottom. The parameters of the experiment: the cell size 4.5×7 μm, the Bessel beam power 150–200 mW, the diameter of the trajectory 17 μm [19].

As an object for experiments with rotation in the light beam the yeast cells have two significant drawbacks:

1. The preparation of these micro-objects for experiment takes several hours;

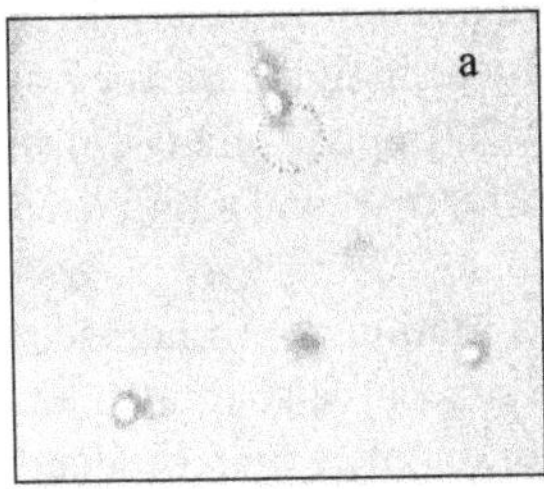

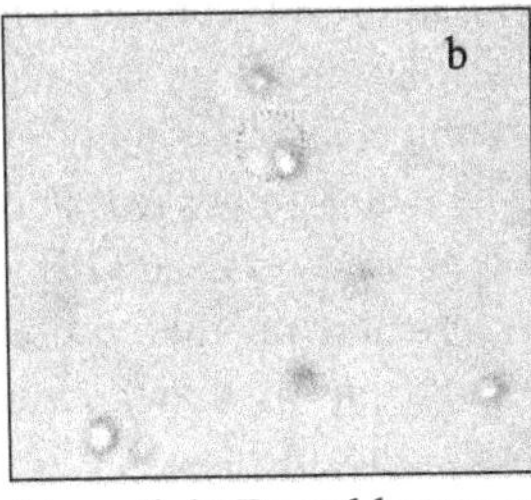

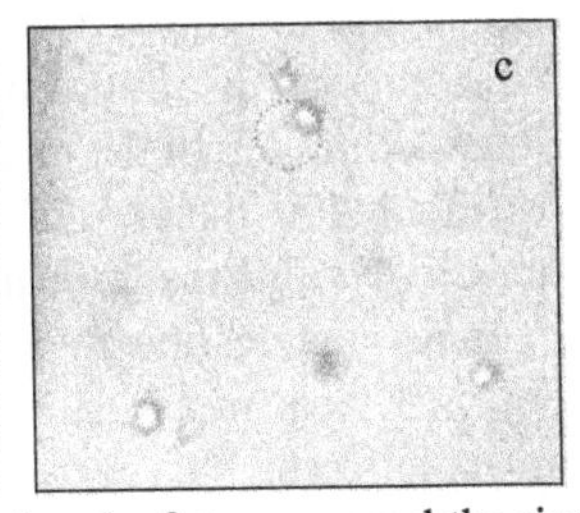

Fig. 8.33. Yeast cells are trapped by a light Bessel beams and make 8 turns around the ring with a diameter 17 μm (the first ring of the Bessel beam), a, b, c – stages of movement by 0.5 s. The trajectory is shown by the contour.

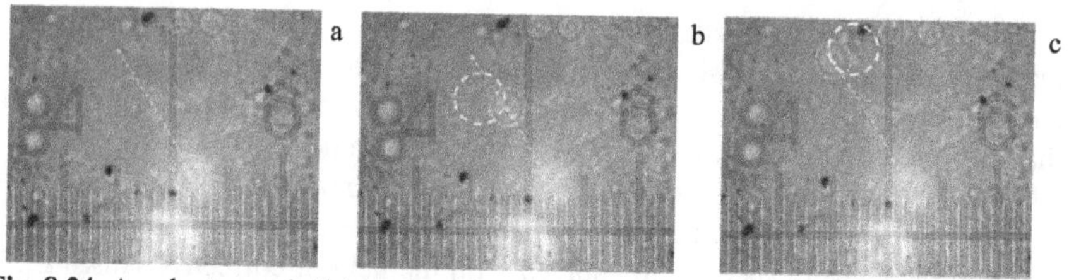

Fig. 8.34. A polystyrene ball is trapped by BB (diameter of the first light ring 3 μm) and moved by 30 μm in a straight line, a, b, c – movement through the stages by 2 s.

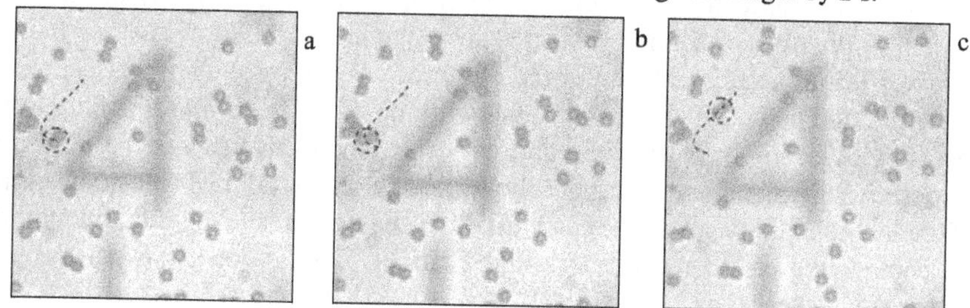

Fig. 8.35. Polystyrene beads are trapped by a Bessel beam and linearly moved by 50 μm, while rotating around the centre of the beam (4 turns), a–c – stages of movement after 1 s.

2. It is impossible to accurately determine the size and shape of yeast cells so it is also difficult to simulate their motion in light beams.

These deficiencies are not found in polystyrene balls, which are manufactured for use in chromatography. In addition, these balls are often used in the work of other experimenters, which facilitates the comparison of experimental results. Experiments were carried out with polystyrene balls with a diameter of 5 μm. They are made with good accuracy of ± 0.1 μm, so are a good target for simulation. It should be noted that the diameter of the balls of 5 μm is very large for experiments with rotation, usually other experimenters used balls with a diameter 1 μm (mass 125 times smaller), but we you think about the future practical use of the experimental results (e.g. in micromechanics), this size is the most suitable, because it is comparable with the characteristic dimensions of the micro-mechanical devices. In the next experiment, BB was focused so that the size of the first ring was 3 μm, which is smaller than the size of the microspheres, and such a beam could carry out the seizure of a micro-object and move it to 30 μm to the side. The micro-object was moved by displacement of the beam by turning the mirror by 1°. Stages of motion of the micro-object are shown in Fig. 8.34 (trapped micro-object is highlighted by the outline). The beam power was about 200 mW, and a microscope objective with a magnification of 90° was used [19].

It was interesting to experiment with the combined motion of a microscopic object. That is, moving the beam to ensure that the micro-object is also rotated. Such an experiment has been done. The phases of the motion of a pair of bonded microspheres are presented in Fig. 8.35. The parameters of this experiment: beam power 250 mW, microscope objective ×20. During the displacement by 50 μm the spheres made 4 turns, rotating as a whole [19].

8.5. Optical rotation using a multiorder spiral phase plate

An experiment with the rotation of microscopic objects [6, 18, 63, 64, 88, 90] was conducted using the optical system shown in Fig. 8.36. The laser light travelled through the collimator to the DOE D, which forms the laser beam with a set of optical vortices. Then, using the optical microscope system (lens L_6 and L_2), the laser beam can be focused into a cell containing an aqueous suspension of polystyrene microspheres. Background light is generated by the lamp I, with the use of lenses L_5. Lens L_2 (microscope objective 16×, 20×) focuses the illumination light and at the same time forms an image of the workspace. The laser beam is focused by the microobjective L6.

The scheme presented in Fig. 8.36 differs from the circuit in Fig. 8.28a by the fact that focusing and observation are carried out through different microscopes. At the same time focusing the laser beam is conducted from the bottom to minimize the friction force of the microscopic object on the bottom of the cell, but, unfortunately, imposes limitations on the power of the light beam (at a specific power the micro-objects are squeezed up and leave the working plane). Experiments with rotation of the micro-objects were carried out using a DOE, forming four optical vortices with the numbers of orders (±3, ±7). The phase of this DOE is shown in Fig. 8.37a. The central part of the relief of this DOE is presented in Fig. 8.37b. Figure 8.37c shows the intensity distribution of the element in the zone of Fraunhofer diffraction.

Figure 8.38 shows the various stages of the movement of polystyrene microspheres trapped by a laser beam in an optical ring (optical vortex of seventh-order). The

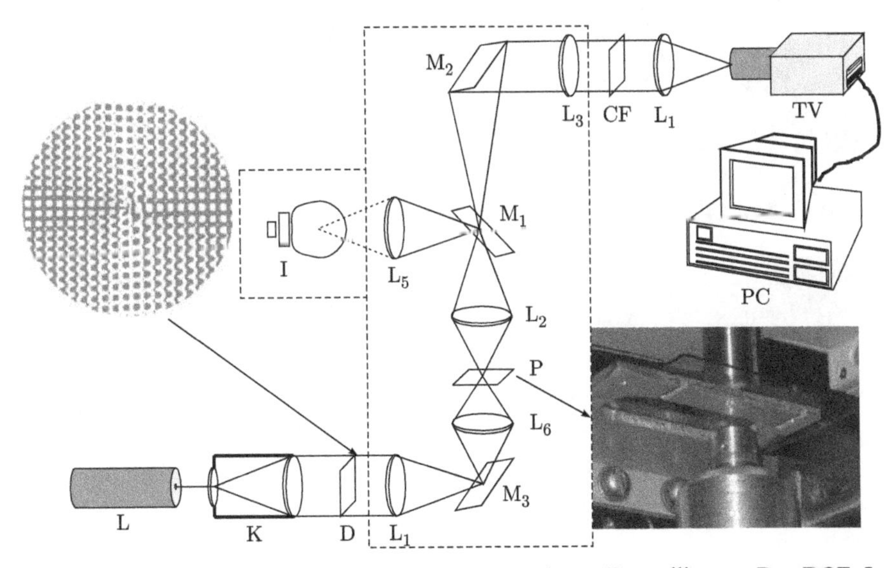

Fig. 8.36. The experimental optical system: L – argon laser, K – collimator, D = DOE, L_1 – corrective lenses, M_1 – semitransparent mirror of the microscope, M_2 – rotating mirror, L2 – microscope objective, P – cell with microspheres, L_3 – the eyepiece of the microscope, CF – red filter, TV – CCD-camera, L_5 – condenser illuminator, and I – light fixture.

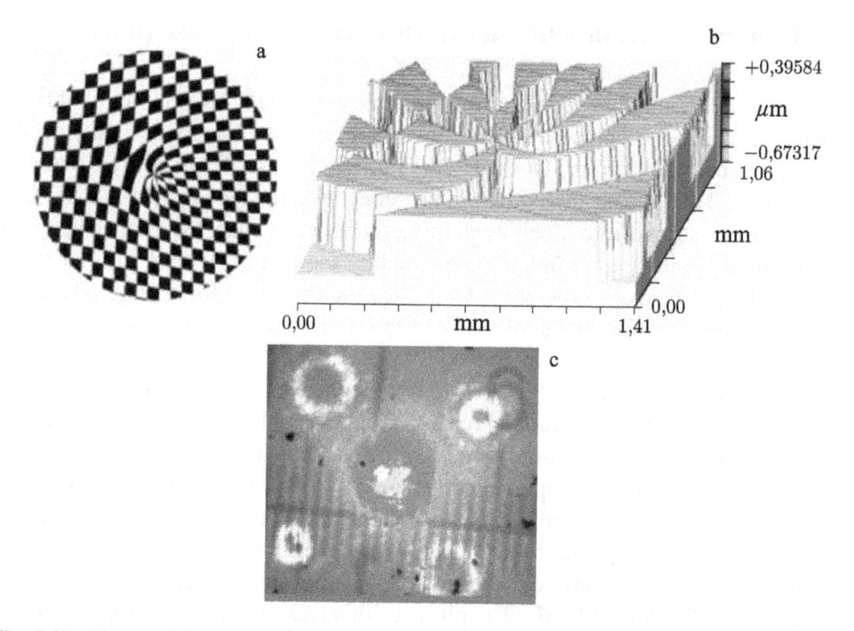

Fig. 8.37. Phase of the binary DOE to create optical vortices with –7, –3, 3, 7 orders of magnitude (a), the central part of the microrelief (b), the intensity distribution in the diffraction pattern (c).

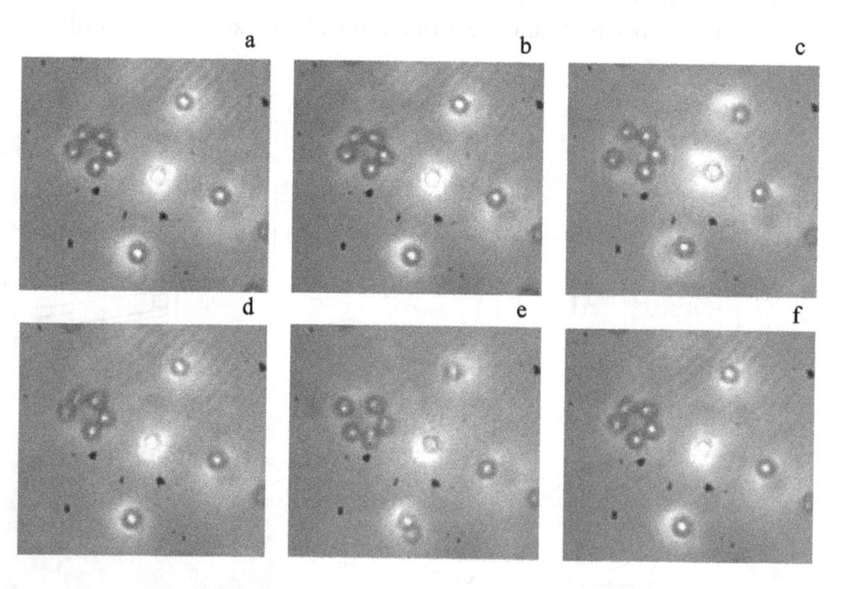

Fig. 8.38. The trapping and rotation of microscopic objects in the optical vortex of order 7, stage movement are shown in 2.5 s.

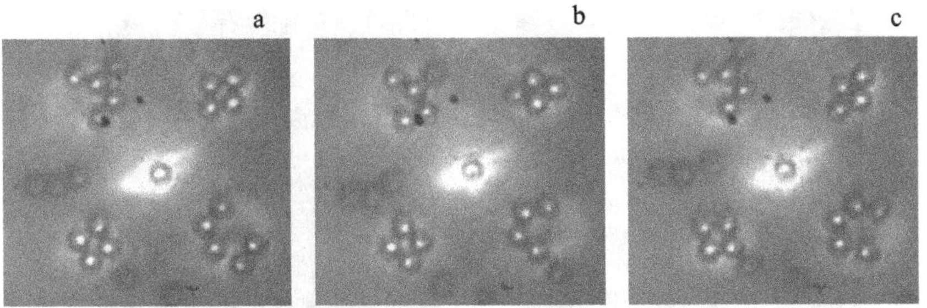

Fig. 8.39. The trapping and rotation of microspheres in optical vortices of 3rd, –3rd and 7th-order, stage of the movement are shown at intervals of 5 s.

diameter of the orbit was 12 μm. Microspheres were in the water. Focusing was carried out with a 20× microscope.

The light beam with the optical vortex simultaneously trapped and rotated a group of micro-objects. It should be noted that the light beams with optical vortices of high orders rotate the micro-objects more efficiently. As can be clearly seen in Fig. 8.38, the microsphere is trapped in a light beam with an optical vortex of order 3 but does not move, while a group of microspheres in the light beam from the optical vortex of seventh order rotates at an appreciable rate. Sophisticated experiments in which micro-objects were rotated in different diffraction orders were carried out.

Figure 8.39 shows the various stages of the movement of polystyrene beads, trapped by several optical vortices. Light beams with optical vortices trapping and at the same time the group of microspheres in different orders. Four microspheres were trapped in optical vortices of the 3rd and –3rd order, four microspheres in an optical vortex of –7th order, and five microspheres in an optical vortex of 7th order were trapped. In the optical vortex of 7th order microspheres did not rotate, apparently due to the fact that some of the microspheres adhered to the bottom of the cell.

Groups of microspheres rotate in different orders of the light beam with optical vortices of the 3rd and 7th orders. Moreover, Fig. 8.39 shows that the microspheres in the optical vortices of opposite sign rotate in opposite directions.

8.6. Rotation of microscopic objects in a vortex light ring formed by an axicon

Lithography technology allows us to produce binary DOEs [55]. However, the helical axicon phase function is not binary. Therefore, it is necessary to use a simple method of converting a grayscale function to a binary one, which is based on use of the carrier frequency. In this case, the transmission function of the binary phase axicon has the form:

$$\tau_{n\beta}(r,\varphi)=\text{sign}\left[\cos\left(\alpha r+n\varphi+\beta r\cos\varphi\right)\right]\text{rect}\left(\frac{r}{R}\right), \tag{8.70}$$

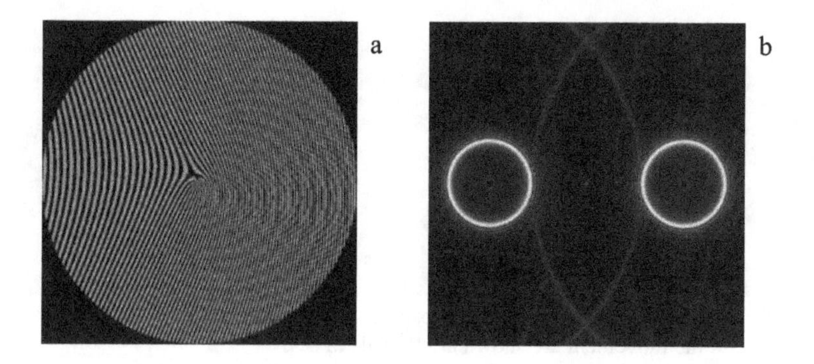

Fig. 8.40. a) the binary phase helical axicon of 10th order with a spatial carrier frequency, b) calculated diffraction pattern of a plane wave on the DOE shown in Fig. 8.40a.

where $\beta = 2\pi/T$ and T is the period of the carrier spatial frequency [69]. As is well known, the binary DOE creates two identical diffraction orders, each of them has an efficiency of about 41% [91]. For the spatial separation of the orders it is necessary to satisfy the condition $\beta > \alpha$. Therefore, the radii of the rings in the Fraunhofer diffraction pattern for the vortex axicon (spiral or helical) are approximated by the expression $\alpha f/k$. Figure 8.40a shows a binary phase DOEs, forming two identical rings with the same numbers $n = 10$, but with different signs. Figure 8.40b shows the calculated Fraunhofer diffraction pattern for the DOE shown in Fig. 8.40a. DOE has a radius $R = 2$ mm, the wavelength $\lambda = 532$ nm.

The axicon parameters $\alpha = 50$ mm^{-1}, the spatial carrier frequency $\beta = 100$ mm^{-1}, the focal length of the spherical lens $f = 420$ mm.

The phase in Fig. 8.40a was used in the manufacture of an amplitude photomask with a resolution of 3 μm using a circular laser writing station CLWS-200. Then, a DOE was produced by etching on a glass substrate 2.5 mm thick, with a refractive index of 1.5.

When illuminating the DOE with a radius $R = 2$ mm (Fig. 8.40) with a plane wave with wavelength $\lambda = 532$ nm, two identical light rings (each with an efficiency of about 41%) form in the focal plane of the lens $f = 460$ mm. The radial section of the ring, shown in Fig. 8.41, was measured with a CCD-camera. The produced DOE was used to trap and rotate polystyrene microspheres with a diameter of 5 μm. The DOE was illuminated by a collimated light beam of a solid-state neodymium laser with a wavelength of 532 nm and 500 mW power. A bright ring of radius 37.5 μm was formed in the focal plane of the microscope objective ($\times 40$). Figure 8.42 shows two successive shots of microspheres, separated by a time interval of ten seconds, ten polystyrene microspheres, moving on a light ring, can clearly seen.

8.7. Optical rotation in a double light ring

The DOEs used to produce a double ring were manufactured in three different ways: by electron-beam lithography with electron beam direct writing on the resist,

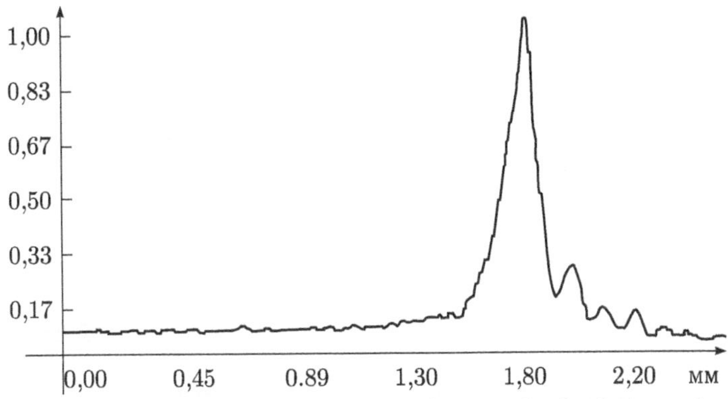

Fig. 8.41. Experimental radial section of the light ring in the focal plane of a lens (f = 460 mm).

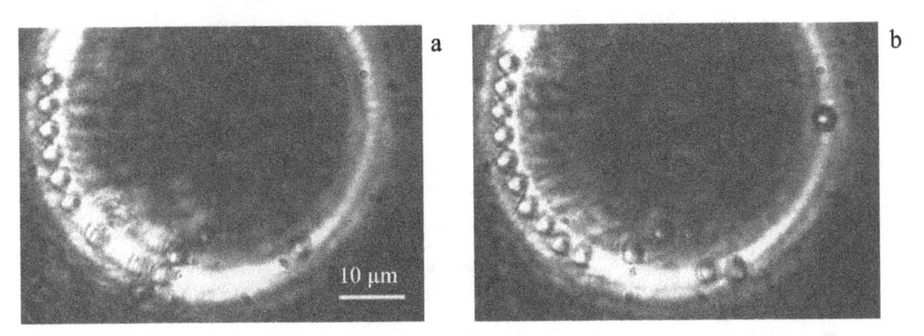

Fig. 8.42. Ten polystyrene beads with a diameter of 5 μm move along the bright ring with a radius of 37.5 μm, formed by a helical axicon with the number $n = 10$, with an average speed of about 4 μm/s. Figures (a) and (b) are separated by an interval of 10 s.

the technology of optical lithography using a binary photomask and wet etching of a glass substrate, and using the liquid crystal display or a dynamic spatial light modulator (SLM) [70].

8.7.1. Production of DOE by electron-beam lithography

The diffraction element was fabricated on a substrate of fused silica with a diameter 1 inch and a thickness of 3 mm. The stages of manufacture are shown in Fig. 8.43 First, a PMMA-resist with a thickness of 200 nm was deposited on the upper surface of the SiO_2 substrate. A 15 nm layer of Cu was then deposited on the substrate (Fig. 8.43a) for ensuring electrical conductivity. A cathode-ray machine Leica Lion LV1 was used for exposure. After exposure, the conductive layer was removed. Then, the resulting relief was coated with a 50 nm layer of Cr (Fig. 8.43c). Etching of the SiO_2 substrate was carried out step by step through reactive ion etching in the atmosphere CHF_3/Ar (Fig. 8.43d). After etching the Cr mask was removed by wet etching (Fig. 8.43f). The measured depth of the relief of the DOE was 578 nm.

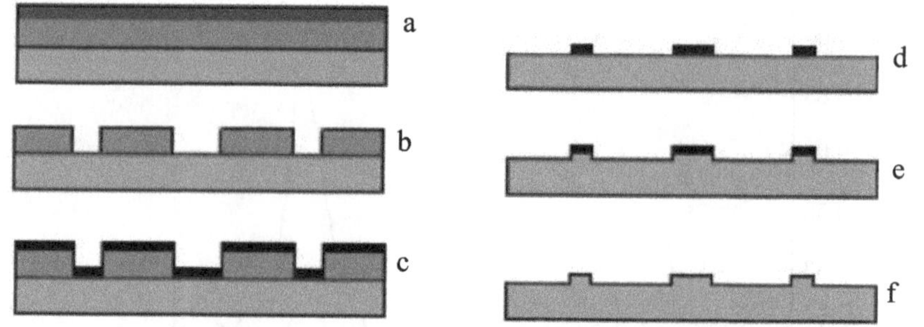

Figure 8.43. Stages of production of binary DOEs with electron-beam lithography and reactive ion etching.

Fig. 8.44. The central part of the relief of the DOE, size 260 × 350 μm.

8.7.2. Production of DOE using photolithography

The synthesized image of the phase of the DOE was used to produce a photomask on a glass substrate with a deposited layer of chromium. Recording on the photomask was done at the laser writing station CLWS-200, with a positioning accuracy of 50 nm and a resolution of 0.6 μm. The DOE was produced by standard methods of photolithography on a glass substrate, thickness 2.5 mm. The depth of etching was 0.5 μm. The error of etching in the height was about 50 nm. Figure 8.44 shows a profile of a binary central part of the DOE with the size of 260 × 350 μm, measured with an NewView 5000 Zygo interferometer.

The cross section for one period of modulation of the surface relief is shown in Fig. 8.45. Figure 8.45 shows that for the wavelength of laser light of 532 nm the depth of etching iof glass was 0.5 μm. The trapezoidal single step did not exceed 25% of its width (approximately 20 μm).

8.7.3 Formation of the DOE with a liquid-crystal display

CRL OPTO SXGA SLM (spatial light modulator) with the active-region of 1316 × 1024 (1280 × 1024, excluding the boundary region) pixels (the size of one pixel

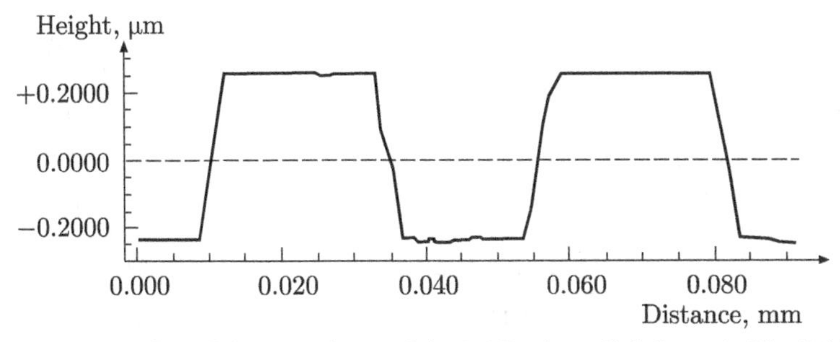

Fig. 8.45. Section of the central part of the DOE microrelief shown in Fig. 8.44.

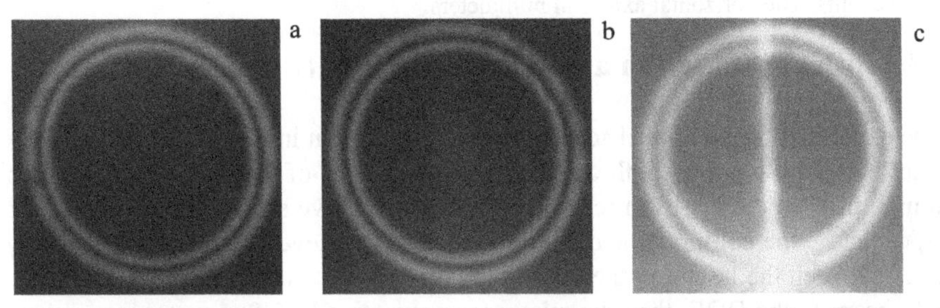

Fig. 8.46. The measured intensity in the Fraunhofer diffraction patterns (double ring) formed using a binary DOE implemented in different ways: by electron-beam lithography (a); optical lithography (b) and with liquid-crystal display (c).

is 15 μm) was used to display the image of the phase of the DOE with the 512 × 512. The microdisplay generates the phase image due to reflection of polarized laser light from the different planes: in the closed state of an individual element of the resolution of the microdisplay light is reflected from the outer surface of the thin film, in the open state – from the inner surface. The result is a binary DOE with a diameter of 6.5 mm. The image formed on the microdisplay is updated with a frequency of 63Hz. At the same time, switching of individual pixels does not exceed 10 μs.

8.7.4. Formation of a double ring of light with different types of DOE

Figure 8.46 shows the distribution of light in a double ring formed in the focal plane of a spherical lens with a focal length f = 138 mm using a binary DOE (8.70), the phase of which is shown in Fig. 8.40a, which was made by a variety of ways: electron lithography (a); optical lithography (b) and with SLM (c). Only the minus first diffraction order is shown. The size of diffraction patterns in Fig. 8.46 is 2 × 2 mm.

Figure 8.47 shows the radial cross section of a double ring of the Fraunhofer diffraction pattern (Fig. 8.46b). It is seen that the ring radius is about 0.8 mm, and width is about 0.3 mm.

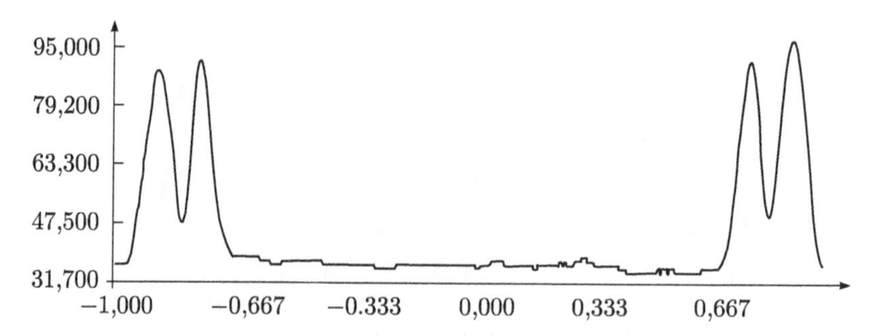

Fig. 8.47. Radial section of the intensity of the Fraunhofer diffraction pattern (Fig. 8.46b) in relative units. The horizontal axis is in millimeters.

8.8. Optical rotation in a double ring of light

|The experiments, the optical scheme of which is shown in Fig. 8.48, used a solid-state laser with a wavelength of 532 nm and a power of 500 mW [20]. In order to minimize power losses in reflections on the refractive surfaces the beam is not expended by the collimator and the desired size is achieved by increasing the distance between the laser and the first turning mirror. Rotating mirror M_1 directs the light beam to the DOE, then the microscope objective L_1 (40×, water immersion, the focal length 4.3 mm) focuses the beam in the work area inside the cell V with micro-objects, microscope objective L_2 (20×) forms an image of the workspace, and

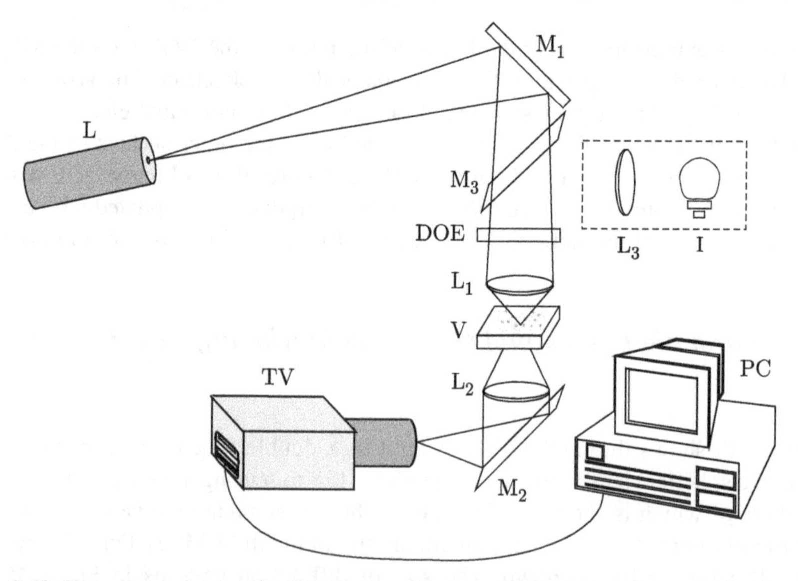

Fig. 8.48. Experimental setup for the rotation of microscopic objects. L – solid-state laser, M_1 – the first mirror, M_3 – semi-transparent mirror to illuminate the work area, M_2 – second rotating mirror, L_1 – focusing microscope objective (40×), L_2 – microscope objective, constructs the image of the working area (20×), L_3 – lens of the illuminator, I – light fixture, V – cell with microobjects, TV – TV camera, PC – a computer.

the mirror M2 turns the light beam in the horizontal direction of the camera. Micro-objects are polystyrene beads with a diameter of 5 μm.

Polystyrene microspheres were trapped by light rings and they move along therm with an approximately constant speed. Different stages of the movement of microspheres at intervals of 2.5 s in a double ring of light are shown in Fig. 8.48.

The radius of the inner ring was 37 μm, the radius of the outer ring 48 μm. As can be seen from Fig. 8.49, there is a steady movement of microspheres along the inner ring of light with an average speed of about 3–4 μm/s and the movement of microspheres along the outer ring of light at a speed of 0.5–0.7 μm/s. This difference in speed is caused by different intensities of the rings. The difference of the intensities of the rings is due to the fact that the DOE is illuminated with a Gaussian beam with the radius smaller than the radius of DOE (to reduce losses during focusing).

8.9. Rotation of micro-objects by means of hypergeometric beams and beams that do not have the orbital angular momentum using the spatial light modulator

Spatial light modulators allow one to generate in real-time the phase DOEs, including for the problems of optical rotation [5, 34, 71, 87, 94]. Unfortunately, the main disadvantage of dynamic modulators, as working for as transmission and reflective, is low diffraction efficiency. This places increased demands on the beams formed. In particular, the need to minimize the number of orders, but it is also clear that it is most efficient to use light beams that have modal properties.

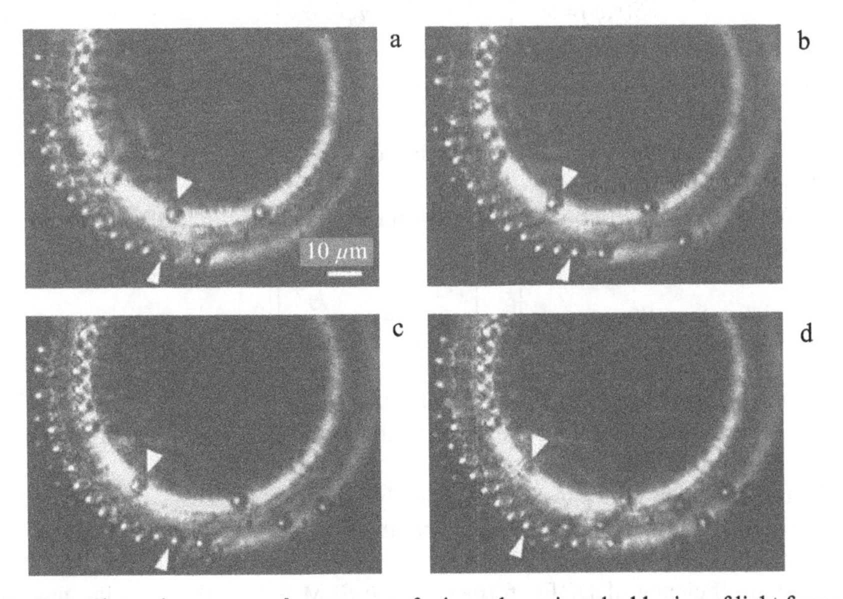

Fig. 8.49. The various stages of movement of microspheres in a double ring of light formed by a compound axicon.

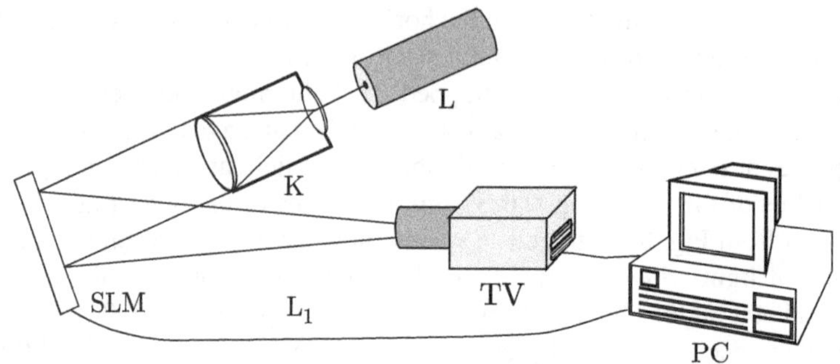

Fig. 8.50. Optical layout for the experiment. L – solid-state laser with a wavelength of 532 nm and a power of 500 mW, K – collimator, TV – TV camera, PC – personal computer.

8.9.1. Rotation of hypergeometric beams

Experimentally the hypergeometric modes were formed with a liquid-crystal microdisplay with an optical arrangement shown in Fig. 8.50 [10].

The laser beam is expanded by a collimator and is incident on the SML of light at an angle close to 90°, and is reflected towards the camera. The DOE is formed on the dynamic CRL OPTO modulator with a resolution of 1316 × 1024, and the physical size of 6.5 mm. The modulator is illuminated by a plane beam from a solid-state laser with a wavelength of 532 nm and a power of 500 mW. At the same time, the binary phase, obtained in the modulator, was encoded in two ways.

1. Binary phase $S(r, \varphi)$ by adding a linear carrier, satisfies the equation:

$$S(r,\varphi)=\operatorname{sgn}\left\{\cos\left[\gamma\ln\frac{r}{w}+n\varphi+\alpha x\right]\right\}, \tag{8.71}$$

where sgn (x) is the sign function, α is the spatial carrier frequency, x is the Cartesian coordinate.

The coded element forms two symmetric orders with hypergeometric modes in the Fraunhofer diffraction plane.

2. With a quadratic radial encoding the binary phase of the DOEs is calculated by the formula

$$S(r,\varphi)=\operatorname{sgn}\left\{\cos\left[\gamma\ln\frac{r}{w}+n\varphi+\frac{kr^2}{2f}\right]\right\}, \tag{8.72}$$

where f is the focal length of the spherical lens.

This means that because of the actual addition of the lens, the hypergeometric modes are formed in a convergent beam.

Figure 8.51 shows the phases of the linearly coded phase DOEs and the intensity distributions at a distance of 2000 mm from the SLM for different n and γ.

Since the energy in linear encoding is divided between two orders, and the diffraction efficiency of DOEs, formed on the SLM is low, the resulting images have insufficient contrast. To get rid of this shortcoming, we used a quadratic

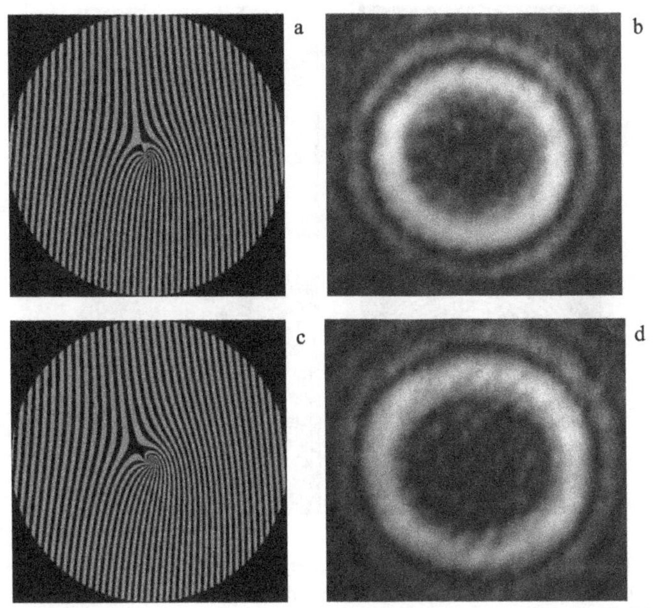

Fig. 8.51. Phase formed in the SLM for n = 10, γ = 5 (a), n = 10, γ = 10 (c) and n = 5, γ = 10 (d) and the corresponding intensity distributions at a distance of 2000 mm from the modulator (b) (d) and (e).

coding (8.72). Figure 8.52 presents the phases of these types of elements (central parts), forming hypergeometric modes, and the intensity distribution at a distance of 700 mm from the DOE.

Figure 8.53 shows the distribution of intensity at different distances from the DOE.

Reducing the size of the hypergeometric mode is due to the fact that the image was shot in the converging beam. As can be seen from Fig. 8.53, the structure of the distribution is preserved, which proves the modal nature of the light beam. Also, it was experimentally determined that the brightest central ring receives 35–40% of the energy beam, which is somewhat higher than in BB. Unfortunately, the low diffraction efficiency of the SLM in combination with relatively low laser power (500 mW) does not allow the hypergeometric modes to be used for the rotation of microscopic objects (in most studies the laser power was greater than 1W). A group of microscopic objects was trapped. Experiments with the rotation of microscopic objects were carried out with the phase DOE shown in Fig. 8.52c. The optical system shown in Fig.8.48 was used in an experiment with the rotation of polystyrene beads with a diameter of 5 μm in the hypergeometric mode with parameters n = 10, γ = 10. Figure 8.54 presents the successive stages of movement (with an interval of 15 s) of trapped polystyrene beads along the brightest light ring of the hypergeometric mode.

The experiment proves the possibility of using the hypergeometric modes in problems of optical trapping and rotation of micro-objects. The presence of an

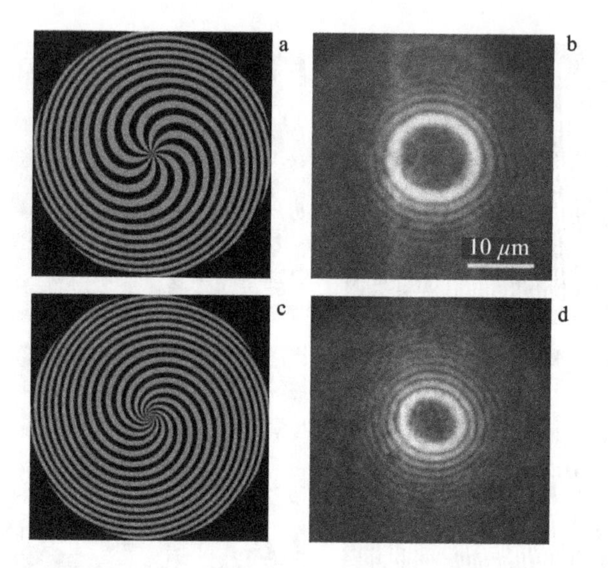

Fig. 8.52. (a) the phase of the DOE for the hypergeometric mode with parameters $n = 10$, $\gamma = 1$, (b) the intensity distribution in the Fresnel diffraction zone, (c) the phase of the DOE for the hypergeometric mode with parameters $n = 10$, $\gamma = 10$, (d) intensity distribution in the zone of Fresnel diffraction.

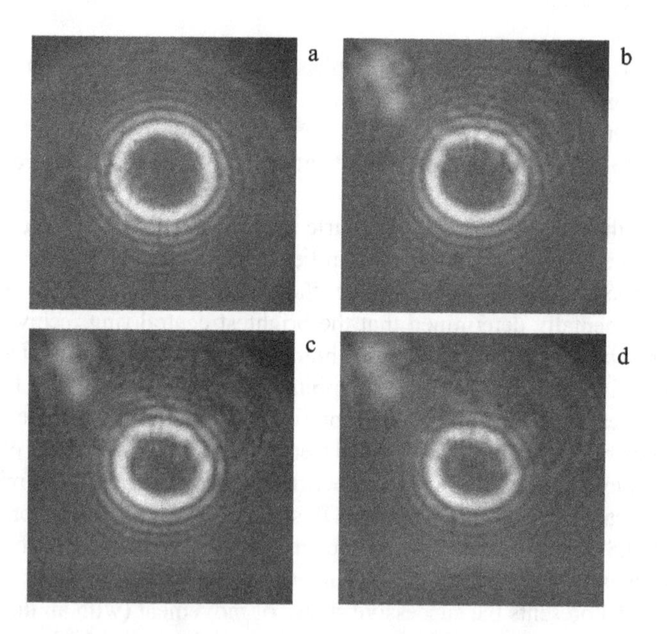

Fig. 8.53. The spread of the beam $n = 10$, $\gamma = 1$ (a) 700 mm, (b) 725 mm (c), 750mm (d) 775 mm.

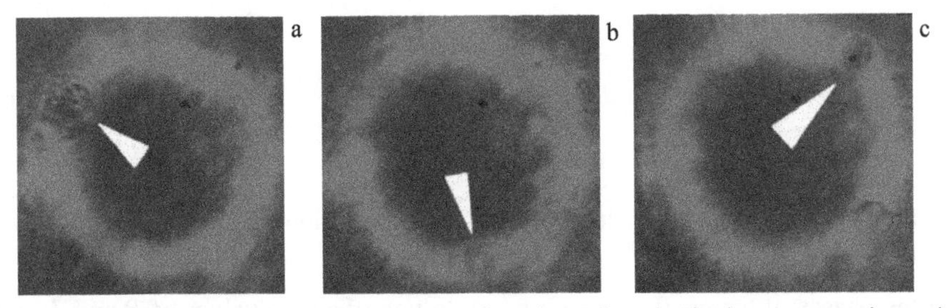

Fig. 8.54. The various stages of movement of a microsphere in the hypergeometric mode with parameters $n = 10$, $\gamma = 10$, formed by a binary DOE, the phase of which is shown in Fig. 8.52c.

additional parameter γ allows to adjust the radius of the brightest ring without changing the parameters of the optical system.

8.9.2. Rotation of the laser beams with no orbital angular momentum

Any paraxial optical field, described by a complex amplitude $E(x, y)$ at $z = 0$ can be decomposed into a number of LG modes in the basis:

$$E(x,y) = \sum_{n,m} C(n,m)\Psi_{nm}(x,y), \tag{8.73}$$

where $C(n, m)$ are the complex coefficients with indices n and m, where m is the azimuthal index,

$$\psi_{nm}(x,y) = a^{-1}\sqrt{\frac{n!}{(n+|m|)!}}\left(\frac{r}{a}\right)^{|m|} L_n^{|m|}\left(\frac{r^2}{a^2}\right)\exp\left[-\frac{r^2}{2a^2}+im\varphi\right], \tag{8.74}$$

where $a = \omega_0/\sqrt{2}$, ω_0 is the Gaussian beam waist radius, (r, ϕ) are the polar coordinates , $L_n^m(x)$ are the associated Laguerre polynomials. In [58] a condition is defined for the number of modes in equation (8.73) in which the intensity in the cross section of a multimode LG beam will rotate as it propagates along the axis z:

$$B = \frac{2(n-n')+|m|-|m'|}{m-m'} = \text{const}, \tag{8.75}$$

where (n, m) and (n', m') are numbers of any two numbers of the linear combination (8.73). Constant $B/4$ is equal to the number of revolutions performed by the multi-mode LG beam $z = 0$ to $z = \infty$. Half of these rotations the beam performs in the distance from $z = 0$ and $z = z_0$, where $z_0 = k\omega_0^2$ is the Rayleigh length, $k = 2\pi/\lambda$ is the wave number of light. In [66] an equation was derived for the projection on the z axis of the linear density of the orbital angular momentum of a linearly polarized laser beam at unit power unit consisting of a superposition of LG modes (8.73):

$$wJ_z = \frac{\sum_{n,m} m|C(n,m)|^2}{\sum_{n,m} |C(n,m)|^2}, \tag{8.76}$$

where w is the angular frequency of light. From equations (8.73)–(8.76) it follows that:

1) the phases of the coefficients $C(n, m)$ do not affect the values of B and J_z, but affect the kind of intensity of the light field from equation (8.73) $I(x, y, z) = |E(x, y, z)|^2$;
2) the number of revolutions according to equation (8.75) during the rotation of the beam (8.73) depends only on the combination of numbers of modes (n, m) and does not depend on the choice of coefficients $C(n, m)$;
3) the orbital angular momentum (8.76) is determined only by the azimuthal numbers m of LG modes and the values of moduli of the coefficients $|C(n, m)|$ and is independent of the number n.Therefore, using a suitable choice of a combination of numbers (n, m) and the moduli of the coefficients $|C(n, m)|$ can be realized by different variants of combinations of values of B and J_z.

Figure 8.55 shows examples. The first row shows the intensity distributions of a four-mode GL beam with coefficients $C(8,0) = 1$, $C(11,2) = 1$, $C(10, -4) = i$, $C(9, 6) = 1$. Such a beam is not rotating ($B = 0$), but has a positive orbital angular momentum ($wJ_z = 1$). In the second row of Fig. 8.55 there are intensity distributions of a 5-mode LG beam with coefficients $C(2, 2) = i$, $C(3,1) = 1$, $C(4,0) = -1$, $C(4, -2) = 1$, $C(4, -4) = 1$. Such a beam is rotated counter-clockwise ($B = -1$) and has a negative orbital angular momentum ($wJ_z = -3/5$). The third row of Fig. 8.55 shows the cross-section of intensity of the three-mode GL beam with coefficients $C(10, -2) = 1$, $C(8,0) = 1$, $C(4,2) = 1$. This beam is rotated counter-clockwise ($B = -3$), but has no orbital angular momentum ($wJ_z = 0$). The fourth row of Fig. 8.55 shows a two-mode LG beam with coefficients $C(1, -1) = 1$, $C(9,1) = 1$. This beam is rotated in a clockwise direction ($B = 8$) and also has an orbital angular momentum ($wJ_z = 0$). Simulation parameters: wavelength $\lambda = 633$ nm, the waist radius of the fundamental LG mode $\omega_0 = 0.1$ mm; the size of each image in Fig. 8.55 1×1 mm, the distance at which the intensities in Fig. 8.55 were calculated (from left to right) $z = 30$ mm, $z = 40$ mm, $z = 50$ mm.

For a superposition of Bessel modes (BM) (8.73) we write:

$$\Psi_{nm}(x,y) = \left[\sqrt{\pi} R J'_m(\gamma_n)\right]^{-1} J_m(k\alpha_n r)\exp(im\varphi), \tag{8.77}$$

where $\alpha_n = \cos \theta n = \gamma_n / kR$, θ_n is the angle of inclination to the z-axis of the conical wave, $J_m(x)$, $J'_m(x)$ is the Bessel function and its derivative, γ_n is the root of the Bessel functions. BMs (8.77) are normalized to unity in the circle of radius R. The laser beam (8.77), consisting of BM, will rotate [66] at a finite distance from the reference plane ($z = 0$), provided that the number of modes (n, m), occurring in the superposition (8.77), will satisfy the condition:

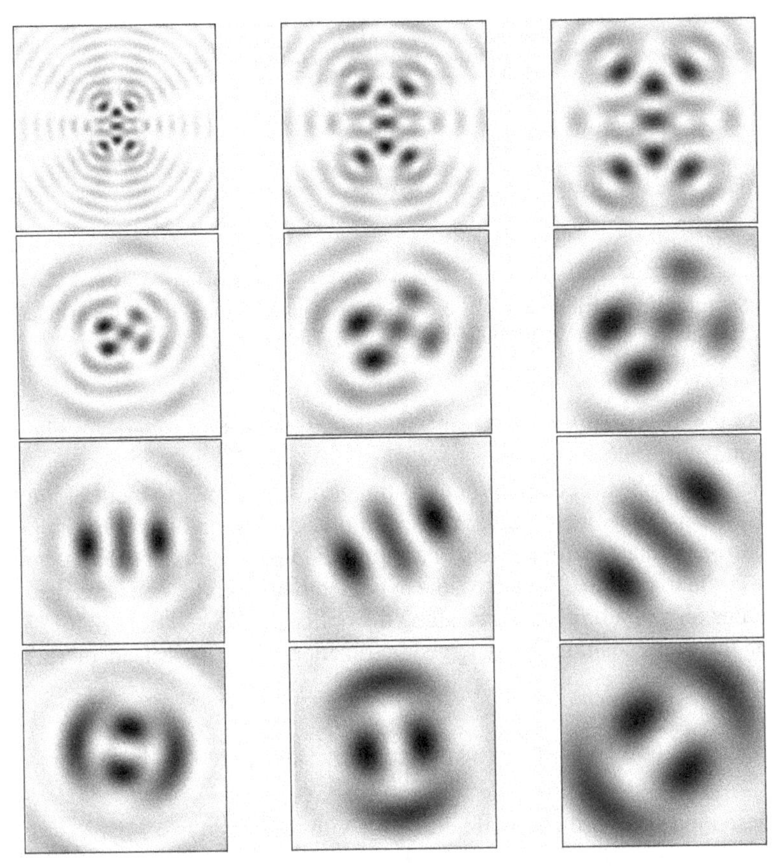

Fig. 8.55. Calculated intensity distributions (negative) in the cross-section of multimode LG beams, calculated at a distance z = 30 mm (column 1), z = 40 mm (column 2), z = 50 mm (column 3) for: 4-mode (row 1), 5-mode (row 2), 3-mode (row 3) and 2-mode (row 4).

$$B_1 = \frac{\alpha_n^2 - \alpha_{n'}^2}{m - m'} = \text{const.} \tag{8.78}$$

The number $B_1/2$ is the number of rotations performed by the intensity in the beam cross section at a distance equal to one wavelength λ.

The projection on the z axis of the linear density of the orbital angular momentum of the laser beam at a power unit consisting of a superposition of BM is calculated using equation (8.76). From equations (8.76) and (8.78) it follows that by selecting the numbers (n, m) we can generate Bessel beams with rotation of intensity in the cross section ($B_1 \neq 0$), but with zero orbital angular momentum ($J_z = 0$). Figure 8.56 shows the intensity distribution of a two-mode Bessel beam with the coefficients $C(\alpha_5, 3) = 1$ and $C(\alpha_{10}, -3) = 1$. Such a beam is rotated counterclockwise ($B_1 = -12.5 \cdot 10^{-8}$) and has no orbital angular momentum ($J_z = 0$). The calculation parameters:

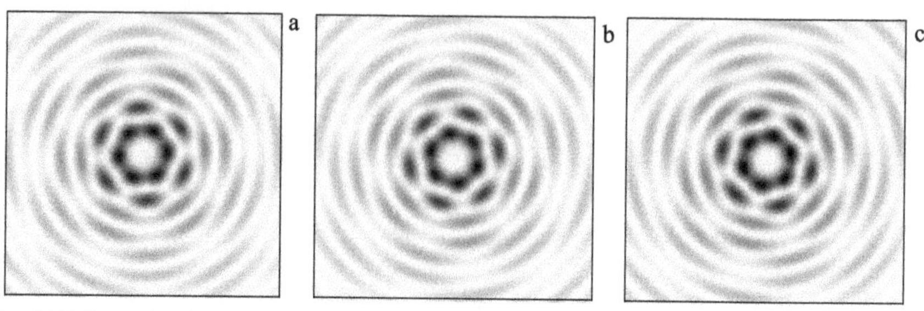

Fig. 8.56. Intensity distribution in the cross section of a two-mode Bessel beam calculated at different distances z from the initial plane: 1 m (a), 2 m (b) and 3 m (c).

$\lambda = 633$ nm, $\alpha_5 = 5\cdot10^{-4}$, $\alpha_{10} = 10\cdot10^{-4}$, the size of diffraction patterns in Fig. 8.56 is 5×5 mm.

The rotating beam can be produced by the hypergeometric mode. Like Bessel modes, the hypergeometric modes have infinite energy, and therefore, in practice, they can be produced with the help of an amplitude-phase filter or a digital hologram with a circular aperture. Therefore, the hypergeometric modes will keep their modal properties only at a finite distance along the optical axis.

For the superposition of hypergeometric modes (8.73) instead of (8.74) and (8.77) we write ($R \gg 1$):

$$\Psi_{nm}(x,y) = \left(2\pi r^2\right)^{-\frac{1}{2}} \exp\left(i\frac{\pi n}{\ln R}\ln r + im\varphi\right),\quad n = 0, \pm1, \pm2, \ldots \tag{8.79}$$

These modes are orthonormal in the ring with radii R and:

$$\int_{R^{-1}}^{R} \Psi_{nm}(r,\varphi)\Psi^*_{n'm'}(r,\varphi)\, r\, dr d\varphi = \delta_{nn'}\delta_{mm'}, \tag{8.80}$$

where $\delta_{nn'}$ is the Kronecker symbol. In the Fresnel diffraction zone the mode (8.79) has the form ($R\to\infty$, $x = kr^2/2z$):

$$\Psi_{nm}(r,\varphi,z) = \frac{1}{2\pi|m|!}\left(\frac{2z}{k}\right)^{\frac{i\gamma-1}{2}} \exp\left[\frac{i\pi}{4}(-|m| + i\gamma - 1) + ix + im\varphi\right]\times$$
$$\times x^{\frac{|m|}{2}}\,\Gamma\left(\frac{|m| + i\gamma + 1}{2}\right) {}_1F_1\left(\frac{|m| + i\gamma + 1}{2}, |m| + 1; -ix\right), \tag{8.81}$$

where $\gamma = \pi n$, ${}_1F_1(a, b, x)$ is the confluent hypergeometric function, $\Gamma(x)$ is the gamma function.

The condition for the rotation of the beam is multimode hypergeometric form:

$$B_2 = \frac{n - n'}{m - m'} = \text{const}, \tag{8.82}$$

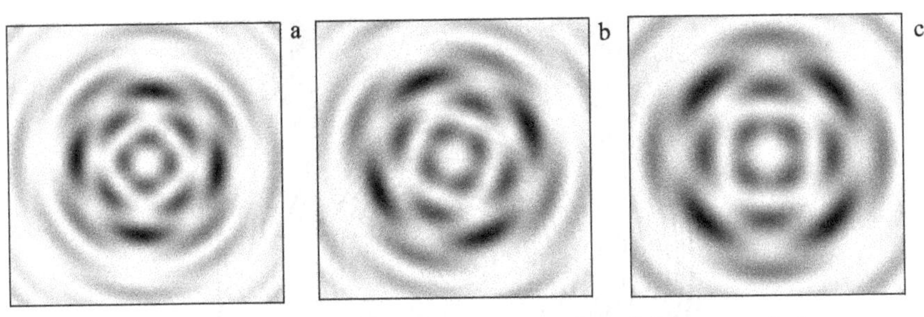

Fig. 8.57. The distribution of intensity in the cross section of the two-mode hypergeometric beam, calculated at different distances z from the initial plane: 1.5 m (a), 2 m (b) and 2.5 m (c).

where $B_2/4$ is the number of rotation which the beam (8.73) with the modes (8.79) makes in the interval from $z = 1$ to $z = R$. Note that in the interval from $z = R^{-1}$ to $z = 1$ the hypergeometric beam makes $B_2/4$ turns in the opposite direction.

The projection on the z axis of the linear density of the orbital angular momentum of the laser beam (8.73) with the modes (8.79) per unit power is described as previously by (8.76). Therefore, using the rotation condition (8.82) and the expression for the orbital angular momentum (8.76) we can generate a laser beam, which, for example, will rotate the transverse distribution of intensity, and there will be a zero orbital angular momentum. Figure 8.57 shows the intensity distribution of a two-mode hypergeometric beam with coefficients C (0, 2) = 1 and C (3, –2) = 1. Such a beam is rotated counterclockwise ($B_2 = -0.75$) and has no orbital angular momentum ($J_z = 0$). The calculation parameters: $\lambda = 633$ nm, $\gamma_0 = 0$, $\gamma_3 = 13.597$, the size of diffraction patterns in Fig. 8.57 is 4×4 mm.

Note that in [66] the authors studied a special case of hypergeometric modes at $\gamma = -i$, which are formed with a spiral phase plate with the transmission exp ($im\varphi$). These modes have the same phase velocities, and therefore their linear combination (8.73) can not rotate during propagation. It also follows from the rotation condition (8.82) at $n = n' = $ const ($B_2 = 0$).

In conclusion, we present some experimental results. The experiments were carrieds out using a binary liquid-crystal SLM CRL Opto SXGA H1 1280 × 1024. Figure 8.58 shows a binary phase intended to generate a light field representing a superposition of two Bessel modes with numbers C (α_1, 3) = C (α_2, –3) = 1 ($\alpha_1 = 1.4 \times 10^{-4}$, $\alpha_2 = 7 \times 10^{-3}$). The size of the formed phases is 7 × 7 mm.

The intensity distribution in the cross section of one of the two beams formed, measured at different distances from the microdisplay with a CCD camera, is shown in Fig. 8.59 [12].

As can be seen from Fig. 8.59, there is qualitative agreement between the experimental and the theoretical data. The experiments were conducted using the optical scheme shown in Fig. 8.48. In this scheme, the mirror M1 was replaced by a spatial light modulator. A rotating multimode BB was formed with C (α_1, 3) = C (α_2, –3) = 1$\alpha_1 = 1.4 \times 10^{-4}$, $\alpha_2 = 0.7 \times 10^{-4}$. Beam power was approximately 5 mW, $\lambda = 0.532$ m.

Fig. 8.58 Binary phase pattern formed on the microdisplay.

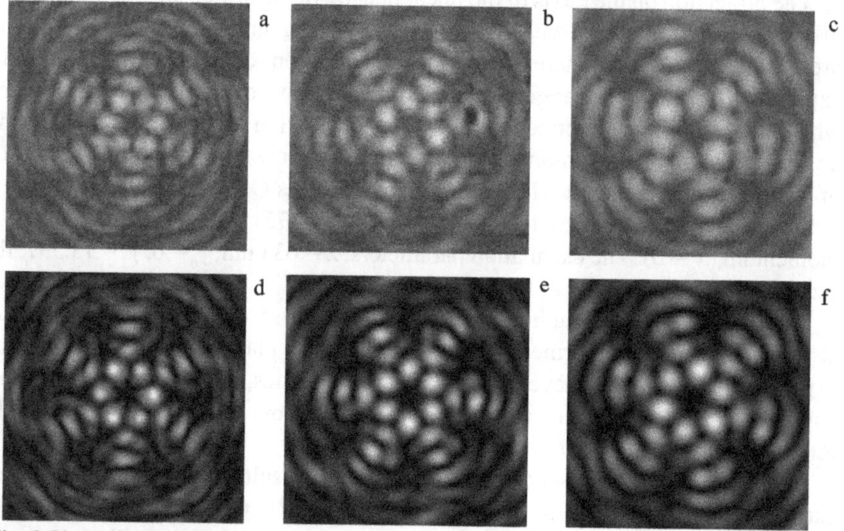

Fig. 8.59. Diffraction pattern of a rotating two-mode Bessel beam at different distances from the microdisplay (a, b, c – experiment, d, e, f – theory): $z = 720$ mm (a, d); $z = 735$ mm (b, e); $z = 765$ mm (c, f).

Figure 8.60 shows the different positions of a rotating beam with zero orbital angular momentum with a trapped polystyrene sphere with a diameter of about 1 μm. Pictures a, b, c were taken at different shifts of the focusing microscope objective (16×) from the initial plane: 0 mm (a), 0.1 mm (b), 0.2 mm (c).

Since the displacement of the micro-object is small we consider separately Fig. 8.60a and c. The dotted line indicates the middle of the beam and was used to construct the cross section of the beam in Fig. 8.61b (Fig. 8.61c).

As can be seen from Fig. 8.61, the microsphere trapped at the maximum intensity is rotated following the rotation of the beam. The beam cross section shows that the

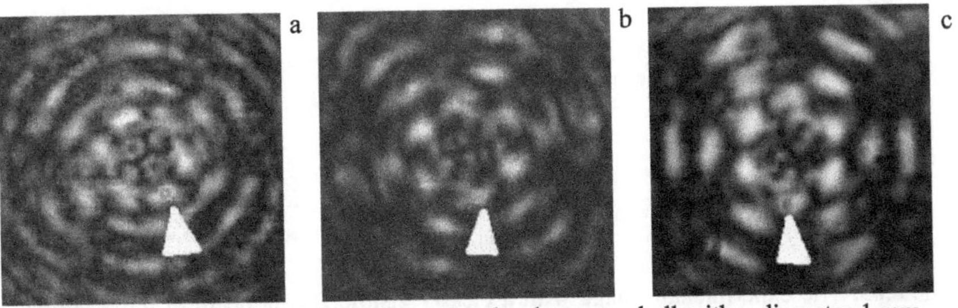

Fig. 8.60. The rotating beam with trapped polystyrene ball with a diameter 1 mm.

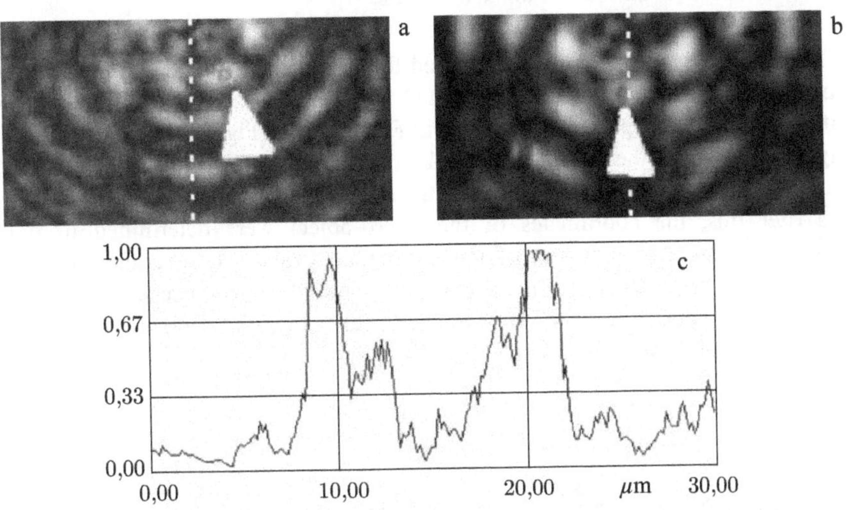

Fig. 8.61 The rotating beam with trapped polystyrene microspheres with a diameter 1 mm, the interval between frames (a) and (b) – 10 s, (c) section along the dotted line of the beam (b).

maxima in Fig. 8.61b are oriented vertically. This experiment shows that using the DOE and a very simple optical arrangement it is possible to control the rotation of the micro-object together with the beam. This effect is usually achieved by using rather complex interferometers.

8.10. Investigation of rotation of micro-objects in light beams with orbital angular momentum

8.10.1. Investigation of rotation of micro-objects in the Bessel beam

The motion of micro-objects in different light beams can be compared most conveniently using the average speed. To determine the average speed, special software was developed allowing processing and separating micro-objects in the image sequence.

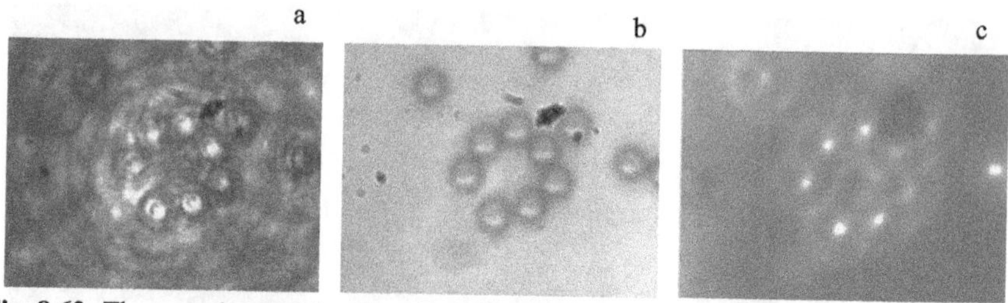

Fig. 8.62. The experimental image with a clearly visible Bessel beam of the 5th order (a), an image with the beam removed and clearly visible microobjects (b), and the image of the correlation peaks at the location of micro-objects.

To automatically determine the speed the correlation function with one of the images of the micro-objects is calculated. Figure 8.62 shows the different stages of processing the experimental images in Fig. 8.62a, in Fig. 8.62b the effect of the light beam due to the separation of colours is completely removed, in Fig. 8.62c correlation peaks are clearly visible on the site of the micro-objects.

After this, the coordinates of the micro-object were determined from the coordinates of the maximum of the correlation peaks. The average speed was defined with respect to both time and the ensemble of micro-objects. The first stage included the determination of the average linear velocity of each micro-object $\langle v_i \rangle$ separately as follows:

$$\langle v_i \rangle = \frac{1}{t}\int_0^t |\mathbf{v}_i(t)|\, dt, \tag{8.83}$$

where t is the time of observation, $\mathbf{v}_i(t)$ is the velocity of the micro-object versus time. We then determine the average velocity V of the ensemble of micro-objects:

$$V = \frac{1}{N}\sum_{i=1}^{N} \langle v_i \rangle. \tag{8.84}$$

The Bessel beam of the fifth-order was formed using a DOE [14] the phase of which is shown in Fig. 8.30a, and the transmission function is given by (8.69). The Bessel beam of the tenth order was produced using a binary helical axicon whose phase is shown in Fig. 8.63, and the transmission function has the form [89]:

$$E_2(r,\varphi) = \operatorname{sgn}\{\exp(in\varphi + iar + iyr\cos\varphi)\}, \tag{8.85}$$

where y is the carrier spatial frequency, a is the parameter of the axicon, $n = 10$ is the order of the helical axicon.

The determining factor for the speed of the micro-objects is the presence of dry and viscous friction forces. But if the force of viscous friction is quite easy to define, and it depends only on the properties of the liquid (as they are the same throughout the volume of the cell) and on the form of a microscopic object, the force of dry friction can greatly vary depending on the location of the micro-object. To minimize

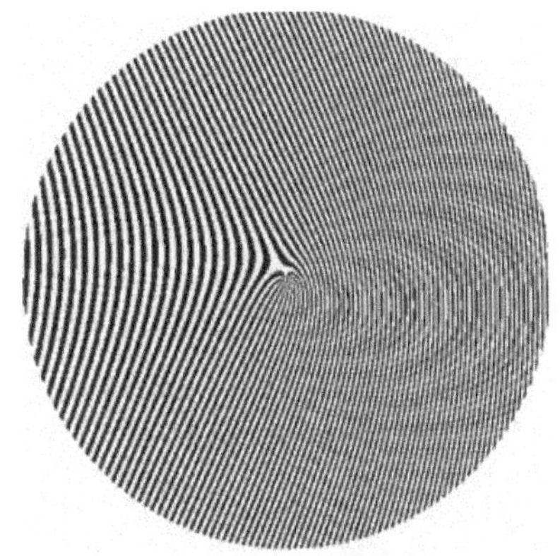

Fig. 8.63. Phase of the binary axicon to form Bessel beam of the tenth order.

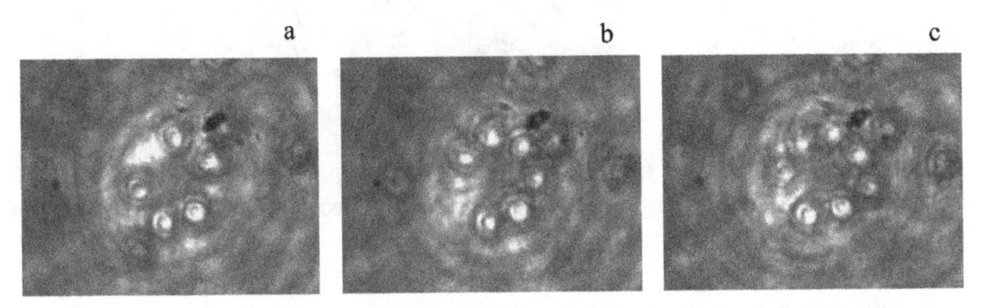

Fig. 8.64. The various stages of the movement of polystyrene beads in a light ring of the Bessel beam formed by the DOE (helical axicon of the 5th order).

the influence of dry friction force on the bottom of the cell the Bessel beam was positioned at the same place with an error of no more than 2 μm, so the nature of the friction of one cell should not be changed by changing the beams. The strength of viscous friction is proportional to the velocity of the micro-object, consequently, the velocity of the micro-object can indicate the magnitude of force of the light beam acting on the micro-object. Computational experiments were carried out prior to each full-scale experiment (using the developed method) to estimate the velocity of microscopic objects in a particular light beam for the given parameters. So simulation was carried out for the fifth-order BB for a 16× focusing microscope objective and the beam power 230 mW (at the output of the laser). The average velocity of the spheres with a diameter of 5 μm was 8 μm/s, which gave reason to believe the success of full-scale experiment with the same parameters.

Initially, the circuit was fitted with a DOE the phase of which is shown in Fig. 8.30a. The DOE formed a fifth-order BB in which the very bright ring trapped as a result seven microspheres. Different stages of the movement of these microspheres at intervals of seven seconds are shown in Fig. 8.64. The diameter of the bright ring of the BB was about 18 μm.

Figure 8.65 presents the processed images with clearly visible micro-objects, and the almost completely 'removed' beam.

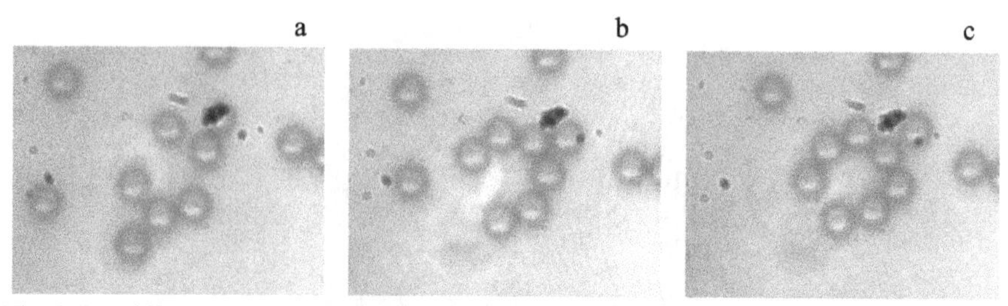

Fig. 8.65. Different stages of the movement of microspheres in a light ring, formed by a helical axicon of the 5th order, after processing to determine the average velocity.

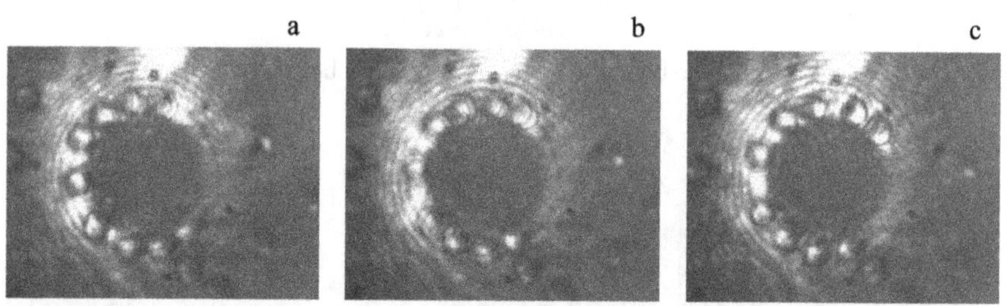

Fig. 8.66. The various stages of the movement of micro-objects in a light ring formed by the binary axicon of the 10th order (Fig. 8.63).

In this experiment the average velocity was determined using seven micro-objects with the observation time of 29 s. A total of 116 images with the successive stages of movement were processed. The average velocity was 1.3 ± 0.1 μm/s.

For comparison, a similar experiment was carried out with the BB of the 10th order. Different stages of movements of the micro-objects at intervals of one second in the BB of the 10th order are shown in Fig. 8.66.

In this experiment the average velocity was determined using nine microparticles with the observation time of 22 s. A total of 88 images with successive stages of movement were processed. The average speed was 2.9 ± 0.3 μm/s. It should be borne in mind that the use of the binary axicon (Fig. 8.63a) reduces by half the beam energy, as the energy is divided between the two (plus and minus first) orders. Table 8.4 presents the parameters of both experiments for comparison.

As shown in Table 8.4, using the BB of the 10th order the velocity of the micro-objects is more than doubled, with half the beam energy. If we assume that the force of viscous friction is proportional to the velocity of micro-objects, then at the same beam energy the force directed along the ring in the BB of the 10th order will be superior to the same effect for the BB of the 5th order four times. It should however be noted that it is difficult to take into account the effect of friction of micro-objects on the bottom of the cell, which increases with the beam power (due to the pressure of light as a result of Fresnel reflection from the micro-object). In particular, because of this force in both experiments there was a complete arrest of movement of some micro-objects (for a short time) in a number of stages. To minimize the influence of friction forces in the same experiments, the average velocity was measured in two

Table 8.4. Comparative experimental parameters

The beam	Power of the beam in the working plane (MW) (including losses)	Average intensity of the bright ring (W/m²)	Diameter of the bright ring (μm)	The average speed of micro-objects (μm/s)	Average speed excluding stopped micro-objects (μm/s)
BP of order 5	230	27 107	18	1.3 ± 0.1	3.4 ± 0.4
PD 10-th order	100	8 107	37	2.9 ± 0.3	3.1 ± 0.4

stages. In the first stage the overall average velocity was calculated from (8.83) and (8.84), in the second stage we determined micro-objects and time intervals, during which their actual velocity was less than half the average speed defined in the first stage. This was followed again by determination of the average velocity from (8.83) and (8.84), but the above-mentioned micro-objects were not considered. The resulting average velocity is indicated in the rightmost column of Table 8.4.

8.10.2. Studies of mechanical characteristics of rotation of micro-objects in optical vortices

In all the above, as a rule special attention was paid to the formation of an optical vortex beam without regard to the effectiveness of this beam in rotation tasks. At the same time, it is obvious that if we consider the task of efficient transmission of the torque to micromechanical systems, it is necessary to investigate how changes of the order number of the optical vortex will change the amount of energy transferred from the beam to a microscopic object. At a qualitative level it was determined that the velocity of the micro-objects increases with increasing numbers of the order of an optical vortex [39, 40]. However, quantification of this relationship was not carried out. To perform such a study, further experiments were carried out by the rotation of polystyrene beads in light beams with the angular harmonics of the 30th and 31st order [17]. The experimental setup for optical rotation is shown in Fig. 8.48.

The experiment used a solid-state laser with a wavelength of 532 nm and a power of 500 mW. The beam was not collimated in order to minimize power losses in reflections on the refractive surfaces, and the desired size is achieved by increasing the distance between the laser and the first rotating mirror. Polystyrene microspheres with a diameter of 5 μm were used as the micro-objects.

In order to form a set of 4 optical vortices (numbers of orders –31, –30, 30, 31) experiments were carried out with a DOE, the binary phase of which is shown in Fig. 8.67a [16, 17]. Figure 8.67b shows the central part of the microrelief. Figure 8.67c shows the distribution of intensity for the DOE in the area of Fraunhofer diffraction.

The scheme included an element, the phase of which is shown in Fig. 8.67a. The optical vortex of 30th orderr trapped 14 micro-objects as a result.

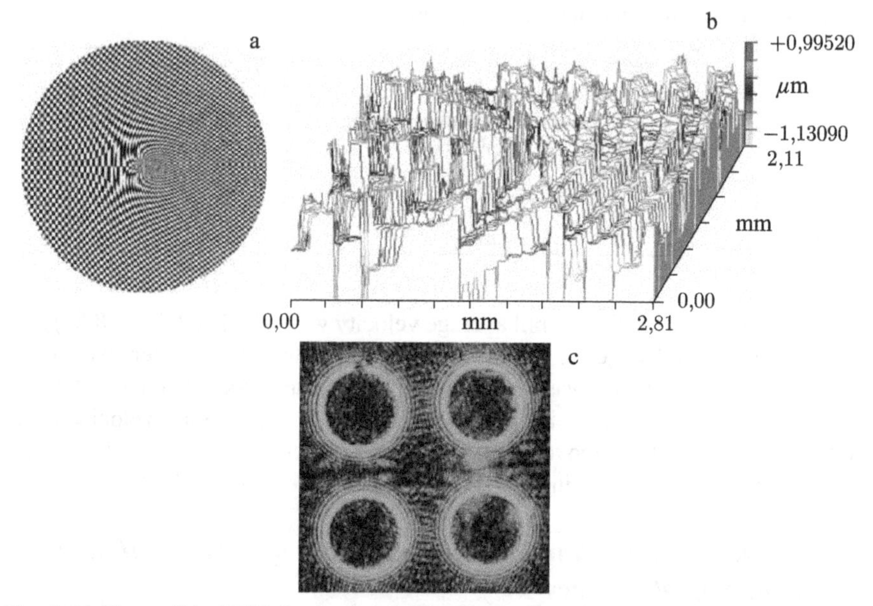

Fig. 8.67. Phase of the DOE for the formation of optical vortices of –31, –30, 30, 31 orders (a), the central part of the DOE microrelief (b), the intensity distribution in the diffraction pattern (c).

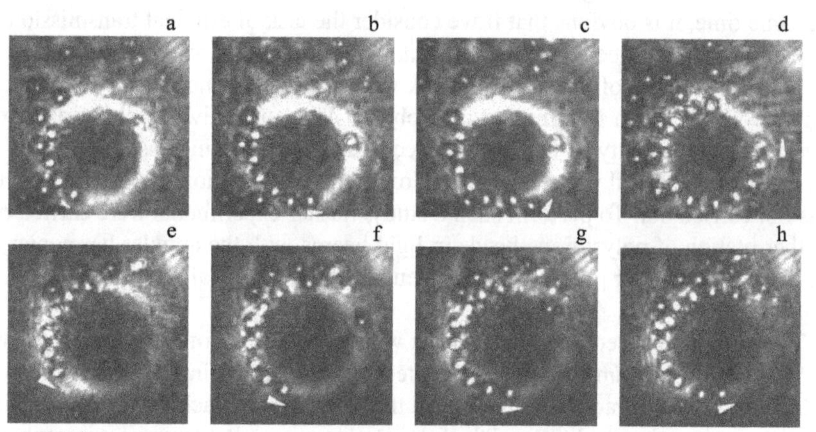

Fig. 8.68. The various stages of motion of micro-objects in the vortex beams, formed by the DOE: a–d) optical vortex of order 30, (e–h) optical vortex of order 31.

Different stages of the movement of these micro-objects the interval of seven seconds are shown in Fig. 8.68 (a–d). Exactly the same experiment was performed for an optical vortex of order 31, the stages of its movement are shown in Fig. 8.68 (e–h).

The motion of micro-objects in different light beams can be compared most conveniently using the average speed. Experimental images were processed to determine the average velocity by the method described in the previous section.

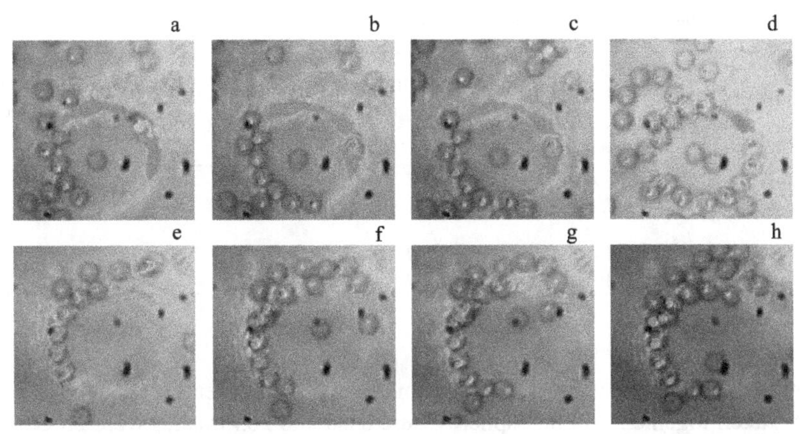

Fig. 8.69. Various stages of motion of micro-objects in the vortex beams, formed by the DOE: (a–d) optical vortex of order 30; (e–h) optical vortex of order 31, after computer processing of images.

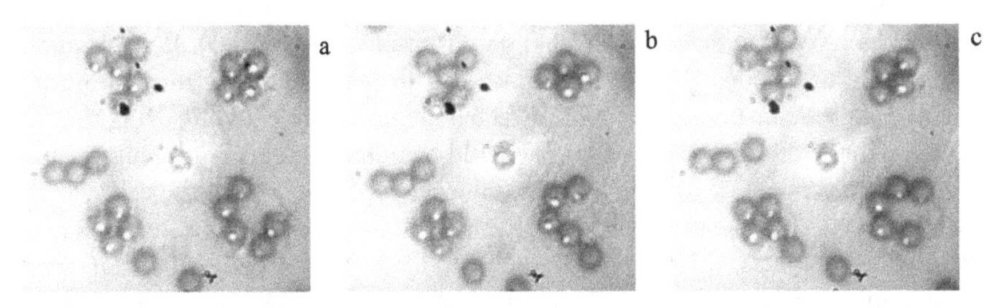

Fig. 8.70. The various stages of motion of micro-objects in the vortex beams, formed by a binary DOE the phase of which is shown in Fig. 8.37a.

Figure 8.69 presents the processed images with clearly visible micro-objects, and an almost invisible beam. The different brightness of the images is due to the change in background illumination, almost invisible on the original images.

In this experiment, the average velocity was determined using fourteen and eleven microspheres (for the 30th and 31st orders, respectively) during 19 s. 76 images of consecutive stages of movement for an optical vortex of the 30th order were processed. The determination of the average velocity did not take into account the moments of time during which the microspheres stayed under the influence of friction. The average velocity was 14 ± 3 μm/s.

175 images were processed for an optical vortex of the 31st order, i.e. the total duration of the experiment was 44 s. The average speed was 11 ± 3 μm/s.

Figure 8.70 shows the images processed to determine the average velocity of various stages of micro-objects in optical vortices in the 3rd and 7th orders.

In this experiment, the average velocity was determined using eight microspheres (for the optical vortex of the 3rd order) during 12 s, Only 48 images with the successive stages of movement were processed. The average velocity was 4 ± 2 μm/s. The average velocity in the micro-objects in the optical vortex of 7th

Table 8.5. Comparative experimental parameters

The order of an optical vortex	Beam power in the working plane (mW) (including losses)	Aaverage intensity of the bright ring (W/m^2) × 10^8	The diameter of the ring (μm)	Average speed of micro-objects (μm/s)
3	50	3.2	9	4 ± 2
7	50	2.1	13	6 ± 2
30	40	0.9	27	14 ± 3
31	40	0.9	28	11 ± 3

order was determined using the results of several experiments (not only in Fig. 8.70). Taken together, 16 microspheres were used and 203 images were processed (total time of four experiments 51 s). The average speed was 6 ± 2 μm/s. These data were used to compile Table 8.5.

As shown in Table 8.5, at increasing numbers of the order the velocity of the micro-objects is initially almost doubled and in the further growth of the number of the order does not change so much (though with reduced intensity). If we assume that the force of viscous friction is proportional to the velocity of micro-objects, then at the same beam energy the force directed along the optical vortex ring should increase with the number of the order. It should be noted that this is difficult to take into account the effect of friction of micro-objects on the bottom of the cell, which increases with the beam power (due to the pressure of light from the Fresnel reflection from the micro-objects). In addition, as shown in Fig. 8.69 and Fig. 8.70, the motion of micro-objects is very uneven (there are short stops), indicating heterogeneities in the bottom of the cell. It is also extremely difficult to accurately determine the power of the specific beam, as available devices allow one to define an integrated beam power (i.e. all four rings simultaneously with the zero-order). This power is then to be divided in proportion to the brightness of each image.

According to [76], the orbital angular momentum of the light field transmitted to the micro-object can be expressed by the formula

$$M = \frac{\lambda n P}{2\pi c}\eta_{\text{abs}}, \tag{8.86}$$

where M is the transmittted moment, λ is wavelength, n is the order (number) of the singularity, P is beam power, η_{abs} is the absorption coefficient of micro-objects. Using this formula and assuming that there is complete coincidence for one of the experimentally obtained points, a theoretical curve of the dependence of the velocity of the micro-objects on the number of singularity was constructed. Figure 8.71 shows this curve with the superimposed experimental points.

As can be seen from the graph in Fig. 8.71, the experimental data are in good agreement with the theoretical ones within the experimental error.

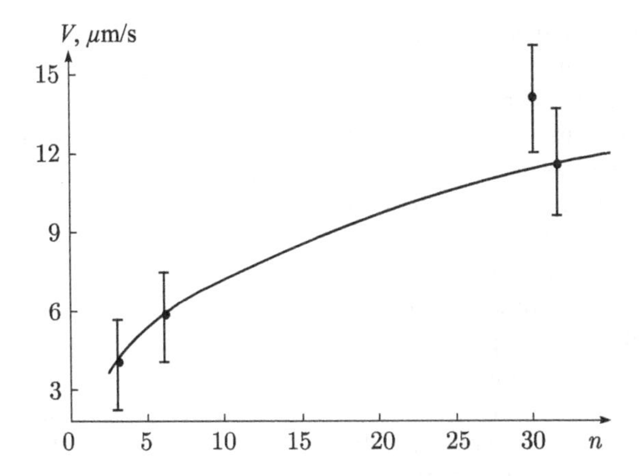

Fig. 8.71. Theoretical dependence of the velocity of polystyrene beads with a diameter of 5 μm on the number of the singularity of the light field (curve), and experimental data on the dependence of the velocity of polystyrene beads with a diameter of 5 μm on the number of the singularity of the light field (points).

8.11. The trapping of micro-objects in Airy beams with ballistic properties

With their compact concentration of light energy in small areas and preservation of this distribution over long distances (i.e. a significant increase in the depth of focus as compared to the Gaussian laser modes), diffractionless beams are widely used in various applications: medical imaging, non-destructive investigation of materials, measurement of the velocity of moving objects – in all of these problems the important property is the invariant length of the diffractionless beam as the measurements of parameters of a moving or extended object are taken.

8.11.1 Airy laser beams

Diffractionless beams, having also an orbital angular momentum, are of particular interest. The effectiveness of their use has already been demonstrated in areas such as optical trapping and multifunctional manipulation of micro- and nano-objects (from biological cells to atoms) in nonlinear optics and plasma physics. One of the promising directions is their application in quantum computing – an infinite number of orthogonal states of the orbital angular momentum significantly increases the amount of information that can be transmitted by a single photon.

Bessel beams were physically formulated in the late 80s of the last century [95, 96] and the effectiveness of different applications was shown. This was followed by other diffractionless solutions of the wave equation: parabolic beams [97], hypergeometric beams [98], Airy beams [99], circular beams [100], hoping to enrich diffractionless beams with new properties.

For example, hypergeometric beams, in contrast to BBs, have a more extended range, allowing them to keep diffractionless properties over larger distance at their physical implementation and the inevitable limitation of the aperture [101].

It was also demonstrated that one- and two-dimensional Airy beams, in contrast to other diffractionless beams, have ballistic properties [102], which opens up new possibilities in optical manipulation, such as the ability to bypass barriers non-transparent for laser radiation.

Interest in the Airy functions, which emerged in the late seventies of the last century in the context of quantum mechanics [103], has been revived recently in connection with the new opportunities offered by diffractive optics in the physical implementation of mathematical abstractions.

As the Bessel function, the Airy function is infinitely extended: exponential decay for positive values of the argument, for negative values an oscillating character with a slowly decaying amplitude [104], so their physical implementation should be truncated. In [105] the authors considered Airy beams with finite energy, which are the product of a classical Airy mode and exponential functions, and in [106] a generalized formula describing the passage of paraxial Airy–Gaussian beams through an optical ABCD-systems was derived.

Although the multiplication by a Gaussian or an exponential function (in this case the Fourier image is proportional to the Gaussian function) allows one to simply generate such beams using a spatial light modulator illuminated by laser radiation, in both cases the generated beams actually cease to be diffractionless, although some retain their form in some distance.

This section describes a way to truncate the infinite Airy mode by a rectangular aperture, from a value of d_0 in the positive part of the argument (for example, the value Ai $(x = d_0 = 3)$ is practically equal to zero) to the nth zero in the negative part. This 'limited' Airy distribution formed in the cross sections of laser modes of the 'whispering gallery' type [107–109]. In [110] the degree of divergence of the three types of truncated Airy beams was compared: exponential, Gaussian and simply limited by a numerical aperture, and it was shown that in the latter case, the oscillating structure of the beam and a narrow well-defined maximum of intensity is preserved much longer than in the first two.

This section also presents the results of experimental formation of one-dimensional Airy beams by a method different from than in [99, 105], namely by means of coded diffractive optical elements. At the same time using the coding parameters it is possible to vary the energy contribution of high-frequency components in the generated beam and demonstrate the generation of the distribution preserving the concentration of energy in a narrow lateral lobe. Such a beam was used to trap polystyrene microparticles.

In [103] in the context of quantum mechanics attention is given to the $(1 +1)D$ Schrödinger equation for a free particle with mass m:

$$\left(-\frac{\hbar^2}{2m}\frac{\partial^2}{\partial x^2}-i\hbar\frac{\partial}{\partial t}\right)\psi(x,t)=0, \tag{8.87}$$

and the solution of this equation in terms of the Airy function Ai(x):

$$\psi(x,t) = \mathrm{Ai}\left[\frac{B}{\hbar^{2/3}}\left(x - \frac{B^3t^2}{4m^2}\right)\right]\exp\left[\frac{iB^3t}{2m\hbar}\left(x - \frac{B^3t^2}{6m^2}\right)\right], \tag{8.88}$$

where $B > 0$ is an arbitrary real constant, $\hbar$ is Planck's constant.

Thus, this paper shows that the probability density function (8.88) $|\psi(x, t)|^2$ does not change its shape depending on the time. If we rewrite equation (8.87) in the form of the paraxial wave equation of propagation in free space:

$$\left(\frac{\partial^2}{\partial x^2} + 2ik\frac{\partial}{\partial z}\right)\psi(x,z) = 0, \tag{8.89}$$

where the wave number $k = 2\pi/\lambda$ (λ is wavelength) obtained by formal replacement $m/\hbar = k$ then a solution of (8.89) is a function of the form:

$$\psi(x,z) = \mathrm{Ai}\left[x - \left(\frac{z}{2k}\right)^2\right]\exp\left\{\frac{iz}{2k}\left[x - \frac{2}{3}\left(\frac{z}{2k}\right)^2\right]\right\}. \tag{8.90}$$

This expression shows that the intensity function (8.90) $|\psi(x, t)|^2$ does not change for different values of z, and is only shifted proportional to the square of this parameter.

However, since the Airy function is defined on the entire numerical line, then it is difficult to realize accurately. Therefore, in [105] the action of (8.89) on the limited energy function is investigated:

$$\psi(x) = \mathrm{Ai}(x)\exp(ax) \tag{8.91}$$

and obtained the following expression:

$$\psi(x,z) = \mathrm{Ai}\left[x - \left(\frac{z}{2k}\right)^2 + \frac{iaz}{k}\right]\exp\left\{a\left[x - 2\left(\frac{z}{2k}\right)^2 + \frac{iaz}{2k}\right]\right\}\exp\left\{\frac{iz}{2k}\left[x - \frac{2}{3}\left(\frac{z}{2k}\right)^2\right]\right\}. \tag{8.92}$$

As can be seen from (8.92), the Airy function modified in this way is no longer a mode, though at a distance it approximately (even at small z the dependence on the apparent parameter exists) retains its shape.

In [106] the generalized one-dimensional Airy–Gauss beams are considered:

$$U_1(x_1;\kappa_1,\delta_1,S_1,q_1) = \mathrm{Ai}\left(\frac{x_1+\delta_1}{\kappa_1}\right)\exp\left[iS_1\left(\frac{x_1+\delta_1}{\kappa_1}\right) + i\frac{S_1^3}{3}\right]\exp\left(\frac{ikx_1^2}{2q_1}\right), \tag{8.93}$$

which in paraxial passage of the optical ABCD system take the following form:

$$U_2(x_2;\kappa_2,\delta_2,S_2,q_2)=\mathrm{Ai}\left(\frac{x_2+\delta_2}{\kappa_2}\right)\exp\left[iS_2\left(\frac{x_2+\delta_2}{\kappa_2}\right)+\frac{iS_2^3}{3}\right]\exp\left(\frac{ikx_2^2}{2q_2}\right)\left(A+\frac{B}{q_1}\right)^{-1/2},\tag{8.94}$$

where $q_2=\dfrac{Aq_1+B}{Cq_1+D}$, $\kappa_2=\kappa_1\left(A+\dfrac{B}{q_1}\right)$, $S_2=S_1+\dfrac{B}{2k\kappa_1\kappa_2}$,

$$\delta_2=\delta_1\left(A+\frac{B}{q_1}\right)-\frac{B}{2k\kappa_1}(S_1+S_2).$$

In [110] the exponential Airy beams (8.91) and the Airy–Gauss beams (8.93) are compared with the limited beams:

$$\psi_n(x,z=0)=\begin{cases}\mathrm{Ai}(x), \ \gamma_n\le x<d_0,\\ 0, \quad \text{otherwise}\end{cases}\tag{8.95}$$

where $d_0=7$, $\gamma_n\approx-\left[\dfrac{3\pi}{8}(4n-1)\right]^{2/3}$ are the roots of the Airy function [111] that are numerically specified in the calculations..

Figure 8.72 shows the distribution of different types of Airy beams in free space: a picture of the intensity distribution at a distance from $z=0$ to $z=0.05z_0=62$ mm, where z_0 is the Rayleigh distance (bottom row images) and the distribution pattern of the structure of the intensity (i.e. the maxima at each distance were reduced to a single value) at a distance from $z=0$ to $z=0.2z_0=248$ mm (top row images). Figure 8.73 clearly shows that the limited Airy beam (8.95) much better approximates the oscillating structure of the ideal function and significantly longer maintains a well-defined narrow intensity peak whose displacement at the beginning of propagation occurs on a curved ballistic trajectory.

Figure 8.73 shows the intensity of the beams at a distance $z=0.05z_0$ and $z=0.1z_0$, from which it can be clearly seen that the limited beam is also much less prone to diffraction than other types of truncated beams. The Airy–Gaussian distribution is kept exponential a little longer, but degenerates still very fast compared to the limited beam.

Figure 8.74 shows the intensity of the Fourier spectrum $|\phi(\xi,f)|^2$ of the truncated Airy beams at $f=z_0/4$. The figure shows that the spectrum of the exponential Airy beam is close to the Gaussian function, and the spectrum of the Airy–Gaussian beam has a more interesting view – 'flat-top' distribution, with an almost flat top, which, for example, can be described by a super-Gaussian function. Such a distribution [112] is often useful in various problems, such as improved print quality, microlithography, materials processing, optical manipulation. The Fourier spectrum of the limited Airy beam (8.95) is approximated by a rectangle.

Another method was proposed for the formation of limited Airy beams (8.95) in comparison with the method described in [99, 105], namely by means of phase coded diffractive optical elements (DOE). Using the method of partial coding [113],

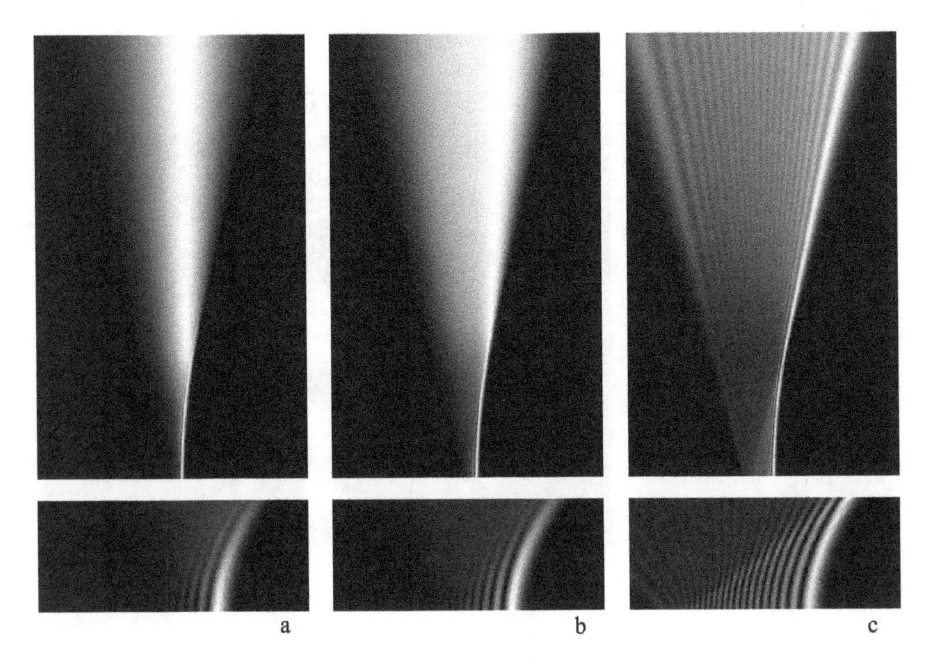

Fig. 8.72. Distribution of the exponential (a) and Gaussian (b) and limited (in) Airy beams in free space: the picture of the intensity at a distance of $z = 0.05z_0$ (bottom row) and a picture of the structure up to $z = 0.2z_0$ (top row).

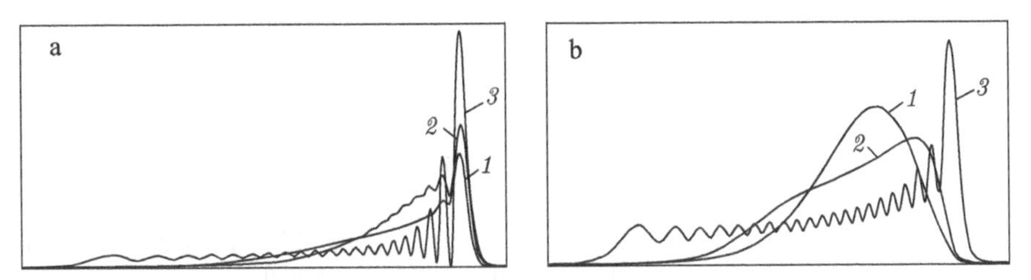

Fig. 8.73. Graphs of the intensity at $n = 21$ of exponential (line 1), Gaussian (line 2) and the limited Airy beams (line 3) at a distance $z = 0.05z_0$ (a) and $z = 0.1z_0$ (b).

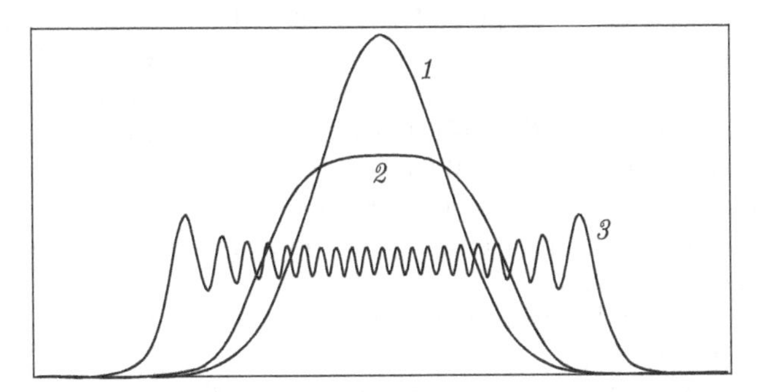

Fig. 8.74. Graphs of the intensity at $n = 21$ of the Fourier spectrum of the exponential (line 1), Gaussian (line 2) and the limited Airy beams (line 3) at $f = z_0/4$.

Table 8.6. Coded DOEs with different encoding parameters α, designed to produce Airy beams (8.95) for $n = 7$, and their spatial spectra

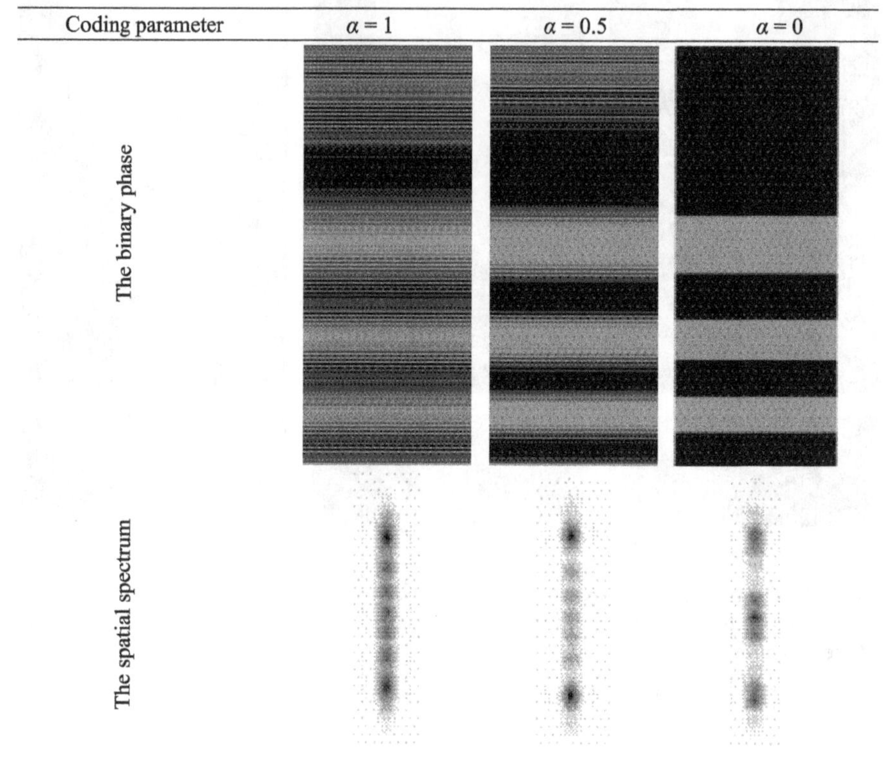

we examine the formation of limited Airy beams with variable energy contribution of high-frequency components. In [110] it was shown that the prediction of the initial distribution allows to concentrate in a narrow lateral lobe the light energy that is retained during propagation.

Table 8.6 shows the binary phase DOEs with different encoding parameters α, designed to produce Airy beams (8.95) for $n = 7$, and their spatial spectra. The parameter of partial coding $\alpha \in [0, 1]$ [113] allows to vary the ratio of two competing characteristics of the DOE – the diffraction efficiency and accuracy of the formation of a given distribution. When $\alpha = 0$ the coding method degenerates into kinoform method that defines high performance, while at $\alpha = 1$ complete coding by the two-phase method is achieved providing accurate formation by discarding part of the energy from the useful area.

Figure 8.75 shows the intensity distribution at different distances generated by each of the above DOEs.

Table 8.6 and Fig. 8.75 show that when $\alpha = 1$ the distribution is close to ideal, persists over a long distance, but has a low diffraction efficiency. At $\alpha = 0$ the distance

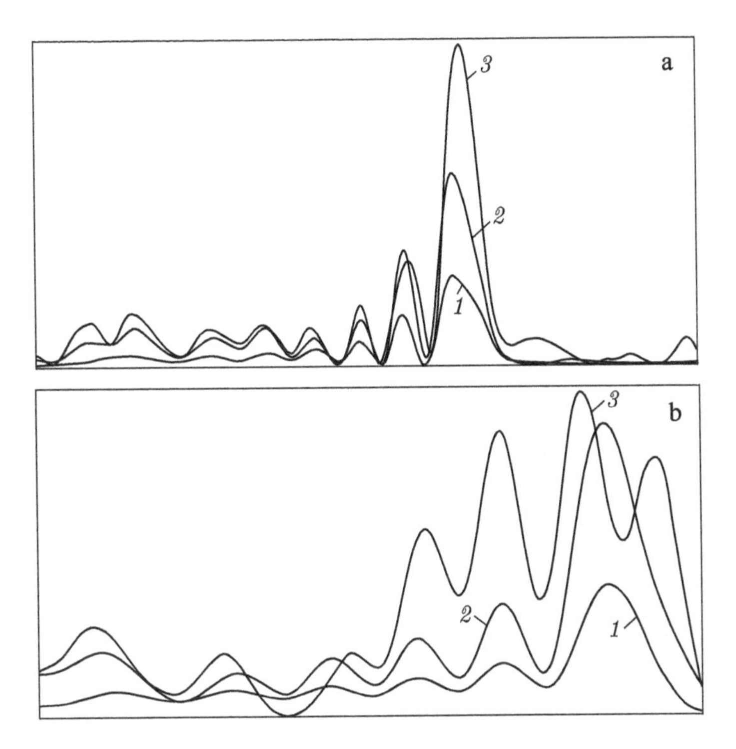

Fig. 8.75. Graphs of the intensity at $n = 7$ for encoding parameters $\alpha = 1$ (line 1), $\alpha = 0.5$ (line 2) and $\alpha = 0$ (line 3) at $z = 0.4z_0$ (a) and $z = z_0$ (b).

at which the Airy beam maintains a distinctive side peak is significantly reduced, although in this case, the useful area gets more than 60% of energy. When $\alpha = 0.5$, the distribution is also close to ideal, except for the high-frequency components – this is clearly seen in Fig. 8.75a and in the reconfiguration of the spectral pattern. This coding parameter value represents the best compromise between energy efficiency and the accuracy of formation.

The experimentally recorded results of propagation of a limited Airy beam, formed by a binary coded DOE with the parameter $\alpha = 0.5$, are shown in Fig. 8.76. It is seen that the beam retains a one-dimensional structure with a small divergence – the initial beam having a longitudinal dimension of 2 mm over a distance of 100 mm has increased by 20%. The transverse beam broadening is due to diffraction on a one-dimensional slit, which was used for the illumination of square DOEs.

8.11.2. Optical trapping of micro-objects in Airy beams

Airy beams allow trapping of microscopic objects. In trapping the micro-objects are grouped into two peaks. For the experiment used optical scheme shown in Fig. 8.48. This scheme included a DOE the phase of which is shown in Table 8.6 for the encoding parameter $\alpha = 0.5$. Polystyrene beads with a diameter of 5 μm were

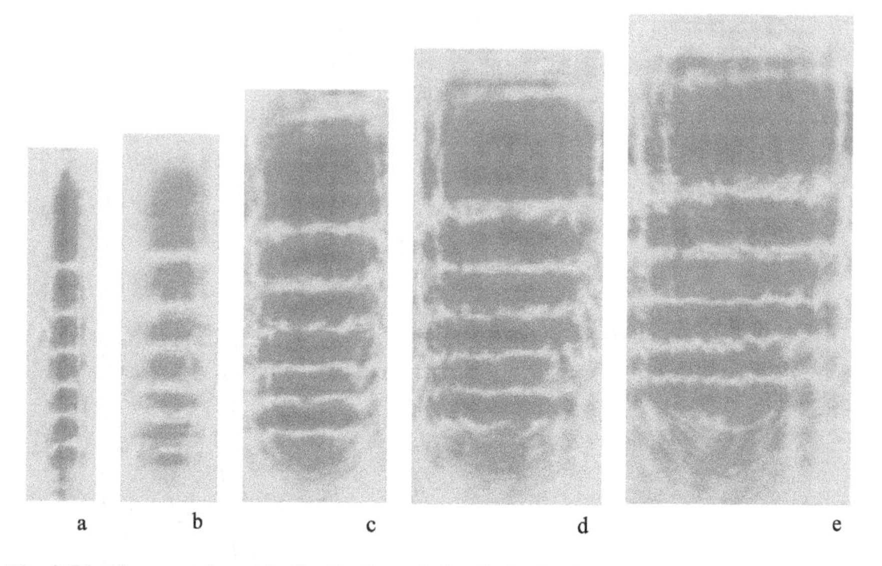

Fig. 8.76. The experimental distribution of the limited Airy beam, formed by a binary partially coded DOE: a) confocal image plane by a system of two spherical lenses, b–e) planes, separated from the image plane at a distance of 25 mm, 50 mm, 75 mm, 100 mm, respectively.

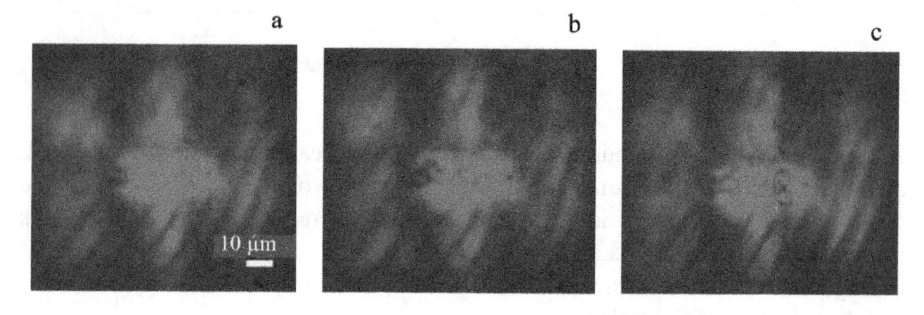

Fig. 8.77. The trapping of micro-objects by an Airy beam.

used for trapping. Figure 8.77 shows the different stages of the trapping of micro-objects in 3 s steps.

As can be seen from Fig. 8.77 there is a stable trapping of several polystyrene beads. The Airy beams may be useful for arranging micro-objects in a row.

The need to work with increasingly small objects gives rise to the need for an appropriate instrument. Nanotechnologies, micromechanics, microbiology require for affordable and convenient tools to move objects ranging in size from several micrometers to several nanometers. The optical trapping using specially-formed beams allows us to solve these problems. Specially shaped beams can trap, position, move and rotate the micro-objects. The rotation of micro-objects using a specially designed DOE can solve the problem of a particular drive for micromechanical systems. Such a task requires to rotate the micro-object with a certain moment of the forces known in advance at the stage of the calculation of the DOE. The possibility of using DOEs to form with good accuracy the required distribution, as well as the ability to accurately calculate the strength of the action of the light field can solve this problem. Using microtrapping laser technology in biology and medicine often requires precise knowledge of the magnitude of the forces acting on the micro-object and in this case it is desirable to minimize the power of the laser beam to prevent damage to biological micro-objects. This problem is also solved in principle using the DOEs. In the future, the DOEs can solve any problem in which it is necessary to move the micro-objects.

References

1. Debye P., Der Lichtdruck and Kugeln von beliebige Material, Ann. Phys. 1908. V. 30. P. 57–136.
2. Navade Y., Asakure T., Radiation forces on a dielectric sphere in the Rayleigh scatterry regime, Opt. Commun., 1996. V. 124. P. 529–541.
3. Pobrem R., Salome C., Radiation force on a nonlinear microsphere by a lightly focused Gaussian beam, Appl. Opt. 2002. V. 41 , No. 36. P. 7694–7701.
4. Rockstuhl C., Herzig H. P. Calculation of the torque on dielectric elliptical cylinders, Opt. Soc. Am. A. 2005. V. 22, No. 1. P. 109–116.
5. Volostnikov V.G., SP Kotova., Rakhmatulin M.P., Izv. Samarsk. Nauch. Tsentra RAN, 2000. V. 2, N. 1. P. 48–52.
6. Kotlyar V.V., et al., Komp. Optika, 2005. N. 29. P. 29–36.
7. Kotlyar V.V., Nalimov P.G., Komp. Optika, 2005. N. 27. P. 105–111.
8. Kotlyar V.V., et al., Komp. Optika, 2003. N. 25. P. 24–28.
9. Kotlyar V.V., et al., Opticheskii Zhurnal. 2005. V. 72, N. 5. P. 55–61.
10. Kotlyar V.V., et al., Komp. Optika, 2006. N. 30. P. 16–22.
11. Kotlyar V.V., et al., Pis'ma Zh. Teor. Fiz.. 1997. V. 23, N. 17. P. 1–6.
12. Kotlyar V.V., et al., Komp. Optika, 2007. N. 31. P. 35–38.
13. Rakhmatulin M.P., Development of methods for manipulating microscopic objects with laser radiation, Disseration, 2003.
14. Skidanov R.V., Komp. Optika, 2006. N. 29. P. 4–23.
15. Skidanov R.V., Komp. Optika, 2005. N. 29. P. 18–22.
16. Skidanov R.V., et al., Izv. SNTs RAN, 2006. N. 4. P. 1200–1211.
17. Skidanov R.V., et al., Komp. Optika, 2007. N. 31. P. 14–21.
18. Soifer V.A., et al., Optical micromanipulation using DOE, Official materials of 2nd Intern. Forum 'Holography EXPO 2005'. P. 69–70.
19. Soifer V.A., et al., Computer Optics. 2005. N. 29. P. 5–17.
20. Soifer V.A., et al., Proceedings of the All Russian seminar 'Yu.I.Denisyuk – founder of optical holography', FTI RAN, St. Peterburg, 2007, 116–123
21. Khonina S.N., et al., Komp. Optika, 1998. N. 18.
22. Abramochkin E.G., Volostnikov V.G. Generalized Gaussian beams, J. Opt. A: Pure APl. Opt. 2004. V. 6. P. 5157–5161.
23. Abramochkin E.G., Kotova S.P., Korobtsov A.V., Losevsky N.N., Mayorova A.M., Rakhmatulin M.A., Volostnikov V.G., Microobject manipulations using laser beams with nonzero orbital angular momentum, Laser Physics. 2006. V. 16 No. 5.
24. Abramochkin E.G., Razueva E.V., Volostnikov V.G., Application of spiral laser beams for beam shaping problem, Proc. of LFNM. Kharkiv, Ukraine, June 29–July 1, 2006. P. 275–278.

25. Abramochkin E.G., Razueva E.V., Volostnikov V.G., Fourier invariant singular wave–fields and beam shaping problem, Proc. of LFNM. Kharkiv, Ukraine, June 29–July 1, 2006. P. 370–373.
26. Abramochkin E.G., Volostnikov V.G., Gaussian beams: new aspects and applications, Proc. of LFNM. Kharkiv, Ukraine, June 29–July 1, 2006. P. 267–274.
27. Allen L., et al., Orbital angular momentum of light and the transformation of Laguerre–Gaussian laser modes, Phys. Rev. 1992. V. 45. P. 8185–8188.
28. Arlt J. et al. Optical micromanipulation using a Bessel light beams, Opt. Commun. 2001. V. 197. P. 239–245.
29. Ashkin A. et al., Observation of a single–beam gradient force optical trap for dielectric particles, Opt. Lett. 1986. V. 11, No. 5. P. 288–290.
30. Ashkin A., Acceleration and trapping of particles by radiation pressure, Phys. Rev. Lett. 1970. V. 24 , No. 4. P. 156–158.
31. Ashkin A., Dziedzic J.M., Observation of resonances in the radiation pressure on dielectric spheres, Phys. Rev. Lett. 1977. V. 38, No. 23. P. 1351–1354.
32. Ashkin A., Dziedzic J.M., Optical levitation by radiation pressure, Appl. Phys. Lett. 1971. V. 19. P. 283–285.
33. Bandres M.A., Gutierrez-Vega J. C., Chavez-Cedra S., Parabolic non-diffracting optical wave fields, Opt. Lett. 2004. V. 29 , No. 1. P. 44–46.
34. Bentley J. B., Devis J.A., Bandres M.A., Gutierrez-Vega J.C., Generation of helical Ince–Gaussian beams with a liquid-crystal display, Opt. Lett. 2006. V. 31, No. 5. P. 649–651.
35. Bretenaker F., Le Floch A., Energy exchange between a rotating retardation plate and a laser beam, Phys. Rev. Lett. 1990. V. 65, No. 18. P. 2316.
36. Chen C., Konkola P., Ferrera J., Keilmann R., Schaffenberg M., Analysis of vector Gaussian beam propagation and the validity of paraxial and spherical approximations, J. of Optical Society of America A. 2002. V. 19, No. 2. P. 404–412.
37. Cojoc D., et al., Design and fabrication of diffractive optical elements for optical tweezer arrays by means of e-beam lithography, Microelectronic Engineering, 2002. V. 61–62. P. 963–968.
38. Curtis J. E., Koss B.A., Grier D.G. Dynamic holographic optical tweezers, Optics Commun. 2002. V. 207. P. 169–175.
39. Davis J.A., Guertin J., Cottrell D.M. Diffraction–free Beam Generated with Programmable Spatial Light Modulators, Appl. Opt. 1993. V. 32, No. 31. P. 6368–6370.
40. Davis J.A., Carcole E., Cottrell D.M., Intensity and phase measurements of non-diffracting beams generated with the magneto-optic spatial light modulator, Appl. Opt. 1996. V. 35, No. 4. P. 593–598.
41. Debye P. Der Lichtdruck and Kugeln von beliebige Material, Ann. Phys. 1908. V. 30. P. 57–136.
42. Dufresne E.R., Grier D.G., Optical tweezer arrays and optical substrates created with diffractive optical elements, Rev. Sci. Instr. 1998. V. 69, No. 5. P. 1974–1977.
43. Dufresne E.R., et al., Computer-generated holographic optical tweezer arrays, Rev. Sci. Instr. 2001. V. 72. P. 1810.
44. Durnin J., Miceli J.J., Eberly J.H., Diffraction–free beams, Phys. Rev. Lett. 1987. V. 58. P. 1499–1501.
45. Durnin J. Exact solution for non-diffracting beams. I. The scalar theory, J. Opt.Soc. Am. 1987. V. 4. P. 651–654.
46. Fällman E., Axner O., Design for fully steerable dual–trap optical tweezers, Appl. Opt. 1997. V. 36. P. 2107.

47. Farafonov V. S., Ilin U.B., Henning T., A new solution of the light scattering problem for axisymmetric particles, J. of Quantitative Spectroscopy and Radiative Transfer. 1998. V. 63. P. 205–215.
48. Fedotowsky A., Lehovec K., Optimal filter design for annular imaging, Appl. Opt. 1974. V. 13, No. 12. P. 2919–2923.
49. Friese M.E.J., et al., Optical alignment and spinning of laser-trapped microscopic particles, Nature. 1998. V. 394. P. 348–350.
50. Friese M.E.J., et al., Optical angular–momentum transfer to trapped absorbing particles, Phys. Rev. 1996. V. 54, No. 2. P. 1593–1596.
51. Ganic D., Gan X., Gu M., Exact radiation trapping force calculation based on vectorial diffraction theory, Opt. Express. 2004. V. 12, No. 12. P. 2670–2675.
52. Garces-Chavez V. et al. Simultaneous micromanipulation in multiple planes using a self-reconstructing light beam, Nature. 2002. V. 419. P. 145–147.
53. Grier D.G., Dufresne E. R. US Patent 6,055,106. The University of Chicago, 2000.
54. Grover S., et al., Automated single-cell sorting system based on optical trapping, J. Biomed. Opt. 2001. V. 6. P. 14.
55. Guo C., Liu X., He J., Wang H., Optimal annulus structures of optical vortices, Opt. Express. V. 12, No. 19. P. 4625–4634.
56. Hahn J., Kim H., Choi K., Lee B., Real-time digital holographic beam-shaping system with a genetic feedback tuning loop, Appl. Opt. 2006. V. 45, No. 5. P. 915–924.
57. Harris M., Hill C.A., Vaughan J.M., Optical helices and spiral interference fringes, Optics Commun., 1994. V. 106, No. 4–6. P. 161–166.
58. He H., et al., Direct observation of transfer of angular momentum to absorptive particles from a laser beam with a phase singularity, Phys. Rev. Lett. 1995. V. 75, No. 5. P. 826–828.
59. Higurashi E., Sawada R., Ito T., Optically induced angular alignment of trapped birefringent microobjects by linear polarization, Appl. Phys. Lett. 1998. V. 73. P. 3034.
60. Holmlin R.E., et al,. Light–driven microfabrication: Assembly of multi-component, three-dimensional structures by using optical tweezers, Angew. Chem. Int. Ed. Engl. 2000. V. 39. P. 3503.
61. Hong Du, Hao Zhang, Ultra-high precision Mie scattering calculations. 2002.
62. Khonina S.N., Skidanov R.V., Kotlyar V.V., Soifer V.A., Rotating microobjects using a DOE-generated laser Bessel beam, Proc. of SPIE. 2004. V. 5456. P. 244–255.
63. Khonina S.N., Skidanov R.V., Kotlyar V.V., Soifer V.A., Turunen J., DOE-generated laser beams with given orbital angular moment: application for micromanipulation, Proc. of SPIE Intern. Soc. Opt. Eng. 5962, 59622W. Optical Design and Engineering II. Jena, Germany, Oct. 2005.
64. Khonina S.N., Skidanov R.V., Kotlyar V.V., Kovalev A.A., Soifer V.A., Optical micromanipulation using DOEs matched with optical vorticies, Proc. SPIE– 2006. V. 6187. P. 61871F.
65. Khonina S.N. et al., The phase rotor filter, J. Modern Optics. 1992. V. 39, No. 5. P. 1147–1154.
66. Kotlyar V.V., Almazov A.A., Khonina S.N., Soifer V.A., Elfstrom H., Turunen J. Generation of phase singularity through diffracting a plane or Gaussian beam by a spiral phase plate, J. Opt. Soc. Am. A. 2005. V. 22, No. 5. P. 849–861.
67. Kotlyar V.V., Khonina S.N., Soifer V.A., Calculation of phase formers of non-diffracting images and a set of concentric rings, Optik. 1996. V. 102, No. 2. P. 45–50.
68. Kotlyar V.V., Soifer V.A., Khonina S.N., Rotation of multimodal Gauss–Laguerre light beans in free space and in a fiber, Optics and Lasers in Engineering. 1998. V. 29. P. 343–350.

69. Kotlyar V.V., Kovalev A.A., Khonina S.N., Skidanov R.V., Soifer V.A., Elfstrom H., Tossavainen N., Turunen J., Diffraction of conic and Gaussian beams by a spiral phase plate, Appl. Opt. 2006. V. 45, No. 12. P. 2656–2665.
70. Kotlyar V.V., Kovalev A.A., Skidanov R.V., Moiseev O.Yu., Soifer V.A., Diffraction of a finite-radius plane wave and a Gaussian beam by a helical axicon and a spiral phase plate, J. Opt. Soc. Am. A. 2007. V. 24, No. 7.
71. Leach J., et al., Interactive approach to optical tweezers control, Appl. Opt. 2006. V. 45. P. 897–903.
72. Lemire T., Coupled-multipole formulation for the treatment of electromagnetic scattery by small dielectric particles of arbitrary sphere, J. Opt. Soc. Am. A. 1997. V. 14. P. 470–474.
73. MacDonald M.P. et al., Creation and manipulation of three–dimensional optically trapped structures, Science. 2002. V. 296. P. 1101–1103.
74. Marston P. L., Chrichton J.H., Radiation torque on a sphere caused by circularly polarized electromagnetic wave, Phys. Rev. A. 1984. V. 30, No. 3. P. 2508–2516.
75. Miller W., Symmetry and separation of variables. Addison-Wesley Publ. Comp., MA, 1977.
76. Mingwei G., Chunqing G., Zhifeng L., Generation and application of the twisted beam with orbital angular momentum, Chinese Optics Lett. 2007. V. 5, No. 2.
77. Navade Y., Asakure T. Radiation forces on a dielectric sphere in the Rayleigh scatterry regime, Opt. Commun. 1996. V. 124. P. 529–541.
78. Nieminen T.A., Rubinsztein-Dunlop H., Heckenberg N.R., Calculation and optical measurement of laser trapping forces on non-spherical particles, J. of Quantitative Spectroscopy & Radiative Transfer. 2001. V. 70. P. 627–637.
79. Reicherter M., et al., Optical particle trapping with computer-generated holograms written on a liquid- crystal display, Opt. Lett. 1998. V. 24. P. 608–610.
80. Rohrbach A., Stelzer E.H., Optical traping of dielectric particles in arbitrary fields, J. Opt. Soc. Am. A. 2001. V. 18, No. 4. P. 813–838.
81. Paterson C., Smith R. Higher–order Bessel waves produced by axicon-type computer-generated holograms, Optics Commun., 1996. V. 124, No. 1–2. P. 121–130.
82. Paterson L. et al., Controlled rotation of optically trapped microscopic particles, Science. 2001. V. 292, No. 5. P. 912–914.
83. Pobre R., Salome C., Radiation force on a nonlinear microsphere by a lightly focused Gaussian beam, Appl. Opt. 2002. V. 41, No. 36. P. 7694–7701.
84. Sasaki K. et al., Pattern formation and flow control of fine particles by lasers canning micromanipulation, Opt. Lett., 1991. V. 16. P. 1463.
85. Sato S., Ishigure M., Inaba H., Optical trapping and rotational manipulation of microscopic particles and biological cells using higher-order mode Nd:YAG laser beams, Electron. Lett., 1991. V. 27, No. 20. P. 1831–1832.
86. Shaohui Y., Baoli Y., Transverse trapping forces of focused Gaussian beam on ellipsoidal particles, J. Opt. Soc. Am. B. 2007. V. 24, No. 7. P. 1596–1602.
87. Schwarz U.T., Bandres M.A., Gutierrez-Vega J., Observation of Ince–Gaussian modes in stable resonators, Opt. Lett. 2004. V. 29, No. 16. P. 1870–1872.
88. Skidanov R.V., Khonina S.N., Kotlyar V.V., Soifer V.A., Optical microparticle trapping and rotating using multi-order DOEs, Proc. of the ICO Topical Meeting on Optoinformatics Information Photonics'2006, St. Peterburg, Russia, Sept. 4–7, 2006. P. 466–468.
89. Skidanov R.V., Kotlyar V.V., Khonina V.V., Volkov A.V., Soifer V.A., Micromanipulation in Higher-Order Bessel Beams, Optical Memory and Neural Networks (Information Optics). 2007. V. 16, No. 2. P. 84.

90. Soifer V.A., Kotlyar V.V., Khonina S.N., Skidanov R.V., Remarkable laser beams formed by computer-generated optical elements: properties and applications, Proc. of SPIE–2006. V. 6252, P. 62521B.
91. Soifer V.A., Kotlyar V.V., Khonina S.N., Skidanov R.V., Optical data processing using DOEs, Methods for Computer Design of Diffractive Optical Elements. Wiley Interscience Publ. John Wiley & Sons, Inc., 2002. Ch. 10. P. 673–754.
92. Turunen J., Vasara A., Friberg A.T., Holographic generation of diffraction-free beams, Applied Optics. 1988. V. 27. P. 959–962.
93. Volke-Sepulveda K., Garces-Chavez V., Chavez-Cerda S., Arlt J., Dholakia K., Orbital angular momentum of a high-order Bessel light beam, J. Opt. B: Quantum Semiclass. Opt. 2002. V. 4. P. 82–88.
94. Xun X.D., Cohn R.W., Phase calibration of spatially nonuniform spatial light modulators, Appl. Opt. 2004. V. 43. P. 6400–6406.
95. Durnin J., Miceli J.J., Eberly J.H., Diffraction-free beams, Phys. Rev. Lett. 1987. V. 58, No. 15. P. 1499–1501.
96. Vasara A., Turunen J., Friberg A.T., Realization of general non-diffracting beams with computer–generated holograms, J. Opt. Soc. Am. A. 1989. V. 6. P. 1748–1754.
97. Bandres M.A., Gutierrez-Vega J. C., Chavez-Cerda S., Parabolic non-diffracting optical wave fields, Opt. Lett. 2004. V. 29 (1). P. 44–46.
98. Kotlyar V.V., Skidanov R.V., Khonina S.N., Soifer V.A., Hypergeometric modes, Opt. Lett. 2007. V. 32. P. 742–744.
99. Siviloglou G.A., Broky J., Dogariu A., Christodoulides D.N., Observation of Accelerating Airy Beams, Phys. Rev. Lett. 2007. V. 99. P. 213901.
100. Bandres M.A., Gutierrez-Vega J. C. Circular beams, Opt. Lett. 2008. V. 33. P. 177–179.
101. Balalaev S.A., Khonina S.N., Comparison of the properties of hypergeometric modes and Bessel modes, Komp. Optika. 2007. V. 31 (4). P. 23–28.
102. Siviloglou G.A., Broky J., Dogariu A., Christodoulides D.N., Ballistic dynamics of Airy beams, Opt. Lett. 2008. V. 33 (3). P. 207–209.
103. Berry M.V., Balazs N.L., Non-spreading wave packets, Am. J. Phys. 1979. V. 47 (3). P. 264–267.
104. Abramowitz M., Stegun I.A., Handbook of Mathematical Functions. Dover, 1972.
105. Siviloglou G.A., Christodoulides D.N., Accelerating finite energy Airy beams. Opt. Lett. 2007. V. 32 (8). P. 979–981.
106. Banders M.A., Gutierrez-Vega J.C.. Airy–Gauss beams and their transformation by paraxial optical systems, Opt. Express. 2007. V. 15 (25). P. 16719–16728.
107. Marhic M.E., Kwan L.I., Epstein M. Whispering Gallery CO_2 Laser, IEEE J. Quant. Electr. 1979. V. QE–15 (6). P. 487–490.
108. Grossman J.G., Casperson L.W., Stafsudd O.M., Radio frequency-excited carbon dioxide metal waveguide laser, Appl. Opt. 1983. V. 22 (9). P. 1298–1305.
109. Al-Mashaabi F.S., Casperson L.W., Direct current-excited cw CO_2 metal waveguide laser, Appl. Opt. 1989. 1989. V. 28 (10). P. 1899–1903.
110. Khonina S.N., Volotovskiy S G., Limited 1D Airy beams : laser fan, Komp. Optika, 2008. V. 32 (2). P. 168–174.
111. Casaubon J.I., Cosentino J P., Buep A.H., Variation Principle for a Linear Potential, Turk. J. Phys. 2007. V. 31. P. 117–121.
112. Khonina S.N., Kotlyar V.V., Skidanov R.V., Soifer V.A., Levelling the focal spot intensity of the focused Gaussian beam, J. of Modern Optics. 2000. V. 47 (5). P. 883–904.
113. Kotlyar V.V., Khonina S.N., Soifer V.A., Method of partial coding for calculation of phase generators of Gauss–Hermite modes, Avtometriya, 1999. V. 6. P. 74–83.

Conclusion

In the book, attention is paid to the diffraction of laser radiation on nanoscale inhomogeneities and optical nanostructures and also focusing of light in nanosized spaces. The authors applied their experience in the field of computer diffractive optics to solving nanophotonics problems. Based on mathematical modelling and numerical solution of inverse problems of the theory of diffraction of light, new devices containing diffractive optical elements (DOE), significantly extending the component base of nanophotonics, were investigated and constructed. The study of the diffraction of light is based on the solution Maxwell's equations in different ways: the difference method (FDTD-method), the beam propagation method (BPM-method), the method of coupled waves (RWCA), integral methods for solving diffraction problems, the method of finite and boundary elements, the method of matched sinusoidal modes. These and other methods are used in the book for the simulation of light diffraction on sub-wavelength gratings profiled metal–dielectric heterostructures, metallic nanorods, two-dimensional and three-dimensional photonic crystals, photonic–crystal fibres and lenses and on other nanophotonics devices. The optical manipulation of micro- and nano-objects with singular vortex laser beams is studied and the problem of sharp sub-wavelength focusing of laser light having radial polarization is also solved.

According to the authors, the following are actual directions of the development of diffractive nanophotonics.

• Research of extraordinary (resonance) of the optical and magneto-optical effects formed in diffraction of electromagnetic waves on heterostructures containing a regular system of curvilinear steps or gaps. The most interesting are radially symmetric heterostructures containing dielectric and metal layers, perforated by a concentric system of annular slits. There is sufficient evidence to suggest the presence in radially symmetric heterostructures of a whole spectrum of resonance effects that exist in two-dimensional structures and including extraordinary transmission, magneto-optical effects associated with resonance changes in the spectra of passage and reflection and rotation of the polarization plane due to the change of magnetization, formation of areas with a high degree of localization of energy due to the interference of evanescent and plasmon waves. The presence of the central zone allows one to expect new optical effects related with increased energy and localization of the light wave energy in the centre of the structure, as well as with focusing in the centre of the structure of plasmon and quasi-plasmon modes.

• An important element in the study of diffractive structures with curved areas is the creation of efficient computational methods of electromagnetic simulation. In this case, it is advisable to solve Maxwell's equations using selected curvilinear coordinates. In particular, the radially symmetric structures have cylindrical symmetry. For efficient simulation of diffraction in such structures it is necessary to develop a modal method for solving diffraction in cylindrical coordinates. In this method, the incident and scattered electromagnetic field and also the field within the layers of the structure will be presented in the basis of conical waves.

• In the field of plasmonics it is interesting to study metal–dielectric heterostructures with magnetized layers in problems of the control of surface electromagnetic waves (SEW) and the formation of SEW interference patterns. In such heterostructures we can expect new magneto-optical effects associated with the change of the form of the interference pattern of SEW depending on the magnetization of the layers. When SEW passes through a magnetized diffraction structure just above the surface of SEW propagation, the change of magnetization can provide new opportunities for the modulation of SEW parameters. Thus, the prospects of this research are related with better control of SEW due to changes in the magnetization of the heterostructure layers. An important area of research is also the study of plasmonic effects in the radial heterostructures. In particular, it is of great interest to study heterostructures consisting of a radial diffractive grating and a uniform metal layer. Such structures can shows the formation of ring interference patterns of surface plasmons and their focusing in the centre of the structure on the surface of the metallic layer.

• Of great interest is the further study of gradient photonic crystals, including photonic–crystal lenses to create nanophotonic devices that allow selective focusing of laser light with a certain wavelength and overcoming the diffraction limit. The main difficulty in modelling gradient photonic crystals is that they have impaired spatial periodicity and, therefore, no complete band gaps. However, the properties of frequency selectivity are retained because their thickness is much greater than the light wavelength. The gradient photonic crystals have a frequency selectivity similar to volume holograms, but unlike them the photonic crystals have higher contrast distribution of the refractive index. The high contrast of changes of the refractive index and large thickness of the gradient photonic crystals allow us to hope for the low threshold for formation of non-linear effects.

• The development of nanotechnologies for structuring the optical materials with nanometer resolution, and the creation of high-performance computing systems open up the possibility of synthesizing full-aperture elements of nanooptics and nanophotonics with complex topology and significant subwave characteristic dimensions of the nanorelief.

The development of the technology of three-dimensional micro- and nanostructuring (such as the two-photon polymerization technique and the technology of refractive index modification by femtosecond laser radiation) allows us to generalize the well-known statement of the problem and the calculation of the two-dimensional function of the microrelief of 'flat' DOEs to three dimensions. This will allow, for example, to produce the desired change in the amplitude–phase distribution in the cross section of the beam without causing energy losses associated with the introduction of spurious diffraction orders to bring the complex transmission function of the 'two-dimensional' DOE to the purely phase form. It is interesting to note that the technology of computer-controlled nanostructuring, successfully used for synthesis of diffractive optical elements, is also used to create optical metamaterials and photonic crystals and quasicrystals with the predetermined spectral properties. In this sense, the advantage of such technologies in comparison with the so-called group technologies (technologies based on self-organization, sol-gel processes, etc.) is obvious, since the computer-controlled nanostructuring allows the creation of optical metamaterials with pre-calculated parameters .

• To adequately describe the sharp focusing of laser radiation with overcoming of the diffraction limit, it is important to develop effective methods of non-paraxial vector modelling of radiation on the basis of the Rayleigh–Sommerfeld and Richards–Wolf integral expressions, or on the basis of an integral expansion of the light fields on the plane waves with the use of the fast Fourier transform algorithm. Improvement of the integrated methods of diffraction theory allows to effectively simulate the propagation of light in the near-field zone with input from the decaying evanescent waves; to simulate the use of DOE in focusing microscopic systems to form distributions in the focal region with the specified spatial and polarization properties and imaging microscopic systems to increase the resolution (and achieving super-resolution) using nanostructured microoptics.

• Development of the theory of vortex laser beams and their use for optical trapping and manipulation of micro- and nano-objects should be aimed at improving the energy efficiency of formation of the annular intensity distribution of the vortex laser beam in the focal plane. It is known that to improve the efficiency of optical trapping and retention of micro-and nano-particles in a light ring it is required to increase the energy density or decrease the diffractive thickness of the light rings. The size of the focal spot or diffractive thickness of the light rings can be reduced by the use of radially polarized light. Radially polarized beams are formed, for example, by subwavelength DOEs the fragments of which are subwavelength diffractive gratings. After passing through these gratings the electric field vector rotates at a certain angle. Therefore current research is focusing on the vector non-paraxial vortex beams with radial or azimuthal polarization.

There remains a pressing task of improving numerical methods of diffractive nanophotonics.

• Development of hybrid computationally efficient schemes numerical solutions of Maxwell's equations, which combine a rigorous approach to solving the problems of diffraction (FDTD method) on optical nanostructures and various approximate methods of solutions in the areas whose characteristic dimensions exceed the wavelength.

• Study of the propagation of electromagnetic pulses through the FDTD method. Calculation of the diffraction relief of optical elements for controlling the amplitude and phase characteristics of the pulses. Application of the method of numerical solution of Maxwell's equations for the study of propagation of radiation in non-linear media, in particular the study of the formation and interaction of solitons in photorefractive crystals under the influence of an external electric field.

• Recording of difference equations, more precisely taking into account the physics of the processes: on the grid areas corresponding to the geometric characteristics of the studied elements and the characteristics of the incident wave; for the areas of sparse on homogeneous areas and dense at the borders sections of media in accordance with the topography of such boundaries; on moving areas, taking into account the spatial localization of the electromagnetic pulse.

• Synthesis of specialized vector algorithms for the calculation of the diffraction pattern on parallel computing devices.

Note that all of the practical results of this book relate to the optical wave range but the methods developed can also be used for X-rays, which certainly requires a large amount of additional research.

The solution of these and other problems of the diffractive nanophotonics is the aim of the efforts of the authors at the present time.

Appendix A

Simulation using FULLWAVE

In the problems of nanophotonics devices often have sizes comparable and even smaller than the wavelength of light. In this case, the geometric optics methods are not applicable, and it is required to investigate the propagation of light within the strict wave vector theory of the electromagnetic field. The analytical solution of such problems is often not possible, and therefore it is necessary to resort to numerical simulations.

The best known universal method of numerical solution of Maxwell's equations is the finite-difference FDTD-method (Finite-Difference Time-Domain).

One of the programs that implements this method is the program FullWAVE of the company RSoft Design Group (http://www.rsoftdesign.com), designed to simulate the propagation of light through a variety of nanophotonic devices, including integrated and fibre optics devices, and photonic crystals.

A.1. Brief description of the FDTD-method

Consider a region of space without charges and currents. In this case, the first two Maxwell's equations take the form:

$$\begin{cases} \dfrac{\partial H_x}{\partial t} = -\dfrac{1}{\mu}\left(\dfrac{\partial E_y}{\partial z} - \dfrac{\partial E_z}{\partial y}\right), \\ \dfrac{\partial E_y}{\partial t} = -\dfrac{1}{\varepsilon}\left(\dfrac{\partial H_x}{\partial z} - \dfrac{\partial H_z}{\partial x}\right). \end{cases} \tag{A.1}$$

The other four equations have a similar form.

In the Maxwell's equations the electric field depends in time on the variation of the magnetic field in the space and vice versa. The FDTD-method solves Maxwell's equations by discretization of the variables occurring in it, by replacing the derivatives by their difference analogues, and by numerical solutions of the resulting systems of linear equations. Time is quantized with a discretization interval of Δt. The components of the electric vector are calculated at moments $t = n\Delta t$, and those of the magnetic vector – a step $t = (n + 1/2)\Delta t$ (n – is an integer). This method leads to six equations, which can be used to calculate the components of the field. For example, two of these equations are:

$$H^{n+1/2}_{x(i,j,k)} = H^{n-1/2}_{x(i,j,k)} + \frac{\Delta t}{\mu\Delta z}[E^n_{y(i,j,k)} - E^n_{y(i,j,k-1)}] - \frac{\Delta t}{\mu\Delta y}[E^n_{z(i,j,k)} - E^n_{z(i,j-1,k)}],$$

$$E^{n+1}_{x(i,j,k)} = E^n_{x(i,j,k)} + \frac{\Delta t}{\varepsilon\Delta y}[H^{n+1/2}_{z,(i,j+1,k)} - H^{n+1/2}_{z,(i,j,k)}] - \frac{\Delta t}{\varepsilon\Delta z}[H^{n+1/2}_{y,(i,j,k+1)} - H^{n+1/2}_{y,(i,j,k)}]. \quad \text{(A.2)}$$

where Δx, Δy, Δz are discretization steps along the axes x, y, z. Discretization steps must satisfy the Courant condition:

$$c\Delta t < \frac{1}{\sqrt{\frac{1}{\Delta x^2} + \frac{1}{\Delta y^2} + \frac{1}{\Delta z^2}}}, \quad \text{(A.3)}$$

where c is the speed of light in vacuum.

To account for the material properties of the medium in the FullWAVE software we used the following equation:

$$\begin{aligned} \mathbf{D} &= \varepsilon_0 \mathbf{E} + \mathbf{P}, \\ \mathbf{B} &= \mu_0 \mathbf{H} + \mathbf{M}, \\ \mathbf{P} &= \varepsilon_0 \left[\chi(\omega)\mathbf{E} + \chi^2\mathbf{E}^2 + \chi^3(\omega)\frac{I}{1+c_{\text{sat}}I}|\mathbf{E}|^2\,\mathbf{E} \right], \\ \mathbf{M} &= \mu_0 \left[\chi_m(\omega)\mathbf{H} + \chi_m^2\mathbf{H}^2 + \chi_m^3(\omega)\frac{I}{1+d_{\text{sat}}I}|\mathbf{H}|^2\,\mathbf{H} \right]. \end{aligned} \quad \text{(A.4)}$$

In the last two equations, the first term takes into account the variance (refractive index dependency on the wavelength of light), the second and third terms correspond to non-linear effects: the non-linearity of order 2 and the frequency-dependent non-linearity of the third order.

A.2. The main components of the program

To perform the numerical simulation by the FDTD-method with the RSoft software package we require the following components to be included therein:

RSoft CAD Layout (file bcadw32.exe) – a program for designing micro-optics elements.

FullWave (file fullwave.exe) – program for the simulation of the electromagnetic fields passing through the device, designed in the program BCADW32.EXE.

WinPlot utility (file winplot.exe) – software for viewing the results stored in simulation by the program FullWAVE.

A.3. Program design elements of micro-optics

The design program allows setting the refractive index distribution in the plane (or in the case of a two-dimensional or axisymmetric three-dimensional modelling) and in space (in the case of three-dimensional modelling). In creating a model the following parameters are chosen:

– Modelling tool (for the nanophotonics tasks we typically require FullWAVE, but in the software package RSoft there are others, for example BeamPROP method for modelling by the beam propagation method).

– The dimension of the model (two-dimensional, three-dimensional or three-dimensional axially symmetric model)

– Type of structure (fibre, planar waveguide, etc)

– Other parameters (wavelength of light in free space, the refractive index of the material used in the model, profiles of the refractive index).

Next we will consider other features of the program.

Using tapers and profiles

When creating microoptics elements in the program BCADW32.EXE we can use profiles and tapers to control the width, height, curvature and refraction index of the waveguide. Figure A.1, *a* shows the distribution of the refractive index of the Mikaelian lens that was created with the profile defined by the formula:

$$n(r)=\frac{n_0}{\mathrm{ch}\left(\pi r/2L\right)}, \tag{A.5}$$

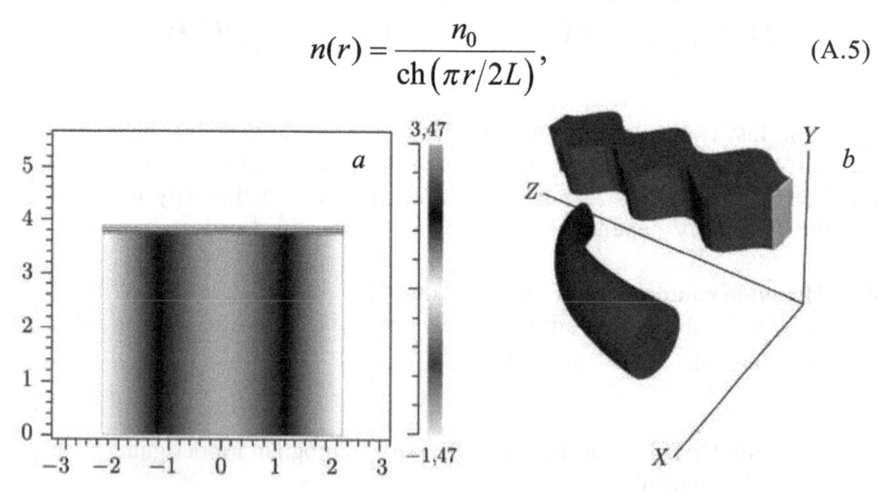

Fig. A.1. The use of profiles and tapers to create models in design program BCADW32. EXE.

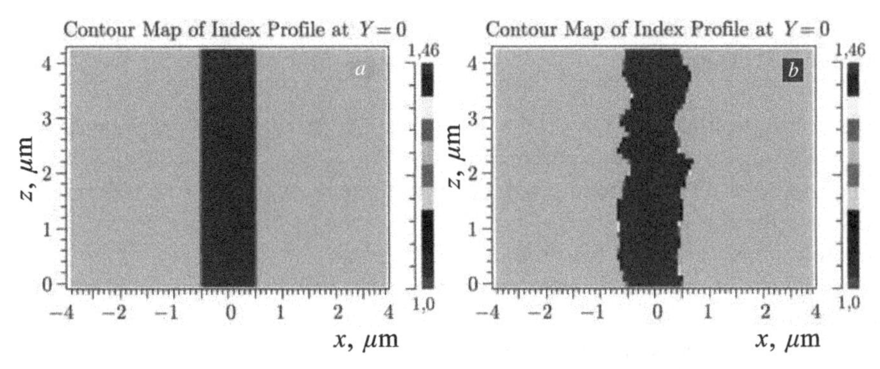

Fig. A.2. Adding of manufacturing errors in modelling in the design program RSoft CAD Layout.

where n_0 and L is the refractive index at the axis and the length of the lens. Fig. A.1, *b* shows a corrugated waveguide and a waveguide with a bend, built using the tapers for coordinates and tapers in width.

Simulation considering manufacturing errors

Modelling can also be carried out taking into account the technological errors of manufacturing of elements which in the packet RSoft are given by the depth and length of roughness correlation. Figure A.2 shows a planar waveguide without or with distortions.

Automatic generation of grids and other periodical devices

For some applications it is required to investigate the devices that are difficult to manually design. For this reason, the program comprises a means for automatic creation of one-, two- and three-dimensional devices comprising periodic elements (Fig, A.3).

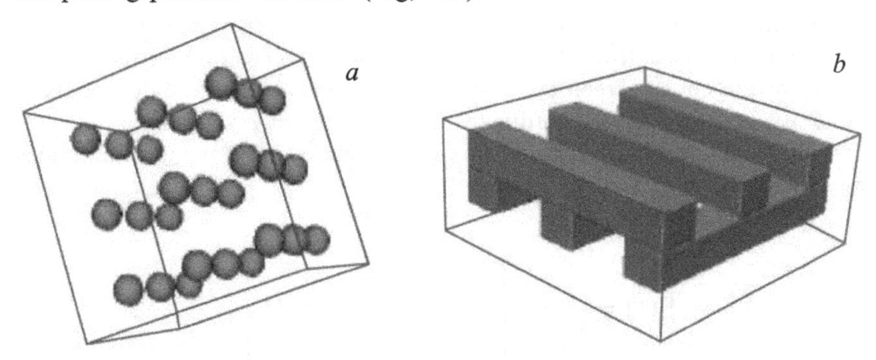

Fig. A.3. Examples of creating periodic structures in the design program RSoft CAD Layout

A.4. The program for the modelling of propagation of the electromagnetic field by FDTD method

The simulation can be run either in the graphic interface of program RSoft CAD Layout, or from the command line. The second option allows for a consistent set of experiments without interference of the user (for example, to leave for the night).

Below are the main features of the program:

– Two-dimensional, three-dimensional and radially symmetrical modelling

– Non-uniform sampling grid

– Taking into account dispersion, non-linear and anisotropic effects

– Assignment of boundary conditions in the form of a completely absorbing layer, as well as periodic, symmetric and asymmetric conditions

– The use of several exciting fields (each has its own spatial–temporal characteristics, such as location, wavelength, direction, polarization)

– A large range of options for measuring and analyzing both the components of the field, as well as quantities such as power flow, the energy density, the overlap integrals, the Umov–Poynting vector

– The ability to automatically determine the modes of the resonators

– The possibility of parallel processing on multiple processors.

Output of results

The FullWAVE software calculates the electromagnetic field as a function of space and time for a given distribution of the refractive index and the given initial conditions and displays this field at the indicated times. Often, however, we must write this information for subsequent processing and analysis. For this the program uses screens, capable of recording both the values of all six components of the electromagnetic field and the propagation of electromagnetic quantities such as the power or energy density. Screens can implement both time and spatial output. Time screens measure the dependence of the given value on time: either some component of the field in the centre of the screen, or the total power flow or the energy density of the entire screen area. Spatial screens measured the dependence of the given value on the coordinates of the screen at given points in time.

With the help of time screens we can measure the following intergrated values:

– The flow of power

$$\mathbf{S}(t)=\frac{1}{S_0}\mathrm{Re}\left\{\int_A[\mathbf{E}(t)\times\mathbf{H}^*(t)]\cdot d\mathbf{A}\right\},$$

where S_0 – input power; A – area of the screen.

– Power density

$$S(t) = \frac{1}{2} \int_V \varepsilon(\mathbf{r}') |\mathbf{E}(t)|^2 \, dV,$$

– Magnetic energy density

$$S(t) = \frac{1}{2} \int_V \mu |\mathbf{H}(t)|^2 \, dV.$$

With the help of spatial screens we can measure the following integral quantities:

– The flow of power along the optical axis

$$S(\mathbf{r},t) = \mathrm{Re}\left[\mathbf{E}(\mathbf{r},t) \times \mathbf{H}^*(\mathbf{r},t)\right]_z,$$

– Power density

$$S(\mathbf{r},t) = \varepsilon(\mathbf{r}') |\mathbf{E}(\mathbf{r},t)|^2,$$

– The magnetic energy density

$$S(\mathbf{r},t) = \mu |\mathbf{H}(\mathbf{r},t)|^2,$$

– The Umov–Poynting vector

$$S(\mathbf{r},t) = \sqrt{\mathrm{Re}[S_x]^2 + \mathrm{Re}[S_y]^2 + \mathrm{Re}[S_z]^2},$$

where

$$S(\mathbf{r},t) = \mathbf{E}(\mathbf{r},t) \times \mathbf{H}^*(\mathbf{r},t).$$

Additional features of the software

The use of non-uniform sampling grid. In most cases, the following calculation parameters are specified:

– The dimensions of the regions on the axes x, y, z.
– Discretization steps along the axes x, y, z.
– The thickness of the absolutely absorbing layer along the axes x, y, z.

The program provides the possibility of forming an adaptive grid of samples, according to the designed model. That is, we can choose a large discretization step in areas with small changes in the refractive index and a small step near the boundaries of objects as shown in Fig. A.4. The use of such a grid can significantly reduce the computation time.

Modelling axially symmetric structures

The program FullWAVE also implements the radial FDTD-method to reduce the three-dimensional problems with axial symmetry for two-dimensional ones (e.g., in the case of waveguides with a round cross-section).

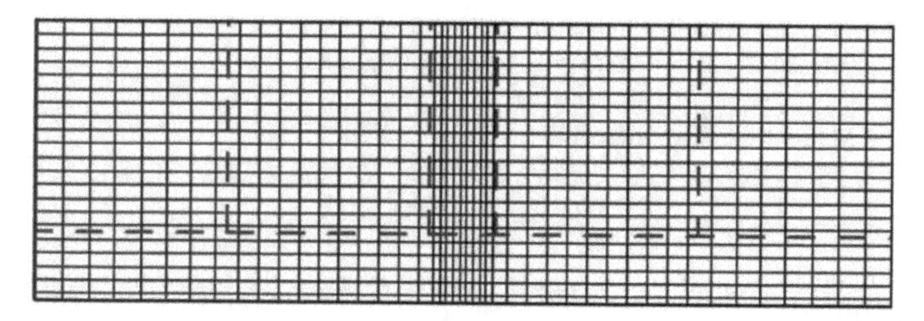

Fig. A.4. The use of non-uniform sampling grid (thin solid line – the border of samples, thick dashed lines – the edges of objects).

Parallel computing

The FDTD algorithm is easy to 'parallelize'. The FullWAVE program is capable of performing on multiple processors using the well-known interface MPI.

A.5. Program charting

After running the FullWAVE program there appears a set of text files containing the output data: the distribution of the electromagnetic field in a predetermined plane, as well as the data stored by the screens. These files are in a format defined by the manufacturer (RSoft), and open the charting programs WinPlot and DataBROWSER (Figs. A.5 and A.6).

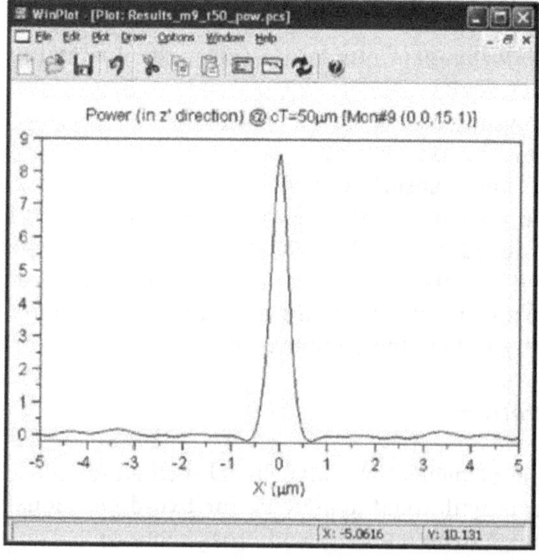

Fig. A.5. Program to visualize graphs WinPlot.

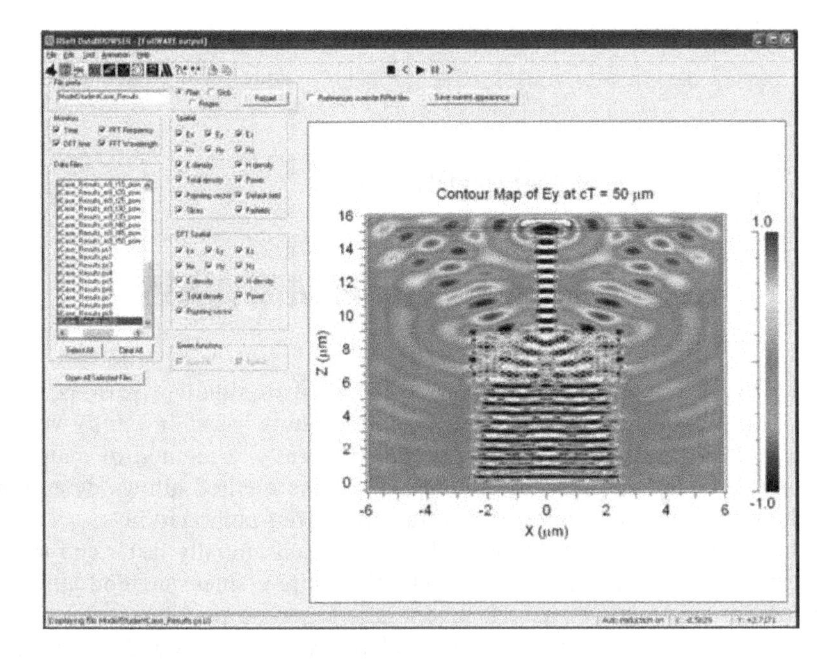

Fig. A.6. DataBROWSER program to visualize graphs.

In these programs, in addition to viewing graphs and images, there is also the ability to export them to common graphics formats. The DataBROWSER is also capable of producing videofiles in the MPEG format from several images.

Appendix B

Simulation using FIMMWAVE

The software product FIMMWAVE is designed to simulate various two-dimensional light-guiding structures. The program is based on a fully vector method of calculating the modes of optical fibres – a method of matched sinusoidal modes authored by Sudbo [1,2]. This method allows detection of almost any natural modes of arbitrary or mixed polarization.

The method of matched sinusoidal modes is substantially faster and more accurate than other competing methods: the finite element method and the finite difference method. Its effectiveness is based on the representation of a cross section of the fibre by a set of relatively large areas with a constant refractive index of the medium. In the case where the fibre has a gradient profile, or for some other reason its description requires a large amount of areas, the advantage of the method of the matched sinusoidal modes front in comparison with the finite-element method decreases.

FIMMWAVE 4.6 is a software implementation of the modified method of matched sinusoidal modes [3], which works significantly faster with complex structures and require fewer RAM resources.

Calculation of optical fibres in FIMMWAVE

B.1. Creation of a project

Before starting work with FIMMWAVE we need to create a working draft. To do this, in the main window, click the icon ⁂ on the new elements panel.

In the window that appears, enter a name for the new project 'Tutorial project' and click 'OK'.

The project appears in the project tree windows, as shown in the figure below. (Fig. B.1).

Elements of other types: models of optical fibres, scanners, etc., will be sub-sites of the created project.

FIMMWAVE allows to work on multiple projects simultaneously. To switch between we using the project tree windows.

Fig. B.1. The main window of the program FIMMWAVE.

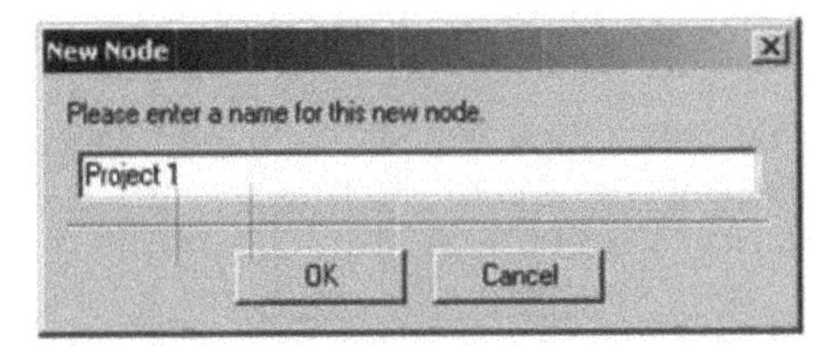

Fig. B.2. Screen for entering the name of the new project.

Fig. B.3. The project tree window with the project added.

B.2. Creating a model of the fibre

FIMMWAVE supports three main types of optical fibres: comb, and fibrous microstructurized. Work with any of these models starts with the determination of the profile of the cross section.

B.2.1. Model of the comb fibre

In the project tree we choose the project to which the model of the comb fibre will be added and press the button on the panel of the new

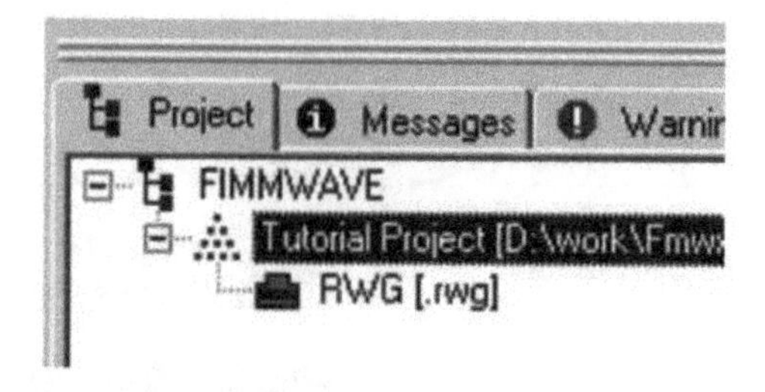

Fig. B.4. The project tree, where the model of the comb fibre is added.

elements. After entering the name of the model, for example 'RWG', the window project tree will look like shown.

The editor of the fibre model is opened by double-clicking on the corresponding node of the tree of the project; its window is as follows.

To set the cross-sectional profile of the created model, we need to click in the editor window, and thus open a window of the profile editor of the appropriate type.

Here, using the 'New slice' we add the required number of columns that will describe the cross-section of the fibre.

Each column has a width defined by the profile (the field 'width'), as well as its own layer structure which is edited by double-clicking on the corresponding item in the column list 'slices (from lhs)'. The editor of the layer structure of a single column allows, with the button 'New', to set the desired number of layers, each having a thickness (field 'thickness') and the refractive index of the medium. The latter can be set manually: options 'rix (iso)' and 'rix (aniso)'; or automatically selected from a database of materials: the option 'mat'. One or more layers may be selected as a domain for the calculation of the concentration coefficient of the mode, the option 'cfreg'.

Adding columns to the structure of the profile can be done by copy (button 'Copy') and paste (button 'Paste') of existing ones, the added columns will have the same structure of layers as the initial one.

In the window of the editor of the model of the fibre, the cross-sectional profile graphically displayed as follows.

To give the created profile the shape characteristic of the model of the comb fibre, it is convenient to perform the 'etching' operation. Its the point is that some part of the column in the height defined in micrometers (field 'etchDepth'), starts to be characterized by the refractive index determined for the uppermost layer of the column in such a way that the original structure in the etching depth of layers is ignored.

The modified cross-sectional profile will look like Fig. B.12, which fully corresponds to the model of the comb fibre.

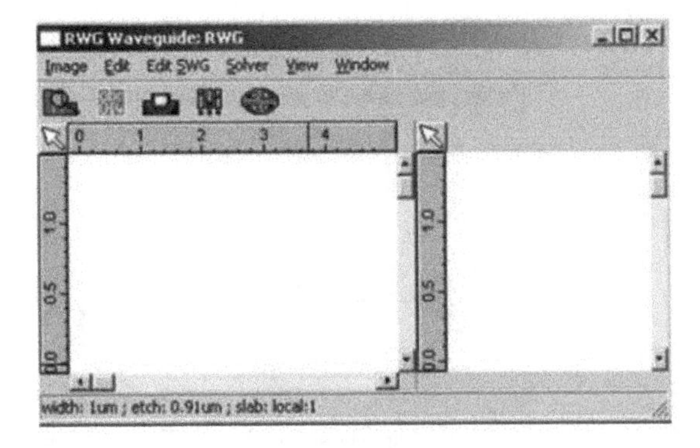

Fig. B.5. Editor window of the model of the comb fibre.

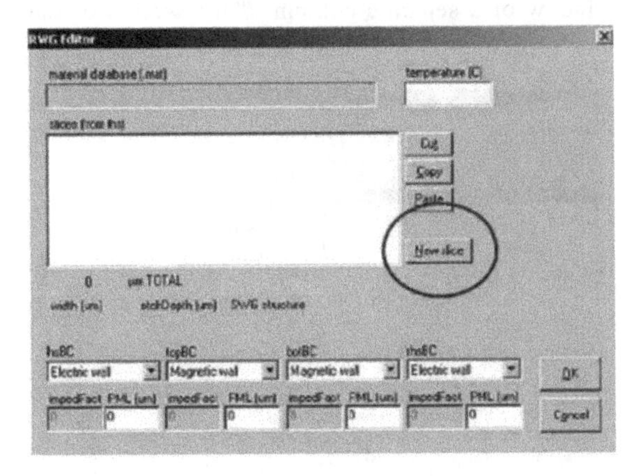

Fig. B.6. Window of the profile editor of the comb fibre.

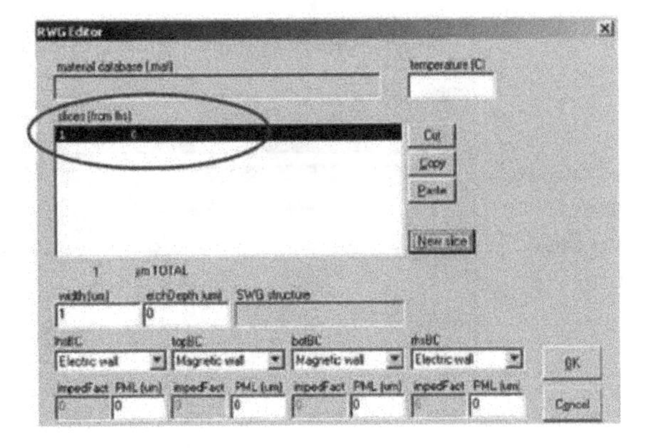

Fig. B.7. Profile editor window of the model of the comb waveguide, where one column is added to the structure of the section.

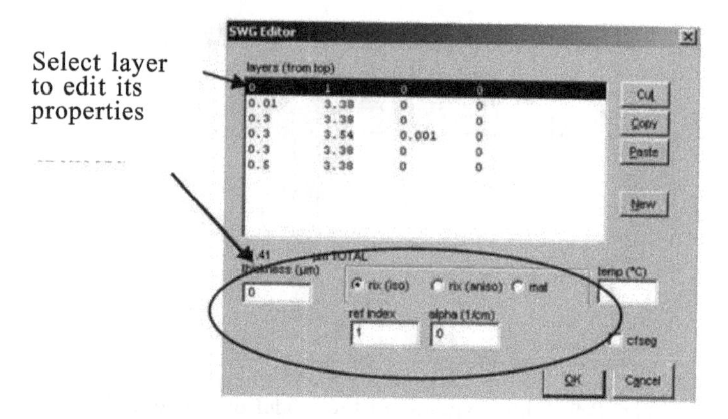

Fig. B.8. Editor window of a separate column of the section of the model of the comb fibre.

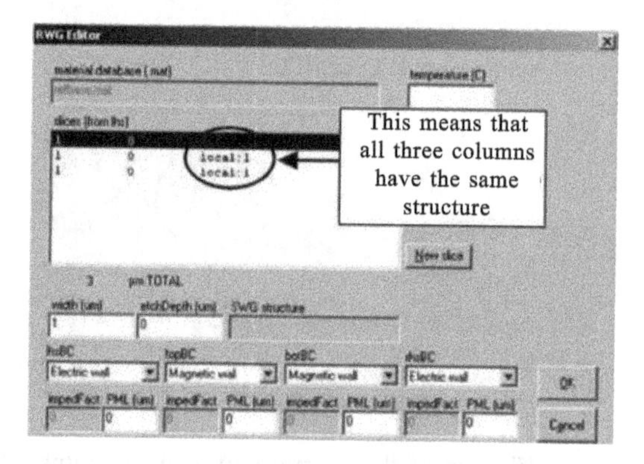

Fig. B.9. Window of the profile editor of the comb fibre model, wherein the second and third columns are the copies of the first column.

Dotted lines indicate that the adjacent columns have the same structure

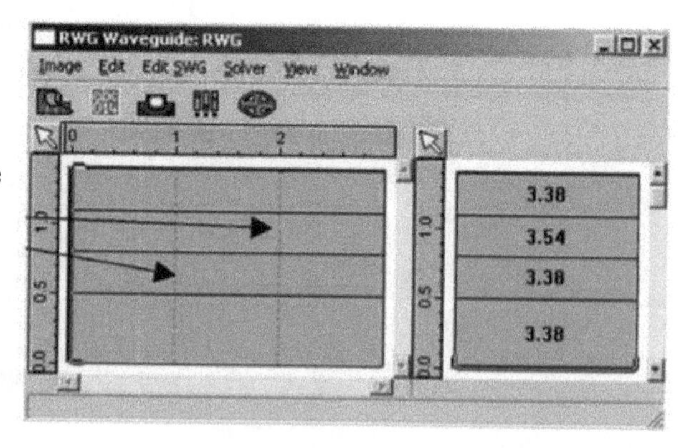

Fig. B.10. Window of the editor of the model of the comb waveguide with three equal columns.

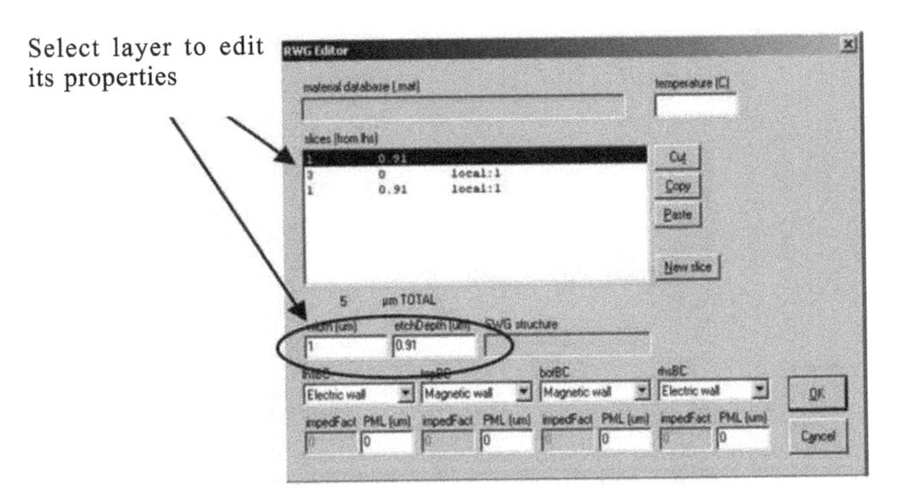

Fig. B.11. Window of the profile editor of the comb fibre model, wherein 'etching' was applied to the first and third column of the section.

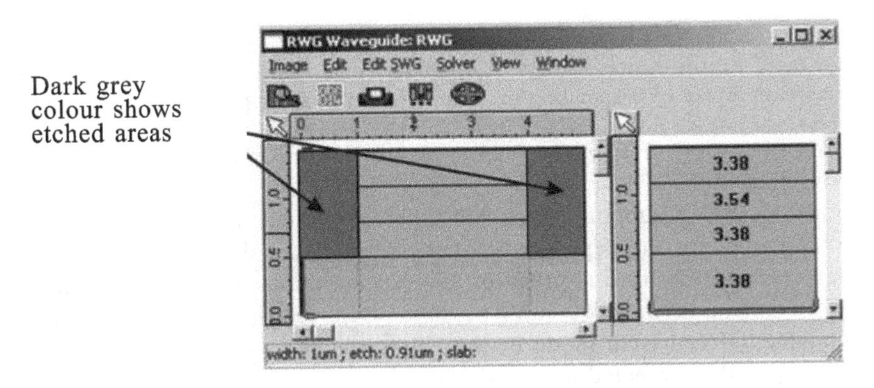

Fig. B.12. Window of the editor of the model of the comb fibre, wherein 'etching' was applied to the first and third column.

For any of the types of fibres supported in FIMMWAVE we can apply several types of boundary conditions. The simplest of these are the 'electric' and 'magnetic' walls ('Electric wall' and 'Magnetic wall'), which is equivalent to the presence on the border of a perfect conductor and ideal magnetic material, respectively. In the first case, the electric field component tangential to the border converts to zero at the boundary, in second – magnetic.

B.2.2. Model of the optical fibre

Creation of a model of the optical fibre should start with adding to the project tree the corresponding object . The editor window of the fibre on the button shows, similar to the previously discussed case of the

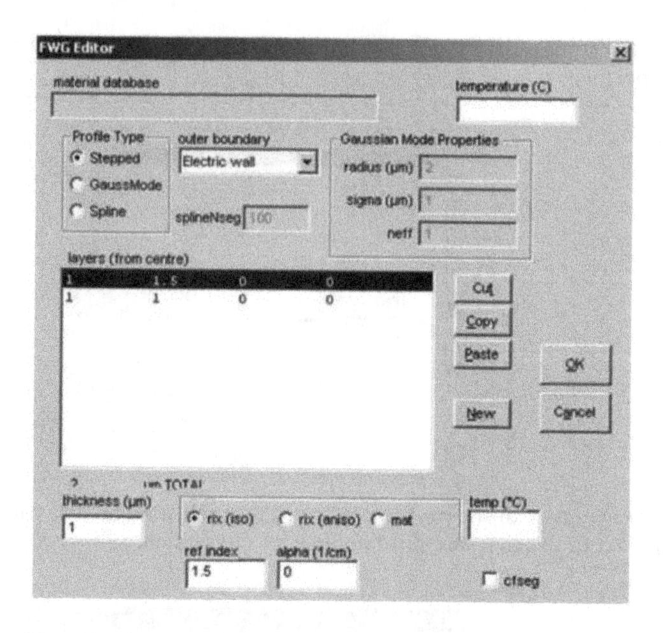

Fig. B.13. Profile editor window of the model of a stepped optical fibre with two concentric layers corresponding to the core and the cladding.

comb fibre, the editor of the profile of the section. In this case the window looks like this:

FIMMWAVE supports three sub-types of optical fibres:

1) Stepped – the dependence of the refractive index of the distance to the centre of the fibre is given by the discontinuous stepped function.

2) Gaussian mode – it is necessary to set the parameters of the fundamental Gaussian mode of the fibre, and the corresponding structure of the profile is created automatically.

3) Spline – it is necessary to add few discrete values of the function of the refractive index of the distance to the centre of the fibre, and FIMMWAVE builds a cubic spline passing through these points.

The layers in the list 'layers (from centre)' refers to the concentric rings of a certain thickness (field 'thickness') with the specified refractive index of the medium.

B.2.3. Model of the microstructured optical fibre

FIMMWAVE allows to model microstructured optical fibres, including photonic crystals. Use the button of the panel of the new elements to add to a project of the model of the microstructured fibre. Immediately after the addition of the windows of the editor of the fibre (Fig. B.14) one sees a homogeneous area with the size of 10 × 10 mm with a refractive index of the medium equal to unity. The context menu is available by right-click, and one can add a variety of geometric shapes (menu 'Add Shape') which

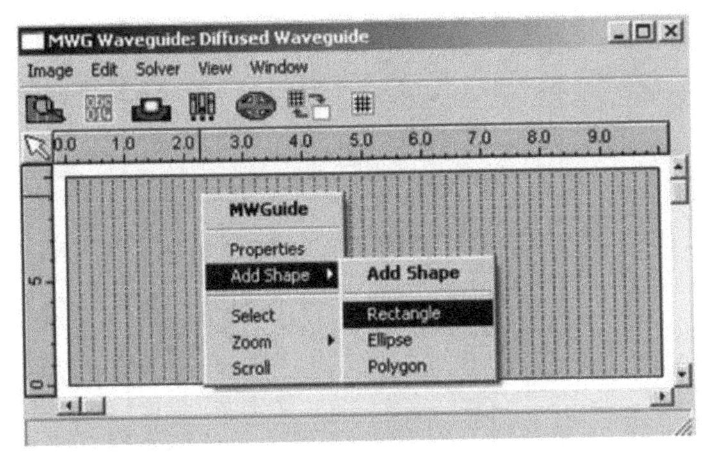

Fig. B.14. Editor window of the model of the microstructured optical fibre with the context menu at the top.

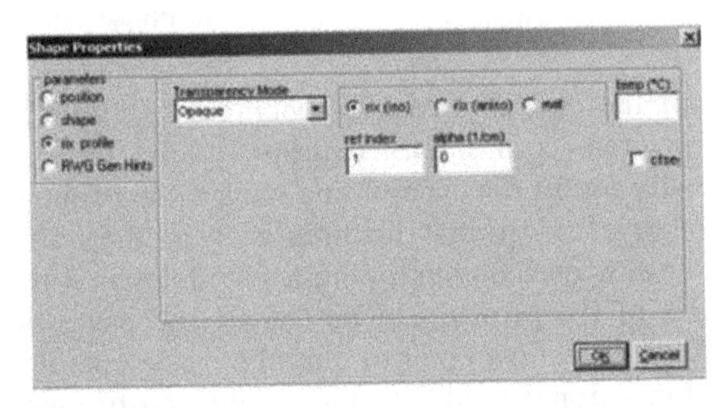

Fig. B.15. Editor window of the element of the profile of the model of a microstructured optical fibre

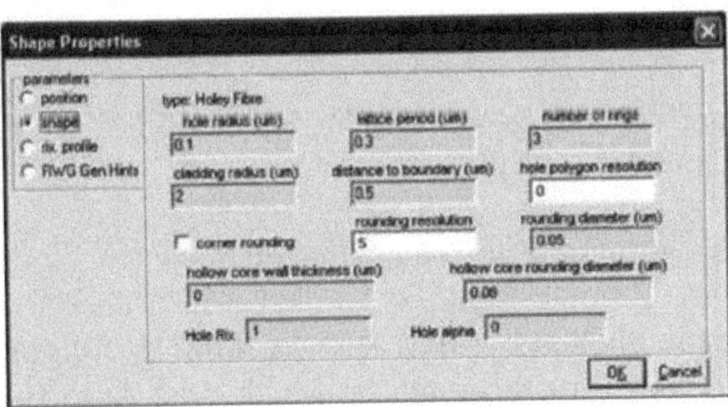

Fig. B.16. Editor window of the element 'Holey Fiber' of the profile of the model of the microstructured fibre.

form the constituent elements of the profile, edit their parameters (menu 'Properties').

Each element of the structure of the profile has a number of adjustable parameters: location, shape, refractive index, absorption coefficient of the material.

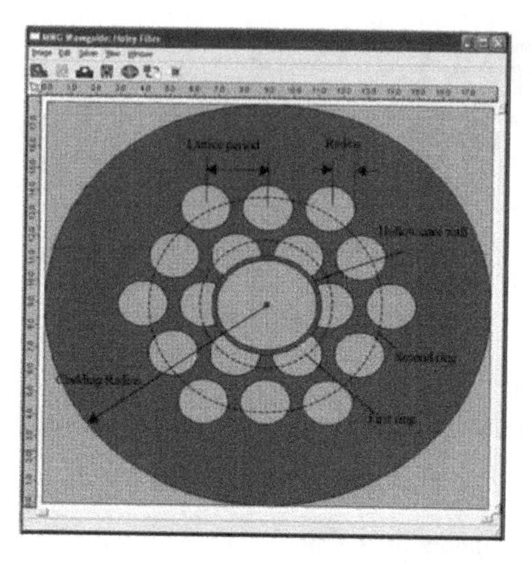

Fig. B.17. Editor window of the model of the microstructured optical fibre the cross-sectional profile of which is formed by 'Holey Fiber'.

The shape of the elements can be rectangular, elliptical, polygonal or constitute a structured set of elements (form 'Holey Fiber'). The tool 'Holey Fiber' is especially useful for determining the section of the perforated fibre, and has a number of additional features such as the polygonal shape of the holes, the hollow core and skipping a few holes in a regular structure.

The elements of the profile can be copied and paste like the columns of the comb fibre that allows one to quickly create complex structures.

The three-dimensional model of the created profile can be viewed by clicking on the toolbar of the editor of the model of the fibre.

B.3. Calculation of modes

After specifying cross-sectional profile of the fibre, we can move to the search its eigenmodes.

B.3.1. MOLAB – automatic search of eigenmodes

This can be accomplished by MOLAB – a reliable, fully automatic search system of searching for eigenmodes ('MOde List Auto Bilder'). MOLAB can calculate almost any natural modes of the fibre, and get a list of events satisfying a predetermined condition, e.g.: five modes with the highest value of the propagation constant of the mode, or modes with the values of the effective index belonging in a certain interval, etc.

To display the preference window of MOLAB, click on in the toolbox in the model editor of the fibre. In the drop-down list of methods of calculation (setting 'Solver') select 'FMM Solver (real)'. Hereinafter we deal only with the method of matched sinusoidal modes (Film Mode

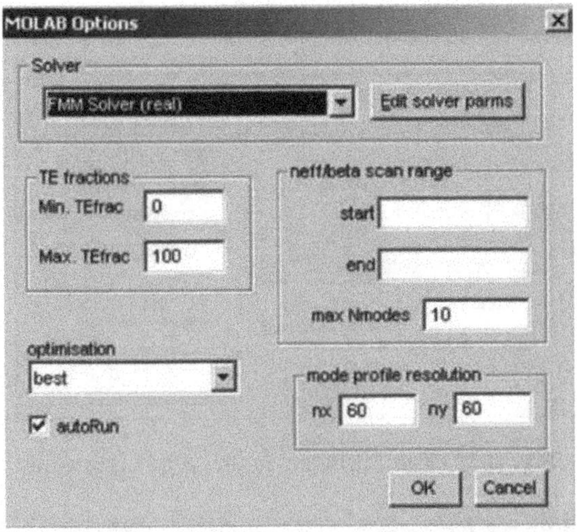

Fig. B.18. The settings window of MOLAB.

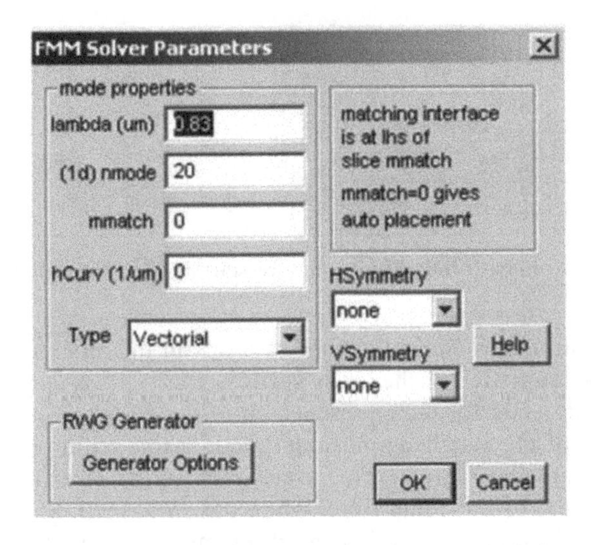

Fig. B.19. The settings window of the method of matched sinusoidal modes.

Matching) in the case of modes with the actual value of the propagation constant of the mode.

In the 'MOLAB Options' we can enter search terms of spatial modes, as mentioned earlier, as well as the resolution ('mode profile resolution') with which the graphs of spatial modes will be displayed. This resolution can be virtually any as the MSM method gives continuous solution for the mode in the form (1.2). In systems of the method 'Edit solver parms' we determine the wavelength λ ('lambda') and other parameters such as the

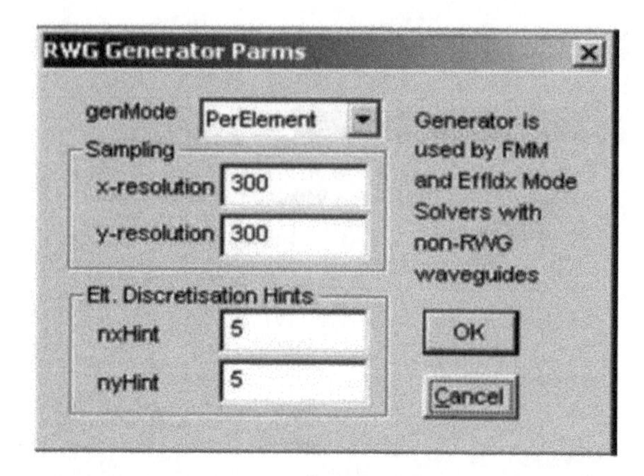

Fig. B.20. The settings window RWG Generator.

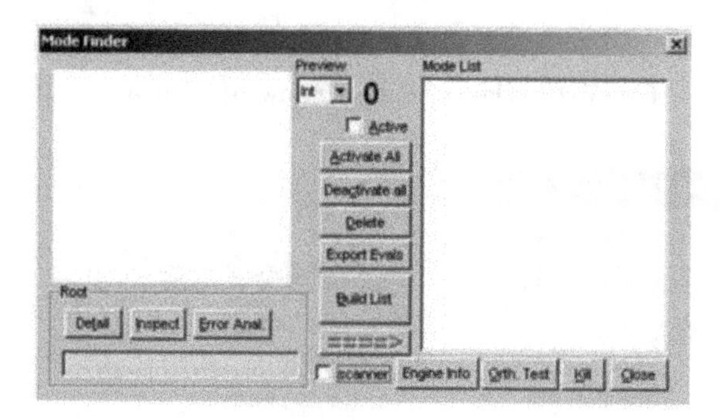

Fig. B.21. Window 'Mode Finder' before starting the calculation modes.

number of local modes (finite number of terms in the inner sum of (1.2)), the symmetry properties of the cross section, etc.

The basic structure for the representation of the cross section of the fibre in the method of the matched sinusoidal mode, as noted in paragraph B.1, is a set of rectangular regions. This description of the section corresponds only to the model of the comb fibre, while the cross sections of round and microstructured optical fibres by an editor must be first converted. For this purpose, FIMMWAVE incorporates RWG Generator, which works in two stages: first it creates a discrete profile of the continuous model with a resolution on the axes 'x-resolution' and 'y-resolution', then the discrete profile is used to produced a model of the comb fibre. Depending on the setting 'genMode' this may be implemented differently. More information about modes of the RWG Generator can be obtained from the documentation accompanying the package FIMMWAVE [4].

The results of the RWG Generator can be checked by pressing the 'Engine Info' in the 'Mode Finder' (see Fig. B.21) and then clicking 'View RWG Waveguide'.

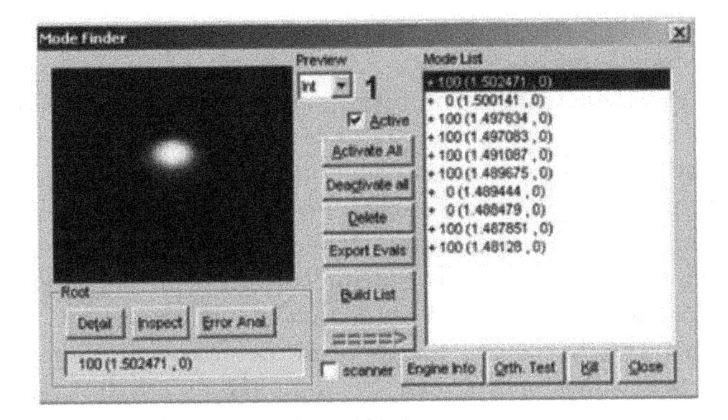

Fig. B.22. Window 'Mode Finder' with the list of calculated modes.

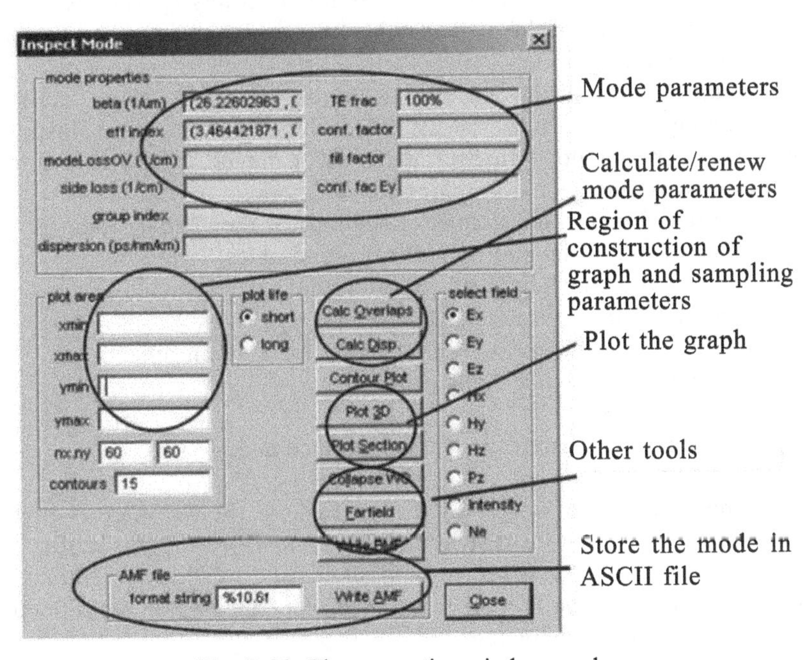

Fig. B.23. The properties window modes.

The calculation algorithm using the method of matched sinusoidal modes consists of three main steps:

– Calculation of the local y-modes for each column of partition of the section.

– The calculation of the matrix of the overlap integrals between the local y-modes of neighbouring columns.

– Search of coefficients, ensuring the satisfaction of the components of the mode for Maxwell's equations and boundary conditions.

To start the calculation, close the preferences window and click [icon] in the window of the editor of the fibre model. After a brief pause, due to the implementation of preliminary calculations, the program will display window 'Mode Finder'.

Press the 'Build List', then press 'Start' in the window 'Build EV List', to start the automatic calculation mode by using MOLAB. The program will perform the calculation in accordance with the introduced parameters. To cancel the process, in the case of incorrect set of parameters or too long time to run, just click 'Kill' in the 'Mode Finder'. Upon completion of the calculation, if the process was not interrupted user, the program displays a list of detected events sorted by descending effective index (the number in brackets). This list is can be stored in the project file.

In the preview window, one can quickly assess the distribution of intensity or of an individual vector component of the mode selected in the 'Mode List'. If the resulting solution appears 'non-physical' it may be necessary to increase the number of local modes (parameter '(1d) nmode') in the settings of the method 'FMM Solver Parameters' and perform calculations again.

The properties of the found events can be viewed in more detail by double clicking on the item in the list 'Mode List' or by pressing 'Inspect'; in both cases, the window 'Inspect Mode' is displayed. It allows you to:

– build two-dimensional graphics spatial mode;

– build one-dimensional graphs of sections of the spatial mode for the given values of the horizontal or vertical coordinates;

– calculate the coefficient of concentration and/or lossed for a given region (optional 'cfreg', see B.2.1);

– calculate the group velocity dispersion for a given mode ;

– calculate the effective mode area.

B.3.2. WG Scanner – parametric scanner of eigenmodes

The scanner ('WG Scanner') allows to see how the parameters of the eigenmodes change in the restructuring of the fibre cross-section. An example is described in which the dependence of the coefficient of concentration of the mode in a predetermined sectional area on the refractive index in the core is investigated.

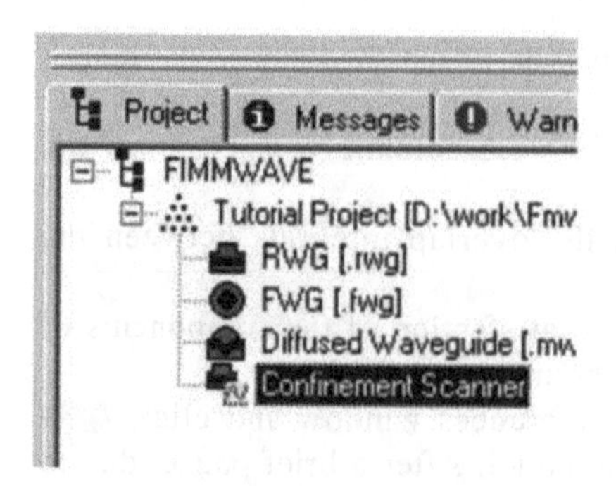

Fig. B.24. The project tree window, with the added scanner object.

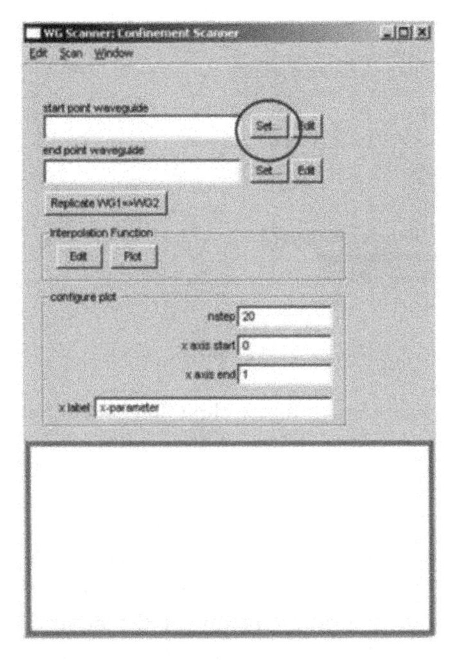

Fig. B.25. Scanner Properties window.

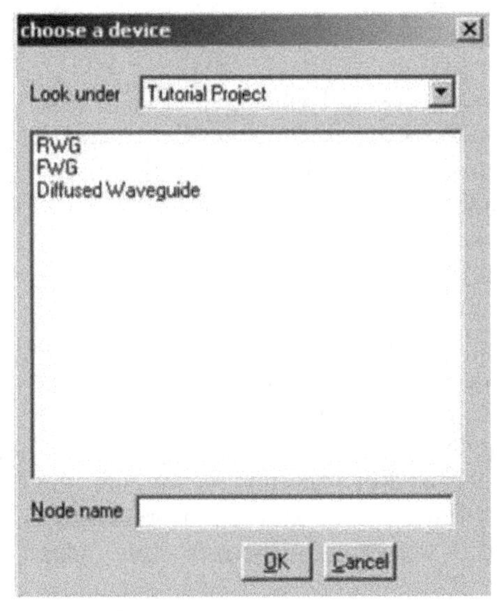

Fig. B.26. A window with a list of models of optical fibres that are available in the working draft.

By pressing the button on the new elements panel we add the object of type 'WG Scanner' to the project, let's call it 'Confinement Scanner' with the project tree will look like Fig. B.24.

Double-clicking on the corresponding node of the project tree opens the property sheet 'Confinement Scanner' (Fig. B.25).

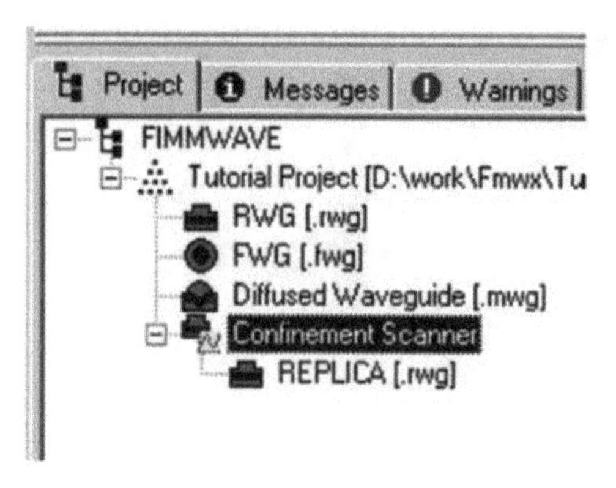

Fig. B.27. The project tree, where the object appeared scanner child node – a copy fibre comb pattern.

Here pressing the 'Set...' as a starting point of scanning we select a previously created model of the comb waveguide RWG (see paragraph B.2.1), where the calculation domain of the coefficient of concentration of the mode (option 'cfreg') we select the fourth layer on top of the central column of the structure of the section (Fig. B.26).

The structure of the model RWG is copied and the copy is assigned the endpoint of scanning at the touch of a button 'Replicate WG1 => WG2', while in the project tree the following changes will occur (see Fig. B.27).

We return to the window of the properties of the element 'Confinement Scanner' and open the rector of the fibre model REPLICA using the button 'Edit'. We change the value the refractive index of the fifth layer of the central column of the structure, and make it equal to 3.54, as shown in Fig. B.28, instead of the original 3.38 model (see Fig. B8).

For the original scanning model ('start point waveguide') we open the editor window of the model by the button 'Edit' and with MOLAB get a list of two TE- and TM-fundamental modes (see paragraph B.2.2). Closing the editor of the original model, we save the resulting list of modes: button 'Yes' in the message window 'Save mode list changes?'.

Scanner 'does not know' the parameter used in scanning, so we enter the appropriate values in the fields 'x axis start' – the value of the refractive index in the original model, 'x axis end' – the value of the refractive index in the final model, 'x label' – the title of the parameter used in scanning, 'nstep' – the number of intermediate models and points for a graph (Fig. B.29).

We are ready to start scanning the coefficients of concentration for two fundamental TE- and TM-modes in dependence on the refractive index of the core, varied between 3.38 and 3.54. The command 'Start' menu 'Scan' opens the window 'Results to plot', where we choose the appropriate chart type, and enter the numbers of the modes '1–2' – two fundamental modes, then press the 'Add' and switch the setting 'plot on same scale' (Fig. B.30).

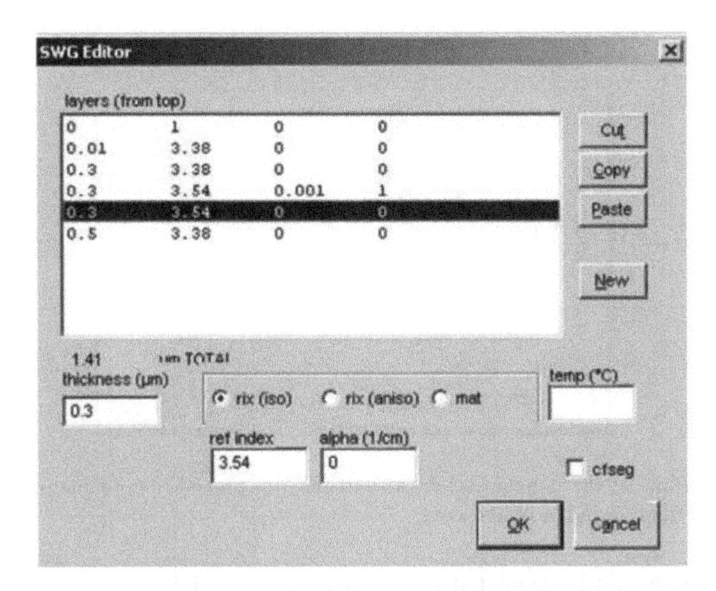

Fig. B.28. The editing window of the central column of the section of the copy comb fibre model.

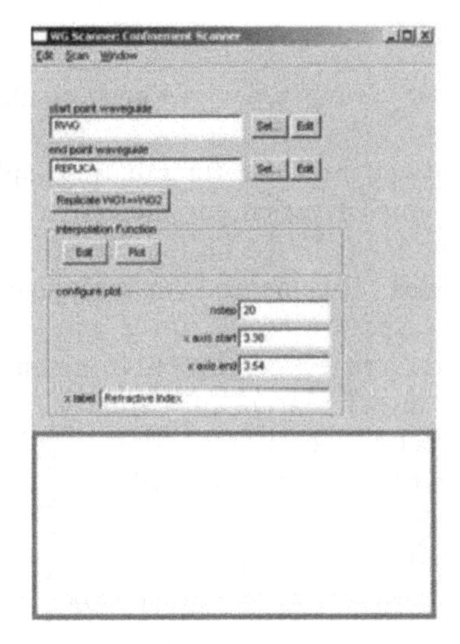

Fig. B.29. The properties window of the scanner before scanning.

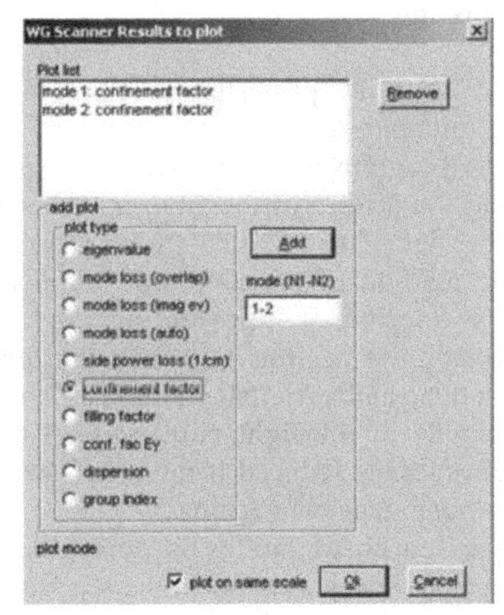

Fig. B.30. The window of the settings of display of the scan results.

After pressing the ‘Ok’ the graph of the following type is displayed (Fig. B.31).

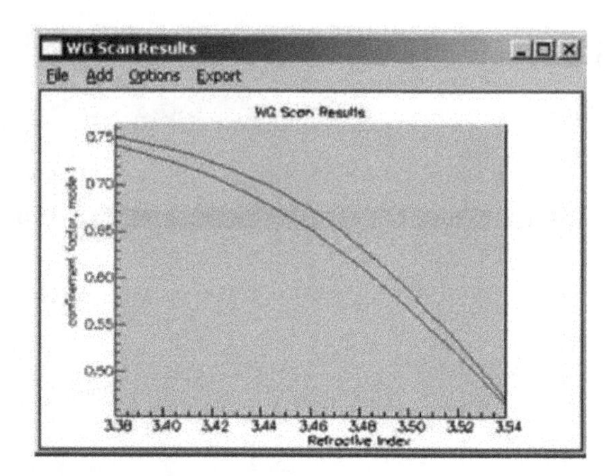

Fig. B.31. Plots of the concentration ratio of the TE- and TM- fundamental modes from the refractive index of the core.

Here, as would be expected in this case, the coefficient of concentration of both modes decreases with increasing refraction index.

Conclusion

FIMMWAVE is a powerful modelling tool for various types of fibres: comb, fibre, microstructured.

The program has a user-friendly graphical interface for creating and editing the profile of the fibre model with lots of varied parameters.

The algorithm of calculation of eigenmodes is based on the formalism of the method of matched sinusoidal modes. The method, in turn, uses the assumption of representing the cross section in the form of a set of rectangular regions with constant characteristics of the environment. This quick and efficient method allows the calculation of virtually any eigenmode of the light guides of all three types of fibres. However, its effectiveness is reduced if the cross-section of the fibre is described by too many homogeneous regions.

The search of modes is possible in the automatic mode for some waveguide model and also in the scanning mode of certain parameters of the mode depending on changes in the parameters of the profile of the section.

References

1. Sudbo A.S., Film mode matching: a versatile numerical method for vector mode field calculations in dielectric waveguides, Pure and Appl. Optics. 1993. V. 2. P. 211–233.
2. Sudbo A.S., Numerically stable formulation of the transverse resonance method for vector mode field calculations in dielectric waveguides, IEEE Phot. Tech. Lett. 1993. V. 5. P. 342–345.

3. Sudbo A.S. Improved formulation of the film mode matching method for mode field calculations in dielectric waveguides. Private communication. Dec. 1993.
4. Photon Design: Supporting documentation for software package FIMMWAVE. file fimmwave.pdf

List of special terms

The method of matched sinusoidal modes
Comb fibres
Optical fibres
Microstructured optical fibres
Eigenmodes of the fibre
Mode propagation constant
Effective mode index
Fundamental mode
'Electric' and 'magnetic' walls'

Appendix C

Simulation using OLYMPIOS program

C.1. The purpose and structure of the program

The software package OlympIOs of the Dutch company Concept To Volume is intended for the development, simulation and formation of templates of integrated optics devices. Development is in the form typical for CAD-systems: the projected device is formed as a combination of optical and control elements (each of which has customizable personal characteristics), and the editing is carried out in a visual form that can easily and effectively change the properties of the apparatus and monitor its performance.

An important architectural feature of the package is its modular structure: functional elements of the program are presented in the form of separate modules, which allows one to compose a package based on user needs. Modules of the package will be considered and their functions and features will be briefly described.

Basic mode solver

This module is designed for the numerical determination of the optical modes of the optical medium. The initial data for its work is a one-dimensional (in planar targets) or two-dimensional (in three dimensions), refractive index profile, determined by the location of the optical elements and their optical properties, the wavelength of coherent radiation, as well as the specific parameters of the specific method.

For the analysis of one-dimensional profiles we have the following numerical methods: the 3-layer method, the transfer matrix method, the complex transfer matrix method, the Usievich method, the Cauchy integral method, the Bakhtazad method and the Bend 1D method.

For two-dimensional modules we use numerical methods: the effective index method, the Marcatili method, the finite difference method).

These methods differ in the number of layers considered when calculating, the maximum value of the imaginary part of the refractive index, and the computing speed. The result of their work is the distribution of the complex amplitudes and the values of the propagation constants of the calculated modes.

Advanced mode solver

This module extends the functionality of the previous one by adding methods for the analysis of two-dimensional refractive index profiles: the multigrid finite difference method, the finite difference generic method, the Bend 2D method and the film mode matching method.

All of these methods allow one to more accurately determine the modes of optical environments, but also have fine setting, which considerably complicates their use.

Beam propagation method

This is the basic unit in the package and includes various implementations of the beam propagation method. The initial data for its work are all parameters of the examined optical elements, optical parameters of the system as a whole and the parameters of the method itself. Implementation of the method for two-dimensional and three-dimensional cases has been proposed. In both cases, BPM versions, based on the Fourier transform (FFT-BPM) and finite-difference methods (FD-BPM), are available

In the two-dimensional case modifications FD-BPM, differing in the following criteria, are available:

- For TE- and TM-polarization ;
- Padé approximation order (0, 2 or 4);
- Order of discretization of transverse derivatives (2 or 4);
- The type of boundary conditions (transmissive or absorbing).

In the three-dimensional case we used the approximation of the smooth envelope and not Padé approximation, and the proposed modifications are distinguished by the following criteria:

- For TE-polarization and for TM-polarization;
- The type of boundary conditions (transmissive or absorbing).

These methods use a uniform discretization of the computing domain, wherein the discretization steps can be specified. It is also possible to specify how to calculate the relative index of refraction.

The electro-optic module

The functionality of this module allows one to take into account the electro-optic effect in the calculations (determining modes and calculation of beam propagation).

Materials designated for optical elements, have, in addition to the optical parameters, also electric parameters (in particular, the electro-optic coefficient). Thus, in addition to the optical elements, sources of the electric field can be added to the modelled area.

On the basis of these data, the module allows the use of the finite difference methods to compute the distribution of the electrostatic field on the basis of which we calculate the refractive index change due to electro-optic effect. The calculated refractive index changes are taken into account in the course of operation of computing modules.

Thermo-optic module

As regards the scheme of operation, this module identical to the previous one, but takes into account the optical effects due to thermal effects on the optical elements.

Similarly, for materials of optical elements we can define thermal properties (thermal conductivity, thermo-optic coefficient etc.), and place heat sources in the simulated domain (in two modes: constant temperature and constant power.)

On the basis of these data we calculated the temperature distribution in the simulated area and the resultant refractive index change, which, in turn, is taken into account in the work of computing modules.

Stress modelling module

This module allows one to take into account in calculations the effects arising from the mechanical effects on the planar optical elements (optical elasticity effect) and the associated thermal expansion of these elements.

Additional settings are reduced to indicating the physical parameters of the materials of the elements that allow the detection of the refractive index and its anisotropy in the case of the optical elasticity effect. When calculating the thermal expansion the calculation is carried out using the data of the thermo-optic module and related parameters.

Mask layout module

Unlike the earlier modules, this module is not connected with the process of calculation, but allows on the basis of project data OlympIOs to get files that are original data for photolithographic devices.

The presence of such a function allows one to move from the development of integrated-optical and MEMS devices to their implementation.

Script editor module

Despite the availability in the package of a range of automation and optimization processes, this may not be enough for integrated studies of devices. Scripts can be used in these cases.

A script is a program in the meta-language of the package OlympIOs. The language features include control of the parameters of the project, additional data processing, etc. The Script editor module is used for writing and running scripts.

Bidirectional eigenmode propagator module

An alternative to BPM methods of calculating field characteristics in the process of propagation of the field in the environment is to use the mode method, which is implemented in this module.

During operation, the simulation area is divided along the propagation axis into sections with a constant refractive index profile. For each of these sections we determine forward- and backward-propagating modes, into which the propagating beam is split, and at the borders of the sections these expansions are merged. The result of the calculation are the coefficients of expansion in the modes in all areas including initial conditions.

Unlike BPM, this approach allows us to take into account the reflection and re-reflections at the boundaries of the media, but it is not applicable in the case of a smoothly changing refractive index profile.

Distributed computing module

Since BPM has sufficiently high computational complexity, simulation of large areas and, consequently, carrying the series of computing experiments for a set of parameters requires considerable computing power and the time required for calculations. The distributed computing method is used to increase the speed of computation.

The distributed computing module allows to solve computing tasks within OlympIOs on multiple computers simultaneously. As distributed computing in the solution of a problem is rather difficult (since it requires the transfer of significant volumes of data in the network), the approach in which one task is transferred to each of the computing nodes during computing experiments is used. Thus, the network transfers only the results of solving problems, and the overall speed of a series of numerical experiments is increased proportionally to the number of nodes.

C.2. Determination of modelling parameters

The initial data for simulations using BPM are the refractive index distribution in the modeled area, physical element parameters and initial conditions in the form of the field distribution.

The advantage of the OlympIOs package is an approach to the determination of the simulation parameters. From the viewpoint of a user interface the editor of the package is similar to the editors of CAD-systems

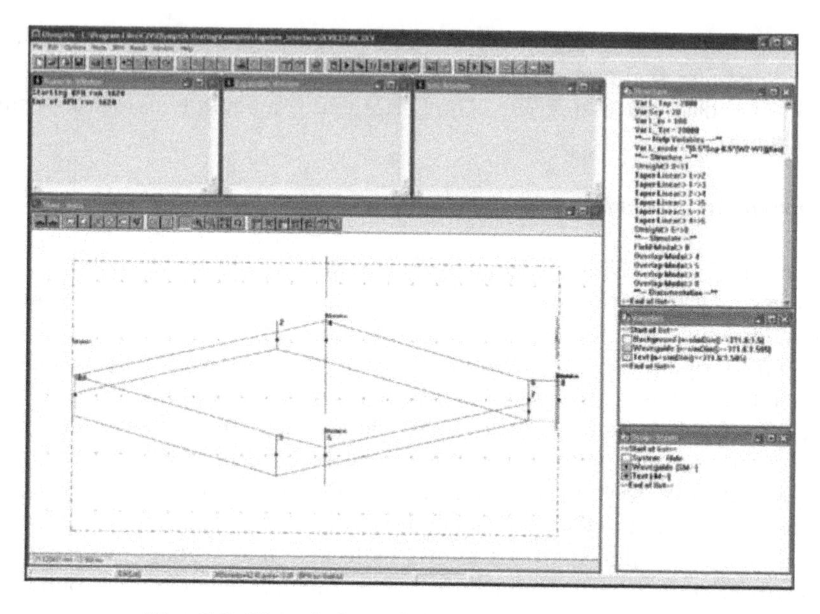

Fig. C.1. The window of the editor of the project.

(e.g. AutoCAD) or three-dimensional image editors (for example, 3D Studio).

The distribution of the refractive index is based on the 'scene' objects. The scene is a collection of objects, their characteristics and relative position. Figure C.1 an example of a scene in the form as it is displayed in the editor.

Each object has a set of characteristics determined by the type of the object (including its shape) and sheet (Design sheet), on which the object is situated.

There are 7 basic types of objects.

1. The waveguide structures (Waveguide). These are objects that define the structure of the modelled optical system. For example, these are waveguide of various types, lenses, tapers, etc.
2. Special items (Special). These objects also determine the structure of the optical system, however, they are more complex than ordinary objects.
3. Control element (Control). Objects of this type are not components of the optical system, but allow one to control it. For example, elements If and For permit elements to be added to the circuit depending on the shape and consistency of similar elements.
4. Elements of two-dimensional simulation (2D Simulate). These objects can control the course of the simulation. The important objects of this form should include objects Field (specify the initial distribution field), Manipulator (one can change characteristics of the beam during

```
DefaultValue(Lambda=1.32,Width=5,ModeNum=0,Polarisation=TE,BpmOrder=3);
field3G(WaistX=5,WaistY=5,Offset=0,Power=1,Phase=0,TiltX=0,TiltY=0);
loop "i" (From=1,To=10,Count=10,loopPlane=1,outPutPlane=0,Format=0) {
  deflayer(layer "layer{$i}");
  straight(Width="16-i",Length="110-i*10");
}
```

a

```
Structure
===Main Level===
   Field3-Gauss:> 0
-  For 'i' :> 0=>0
      Use DesignSheet:> layer{$i}
      Straight:> 0=>0
==End of list==
```

b

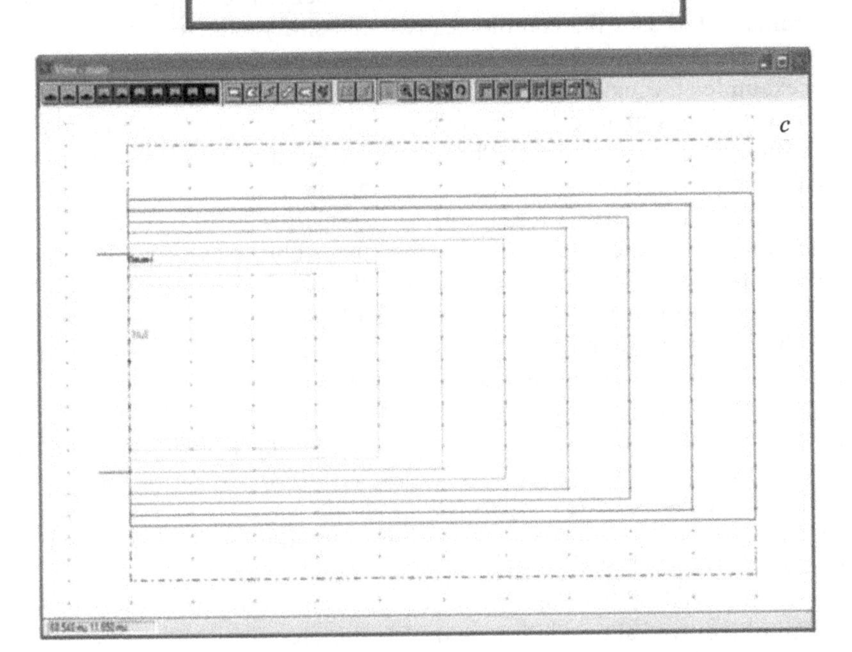

c

Fig. C.2. Project code using control elements (*a*), elements of the project in the editor of the structure (*b*) and the appearance of this project in the visual editor (*s*).

its propagation), Overlap (computes overlap integrals with the modes and set fields).

5. Elements of three-dimensional simulation (3D Simulate). Similar facilities as for the two-dimensional case.
6. Documentation elements (Documentation). These objects are also not involved in the scene in an explicit form, but allow one to add to the project text information describing the project.
7. Mask elements (Mask). Also they do not affect the process of modelling, but are displayed in the formation of mask files.

The sequence of describing objects of different types determines the project OlympIOs and the structure of the simulated optical system.

Despite the fact that the editing of the project is carried out by visual means, the project itself is a sequence of commands that describe the objects. The fact that the descriptions of the objects in this case are ordered, allows the efficient use of control elements. In particular, powerful means are variables that can be used in parameter description of other objects. Due to their use, the parameters of the optical system and even its structure may be changed by editing values of the variables. Figure C.2 shows the code for the project, describing a set of straight waveguides with the given parameters, and also the view of the scene for this project.

One of the features of the OlympIOs package is a way of specifying the relative position of the elements in the scene: it does not use absolute coordinates but uses the coordinates (tilt and shift) relative to the reference points. The reference points are determined by the structural elements: for example, the output of the waveguide forms a reference point. Such approach allows us to easily change the whole scene when the parameters of structural elements change. The scene generally has a two-dimensional character, i.e. describes the location of objects in the plane.

Volume characteristics and optical parameters of objects are in turn, determined by the canvas to which the given object belongs. The canvas defines the media (and, as a consequence, the physical properties of objects) and cross-sectional profile of the objects of this canvas.

The following forms of cross-section are allowed for three-dimensional objects: rectangular, elliptical and trapezoidal. From this point the possibility of the package are considerably reduced (as it is not possible to construct an arbitrary three-dimensional structure), but such a set is sufficient for most typical problems.

Each canvas is edited separately, and when an object is formed it is specified to what kind of canvas it belongs. In addition, each canvas forms its own set of reference points.

The material of the object, in turn, determines the physical properties of the object: refractive index, and also the electrical, thermal and other properties. The material editor allows one to create materials with the desired characteristics. In addition, there is a library of completed materials and also the possibility of formation of own user libraries.

In addition to the parameters of objects and scenes, the project has a set of global characteristics. These include the dimension of the simulation, the radiation wavelength, the kind of polarization, as well as several others.

Thus, the simulation parameters are determined by global characteristics of the project, a set of materials, the canvas and a set of objects.

C.3. Modelling and analysis of results

After determining the parameters of the model we can carry out simulations

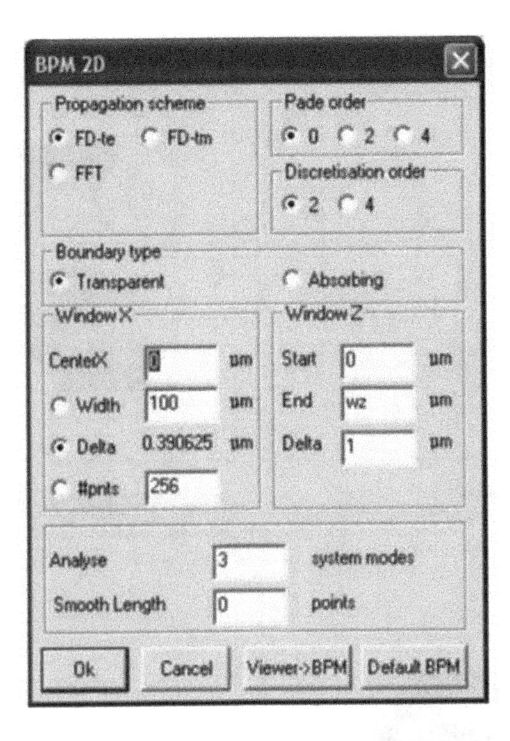

Fig. C.3. Dialog window of setting modelling by the BPM method for a two-dimensional case.

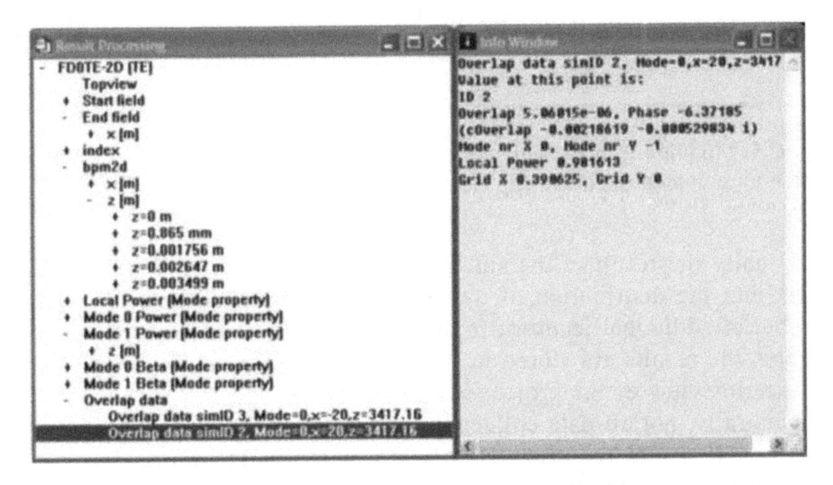

Fig. C.4. View of the window of simulation results.

of the propagation of electromagnetic radiation by one of the methods available in the package. It is necessary to specify the parameters of a particular run of the modelling process. Figure C.3 shows the dialog window of setting simulations in two dimensions.

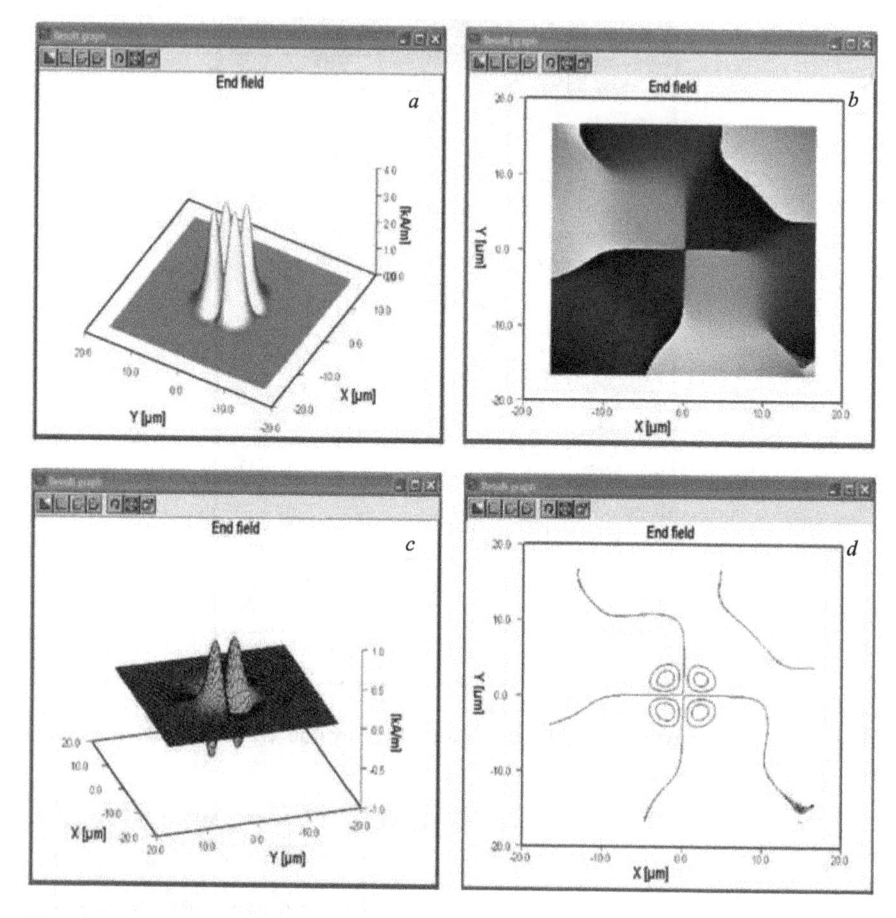

Fig. C.5. Module (*a*), phase (*b*), real (*c*) and imaginary (*d*) components of the beam with a high content of mode LP-21.

Usually the result of the simulation is the distribution of the refractive index and the distribution of the field in the modelled area, as well as a number of additional features. In the case of multiple runs of the simulation process the results are stored in the project for all runs. Figure C.4 shows an exterior view of the project window.

Naturally, not all data collected during the simulation are stored as the results – it would require a significant memory. Therefore, some of the characteristics of the project are the parameters of storing data defining the number of stored points in all measurements.

The results can be viewed in several ways: in the form of graphs of two-dimensional dependences and in the form of drawings and diagrams for the three-dimensional relationships. Furthermore, in the case of study of fields the amplitude, phase, real and imaginary components of the field

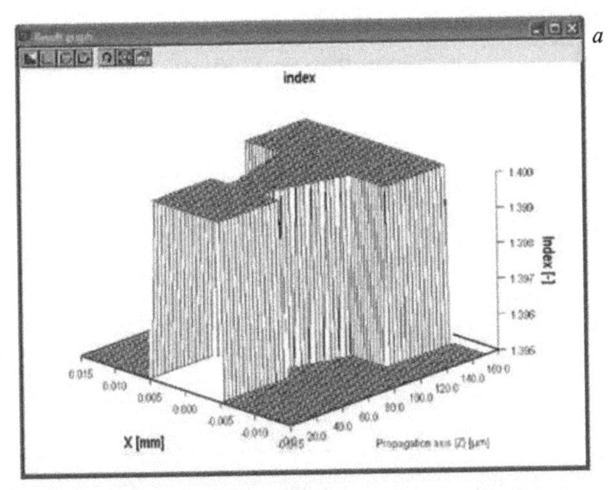

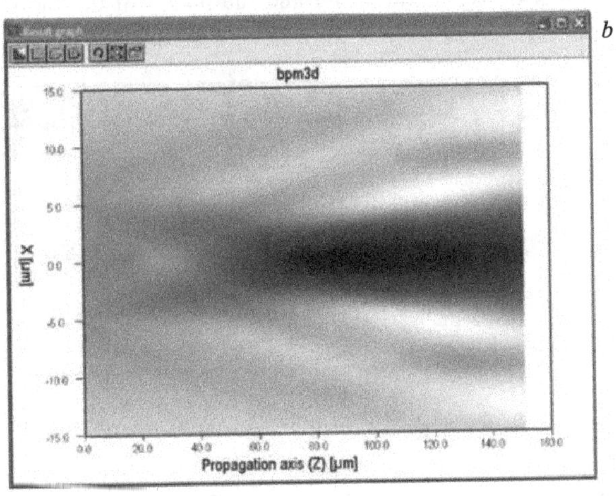

Fig. C.6. Distribution of the refractive index (*a*) and the modulus of the complex amplitude in the longitudinal section of the three-dimensional waveguide transition (*b*).

can be displayed. As an example, Fig. C.5 shows the module, phase, real and imaginary components of the field imaged using various diagrams.

The advantages of the package also include the ability to analyze not only transverse 'slices' of the fields, but also longitudinal ones. Thus, Fig. C.6 shows the distribution of the refractive index in the longitudinal section of a three-dimensional waveguide transition (taper), as well as the distribution of the complex amplitude modules in the same section.

Furthermore, all the results may be exported from the programs in the text (available for opening in a spreadsheet), or in graphic form (one of the standard image formats). This feature greatly facilitates the handling and use of results.

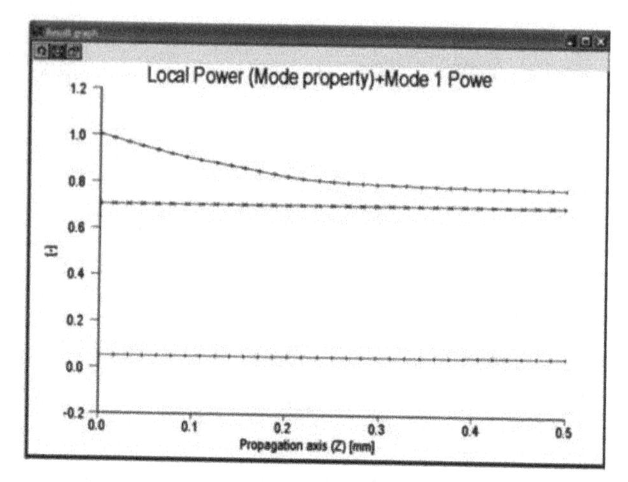

Fig. C.7. The total power of the beam and the power of the individual modes in the area of the transition process in the planar waveguide.

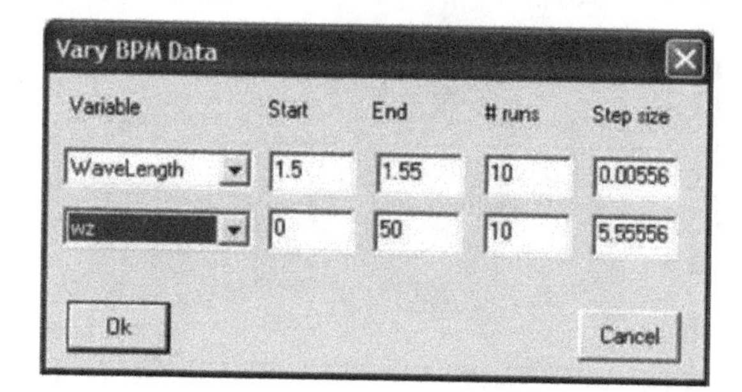

Fig. C.8. Appearance of the window for starting variant modelling.

The OlympIOs package also provides a means of analyzing the energy characteristics of the propagating beams. In particular, the package facilities make it possible to analyze the total power of the propagating radiation, the distribution of power in modes and compute the value the overlap integral for the predefined fields. As an example, Fig. C.7 shows the dependence beam power and power modes in a planar wave waveguide in the area of transition.

A powerful tool for the study of optical systems implemented in the package is variant modelling (Vary run). The capabilities of this tool are based on the use of variables (control objects) to describe the simulation parameters: the change in the value of the variable that leads to a change in the system being modeled. The variant modelling tool, in turn, allows one to perform a series of numerical experiments with different values of one or two variables. The range and step of variation is specified for each of the variables. Figure C.8 shows the exterior view of the startup screen of variant modelling.

Thus, the simulation can be performed for the whole set of various variables, followed by an analysis of the dependences of calculated values on the values of variables.

Due to the range of available simulation algorithms, visualization facilities and data export, as well as built-in functions that define the field, the OlympIOs package is a powerful tool for modelling and analysis of processes in waveguide structures.

Index